国外优秀物理著作
原 版 系 列



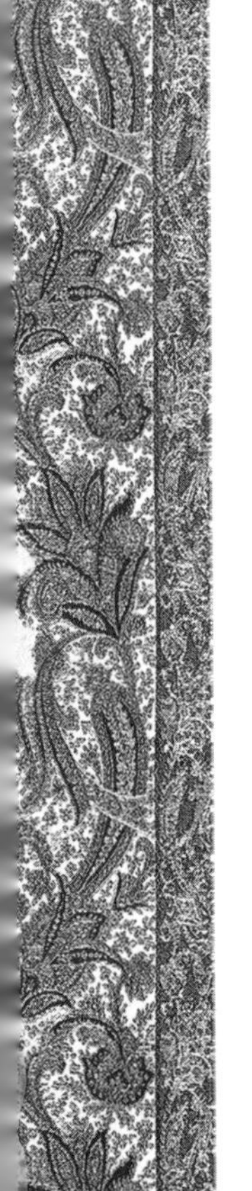

# The Scientific Papers of Sir Geoffrey Ingram Taylor
## —Volume I. Mechanics of Solids

# 杰弗里·英格拉姆·泰勒科学论文集
## ——第1卷，固体力学

（英文）

[英] G. K. 巴彻勒（G. K. Batchelor） 主编

哈爾濱工業大學出版社
HARBIN INSTITUTE OF TECHNOLOGY PRESS

**黑版贸审字 08-2020-077 号**

**图书在版编目(CIP)数据**

杰弗里·英格拉姆·泰勒科学论文集.第1卷,固体力学=The Scientific Papers of Sir Geoffrey Ingram Taylor: Volume Ⅰ. Mechanics of Solids:英文/(英)G. K. 巴彻勒(G. K. Batchelor)主编.—哈尔滨:哈尔滨工业大学出版社,2021.8

ISBN 978-7-5603-9403-9

Ⅰ.①杰… Ⅱ.①G… Ⅲ.①固体力学-文集-英文
Ⅳ.①N53 ②O34-53

中国版本图书馆 CIP 数据核字(2021)第 078714 号



策划编辑 刘培杰 杜莹雪
责任编辑 李 鹏 张嘉芮
封面设计 孙茵艾
出版发行 哈尔滨工业大学出版社
社 址 哈尔滨市南岗区复华四道街 10 号 邮编 150006
传 真 0451-86414749
网 址 http://hitpress.hit.edu.cn
印 刷 哈尔滨博奇印刷有限公司
开 本 720 mm×1 000 mm 1/16 印张 40.75 字数 662 千字
版 次 2021 年 8 月第 1 版 2021 年 8 月第 1 次印刷
书 号 ISBN 978-7-5603-9403-9
定 价 78.00 元

(如因印装质量问题影响阅读,我社负责调换)

# EDITOR'S PREFACE

These volumes contain all the scientific papers by Sir Geoffrey Taylor that he and I have been able to locate, apart from some that duplicate work herein, and apart from a number of minor unpublished notes thought to be no longer of interest. The papers have been prepared for the present volumes with as little change of their original form as possible. Printing errors and small mathematical slips in the first publication have been rectified without comment, but any substantial change in the text as originally published has been disclosed in a footnote. All the figures have been redrawn in a uniform style. Several papers that were written for Government departments or advisory committees, mostly during the last war, are here published for the first time; Sir Geoffrey Taylor has looked through them and has made a number of minor changes to meet the needs of more general distribution. Purely expository articles prepared for delivery as lectures or for publication in collective works have not been included in these volumes.

The papers have been grouped into volumes according to subject, although the divisions between the different aspects of fluid mechanics in Volumes II, III and IV are not sharp. Within each volume the papers have been arranged in chronological order, and each volume has its own numbering system based on that order. A complete list of the papers in all four volumes will be provided at the end of Volume IV.

The publication of these volumes was made possible by a financial guarantee received from the Ministry of Supply (through the good offices of the Fluid Motion Sub-Committee and its parent body, the Aeronautical Research Council), supplemented by a gift from the Master and Fellows of Trinity College. My personal thanks are due to Dr J. W. Maccoll and his staff at the Armament Research and Development Establishment for the provision of an invaluable list of nearly all Sir Geoffrey Taylor's papers, both published and unpublished, and to them and Mr J. L. Nayler and Mr R. W. Gandy, past and present secretaries of the Aeronautical Research Council, and the staff of the A.R.C. office, for their help in matters relating to the papers hitherto unpublished.

Papers 8, 25 and 30–34 inclusive are crown copyright material; the Controller of H.M. Stationery Office has given permission for their inclusion in this volume. Paper 28 is reprinted by permission of The Macmillan Company of New York, and Paper 41 by permission of Springer-Verlag, Berlin. Acknowledgement is made also to the President of the Executive Committee of the International Congress for Applied Mechanics for permission to reprint Paper 11.

G. K. BATCHELOR

*January 1957*

# CONTENTS

# 1

# THE USE OF SOAP FILMS IN SOLVING TORSION PROBLEMS*

REPRINTED FROM

*Reports and Memoranda of the Advisory Committee for Aeronautics*, no. 333 (1917), and *Proceedings of the Institution of Mechanical Engineers* (1917), pp. 755–89

## INTRODUCTION

The equations which represent the torsion of an elastic bar of any uniform cross-section are of exactly the same form as those which represent the displacement of a soap film, due to a slight pressure acting on its surface, the film being stretched over a hole in a flat plate of the same shape as the cross-section of the bar. The theory of this relationship is briefly outlined, and it is shown that advantage may be taken of the analogy to find the stresses and torsional stiffness of a twisted bar or shaft of any cross-section whatever, by making appropriate measurements of soap films. The method is technically useful because there is no restriction on the shape of sections with which it is capable of dealing, whereas the number of cases in which the equations can be solved analytically is extremely limited.

The apparatus used for measuring films is described and illustrated, and examples of its use are given. These include simple geometrical figures, for which the results of the soap-film method may be checked by calculation, and also two instances of technically important sections which are not amenable to mathematical treatment. In the first of these, the magnitude of the stress in an internal corner, and its dependence on the radius of the fillet, are investigated, while in the second the stresses and torque of a twisted aeroplane wing spar, of I section, are discussed, and comparisons between the results of the method and those of some direct torsion experiments are given.

Finally, a number of general theorems relating to, and approximate formulae for, the stiffness and strengths of shafts and beams, are obtained with the help of the soap-film analogy. It is shown, by comparison with other results, that it is possible to deduce thus, in nearly every case, figures for those torsional data usually required in practice, which are within a small percentage of the exact values. The superiority of these formulae over those now in use appears to be due to the introduction, it is believed for the first time, of the length of the perimeter of the cross-section as a factor. This was suggested almost immediately by the soap-film analogy, and is an instance of the value of the latter as a means of forming a clear idea of the nature of the torsion problem.

* With A. A. GRIFFITH.

## General Considerations

In the old theory of the torsion of shafts or beams of uniform cross-section, which was originated by Coulomb, it was assumed that sections of the bar, initially plane and at right angles to the axis of torsion, remained so when the bar was twisted, and that the only strains set up were those due to the relative rotation of adjacent sections about the axis.

In his classical memoir on the mathematical theory of torsion, Saint-Venant showed that the assumptions made by Coulomb were valid only in the case of circular shafts, either solid or having concentric circular holes. In every other instance the initially plane section is distorted into a curved surface, and the stresses and strains set up in the bar cannot be calculated until the shape of this curved surface has been found.

A complete discussion of the theory of torsion put forward by Saint-Venant would be out of place in the present Paper. It is fully dealt with in books on the mathematical theory of elasticity, among which the treatise of Professor Love* may be mentioned. It is necessary to remark, however, that he showed how to reduce the problem to that of finding a function of the co-ordinates of points on the cross-section, which satisfies a certain partial differential equation. There is, however, no known general analytical method of finding this function for any assigned cross-section, and therefore the torsion problem cannot be solved mathematically for the great majority of technically important sections.

A simple method of determining these stresses would be of the very greatest assistance in general engineering work, and even more so in the many fresh problems which have to be dealt with in aeronautical calculations. In the very complex sections which occur in this work, such as those of airscrew blades, and the many forms of spars and struts, etc., used, it is of the highest importance that correct knowledge should be available, and therefore the authors have carried out work at the Royal Aircraft Factory, Farnborough, with a view to solving the problem by means of a simple experimental method. The following is a very brief description of the method which has been developed.

A hole is cut, in a thin plate, of the section required to be investigated, and a circular hole of a predetermined diameter is cut alongside it. The plate is placed in a box and soap films are stretched across the holes. The films are blown out slightly by reducing the air pressure on one side of them. By making suitable measurements of the shape of the resulting film surfaces, as will be explained later, it is possible to find the stresses in a bar of the given section, in terms of the stresses in a circular bar of the same diameter as the circular hole, when the two bars are twisted through the same angle per unit length. It is equally easy, by means of other measurements, to find the ratio of the torques which must be applied to the two bars in order to produce the same twist in each. It will readily be seen that by this means the most complicated sections can be dealt with.

* A. E. H. Love, *Mathematical Theory of Elasticity*, 2nd ed., chap. XIV.

The experimental work is described in the body of the paper, while the mathematical theory of the method is discussed in an Appendix.

## Experimental Methods

It is seen from the mathematical discussion given in the Appendix (p. 20), that, in order that full advantage may be taken of the information on torsion which soap films are capable of furnishing, apparatus is required with which three kinds of measurements can be made, namely:

(A) Measurements of the inclination of the film to the plane of the plate at any point, for the determination of stresses.

(B) Determination of the contour lines of the film.

(C) Comparison of the displaced volumes of the test film and circular standard for finding the corresponding torque ratio.

The available means of measurement will now be enumerated under these three heads.

(A) For this purpose optical reflection methods naturally suggest themselves. In the apparatus used by the authors, the image of an electric-lamp filament is viewed in the film in such a way that the reflected ray is coincident with the incident one, so that their common direction gives the inclination of the normal to the surface of the film. This experiment may conveniently be referred to as the measurement of angles by auto-collimation.

(B) For mapping contour lines, a steel needle point, moistened with soap solution, is arranged to move about over the plate carrying the film, its distance therefrom being adjustable by means of a micrometer screw. The point is made to approach the film till the distortion of the image in the latter shows that contact has occurred. This position is remarkably definite, so much so, indeed, that it is possible, with ordinary care, to limit the error in the measurement of the normal co-ordinate to $\pm 0{\cdot}001$ in. This method of mapping contours will be referred to as the 'spherometer' method. Another method, which was suggested to the authors by Mr Vernon Boys, F.R.S., though not so convenient as the one already described, is, nevertheless, useful in affording a ready means of exhibiting the shape of the contour lines to the eye. If a film be left undisturbed for, say, 15 min., owing to drainage and consequent thinning of the film, a black spot appears at the highest point and gradually increases in size till, after the lapse of several hours, it may include the whole surface of the film. Its edge is quite sharply defined and is horizontal. Hence, if the plate has been levelled up beforehand, the edge of the black spot coincides at any moment with a contour line of the film.

(C) The most obvious way of measuring the displaced volume of the films is to blow them up by running a known volume of water, or, preferably, soap solution, into the apparatus from a pipette or burette. The volume of the circular film may be calculated from the observed value of the inclination at its boundary, since its surface is a portion of a sphere, and hence the volume of the other film may be obtained by difference. The most accurate results are obtained by giving the film a slight initial displacement before running in the known volume of liquid, and measuring the difference of the inclinations at the boundary of the circular film.

Another method, which requires a certain amount of practice, but which has the advantage of great simplicity, is to blow up the two films, observe the angle at the edge of the circular one, and then carefully place a flat plate, moistened with soap solution, on the test film, so as to cover it completely, until the flat plate is in contact with the test-plate. The total volume is then contained in the circular film, and it can be determined in the ordinary way by again observing the inclination. Hence the volume of the test film may be found.

## Description of Apparatus

In the apparatus used by the authors, the films are formed on holes cut to the required shapes in flat aluminium plates, of no. 18 s.w.g. thickness. The plates are held in a horizontal position during the experiment, and the edges of the holes are chamfered off on the underside, to an angle of about 45°, in order to fix the plane of the boundary. The soap solution used is that recommended by Mr Boys, namely, pure sodium or potassium oleate, glycerine and distilled water. It may be obtained ready for use from Messrs Griffin, Kingsway, London.

The photograph (plate 1) shows the apparatus in which the films are formed, and also illustrates the construction of the spherometer. The test-plate is clamped between the two halves of the cast-iron box *A*. The lower part of this box takes the form of a shallow tray $\frac{1}{4}$ in. deep, blackened inside and supported on levelling screws, while the upper portion is simply a square frame carrying the clamping studs and enamelled white inside. A three-way cock communicates with the former and a plain tube with the latter. The film shown in the photograph represents a section of an airscrew blade. It will be noticed that a black spot has commenced to form at the top of the bubble.

The spherometer apparatus consists of a screw *B*, of 1 mm. pitch, passing through a hole in a sheet of plate glass $\frac{1}{8}$ in. thick and sufficiently large to cover the box in any possible position. It slides about on the flat upper face of the latter. The lower end of the screw carries a hard steel point *C*, tapering about 1 in 4, and its divided head moves beside a fixed vertical scale. Fixed above the screw and in its centre line is the steel recording point *D*. The record is made on a sheet of paper fixed to the board *E*, which can swing about a horizontal axis at the same height as *D*. To mark any position of the screw, it is merely necessary to prick a dot on the paper by bringing it down on the recording point.

In the auto-collimator (plate 1), light from the straight filament of the 2 V. bulb *A* is reflected from the surface of the film through a V-nick *B* and a pin-hole eyepiece *C*, placed close to the lamp and shaded from direct light by a small screen. The inclinometer *D*, which measures the angle which the optical axis makes with the vertical, consists of a spirit level, of 6 ft. radius, fixed to an arm which moves over a quadrant graduated in degrees. The apparatus is mounted, by means of a stiff-jointed link, on a tripod stand weighted with lead. Fine adjustment of angle is made with a screw.

### Method of using Apparatus

In using the soap-film apparatus, the test-plate and lower half of the test-box, which must both be perfectly clean, are moistened with soap solution and clamped together by means of the upper frame. The soap solution not only forms an airtight joint between the plate and box, but also serves to saturate the air within the apparatus, so that evaporation from the surface of the film is minimized. The edges of the holes are now tested with the spherometer point; if they are not parallel to the plane of motion of the glass plate they must be adjusted. A film is then drawn across the holes by means of a strip of celluloid wetted with soap solution fresh from the stock bottle and the glass cover immediately replaced. The blowing up should be done by suction from the tube in the upper frame, and not by blowing through the stopcock, as the carbon dioxide introduced by the latter method might affect the life of the film adversely. Measurements may now be made as desired. It should be remembered that if the auto-collimator is used, the apparatus must be levelled up beforehand.

In the case of the spherometer, the point, previously moistened with fresh solution, is set to a given height and made to touch the film at a number of positions, which are marked on the paper. This is repeated for as many contour lines as may be required. The plate need not be levelled. A contour map taken in this way is to be seen on the board (plate 1).

Usually, the use of the auto-collimator is confined to the determination of inclinations at given points on the boundary, which are marked by scratches on the plate. It is better for stress measurements than the contour-line method, since it gives the inclination directly, whereas in the other case the latter can only be found by a graphical differentiation. The use of the optical method may be extended to the finding of inclinations at points other than those on the boundary, with the help of the spherometer, in the following manner. The outline of the experimental hole is marked on the paper by means of the recording point, and the position of the point for which the stress is desired is added. The glass plate is adjusted until the recording point coincides with it. The needle is screwed down till it just touches the film, and its height is noted. It is then screwed back till the film breaks away and finally brought down again to within one- or two-thousandths of an inch of its former height. The auto-collimator is now adjusted till the image of the filament is seen in the film just below the needle-point. The reading of the inclinometer then gives the required angle.

### Accuracy of Results

Strictly speaking, the soap-film surface can only be taken to represent the torsion function if its inclination $\gamma$ is everywhere so small that $\sin\gamma = \tan\gamma$ to the required order of accuracy. This would mean, however, that the quantities measured would be so small as to render excessive experimental errors unavoidable. A compromise must therefore be effected. In point of fact, it has been found from experiments on sections for which the torsion function can be calculated, that the ratio of the stress at a point in any section to the stress at a point in a circular shaft, whose radius equals the value of $2A/P$ for the section, is given quite satisfactorily by the value of $\sin\gamma/\sin\mu$, where $\gamma$ and $\mu$ are the respective inclinations of the corresponding films, even when $\gamma$ is as much as 35°. Similarly, the volume ratio of the films has been found to be a sufficiently good approximation to the corresponding torque ratio, for a like amount of displacement.

In contour mapping, the greatest accuracy is obtained, with the apparatus at present in use, when $\mu$ is about 20°. That is to say, the displacement should be rather less than for the other two methods of experiment.

Table 1. *Showing experimental error in determining stress by means of soap films*

| | Section | Radius of circle (in.) | $\alpha$ (deg.) | $\beta$ (deg.) | $\alpha/\beta$ | $\sin\alpha/\sin\beta$ | True value | Error $\alpha/\beta$ (%) | Error $\sin\alpha/\sin\beta$ (%) |
|---|---|---|---|---|---|---|---|---|---|
| 1 | Equilateral triangle: height, 3 in. | 1·00 | 32·55 | 21·19 | 1·536 | 1·490 | 1·500 | +2·4 | −0·7 |
| 2 | Square: side, 3 in. | 1·5 | 29·11 | 21·34 | 1·364 | 1·337 | 1·350 | +1·0 | −1·0 |
| 3 | Ellipse: semi-axes, 2 × 1 in. | 1·296 | 30·71 | 24·32 | 1·263 | 1·240 | 1·234 | +2·4 | +0·5 |
| 4 | Ellipse: 3 × 1 in. | 1·410 | 31·10 | 24·00 | 1·296 | 1·270 | 1·276 | +1·6 | −0·5 |
| 5 | Ellipse: 4 × 0·8 in. | 1·196 | 35·35 | 26·58 | 1·331 | 1·293 | 1·286 | +3·5 | +0·5 |
| 6 | Rectangle: 4 × 2 in. | 1·333 | 31·70 | 22·36 | 1·418 | 1·380 | 1·395 | +1·6 | −1·1 |
| 7 | Rectangle: 8 × 2 in. | 1·60 | 34·83 | 27·23 | 1·279 | 1·247 | 1·245 | +2·7 | +0·2 |
| *8 | Infinitely long rectangle: 1 in. wide | 1·00 | 36·42 | 36·19 | 1·006 | 1·005 | 1·000 | +0·6 | +0·5 |

* On 4 in. length.

In all soap-film measurements the experimental error is naturally greater the smaller the value of $2A/P$. Reliable results cannot be obtained, in general, if $2A/P$ is less than about half an inch, so that a shape such as a rolled I beam section could not be treated satisfactorily in an apparatus of convenient size. As a matter of fact, however, the shape of a symmetrical soap film is unaltered if it be divided by a septum or flat plate which passes through an axis of symmetry and is normal to the plane of the boundary. It is therefore only necessary to cut half the section in the test-plate and to place a normal septum of sheet metal at the line of division. This device, for the suggestion of which the authors are indebted to the late Dr C.V. Burton, may also be employed in many other cases where contour lines are so nearly normal to the septum that they are not sensibly altered by its introduction. An I beam, for instance, might be treated by dividing the web at a distance from the flange equal to two or three times the thickness of the web. It has been found advisable to carry the septum down through the hole so that it projects about $\frac{1}{8}$ in. below the underside of the plate, as, otherwise, solution collects in the corners and spoils the shape of the film.

The values set down in table 1 indicate the degree of accuracy obtainable with the auto-collimator in the determination of the maximum stresses in sections for which the torsion function is known. They also give an idea of the sizes of holes which have been found most convenient in practice. The angles given are ($\alpha$) the maximum inclination at the edge of the test film, and ($\beta$) the inclination at the edge of the circular film of radius $2A/P$. They are usually the means of about five observations and are expressed in decimals of a degree.

The last two columns show the errors due to taking the ratio of angles and the ratio of sines respectively as giving the stress ratio.

The error is always positive for $\alpha/\beta$, and its mean value is 1·98 %. In the case of $\sin\alpha/\sin\beta$ the average error is only 0·62 %. In only two instances does the error reach 1 %, and in both it is negative. The presence of sharp corners seems to introduce a negative error which is naturally greatest when the corners are nearest to the observation point. Otherwise, there is no evidence that the error depends to any great extent on the shape. Nos. 4, 5, 7 and 8 in the table are examples of the application of the method of normal septa.

Table 2 shows the results of volume determinations made on each of the sections 1–8 given in the previous table.

Table 2. *Showing experimental error in determining torques by means of soap films*

| No. | Section | Maximum inclination (deg.) | Observed volume ratio | Calculated torque ratio | Error (%) |
|---|---|---|---|---|---|
| 1 | Equilateral triangle: height, 3 in. | 32·06 | 1·953 | 1·985 | −1·6 |
| 2 | Square: side, 3 in. | 30·39 | 1·416 | 1·432 | −1·1 |
| 3 | Ellipse: semi-axes, 2 × 1 in. | 30·50 | 1·143 | 1·133 | +0·9 |
| 4 | Ellipse: 3 × 1 in. | 31·01 | 2·147 | 2·147 | 0 |
| 5 | Ellipse: 4 × 0·8 in. | 36·12 | 3·041 | 3·020 | +0·7 |
| 6 | Rectangle: sides, 4 × 2 in. | 31·33 | 1·456 | 1·475 | −1·3 |
| 7 | Rectangle: 8 × 2 in. | 35·28 | 1·749 | 1·744 | +0·3 |
| *8 | Infinitely long rectangle | 36·00 | 0·858 | 0·848 | +1·2 |

* On 4 in. length.

The average error is 0·89 %. In four of the eight cases considered the error is greater than 1 %, and in three of these it is negative. One may conclude that the probable error is somewhat greater than it is for the stress measurements, and that it tends to be negative. Its upper limit is probably not much in excess of 2 %. The remarks already made regarding the dependence of accuracy on the shape of the section apply equally to torque measurements.

As additional confirmation of the correctness of solutions of the torsion problem obtained by the soap-film method, some experiments on wooden beams may be cited. In the first of these, a walnut plank was shaped so that its section was exactly the same as the hole in one of the test-plates, which represented a section of an air-screw blade, of fineness ratio 10·55, having its thickest part about a third of the way from the leading edge. The value of the modulus of rigidity, $N$, was found by performing a torsion test on this plank, using the expression for the torque given by a soap-film experiment on the plate which was used in shaping the plank. $N$ was found to be $0{\cdot}1355 \times 10^6$ lb. per sq.in. Five circular rods were then cut from the plank and their rigidities were measured. The mean value of $N$ found in this way was $0{\cdot}1387 \times 10^6$, a difference of only 2·3 %.

Similar experiments were made on three lengths of spruce wing-spar, of I section. The results are set down in table 3. Column A shows the value of $N$, obtained by twisting the spar, using the figure for torque obtained by soap films. Columns B and C show the values of $N$ found from round specimens cut from the thickest part of the two flanges, while column D gives the percentage difference between A and the mean of B and C.

Table 3. *Comparison of soap-film results with those of direct torsion experiments*

| Spar no. | A<br>lb. per sq.in. | B<br>lb. per sq.in. | C<br>lb. per sq.in. | D<br>(%) |
|---|---|---|---|---|
| 1 | $0{\cdot}1091 \times 10^6$ | $0{\cdot}1172 \times 10^6$ | $0{\cdot}1063 \times 10^6$ | 2·5 |
| 2 | 0·0873 | 0·0640 | 0·0966 | 8·7 |
| 3 | 0·1156 | 0·1200 | 0·1151 | 1·7 |

The comparatively large discrepancy in no. 2 is probably due to the extraordinarily large variation of $N$ over this particular spar.

When contour lines have been mapped, the torque may be found from them by integration. If the graphical work is carefully done, the value found in this way is rather more accurate than the one obtained by the volumetric method. Contours may also be used to find stresses by differentiation, that is, by measuring the distance apart of the neighbouring contour lines; but here the comparison is decidedly in favour of the direct process, owing to the difficulties inseparable from graphical differentiation. The contour map is, nevertheless, a very useful means of showing the general nature of the stress distribution throughout the section in a clear and compact manner. The highly stressed parts show many lines bunched together, while few traverse the regions of low stress, and the direction of the resultant stress is shown by that of the contours at every point of the section. Furthermore, the map solves the torsion problem, not only for the boundary, but also for every section having the same shape as a contour line.

## Experimental Results

The two examples which follow serve to illustrate the use of the soap-film apparatus in solving typical problems in design:

(1) It is well known that the stress at a sharp internal corner of a twisted bar is infinite or, rather, would be infinite if the elastic equations did not cease to hold when the stress becomes very high. If the internal corner is rounded off the stress is reduced; but so far no method has been devised by which the amount of reduction in strain due to a given amount of rounding can be estimated. This problem has been solved by the use of soap films.

An L-shaped hole was cut in a plate. Its arms were 5 in. long by 1 in. wide, and small pieces of sheet metal were fixed at each end, perpendicular to the shape of the hole, so as to form normal septa. The section was then practically equivalent to an angle with arms of infinite length. The radius in the internal corner was enlarged step by step, observations of the maximum inclination at the internal corner being taken on each occasion.

The inclination of the film at a point 3·5 in. from the corner was also observed, and was taken to represent the mean boundary stress in the arm, which is the same as the boundary stress at a point far from the corner. The ratio of the maximum stress at the internal corner to the mean stress in the arm was tabulated for each radius on the internal corner. The results are given in table 4.

Table 4. *Showing the effect of rounding the internal corner on the strength of a twisted L-shaped angle beam*

| Radius of internal corner (in.) | Ratio: $\frac{\text{maximum stress}}{\text{stress in arm}}$ | Radius of internal corner (in.) | Ratio: $\frac{\text{maximum stress}}{\text{stress in arm}}$ |
|---|---|---|---|
| 0·10 | 1·890 | 0·70 | 1·415 |
| 0·20 | 1·540 | 0·80 | 1·416 |
| 0·30 | 1·480 | 1·00 | 1·422 |
| 0·40 | 1·445 | 1·50 | 1·500 |
| 0·50 | 1·430 | 2·00 | 1·660 |
| 0·60 | 1·420 | | |

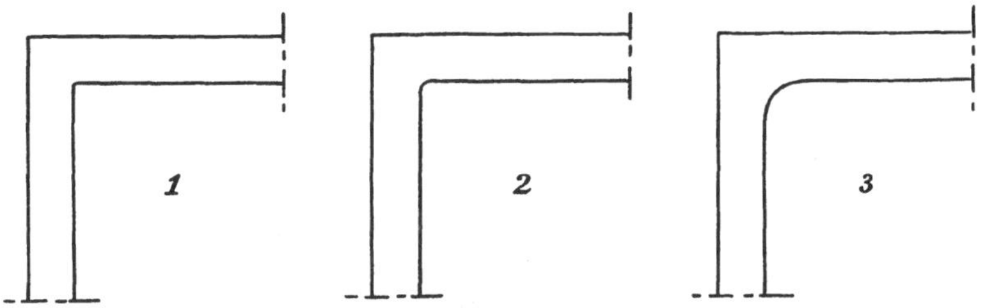

Fig. 1. Stress in internal corner.

It will be seen that the maximum stress in the internal corner does not begin to increase to any great extent till the radius of the corner becomes less than one-fifth of the thickness of the arms. A curious point which will be noticed in connection with the table is the minimum value of the ratio of the maximum stress to the stress in the arm which occurs when the radius of the corner is about 0·7 of the thickness of the arm.

In fig. 1 is shown a diagram representing the appearance of these sections of angle-irons.

No. 1 is the angle-iron for which the radius of the corner is one-tenth of the thickness of the arm. This angle is distinctly weak at the corner.

In no. 2 the radius is one-fifth of the thickness. This angle-iron is nearly as strong as it can be. Very little increase in strength is effected by rounding off the corner more than this. No. 3 is the angle with minimum ratio of stress in corner to stress in arm.

A further experiment was made to determine the extent of the region of high stress in angle-iron no. 1. For this purpose contour lines were mapped, and from

Table 5. *Showing the rate of falling off of the stress in the internal corner of the angle-iron*

| Distance from boundary (in.) | Ratio: $\frac{\text{stress at point}}{\text{boundary stress in arm}}$ | Distance from boundary (in.) | Ratio: $\frac{\text{stress at point}}{\text{boundary stress in arm}}$ |
|---|---|---|---|
| 0·00 | 1·89 | 0·30 | 0·49 |
| 0·05 | 1·36 | 0·40 | 0·24 |
| 0·10 | 1·12 | 0·50 | 0·00 |
| 0·20 | 0·77 | | |

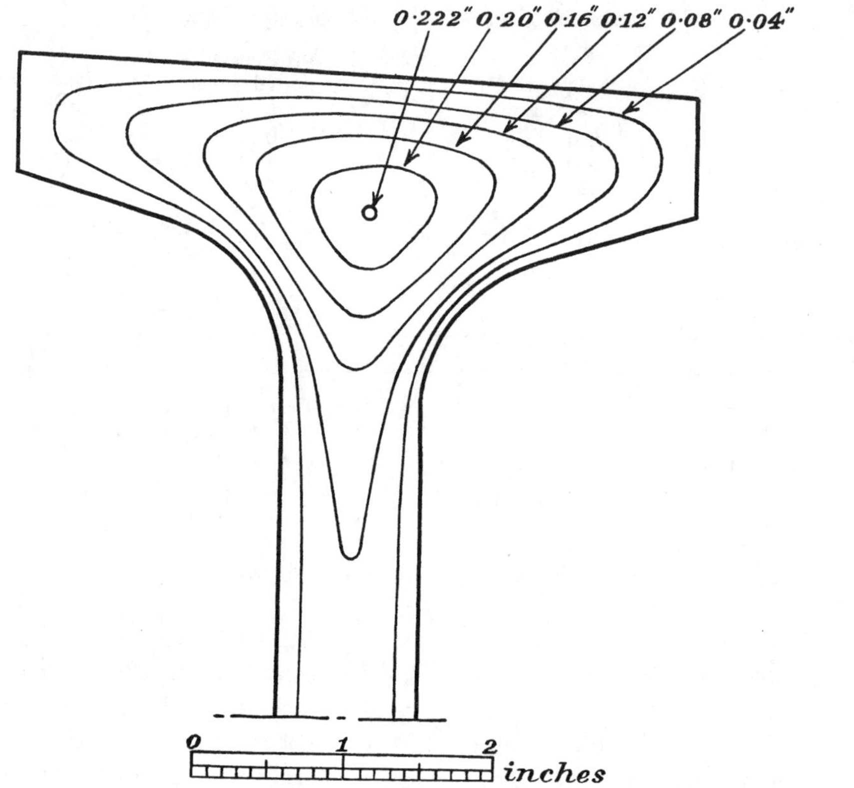


Fig. 2. Lines of shearing stress in the torsion of a wooden spar to scale. The figures give the heights of the contour-lines of the corresponding soap film. Stress at any point $= 2{\cdot}70\ N\tau \sin \nu$ lb. per sq.in., where $\nu$ is the inclination of the film. Torque on half section $= 4{\cdot}09\ N\tau$ in.lb.

these the slope of the bubble was found at a number of points on the line of symmetry of the angle-iron. Hence the stresses at these points were deduced. The results are given in table 5.

It will be seen that the stress falls off so rapidly that its maximum value is to all intents and purposes a matter of no importance, if the material is capable of yielding. If the material is brittle and not ductile a crack would, of course, start at the point of maximum stress and penetrate the section.

(2) The diagram shown in fig. 2, which represents the half-section of a wooden wing spar, is a good example of the contour-line method. The close grouping of the lines near the internal radii, denoting high stress, is immediately evident. The projecting parts of the flange are lightly stressed and contribute little to the torsional stiffness. The stress at the middle of the upper face is, however, considerable, being in fact next in order of magnitude below that in the radii. The stress near the middle of the web is practically constant and equal to that in a very long rectangular section of the same thickness under the same twist.

A further point of interest is that the 'unstressed fibre' is very near the centre of the largest circle which can be inscribed in the section. It will also be observed that the three points of greatest stress are almost coincident with the points of contact of the circle. The maximum stress is about 1·89 times the mean boundary stress.

The figures given below the diagram for the values of the stress and torque on the section fully confirm the generally accepted notions regarding the extreme weakness of I beams in torsion.

## General Deductions from the Soap-film Analogy

One of the greatest advantages of considering the torsion problem from the soap-film standpoint arises from the circumstance that it is very much easier to form a mental picture of a soap bubble than it is to visualize the complicated system of shear-strains in a twisted bar. It cannot be too strongly urged that the surest way of forming a clear idea of the nature of the torsion problem is to blow a few soap films on boundaries of various shapes. This can be done with the simplest of apparatus; the holes may be cut in plates of thin sheet metal, which can be luted on to the top of a biscuit tin with vaseline or soft soap. To blow the films up it is only necessary to bore a hole in the bottom of the tin and stand it in a vessel containing water. Two sections may readily be compared by cutting them in the same plate. A simple way of estimating inclinations is to view the image of the eye in the film and adjust the arm of a clinograph so that it lies along the line of sight. Black spots, as previously mentioned, may be observed if arrangements are made to cover the films with a sheet of glass, in order to exclude dust and air currents.

With the aid of simple apparatus of this kind the truth of theorems, such as those contained in the following list, may be readily demonstrated:

(*a*) The stress distribution (and therefore the torque) for any section is independent of the axis of twist. This is easily seen, since the shape of the soap film is completely determined by the boundary and the value of $4S/p$. Hence the torque on a number of bars clamped together at their ends may be found by adding the separate torques which would be necessary to twist each through the same angle. This, as in other cases, applies to torsion only. It will be realized that in practice there will be bending stresses which must be taken account of in the usual way.

(*b*) Any addition of material to a section must increase the torque, and vice versa, so long as the distribution of material in the original section is unaltered.

(*c*) Any cut made in a section, whether it decreases the area or not, must decrease the torque.

(*d*) The stress at any point of the boundary of a section is never less than the boundary stress in a circular bar under the same twist, whose radius is equal to that of the circle inscribed in the section, which touches the boundary at the point in question.

More generally, if one section lie entirely inside another, so as to touch it at two or more points, the stresses in the inner figure are less than those in the outer one at the points of contact; if the two figures are approximately congruent in the neighbourhood of the points of contact, the difference between the stresses is small. The maximum stress in a section is not greater than $2aN\tau$, where $a$ is the radius of the largest inscribed circle, unless the boundary is concave, that is, re-entrant.

(*e*) If a concave part of the boundary approximates to a sharp corner, the stress at this point may be very high, and if the curvature is infinite then the stress is also theoretically infinite, whatever be the situation of the corner with respect to the rest of the section. Actually, of course, if the material is ductile, we can only deduce that the stresses at such a corner are above the elastic limit.

(*f*) It is a consequence of (*e*) that it does not necessarily follow that the making of a cut in a section will reduce its strength, whether material is removed or not. As an example of this, one may quote the case of an angle-iron in which the internal corner is quite sharp. It is well known in practice that this will often fracture. It may be strengthened, however, by reducing the section, planing out a semicircular groove at the root of the angle-iron.

(*g*) There can be no discontinuous changes of stress anywhere in a section, excepting only those parts of the boundary where the curvature is infinite (concave or convex sharp corners).

(*h*) The maximum stress occurs at or near one of the points of contact of the largest inscribed circle, and not, in general, at the point of the boundary nearest the centroid, as has been hitherto assumed. An exception may occur if, at some other part of the boundary, the curvature is (algebraically) considerably less (that is, the boundary is more concave) than it is at this point.

(*i*) If a section which is long compared with its greatest thickness be bent so that its area and the length of its median line are unchanged, its torque will not be greatly changed thereby. For instance, the torsional stiffness of a metal plate is practically unaltered by folding or rolling it up into the form of an L or a split tube. Soap-film experiments show, in fact, that there is a diminution of less than 5 % when the inner radius of the boundary is not less than the thickness at the bend.

(*j*) The 'unstressed fibre', which is situated at the point corresponding with the maximum ordinate of the soap film, is near the centre of the largest circle which can be inscribed in the section.

In general, the inscribed circle has a maximum value wherever it touches the boundary at more than two points, and there is usually an unstressed fibre near the centre of each of these circles. Between each pair of maximum ordinates on the soap film, however, there is a 'minimax' point, which is near the centre of the corresponding minimum inscribed circle. This fibre in the bar is also unstressed.

(*k*) The 'lines of shearing stress' round the unstressed fibres of the first sort are initially ellipses, and round those of the second sort hyperbolae, from which shapes they gradually approximate to that of the boundary. Notions of this sort are useful in practice, because it is possible, with their help, to sketch in the general nature of the lines of shearing stress for any section.

## Approximate Formulae for Torques and Stresses

The torque on any section is given by

$$T = N\tau C,$$

where $C$ is a quantity of the fourth degree in the unit of length, which may be called the torsional stiffness of the section.

In the case of a circular shaft, in which there is no distortion of cross-sections, $C$ is equal to the polar moment of inertia, so that we have

$$C = \tfrac{1}{2}Ar^2,$$

where $r$ is the radius of the circle.

In the general case we may put

$$C = \tfrac{1}{2}Ak^2.$$

$k$ is a length, which, by analogy with the circle, may be called the 'equivalent torsional radius' of the section.

It is seen (see Appendix, p. 20) that the mean stress round the boundary of any section is equal to the boundary stress in a circular shaft whose radius equals the quantity $2A/P$, which we have called $h$. This result suggested that some fairly simple approximate relation might be found between $h$ and $k$.

When this idea was tested by application to known results, it became immediately evident that the fraction $k/h$ was not very different from unity for a large number of sections. It was observed, however, that the presence of sharp outwardly projecting corners tended to make $k$ greater than $h$, while the opposite effect was noticed in the case of sections whose length was great compared with their greatest thickness. For instance, $h$ for the square is equal to $a$, the radius of the inscribed circle, whereas $k$ is about 6 % greater. In the equilateral triangle $h$ is still equal to $a$, while $k$ is 9 % greater. For long rectangles and ellipses, however, $k$ is considerably less than $h$.

At first sight, since, in many sections, these two effects are operating simultaneously, it might be thought that their separation, with a view to formulating a method of finding $k$ empirically, would be a matter of some difficulty. It has been accomplished, however, by a process of successive trial, with the result that the empirical treatment about to be described has been evolved. The curves giving the values of the constants were found by plotting the values they should have for all the sections, for which a solution has been obtained, in order to get the correct result, and then drawing the best curves through these points.

If the figure contains sharp, outwardly projecting corners, construct a new figure by rounding off each corner with a radius $r$, which is a certain fraction of $a$, the radius of the largest inscribed circle. The value of this fraction depends on the angle $\theta$, turned through by the tangent to the boundary in passing round the corner in question. In fig. 3 (p. 15), $r/a$ is shown graphically as a function of $\theta/\pi$, and, in addition, a table of values is subjoined:

| $\theta/\pi$ | $r/a$ | $\theta/\pi$ | $r/a$ |
|---|---|---|---|
| 0·0 | 1·00 | 0·6 | 0·375 |
| 0·1 | 0·93 | 0·7 | 0·270 |
| 0·2 | 0·85 | 0·8 | 0·210 |
| 0·3 | 0·75 | 0·9 | 0·170 |
| 0·4 | 0·625 | 1·0 | 0·155 |
| 0·5 | 0·500 | | |

If the area of this new figure be called $A_1$, and its perimeter $P_1$, the value of $2A_1/P_1$ is a close approximation to the $k$ of the original boundary, subject to the second modification, which must be made for long sections.

It is not difficult to see that a certain amount of common sense may be required in applying the above rule. For instance, if the figure has a projection which is slightly rounded instead of being quite sharp, the value of $r$ is that which would be used if the projection did run out to a sharp point. In most cases of this sort, however, it is found that the correction makes little difference.

The criterion, which has been adopted for fixing the value of the correction factor for long sections, is the fraction $a/h$. Where this is appreciably less than unity, the stiffness calculated by the process already described should be multiplied by the correction factor $K$, which is given in table 6, and which is also shown graphically in fig. 4.

The expression for $C$ now takes the form

$$C=\tfrac{1}{2}KA\left(\frac{2A_1}{P_1}\right)^2.$$

This formula is quite satisfactory for figures such as triangles, squares, ellipses, etc., in which $a$ has one maximum value only, which may be called 'simple sections', but if $a$ has more than one maximum the solution in its present form is ambiguous, and it is necessary to split the section up into two or more parts, which will be referred to as the 'components' of the original figure. The stiffness of each component must be found separately and the total stiffness obtained by addition.

In order to evolve a method of division, it is to be noted that the process already described is based on the equality of the resultant air pressure and surface-tension forces acting on the analogous soap film. If the film is divided by a series of 'normal septa', which are so arranged that they are everywhere at right angles to the contour lines which they cut, the equilibrium equations are in no way altered, and the

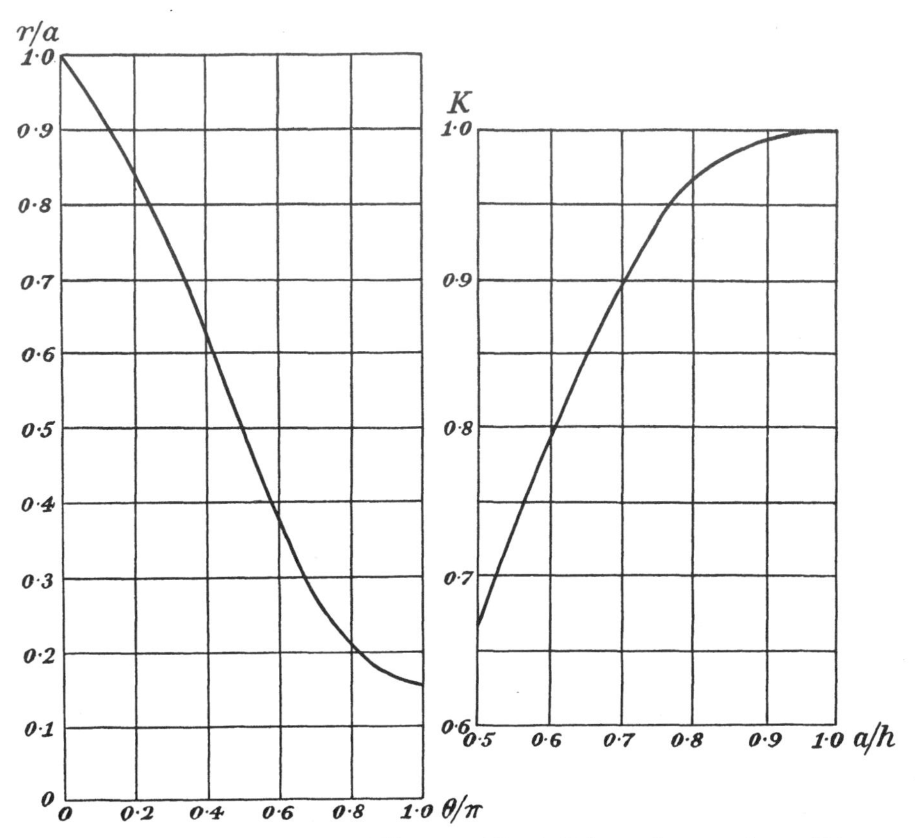


Fig. 3. Values of $r/a$ in terms of $\theta/\pi$.

Fig. 4. Values of torque factor $K$ in

$$T = KN\tau\frac{2A_1^2}{P_1^2}A.$$

Table 6

| $a/h$ | $K$ | $a/h$ | $K$ |
|---|---|---|---|
| 1·00 | 1·000 | 0·70 | 0·897 |
| 0·95 | 0·998 | 0·65 | 0·848 |
| 0·90 | 0·994 | 0·60 | 0·793 |
| 0·85 | 0·984 | 0·55 | 0·732 |
| 0·80 | 0·966 | 0·50 | 0·667 |
| 0·75 | 0·938 | | |

theorem is still true of each separate part of the film. Hence, if the section is divided in this way, the empirical treatment explained above should be applicable to each component. It is to be noted, however, that the term 'perimeter' must be taken to mean that part of the boundary of the component which formed part of the perimeter of the original figure. The remainder is not, strictly speaking, part of the boundary at all. It remains to formulate rules for the division of these 'compound' sections.

Imagine a circle to be drawn in the section so as to *touch* the boundary at two points. Now let the centre of this circle move through the figure, the radius being varied simultaneously so that there is always contact at two points. At some places

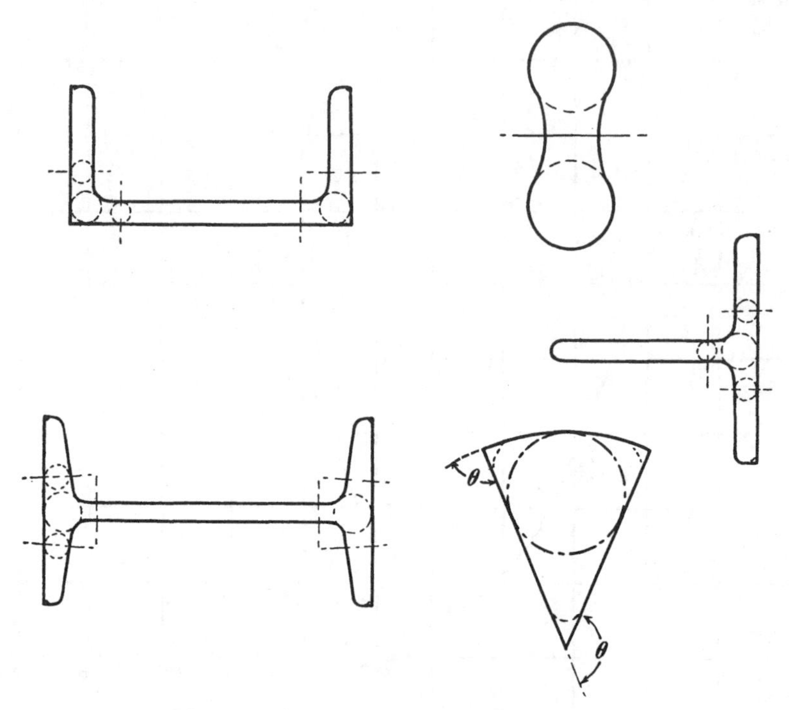


Fig. 5. Subdivision of compound sections.

the circle will touch at three (or more) points. It is then an 'inscribed circle of maximum radius', and between every such pair of maxima there must be a position where the radius is a minimum. The section should be divided by straight lines passing through the points of contact of these minimum circles.

In some cases, such as the web of an I beam, there is a long thin parallel portion, and the position of the minimum circle is indeterminate. Here the line of division should be at a distance from the commencement of the parallel part equal to half its thickness. The portion of the web cut out may be treated separately as part of an infinitely long thin rectangle, the torsional properties of which are well known. If the piece cut off is 'closed' at the other end (e.g. the arm of an angle), it may be treated as a separate component. It is advisable to cut off all long, thin, projecting parts in the same way, even though the sides are not quite parallel. The tapering flanges of I beams may be cited in illustration.

Fig. 5 shows some typical examples of the subdivision of compound sections, and also illustrates the rounding off of sharp corners. The I beam, for instance, has seven

PLATE 1

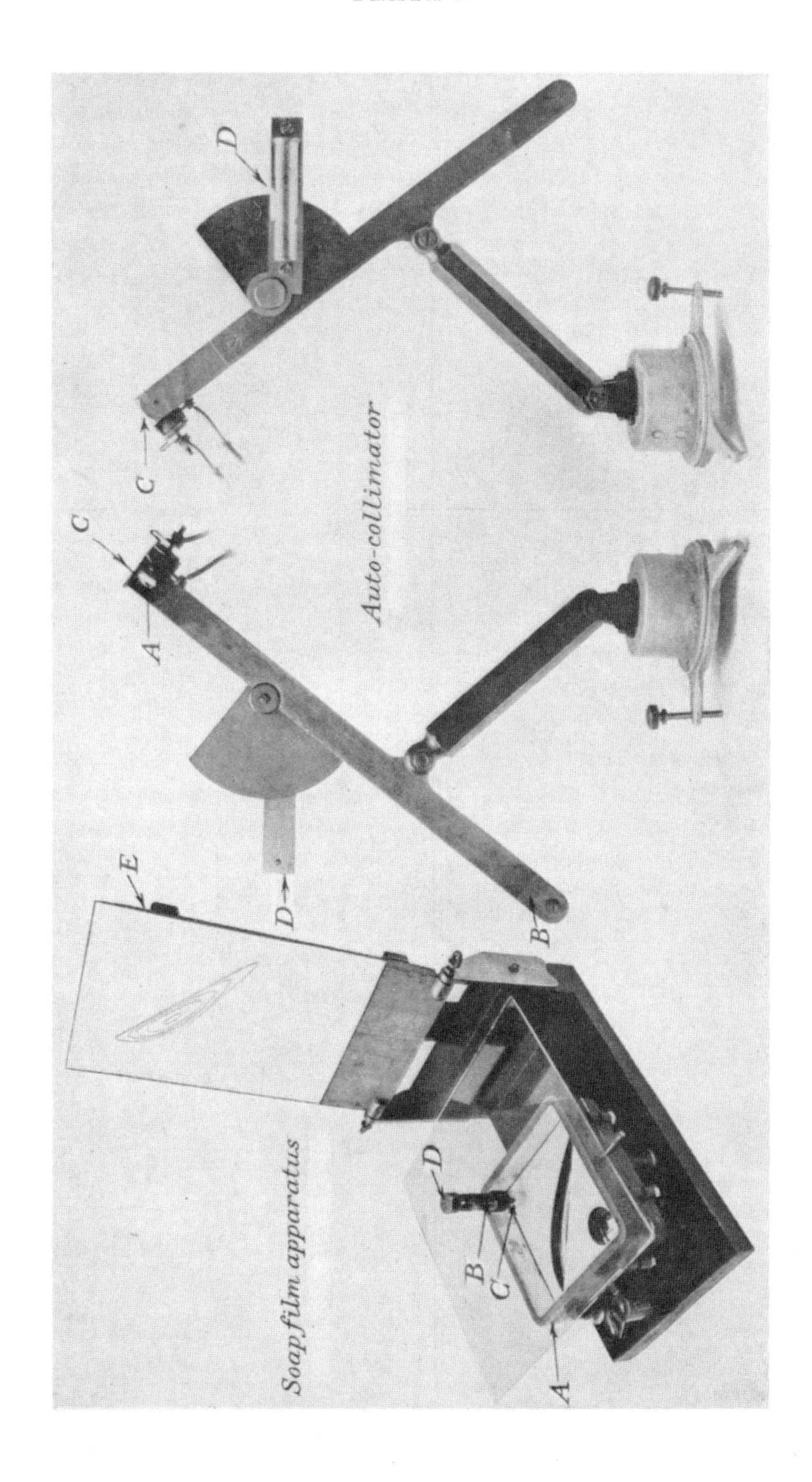

components, the channel five, and the tee four. In the 45° sector there is only one component. The angle turned through at the apex is 135°, so that $\theta/\pi = 0{\cdot}75$. Hence, from table 5 (p. 10), $r = 0{\cdot}24a$ ($a$ is the radius of the chain-dotted circle). At the other two corners $\theta/\pi = 0{\cdot}5$, and hence $r = 0{\cdot}5a$.

In the case of certain sections, another form of expression may be arrived at by a more direct method. Consider a soap film on a long narrow slit of varying width. If the rate of change of width with length is nowhere large, we may neglect the longitudinal curvature $\delta^2 z/\delta x^2$ of the film, and put the transverse curvature $\delta^2 z/\delta y^2$

Table 7

| Section | $C$ (formula) | $C$ (calculated) | Error (%) |
|---|---|---|---|
| Square: side $2s$ | $2{\cdot}249S^4$ | $2{\cdot}249s^4$ | 0 |
| Rectangle: sides $2b$, $3b$ | $4{\cdot}710b^4$ | $4{\cdot}698b^4$ | $+0{\cdot}26$ |
| sides $2b$, $4b$ | $7{\cdot}320b^4$ | $7{\cdot}318b^4$ | $-0{\cdot}03$ |
| sides $2b$, $10b$ | $23{\cdot}15b^4$ | $23{\cdot}31b^4$ | $-0{\cdot}68$ |
| sides $2b$, $2l$ ($l \to \infty$) | $\frac{16}{3}lb^3$ | $\frac{16}{3}lb^3$ | 0 |
| Ellipse: axes $2b$, $3b$ | $3{\cdot}250b^4$ | $3{\cdot}260b^4$ | $-0{\cdot}31$ |
| axes $2b$, $4b$ | $5{\cdot}035b^4$ | $5{\cdot}025b^4$ | $+0{\cdot}20$ |
| axes $2b$, $10b$ | $15{\cdot}14b^4$ | $15{\cdot}10b^4$ | $+0{\cdot}26$ |
| axes $2b$, $2l$ ($l \to \infty$) | $3{\cdot}235lb^3$ | $\pi lb^3$ | $+3{\cdot}00$ |
| Equilateral triangle: side $2s$ | $0{\cdot}3476S^4$ | $0{\cdot}3464S^4$ | $+0{\cdot}27$ |
| 45° sector: radius $R$ | $0{\cdot}01810R^4$ | $0{\cdot}01815R^4$ | $-0{\cdot}27$ |
| 90° sector: radius $R$ | $0{\cdot}0830R^4$ | $0{\cdot}0824R^4$ | $+0{\cdot}73$ |
| Curtate sector: 180°, $R_1 = 2R_0$ | $1{\cdot}355R_0^4$ | $1{\cdot}369R_0^4$ | $-1{\cdot}02$ |

equal to a constant $R^{-1}$, say. If the width at a distance $x$ from one end be $y$, and the total length $l$, we readily obtain the volume, $V$, of the film in the form

$$V = \frac{1}{12R}\int_0^l y^3\,dx.$$

This result must be exact for an indefinitely long rectangle, whence we have, by comparison with the known stiffness of the latter,

$$C = \frac{1}{3}\int_0^l y^3\,dx = I, \quad \text{say},$$

for the torsional stiffness of any long thin section.

A consideration of the case of ellipses suggests the modification

$$C = \frac{I}{1 + 4\dfrac{I}{Al^2}}$$

to allow for the longitudinal curvature of the figure.

The expression is now exact for all ellipses, whatever their fineness ratio, and, as will be seen, its error is within the limits of accuracy of soap-film measurements, for sections such as those of airscrew blades, down to a fineness ratio of two, at least.

The formula may also be applied, though with somewhat less accuracy, to thin sections having a curved median line, provided that $x$ is measured along the latter and $y$ at right angles thereto.

Tables 7, 8 and 9 have been prepared to indicate the degrees of accuracy which may be expected in the application of the preceding formulae.

Table 8

| Section | $C$ (formula) | $C$ (soap film) |
|---|---|---|
| Wing spar: $2\frac{1}{2} \times 1\frac{1}{4}$ in. (I section) | 0·0678 in.$^4$ | 0·0680 in.$^4$ |
| $3 \times 1\frac{1}{2}$ in. (I section) | 0·1042 in.$^4$ | 0·1051 in.$^4$ |
| Angle: $3 \times 3$ in. | 0·478 in.$^4$ | 0·487 in.$^4$ |
| Airscrew section: A | 11·70 in.$^4$ | 11·72 in.$^4$ |
| B | 7·44 in.$^4$ | 7·50 in.$^4$ |
| C | 2·42 in.$^4$ | 2·38 in.$^4$ |
| D | 0·846 in.$^4$ | 0·835 in.$^4$ |

Table 9

| Section | $C$ (formula) | $C$ (experiment) |
|---|---|---|
| Angle: $1·175 \times 1·175$ in. | 0·01234 in.$^4$ | 0·01284 in.$^4$ |
| $1·00 \times 1·00$ in. | 0·00440 in.$^4$ | 0·00455 in.$^4$ |
| Tee: $1·58 \times 1·58$ in. | 0·01451 in.$^4$ | 0·01481 in.$^4$ |
| I beam: $5·01 \times 8·02$ in. | 1·160 in.$^4$ | 1·140 in.$^4$ |
| $3·01 \times 3·00$ in. | 0·1179 in.$^4$ | 0·1082 in.$^4$ |
| $1·75 \times 4·78$ in. | 0·0702 in.$^4$ | 0·0635 in.$^4$ |
| Channel: $0·97 \times 2·00$ in. | 0·0175 in.$^4$ | 0·0139 in.$^4$ |

In table 7, comparison is made with the results of Saint-Venant's exact analysis; in table 8 the second column of values has been obtained from soap-film measurements; while in table 9 the results of the method are compared with those of some direct torsion experiments on rolled beams, carried out by Mr E. G. Ritchie.*

It will be seen that all the figures in table 9 show good agreement with the exception of those referring to the last three beams. In view of the remarks made by the author cited in regard to the want of homogeneity of rolled beams, and more particularly the comparative weakness of the metal in the internal radii, the discrepancy in these cases cannot be considered unsatisfactory.

The method of calculating $C$ should be chosen according to the nature of the section.

If there is only one maximum inscribed circle, and the section is not a long thin one, proceed by the method of rounding off sharp corners and finding $2A_1/P_1$, etc.

If the section is compound, divide it into its components and then proceed as before. Alternatively, if some of the components are thin compared with their length, they may be dealt with by finding $\int y^3 dx$.

If the median line of the section is long in comparison with the greatest thickness, straighten out the median line where necessary and use the $\int y^3 dx$ method.

* *A Study of the Circular Arc Bow Girder*, by Gibson and Ritchie (Constable and Co. 1914).

## Estimation of Stresses

The empirical calculation of the stress at any given point of a section is naturally a matter of greater difficulty than the determination of torques. If the section contains no re-entrant angles, the stresses at the three points of contact of the inscribed circle of maximum radius $a$ are usually given sufficiently well by the expression

$$\frac{2a}{1+m^2}\left[1+0{\cdot}15\left(m^2-\frac{a}{\rho}\right)\right],$$

where $m$ is the quantity $\pi a^2/A$ and $\rho$ is the radius of curvature of the boundary.

In the case of a 'compound' section, the formula may be applied to each component separately.

Table 10

| Section | Stress/$N\tau$ (formula) | Stress/$N\tau$ (true) |
|---|---|---|
| Ellipse: axes $2a$, $2b$ | $\frac{2ab^2}{(a^2+b^2)}$ | $\frac{2ab^2}{(a^2+b^2)}$ |
| Square: side $2s$ | $1{\cdot}35s$ | $1{\cdot}35s$ |
| Rectangle: sides $2s$, $3s$ | $1{\cdot}64s$ | $1{\cdot}69s$ |
| sides $2s$, $4s$ | $1{\cdot}77s$ | $1{\cdot}86s$ |
| sides $2s$, $8s$ | $1{\cdot}94s$ | $1{\cdot}99s$ |
| sides $2s$ | $2{\cdot}00s$ | $2{\cdot}00s$ |
| Equilateral triangle: sides $2\sqrt{3}s$ | $1{\cdot}53s$ | $1{\cdot}50s$ |
| Wing spar (I) $a=1{\cdot}05$ in. | 2·14 | 2·13* |
| Wing spar (I) $a=1{\cdot}27$ in. | 2·60 | 2·58* |

* Determined by soap films.

The mean value of the stress round the boundary of any component is accurately equal to $2N\tau A/P$. By combining this value with those obtained for the maximum stresses, and bearing in mind the general properties of soap films, it is possible to sketch in a boundary stress diagram for the component, with sufficient accuracy for most purposes.

Obviously the formula cannot be expected to apply to points where the boundary is concave—that is, re-entrant angles, since it fails to differentiate between an acute re-entrant angle and an obtuse one. It is possible to devise a formula which will take account of this angle and which will fit any assigned number of observed results within, say, 4 or 5 %, but such a formula naturally becomes more complicated as its range of application is increased, and hence the practical utility of such generalization is doubtful. Probably the most satisfactory way of dealing with re-entrant sections is to make soap-film measurements and to deduce, from these, formulae or curves which apply to one particular class of figure only.

It should be mentioned, however, that the formula given has been found to agree with soap-film measurements on a number of re-entrant sections, in which the angle is approximately a right angle, when $\rho$ is not very small. I beams, channels and tees are examples of such sections, to which the formula may be applied. It should be borne in mind, however, that $\rho$ is now negative.

2-2

The stress at any point of a rolled standard section may be taken to be $2aN\tau$, where $a$ is the radius of the inscribed circle which touches at that point, except at places near the end of a flange, where the stress is smaller. The same thing holds for figures such as airscrew sections, when the fineness ratio is greater than about eight.

## MATHEMATICAL APPENDIX

The solution of the problem of torsion can be made to depend (see the book referred to in the introduction) on the finding of a function, $\psi$, of $x$ and $y$, the co-ordinates of points on the cross-section, which satisfies the partial differential equation

$$\frac{\delta^2\psi}{\delta x^2}+\frac{\delta^2\psi}{\delta y^2}+2=0 \tag{1}$$

at all points of the cross-section, and is zero at all points on the bounding curve.

Consider the equations which represent the surface of a soap film stretched over a hole of the same size and shape as the cross-section of the twisted bar, cut in a flat plate, the film being slightly displaced from the plane of the plate by a small pressure $p$.

If $S$ be the surface tension of the soap solution, the equation of the surface of the film is

$$\frac{\delta^2 z}{\delta x^2}+\frac{\delta^2 z}{\delta y^2}+\frac{p}{2S}=0, \tag{2}$$

where $z$ is the displacement of the film and $x$ and $y$ are the same as before. Round the boundary, of course, $z=0$.

It will be seen that if $z$ is measured to such a scale that $\psi=4Sz/p$, then the two equations are identical. It appears, therefore, that the value of $\psi$ corresponding with any values of $x$ and $y$ can be found by measuring the quantities $p/S$ and $z$ on the soap film.

To put the matter in another light, the soap film is a graphical representation of the function $\psi$ for the given cross-section. Actual values of $\psi$ can be obtained from it by multiplying the ordinates by $4S/p$.

If $N$ is the modulus of rigidity of the material and $\tau$ the twist per unit length of the bar, the shear stress at any point of the cross-section can be found by multiplying the slope of the $\psi$ surface at the point by $N\tau$, so that, if $\gamma$ is the inclination of the bubble to the plane of the plate, the stress is

$$f_s=\frac{4S}{p}N\tau\gamma. \tag{3}$$

The torque $T$ on the bar is given by

$$T=2N\tau\iint\psi\,dx\,dy$$

or

$$T=\frac{8S}{p}N\tau V, \tag{4}$$

where $V$ is the volume enclosed between the film surface and the plane of the plate.

The contour lines of the soap film in planes parallel to the plate correspond to the 'lines of shearing stress' in the twisted bar, that is, they run parallel to the direction of the resultant shear stress at every point of the section.

It is evident that the torque on and stresses in a twisted bar of any section whatever may be obtained by measuring soap films in these respects.

In order to obtain quantitative results, it is necessary to find the value of $4S/p$ in each experiment. This might be done by measuring $S$ and $p$ directly, but a much simpler plan is to determine the curvature of a film, made with the same soap solution, stretched over a circular hole and subjected to the same pressure difference, $p$, between its two surfaces, as the test film.

The curvature of the circular film may be measured by observing the maximum inclination of the film to the plane of its boundary.

If this angle be called $\mu$, then

$$\frac{4S}{p}=\frac{h}{\sin\mu}, \tag{5}$$

where $h$ is the radius of the circular boundary.

The most convenient way of ensuring that the two films shall be under the same pressure is to make the circular hole in the same plate as the experimental hole.

It is evident that, since the two films have the same constant $4S/p$, we may, by comparing inclinations at any desired points, find the ratio of the stresses at the corresponding points of the cross-section of the bar under investigation to the stresses in a circular shaft of radius $h$ under the same twist. Equally, we can find the ratio of the torques on the two bars by comparing the displaced volumes of the soap films. This is, in fact, the form which the investigations usually take.

As a matter of fact, the value of $4S/p$ can be found from the test film itself by integrating $\gamma$, its inclination, round the boundary. If $A$ be the area of the cross-section, then the equilibrium of the film requires that

$$\int 2S\sin\gamma\, ds=pA. \tag{6}$$

This equation may be written in the form

$$\frac{4S}{P}=2\times\frac{\text{area of cross-section}}{(\text{perimeter of cross-section})\times(\text{mean value of }\sin\gamma)}. \tag{7}$$

By measuring $\gamma$ all round the boundary the mean value of $\sin\gamma$ can be found, and hence $4S/p$ may be determined. This is, however, more laborious in practice than the use of the circular standard.

It is evident that if the radius of the circular hole be made equal to the value of $2A/P$, where $A$ is the area and $P$ the perimeter of the test hole, then $\sin\mu=$ mean value of $\sin\gamma$. It is convenient to choose the radius of the circular hole so that it satisfies this condition, in order that the quantities measured on the two films may be of the same order of magnitude.

The corresponding theorem in the torsion problem states that the mean stress round the boundary of a twisted bar is equal to the stress at the boundary of a circular shaft of radius $2A/P$. It is shown in the text that this property can be made the basis of a method of approximating to the torsional stiffness of any bar by calculation.

### Symbols and Formulae used in the Paper

$N$ = modulus of rigidity of material.

$\tau$ = twist of bar in radians per unit of length.

$A$ = area of cross-section of bar.

$P$ = length of perimeter of cross-section.

$h = 2A/P$.

$f_s$ = shear stress in bar.

$f_c$ = shear stress in circular bar of radius $h$ under twist $\tau$.

$T$ = torque applied to bar.

$T_1$ = torque applied to circular bar to give twist $\tau$.

$\gamma$ = inclination of soap film blown on a hole of the same shape as the twisted bar.

$\mu$ = inclination of film blown on a circular hole of radius $h$.

$V$ = displaced volume of the test-film.

$V_1$ = displaced volume of the circular film.

$S$ = surface tension of soap solution.

$p$ = pressure difference causing displacement.

$C = T/N\tau$.

$k$ = 'equivalent torsional radius.'

$a$ = radius of inscribed circle.

$r$ = radius for rounding projecting corners.

$\theta$ = angle turned through at a corner by the tangent to the boundary.

$A_1$ = area of modified section, when the corners have been rounded off.

$P_1$ = perimeter of modified section.

$K$ = torque correction factor.

$I = \frac{1}{3}\int_0^l y^3\,dx$, the integration being taken along the median line of the section ($l$ = length of median line).

$m = \frac{\pi a^2}{A}$.

$\rho$ = radius of curvature of boundary of section.

(1) $f_s = \frac{4S}{p} N\tau\gamma$.

(2) $T = \frac{8S}{p} N\tau V$.

(3) $\frac{4S}{p} = \frac{h}{\sin\mu}$.

(4) $\dfrac{f_s}{f_c}=\dfrac{\sin\gamma}{\sin\mu}$, for any pair of points on the sections.

(5) $\dfrac{T}{T_1}=\dfrac{V}{V_1}$.

(6) $C=\frac{1}{2}KA\left(\dfrac{2A_1}{P_1}\right)^2$ for a simple section or for any component of a compound section.

(7) $C=\dfrac{I}{1+\dfrac{4I}{Al^2}}$ for a long thin section.

(8) Stresses at points of contact of inscribed circles of maximum radius $a$

$$f_s=\frac{2aN\tau}{1+m^2}\left[1+0{\cdot}15\left(m^2-\frac{a}{\rho}\right)\right].$$

(9) Mean stress round the boundary of any section

$$f_s=\frac{2A}{P}N\tau.$$

(10) Stress at any point of the boundary of a rolled standard section

$$f_s=2aN\tau$$

($a$ is the radius of the inscribed circle which touches at the point in question).

# 2

# THE PROBLEM OF FLEXURE AND ITS SOLUTION BY THE SOAP-FILM METHOD*

REPRINTED FROM

*Reports and Memoranda of the Advisory Committee for Aeronautics*, no. 399 (1917)

In previous papers† the authors described a method of solving, by means of soap films, certain differential equations which occur in the theory of the torsion of cylindrical bars, which are not otherwise generally soluble. In the present paper it is shown that the same method may be extended so as to include equations of a similar but more general type, which arise in the consideration of the bending of a uniform cantilever by a single concentrated load.

The technical importance of this application arises from the circumstance that the accurate evaluation of the shearing stresses in the beam is impossible unless these equations can be solved, and, so far, no other general method of solution has been proposed.

The apparatus required for forming and measuring the soap films is described and illustrated. The examples of the technical data which may be obtained by its use comprise diagrams of shearing stress for propeller, wing spar I and T sections, and also a brief discussion of the effect, on the strength of beams, of rounding sharp internal corners.

In the case of certain sections, for instance, rolled beams and tubes, the equations may be solved, without forming soap films, by making a certain assumption, which is shown to be justified both by soap-film observations and the calculated results which it yields. Among other examples of this method, the case of a circular tube split longitudinally is discussed.

There is one point of the section of a beam through which the load must pass in order that the beam shall bend without twisting. If the section is unsymmetrical, this point does not coincide with the centroid, but its position may be found by means of the soap-film measurements or the approximate calculation method. Several examples are given, and it is shown that the results agree with those of direct flexure experiments on beams of various shapes. The mathematical basis of the work is discussed in the appendix.

## 1. INTRODUCTORY

The problem of estimating the strength of a beam under given conditions of loading and support may be divided into two parts. In the first place it is necessary to determine the magnitude of the stresses arising from the longitudinal extension and compression of the fibres, which are due to the bending moment on the beam, and, secondly, to find the shear stresses, due to the relative sliding of the adjacent fibres, which balance the load.

* With A. A. GRIFFITH.

† *Reports and Memoranda*, nos. 333 and 334 (1917).

A complete solution of the elastic equations for a beam under any assigned conditions has not yet been obtained. For technical purposes it is usual to assume that the state of stress at a given section of any beam is the same as that in a uniform cantilever of the same shape, bent by a single concentrated load, at a section where the bending moment and the shear force are the same as at the given section of the beam under consideration. It has been found, in all cases where solutions of the flexure problem have been obtained for more general conditions of loading, that this assumption is substantially justified. The problem of the bending of the cantilever of uniform cross-section by a transverse load acting at the free end may be considered to be the standard flexure problem, to which all others may be referred.

The exact elastic equations of the cantilever were solved by Saint-Venant. He showed how to calculate the longitudinal extensions and compressions of beams of any sections whatever; he also deduced the equations which determine the distribution of shear stress, and solved them in special cases where the cross-sections had certain simple mathematical forms. He did not succeed in finding the shear stresses in the general case of a beam of arbitrary cross-section.

Usually, beams occurring in structural work are of metal, for which Young's modulus is not more than two or three times the modulus of rigidity. It is known that beams made of such material generally fail in direct tension or compression under much lower loads than would be necessary to cause the shear stress to reach the elastic limit, so that, so far as an estimate of strength is concerned, our inability to calculate the precise value of the maximum shear stress is not a matter of great moment.

In aeroplane construction, however, highly stressed beams are made of wood, whose rigidity may be one-fifteenth to one-twentieth of its direct modulus.

Without exact knowledge, it is unsafe to predict that the shear stresses in such a beam are of less importance than the direct stresses. Hence it is of interest to investigate means of solving Saint-Venant's equations when the cross-section is of any assigned shape.

A further problem, which has attained great importance in the development of aeronautical engineering, is the determination of the twist of a beam of unsymmetrical section due to the application of a load at a given point. This twist cannot be found unless the Saint-Venant equations can be solved.

## 2. Experimental Method of Solution by means of Soap Films

It is shown in the appendix (see Appendix, §1) that the equations can be expressed in forms which are mathematically identical with those which determine the shape of a soap film, having no pressure difference between its two surfaces, and stretched over a hole cut in a curved sheet of metal or other suitable material. The sheet must be formed in such a way as to make the projection of the hole on a given plane the same shape as the cross-section to be investigated, while the perpendicular distances of points on the edge of the hole from the plane represent, to some scale, a given function of the shape of the section. If the slopes of the soap-film surface in directions parallel and perpendicular respectively to the direction of bending can be found, the shear stresses in the beam can be deduced by means of equations (15) in the appendix. Equation (21), which determines the shape of the edge of the hole, readily lends itself to graphical treatment, and hence the method may be applied to any arbitrary section.

Methods of measuring the slope of soap films were described in a previous paper (R. & M. no. 333) (see Summary), on 'The use of soap films in solving torsion problems', to which reference should be made. In the case of flexure, we generally wish to find the stresses at every point of the section. As this involves a complete knowledge of the film surface, the most satisfactory method for our present purpose is that of mapping contour lines, which we have termed the 'spherometer' method. Slopes may be found from the contour map by drawing straight lines across it parallel to the axis and constructing curves representing the corresponding sections of the soap film. These curves must then be differentiated graphically and the slopes so found inserted in the expressions for the shear stresses. The most convenient way of exhibiting the results is to construct diagrams showing lines of equal shear stress on the section.

The resultant of all the shear stresses must be a single force whose line of action has a certain position. For equilibrium, this must also be the position of the load acting on the beam. Hence the line of action of the load may be found from the diagrams of the shear stresses by calculating their total moment about the central line and making the moment of the load balance it.

The equations deduced in the appendix have been adjusted so as to make the position found, the point at which the load must be applied in order to bend the beam without twisting it. It is proposed to call this point the 'flexural centre' of the section. In general, it coincides with the centroid only when the section is symmetrical about an axis perpendicular to the axis of bending.

If the load be not applied through this point, there is a twisting couple on the beam equal to the product of the load into the distance of its line of action from the flexural centre. Since it is known that many beams commonly used to resist bending are excessively weak in torsion, the importance of finding accurately the position of the flexural centre is obvious.

### 3. Methods of Constructing Boundaries

In one method which has been found generally satisfactory, the hole is cut in a sheet of brass 22–24 s.w.g. thickness, which is afterwards bent to the right shape and soldered on to sheet-brass supporting walls which conform approximately to the shape of the hole. Their upper edges are cut to fit the curve of the plate and they are soldered on at a distance of about $\frac{1}{4}$ in. from the edge of the hole. The shape of their lower edges depends on the manner in which the boundary is to be fixed in the measuring apparatus; in the form used by the authors they are made straight and soldered on to a brass levelling plate. The edges of the holes are filed sharp.

To design a boundary, the section is set out to the scale selected for the horizontal or '$x$, $y$' dimensions, and the position of the centroid and the directions of the principal axes of inertia are found. The values of expression (21) in the appendix are then found for all points on the boundary by the usual graphical methods, and a view of the boundary in elevation is constructed on one of the axes (usually the longer one) to the scale it is proposed to use for the vertical, or '$z$', dimensions. A curve is then drawn round the boundary, approximately parallel to it and about

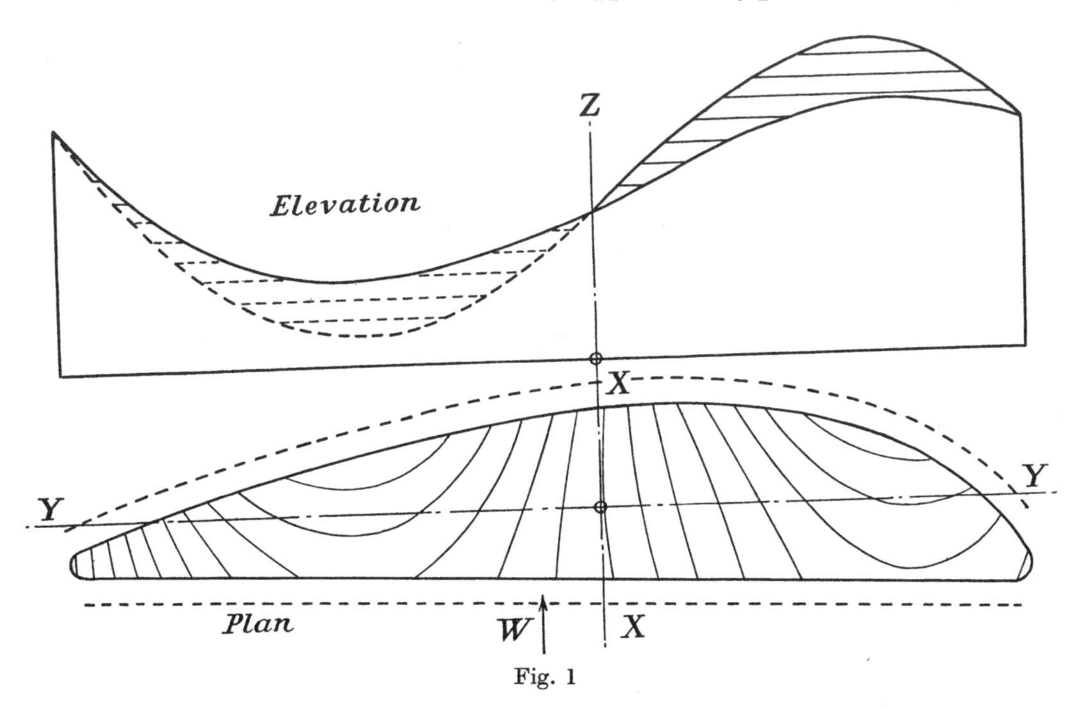


Fig. 1

$\frac{1}{4}$ in. from it, to represent the plan view of the brass supporting walls. The plan and elevation of the boundary being thus determined, developments of the several parts may be constructed. It is not necessary to make the walls in one piece—they may be made in as many sections as ease of manufacture may require.

The process is quite straightforward if the section is symmetrical about the axis on which the elevation is drawn, but if it is unsymmetrical, the edge of the hole is not developable to a plane curve. In this event the development may be carried out in two parts, and a cut made in the plate at the place where the developed boundary fails to join up. After the plate has been joint bent, the two ends can be brought together and butt-jointed.

When the various parts have been cut out in the flat as carefully as possible, they may be bent and soldered together. Every precaution should be taken to avoid tinning the upper surface of the plate or the edge of the hole, as soft solder acts comparatively rapidly on soap solution. For the same reason it is advisable, though not absolutely necessary, to coat all soldered joints with lacquer or celluloid varnish.

The finished boundary may now be set up on a marking-off table and tested. Adjustments can be made where necessary by bending the plate near the edge of the hole, care being taken to avoid irregularities. Extreme accuracy is not necessary, save in very exact work, for a reason which will be explained later.

The description may be made clearer by reference to plate 1*a* and fig. 1. Plate 1*a* is a photograph of a plaster model of the soap film used in solving the flexure problem for a propeller blade. The white lines represent sections of the film by planes parallel respectively to the two axes. Fig. 1 is a diagram showing the plan and elevation of the same soap film. The dotted lines indicate the positions of the brass supporting walls. This is an example of an unsymmetrical section. The upper and lower parts of the boundary were developed separately and the junction was made at the trailing edge.

In an alternative method, the plate is cut out of a sheet of white celluloid, to the correctly developed shape, and two rows of holes are drilled in it, at distances from each other corresponding with equal spacing of, say, half an inch to an inch on the plan view. The plate is then fixed on two rows of vertical studs, having the same spacing, and adjusted, by means of nuts and washers on the studs, until the boundary is correct. In use, the bars carrying the studs are bolted down on to the levelling plate. The free outer edges of the celluloid sheet, at right angles to the studs, are stiffened by means of metal end-pieces. The photograph (plate 1*b*) illustrates this method of construction.

## 4. Considerations which Determine the Scale of the Boundary Heights

As in the case of torsion, the analogy between the soap film and the elasticity equations does not hold rigorously unless the slope of the film is everywhere infinitesimal, but the method is generally sufficiently accurate in practice if the maximum inclination is not greater than about 30°. It is not possible, however, to lay down any definite rule in regard to this point in the case of flexure, since the actual value of the error which occurs when the slope has any assigned value depends on the magnitude of the principal curvatures. In particular, the error is obviously zero if the film is flat.

A working rule, which has been found to be generally satisfactory, is to make the total range of the vertical co-ordinate not more than one-tenth of the maximum horizontal dimension in plan. In the apparatus used by the authors, the latter is about 11·5 in. and the travel of the spherometer point 1·1 in.

It is often possible to reduce the range of the boundary function by making a substitution such as that given in the appendix in equation (23). Effectively, this amounts to increasing the scale of boundary heights without increasing the maximum inclination, and hence it conduces to greater accuracy.

### 5. Description and Use of the Apparatus (see Plate 2)

The boundary is fixed to the brass levelling plate $A$ (plate 2), which is placed inside the cast-iron pot $B$. The function of the levelling plate is to enable the boundary to be placed in its proper position relatively to the spherometer. The pot is 12·5 in. diameter and is enamelled white inside. The upper edge is machined and scraped flat to a surface plate, and on this edge slides a disc of plate glass $C$, 25 in. diameter and $\frac{3}{8}$ in. thick. In the centre of the disc is a hole about $\frac{5}{8}$ in. diameter. The spherometer $D$ is supported on three adjusting screws, which rest on the upper surface of the glass disc. The spherometer point is a hardened tool-steel rod $\frac{1}{4}$ in. diameter, tapering 1 in 4 at the end; it passes through the hole in the plate. A washer fits the rod loosely and rests on the glass so as to exclude dust and air currents. Coarse adjustment of the position of the spherometer point is obtained by moving the disc as a whole; fine adjustment is made by moving the spherometer itself. The divided head of the spherometer screw reads hundredths of a turn and the screw has forty threads to the inch.

The record is made on a sheet of paper or millboard, clamped to the circular recording board $E$. The frame $F$, on which $E$ is pivoted, is a very rigid structure built up of mild steel bars.

The method of using the apparatus is, in the main, the same as that already described in the paper on torsion; a few additional points must, however, be mentioned.

There is sometimes a tendency for solution to collect at the lowest part of the film, causing it to sag. When this occurs, a needle should be placed in a vertical position under the film at this place, with its point just touching the film. The needle drains away the surplus liquid and so prevents the deformation of the film surface. It may be conveniently mounted on a cork. Since the film is not blown up, as it is in solving the torsion problem, it may be renewed any number of times during an experiment. Hence it is not necessary to protect the film so carefully against deleterious influences, and the soap solution need not be of such high quality.

Theoretically, the contour lines mapped by the spherometer should pass through the points on the boundary where the heights are the same as those of the contours. In practice, owing to edge effects and slight errors in the construction of the boundary, they do not, in general, pass through these points. We may take advantage of a well-known device of graphical mathematics to adjust the contour lines by drawing new curves, of the same shape as the originals, but passing through the correct points on the boundary. Substantially, this amounts to the practical elimination of edge effect and constructional errors. It is evident, however, that it is not possible in this way to smooth out *irregularities* in the boundary, since they must have a sensible influence on the shape of the contours.

## 6. Experimental Results

(A) Fig. 1 illustrates the boundary and contour lines of the soap film for the bending of a beam of propeller section by a load acting perpendicular to the longer axis of inertia, while in fig. 2 is shown the diagrams of the two components of shear stress which were deduced from it. $\sigma N/E$ was taken to be 0·03, which is an average value for walnut. The shear stresses are expressed in terms of $W/A$, the mean value of the $X_z$ component.

Theoretically, of course, the integral of $X_z$ over the surface of the section should equal $W$, while the integral of $Y_z$ should be zero. In the present case they were actually found to be $1{\cdot}015W$ and $0{\cdot}02W$ respectively. These figures give an idea of the order of accuracy which may be expected from the soap-film method in solving the flexure problem.

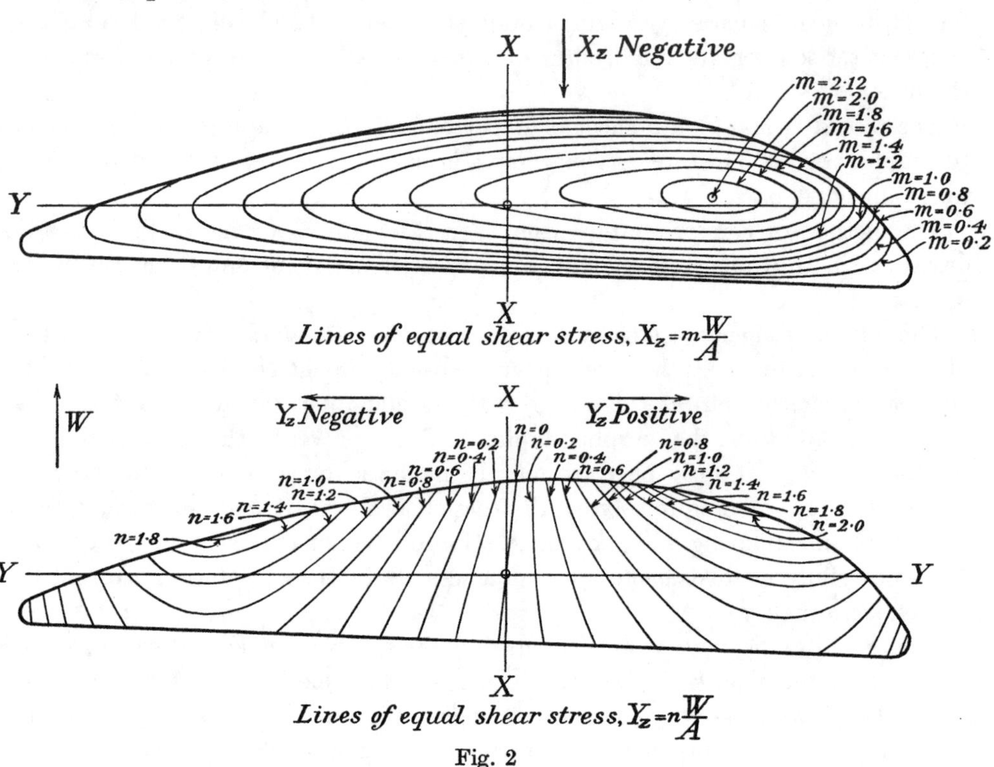


*Lines of equal shear stress*, $X_z = m\frac{W}{A}$

*Lines of equal shear stress*, $Y_z = n\frac{W}{A}$

Fig. 2

The maximum value of $X_z$, which occurs between the centroid and the leading edge, near the axis of $y$, is $2{\cdot}12\ W/A$. The maximum $Y_z$ is to be found at a near point on the boundary and its magnitude is about the same. The greatest total shear stress occurs at a point between these two. It is about $2{\cdot}4\ W/A$.

The line of action of the load was obtained by finding the moment of the $X_z$ and $Y_z$ components about the central line. It is between the centroid and the leading edge, at a point distant 0·0628 of the chord from the former.

(B) In fig. 3 we give the contour lines of the soap films and the shear-stress diagrams for the flexure of a wooden wing spar of I section. Our method of dealing with this section is typical of the treatment which may be applied to symmetrical boundaries. The constant being adjusted so as to make it zero at the origin, $\psi$ is an

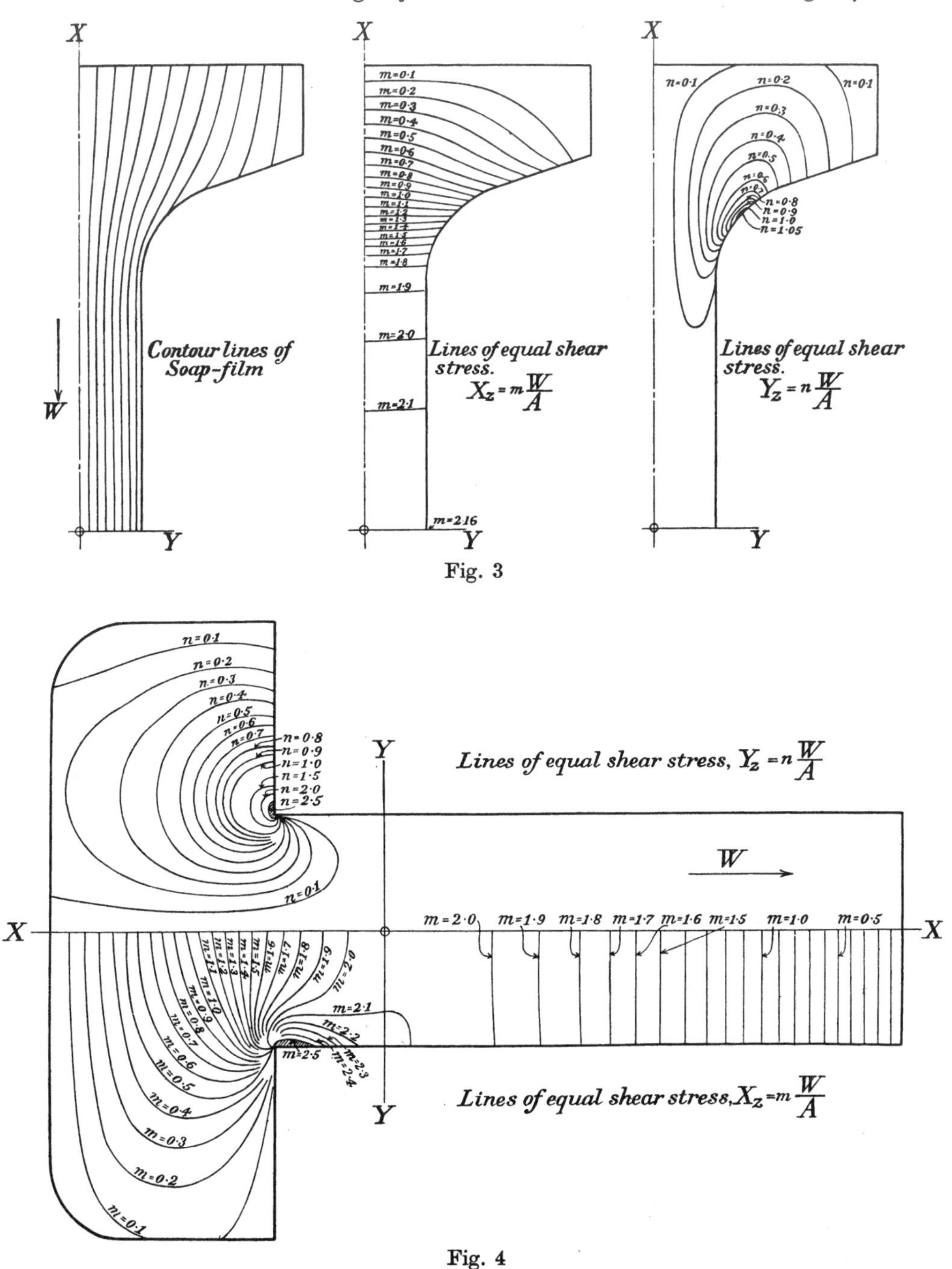


Fig. 3



Fig. 4

even function of $x$ and an odd function of $y$. Hence, if $ox$ is an axis of symmetry, as it is in the present case, it is also a contour line of $\psi$. $oy$ is not a contour line but, since it is an axis of symmetry it cuts the contours orthogonally. It follows that we need only construct the boundary of the quarter section, and place a straight edge along $ox$ and parallel to the plane of $xy$ and a 'normal septum' along $oy$.

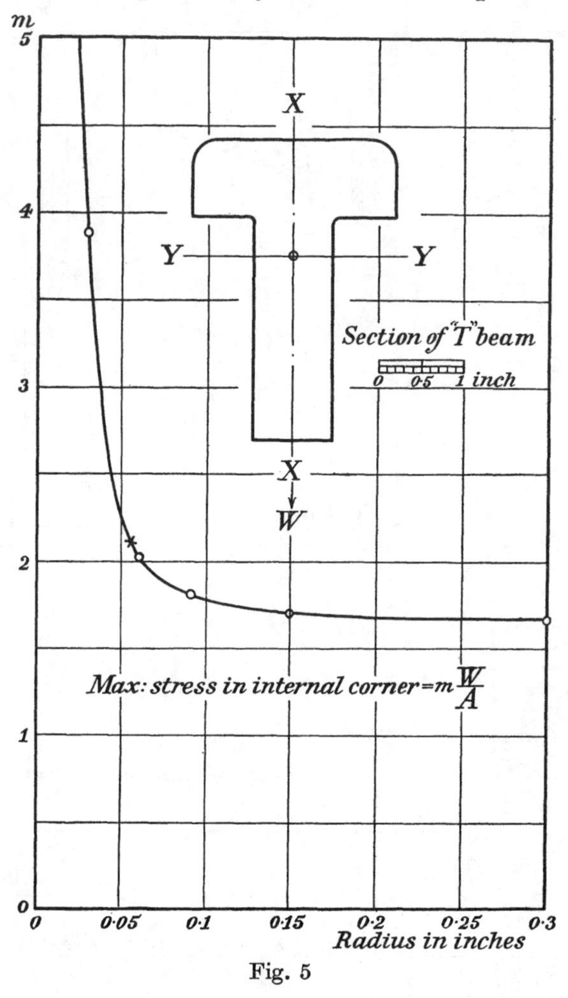


Fig. 5

This method of treating sections which have axes of symmetry is perfectly general.

To return to the diagrams, it will be seen that the accepted belief, that the web of an I beam takes most of the $X_z$ shear, is fully confirmed. Its maximum value, at the middle of the web, is almost exactly that which would have been calculated by the approximate methods at present in use. The maximum $Y_z$ shear occurs in the internal radius. It is only about half the maximum of the $X_z$ component. The greatest total shear in the radius is $1{\cdot}74\ W/A$, or about 80 % of the maximum shear in the web. This might be of importance if the beam were subjected to torsion as well as to bending, since the greatest torsional shear stress occurs in the internal radius.

PLATE 1

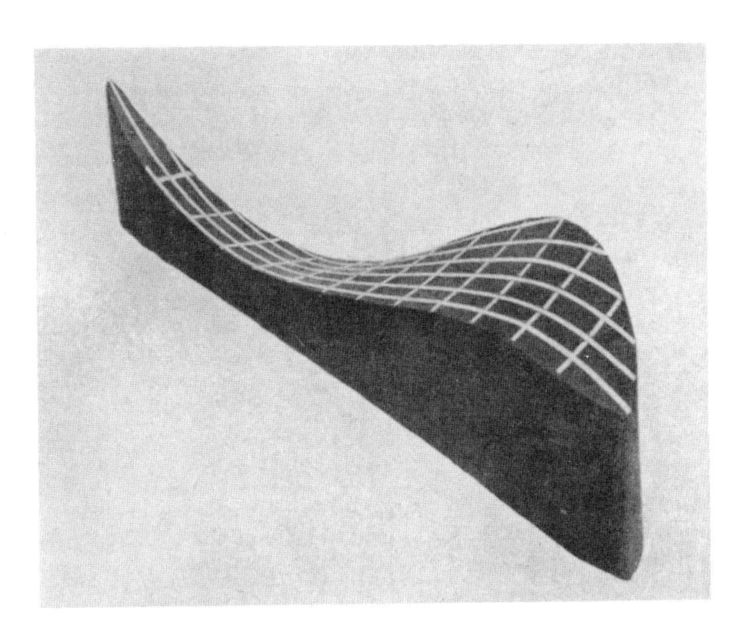

$(a)$

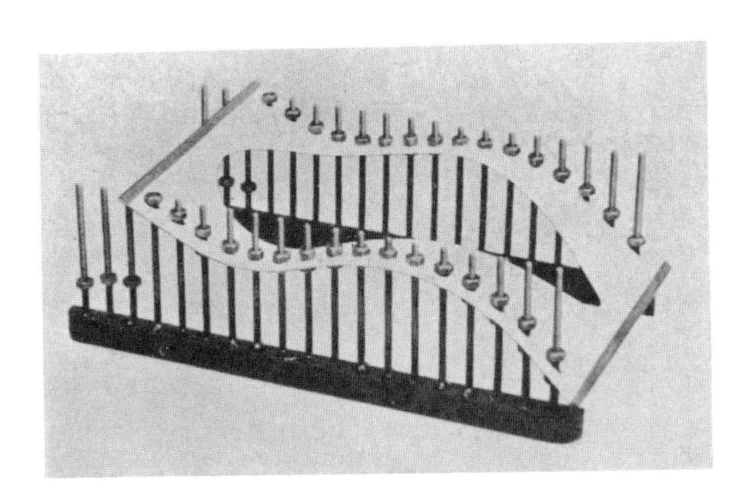

$(b)$

PLATE 2

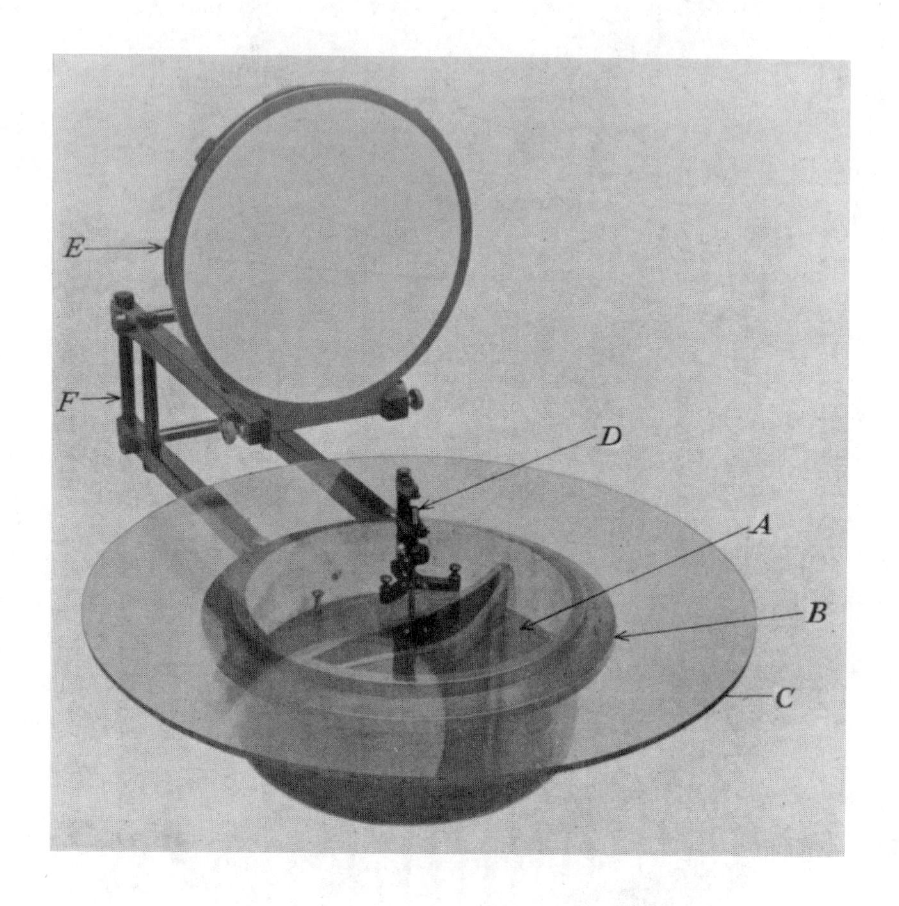

(C) Fig. 4 shows the lines of equal shearing stress in the bending of a T beam by a load acting along the line of symmetry. Here the chief point of interest is the sharp internal corner, where the stress is theoretically infinite. Naturally it is not possible to stretch a film on a boundary having a perfectly sharp corner of this sort, but, actually, it was observed that the bubble adhered to the plate at a distance from the corner which was so small as to be practically negligible.

It is immediately evident that, although the stress appears to tend towards infinity at the corner, the area of the region in which it is dangerously high is remarkably small. To bring out this point more clearly, we have shaded the portions in which the component stresses exceed $2{\cdot}5\ W/A$.

A further experiment was made on this section to determine the effect of rounding the internal corner on the value of the maximum shear stress. The radius of the corner was increased in a series of steps and contour lines were mapped on each occasion, after the necessary adjustments to the height of the boundary had been made. From the contour maps the maximum stresses were deduced.

The results of this research are exhibited graphically in fig. 5. The abscissae represent the internal radius in inches, on the same scale as the outline of the section which is given above the curve. The ordinates give the maximum shear stress at the internal corner in terms of $W/A$. The circles are the experimental results, while the cross indicates the value of the radius at which the stress in the corner equals the maximum stress in the web. The view of the section shown on the diagram has been drawn with this radius. So far as flexure is concerned, no increase of strength is obtained by making the radius larger. It will be observed that this 'minimum radius for maximum strength' is about one-sixteenth of the thickness of the web.

## 7. Method of Approximate Calculation for Thin Sections

It has often been observed, in the course of experiments on soap films with undulating boundaries, that contour lines joining portions of the boundary which are near together and approximately parallel in plan, are very nearly straight.

If it be granted that this is a general property of the flexure functions of sections whose boundaries consist of two nearly parallel curves, separated by a small distance, the shear stresses in beams of such forms may be found by direct calculation. The details of the work are given in §§ 2 and 3 of the appendix.

The following examples serve to illustrate the application of the general expressions to particular problems. Some experimental verifications of the results deduced will be described later.

(A) An angle-iron having arms of equal length $a$ and thickness $t$, bending about its axis of symmetry $(oy)$.

We have

$$A = 2at$$

and

$$k^2 = \frac{a^2}{6}.$$

Taking our origin of $s$ at the end of the upper arm (where the two bounding lines join), and noting that $x=(a-s)/\sqrt{2}$, we see, by (41), that the stress in this arm is given by

$$S_z=-\frac{W}{A}\frac{6}{\sqrt{(2a^2)}}\int(a-s)\,ds$$

$$=-\frac{W}{A}\frac{6}{\sqrt{(2a^2)}}(as-\tfrac{1}{2}s^2).$$

By taking our origin of $s$ at the end of the lower arm, we find for the stress in that arm the same expression with the sign reversed, as is obvious from considerations of symmetry.

The maximum stress occurs at the junction of the arms—its value is $3/\sqrt{2}(W/A)$ or $2{\cdot}12W/A$.

The position of the flexural centre may be found from equation (42), thus

$$\bar{y}=\frac{3}{a^3t}2\int_0^a t(as-\tfrac{1}{2}s^2)\,ds$$

$$=\frac{a}{\sqrt{2}},$$

or the flexural centre is at the junction of the arms.

(B) A circular tube of mean radius $a$ and thickness $t$, with the wall of the tube cut by a slit, parallel to the central line, at an angle $\alpha$ with the axis of $x$.

Here

$$A=2\pi at,$$

$$k^2=\frac{a^2}{2}.$$

Calling $\theta$ the angle measured from $ox$, we see that $x=a\cos\theta$, and (41) becomes

$$S_z=-\frac{W}{A}\frac{2}{a^2}\left\{\int a^2\cos\theta\,d\theta+C\right\}$$

$$=-\frac{2W}{A}(\sin\theta-\sin\alpha),$$

on adjusting $C$ so as to make the integral zero at $\theta=\alpha$, where the inner and outer bounding circles meet.

The maximum value of $S_z$ is

$$2\frac{W}{A}(1+\sin\alpha).$$

(47) becomes

$$\bar{y}=\frac{1}{\pi a}\int_0^{2\pi}a^2(\sin\theta-\sin\alpha)\,d\theta$$

$$=-2a\sin\alpha,$$

or the flexural centre is at a distance from the centre of the tube equal to $2a\sin\alpha$ on the side of $ox$ opposite the slit.

A method of calculating the torsional stresses and stiffness of a section such as this was described in a previous paper (R. & M. no. 334).

In the present case, the torsional stiffness

$$G=\frac{2\pi}{3}at^3 \text{ approximately.}$$

Hence the twist of the beam due to the application of the load at the centroid instead of the flexural centre is given by

$$\tau=\frac{2Wa\sin\alpha}{\frac{2\pi}{3}at^3N}=\frac{3W\sin\alpha}{\pi Nt^3}.$$

The stress due to this twist is $N\tau t$ or

$$6\frac{W}{A}\frac{a}{t}\sin\alpha.$$

Hence the maximum total shear stress is given by

$$S_z(\max)=2\frac{W}{A}\left\{1+\sin\alpha\left(1+3\frac{a}{t}\right)\right\}.$$

We see that if $a/t=10$ and $\alpha=\frac{1}{2}\pi$, the maximum shear stress is 32 times greater than if there were no slit at all or, which comes to the same thing, if $\alpha$ were zero.

(C) An I beam, having flanges of uniform thickness $t_1$, and width $2a$, and a web of thickness $t_2$ and length $2b$.

Integrating, in (41), from the end of a flange, we see that

$$s=a-y,$$

and we get

$$S_z=-\frac{W}{Ak^2}b(a-y)$$

for the stresses in the flanges where $y$ is positive, and, similarly,

$$S_z=\frac{W}{Ak^2}b(a+y),$$

when $y$ is negative.

For the integral along the web, we note that $C$ is given by the difference of the flange integrals at the junction, or it is

$$-\frac{2W}{Ak^2}t_1ab.$$

Hence

$$S_z=\frac{W}{Ak^2}\left\{\tfrac{1}{2}(b^2-x^2)+\frac{t_1}{t_2}ab\right\}.$$

It may be pointed out here that, in the present theory, the increase of stress at the junction of two or more branches, due to the internal corners, is neglected. In view

3-2

of the results already quoted, for the bending of a T beam, it is probable that the increase so neglected is quite small, when the radius of the corner is of the order of magnitude common in engineering practice.

(D) A thin plank or blade, of trapezoidal section, width $a$, thickness at the ends $t_1$ and $t_2$ respectively.

In this case

$$t=t_1+\frac{y}{a}(t_2-t_1),$$

so that

$$\int_0^a t^3 y\,dy=\frac{a^2}{20}(t_1^3+4t_1^2t_2+3t_1t_2^2+4t_2^3)$$

and

$$\int_0^a t^3\,dy=\frac{a}{4}(t_1^3+t_1^2t_2+t_1t_2^2+t_2^3).$$

Hence

$$\bar{y}=\frac{\int_0^a t^3 y\,dy}{\int_0^a t^3\,dy}$$

$$=\frac{a}{5}\left\{4-\frac{t_1(3^3t_1^2+t_2^2)}{(t_1+t_2)(t_1^2+t_2^2)}\right\}.$$

If $t_1=0$, the figure is approximately equivalent to a circular sector of small angle. In this case

$$\bar{y}=\tfrac{4}{5}a \text{ (cf. § 8 A, below).}$$

## 8. Experimental Verifications

The direct measurements of the shear strains in a bent beam is a matter of considerable difficulty, but a very easy experiment, requiring the simplest of apparatus, enables us to find the position of the flexural centre. If this experiment be performed on beams whose sections are not symmetrical about the axis of $x$, it constitutes a check on the soap-film results which is quite as satisfactory as the direct measurement of shear strains would be.

Several experiments of this sort have been made; the results are given below.

(A) Soap-film measurements on a circular sector of 10° angle showed that the flexural centre of this figure was on the line of symmetry at a distance from the point equal to 0·78 of the radius.

A wooden beam of this section was set up as a cantilever with the line of symmetry horizontal. Two mirrors were fixed to the beam so that the twist could be observed by means of a telescope and scale in the usual manner. A weight was hung on the free end and its position adjusted till the twist was zero. It was then found that the line of action of the load passed through a point distant 0·795 of the radius from the thin end of the sector.

(B) The soap-film method indicated that the flexural centre of an airscrew section, of chord $b$, was distant 0·396$b$ from the leading edge. Two cylindrical wooden beams of the same section were set up, with their principal axes of inertia horizontal and vertical respectively, and loaded as before. The values obtained in this way were $0{\cdot}44b$ and $0{\cdot}36b$, so that the mean was $0{\cdot}40b$.

(C) (see § 7 A above). We have seen that on the theory of the bending of thin sections, which is discussed in the appendix, an angle beam having arms of equal length should not twist when loaded at the junction of the arms. This was fully confirmed by an experiment on an angle beam made of sheet brass, the load being applied at right angles to the axis of symmetry.

(D) (see § 7 B). A brass tube was split by a saw-cut parallel to its central line. The saw-cut was closed and brass plugs were soldered into the ends. The beam was set up as a cantilever in such a way as to permit the cut to be placed at any angle with the vertical, and arrangements were made for loading at any position along the horizontal diameter. The flexural centre was determined for several different values of the angle.

The results are given in the table below, together with the values obtained by calculation. The mean diameter of the tube was 0·704 in., the thickness of the material 0·036 in. and the length of the beam 9 in.

| | $\bar{y}$ (experiment) (in.) | $\bar{y}$ (calculation) (in.) | Error (%) |
|---|---|---|---|
| 0 | 0 | 0 | 0 |
| 30° | 0·35 | 0·352 | 0·5 |
| 42° | 0·45 | 0·471 | 4·5 |
| 61° | 0·59 | 0·615 | 4·2 |
| 90° | 0·68 | 0·704 | 3·5 |

(E) (see § 7 D). Two walnut planks were made of trapezoidal section, 48 in. long, 6 in. wide and $\frac{5}{8}$ in. thick at one side, tapering to $\frac{1}{8}$ in. at the other. They were cut from a single board $1\frac{1}{2}$ in. thick, with the thin edge of one opposite the thick edge of the other, in order that the effect of variations in the wood might be eliminated, by taking the mean result from the two planks.

The planks were set up and loaded as in the previous experiments, care being taken to make the line of symmetry of the section horizontal.

The flexural centres were found to be at 5·05 in. and 3·90 in. respectively from the thin edge, the mean value being 4·47 in. The figure obtained by the approximate calculation method was 4·58 in. In view of the comparatively large difference between the two planks, the agreement of the mean value with the calculated result appears to be quite satisfactory.

## 9. Cantilever Bent by any End Load

In the most general possible case of a cantilever deformed by a load applied at its free end, the load may be resolved into the following components:

(*a*) A force acting along the central line of the beam.

(*b*) Two forces acting through the flexural centre of the cross-section, parallel respectively to its two principal axes of inertia.

(*c*) A couple about the central line.

(*d*) Two couples about the principal axes of the section.

The strain due to (*a*) is a uniform extension; those due to (*d*) correspond to uniform flexure. The stresses set up by these strains are well known.

(*b*) and (*c*) involve non-uniform flexure and torsion respectively. The stresses to which they give rise can be completely determined if we know three functions of the shape of the cross-section of the beam, namely, the two flexure functions corresponding to bending about the two axes of inertia, and the torsion function. All these functions can be found, with the help of soap films, for any section whatever, and hence the stress-distribution due to any end load may be determined.

## APPENDIX

### (1) *General Theory* (see diagram *a*, fig. 6)

It is required to find the shearing stresses in a cantilever of uniform cross-section, bent by a concentrated transverse load $W$ acting parallel to the axis of $x$, the axis of $z$ being coincident with the line of centroids of the sections of the beam in its unstrained state. $Ox$ and $Oy$ are taken parallel respectively to the principal axes of inertia of the sections and $Oz$ is assumed to be an axis of symmetry of elastic structure.

It will be premised that the load is applied at such a point that the beam does not twist. We have to determine the two components of shear stress $X_z$ and $Y_z$, acting in the directions $xz$ and $yz$ respectively and, from them, the position of the line of action of the resultant shear force on the terminal cross-section.

Let $E$ be the Young's modulus parallel to $Oz$ and $\sigma$ the corresponding Poisson's ratio; and let $N$ be the modulus of rigidity for shears in the direction $xz$ and $yz$; further let $A$ be the area of the section, and $k$ its radius of gyration about $Oy$.

It is shown in books on the mathematical theory of elasticity* that the stresses and strains in the beam must be such as to satisfy three conditions, namely:

(A) The elements of the beam must be in equilibrium under the force acting on them, whence we deduce the equation

$$\frac{\partial X_z}{\partial x}+\frac{\partial Y_z}{\partial y}+\frac{W_x}{Ak^2}=0, \tag{1}$$

which must be satisfied at all points of the cross-section.

(B) The resultant shear stress at any point of the cylindrical bounding surface must act parallel to the tangent to the boundary of the cross-section at that point so that

$$\frac{Y_z}{X_z}=\frac{dy}{dx}, \tag{2}$$

or

$$Y_z\frac{dx}{ds}-X_z\frac{dy}{ds}=0, \tag{3}$$

where $ds$ is the element of the bounding curve. Equation (3) must hold at every point of the boundary.

* A. E. H. Love, *Mathematical Theory of Elasticity*, 2nd ed., chap. xv.

(C) In order that the material of the bar may be continuous in the strained state (which is the same thing as saying that rupture does not take place), we must have, between the strain components $e_{xz}$ and $e_{yz}$, the two relations

$$\frac{\partial}{\partial x}\left(\frac{\partial e_{yz}}{\partial x}-\frac{\partial e_{xz}}{\partial y}\right)=0, \tag{4}$$

$$\frac{\partial}{\partial y}\left(\frac{\partial e_{yz}}{\partial x}-\frac{\partial e_{xz}}{\partial y}\right)=-\frac{2\sigma W}{EAk^2}. \tag{5}$$

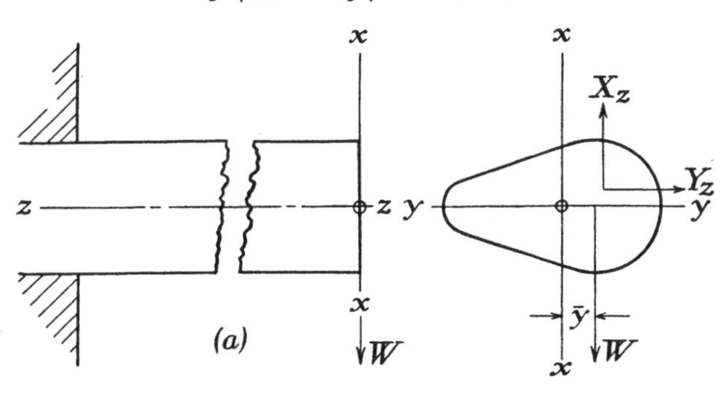


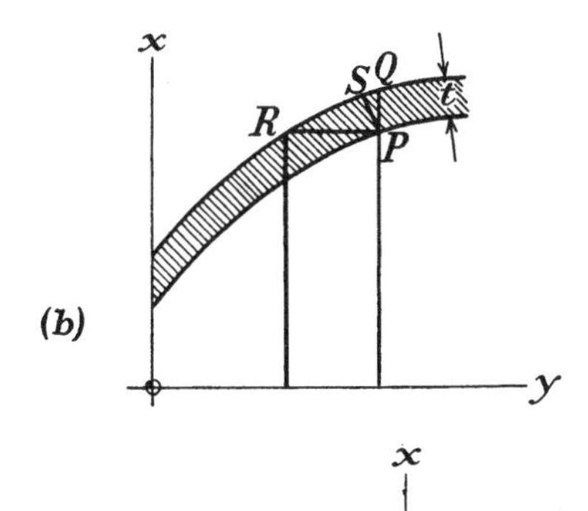


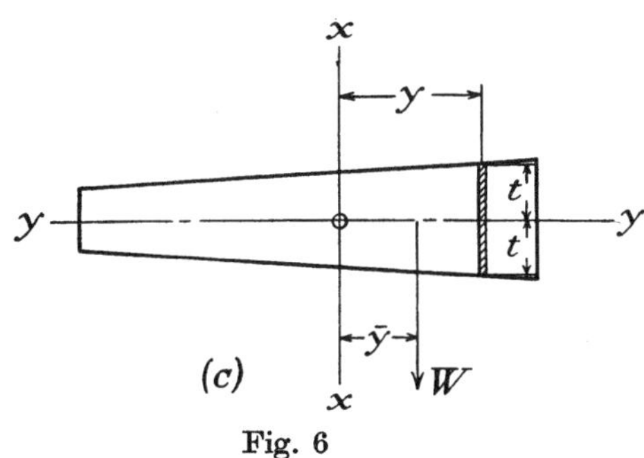


Fig. 6

From (4) and (5) we deduce the equation

$$\frac{\partial e_{yz}}{\partial x}-\frac{\partial e_{xz}}{\partial y}=2\tau-\frac{2\sigma W}{EAk^2}y, \tag{6}$$

where $\tau$ is a constant of integration.

Evidently, $\tau$ represents a relative rotation of adjacent sections about $Oz$, which, by hypothesis, is zero.

It follows that $e_{yz}, e_{xz}$, can be expressed in the forms

$$e_{yx}=\frac{W}{NAk^2}\frac{\partial\phi_0}{\partial y},\quad e_{xz}=\frac{W}{NAk^2}\frac{\partial\phi_0}{\partial x}+\frac{\sigma W}{EAk^2}y^2, \tag{7}$$

where $\phi_0$ is some function of $x$ and $y$ only. Hence

$$Y_z=\frac{W}{Ak^2}\frac{\partial\phi_0}{\partial y},\quad X_z=\frac{W}{Ak^2}\left(\frac{\partial\phi_0}{\partial x}+\frac{\sigma N}{E}y^2\right). \tag{8}$$

Therefore, substituting in (1),
$$\frac{\partial^2\phi_0}{\partial x^2}+\frac{\partial^2\phi_0}{\partial y^2}+x=0, \tag{9}$$

while (3) becomes
$$\frac{\partial\phi_0}{\partial y}\frac{dx}{ds}-\frac{\partial\phi_0}{\partial x}\frac{dy}{ds}-\frac{\sigma N}{E}y^2\frac{dy}{ds}=0. \tag{10}$$

Take a new function, $\phi$, such that

$$\phi_0=\phi-\frac{x^3}{6}. \tag{11}$$

Then, from (9),
$$\frac{\partial^2\phi}{\partial x^2}+\frac{\partial^2\phi}{\partial y^2}=0. \tag{12}$$

The boundary condition (10) takes the form

$$\frac{\partial\phi}{\partial y}\frac{dx}{ds}=\frac{\partial\phi}{\partial x}\frac{dy}{ds}+\left(\frac{x^2}{2}-\frac{\sigma N}{E}y^2\right)\frac{dy}{ds}=0, \tag{13}$$

since, by (12), $\phi$ is a plane harmonic function of $x$ and $y$, there must exist a conjugate function, $\psi$, such that

$$\left.\begin{aligned}&\frac{\partial^2\psi}{\partial x^2}+\frac{\partial^2\psi}{\partial y^2}=0,\\ &\frac{\partial\phi}{\partial x}=\frac{\partial\psi}{\partial y},\\ &\frac{\partial\phi}{\partial y}=-\frac{\partial\psi}{\partial x},\end{aligned}\right\} \tag{14}$$

at all points where (12) holds.

Hence, by (8), (11) and (14),

$$\left.\begin{aligned}X_z&=\frac{W}{Ak^2}\left(\frac{\partial\psi}{\partial y}-\frac{x^2}{2}+\frac{\sigma N}{E}Y^2\right),\\ Y_z&=-\frac{W}{Ak^2}\frac{\partial\psi}{\partial x}.\end{aligned}\right\} \tag{15}$$

The moment of these forces about $Oz$ is given by

$$\iint(xY_z-yX_z)\,dx\,dy, \tag{16}$$

the integration extending over the section, so that if the line of action of $W$ pass through $\bar{y}$, we must have

$$W\bar{y}+\iint(xY_z-yX_z)\,dx\,dy=0, \tag{17}$$

or, by (15),

$$\bar{y}=\frac{1}{Ak^2}\iint\left(x\frac{\partial\psi}{\partial x}+y\frac{\partial\psi}{\partial y}-\frac{x^3}{2}+\frac{\sigma N}{E}xy^2\right)dx\,dy. \tag{18}$$

We see from (15) and (18) that the problem is solved if we can find $\psi$ at all points of the section.

The boundary condition (13) takes the form

$$\frac{\partial\psi}{\partial x}\frac{dx}{ds}+\frac{\partial\psi}{\partial y}\frac{dy}{ds}-\left(\frac{x^2}{2}-\frac{\sigma N}{E}y^2\right)\frac{dy}{ds}=0, \tag{19}$$

or

$$\frac{\partial\psi}{\partial s}=\left(\frac{x^2}{2}-\frac{\sigma N}{E}y^2\right)\frac{dy}{ds}. \tag{20}$$

Integrating along $s$, we obtain the expression

$$\psi=\frac{1}{2}\int x^2dy-\frac{\sigma N}{E}\frac{y^3}{3}+\text{const.}, \tag{21}$$

which gives the value of $\psi$ for every point of the bounding curve.

Now imagine a hole cut in a sheet of metal so shaped that the projection of the edge of the hole on the plane of $xOy$ (i.e. its plan), has the same shape and size as the boundary of the cross-section of the beam, while the $z$ co-ordinates of the edge represent, to some scale, the values of $\psi$ given by (21) for corresponding points on the boundary.

It will be noticed that, since the origin is at the centroid, $\int x^2dy$ vanishes when taken round the boundary. Hence the edge of the hole is a closed curve.

Let the soap film be stretched over the hole with the difference of pressure between the two sides of the film equal to zero. If the $z$ co-ordinates be sufficiently small compared with the $x$ and $y$ dimensions, so that the curvature of the film in any one direction is small, the equation of the surface of the film is

$$\frac{\partial^2 z}{\partial x^2}+\frac{\partial^2 z}{\partial y^2}=0. \tag{22}$$

Hence, by the first equation of (14) the $z$ co-ordinates of the soap film represent the flexure function $\psi$ at all points of the cross-section, to the scale on which the $z$ co-ordinates of the edge of the hole represent the values of $\psi$ at the boundary.

It follows that if we can obtain a diagram of the contour lines of the soap film, we can deduce the values of $X_z$ and $Y_z$ at all points, by measuring the slopes of the film surface in directions parallel to $Ox$ and $Oy$ respectively and substituting in (15).

In addition, the position of the load may be found from (18) or, alternatively, by finding the moment (16) directly from the shear stress curves.

It will be observed that if we put

$$\psi=\psi'+Bx+Cy, \tag{23}$$

where $B$ and $C$ are arbitrary constants, then the equation

$$\frac{\partial^2\psi'}{\partial x^2}+\frac{\partial^2\psi'}{\partial y^2}=0 \tag{24}$$

is satisfied, and $\psi'$ may be found by the soap-film method. Expressed in terms of $\psi'$, equation (15) becomes

$$\left.\begin{aligned}X_z&=\frac{W}{Ak^2}\left(\frac{\partial\psi'}{\partial y}+C-\frac{x^2}{2}+\frac{\sigma N}{E}y^2\right),\\ Y_z&=-\frac{W}{Ak^2}\left(\frac{\partial\psi'}{\partial x}+B\right).\end{aligned}\right\} \tag{25}$$

The values of $\psi'$ at the boundary are given by

$$\psi'=\frac{1}{2}\int x^2dy-\frac{\sigma N}{E}\frac{y^3}{3}-Bx-Cy+\text{const.} \tag{26}$$

It is often useful to make substitutions of this sort, for reasons which are explained in the text.

(2) *Approximate Theory of Thin Sections* (see diagram *b*, fig. 6)

It is assumed that the boundary consists of two nearly parallel curves separated by a small distance $t$ (measured normally to the curves).

Referring to the diagram, and remembering equation (21), if we put

at $P$ $$\psi_1=\frac{1}{2}\int x^2dy-\frac{\sigma N}{E}\frac{y^3}{3}, \tag{27}$$

then at $Q$ $$\psi_2=\frac{1}{2}\int\left(x+t\frac{ds}{dy}\right)^2dy-\frac{\sigma N}{E}\frac{y^3}{3}+C, \tag{28}$$

at $R$ $$\psi_3=\frac{1}{2}\int\left(x+t\frac{ds}{dy}\right)^2dy-\frac{\sigma N}{E}\frac{y^3}{3}+C-\left(\frac{x^2}{2}-\frac{\sigma N}{E}y^2\right)t\frac{ds}{dx}, \tag{29}$$

and at $S$ $$\psi_4=\frac{1}{2}\int\left(x+t\frac{ds}{dy}\right)^2dy-\frac{\sigma N}{E}\frac{y^3}{3}+C-\left(\frac{x^2}{2}-\frac{\sigma N}{E}y^2\right)t\frac{dx}{ds}. \tag{30}$$

$PQ$, $PR$ and $PS$ are drawn parallel to $Ox$, $Oy$ and $\nu$, the normal, respectively, and $C$ is a constant, at present unknown.

Experience with soap films shows that $PQ$, $PS$ and $PR$ may be taken to be straight lines when $t$ is sufficiently small. Hence we may write

$$\left.\begin{aligned}\frac{\partial\psi}{\partial x}&=\frac{\psi_2-\psi_1}{t\dfrac{ds}{dy}},\\ \frac{\partial\psi}{\partial y}&=\frac{\psi_1-\psi_3}{t\dfrac{ds}{dx}},\\ \frac{\partial\psi}{\partial\nu}&=\frac{\psi_4-\psi_1}{t}.\end{aligned}\right\} \tag{31}$$

Substituting in (15) we get

$$\left.\begin{aligned} X_z &= -\frac{W}{Ak^2}\frac{1}{t}\frac{dx}{ds}\left[\int t\left(x+\frac{t_1}{2}\right)ds+C\right],\\ Y_z &= -\frac{W}{Ak^2}\frac{1}{t}\frac{dy}{ds}\left[\int t\left(x+\frac{t_1}{2}\right)ds+C\right],\end{aligned}\right\} \tag{32}$$

where $$t_1 = t\frac{ds}{dy},$$

while, from (17), $$\bar{y} = \frac{1}{Ak^2}\int\left[\int t(x+\tfrac{1}{2}t_1)\,ds+C\right]\left(x\frac{dy}{ds}-y\frac{dx}{ds}\right)ds. \tag{33}$$

In many cases (32) and (33) may be expressed more conveniently in the forms

$$S_z = -\frac{W}{Ak^2}\frac{1}{t}\left(\int t\left(x+\frac{t_1}{2}\right)ds+C\right), \tag{34}$$

$$\bar{y} = \frac{1}{Ak^2}\int\left[\int t\left(x+\frac{t_1}{2}\right)ds+C\right]p\,ds, \tag{35}$$

where $S_z$ is the resultant shear stress and $p$ is the perpendicular from the origin on to the tangent to the boundary.

To complete the solution it is necessary to determine $C$.

If the two bounding curves meet at any point, $C$ must be adjusted so as to make the two values of $\psi$ equal at this place. In other words, if the integration of $\int t\left(x+\frac{t_1}{2}\right)ds$ be started from this junction, $C$ is zero.

If the two boundaries do not meet, as in the case of a tube, $C$ cannot be found thus. We note, however, that the function $\phi$, conjugate to $\psi$, represents a displacement of the elements of the beam, and must, therefore, be single-valued.

Hence

$$\int\frac{\partial\phi}{\partial s}ds = 0, \tag{36}$$

the integration being taken round any closed curve in the section. Whence, by the second and third equations of (14),

$$\int\frac{\partial\psi}{\partial\nu}ds = 0, \tag{37}$$

$\nu$ being the normal to the curve.

In the present case, from the third equation of (31), we have

$$\frac{\partial\psi}{\partial\nu} = \frac{1}{t}\left[\int t\left(x+\frac{t_1}{2}\right)ds+C-\left(\frac{x^2}{2}-\frac{\sigma N}{E}y^2\right)\frac{dx}{ds}\right]. \tag{38}$$

Therefore, from (37),

$$\int\frac{1}{t}\left[\int t\left(x+\frac{t_1}{2}\right)ds+C\right]ds+\frac{2\sigma N}{E}A\,g_y = 0, \tag{39}$$

where $g_y$ is the $y$ co-ordinate of the centroid of the area enclosed by the boundary round which the integration is taken.

$C$ is the only unknown in this equation; hence it may be determined.

In the general case, if there are $n$ separate boundaries, so that the thin walls form a region of $n$ connections, we may integrate round each of the $(n-1)$ internal boundaries and so obtain $n-1$ equations of the type (39) containing the necessary constants; hence all the constants may be found.

It must be noted, however, in connection with hollow sections, that the general equations were deduced for a section bounded by a single closed curve, bending about an axis passing through the centroid of the area. In general, if the section bend about the axes of $y$, which, however, does not pass through the centroid, then $x$ must be replaced by $x-g_x$ throughout the equations, $g_x$ being the $x$ co-ordinate of the centroid. Where the section has more than one boundary, the $x$ co-ordinates of each must be measured from an axis passing through the centroid of the area enclosed by that particular boundary.

In the majority of cases, the expressions are simplified by putting

$$x+\tfrac{1}{2}t_1=x_1. \tag{40}$$

Equations (34), (35) and (39) then become

$$S_z=-\frac{W}{Ak^2}\frac{1}{t}\left(\int x_1 t\,ds+C\right), \tag{41}$$

$$\bar{y}=\frac{1}{Ak^2}\int\left(\int x_1 t\,ds+C\right)p\,ds, \tag{42}$$

$$\int\frac{1}{t}\left(\int x_1 t\,ds+C\right)ds-\frac{2\sigma N}{E}A\,g_y=0, \tag{43}$$

and $x_1$ is measured to the line midway between the two bounding curves.

### (3) *Bending of a Blade* (see diagram *c*, fig. 6)

If the section is a thin 'blade' or plank, symmetrical about $Oy$, problems relating to its flexure may be solved in a similar manner.

Assuming, as before, that the contour lines of the film are straight lines parallel to $Ox$ and calling $2t$ the thickness at a distance $y$ from the origin, we have, by (20),

$$\left.\begin{aligned}\frac{\partial\psi}{\partial y}=\frac{\partial\psi}{\partial s}&=\frac{t^2}{2}-\frac{\sigma N}{E}y^2,\\ \frac{\partial\psi}{\partial x}&=0.\end{aligned}\right\} \tag{44}$$

Hence, by (15),

$$\left.\begin{aligned}X_z&=\frac{W}{2Ak^2}(t^2-x^2),\\ Y_z&=0.\end{aligned}\right\} \tag{45}$$

Substituting in (17)

$$\bar{y}=\frac{1}{2Ak^2}\iint y(t^2-x^2)\,dx\,dy. \tag{46}$$

Integrating with respect to $x$ and putting in the value of $Ak^2$, we see that (46) becomes

$$\bar{y}=\frac{\int t^3 y\,dy}{\int t^3\,dy}. \tag{47}$$

(4) *Effect of Changing the Origin of y*

In discussing the general theory, we have placed the origin of $y$ at the centroid of the section. In many cases it would be more convenient to choose some other point as origin. It is therefore of interest to investigate the influence of such a change on the stress distribution.

If the curvature of the central line of the beam be $1/\rho$, any line in a section, initially straight and parallel to $Oy$, receives a curvature $\sigma/\rho$ when the beam is bent. Hence a particle at a distance $y$ from $Ox$ receives an angular displacement $\sigma y/\rho$ about an axis parallel to the central line, so that there is a twist on the particle of magnitude

$$\tau_1=\frac{\partial}{\partial z}\left(\frac{\sigma y}{\rho}\right)$$

or

$$\tau_1=\frac{\sigma W}{WAk^2}y. \tag{48}$$

We see from this equation that we cannot adjust the line of application of the load so as to make the twist zero over the whole section, but we can make it zero along one line in the section, parallel to $Ox$. In putting $\tau=0$ in equation (6), we have made this line coincide with $Ox$. If we take any other line as our origin of $y$, we also make it the line which does not twist. The change in the position of the load due to a removal of the origin to $y_1$ is given by

$$\sigma\frac{NG}{EAk^2}y,$$

where $G$ is the torsional stiffness of the section.

If we know the torsional properties of the section we can deduce, from any solution of the flexure problem, the solution corresponding with any other position of the origin.

# 3

# THE APPLICATION OF SOAP FILMS TO THE DETERMINATION OF THE TORSION AND FLEXURE OF HOLLOW SHAFTS*

REPRINTED FROM

*Reports and Memoranda of the Advisory Committee for Aeronautics*, no. 392 (1918)

The method of solving by means of soap films the equations of Saint-Venant for the shearing stresses set up in bent and twisted beams, which was discussed in the other reports (R. & M. nos. 333 and 399), is here extended to hollow beams—that is, to beams whose cross-sections are multiply-connected regions.

By the method which has been developed it is necessary to make measurements on as many films as the section has boundaries. In nearly all practical cases there are two boundaries, but the process is not intrinsically more difficult when applied to problems involving a greater number.

The examples given include the case of a hollow circular shaft with eccentric bore and also a description of a research undertaken with a view to improving the design of aircraft engine propeller shafts, in which the propeller is fixed in place either by means of a key or serrations cut in the periphery of the shaft.

The mathematics of the method is discussed in an Appendix.

## 1. Preliminary

In previous reports the authors described an experimental method of solving by means of soap films Saint-Venant's equations for the shearing stresses in beams, due to torsion and flexure. In those reports the discussion was limited to beams whose cross-sections were bounded by single curves.

The equations can be solved by the same method for hollow beams—that is, beams whose cross-sections have two or more boundaries—but the problem is then of a more general type, and an extension of the theory is necessary.

The nature of the requisite extension may be seen by referring to the boundary condition which must be satisfied by $\psi$. This takes the form $\psi$ = a known function of $x$ and $y$, + a constant. In the case of sections having multiple boundaries, it is not legitimate to assume, without proof, that the additive constant is the same for each boundary, and since the soap film cannot now be formed until the constants have been found it is necessary to seek a method of evaluation.

This may be done by observing that the function $\phi$, which represents a distortion of the cross-sections of the beam, must be single-valued. Hence by means of the relations connecting $\phi$ and $\psi$, it is possible to obtain a series of equations which must be satisfied by the slopes of the latter function. The details of the work are given in

* With A. A. Griffith.

the appendix (see Appendix, § 1), where it is shown that the equations reduce to the condition that the total force acting on each boundary, due to the components of the surface tension, normal to the $xy$-plane, must balance the force due to the air pressure acting on the area enclosed by the boundary. In the case of 'zero-pressure' films, which satisfy

$$\frac{\partial^2\psi}{\partial x^2}+\frac{\partial^2\psi}{\partial y^2}=0,$$

the condition is simply that the resultant normal surface-tension force on each boundary is zero.

At first it was sought to realize this condition practically by suspending each of the interior boundaries from a balance arm and so letting it take up its own position of equilibrium. This idea, however, was finally abandoned on account of experimental difficulties and the very small magnitude of the forces involved, and another method, which is discussed in the appendix, was devised. In this method of finding the constants, it is generally necessary to make measurements on a series of different films, whose number is equal to the number of boundaries. The films differ in the additive constants only, in a manner which is fully explained in the appendix. It is then possible to determine the correct values of the constants and so to plot the surface of the film which represents the required function.

## 2. Methods of Experiment

Theoretically the torsion of hollow sections could be investigated by means of 'pressure' films in a manner similar to that employed for solid ones. As a matter of fact, however, this method suffers from several serious practical disadvantages when applied to multiple boundaries, and it has been found as a consequence that the initially greater difficulty of building up the boundaries of zero-pressure films is more than compensated for by the considerable saving effected in the subsequent operations.

In the first place it is much more troublesome to stretch a film over a number of boundaries than it is to stretch it over a single boundary, and it is a fact of experience that the life of the former film is very much shorter than that of the latter, so much so that, under ordinary working conditions, the time taken to adjust the pressure acting on the film is usually an important fraction of its life. In addition, it is to be noted that the necessary use of two films in the pressure method unavoidably doubles the chance that one will break prematurely.

A further consideration of no little weight is that the zero-pressure method may be made considerably more accurate than its rival. This is partly due to the possibility of smoothing out edge-effect errors by adjusting the measured contour lines, as explained in R. & M. no. 399, but also to the fact that it is usually possible, in reducing the problem to the zero-pressure form, to choose a function which can be represented on a much larger scale than would be possible with the original blown-up bubble. This point is discussed in detail in the appendix.

Lastly, it is easier to make the dispositions for altering the boundary constants in the case of the zero-pressure bubbles, as they need not be airtight.

In view of these considerations, it is advisable, in all applications of the soap-film method to problems involving multiple boundaries, to use zero-pressure films.

The general procedure is similar to that already described in the previous reports, the film being stretched over the boundaries with a piece of thread or twine dipped in soap solution. It is frequently necessary to resort to such devices as the use of temporary wire bridges to lead the twine from one boundary to the next, but these naturally suggest themselves during the manipulation of the films, whereas it is difficult to describe them without recourse to practical demonstration.

There are several obvious methods of securing the necessary adjustment of the heights of the interior boundaries. One is to mount each on a vertical sliding column. Alternatively, they may be supported on levelling screws, which may be located on the main levelling plate by means of geometric clamps.

The spherometer apparatus illustrated in R. &. M. no. 399 has been used exclusively in the work relating to hollow sections.

## 3. Experimental Results

(*a*) Fig. 1 shows the contour-line map obtained in the application of the soap-film method to the problem of the torsion of a circular tube having an eccentric bore.

Here the origin was taken at the centre of the outer circle, and the particular integral which was used in transforming the problem to the zero-pressure form was $\frac{1}{2}(x^2+y^2)$, or $\frac{1}{2}r^2$. Hence the outer boundary lies in a horizontal plane, while the inner one lies in a plane inclined at an angle to the horizontal.

The figures attached to the contours represent spherometer readings, the unit being 0·0025 in. To transform them to absolute units they must be divided by the 'scale constant' of the boundary, which is 67·2.

The more convenient torsion diagram of the contour lines of the constant-pressure film may be deduced from fig. 1 by adding to it the values of $\frac{1}{2}r^2$ at all points, due regard being paid to the scale factor.

It is found from the diagram that this tube has 0·924 time the torsional stiffness of a similar tube with concentric bore, and that the ratio of the maximum stresses, when both tubes undergo the same twist, is 1·373. Hence it is inferred that the ratio of the strength of this tube to the strength of the concentric one, under given torque, is 0·673. The maximum stress occurs, of course, on the outer boundary at the thinnest part.

As an illustration of the greater accuracy of the results obtainable by this method, as compared with the constant-pressure method, it may be mentioned that the volume which was measured, in this instance, in order to find the torsional stiffness, was only about one-twelfth of the volume representing the total stiffness which would have to be measured in the other method; so that, other things being equal, the experimental error should be only one-twelfth as great. Similarly, the maximum observed inclination was little more than a quarter of the inclination representing the maximum stress.

This torsion problem can also be solved analytically by transforming from Cartesian co-ordinates to the double system of orthogonally intersecting coaxial circles, cutting the common radical axis in real and imaginary points respectively. The two bounding circles are members of the latter family.

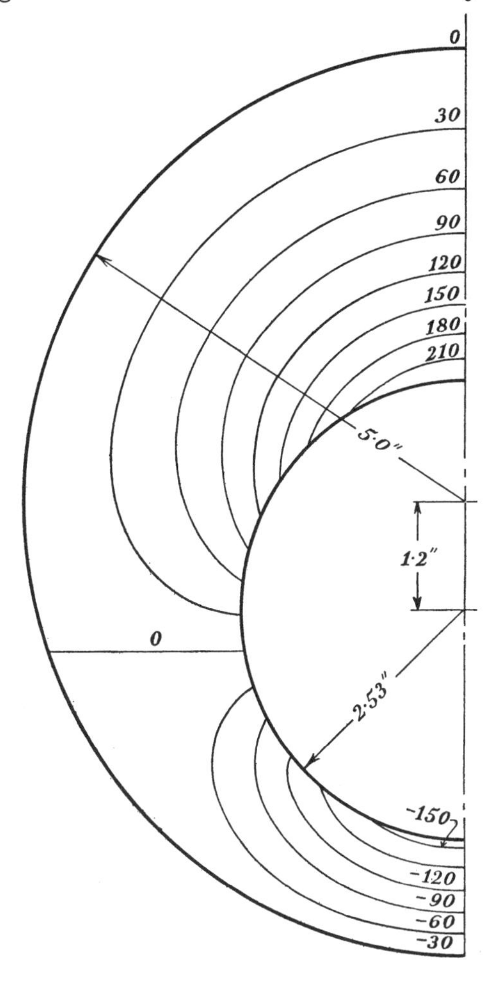


Fig. 1. Torsion function of a tube.

It was hoped to include the results of the mathematical analysis in the present paper, for the purpose of comparison with the experimental results; it is regretted, however, that the arithmetic could not be completed in time for inclusion. It is proposed to issue a supplementary report containing these results, together with a comparison of the soap-film and analytical solutions of the flexure problem for this section. It may be mentioned, however, that it has been established mathematically that the zero contour line of the diagram shown in fig. 1 is a straight line passing through the limiting point, or pole, of the boundary circles with respect to the common radical axis. The same result is obtained from the experimental contour map. This is a proof that the method used for determining the boundary constants does in fact give the correct results, which is quite independent of the direct one obtainable by finding the line integral of the normal inclinations of the soap film round the inner boundary.

(*b*) The following description of a research which has been carried out by the above method, whereby troubles which had been actually experienced in practice were explained and remedies suggested, affords an excellent example of the usefulness of soap films in engineering design.

The propellers of aircraft are usually mounted on hollow shafts and are prevented from rotating on those shafts by means of keys, or, alternatively, by cutting in the shaft axial teeth which engage with corresponding teeth in the sleeve of the propeller boss. In the former case the keyway which was cut in the shaft formed re-entrant angles in the outer boundary of the section, and the corners were commonly left sharp. These shafts often failed, owing to the torsional cracks which started at the corners. It was proposed to mitigate this evil by putting radii or fillets in the corners, and it was required to know what amount of rounding, if any, would make the shafts safe. Figures were also required for the comparative torsional strengths of the serrated and keywayed forms of shaft.

The necessary data were obtained by means of soap films. Before describing the work, however, it should be remarked that, although the soap-film results are only accurately true of steel members when the latter are everywhere stressed below the yield point, yet it is to be expected that they will afford a fairly good indication of the strength of the shafts at present under consideration, since the vibration and consequent stress variation, which are unavoidable under the extraordinary working conditions of aeroplane engines, render the ductility of the steel of less account in minimizing the importance of localized high stresses.

In all cases the shafts investigated were 10 in. outside and 5·8 in. inside diameter.

It is customary, in current engineering practice, to express the strength of a keywayed shaft in terms of the strength of the corresponding uncut shaft, whereas a serrated shaft is thought of in relation to one whose diameter equals the diameter of the serrated shaft at the bottom of the teeth. This is unfortunate, since it is evident that the only rational figure of torsional merit is the strength of a shaft of given weight and given maximum diameter. If no limit is set to the latter, we can, theoretically at least, make the figure of merit anything we please.

In the subjoined results the term 'stiffness ratio' is used to denote the ratio of the stiffness of the shaft under consideration to the stiffness of an uncut shaft 10 in. outside diameter, having a (concentric) bore 5·8 in. diameter. 'Stress ratio' means the ratio of the maximum (or other specified) stress to the maximum stress in the standard, when both undergo the same twist, while the ratio of the torques required to cause the same stress in each is called the 'strength ratio'.

All the results are, of course, true for similar shafts of different absolute dimensions.

The subjoined table (table 1) shows the data obtained for a shaft, having a keyway 2·5 in. wide and 1 in. deep, for a series of values of the radius in the corner. With a sharp corner the stress is, of course, theoretically infinite. The stress and strength ratios are given (*a*) on the basis of the stress in the radius, (*b*) on the basis of the stress at the middle of the keyway.

Table 1. *Torsion of keywayed hollow shafts*

| Radius in inches | Stiffness ratio | Stress ratio | | Strength ratio | |
|---|---|---|---|---|---|
| | | *a* | *b* | *a* | *b* |
| 0·1 | 0·900 | 4·84 | 2·05 | 0·186 | 0·438 |
| 0·2 | 0·905 | 3·05 | 2·00 | 0·297 | 0·452 |
| 0·4 | 0·915 | 2·11 | 1·97 | 0·435 | 0·465 |
| 0·7 | 0·930 | 1·75 | 1·85 | 0·531 | 0·502 |

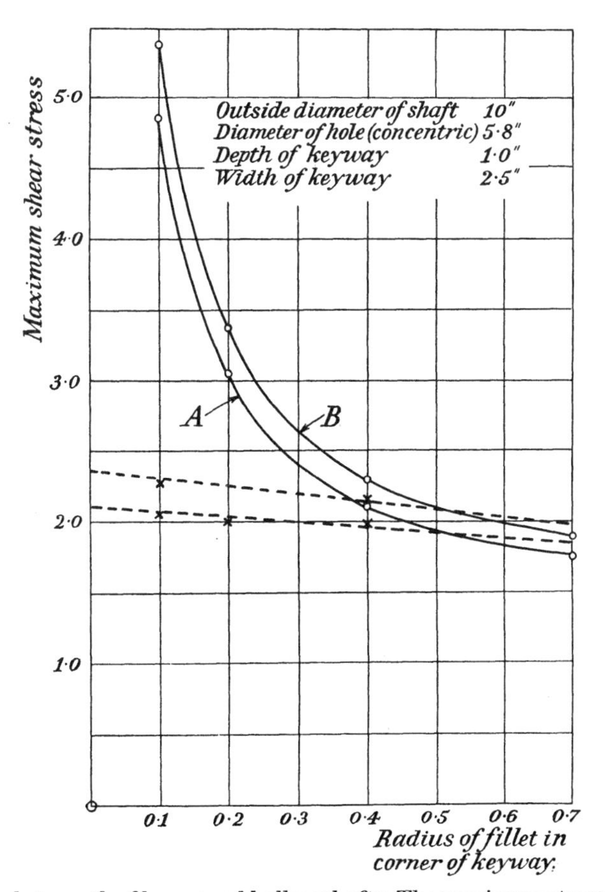


Fig. 2. Torsional strength of keywayed hollow shafts. The maximum stress is given as a multiple of the maximum stress in a similar shaft without keyway: (A) When the two shafts are twisted through the same angle. (B) When they are subjected to the same torque. The radius of the fillet is given as a fraction of the depth of the keyway. The dotted curves show the respective stresses in the middle of the keyway.

4-2

These figures are shown graphically in fig. 2 (in the case of the strength ratios, the reciprocals have been plotted, as is indicated below the diagram).

It is evident that if the radius of the fillet is less than one-fifth of the depth of the keyways the stress in it is dangerously high. For fillets greater than 0·3 of the depth

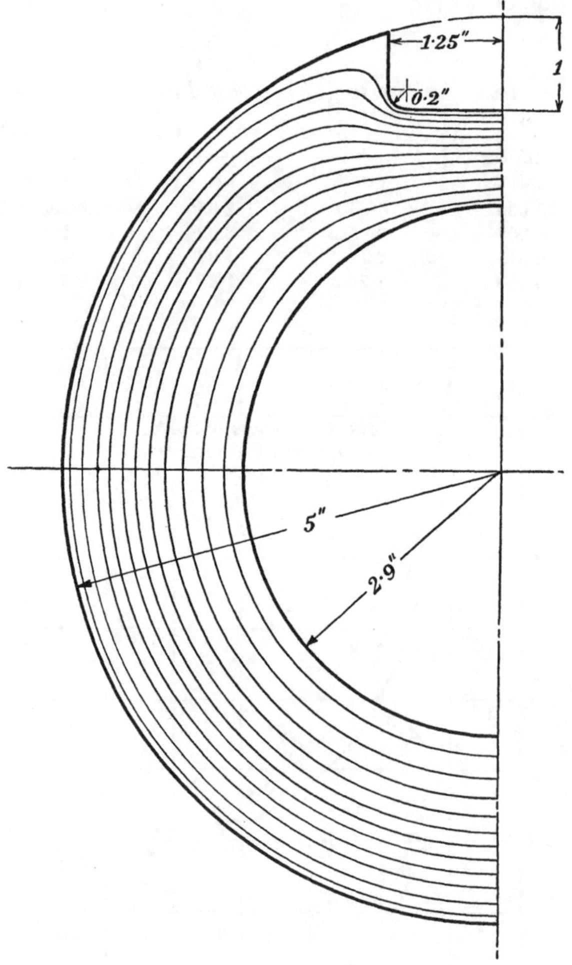


Fig. 3. Lines of shearing stress in the torsion of a keywayed hollow shaft.

of the slot, the rate of increase of strength with increase of radius is comparatively small. A curious point which will be noted is that for radii greater than 0·5 in. the stress in the radius is less than the stress at the middle of the keyway.

The torsional weakness of hollow keywayed shafts is due only in part to the local region of high stress at the corner—in all shafts where the bore is concentric with the outside there is, necessarily, a wide area of high stress in the thin part of the wall between the keyway and the bore. This is brought out very clearly by the diagram of lines of shearing stress in fig. 3. The example selected is that where the radius of the fillet is 0·2 in. The close grouping of the lines round the slot, connoting weakness, is immediately evident. This shaft is, in fact, less than half as strong as the uncut shaft, even under a steady torque where the ductility of the material would render regions of local high stress unimportant.

The appearance of the soap films suggested that this source of weakness could be avoided by throwing the bore out of centre, so as to make the wall thicknesses at the keyway and on the opposite side of the shaft the same.

Table 2. *Torsion of keywayed shafts with eccentric bores*

| Radius in inches | Stiffness ratio | Stress ratio | | Strength ratio | |
|---|---|---|---|---|---|
| | | *a* | *b* | *a* | *b* |
| 0·1 | 0·890 | 1·965 | 1·320 | 0·452 | 0·675 |
| 0·2 | 0·892 | 1·471 | 1·310 | 0·606 | 0·680 |
| 0·4 | 0·902 | 1·317 | 1·310 | 0·685 | 0·689 |
| 0·7 | 0·916 | 1·291 | 1·305 | 0·709 | 0·689 |

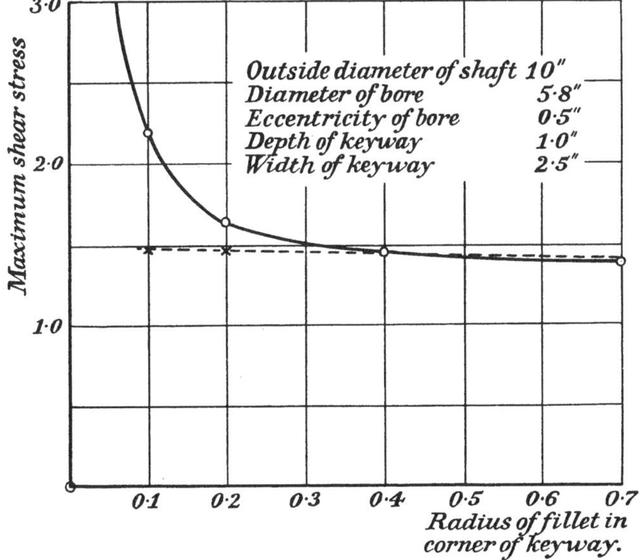


Fig. 4. Torsional strength of keywayed hollow shafts. The maximum stress is given as a multiple of the maximum stress in a shaft, of the same size, without keyway, having a concentric bore of the same diameter, when the two shafts are subjected to the same torque. The dotted curve gives the stress at the middle of the keyway. The radius of the fillet is given as a fraction of the depth of keyway.

Soap-film measurements on shafts with eccentric bores showed that this inference was perfectly correct. The following table (table 2) gives the results obtained when the bore of the shaft was thrown 0·5 in. out of centre.

The reciprocal of the strength ratio is plotted in fig. 4, while fig. 5 shows the contour lines for the case where the radius equals 0·2 in.

It is evident that there is a very large reduction, not only of the stress in the middle of the keyway, but also of the local stress in the corner. The contour map illustrates very clearly the great improvement in the stress distribution which is obtainable by throwing the bore out of centre.

The figures for the serrated shaft, when the radius at the root of the teeth has a series of values, are given below in table 3.

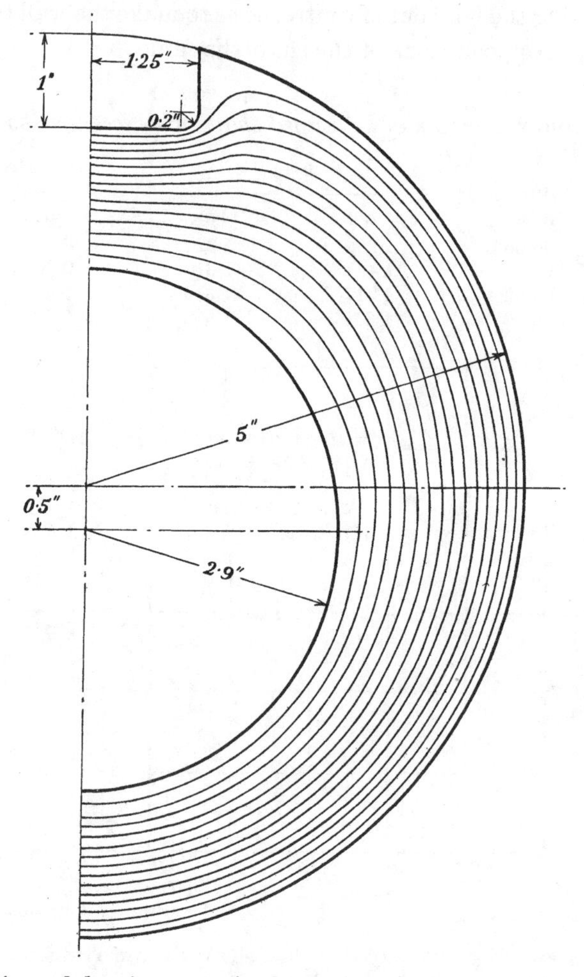


Fig. 5. Lines of shearing stress in the torsion of a keywayed hollow shaft.

Table 3. *Torsion of serrated shafts*

| Radius in inches | Stiffness ratio | Stress ratio | Strength ratio |
|---|---|---|---|
| 0·10 | 0·618 | 2·11 | 0·293 |
| 0·15 | 0·642 | 1·70 | 0·377 |
| 0·20 | 0·670 | 1·67 | 0·401 |
| 0·25 | 0·700 | 1·64 | 0·425 |

A section of the teeth, for the case where the radius is 0·1 in., together with the lines of shearing stress, is shown in the diagram (fig. 6). This value of the radius is representative of current practice.

It would appear that the shaft could be strengthened quite appreciably by increasing the radius 50 %, but the saving effected by enlarging it still further would hardly justify such a course.

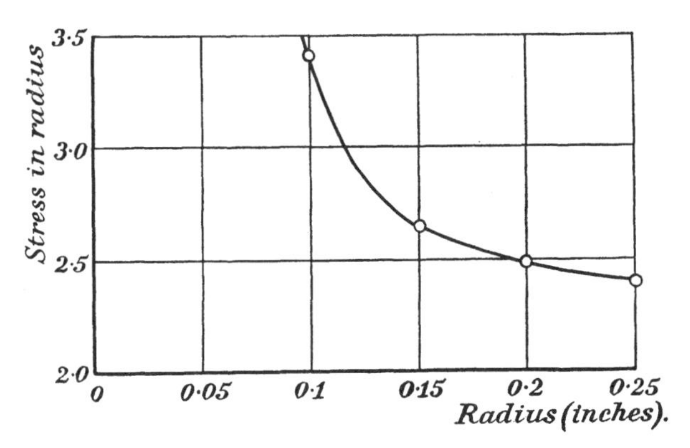


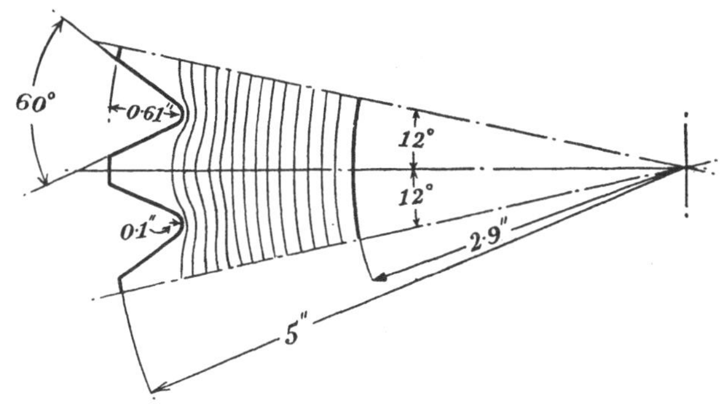


*View of section and lines of shearing stress.*

Fig. 6. Torsional strength of serrated shafts.

The stiffness of the shaft is seriously reduced by the cutting of teeth in it. The reduction is, in fact, nearly double the diminution in the polar moment of inertia, as is evident alike from the table and the diagram of shearing stress.

## 4. Application to Flexure Problems

The method of applying the present theory to the problem of finding flexural shear stresses differs from the foregoing only in the method of calculating the shape of the boundary, and hence no separate examples need be given. The torsion problem has been chosen for the purpose of illustration on account of its greater practical importance. As has been mentioned, it is hoped to give an example of the flexure problem in a supplementary paper.

## MATHEMATICAL APPENDIX

### (1) *General Theory*

Both the torsion and flexure problems can be reduced to the finding of a function $\psi$, which satisfies the differential equation

$$\frac{\partial^2\psi}{\partial x^2}+\frac{\partial^2\psi}{\partial y^2}=0 \tag{1}$$

at all points of the cross-section of the bar, and the equation

$$\psi=F+C \tag{2}$$

round the boundary, $F$ being a known function of $x$ and $y$ and $C$ an unknown constant.

The addition of a constant to the function $\psi$ does not affect the stress distribution, which it is the object of the investigation to obtain, and hence the boundary equation is, for our purpose, determinate, provided that the boundary is a single curve. It follows that the problem can be solved directly by the soap-film method.

If, however, the boundary is composed of $n$ separate curves, or, in other words, if the section is $n$-tuply connected, there must, in general, be $n$ undetermined constants, one for each boundary. One only of these can be eliminated by means of the additive constant, and it is therefore necessary to seek a method of finding the other $n-1$ constants.

To do this we note that the function $\phi$ conjugate to $\psi$ represents a distortion of the cross-sections of the beam, and hence it must be single-valued. The solution of the equations for $\psi$ with any arbitrary values of the constants leads to a corresponding solution of the equations for $\phi$, but in only one of these is $\phi$ a single-valued function.

The mathematical expression of the fact that $\phi$ is single-valued is that

$$\int\frac{\partial\phi}{\partial s}ds=0, \tag{3}$$

the integration being taken round any closed curve in the section. But

$$\frac{\partial\phi}{\partial s}=\frac{\partial\psi}{\partial\nu}, \tag{4}$$

$\nu$ being the normal to the curve, and hence

$$\int\frac{\partial\psi}{\partial\nu}ds=0 \tag{5}$$

round any closed curve.

In order that $\phi$ may be single-valued, this equation must be satisfied when the integration is taken round each of the boundaries of $\psi$.

It will be seen that (5) is equivalent to the condition that the resultant of the components of surface tension acting on the boundary, normal to the $xy$ plane, is zero.

Since the condition must hold for each boundary we apparently get $n$ equations to determine the $n-1$ constants, but it is obvious that one of the equations must be satisfied identically.

It is easy to see that similar conditions must hold for 'constant-pressure' films, which represent $\psi$ in

$$\frac{\partial^2\psi}{\partial x^2}+\frac{\partial^2\psi}{\partial y^2}+2=0. \tag{6}$$

In this case it is found that the constants must be determined in such a way as to make the normal components of surface tension balance the resultant forces due to the air pressure acting on the areas enclosed by the interior boundaries.

The correct values of the $n-1$ constants may be found, in both instances, by making suitable measurements on $n$ films whose boundaries satisfy equation (2), the constants being chosen arbitrarily, save for one condition which will be discussed later, in such a way that no two of the films are alike.

The required observations comprise measurements of normal inclinations round all the boundaries of each of the $n$ films.

If $\psi_1\psi_2\psi_3\ldots\psi_n$ denote the $n$ films, the quantities $a, b, c, \ldots, n$, the boundary constants, and $p, q, r, \ldots, z$, the observed values of $\int\frac{\partial\psi}{\partial\nu}ds$ round the boundaries of the films, the notation may be represented by the scheme:

| | Boundary (1) Values of | | Boundary (2) Values of | | ... | Boundary ($n$) Values of | |
|---|---|---|---|---|---|---|---|
| Film | $\psi$ | $\int\frac{\partial\psi}{\partial\nu}ds$ | $\psi$ | $\int\frac{\partial\psi}{\partial\nu}ds$ | | $\psi$ | $\int\frac{\partial\psi}{\partial\nu}ds$ |
| $\psi_1$ | $F_1+a_1$ | $p_1$ | $F_2+a_2$ | $p_2$ | ... | $F_n+a_n$ | $p_n$ |
| $\psi_2$ | $F_1+b_1$ | $q_1$ | $F_2+b_2$ | $q_2$ | ... | $F_n+b_n$ | $q_n$ |
| $\psi_3$ | $F_1+c_1$ | $r_1$ | $F_2+c_2$ | $r_2$ | ... | $F_n+c_n$ | $r_n$ |
| — | — | — | — | — | ... | — | — |
| $\psi_n$ | $F_1+n_1$ | $z_1$ | $F_2+n_2$ | $z_2$ | ... | $F_n+n_n$ | $z_n$ |

Now the function
$$\psi=m_1\psi_1+m_2\psi_2+\ldots m_n\psi_n, \tag{7}$$
where
$$m_1+m_2+\ldots+m_n=1 \tag{8}$$
is a solution of equations (1) and (2), and the values of

$$\int\frac{\partial\psi}{\partial\nu}ds$$

round the $n$ boundaries are, in this case,

$$m_1p_1+m_2q_1+\ldots+m_nz_1,$$
$$m_1p_2+m_2q_2+\ldots+m_nz_2,$$
$$\ldots\ldots\ldots\ldots\ldots\ldots\ldots\ldots\ldots$$
$$m_1p_n+m_2q_n+\ldots+m_nz_n.$$

Hence, if we choose $m_1, m_2, \ldots, m_n$ so that

$$\left.\begin{aligned} m_1p_1+m_2q_1+\ldots+m_nz_1&=0,\\ m_1p_2+m_2q_2+\ldots+m_nz_2&=0,\\ \ldots\ldots\ldots\ldots\ldots\ldots\ldots\ldots&\ldots\\ m_1p_n+m_2q_n+\ldots+m_nz_n&=0,\end{aligned}\right\} \tag{9}$$

the condition (5) is satisfied for each boundary, and $\psi$ is the required solution of equations (1) and (2).

The values of $m_1, m_2, \ldots, m_n$ are, of course, given by

$$\frac{m_1}{\begin{vmatrix} q_2r_2 & \ldots & z_2\\ q_3r_3 & \ldots & z_3\\ \ldots & \ldots & \ldots\\ q_nr_n & \ldots & z_n\end{vmatrix}}=\frac{m_2}{\begin{vmatrix} r_2 & \ldots & z_2p_2\\ r_3 & \ldots & z_3p_3\\ \ldots & \ldots & \ldots\\ r_n & \ldots & z_np_n\end{vmatrix}}=\text{etc.} \tag{10}$$

together with (8), and the correct values of the constants, $c_1c_2c_3\ldots c_n$ are given by

$$\left.\begin{aligned} c_1&=m_1a_1+m_2b_1+\ldots+m_nn_1,\\ c_2&=m_1a_2+m_2b_2+\ldots+m_nn_2,\\ \ldots&\ldots\ldots\ldots\ldots\ldots\ldots\ldots\ldots\\ c_n&=m_1a_n+m_2b_n+\ldots+m_nn_n.\end{aligned}\right\} \tag{11}$$

The boundaries may now be adjusted so as to conform to these values of the constants and the function $\psi$ plotted.

The constants $a, b, c, \ldots, n$ cannot be chosen entirely arbitrarily. Let $\psi_0$ be the function satisfying the body equation

$$\frac{\partial^2\psi_0}{\partial x^2}+\frac{\partial^2\psi_0}{\partial y^2}=0, \tag{12}$$

and the boundary equations

$$\psi_0=F_1,\quad \psi_0=F_2,\quad \text{etc.}, \tag{13}$$

must be satisfied on the $n$ bounding curves.

The $n$ functions

$$\left.\begin{aligned} &\psi_1-\psi_0,\\ &\psi_2-\psi_0,\\ &\ldots\ldots\\ &\psi_n-\psi_0,\end{aligned}\right\} \tag{14}$$

satisfy body equations of the type (1) and have at the boundaries the constant values,

$$\begin{gathered} a_1, a_2, a_3, \ldots, a_n,\\ b_1, b_2, b_3, \ldots, b_n,\\ \ldots\ldots\ldots\ldots\ldots\\ n_1, n_2, n_3, \ldots, n_n.\end{gathered}$$

Now, suppose that it is possible to find $n$ quantities $\alpha, \beta, \gamma, \ldots, \mu$ such that

$$\left.\begin{aligned} a_1\alpha+b_1\beta+c_1\gamma+\ldots+n_1\mu=0, \\ a_2\alpha+b_2\beta+c_2\gamma+\ldots+n_2\mu=0, \\ \ldots\ldots\ldots\ldots\ldots\ldots\ldots\ldots\ldots \\ a_n\alpha+b_n\beta+c_n\gamma+\ldots+n_n\mu=0, \end{aligned}\right\} \tag{15}$$

then we can form any one of the functions (14) by adding together suitable multiples of the other $n-1$ functions.

It appears from (9) that we can get, from the $n$ films, $n$ relations from which to find the $n-1$ unknown constants. Actually, however, one of the equations (9) is identically satisfied, and we really get the right number of equations from the $n$ films. If, however, equations (15) can be satisfied, one of our films is merely a surface derived from the others, and we have only $n-1$ independent films. Hence we have only $n-2$ equations of the type (9), and the constants cannot be found.

The condition that this shall not be so is

$$\begin{vmatrix} a_1, & a_2, & a_3, & \ldots, & a_n \\ b_1, & b_2, & b_3, & \ldots, & b_n \\ \ldots & \ldots & \ldots & \ldots & \ldots \\ n_1, & n_2, & n_3, & \ldots, & n_n \end{vmatrix} \neq 0, \tag{16}$$

which must accordingly be satisfied by these otherwise arbitrary constants.

The simplest scheme which conforms to (16) is to change the boundaries successively by a uniform amount '$a$', all boundaries but one being left unaltered on each occasion. (16) then becomes

$$\begin{vmatrix} a, & 0, & 0, & \ldots, & 0 \\ 0, & a, & 0, & \ldots, & 0 \\ 0, & 0, & a, & \ldots, & 0 \\ \ldots & \ldots & \ldots & \ldots & \ldots \\ 0, & 0, & 0, & \ldots, & a \end{vmatrix} = a^n, \tag{17}$$

which cannot be zero. Equations (11) then take the form

$$\left.\begin{aligned} c_1=m_1a, \\ c_2=m_2a, \\ c_3=m_3a, \end{aligned}\right\} \tag{18}$$

etc.

### (2) *Reduction of Torsion Function to Zero-pressure Form*

The solution of the torsion problem depends on the determination of a function $\psi'$ which satisfies the body equation

$$\frac{\partial^2\psi'}{\partial x^2}+\frac{\partial^2\psi'}{\partial y^2}+2=0, \tag{19}$$

and the boundary condition

$$\psi' = C, \text{ a constant.} \tag{20}$$

Choose a function $\psi$, such that

$$\psi' = \psi - (Ax^2 + 2Hxy + By^2 + Gx + Fy + D)/(A + B), \tag{21}$$

where $A$, $B$, $D$, $F$, $G$, $H$ are arbitrary constants.

Then $\psi$ satisfies the body equation

$$\frac{\partial^2 \psi}{\partial x^2} + \frac{\partial^2 \psi}{\partial y^2} = 0, \tag{22}$$

and the boundary condition

$$\psi = C + (Ax^2 + 2Hxy + By^2 + Gx + Fy + D)/(A + B), \tag{23}$$

and the problem is reduced.

Equation (23) gives the boundary heights, and it is evident that if the constants be suitably chosen, so as to make the difference between the maximum and minimum heights as small as possible, the film can be made to represent the torsion function on the largest possible scale, and hence experimental errors may be minimized. Take, for example, the section of a propeller blade. Suppose the chord to be 10 in. and the maximum thickness 1 in. Take the origin $\frac{1}{2}$ in. from the flat face and 5 in. from either end, and let the axis of $x$ be parallel to the chord.

Let $B = 1$ and $A = H = F = G = D = 0$.

Then the difference between the greatest and least values of $\psi$ on the boundary represents $\frac{1}{4}$ sq.in. Had we employed the usual particular integral wherein $A = B = \frac{1}{2}$, the greatest difference would have represented 12·56 sq.in., and the experimental errors would have been 50 times as great.

It is of interest to note that the boundary of the film represents the section of the quadric surface

$$(A + B)z = Ax^2 + 2Hxy + By^2 + Gx + Fy + D \tag{24}$$

by a cylinder whose axis is parallel to $oz$ and whose cross-section is the figure under investigation. The choice of appropriate constants resolves itself into a question of selecting a quadric whose intersection with the cylinder lies as nearly as possible in a plane parallel to $xoy$.

# 4

# A RELATION BETWEEN BERTRAND'S AND KELVIN'S THEOREMS ON IMPULSES

REPRINTED FROM

*Proceedings of the London Mathematical Society*, ser. 2, vol. XXI (1922), pp. 413–14

The theorems known as Bertrand's and Kelvin's theorems on impulses relate to the effect of constraints on the energy of a dynamical system when it is set in motion by impulses. Bertrand's theorem states that if a system is set in motion by given impulses applied to given co-ordinates, the effect of a constraint is to reduce the energy which the impulses give to the system. Kelvin's theorem states that if, instead of the impulses, the velocities imparted by those impulses to the corresponding co-ordinates are given, the effect of a constraint is to increase the energy of the motion.

The following simple connection between these two theorems does not seem to have been pointed out before. 'The reduction in energy due to the imposition of constraints in the Bertrand case is less than the increase in energy due to the imposition of the same constraints in the Kelvin case.'

The proof of this theorem is very simple. Let $T_o$ represent the energy of a dynamical system due to a set of impulses $p_r$ applied to co-ordinates $q_r$, another set of co-ordinates $q_s$ being unconstrained, and let the velocities due to these impulses be $\dot{q}_r$. This motion will be represented by the symbol $(o)$.

Suppose now that the system is constrained by fixing certain co-ordinates $q_s$. Two cases will be considered. $(a)$ Bertrand's case where the system is set in motion by impulses $p_r$ applied to co-ordinates $q_r$. The energy $T_a$ of the constrained system is less than $T_o$. This is proved by showing that $T_o - T_a$ is equal to the energy of the system whose velocity is equal to the velocity of the motion which is the difference between $(o)$ and $(a)$. This motion, which will be denoted by $(o-a)$, is that produced by the action of a set of impulses $p_s$* which act on the co-ordinates $q_s$ producing velocities $\dot{q}_s$,† the co-ordinates being *unconstrained.*

The next constrained motion, $(b)$, to be considered is Kelvin's case when the constrained system is set in motion by the application of a set of impulses $p'_r$, say, acting on $q_r$ so as to produce velocities $\dot{q}_r$. The energy $T_b$ of this motion is greater than $T_o$. This is proved by showing that $T_b - T_o$ is equal to the energy of the motion $(o-b)$ which is the difference between motions $(o)$ and $(b)$. This derived motion is that produced by the action of a set of impulses $p'_s$‡ which produce the same velocities $\dot{q}_s$

* $p_s$ are the impulsive reactions at the constraints in case $(a)$.

† $\dot{q}_s$ are the velocities of the co-ordinates $q_s$ in case $(o)$.

‡ $p'_s$ are the impulsive reactions at the constraints in case $(b)$.

when they act on $q_s$ as were produced in motion $(o-a)$ by the action of the impulses $p_s$, but in this case the co-ordinates $q_r$ are *fixed*.

It will be seen that the relation between motions $(o-a)$ and $(o-b)$ is exactly that contemplated in Kelvin's theorem. Hence $T_b - T_o$ is greater than $T_o - T_a$, which is the theorem that was to be proved.

The difference between $T_b - T_o$ and $T_o - T_a$, that is, $T_a + T_b - 2T_o$, is, in fact, the energy of the system which is the difference between the motions $(a)$ and $(b)$, that is to say, it is the energy of the system set in motion by the application of impulses $p'_r - p_r$ to the co-ordinates $q_r$, when the system is constrained by fixing the co-ordinates $q_s$.

The physical bearing of these results may be seen at once from the consideration of a simple case. Consider the case of a rigid rod which is set in motion by an impulse applied at some point in a direction perpendicular to its length. If the constraint consists in the fixing of a point in the rod, it is clear that the energy of the constrained motion tends to zero in the Bertrand case when the fixed point is very near to the point of application of the impulse. In the Kelvin case, however, the energy of the constrained motion tends to infinity. The reduction in energy due to the constraint in the Bertrand case can never be greater than the energy of the unconstrained motion, whereas the increase in energy due to the constraint in the Kelvin case may tend to become infinite.

# 5

# THE DISTORTION OF AN ALUMINIUM CRYSTAL DURING A TENSILE TEST*

(Bakerian Lecture to the Royal Society, delivered 22 February 1923)

REPRINTED FROM

*Proceedings of the Royal Society*, A, vol. CII (1923), pp. 643–67

The work described in the following pages was inspired by a paper in which Professor Carpenter† and Miss Elam described the result of applying tensile tests to specimens of aluminium which had been treated in such a way that they appeared to turn into single crystals. The resulting distortions of the test pieces were very remarkable and clearly suggested that the crystal axes were not orientated in the same direction in different specimens. The uniformity of the distortion in different parts of the same specimen made it seem likely that it would be a straightforward, though possibly laborious, matter to determine the relationship between the orientation of the axes and the distortion produced in a tensile test. And it seemed possible that by examining a number of specimens some general results might be obtained about the forces necessary to produce distortions of this type.

On discussing the matter with Professor Carpenter and Miss Elam it was found that it would not be possible to determine the distortion from the measurements they had already made. Moreover, no measurements of the orientation of the crystal axes had been made, though Sir W. Bragg had made a few observations indicating that the material retained its crystalline character after it had been distorted. Under these circumstances, it was decided to carry out a test, making all the necessary measurements at various stages during the extension of a specimen.

Before describing the test, however, it is necessary to refer to the work of previous experimenters on the subject.

## PREVIOUS WORK‡

When a metal is strained beyond the elastic limit a microscopic examination of the surfaces of crystals in it frequently shows the existence of lines known as slip bands.

* With C. F. ELAM.

† *Proc. Roy. Soc.* A, C (1921), 329.

‡ *Note added 8 February 1923.* Since this paper was communicated, an account of some similar work by Mark, Polanyi and Schmid has appeared in the *Zeitschrift für Physik*, December 1922. The distortion of a zinc crystal is discussed. The method used is based chiefly on measurements of slip bands (which in the case of aluminium were found to be nearly useless as a basis for distortion measurements). The results obtained are similar to those described in the present work, but, owing to the fact that zinc has hexagonal instead of cubic symmetry, the complication introduced by the existence of a large number of crystallographically similar slip-planes does not arise.

These bands have been shown to consist of small steps, and the conclusion is naturally drawn that there are planes inside the crystal, presumably crystal planes, on which slipping takes place. The bands would then mark the intersection of the face of the specimen with these crystal planes.

Up to the present, however, the evidence on slipping is purely qualitative. It has not been shown that the deformation of a metallic crystal when the material is strained is such as could be produced by slipping, nor has the relationship between the crystal axes and the slip planes been determined.

A number of experiments have been made on the direction of the crystal axes in drawn wires, and certain conclusions have been reached by Polanyi* in regard to the orientation of the crystal axes with respect to the axis of the wire. References are given in Polanyi's paper.

## General Description of Test

Before the test contemplated could be carried out, it was necessary to enlist the help of an expert in crystal analysis by X-rays. Fortunately, Dr Alex Müller took up the work and succeeded in devising a satisfactory method of determining the orientation of the crystal axes. This method and some of his results will be described elsewhere.

The results of his and Miss Elam's X-ray analysis of the specimen with which the present work is concerned are given in table 5.

Most of the specimens with which Professor Carpenter and Miss Elam's experiments had been carried out were flat strips, about $\frac{1}{8}$ in. thick $\times$ 1 in. broad. These strips were unsuitable for the present purpose, partly because of the difficulty of making accurate measurements on the narrow faces and partly because they were so broad in comparison with their length that it was not possible to make sure of getting an evenly stretched parallel piece in the middle.

Specimens could be produced with circular, but not with square sections. On the other hand, there seemed to be no very simple way of making on a round specimen† the measurements which are necessary for calculating its distortion. For this reason, therefore, a round specimen was machined down till its section was square. Its dimensions were then approximately $1 \cdot 0 \times 1 \cdot 0 \times 20 \cdot 0$ cm. Each face was marked by a scratch parallel to the length of the specimen or axis, as it will be called, and by cross-scratches. The appearance of the specimens so marked is shown in fig. 1. The faces were numbered 1, 2, 3, 4, so that when the specimen was placed upright in the testing machine, the faces appeared in this order when the observer moved round the machine in an anti-clockwise direction. Fig. 1 represents the specimen lying with its top end to the right.

At each successive stage of the test the extension between each pair of cross-marks

* Polanyi, *Die Naturwissenschaften*, April 1922.

† This difficulty has now been overcome.

was measured on each face. The ratio of the length at any stage to the initial length will be denoted by the symbol $\epsilon$. At the same time the angles between these cross-scratches and the longitudinal scratch were measured in each case. These will be denoted by $\beta$ on face 1 and $\gamma$ on face 4. The thickness of the specimen between pairs of opposite faces, $t_{13}$ and $t_{24}$, and the angles, $\lambda$, between neighbouring faces were also measured. These measurements are sufficient to determine the nature of the distortion. The scheme is illustrated in fig. 1.

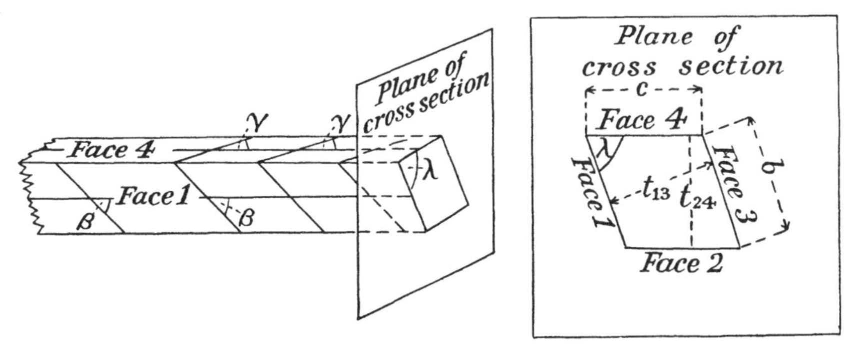


Fig. 1. Scheme for marking and measuring specimen.

## Methods of Measurement

The measurements of extension were made with a reading microscope, and the measurements of thickness between opposite faces with a micrometer. The angles between the cross-lines and the axis of the specimen were measured with a crystallographer's microscope, with rotating eyepiece, containing cross-wires, which was kindly lent us by Mr A. Hutchinson, F.R.S. The angles between the faces were measured by sticking small pieces of cover-slip glass to them with gum, setting the specimen upright in a goniometer, and observing the reflection in the glass of a distant source of light.

## Degree of Uniformity of Measurements

If the distortion of the specimen had been uniform throughout its entire volume, and if the section had been accurately rectangular to begin with, the angles of the cross-scratches measured on pairs of opposite faces would have been the same, the extensions measured on all four faces would have been the same, and the angles between any pair of faces would have been the same as that between the opposite pair.

It turned out that near the ends of the specimen, where it was held in the grips of the testing machine, the measurements were not quite the same as those near the middle, but that the central portion was nearly uniformly strained. The degree of uniformity can be judged from the figures given in table 1, which represent the ratio of the extended to the initial lengths in each of the five compartments into which the faces were divided by the six cross-lines. The ratio will be represented by the symbol $\epsilon$, $\epsilon - 1$ being the extension.

In table 1 the figures in each vertical column represent the values of $\epsilon$ for one of the eight stages of the test at which the specimen was measured. These stages correspond roughly with 5, 10, 15, 20, 30, 40, 60 % extension, and the last column represents the extension after the specimen had broken at 78 % extension.

Similar tables were prepared for the other measurements, but it seems unnecessary to print them in detail as they show a similar type of uniformity among the measurements taken in the central part of the specimen.

In fig. 2 is shown a photograph* of part of the specimen after it had broken. It will be seen that its appearance suggests that the distortion is due to a uniform strain at all points, except those which lie in a small region close to the point where the breakage occurred.

Fig. 2

### Dimensions of Mean Parallelepiped

On looking at table 1 it will be seen that there is a high degree of uniformity among the measurements of extension in the three middle compartments on each face, especially in the earlier stages of the test.

Similar conclusions were reached by inspection of the other measurements; accordingly, the mean has been taken in each case of the measurements made in each of the three central compartments. The mean extension, for instance, at any stage is the mean of the twelve figures obtained on all four faces. These are given at the bottom of table 1.

The mean angle of the cross-line with the axis of the specimen is found by taking the mean of the eight measurements made on the four middle cross-marks on two opposite faces.

In this way the dimensions and angles of a series of mean parallelepipeds have been drawn up, and these have been used in calculating the strain, or distortion, of the specimen at any stage. They are given in table 2.

It is known that when aluminium is worked the change in density is small, at any rate it is less than 1 %. The volumes of the strained figures should differ by less than 1 % from the volumes of the unstrained figure. The ratio of the strained to unstrained volumes has been calculated from the figures in table 2. They are given at the bottom of table 2, and on inspecting them it will be seen that none of them differ from 1 by more than 1 %, except the figure in the last column but one corresponding to an extension of 62 %. This differs from 1 by nearly 8 %. It seems probable, therefore, that some mistake has been made in this case, and all the measurements for this stage have accordingly been rejected.

* Reproduced here as a line drawing.

Table 1. *Ratio of extended to initial length in each of the five compartments into which each face was divided*

| Approximate extension (%) ... | 5 | 10 | 15 | 20 | 30 | 40 | 60 | 78 |
|---|---|---|---|---|---|---|---|---|
| Face 1 | 1·049 | 1·103 | 1·157 | 1·196 | 1·300 | 1·401 | 1·623 | — |
| | 1·047 | 1·108 | 1·160 | 1·199 | 1·309 | 1·414 | 1·645 | 1·834 |
| | 1·058 | 1·116 | 1·164 | 1·200 | 1·303 | 1·398 | 1·604 | 1·756 |
| | 1·052 | 1·109 | 1·160 | 1·199 | 1·300 | 1·394 | 1·595 | 1·739 |
| | 1·052 | 1·109 | 1·153 | 1·278* | 1·297 | 1·390 | 1·582 | 1·692 |
| Face 2 | 1·043 | 1·100 | 1·141 | 1·185 | 1·283 | 1·384 | 1·599 | — |
| | 1·048 | 1·106 | 1·160 | 1·199 | 1·300 | 1·403 | 1·628 | 1·805 |
| | 1·053 | 1·112 | 1·162 | 1·201 | 1·311 | 1·412 | 1·645 | 1·845 |
| | 1·051 | 1·107 | 1·161 | 1·198 | 1·295 | 1·393 | 1·649 | 1·720 |
| | 1·052 | 1·109 | 1·161 | 1·202 | 1·308 | 1·406 | 1·561 | 1·735 |
| Face 3 | 1·046 | 1·099 | 1·153 | 1·191 | 1·296 | 1·392 | 1·618 | 1·828 |
| | 1·051 | 1·111 | 1·162 | 1·202 | 1·311 | 1·424 | 1·650 | 1·845 |
| | 1·059 | 1·112 | 1·163 | 1·201 | 1·304 | 1·401 | 1·613 | 1·760 |
| | 1·054 | 1·110 | 1·163 | 1·201 | 1·298 | 1·397 | 1·596 | 1·800 |
| | 1·056 | 1·111 | 1·150 | 1·198 | 1·300 | 1·390 | 1·581 | 1·611 |
| Face 4 | 1·051 | 1·103 | 1·157 | 1·200 | 1·304 | 1·410 | 1·641 | — |
| | 1·050 | 1·111 | 1·169 | 1·203 | 1·316 | 1·420 | 1·648 | 1·830 |
| | 1·056 | 1·112 | 1·152 | 1·200 | 1·300 | 1·400 | 1·615 | 1·772 |
| | 1·054 | 1·109 | 1·159 | 1·198 | 1·299 | 1·394 | 1·594 | 1·720 |
| | 1·058 | 1·111 | 1·080 | 1·201 | 1·300 | 1·393 | 1·591 | 1·705 |
| Means of middle three | 1·053 | 1·110 | 1·161 | 1·200 | 1·304 | 1·404 | 1·623 | 1·785 |

* Probably an error in measurement.

Table 2. *Dimensions of mean parallelepiped*

| Extension | 0 | 1·053 | 1·110 | 1·161 | 1·200 | 1·304 | 1·404 | 1·623 | 1·785 |
|---|---|---|---|---|---|---|---|---|---|
| Thickness in mm.: $t_{13}$ | 10·16 | 9·68 | 9·17 | 8·82 | 8·55 | 7·90 | 7·38 | 6·57 | 6·03 |
| $t_{24}$ | 10·35 | 10·33 | 10·31 | 10·28 | 10·26 | 10·12 | 10·00 | 10·17 | 9·29 |
| Angle between faces 1 and 4, $\lambda^\circ$ | 90·6 | 88·8 | 87·2 | 86·7 | 84·9 | 81·8 | 79·2 | 73·1 | 70·8 |
| Angle between cross-marks and axis: | | | | | | | | | |
| Face 1, $\beta^\circ$ | 90·0 | 88·5 | 87·3 | 86·7 | 85·7 | 84·2 | 83·0 | 80·2 | 78·0 |
| Face 4, $\gamma^\circ$ | 90·0 | 90·3 | 89·7 | 89·3 | 88·5 | 85·2 | 82·1 | 73·2 | 65·5 |
| Volumes | 1·000 | 1·002 | 0·990 | 1·003 | 1·005 | 1·002 | 1·003 | 1·078 | 1·007 |

## Analysis of Strain in Mean Parallelepiped

It is obvious that the measurements given in table 2 are sufficient to determine the strain completely. A uniform strain can be specified by giving the directions and magnitudes of the axes of the ellipsoid into which a sphere of unit radius in the unstrained material is transformed by the strain. This ellipsoid, known as the 'strain ellipsoid', can be found from the measurements of table 2, but it is not at once obvious how the axes of the strain ellipsoid would be related to the crystal axes, though there would probably be some indirect connection between the two. On the other hand, there are other possible ways in which a uniform strain could be specified, and some of these may be more likely to throw light on the present problem than others.

For this reason it seems desirable to consider what types of strain can be conceived which would satisfy the conditions indicated by experiment, namely, that the material remains a crystal, so that all molecules are orientated in the same direction at any stage in the strain, and that the density is practically unchanged. In the first place, it seems clear that the relative displacements of neighbouring molecules cannot be the same as the relative displacements of particles in a similar position in the material in bulk. If they were the same, that is, if the strain were uniform even when portions of matter of molecular dimensions were examined, the material would remain crystalline when subjected to any uniform strain (it is true), but the crystal symmetry would be altered by the strain. A plane of molecules which lay in a crystal plane in the unstrained material would continue to do so in the strained material, and the angle between two such planes of molecules would alter continuously during any continuously varying strain. This state of affairs is mathematically conceivable, but it is contrary to all physical experience; moreover, measurements of the inclination of two different crystal planes during the present experiments showed that the angle between them remained practically constant during the whole course of the distortion (see last column table 5).

The fact that the material preserves its cubic crystal symmetry during the distortion leads to the conception that the apparently uniform strain of the material in bulk must be made up of a large number of non-uniform strains or relative displacements between neighbouring molecules.

In order that crystal symmetry may be preserved, the displacement of any molecule relative to a neighbouring molecule must be at least as great as the distance between neighbouring molecules in the direction of its displacement. A small strain in the material in bulk must therefore be due in some way to the occurrence of a small number of these relatively large displacements, the greater part of the molecules preserving their relative positions unchanged.

When a molecule is displaced along any line into the position of a neighbouring molecule, the molecule which previously occupied that space must displace in its turn a third molecule and so on. Owing to the fact that the stress in a material under a uniform tension is constant along any line, it seems likely that all the molecules in a line would be displaced together. By making appropriate combinations of displacements of lines of molecules parallel to a given line in the crystal, the material in bulk might be conceived to assume a great variety of shapes, but they would all be characterized by the fact that lines of particles parallel to this direction would be unchanged in length during the distortion.

These considerations, though admittedly hypothetical, suggest that it would be more promising to analyse the strain, with a view to finding the directions in the material which are unchanged in length, than to find the strain ellipsoid. In a uniform strain, the directions which remain unchanged in length lie on a quadric cone in the strained material. This cone evidently passes through the curved line of intersection of the strain ellipsoid and the unit sphere when these two are placed concentrically. The principal axes of this cone therefore coincide in direction with those of the strain ellipsoid.

The particles of this 'unstretched cone' have evidently two positions corresponding with the unstrained and the strained material respectively. It is necessary to determine both of them.

### Determination of the Cones of Unextended Directions

The method adopted was to find first a series of corresponding positions in the strained and unstrained material respectively of planes of particles which pass through one edge of the specimen; actually the intersection of faces 1 and 4 was chosen. This was accomplished as follows:

The cross-sections of the strained and unstrained specimen were set out on a piece of paper in such a way that the lines representing face 1 coincided in direction, while the points representing the edge of intersection of faces 1 and 4 coincided. The figure produced in this way is shown in fig. 3; $OPQR$ is the cross-section of the unstretched specimen, while $OSTU$ is that of the stretched specimen. Evidently, $OR$, $OU$, which are the traces of face 1, and $OP$, $OS$, which are the traces of face 4, are the traces of two pairs of corresponding planes; the diagonals $OQ$, $OT$ are another pair. To find others, take any point $V$ in $QR$, and let $n$ represent the ratio $RV:RQ$. Divide $UT$ so that $UW = nUT$. Then $OV$ and $OW$ are the traces of corresponding planes.

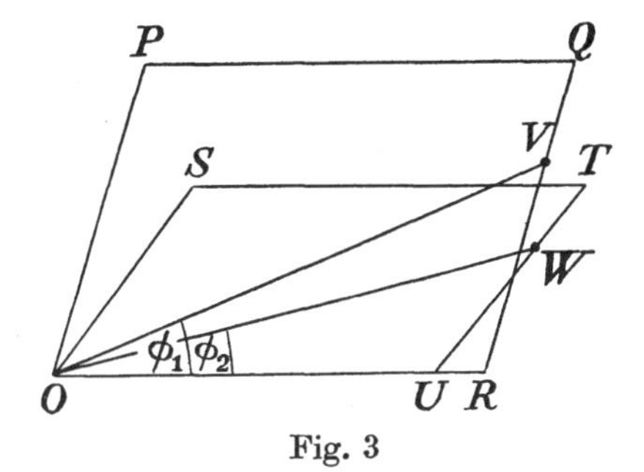

Fig. 3

Let $\lambda_1$ and $\lambda_2$ be the angles between face 1 and face 4 before and after straining, namely, $POR$ and $SOU$ in fig. 3, let $b_1$, $b_2$ represent the breadth of face 1 before and after straining, so that $OR$ is $b_1$ and $OU$ is $b_2$, and let $c_1$, $c_2$ represent the corresponding breadths of face 4. If $\phi_1$ and $\phi_2$ are the angles between corresponding planes in the unstrained and strained material and face 1 in each case, and if $d_1$, $d_2$ represent $OV$, $OW$, the breadths of the corresponding longitudinal sections, then

$$\left.\begin{aligned} d_1^2 &= b_1^2 + n^2c_1^2 + 2nb_1c_1\cos\lambda_1, \\ d_2^2 &= b_2^2 + n^2c_2^2 + 2nb_2c_2\cos\lambda_2, \\ \sin\phi_1 &= nc_1\sin\lambda_1/d_1, \\ \sin\phi_2 &= nc_2\sin\lambda_2/d_2. \end{aligned}\right\} \tag{1}$$

Varying $n$ from $-\infty$ to $+\infty$, all possible pairs of corresponding planes through the edge where faces 1 and 4 intersect are found.

The next step is to find the directions in corresponding planes which are unchanged in length by the strain. These are evidently the intersections of the cones of unstretched directions with this pair of corresponding planes.

At this stage it is necessary to make use of the observed angles between the cross-lines on faces 1 and 4 and the axis of the specimen. If corresponding values of these are $\beta_1$, $\beta_2$ on face 1 and $\gamma_1$, $\gamma_2$ on face 4, it is possible to find, geometrically or analytically, the angles $\alpha_1$, $\alpha_2$ between the axis of the specimen and the lines of intersection of the planes containing the cross-lines on faces 1 and 4 and any pair of corresponding planes through the edge where faces 1 and 4 intersect. The relation between $\alpha$, $\beta$, $\gamma$, $\phi$ and $\lambda$ is shown in fig. 4.

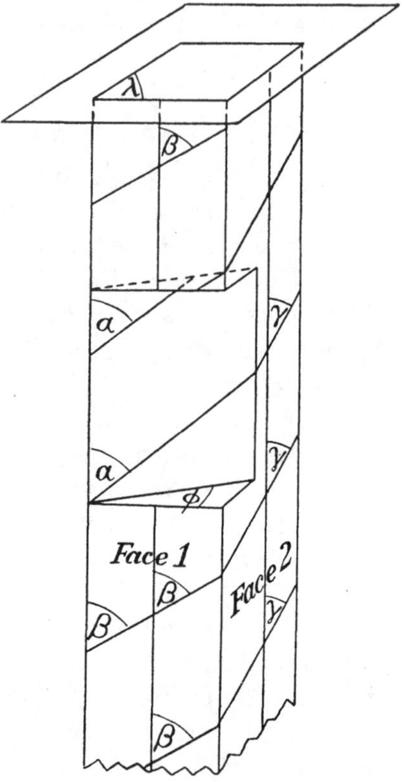

Fig. 4. Diagram showing how $a$ and $\phi$ are related to the faces of the specimen.

Next, by rotation about the line of intersection of faces 1 and 4, bring the two corresponding planes into coincidence. On these two coincident planes construct two corresponding parallelograms, $ABCD$, $AEFG$ (see fig. 5), as follows:

Draw two lines, $BC$, $EF$, parallel to $AD$ (which is on the line of intersection of faces 1 and 4), and at distances $d_1$ and $d_2$ from it. At the point $A$ draw two lines, $AB$, $AE$, at angles $\alpha_1$ and $\alpha_2$, to $AD$, so that $BAD = \alpha_1$, $EAD = \alpha_2$. On $AD$ take any length $AD$ in the unstrained plane. The length $AG$ of the corresponding side of the strained rectangle is found by taking $AG = \epsilon(AD)$, where $\epsilon$ is the ratio of the lengths of the specimen after and before stretching. $ABCD$ and $AEFG$ are the unstrained and strained shapes of a parallelogram of particles in the material.

To find the pairs of lines which are unstretched by the strain:

Join $BE$ and $CF$ (fig. 5), and let them cut in $H$. Any straight line through $H$ cuts the lines $BC$ and $EF$ in two points, $K_1$ and $K_2$ say, which are such that $AK_1 AK_2$ are corresponding lines of particles in the unstrained and strained figures respectively.

To find the positions of $K_1$ and $K_2$ which are such that $AK_1 = AK_2$, draw a line mid-way between $BC$ and $EF$ and parallel to them; let it cut a circle described on $HA$ as diameter in $I$ and $J$. Join $HI$, $HJ$, and let these lines cut $BC$, $EF$ in $L_1$, $L_2$, and $M_1$, $M_2$. Evidently $AL_1 = AL_2$ and $AM_1 = AM_2$, so that these lines are in the required directions.

These operations can be performed analytically, and it is sometimes more convenient to do so. Determining $d_1$, $d_2$, $\phi_1$, $\phi_2$ from the measured quantities $b_1$, $b_2$, $c_1$, $c_2$, $\lambda_1$, $\lambda_2$ by means of equation (1), the angles $\alpha_1$, $\alpha_2$ are given by

$$\left.\begin{aligned} d_1 \cot\alpha_1 &= b_1 \cot\beta_1 + nc_1 \cot\gamma_1, \\ d_2 \cot\alpha_2 &= b_2 \cot\beta_2 + nc_2 \cot\gamma_2. \end{aligned}\right\} \qquad (2)$$

If the angles $L_1AD$, $M_1AD$ of the two unstretched lines of particles in the unstrained material be $\theta_1, \theta_1'$, and if $\theta_2, \theta_2'$ represent the corresponding angles $L_2AG$, $M_2AG$ in the strained material, $\theta_1, \theta_1'$ are the two values of $\theta_1$ given by

$$\cot\theta_1 = \frac{\epsilon}{\epsilon^2-1}(\epsilon\cot\alpha_1 - f\cot\alpha_2) \pm \frac{1}{\epsilon^2-1}\sqrt{\{(\epsilon^2-1)(1-f^2)+(\epsilon\cot\alpha_1-f\cot\alpha_2)^2\}}, \quad (3a)$$

and $\theta_2, \theta_2'$ are the two values of $\theta_2$ given by

$$\cot\theta_2 = \frac{1}{\epsilon f}\left\{\frac{\epsilon}{\epsilon^2-1}(\epsilon\cot\alpha_1 - f\cot\alpha_2) \pm \frac{\epsilon^2}{\epsilon^2-1}\sqrt{\{(\epsilon^2-1)(1-f^2)+(\epsilon\cot\alpha_1-f\cot\alpha_2)^2\}}\right\}, \quad (3b)$$

where $f$ represents the ratio $d_2/d_1$.

Thus we can determine the spherical polar co-ordinates $(\theta_1, \phi_1)$ and $(\theta_1', \phi_1)$ of two directions in the unstrained material which after straining are unextended. Their new co-ordinates are then $(\theta_2, \phi_2)$ and $(\theta_2', \phi_2)$. By taking all possible values of $n$ in formulae (1) and (2) we can map out the two corresponding cones of unstretched directions.

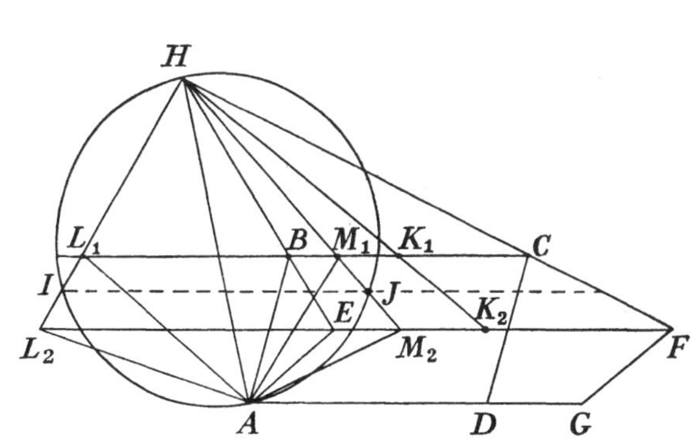


Fig. 5. Construction for finding directions which are unchanged in length by strain.

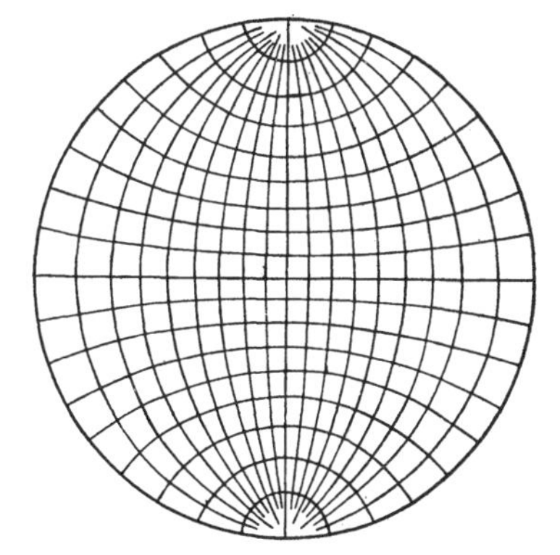

Fig. 6. Stereographic net.

## Representation of Unstretched Cones

The natural way to represent these cones is on a sphere, but for convenience it is necessary to use some flat representation of the sphere. The most convenient of these is the stereographic projection, which is the projection from a point on the surface of the sphere on to the plane through the centre parallel to the tangent plane at the point of projection.

The plane of projection is divided by the sphere into two parts. The inner part represents the hemisphere which lies on the opposite side of the plane of projection to the point of projection. As a rule, a convention is adopted which enables the whole sphere to be represented by the area inside the circle of intersection of the sphere and plane of projection. Assuming the plane to be horizontal, all points of the hemisphere above this are projected from the lowest point of the sphere, while all points of the lower hemisphere are projected from the highest point of the sphere.

The representations of the two hemispheres then cover the same area, and they are distinguished by representing points of the upper hemisphere with a dot, and points of the lower hemisphere with a cross. In the diagrams which follow, curves on the upper hemisphere are represented by full lines through the dots, while curves on the lower hemisphere are represented by dotted lines through the crosses. With these conventions a great circle on the sphere is represented by a lens-shaped figure, consisting of two equal circular arcs, which pass through opposite ends of a diameter of the bounding circle of the projection, and lie on opposite sides of it. One of these arcs is a full line and the other is dotted. Several of them will be seen in figs. 7 and 8.

The usefulness of the stereographic projection is greatly increased if a figure known as a 'stereographic net' is used. This consists of the representation of a series of meridian circles and parallels of latitude projected from a point on the equator, so that the two poles of the system lie on the bounding circle of the projection. Such a net is shown in fig. 6.

To set out a point on a stereographic projection when its spherical polar co-ordinates $(\theta_1, \phi_1)$ are given, let the pole of the spherical co-ordinates, i.e. the point $\theta = 0$, be represented by the centre of the sterographic figure. Draw a circle on a piece of tracing paper equal to the bounding circle of the stereographic net, and mark a radial line on it to represent the meridian circle $\phi = 0$. Next draw another radial line at angle $\phi_1$ with the line to represent the meridian $\phi = \phi_1$. Then lay the tracing paper over the stereographic net so that the line $\phi = \phi_1$ coincides with one of its two principal diameters, and mark off a point to represent the angle $\theta = \theta_1$, using a dot if $\theta_1$ is less than $\frac{1}{2}\pi$ and a cross if it is greater. The points in figs. 7 and 8 are plotted in this way.

## Application of the Method in the Present Case

To analyse a strain by this method, two stereographic figures must be made, showing the undistorted and the distorted positions of the unstretched cone. In the first instance the distortion produced by a 30 % extension of the specimen was analysed. Using the following values from table 2,

$$b_1 = t_{24}/\sin\lambda_1 = 1{\cdot}035, \quad b_2 = t_{13}/\sin\lambda_2 = 1{\cdot}022, \quad c_1 = 1{\cdot}016, \quad c_2 = 0{\cdot}798,$$
$$\lambda_1 = 90°\,35', \quad \lambda_2 = 81°\,50', \quad \beta_1 = 90°, \quad \beta_2 = 84°\,10', \quad \gamma_1 = 90°, \quad \gamma_2 = 85°\,10',$$

and taking $n$ successively as $0, \frac{1}{2}, 1, 2, \infty, -2, -1, -\frac{1}{2}$, the values given in table 3 were calculated for $\phi_1, \phi_2, \theta_1, \theta_1', \theta_2, \theta_2'$. These values were then set out on the two stereographic diagrams shown in figs. 7 and 8.

It was noticed that the points of these diagrams appeared to lie on two great circles, and on placing the tracing paper on which the diagrams were drawn on the stereographic net it was found that two great circles could be described to pass very close to all the points. These great circles are marked in figs. 7 and 8. The unextended cone is therefore a degenerate form, consisting of two planes in this case.

This is very different from the form which would arise when a bar was stretched which consisted of a number of small crystals orientated at random. The 'un-

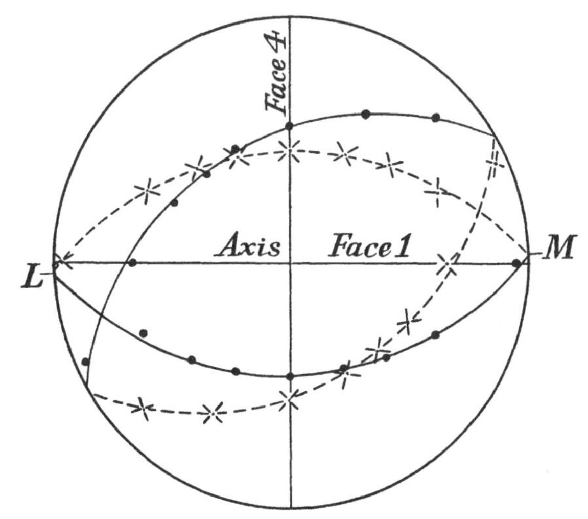


Fig. 7. Stereographic projection of positions in *unstrained* material of cone of directions which are unstretched after the material has stretched by 30 %.

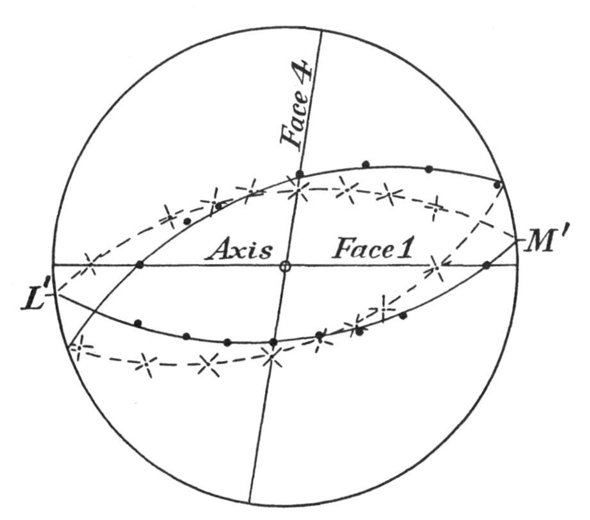


Fig. 8. *Strained* position of the same cone.

Table 3. *Spherical polar co-ordinates of directions which are unextended during an increase of 30 % in length of the specimen*

| | Unstrained material | | | Strained material | | |
|---|---|---|---|---|---|---|
| $n$ | $\phi_1^\circ$ | $\theta_1^\circ$ | $\theta_1'^\circ$ | $\phi_2^\circ$ | $\theta_2^\circ$ | $\theta_1'^\circ$ |
| −2 | −62·6 | 52·0 | 126·2 | −63·3 | 37·4 | 141·4 |
| −1 | −44·2 | 57·8 | 125·9 | −42·1 | 44·3 | 138·0 |
| −½ | −25·7 | 67·7 | 122·5 | −22·7 | 56·5 | 130·5 |
| 0 | 0 | 87·3 | 112·9 | 0 | 80·7 | 114·5 |
| ½ | 25·8 | 111·6 | 93·0 | 20·5 | 111·6 | 87·0 |
| 1 | 44·7 | 121·0 | 81·1 | 35·2 | 124·7 | 71·4 |
| 2 | 63·4 | 127·5 | 69·4 | 51·6 | 135·3 | 56·1 |
| ∞ | 90·6 | 131·0 | 58·1 | 81·8 | 143·7 | 41·8 |

extended cone' would in that case be a circular cone whose axis coincided with the axis of the specimen.

Before proceeding further it may be well to notice the nature of the strain which causes the unextended cone to take the form of two planes. All figures described on either of these planes are unchanged both in dimensions and in shape by the strain; that is to say, the strain does not affect the relative positions of particles in either of them. Since the density of the material is practically unchanged by the strain, the distance apart of any two planes of particles which are parallel to either of these planes is also unchanged by the strain. The strain brought about by the extension of 30 % can therefore be regarded as being due to a simple shear parallel to either of the two unstretched planes.

At a later stage we shall see that the X-ray analysis shows that the orientation of the molecules remains constant with respect to one of them, but not with respect to the other. From the external measurements taken at only two stages of the extension, however, we cannot distinguish between them and say that the shear is due to slipping on one plane rather than on the other.

## Application of the Method to other Stages of the Extension

The same method was applied to the other stages in the extension. It was found that in the cases 0–10, 0–15, 0–20 and 0–40 % extension, the same result was obtained as in the case of 0–30 %, which has already been studied. In each case the cone of unstretched directions was found to consist of two planes. In the case of the final extension 78 %, however, it was found that this was not the case. We shall therefore study first all the stages of the extension up to 40 %.

Table 4. *Spherical polar co-ordinates of directions in unstrained material which remain unstretched at four stages of the test, namely,* 0–10, 0–20, 0–30, 0–40 % *extension*

| | | $\theta_1°$ | | | | $\theta_1'°$ | | | |
|---|---|---|---|---|---|---|---|---|---|
| $n$ | $\phi_1°$ | 0–10 | 0–20 | 0–30 | 0–40 | 0–10 | 0–20 | 0–30 | 0–40 |
| 0 | 0 | 86·5 | 87·5 | 87·3 | 87·0 | 114 | 113 | 113 | 111 |
| 1 | 45 | 122·3 | 120·5 | 121·0 | 120·2 | 74·3 | 78·2 | 81·3 | 82·2 |
| ∞ | 90·6 | 132·5 | 131·3 | 131·3 | 130·0 | 48·8 | 53·0 | 58·0 | 60·2 |
| −1 | −45 | 57·2 | 58·5 | 58·2 | 58·5 | 133·3 | 129·2 | 125·8 | 123·2 |

On comparing the stereographic diagrams showing the positions in the unstrained material of the planes which are unstrained at the various stages of the extension up to 40 %, it was found that one of the planes in each case consisted always of the same particles.

In table 4 are given the spherical polar co-ordinates of eight directions which remain unextended at each of the four stages, 10, 20, 30 and 40 % extension. It will be seen that the figures under heading $\theta_1$, which give the position before straining of one of the unextended planes, are practically identical for all four stages of the test. The figures under the heading $\theta_1'$, which gives the positions of the other planes, vary according to the particular stage concerned. It appears, therefore, that if the shear which produces the extension be considered as due to slipping on the plane given by the columns under heading $\theta_1$ (table 4), this plane of particles remains undeformed and unextended throughout the whole extension from 0 to 40 %. If, however, the other planes determined by the figures under the heading $\theta_1'$ be considered as planes of slipping, these planes will consist of different particles at different stages of the extension.

It is evident that the former is a simpler physical conception, but one cannot be sure that it is the correct one till one studies the directions of the crystal axes. Before proceeding to do this, however, there is still some more information to be got out of the measurements of the external shape of the specimen—the direction of the shear on the slipping plane can be determined.

## Direction of Slipping on Unextended Planes

Let us now consider how the unextended planes would be situated in the case of a simple shear on a given plane of particles. Let $ABCD$ (fig. 9) be a rectangle in the plane perpendicular to the plane of slip, and let $AD$ be in the plane on which the slipping takes place. Let the sheared position of $BC$ be $B'C'$. All particles in the line $BB'CC'$ shift through distance equal to $BB'$. One of the unextended lines of particles is evidently $AD$. The other is found by taking two points $E$, $E'$ at distances equal to $\frac{1}{2}BB'$ on opposite sides of $B$. The line $EA$ in the unstrained material moves to $E'A$ in the strained material and $AE = AE'$, so that $AE$, $AE'$. are the traces of the second unextended planes.

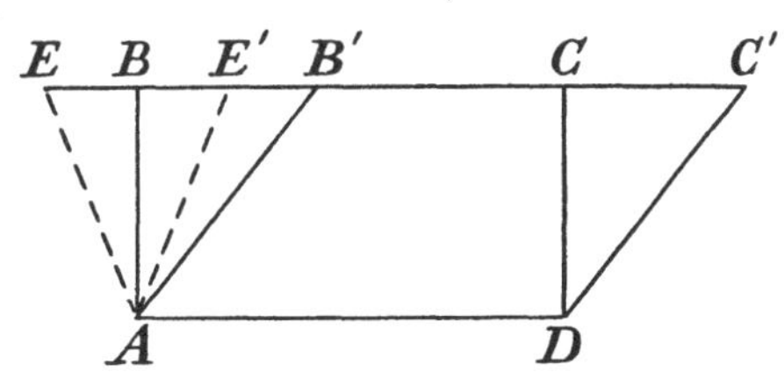

Fig. 9. Section of material shearing on a slipping plane.

From these considerations it will be seen that:

(*a*) The direction of shear is in the line at right angles to the line of intersection of the two unstretched planes.

(*b*) In a finite shear, due to slipping on one plane, the strained and unstrained positions of the other plane of particles, which is undeformed, lie at equal angles on opposite sides of the plane, which is perpendicular to the plane of slip and the direction of slip, i.e. the plane whose trace is $AB$. If the slip be regarded as being measured by $\tan BAB'$ (fig. 9), this is equal to $2 \tan BAE$, i.e. where $\frac{1}{2}\pi - BAE$ is the angle between the two undeformed planes.

(*c*) The only plane which remains undeformed during the whole course of a finite shear is the plane of slip.

To determine the direction of shear, therefore, it is necessary to find the line of intersection of the two planes which constitute the 'cone of unextended directions'. In the present case this was accomplished by constructing a stereographic diagram in which the plane of slipping, i.e. the plane which was undistorted throughout the extension from 0 to 40 %, was represented by the outer circle of the diagram, the normal to the plane of slipping being the centre.

In the case of the figure corresponding with the unstrained material this amounts, in the present case, to rotating the figure through 40° about the line of intersection of the slip-plane with the plane perpendicular to the axis of the specimen. This operation can easily be performed with the stereographic net. In the case of the 30 % extension diagram, for instance, the net is placed with its poles on the points *L*, *M* (fig. 7). The new positions of the points of the diagram are then found by moving them 40° round on a latitude circle of the net.

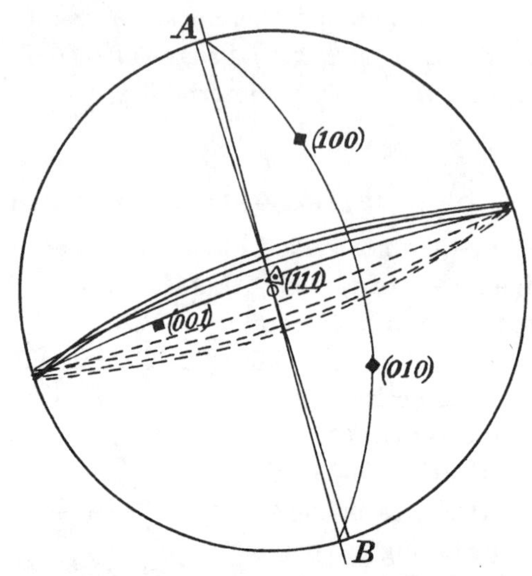


Fig. 10. Projection on to the plane of slip of crystal axes and of second unstretched planes for extensions 0–10, 0–20, 0–30, 0–40 %. *Unstretched* material.

In the case of the figure for the strained material the same procedure is adopted, but in this case the angle through which the diagram has to be rotated is different for different extensions. In each case the boundary circle of the figure represents the same plane of particles, and in order to make it easy to superpose the different figures, the directions of the lines of intersection of the slip plane with the faces and with the axial planes which are parallel to the diagonals of the cross-sections are marked as dots on the edge of the stereographic diagram for each extension. The fact that these dots can be superposed with diagrams obtained from measurements taken at different stages of the extension is confirmatory evidence that the slip-plane is undistorted.

In fig. 10 is shown the position in the unstrained material of the second set of planes, which are undistorted at four stages of the extension, namely, 10, 20, 30 and 40 %. In fig. 11 is shown the positions of three of these same planes in the strained material.* It will be seen that in both cases all the great circles which represent the

* The position of the second undistorted plane for 10 % extension has not been shown in this figure because it confused the central part of the diagram.

second set of planes cut the boundary of the diagram near the same two points. The direction of slip, which, as can be seen from consideration (*a*) above, is represented by the point on the boundary at 90° from these points, is therefore *practically the same for all the stages of the test up to* 40 % *elongation of the specimen.*

It will be noticed that the relation between figs. 10 and 11 is what we should expect from consideration (*b*) above. The fact that the dotted lines in fig. 10 correspond with the full lines in fig. 11 and vice versa, shows that the second set of planes make equal angles on opposite sides of the plane perpendicular to the direction of shear.

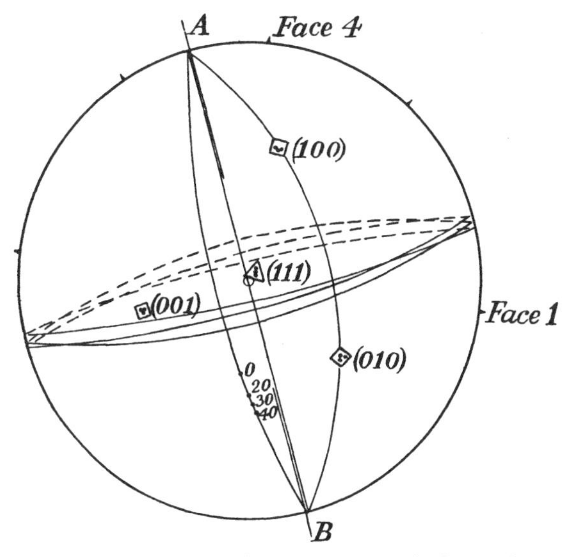


Fig. 11. Positions in *strained* material of the axes and planes shown in fig. 10. The positions of the axis at 0, 20, 30 and 40 % extension are also shown.

To summarize, we have shown that, up to 40 % elongation, the distortion of the specimen is due to simple shear in one direction on a certain plane. We have determined the plane and the direction. It now remains to determine the relation between this plane and the crystal axes.

## Relation between Plane of Slip and Crystal Axes

The co-ordinates of the crystal axes were measured by Dr Müller and Miss Elam at each stage during the extension. They are given in table 5.

It will be seen that one of the (111) planes and one of the (100) planes were determined. In order to test the accuracy of the determinations, the angles between these two directions, which should be 54° 44′, were worked out. These angles, which are represented by the symbol *i* in table 5, are given in the last column. It was found that in some cases there was an error of 2°, and in one of them an error of 4°.

When the positions of two of the axes are found, it is possible to use our knowledge of the type of symmetry which the crystal possesses to determine the positions of the others, but when the measured angle between these two axes is not what we should expect from the known symmetry of the crystal, there is a considerable degree of uncertainty in placing the other axes. In placing the axes, I have accordingly adopted the following system: I have taken the great circle through the two measured axes as being correct. I have then taken the mid-point between the two measured axes as correct, and have shifted both axes by equal amounts in opposite directions along the great circle joining them till their distance apart is 54° 44′. This is equivalent to shifting each of them through an angle equal to half the error in the determination of the angle between them. It is then possible to find the positions of the remaining axes. Such a proceeding is very arbitrary, but something of the

Table 5. *Spherical polar co-ordinates of* (111) *and* (100) *axes at various stages of the test*

| $\epsilon$ | $\theta_{111}$° | $\phi_{111}$° | $\theta_{100}$° | $\phi_{100}$° | $i$° |
|---|---|---|---|---|---|
| 1·000 | 131 | 269·3 | 79·0 | 262·2 | 52·3 |
| 1·053 | 127·4 | 270·2 | 75·0 | 263·8 | 52·5 |
| 1·110 | 126·0 | 270·3 | 73·3 | 263·2 | 53·2 |
| 1·161 | 123·5 | 270·9 | 70·2 | 263·8 | 53·6 |
| 1·200 | 123·2 | 270·5 | 69·7 | 263·2 | 54·0 |
| 1·304 | 118·9 | 271 | 69·5 | 262·2 | 50·2 |
| 1·404 | 118·2 | 272·2 | 66·0 | 260·5 | 53·4 |
| 1·623 | 113·9 | 274·0 | 62·0 | 260·6 | 52·8 |
| 1·785 | 114·7 | 272·8 | 62·4 | 257·7 | 54·5 |

*Note.* $i$ is angle between (111) and (100) axes. It should be 54·7°.

kind is necessary, and any system adopted must be arbitrary. The maximum angle through which either of the axes is shifted by this scheme of averaging is 2·2° in the case of the 30 % extension. The shifts in all other cases are less than this, the average being 1·0°. Fortunately, the distortion of the material is so great that very large changes in the orientation of the crystal axes relative to *some* planes of particles must take place.

To find the relative positions of the axes and the slip-plane, the particular triad (111) axis which was measured and the three tetrad (100) axes have been marked on the stereographic diagrams (figs. 10, 11). Fig. 10, which applies to the unstrained material, of course, has only one set of axes marked on it, but fig. 11 has a set of axes for each stage of the extension.

On looking at these diagrams the following facts will be noticed:

(1) The (111) axis is close to the pole of the diagram. That is, the plane of slip is nearly coincident with a (111) plane of the crystal.

(2) The positions of all the axes are fixed relatively to the plane of slip during the whole course of the distortion, though, of course, the lines of particles which originally occupied positions along the axes move through very considerable angles with respect to it. The variation in the position of the normal to the (111) plane, for instance, is only about 4°.

(3) The great circle through two of the (100) planes is very nearly parallel to the

direction of slip. The angle which the line of intersection of the (100) plane containing these two (100) axes and the plane of slip makes with the direction of slip is less than the margin of experimental error.

If the planes represented by the symbols $(\pm 1 \pm 1 \pm 1)$ be regarded as forming a regular octahedron, the slip is then on a plane which is very nearly parallel to one face, and is almost exactly in the direction of one edge.

The fact that the position of the slip-plane determined by external measurements is so close to the (111) plane determined by X-ray analysis makes one suspect that they really coincide, and that the difference between their measured positions is due to experimental error. This idea is confirmed by the fact that the direction of slip is so exactly parallel to one of the three principal lines of atoms in the (111) plane. It may be, however, that the difference between the measured positions of the slip-plane and the (111) plane is due to some slight slipping on some other plane. During the test, two sets of slip bands appeared on the faces of the specimen. It is hoped to settle this point in another test, in which improved methods of measurement will be used.

In the remaining part of this paper we shall assume that the slip was actually on the (111) plane, and shall discuss some possible consequences of this assumption.

## Changes in Orientation of Specimen relative to Crystal Axes during Extension

The effect of a shear in any direction on a fixed plane of particles is to move any given line of particles in such a way that it remains parallel to the plane containing its original direction and the direction of slip. In fig. 11, for instance, the positions of the axis of the specimen relative to the slip plane at successive stages of the extension is shown. These are marked 0, 20, 30, 40. It will be seen that a great circle can be drawn to pass through them and through the direction of slip.

As the point representing the direction of the axis moves along this circle, the angles between this direction and all the four octahedral planes vary continuously. After a time, it seems likely that the axis of the specimen will move into a position in which it makes an angle with one of the other octahedral planes which is equal to the angle it makes with the particular octahedral plane on which the slipping is taking place. At this stage the force tending to produce slipping on one plane is equal to that tending to produce slipping on the other. It is possible, therefore, that at this stage slipping may take place on both planes. Possibly this is the reason why the cone of unstretched directions in the case of the 78 % extension no longer consists of two planes.

To examine this point, first turn the axes round so that one of the cubic or tetrad axes is in the centre of the figure. The two remaining cubic axes will then be found on the circumference, and the four octahedral axes will be disposed symmetrically round the central cubic axis.

In fig. 12 the axes have been rotated so that the cubic axis denoted by (010) in fig. 11 is in the centre. The slip-plane is then represented by the great circle *DBE* (fig. 12), and the point *B* represents the direction of slip, corresponding with the point *B* in fig. 11.

We have already shown that the points representing the axis of the specimen lie on a great circle through *B*. An arc of this circle is shown in fig. 12 as *FGB*, where *G* is on the line *DE* which represents the (101) plane.

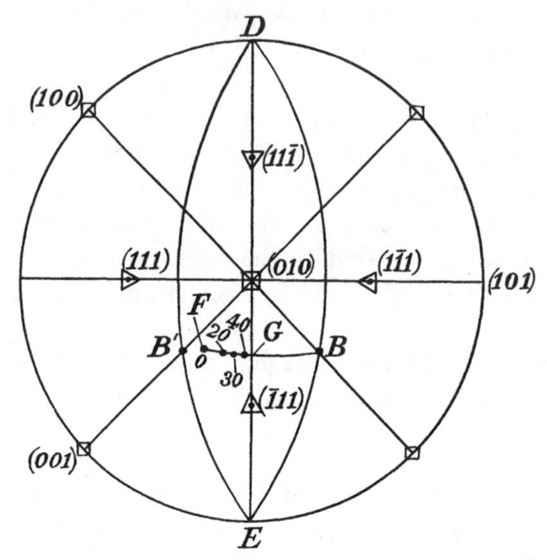


Fig. 12. Figure showing position of axis of specimen relative to crystal axes at 0, 20, 30, 40 % extension.

As the slipping on the plane *DBE* proceeds, the point representing the axis of the specimen moves along the arc *FGB* till, at about 50 % extension, it gets to the point *G*. At this point, *G*, it is in such a position that it is equally inclined to the two octahedral planes *DBE*, *DB′E*, which are perpendicular to the two axes (111) and (1$\bar{1}$1).*

It should now be noticed that up to this stage of the extension the slipping has always been along that one of the twenty-four crystallographically similar possible directions of shear which has the maximum component of shearing force tending to set it going. If this condition continues to hold, then directly the point representing the axis has moved beyond *G* along its path *FGB*, the force tending to produce a slip on the plane *DB′E* in the direction of *B′* is greater than the force tending to continue the slip on *DBE*. Slipping will then begin on *DB′E*, and the axis will start moving along a great circle towards *B′*.

It can easily be seen that the result of successive slips towards *B* and *B′* will be that the point representing the axis will move along the line *DGE* towards the point where the great circle through *B* and *B′* cuts it. At this point no further movement of the axis will take place, however much the specimen extends.

* No distinction is made here between the planes (1$\bar{1}$1) and ($\bar{1}$1$\bar{1}$).

These considerations lead to the prediction that there will be a tendency for the axes to orientate themselves, so that two of the octahedral planes make equal angles of 61° 52′ with the axis of the specimen, while their line of intersection makes an angle of 54° 44′ with the axis.

It is hoped to examine this point later, when the distortions and crystal axes of more specimens have been determined.

### Verification that Slipping in the Last Stage of the Distortion occurs simultaneously on Two Octahedral Planes

Though no very satisfactory method has yet been devised for determining the distortion due to simultaneous slipping on two planes, it is possible to apply the methods used in the early part of this paper to test whether the distortion of the specimen at its breaking point (78 % extension) could have been produced by slipping on the two octahedral planes which are at equal angles to the axis when the extension is about 50 %.

The cones of unextended directions in the distortion of the specimen during its extension from 40 to 78 % were first determined by the equations (1), (2) and (3). The two positions of this cone are plotted on the stereographic figures 13 and 14. Fig. 13 shows the position of the cone in the material at 40 % extension, while fig. 14 shows its position at 78 % extension. In both these figures the centre spot represents the axis of the specimen, as it does in figs. 7 and 8. It will be seen that the cone is now very different from the pair of planes which were characteristic of the first 40 % of extension. The distortion is, therefore, no longer due to slipping on one plane.

We have seen that during the last stage of the extension it seems likely that slipping might take place on the second octahedral plane, *DB′E* (fig. 12). If this is actually the case, then there is one line of particles in the specimen which will not stretch during the whole extension from 0 to 78 %, namely, the line of intersection of the two octahedral planes (111) and $(1\bar{1}1)$. The position of this crystal axis, which is represented by the symbol $(10\bar{1})$, can be found from the crystal measurements. It is a line which makes 45° with the plane (100) and 90° with the plane (111). Using the spherical polar co-ordinates $(\theta, \phi)$, previously used in table 5 for giving the positions of the crystal axes, it was found that the co-ordinates of this crystal axis were $(\theta = 56{\cdot}5°,\ \phi = 209{\cdot}6°)$ in the material at 40 % extension, and $(\theta = 51{\cdot}5°,\ \phi = 204{\cdot}8°)$ in the material at 78 % extension.

To find the directions which were unextended during the extension 0–40 %, and remained unextended at 78 % extension, the positions in the material at 40 % extension of both the slip-plane for the extension 0–40 %, and the 'unextended cone' for the extension 40–78 %, were marked on the same diagram (see fig. 13). The points *P* and *P′* where these cut then represent the directions required. It is clear that one of them must have remained unextended during the whole of the stretching

6 BSP

from 40 to 78 %, whereas the other is merely a line of particles which happens to be unextended at the particular stage, 78 % of extension; but it is not possible to prove that this is the case, because there are no reliable measurements available for stages intermediate between 40 and 78 %.

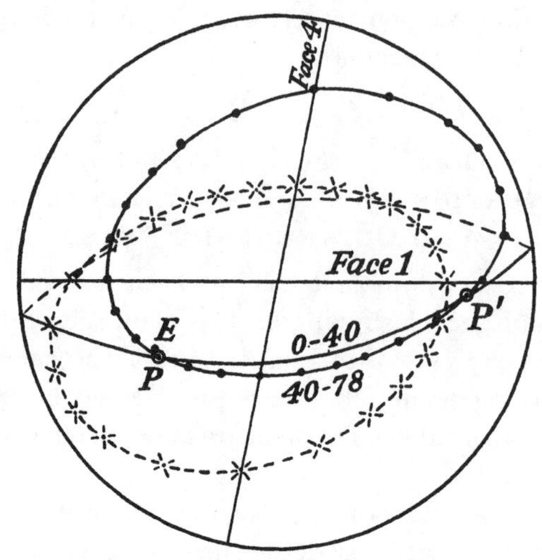


Fig. 13. Positions in material extended 40 % of unstretched cone for extension from 40 to 78 %, and of slip plane for extension 0–40 %.

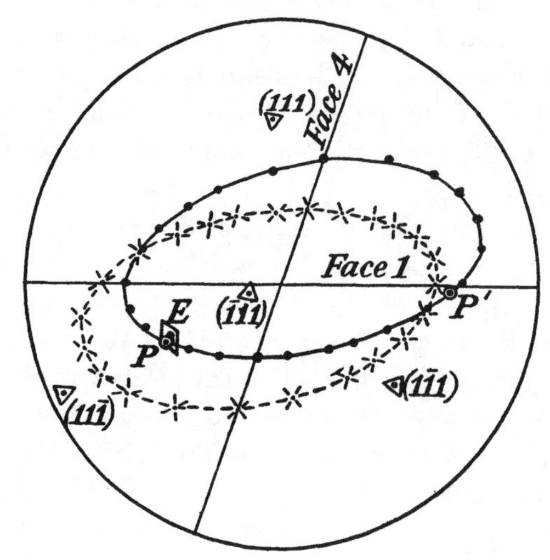


Fig. 14. Position in material extended 78 % of unstretched cone shown in fig. 13.

One of these two directions should therefore, if our theory is correct, coincide with the diad axis whose position in the 40 % material is ($\theta = 56{\cdot}5°$, $\phi = 209{\cdot}6°$).

On measuring up fig. 13 it was found that the co-ordinates of $P$ were ($\theta = 57°$, $\phi = 213°$), and the co-ordinates of $P'$ were ($\theta = 73°$, $\phi = 355{\cdot}5°$).

The diad axis $(10\bar{1})$ is marked at the point $E$ (fig. 13). It is clear that it is very near to its predicted position.

## Verification of Relation between Unstretched Lines of Particles and Crystal Axes in Material at Breaking Point

The positions in the 78 % material of the unextended lines of particles $P$ and $P'$ are marked on the diagram (fig. 14). The co-ordinates of the diad axis, which lies on the intersection of the (111) and the ($1\bar{1}1$) planes, were calculated from the crystal measurements given in the last line of table 5; they are ($\theta = 51{\cdot}5°$, $\phi = 204{\cdot}8°$). This diad axis is marked on the diagram (fig. 14) at the point $E$.

The co-ordinates of the point $P$ (fig. 14) were determined by measuring the diagram; they are ($\theta = 53°$, $\phi = 206{\cdot}6°$).

The co-ordinates of the diad axis and the unstretched lines $P$ are given in table 6.

Table 6. *Spherical polar co-ordinates of diad axis and unextended direction*

| Extension of material ... | 40 % | | 78 % | |
|---|---|---|---|---|
| Co-ordinates ... ... | $\theta$ | $\phi$ | $\theta$ | $\phi$ |
| Diad axis | 56·5° | 209·6 | 51·5° | 204·8° |
| Unextended direction | 57° | 213·0 | 53° | 206·6° |

The agreement between these two is striking, when it is remembered that the diad axis was determined in each case by X-ray analysis of the crystal at one stage only, while the direction of the unextended line of particles was found, using only external measurements of the specimen at a series of successive stages of the distortion.

It will be noticed that this agreement depends only on the fact that the slipping is on the two octahedral planes (111) and ($1\bar{1}1$). If the slipping on the two planes were of equal amounts (as contemplated on p. 81 above) we should expect the two octahedral planes on which the slipping takes place to remain at equal angles to the axis of the specimen during the last part of the stretching. On calculating these angles from the X-ray measurements it was found that they were 66·1° and 63·5° in the material at 60 % extension, and 65° and 61·5° in the material at 78 % extension. The limiting value should, according to the theory, be 61° 52′. It will be seen that the agreement is as good as could be expected in view of the uncertainty of the X-ray measurements. The symmetry of the octahedral planes with respect to the axis of the specimens during the last stages of the extension is shown in fig. 14, where the four triad axes, which are normal to the four octahedral planes, have been marked as spots surrounded by triangles.

It will be seen that the two planes (111) and ($1\bar{1}1$) are at nearly equal angles to the axis, and that the other two triad axes lie on a plane which very nearly passes through the axis of the specimen.

The cone of unextended directions for the extension 40–78 % is not symmetrical with respect to this plane, because in the early part of this range of extension the

6-2

slipping was only on the original slip-plane, but if the external measurements at 62 % extension had been satisfactory I should have expected the cone of unextended directions for the extension 62–78 % to be symmetrical with respect to the (101) plane. In this way it is hoped that it will be possible to test whether there is an equal amount of slip on the two slip-planes during the final stage of the extension. The fact that the (111) plane and the $(1\bar{1}1)$ plane are nearly equally inclined to the axis both in the case of 60 % extension and in the case of 78 % extension suggests that this must be the case, but it cannot be proved without further external measurements.

# 6

# THE HEAT DEVELOPED DURING PLASTIC EXTENSION OF METALS*

REPRINTED FROM

*Proceedings of the Royal Society*, A, vol. CVII (1925), pp. 422–51

## 1. Introduction

When a soft metal, such as annealed copper or aluminium, is deformed while cold, either by stretching, hammering, rolling, or other method of 'cold working', it hardens; that is, the forces necessary to deform it increase as the amount of plastic deformation increases. The physical state of 'cold-worked' metal is undoubtedly different from that of the metal in its original soft or annealed state, and various explanations have been put forward to account for the difference. Some of these explanations involve the hypothesis that the process of hardening is associated with the formation of amorphous material at the crystal planes where slipping occurs during the deformation. The formation of amorphous material from a crystalline mass would involve a phase change, which would in general be accompanied by a change in the internal energy of the material. It has been suggested that the phase change could be detected by measuring the heat evolved during a deformation, and comparing with the heat equivalent of the work done on the metal by the forces producing the deformation. Any difference between the two would imply a change in the internal energy of the metal.

It is curious that very few measurements of this type appear to have been made. The only reference which we have been able to find occurs in Dr Rosenhain's article on 'Metals', in the *Dictionary of Physics*,† where he quotes some previously unpublished observations made by Dr Sinnat.

According to these observations, only one-tenth of the work done reappears in the form of heat, the remaining 90 % being presumably used up in changing the phase of the material. This result, if true, would be of very great interest, but so far no further details have been published. Dr Rosenhain has informed us that he is carrying out further experiments on the subject.

On the other hand, as will be seen later, the experimental difficulties in making measurements of this kind are considerable, and if several independent workers performed such experiments along parallel lines their time would not be wasted, even if they got identical results.

* With W. S. Farren.

† Vol. v, p. 398.

## 2. Methods of Measurement

The simplest way to deform a metal is to take a straight bar or 'test-piece' and stretch it by a longitudinal pull. The work done in any portion of the bar during an operation of this kind can be determined by measuring the pull $P$ and the length $l$ of the portion concerned at successive stages of the test. The work done, namely, $\int P\,dl$, is represented by the area of the stress-strain curve. The stress-strain relation is usually determined in engineering laboratories by means of a testing machine in which the stress is measured by balancing the tension in the specimen against a weight on a lever, and the strain by direct measurement on the specimen.

This method was not suitable for the experiments described here, because it was necessary to do the whole of the stretching in a few seconds of time, in order that the rise in temperature of the metal might be measured before any appreciable cooling had taken place. For this reason a self-recording testing machine was made which automatically produced a stress-strain curve as the specimen was being stretched. This machine is described later in some detail, because the value of our results depends entirely on the accuracy of the measurements, and it is necessary to show that the accuracy of our stress-strain curves is at least equal to that of our temperature measurements.

To measure the heat evolved during the stretching, the use of a calorimeter naturally suggests itself, but the experimental difficulties in the way of getting accurate results in this manner appear insuperable; and even if they could be overcome the result obtained would give only the total heat evolved throughout the whole specimen, while what is wanted is the heat evolved in the middle of the bar, where the stretching is uniform.

The method adopted was to measure the rise in temperature which occurs during a rapid plastic extension of the bar. For this purpose a thermocouple was used and the temperature was recorded photographically on a moving plate.

Preliminary experiments showed that, when fixed in the testing machine and exposed to the air, the specimens first used cooled through about 4 % of their excess of temperature over that of the room in 1 sec. It was clear, therefore, either that large cooling corrections would have to be introduced or that the temperature measurements would have to be made rapidly. The second alternative is preferable, if possible, and the first part of the work was devoted to a study of the heat losses immediately after the extension of the specimen, and to a search for the method of measuring temperature which suffered least from lag.

The cooling of the specimen was due almost entirely to conduction of heat through the ends, only a very small fraction of the heat being radiated or carried away by convection. This might have been guessed beforehand, but it was proved by observing the rate of cooling of the specimen when hung up horizontally by a silk thread, and comparing it with the rate of cooling of the specimen when fixed on the testing machine. In the former case the excess of temperature of the specimen over that of the atmosphere was reduced in the ratio 1 to $e$* in 18 min., while in the latter the time taken was 28 sec. It appears, therefore, that only $\frac{28}{18\times 60}$, i.e. $\frac{1}{40}$ of the heat loss, was due to convection and radiation, and that the remaining $\frac{39}{40}$ must have been due to conduction of heat from the central uniform part of the specimen to the ends and grips of the testing machine.

The coldness of the ends takes some time to reach the middle, and it was during the interval between the time of generation of the heat and the time of arrival of the cold waves from each end of the bar that the temperature measurements were made.

To find how long this interval might be expected to be it is sufficient to idealize the condition slightly. Consider a uniform bar of length $l$, coefficient of conductivity $k$, specific heat $\sigma$, density $\rho$, and suppose that at time $t=0$ the temperature of the whole bar is suddenly raised by an amount $T_0$, the ends being maintained subsequently at their original temperature, which may be taken as $T=0$. The subsequent temperature at distance $x$ from either end is represented by the series

$$T=\frac{4}{\pi}T_0\left\{e^{-at}\sin\frac{\pi x}{l}+\tfrac{1}{3}e^{-3^2at}\sin\frac{3\pi x}{l}+\tfrac{1}{5}e^{-5^2at}\sin\frac{5\pi x}{l}+\dots\right\},\tag{1}$$

where $$a=\frac{\pi^2 k}{l^2\rho\sigma}.$$

When $t=0$, $T=T_0$ for all values of $x$.

When $t$ is large the series reduces to the first term

$$T=\frac{4T_0}{\pi}e^{-at}\sin\frac{\pi x}{l}.\tag{2}$$

When $t$ is small this series is inconvenient for numerical calculation and the expression

$$T=T_0\left\{2-\frac{2}{\sqrt{\pi}}\int_0^{\zeta_1}e^{-\mu^2}d\mu-\frac{2}{\sqrt{\pi}}\int_0^{\zeta_2}e^{-\mu^2}d\mu\right\}\tag{3}$$

may be used, where $\zeta_1=\frac{x}{2}\sqrt{\frac{\rho\sigma}{\kappa t}}$ and $\zeta_2=\frac{l-x}{2}\sqrt{\frac{\rho\sigma}{\kappa t}}$.

The expression (3) is approximately the same as (1) for small values of $t$. The maximum errors occur first at the ends, $x=0$ and $x=l$. When $t=0{\cdot}077\rho\sigma l^2/k$ the error is 1 % at the ends but is still quite inappreciable in the middle, where (3) gives $T=0{\cdot}594T_0$, while (2) gives $T=0{\cdot}595T_0$. It appears that the error in using (3) from $t=0$ to $t=0{\cdot}077\rho\sigma l^2/k$ and (2) from this value to $t=\infty$ is never more than a fraction of 1 %.

The theoretical cooling curve for points in the middle of a bar, cooled by conduction of heat to both ends, is shown in fig. 1. It will be seen that from

$$t=0 \quad \text{to} \quad t=0{\cdot}014\rho\sigma l^2/k\tag{4}$$

the temperature is practically constant. At $t=0{\cdot}014\rho\sigma l^2/k$ it has fallen only $\frac{1}{166}$th of its initial value. After this the temperature begins to fall rapidly, and at $t=0{\cdot}04\rho\sigma l^2/k$

* $e$ the base of Napierian logarithms.

the curve becomes practically indistinguishable from the ordinary exponential curve (shown as a dotted line in fig. 1), which would result from Newton's law of cooling.

In the second part of the curve, when the temperature of the middle of the bar is falling exponentially, the time $t_e$ taken to fall to $1/e$ of its value is $\rho\sigma l^2/\pi^2 k$ or $1/a$. It appears, therefore, that if this time $t_e$ is observed experimentally the temperature will not fall through more than $\frac{1}{166}$th of its initial value during the time $0{\cdot}014\pi^2 t_e$ after the instant when the heat is generated. If the temperature measurements can be made within this interval of time the error due to neglecting cooling corrections will not amount to 0·6 %.

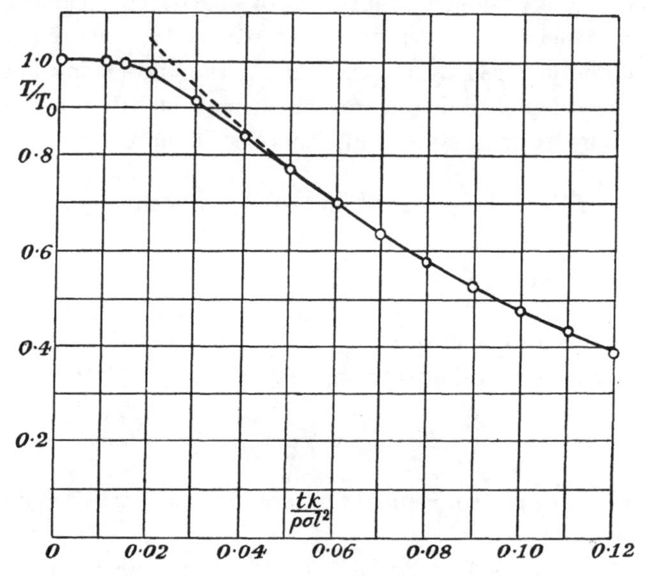


Fig. 1. Theoretical cooling curve for the middle point of a bar initially at a uniform temperature and subsequently cooled by maintaining the ends at a steady temperature.

In the first experiment with aluminium bars having a parallel portion in the middle 11 cm. long before the stretching began, the time $t_e$ was observed to be 28 sec., so that the time available was $0{\cdot}014 \times 28 = 4$ sec., whereas using the expression (4) and inserting the known physical constants for aluminium, namely, $\rho = 2{\cdot}7$, $\sigma = 0{\cdot}212$, $k = 0{\cdot}5$, and putting $l = 11$, the time comes out to be 1·9 sec. The difference is due no doubt to the fact that the ends of the parallel middle portion are not in practice maintained at the temperature of the atmosphere, and it seems clear that the time during which the temperature of the middle of the bar might be expected to remain practically constant is considerably underestimated by the expression (4). This is confirmed by measurements of temperature records taken in the course of the experiments.

In some of the experiments here described it was possible to use specimens containing as much as 30 cm. of parallel part in the middle, and with these there was ample time for measuring the temperature. It was not possible, however, to obtain single-crystal bars of aluminium with a parallel middle portion greater than 16 cm. For this reason it was necessary to have a temperature-recording system which could be relied on to give a correct reading within 2 or 3 sec. of the time of application of the strain.

For this purpose an iron-constantan thermojunction was used. This was connected directly to a galvanometer, the moving part of which reflected light from a vertical slit on to a fine horizontal slit placed close in front of a photographic plate which was made to move vertically at a constant speed by means of an electric motor fitted with a speed governor.

The galvanometer used was of the moving-magnet type, but the suspended system was specially designed and made for us by Dr P. Kapitza, with a view to reducing the period to the minimum value consistent with the required sensitivity. It consisted of an astatic pair of cobalt-steel magnets, each $2\frac{1}{2}$ mm. long × 0·02 mm. thick. These were fixed to a fine straight glass thread which was hung vertically by a quartz fibre. The reflecting mirror was made of thin cover-slip glass, 2 mm. square. A magnet was used to control the sensitivity of the system, and in the present experiments the period of oscillation was 2·3 sec., and the damping per half-period was 0·44. The records obtained with this apparatus could be reduced to eliminate the effects of periodicity and damping in the galvanometer; in most cases, however, this was found to be unnecessary, the temperature of the middle of the bar remaining constant for a time which was long enough to allow the oscillation to die away.

### 3. Method of Using a Thermojunction

The usual method of using a thermojunction for measuring the temperature of solid bodies is either to insert it in a small hole in the body, or, if possible, to solder it to the surface. Neither of these methods was available in the present instance. To bore a small transverse hole in the specimen would alter the type of distortion which the metal would experience in the immediate neighbourhood of the hole. This would give rise to an error whose magnitude it would be impossible to estimate. Solder could not be used because many of the experiments were made with aluminium. It was, therefore, necessary to devise some method of attaching or pressing the thermojunction to the surface of the metal.

When a thermojunction touches a specimen on one side only, unless it can be soldered on, some other material, preferably a non-conductor, must be in contact with its other side in order to press it against the specimen. The temperature of the thermojunction will, therefore, be intermediate between that of the specimen and that of the non-conductor, and if the resistance to the passage of heat at the contact is small, presumably the temperature recorded will be close to that of the specimen, but there will be an error whose magnitude must be determined. In the experiments here described this source of error was reduced to a minimum by pressing the thermojunction against the inside of a $\frac{3}{16}$ in. hole bored symmetrically from end to end of the specimen. The total amount of non-conducting material used for pressing the thermojunction against the inside of the hole was very small, and in any case it was certain that everything inside the hole would warm up to the temperature of the metal, whereas it was by no means certain that a thermojunction pressed on the outside of the specimen would do so.

The chief difficulty in the way of getting an accurate record of the temperature of the metal immediately after stretching was due to the lag in temperature between the thermojunction and the specimen. To test the efficiency of different ways of fitting the thermojunction a very thin-walled metal tube with a bore of $\frac{3}{16}$ in. was made. The thermojunction was fitted into this, and a sudden change in temperature was produced by removing the tube from a jar of water and suddenly plunging it in a jar of water whose temperature was about 2° higher, and stirring vigorously. The temperature record produced in this way had to be analysed to allow for the lag in the galvanometer system. The method adopted for this purpose will next be described.

If there were no lag in the heating of the thermojunction, the record would show a damped harmonic oscillation, the characteristics of which would be determined solely by the galvanometer. In the present case the galvanometer had a period of 2·3 sec. and a damping of 0·44 per half-period. It is clear, therefore, that if the temperature of the thermojunction were suddenly raised from $T=0$ to $T=T_0$, and then kept at temperature $T_0$, the record would overshoot the true temperature by an amount $0{\cdot}44\,T_0$, so that the maximum temperature shown on the record would be $1{\cdot}44\,T_0$. The next minimum would be $\{1-(0{\cdot}44)^2\}\,T_0=0{\cdot}81T_0$, and so on.

If the lag in the temperature of the thermojunction were large, the recorded temperature would be close to the true temperature and there would be no maximum on the record till the true steady temperature had been attained. For intermediate cases, where the lag in the temperature and the period of the galvanometer were of the same order of magnitude, there might be a first maximum on the record, which would necessarily be less than $1{\cdot}44\,T_0$, but might be greater than $T_0$ by an amount which would depend on the temperature lag in the thermojunction.

If $T_m$ represents the first maximum temperature shown on the record, it is possible by measuring $T_m/T_0$ to measure the amount of the temperature lag in the thermojunction.

If $y$ represents the reading of the galvanometer record when multiplied by the calibration factor, so as to reduce it to a measure of temperature, the equation which represents the galvanometer record, in the case when there is no lag in the temperature of the thermojunction, is

$$\frac{d^2y}{dt^2}+\kappa\frac{dy}{dt}+\mu y=0, \tag{5}$$

and the solution of this is

$$y=T_0(1-e^{-\frac{1}{2}\kappa t}\cos nt), \tag{6}$$

where $n^2=\mu-\frac{1}{4}\kappa^2$.

$2\pi/n$ is the period of the galvanometer, while $e^{-\frac{1}{2}\kappa}$ is the damping factor.

When there is a lag, (5) becomes

$$\frac{d^2y}{dt^2}+\kappa\frac{dy}{dt}+\mu(y-T)=0, \tag{7}$$

where $T$ is the temperature of the thermojunction.

If $1/\lambda$ represents the lag in the temperature of the thermojunction when the temperature outside the case containing it is suddenly changed, the equation for $T$ may be assumed to be

$$T=T_0(1-e^{-\lambda t}). \tag{8}$$

This equation in fact defines the 'lag'.

Table 1

| $\lambda$ | $\infty$ | 2 | 1 | 0·71 |
|---|---|---|---|---|
| $T_m/T_0$ | 1·44 | 1·157 | 0·932 | 0·759 |

Substituting for $T$ in (7), solving, and inserting the condition, that when $t=0$, $y=0$ and $dy/dt=0$, an equation is formed for $y$. Using the values for $n$ and $\frac{1}{2}k$ obtained from the experiments on the period and damping of the galvanometer, the solution of (7) which fits the conditions is found to be

$$\frac{y}{T_0}=1+\frac{-8{\cdot}17e^{-\lambda t}+(1{\cdot}42\lambda-\lambda^2)\,e^{-0{\cdot}71t}\cos 2{\cdot}77t-(0{\cdot}256\lambda^2+2{\cdot}59\lambda)\,e^{-0{\cdot}71t}\sin 2{\cdot}77t}{\lambda^2-1{\cdot}42\lambda+8{\cdot}17}.$$

Taking a series of different values of $\lambda$, it is found that if $\lambda$ is less than 0·5 there is no maximum value for $y$, but that when $\lambda$ is greater than 0·71, $y$ has a maximum $T_m$, the values of which are given in table 1.

The value $T_m/T_0$ was measured for a number of different types of thermojunction and the results given in table 1 were used to estimate the lag. It was found that the lag varied very greatly, according to the method used for pressing the thermojunction against the heated surface. In the first place the lag was always far too great unless there was actual metallic contact between the junction and the heated metal. For this reason it would be impossible to obtain accurate results if a multiple junction were used, because of the necessity in that case for having insulating material between the metal and the junction.

Various methods were tried for pressing the thermojunction directly against the inside of the hole through the specimen, but they were all failures. In every case a lag of several seconds occurred, though the lag was not so great as when there was no metallic contact.

The method which proved most successful consisted in soldering the junction to a plate of silver foil 0·1 mm. thick, 7 mm. long and 9 mm. wide. This plate was bent round a small cylindrical piece of cork which could be forced into the hole. In this way the silver was pressed hard up against the inside of the specimen.

This method had the advantage that when the hole contracted as the specimen lengthened the only effect was to make the cork press the silver plate still harder against the metal, and even when the hole became elliptical, as it did when single crystal specimens were used, the silver foil was still pressed against the metal all over its surface.

Using this method the lag was reduced to 0·4 sec. (i.e. $1/\lambda = 0{\cdot}4$). It will be noticed that this includes any lag which may have existed between the temperature of the copper tube and that of the water into which it was plunged, so that the lag due to the passage of heat to the thermojunction must have been less than this. In order to ensure that the temperature of the thermojunction was within 1 % of that of the specimen with which it was in contact it was therefore necessary to read the record at a time greater than $-0{\cdot}4 \log_e 0{\cdot}01$, or 1·8 sec., after the stretching was finished.

### 4. Method of Carrying Out an Experiment

The thermojunction was first fitted inside the middle of the specimen. The iron and constantan wires passed out through one end to the cold junction which was contained in an oil-filled tube dipping into a thermos flask full of water. The other end of the axial hole through the specimen was made watertight with wax, and the thermojunction was calibrated by moving the specimen from one large jar of water at the temperature of the room to another containing water about 2° C. higher. The temperature of the water was found by means of Beckmann thermometers graduated in hundredths of a degree Centigrade. The water in both jars was stirred vigorously by means of small fans, run by an electric motor, but even then the specimen took some 18 sec. to attain the temperature of the surrounding water.

An actual stretching experiment lasted a few seconds only, so the speed of the falling plate on which the records were taken was reduced when the calibrations were being done (see fig. 2). In order to provide a base-line from which measurements of the plate could be made, an image of the same vertical slit which was used to illuminate the galvanometer mirror was cast by a fixed mirror on the horizontal slit of the recording apparatus, and the light was cut off by a shutter from both images, at regular intervals of 6 sec., so that small simultaneous gaps appear in the temperature record and the base line.

In most of the experiments two records were taken on each plate. In fig. 2 the outer and thicker line, marked $A$, is the calibration record.

The next step was to put the specimen in the machine and stretch it by winding the handle as rapidly as possible, till the temperature of the specimen had risen through about 2° C. The machine was then left for about 10 min., with the specimen still under tension, in order that it might have time to cool. It was then again stretched, till its temperature rose again by about 2° C. With the steel, copper and aluminium specimens three such extensions were made. With the single crystals of aluminium there were five.

The load was then released suddenly and a record taken of the reversible adiabatic rise in temperature which accompanies the sudden elastic contraction. This usually amounted to about 0·3° C., or about 3 % of the total rise which would have occurred if the whole extension had been done at once.

The main object of the experiments was to determine how much of the energy put into the metal by external forces was used in increasing the internal energy of the material. For this reason it was thought that the ideal way to carry out an experiment would be to stretch the material and then immediately to release the stress, so that the material would be unstressed as a whole both before and after the experiment. This was inconvenient, and it was decided to divide up the adiabatic heating which occurred when the stress was finally released, and to distribute it among the successive stages of the stretching in proportion to the change in stress which occurred in each stage. Thus if the stresses at the ends of the first three stages were $p_1$, $p_2$ and $p_3$, and if the change in temperature due to adiabatic heating on releasing the stress $p_3$ was $T_a$, an amount $p_1 T_a/p_3$ was added to the observed rise of temperature during the first stretching, $(p_2-p_1)T_a/p_3$ for the second stretching, and $(p_3-p_2)T_a/p_3$ for the third. Except for the first extension in each experiment these corrections are small, but they are founded on experimental evidence that the elastic strain, and hence the adiabatic heating, is proportional to the change of stress and does not depend on the hardness of the material.

## 5. Temperature Records

Some of the temperature records are shown in figs. 2–5. In each the stretching of the specimen was complete before the spot of light had travelled its maximum distance, but since the stretching was not instantaneous (it occupied about 2 sec.) the amount by which the galvanometer mirror overshot the mark was small. It is a comparatively simple matter to analyse the record so as to make allowance for the galvanometer lag and inertia, but an inspection of the records shows that the oscillation has practically died away before the specimen begins to cool, so that in most cases no correction is required. In each of the records the temperature curve has the flat top which was predicted on p. 88 (fig. 1), but it is most obvious in fig. 3 which is the record of the stretching of a single-crystal specimen of aluminium. It will be seen that the temperature did not begin to decrease till at least 5 sec. after the stretching was complete, and by that time the galvanometer oscillation had practically disappeared. In figs. 4 and 5, which are the records for annealed steel and copper, respectively, the specimens were 30 cm. long, and the rate of cooling was much slower than for the aluminium specimen, which had only 16 cm. of parallel part in the middle. The flat top in these records extends over about 8 sec.

## 6. Discussion of Results

In order to compare the observed rise in temperature with the heat evolved the heat equivalent of the work done in each experiment has been calculated. If $l_0$ is the original length of the specimen between its marks, and $\epsilon$ its extension at any stage, its length is then $l_0(1+\epsilon)$. If $P$ is the pull in the testing machine and $a_0$ the original cross-section of the specimen, the work done on unit volume of the material during

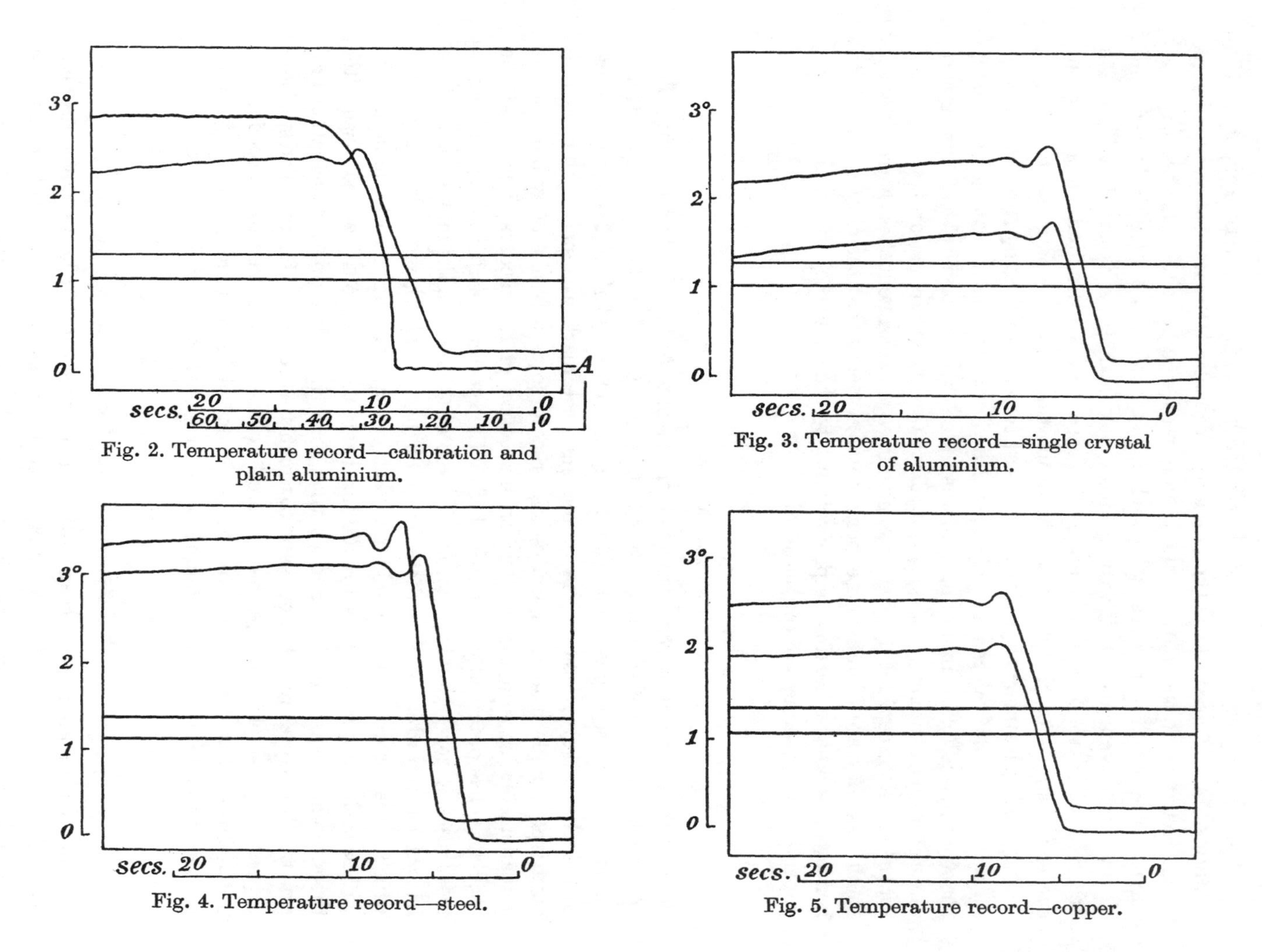


Fig. 2. Temperature record—calibration and plain aluminium.

Fig. 3. Temperature record—single crystal of aluminium.

Fig. 4. Temperature record—steel.

Fig. 5. Temperature record—copper.

a stretching between extensions $\epsilon_1$ and $\epsilon_2$ is $\frac{1}{l_0 a_0}\int_{\epsilon_1}^{\epsilon_2} P l_0 d\epsilon$. This integral is evaluated from the stress-strain records (see p. 107).

The cross-section of the specimen was measured in square inches and the work done in foot-pounds. Using for Joule's equivalent $J$ the value $4{\cdot}18\times 10^7$ ergs per calorie, the factor for converting the above integral to the equivalent rise in temperature is $\frac{1{\cdot}65\times 10^{-3}}{\rho\sigma}$.

The densities of the materials used were determined by weighing in water. The specific heats were not measured, but their values were taken from the latest edition of Landoldt and Börnstein's tables. The values adopted for $\rho$ and $\sigma$ are shown in table 2 below.

Table 2

| Material | $\rho$ | $\sigma$ |
|---|---|---|
| Aluminium | 2·70 | 0·212 |
| Steel | 7·80 | 0·106 |
| Copper | 8·93 | 0·092 |

The results of measurements on eleven specimens, namely, three each of annealed copper, steel and plain aluminium, and two single-crystal specimens of aluminium are given in table 3.

In this table the number of the specimen is given at the top so that the figures may be compared with the reproductions of some of the temperature records (figs. 2–5, p. 94) and stress-strain records (fig. 17, p. 107). In the first column is given the extension $\epsilon$ of the material (percentage of its unstrained length) at the end of each experiment. The second gives the observed rise in temperature $T_0$. The third gives the correction $T_a$ for adiabatic heating which must be added in order to give the rise in temperature which would be observed if the material were unstressed at the beginning and end of each experiment. The fourth column gives the corrected temperature $T_1 = T_0 + T_a$. The fifth column gives the temperature rise $T_2$ corresponding to the heat equivalent of the work done. The sixth column gives the final result, namely, the ratio $T_1/T_2$, i.e. the proportion of the work done which is converted into heat. The bottom line of each table gives the total results for each specimen.

It will be seen that in every case the observed evolution of heat falls short of the heat equivalent of the work done. The difference which represents increase in the internal energy of the material amounts to $13\frac{1}{2}$ % of the work done in the case of steel, $8$–$9\frac{1}{2}$ % in the case of copper, 7–8 % in the case of aluminium, and $4\frac{1}{2}$–5 % in the case of the single-crystal specimens of aluminium.

One remarkable feature of the table is the constancy of the ratio of the increase in internal energy to the work done on the specimens during different stages of the test. It is the same during the first extension when the metal is very soft and hardening

## Table 3

### Steel

**No. 81**

| | $\epsilon$ | $T_0$ | $T_a$ | $T_1$ | $T_2$ | $T_1/T_2$ |
|---|---|---|---|---|---|---|
| | 5·92 | 3·41 | 0·28 | 3·69 | 4·20 | 0·88 |
| | 9·70 | 2·77 | 0·06 | 2·83 | 3·36 | 0·84 |
| | 13·10 | 2·82 | 0·03 | 2·85 | 3·28 | 0·87 |
| Total | 13·10 | 9·00 | 0·37 | 9·37 | 10·84 | 0·865 |

**No. 94**

| | $\epsilon$ | $T_0$ | $T_a$ | $T_1$ | $T_2$ | $T_1/T_2$ |
|---|---|---|---|---|---|---|
| | 3·99 | 2·14 | 0·25 | 2·39 | 2·79 | 0·86 |
| | 8·13 | 3·12 | 0·06 | 3·18 | 3·72 | 0·855 |
| | 11·83 | 3·23 | 0·04 | 3·27 | 3·72 | 0·88 |
| Total | 11·83 | 8·49 | 0·35 | 8·84 | 10·23 | 0·865 |

**No. 95**

| | $\epsilon$ | $T_0$ | $T_a$ | $T_1$ | $T_2$ | $T_1/T_2$ |
|---|---|---|---|---|---|---|
| | 6·11–10·50* | 3·65 | 0·06 | 3·71 | 4·29 | 0·865 |
| | 13·37 | 2·54 | 0·03 | 2·57 | 3·00 | 0·855 |
| | 16·22 | 2·66 | 0·02 | 2·68 | 3·03 | 0·88 |
| Total | 16·22* | 8·85 | 0·11 | 8·96 | 10·32 | 0·865 |

### Copper

**No. 91**

| | $\epsilon$ | $T_0$ | $T_a$ | $T_1$ | $T_2$ | $T_1/T_2$ |
|---|---|---|---|---|---|---|
| | 7·66 | 1·91 | 0·22 | 2·13 | 2·31 | 0·92 |
| | 12·67 | 2·06 | 0·05 | 2·11 | 2·29 | 0·92 |
| | 17·45 | 2·32 | 0·04 | 2·36 | 2·57 | 0·92 |
| Total | 17·45 | 6·29 | 0·31 | 6·60 | 7·17 | 0·92 |

**No. 92**

| | $\epsilon$ | $T_0$ | $T_a$ | $T_1$ | $T_2$ | $T_1/T_2$ |
|---|---|---|---|---|---|---|
| | 7·78 | 1·79 | 0·22 | 2·01 | 2·25 | 0·895 |
| | 13·63 | 2·38 | 0·07 | 2·45 | 2·69 | 0·91 |
| | 19·90 | 3·09 | 0·04 | 3·13 | 3·44 | 0·91 |
| Total | 19·90 | 7·26 | 0·33 | 7·59 | 8·38 | 0·905 |

**No. 93**

| | $\epsilon$ | $T_0$ | $T_a$ | $T_1$ | $T_2$ | $T_1/T_2$ |
|---|---|---|---|---|---|---|
| | 8·76 | 2·18 | 0·23 | 2·41 | 2·72 | 0·885 |
| | 15·35 | 2·88 | 0·06 | 2·94 | 3·23 | 0·91 |
| | 20·20 | 2·50 | 0·03 | 2·53 | 2·76 | 0·915 |
| Total | 20·20 | 7·56 | 0·32 | 7·88 | 8·71 | 0·905 |

### Aluminium (plain)

**No. 96**

| | $\epsilon$ | $T_0$ | $T_a$ | $T_1$ | $T_2$ | $T_1/T_2$ |
|---|---|---|---|---|---|---|
| | 11·12 | 2·16 | 0·25 | 2·41 | 2·59 | 0·93 |
| | 16·88 | 1·67 | 0·02 | 1·69 | 1·82 | 0·93 |
| | 23·06 | 1·85 | 0·01 | 1·86 | 2·01 | 0·93 |
| Total | 23·06 | 5·68 | 0·28 | 5·96 | 6·42 | 0·93 |

**No. 97**

| | $\epsilon$ | $T_0$ | $T_a$ | $T_1$ | $T_2$ | $T_1/T_2$ |
|---|---|---|---|---|---|---|
| | 10·41 | 1·99 | 0·21 | 2·20 | 2·42 | 0·91 |
| | 16·70 | 1·80 | 0·01 | 1·81 | 1·96 | 0·925 |
| | 21·95 | 1·58 | 0·01 | 1·59 | 1·71 | 0·93 |
| Total | 21·95 | 5·37 | 0·23 | 5·60 | 6·09 | 0·92 |

**No. 98**

| | $\epsilon$ | $T_0$ | $T_a$ | $T_1$ | $T_2$ | $T_1/T_2$ |
|---|---|---|---|---|---|---|
| | 9·16 | 1·64 | 0·24 | 1·88 | 2·03 | 0·925 |
| | 15·91 | 1·94 | 0·03 | 1·97 | 2·09 | 0·945 |
| | 21·88 | 1·82 | 0·01 | 1·83 | 1·95 | 0·94 |
| Total | 21·88 | 5·40 | 0·28 | 5·68 | 6·07 | 0·935 |

### Aluminium (single crystal)

**No. 99**

| | $\epsilon$ | $T_0$ | $T_a$ | $T_1$ | $T_2$ | $T_1/T_2$ |
|---|---|---|---|---|---|---|
| | 14·58 | 1·58 | 0·13 | 1·71 | 1·80 | 0·95 |
| | 24·10 | 1·59 | 0·02 | 1·61 | 1·67 | 0·965 |
| | 35·25 | 2·07 | 0·02 | 2·09 | 2·20 | 0·95 |
| | 43·20 | 1·57 | 0·01 | 1·58 | 1·68 | 0·94 |
| | 52·72 | 1·96 | 0·01 | 1·97 | 2·09 | 0·94 |
| Total | 52·72 | 8·77 | 0·19 | 8·96 | 9·44 | 0·95 |

**No. 100**

| | $\epsilon$ | $T_0$ | $T_a$ | $T_1$ | $T_2$ | $T_1/T_2$ |
|---|---|---|---|---|---|---|
| | 16·19 | 2·12 | 0·16 | 2·28 | 2·37 | 0·96 |
| | 25·75 | 1·66 | 0·01 | 1·67 | 1·77 | 0·945 |
| | 37·70 | 2·24 | 0·01 | 2·25 | 2·38 | 0·945 |
| | 46·52 | 1·73 | 0·00 | 1·73 | 1·82 | 0·95 |
| | 55·72 | 1·82 | 0·00 | 1·82 | 1·93 | 0·945 |
| Total | 55·72 | 9·57 | 0·18 | 9·75 | 10·27 | 0·945 |

$\epsilon$ = percentage extension at end of each experiment.
$T_0$ = observed temperature rise.
$T_a$ = 'adiabatic' correction.
$T_1 = T_0 + T_a$.
$T_2$ = temperature rise equivalent to observed work in extension $\epsilon$.

* Owing to an accident, no temperature record was obtained for the first extension, 6·11%, of this specimen.

PLATE 1

rapidly as it is in the last extension when it is quite hard and only hardening slowly. It seems therefore that the increase in internal energy does not bear any very direct relationship to the hardening.

## 7. Description of Machine for Extending Specimen, and Recording Extensometer

The mechanism by which the specimens were stretched was designed specially for these experiments, particularly for use with single crystals of aluminium. These have, at the breaking load of about 1 ton on an initial cross-section of $\frac{1}{4}$ sq.in., an elongation of the order of 70 % on a length of 6 in. Standard tensile testing machines will not deal satisfactorily with large and relatively weak specimens. The wedge grips deform the ends unduly, and require a large initial load. The force is generally applied by a weight, through a lever, and is difficult to control with such large extensions at almost constant loads. The vertical position of the specimen is inconvenient. Finally, a rapid time rate of loading is almost out of the question.

In the machine described in detail below the specimen is horizontal. Its ends are screwed into steel end-pieces with spherical seatings, which fit into corresponding recesses in two moving carriages. Of these, one rests against a pair of steel springs, whose compression measures the applied load. These springs are very short and stiff, so that the energy stored is small and there is no tendency towards instability with plastic specimens. The second carriage is moved by a square thread screw, the nut being rotated by hand through a reduction gear. Loads of 2500 lb. can be conveniently applied in about 2 sec. The recording mechanism consists of a drum on which a celluloid film is stretched, the trace being made on it by a fine steel point. This method (due originally to Mr Collins of the Cambridge Instrument Company) has proved very satisfactory. The record is immediately available, requires no treatment, is permanent, and can be measured with great accuracy. The drum has both axial and rotational motion, the former being proportional to the compression of the springs referred to above, and the latter to the absolute movement of a point on the specimen. The scribing point has a rotational motion only, proportional to the absolute motion of another point on the specimen, situated similarly to, and some 10 cm. from, the first point.

Providing certain geometrical conditions are fulfilled the co-ordinates of the resulting diagram are proportional to the compression of the spring and elongation of the specimen. Actually the abscissae are $1/3{\cdot}59 \times$ the elongation of the specimen between the selected points, the length of the diagram in this direction varying from 6 to 15 mm. (see fig. 17). The ordinates (2000 lb. is represented by about 9 mm.) are approximately twice the compression of the springs, but as the springs do not obey Hooke's law exactly, areas are not exactly proportional to work done. The consequent distortion of the diagram from a true stress-strain figure is unimportant so far as concerns deductions from the shape of the curves, as the departure from Hooke's law is mainly at low loads. For estimation of work done the diagrams are measured on a double-stage microscope and areas of the true stress-strain curves calculated.

The calibration of the mechanism was done directly. By a system of levers described below, loads were applied to the spring in the same way and through the same members as when the specimen is in place. The extension co-ordinate was calibrated by substituting for the specimen an arrangement of telescopic brass tubes incorporating a vernier reading to $\frac{1}{1000}$ in.

The whole of the testing machine and recorder is self-contained and occupies a space approximately 4 ft. × 1 ft. 8 in. The weight is sufficient to enable the maximum load to be applied in a few seconds without the necessity for fixing to the ground. On the other hand, the whole apparatus can be lifted by four men with ease.

## 8. TENSILE TESTING MACHINE (fig. 6)

The main frame $A$ is rectangular, approximately 4 ft. × 1 ft. 8 in., and is made of 4 in. × 3 in. × $\frac{1}{2}$ in. channel steel, bolted together with angle pieces. The longer sides $A_2$ are extended, carrying the recording instrument at one end and the gearing for

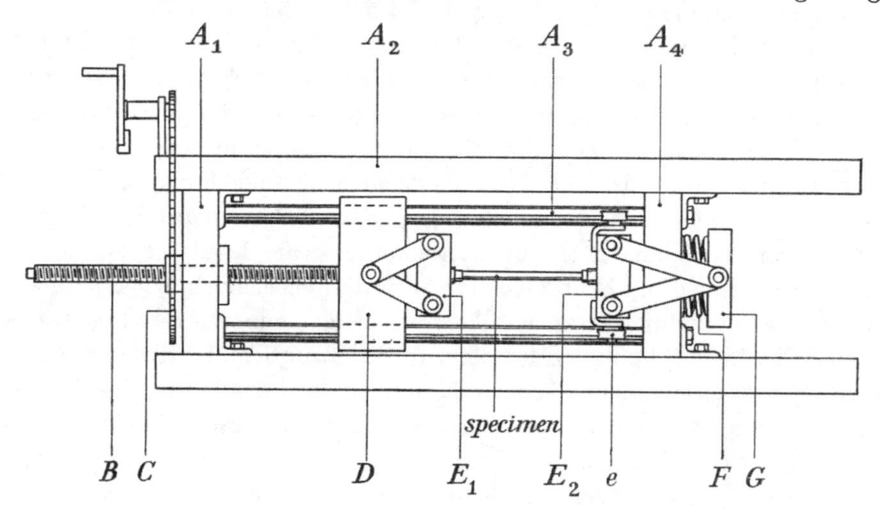


Fig. 6. Arrangement of testing machine.

extending the specimen at the other. To the shorter sides $A_1$, $A_4$, are bolted, parallel to the longitudinal axis of the machine, two bars of $1\frac{1}{2}$ in. round steel $A_3$, which act as guides for the cross-head $D$. This cross-head is moved by means of the screw $B$ (1 in. diameter × $\frac{1}{4}$ in. pitch). The phosphor-bronze nut through which this screw passes is mounted in bearings attached to the cross-member $A_1$, and is rotated by the 3 to 1 chain-gearing $C$. The end thrust is taken by a ball bearing.

On the cross-member $A_4$ rest two large helical springs $F$, held in position by short circular pieces of steel fitting freely inside them. Across their outer ends is a piece of channel steel $G$ provided with two similar discs. To this cross-bar, and to the cross-head $D$, are attached two pieces of channel steel $E_1$, $E_2$, forming carriages for the ends of the specimen. $E_1$ is sufficiently supported by its attachment to $D$. To $E_2$ are fixed two wheels $e$ which roll on the guide bars $A_3$.

The ends of the specimen are screwed into end-pieces $H$ (fig. 7), which form, together with the parts $E_1$, $E_2$, a species of bayonet-joint. In the face of $E_1$ is a rectangular hole through which the end-piece $H$ passes freely. On then rotating the latter through approximately $90^\circ$ the spherical seating on it bears on a corresponding recess in the back of $E_1$. The arrangement at the other end $E_2$ is similar, with an additional complication described below.

The spiral springs $F$ have each three coils of $\frac{1}{2}$ in. diameter wire, wound on a 2 in. mandrel, with a pitch of 0·4 in. The pitch of the end coils decreases to zero and the ends of the spring are ground flat. This forms a suitable bearing surface and gives the spring, regarded as a strut, some stability. But it causes the law of the spring

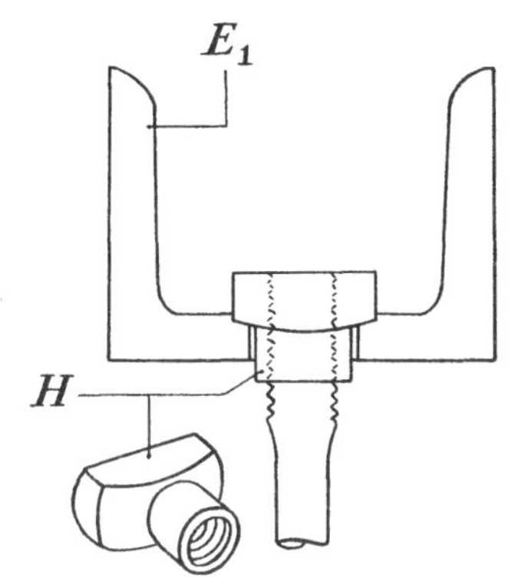


Fig. 7. Grip for end of specimen.

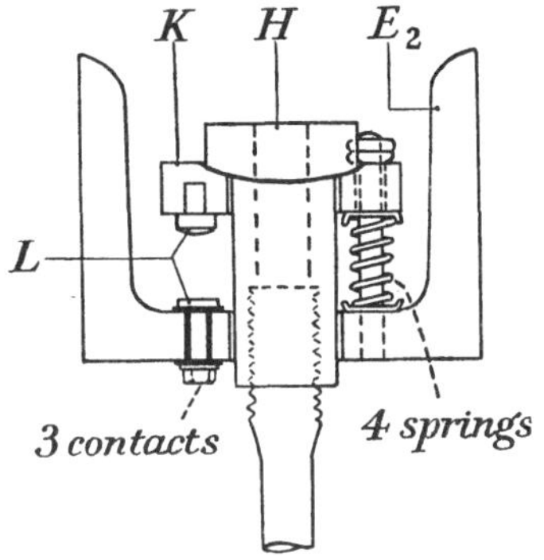

Fig. 8. Grip for end of specimen with spring link.

to be non-linear on account of the closing of the end coil in the early stages of compression. Since there is no actual fixture of the ends either of the specimen or of the springs, a small initial tension was necessary both in the actual experiment and in calibration. To ensure that the same amount was used always, a spring link was introduced which closed hard up at a known load, its closing being indicated electrically. In addition, in order to avoid any uncertainty which might have arisen owing to the springs seating themselves differently at different times, stops were inserted between $E_2$ and $A_4$ which prevented the springs from ever becoming slack. The initial tension (72 lb., of the order of one-fifth of the force required to produce permanent set in the weakest specimens) was sufficient to pull $E_2$ away from these stops.

The spring link is shown in fig. 8. The spherical seating $K$, at the end nearer the springs $F$, is separate from the carriage $E_2$, being supported on four light springs, suitably guided. Three pairs of stops $L$ make contact when the load of 72 lb. is applied to $H$, causing three lamps to light. Provided the load is applied axially, so that the lamps light nearly simultaneously, this device is very sensitive, a decrease of 2 lb. being enough to extinguish the lamps. It was calibrated directly, the carriage $E_2$ being removed and suspended.

7-2

9. MECHANISM FOR CALIBRATING SPRINGS (see fig. 9)

The springs were calibrated in place in the machine by the application of weights through the system of levers shown diagrammatically in fig. 9. A vertical lever $N$, suspended by an adjustable wire, is pivoted at its lower end on a hook-piece $O$, which engages with the lower flange of the carriage $E_1$. To this lever are attached the links $M$ and $P$. The former screws into the standard end-piece $H$ and engages, via the spring link (fig. 8), with the carriage $E_2$, and so to the springs $F$. To $P$ is attached a bell-crank lever $Q$, pivoted at $R$ and carrying weights at $S$. All pivots (indicated by $\otimes$) are hardened knife-edges.

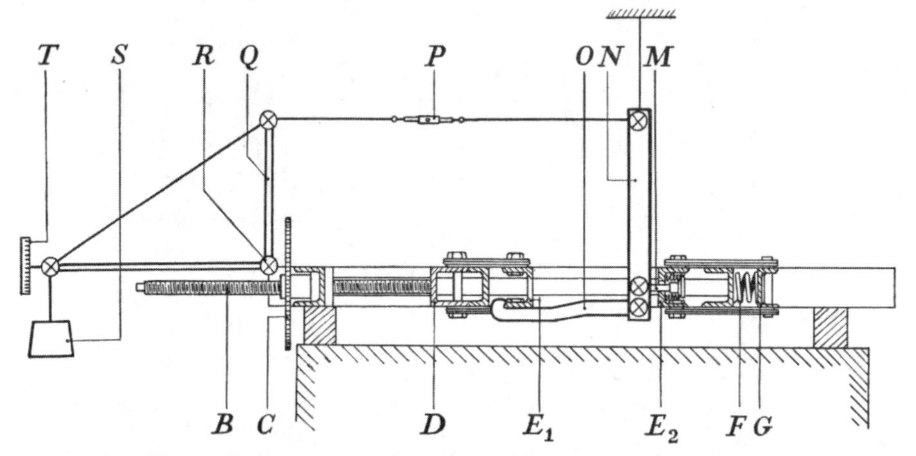


Fig. 9. Arrangement of mechanism for calibrating springs.

The arms of the levers are so proportioned that the force at $M$ is approximately 10 times that at $P$, which is approximately double the weight $S$. The overall ratio is thus 20 to 1, enabling forces of 2500 lb. to be applied to the springs by a weight of 125 lb. at $S$.

The geometry of the mechanism was restored to standard by the following adjustments. Weights being hung at $S$, the cross-head $D$ was moved, carrying with it the hook $O$, until the lever $N$ was parallel to a plumb line. The turn-buckle in $P$ was then adjusted till the bell-crank lever was in the standard position, as indicated by the index and scale $T$.

The accuracy of the calibration is dealt with below (§ 13).

10. EXTENSOMETER (see figs. 10–12)

The essential problem of ordinary extensometers is the multiplying mechanism by which the minute extensions occurring within the elastic range are made visible. In this instrument, with extensions of the order of 50 % on 10 cm., it was found desirable to use a reducing gear (actually 3·59 to 1, in all) in order to produce a convenient diagram. On account of the inconvenience of very long levers, a mechanism was developed which kept the ratio of reduction constant to a high degree of accuracy with levers of reasonable length. This is shown in fig. 10. Two levers, $A_1$, $A_2$, of special construction to avoid errors due to deflection, engage at their lower ends with the selected points on the specimen (the method of attachment is described below). Their upper ends pivot in cross-heads sliding in guides $B_1$, $B_2$. From a point on each of these levers a connection is taken to the recorder. Provided the following conditions are fulfilled, the ratio is invariable:

(1) The points of attachment of the connections must divide the levers in the same ratio.

(2) The guides for the upper ends of the levers must be perpendicular to the axis of the specimen and must be fixed in relation to one another.

(3) The connections to the recorder must be parallel to the axis of the specimen.

The levers may be, and in fact are, of different lengths.

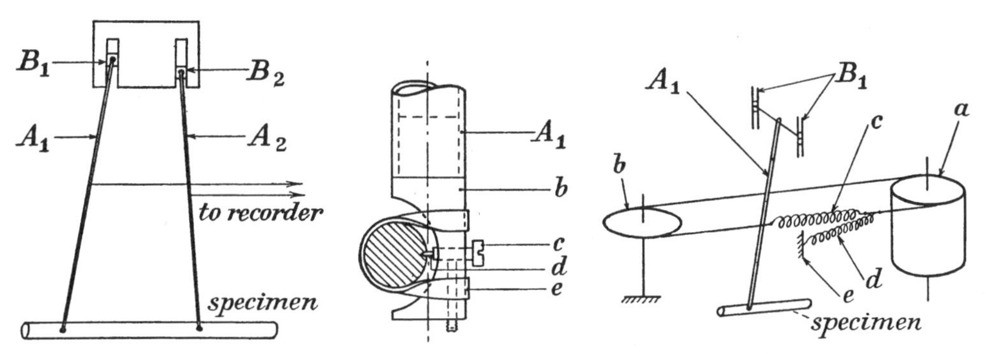


Fig. 10. Arrangement of extensometer mechanism.

Fig. 11. Method of attachment of extensometer to specimen.

Fig. 12. Arrangement of connections to recorder.

Motion of the specimen, as a whole, in relation to the two guides $B_1$, $B_2$ produces no movement of the tracing-point on the drum, and hence it is not essential, from this point of view, that the guides should be fixed, except relatively. As, however, they experience the reaction due to the force required to move the recorder, they are, in fact, fixed to the framework shown in the accompanying photograph (plate 1).

The levers $A_1$, $A_2$ are tubular, stiffened by kingpost bracing tensioned up by adjusting screws. The general arrangement can be seen in the photograph.

One of the problems of the experiment was the attachment of the lower ends of the levers $A_1$, $A_2$ to the selected points in the specimen. Since normal plane sections of a single crystal do not remain normal when the specimen is stretched, it is impossible to use the conventional form of attachment—namely, two screws engaging in punch-marks at opposite ends of a diameter. On the other hand, the deformation of a single crystal is so closely uniform along its whole length that the extension of any generating line of the original cylinder may be taken as the extension of the whole specimen. Two punch-marks are made about 10 cm. apart, on a generating line, the punch used being a gramophone needle. Into each of these fits a similar needle fixed into the lower end of the lever $A_1$ or $A_2$, the details being shown in fig. 11. In the end of the tube forming the lever is soldered a piece of brass *b*, shaped as shown, carrying the brass plug *c*, in which the needle-point *d* is soldered. The point is held firmly in the punch-mark by a rubber band *e*. It will be seen that the position of the point, practically on the axis of the tube, forming the lever $A_1$, reduces the torsion on the lever to a negligible amount, and any tendency of the lever to twist as a whole is reduced by arranging the trunnions which slide in the guides $B_1$, $B_2$ at some distance from the axis of the tube (see fig. 12, $B_1$).

The intermediate points on the lever $A_1A_2$ are connected to the recorder by steel bands 0·1 in. wide by 0·004 in. thick. In order to avoid errors due to the slight flexural rigidity of this band, a certain minimum tension is needed. This is obtained by making each band a complete loop (see fig. 12), passing from the attachment to the lever $A_1$, round the pulley *a*, by which the recorder is driven, back to a free pulley *b* at the other end of the machine, and finally returning to the lever $A_1$, which therefore experiences no force due to any tension in the band. A long spiral spring *c* is inserted in each loop between the two pulleys mentioned above. Finally, in order to ensure that there is no lost motion in the system a light adjustable spiral spring *d* connects a point on each band to the framework *e*.

It will appear from the calibrations referred to in § 13 below that these precautions had the desired result, the *x* co-ordinate of the records being a constant fraction of the extension between the selected points on the specimen to a high order of accuracy.

## 11. Recorder (see figs. 13–15)

A diagrammatic sectional view of the recorder is shown in fig. 13, and its external appearance in plate 1. It is mounted on a cross-bar of channel steel attached to the extended sides of the main frame of the testing machine. On this is a bearing *A*, in which turns the external drum *B*, which carries the scribing point $b_1$. On a parallel cross-bar of flat steel, firmly supported by arms which are clearly visible in the photograph, is a second bearing *C*.

On the upper end of the drum *B* is formed a groove for the steel band $e_2$ from which it derives its motion. Three large oval panels are removed from the drum in order to enable the celluloid film to be mounted on the internal drum *F*. The scribing point $b_1$ is fixed in a triangular frame which pivots on pointed screws $b_2$, attached to the drum *B*. It is pressed on the film by a flat spring (see photograph), which is so arranged that the point can be swung back while the film is put in place, and restored, with the same pressure, when required.

The inner drum *F* carries the film on which the record is made. This is ordinary cinematograph film, the gelatine being first removed. The axial motion of the drum, approximately twice the compression of the springs which measure the force on the specimen, is obtained as follows. To the cross-bar by which these springs are compressed (see fig. 14) is attached a compensating lever *a*, from whose ends two steel bands pass to two equal pulleys $L_1$, mounted on a spindle fixed below the recorder. In between, and integral with, these pulleys is a larger one $L_2$, from whose circumference a steel band passes up to the recorder. Each band is attached to the circumference of its pulley, the compensating lever $a$ eliminating the effect of any slight inequality in the diameters of the smaller ones. The bands are kept taut by a spring $K$ (fig. 13) so arranged that it does not contribute to the friction of the recorder.

The band from the larger pulley $L_2$ is attached to the lower end of a tube $H$ which passes up through both drums and has on its upper end a collar, against which the inner drum is pressed by a spring $G$, a ball thrust bearing reducing the friction. The spring $G$ also serves to press the outer drum $B$ down on to the bearing $A$.

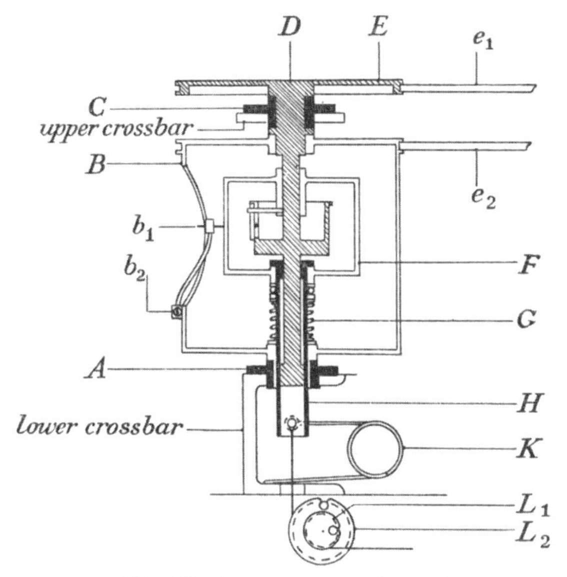


Fig. 13. Arrangement of recorder.

The most difficult problem of the instrument was the rotational motion of the inner drum $F$. This is derived from the upper pulley $E$ and steel band $e_1$. The spindle $D$ of this pulley passes down through the upper bearing $C$, both drums, $B$ and $F$, and the tube $H$, the bearing surfaces being arranged as shown in fig. 13 in order that the tension in the bands may not cause the mechanism to bind.

On the part of this spindle $D$ which is inside the drum $F$ is mounted a cylinder $d_1$ (fig. 15) with a diameter approximately half that of the drum. This cylinder is open at the upper end and is slotted parallel to its axis, as shown in the sketch. To its outer surface is soldered a piece of silver steel rod $d_2$ which was carefully tested for straightness. Its position is indicated in fig. 15. It was adjusted to be as nearly as possible parallel to the spindle $D$. Through the slot in $d_1$ there projects a similar piece of rod $f$, which is fastened to a projection of the end of the drum $F$. The rods $f$ and $d_2$ are kept in contact by springs (not shown) to avoid backlash, and form a 'key' and 'keyway' connecting $D$ and $F$ rotationally.

This construction reduces the errors in the axial motion of the drum $F$ to a minimum. As the result of the precautions described it is not possible to detect, on a Hilger measuring microscope, any departure from straightness in fiducial lines on the records parallel to either $x$- or $y$-axis (see fig. 17 below). The angle between the axes is almost exactly 90°, showing that the 'keyway' is practically parallel to the axis of the drum. As, however, the angle between the stages of the microscope could be adjusted so that the movements were parallel to the fiducial lines, the exact magnitude of this angle is not important.

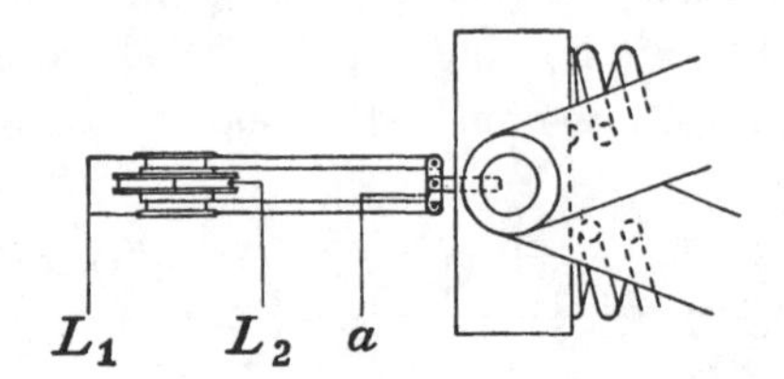


Fig. 14. Connection of extensometer to springs.

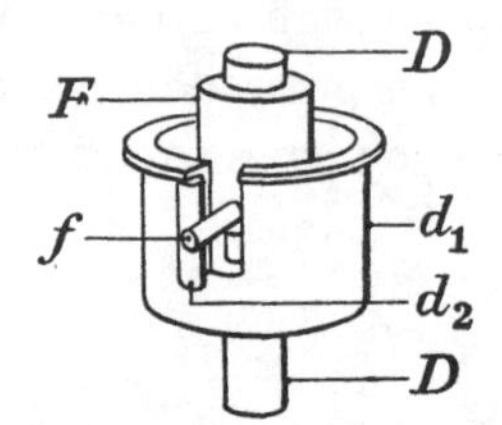


Fig. 15. Rotational drive to drum of recorder.

## 12. Instrumental Errors

The errors to be anticipated in such a mechanism as has been described may be classified as follows:

(1) *Due to strain.* The supports of all parts are very firm. The stresses in the main frame are very small, and any deflection which takes place when the force is applied to the specimen is certainly elastic and, since the springs are calibrated *in situ*, is absorbed in the calibration. The steel bands driving the recorder are of ample size, and are under sufficient initial tension to ensure that their flexural rigidity is not a source of error.

(2) *Due to friction.* There is very little frictional resistance to the rotational movements of the recorder, but there is an appreciable frictional resistance to the axial motion of the inner drum. Since, however, the forces to be measured are of the order of 1000 lb. and this friction is of the order of ounces, no appreciable error arises. All the surfaces on the recorder are copiously lubricated with a fairly viscous oil, and no tendency to stick has ever been observed. It may be noted that in all experiments the motion of every part of the mechanism is in one direction only (the fall of stress at the elastic limit with steel is the only exception and cannot give rise to any appreciable error).

(3) *Due to slack.* The precautions taken to avoid slack at every point have been described. The points which engage in the punch marks on the specimen fit very firmly, and though the latter must extend into an oval shape when the specimen is stretched, it appears from separate measurements made on the stretched specimen that the points remain in the centre of the punch marks.

(4) *Due to change of conditions between calibration and experiments.* As has been mentioned in describing the methods of calibrating, elaborate precautions were taken to ensure that, so far as the condition of the machine and recorder were concerned, the calibrations and experiments were identical.

## 13. Order of Accuracy of Results

The magnitude of the probable error of the final result (the area of the stress-strain figure) is influenced chiefly by the calibration of the springs. The levers used in this calibration (see § 10) were found to have ratios of 1·959 and 10·21 as the mean of several calibrations by weights, giving a combined ratio of 20·00. The variations in the figures for the individual levers were of the order of 1 in 1000 and the combined result may be relied on to 1 in 500. The 'out of balance' of the bellcrank lever was determined directly to be equivalent to 21·5 lb. at the springs.

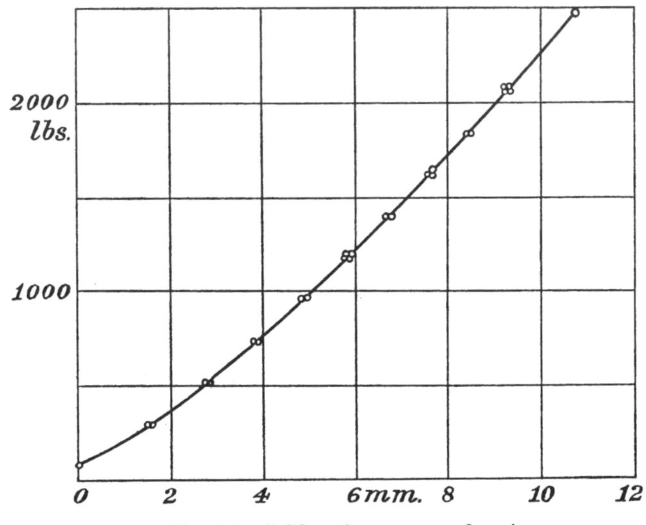


Fig. 16. Calibration curve of springs.

The calibration curve of the springs is shown in fig. 16. It includes the results of calibrations taken both before and after the experiments, showing that the characteristics of the springs did not alter appreciably with time. The points nowhere depart from the curve adopted by more than 1 %, and it is considered that this represents the maximum error in a force measurement.

The calibration of the extension co-ordinate by the method described in § 7 (p. 98) gave a ratio of 3·59 within 1 in 500, the ratio calculated from the measurements of the mechanism involved being 3·58. The former figure was used as it corresponded to a direct comparison of specimen and record, whereas the latter involved several separate measurements, one being of a type in which great accuracy could not be attained. The constancy of this ratio for extensions of both large and small amounts was within the figure given: 'large' meaning of the order of 5 cm. on the specimen, and 'small' of the order of 5 mm. In the experiments the actual extensions corresponding to each temperature measurement had the following average values for the materials tested:

| | mm. | | mm. |
|---|---|---|---|
| Copper | 6 | Aluminium (plain) | 7 |
| Steel | 4 | Aluminium (single crystal) | 10 |

The *total* extensions of each specimen averaged 18, 12, 22 and 50 mm. respectively.

The actual records were measured on a Hilger comparator, on which readings could be repeated within one-hundredth of a millimetre. The final areas were computed from measurements of ordinates sufficiently closely spaced.

It is considered that the above figures enable it to be claimed that the final areas, representing the work done, are determined within 1 %.

## 14. Stress-strain Records

Some examples of these are shown, enlarged 3·1 times, in fig. 17.* The upper three are from steel, copper and plain aluminium, respectively, and the lower two from single-crystal specimens of aluminium. The numbers correspond to those in the table of results (table 3) and in the temperature records, figs. 2–5.

The approximate scales of the diagrams as reproduced are:

*Extension* ($x$-axis) 1 cm. = 11·6 % on the original length of 10 cm. between marks.

*Forces* ($y$-axis) 1 cm. = 715 lb. approximately. The force scale is not exactly linear, for reasons which have been described above.

The original dimensions of the specimens are given in table 4.

The approximate stress scales are therefore as follows:

| | |
|---|---|
| Steel | 1 cm. = 17,000 lb./sq. in., |
| Copper | 1 cm. = 18,600 lb./sq. in., |
| Aluminium | 1 cm. = 4,200 lb./sq. in., |
| Aluminium (single crystal) | 1 cm. = 3,300 lb./sq. in., |

referred, in all cases, to the original cross-sectional area.

The short vertical marks on the records represent the end of each experiment. With steel and aluminium they appear on the record automatically and were thought at first to be due to a defect in the instrument, but as they did not appear in experiments on copper (the marks on the record for copper were made deliberately in order to give greater precision to the measurements) they are presumably characteristic of certain materials only. Their significance is discussed below.

The interval between two extensions of any one specimen was about 10 min., and during this time the specimen was under tension, constant except in so far as any 'creep' of the specimen allowed the springs to extend. The records show that this creep occurred with steel, and (to a much smaller extent) with plain aluminium, but not with copper or single crystals of aluminium. It should be noted that the circumstances during this 'creep' are very different from those which occur when a specimen is left under load in an ordinary testing machine. The load is then independent of any extension which may occur, whereas here extension of the specimen relieves the load rapidly. In fact, the springs extend the same amount as

* The definition of the lines has suffered in reproduction. As seen by transmitted light the trace of the scribing point appears as two dark lines, separated by a white line. The overall width is about 0·04 mm. In measuring the records, the cross-wires of the microscope were focused on the outer edge of one of the dark lines.

the specimen. The recorder magnifies extensions of the spring twice (approximately) and reduces extensions of the specimen, between the selected points, 3·59 times. Hence, if the whole length of the specimen is $n$ times the length between the points, the line on the record representing the 'creep' referred to above should slope down

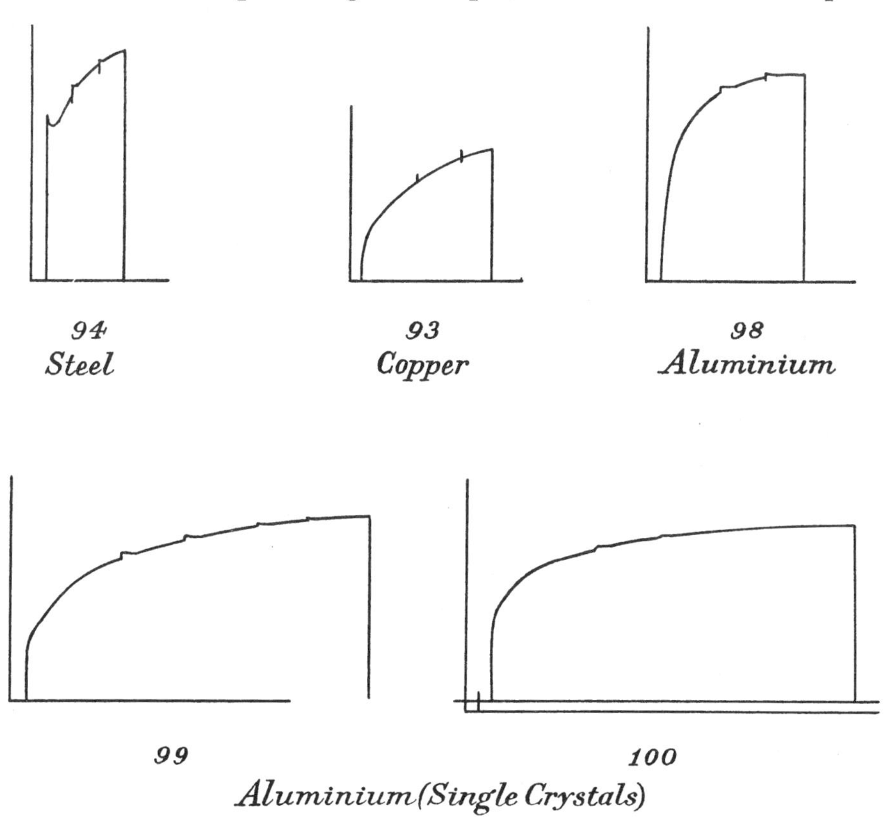


Fig. 17. Stress-strain records.

Table 4

| Material | Length of parallel part (cm.) | Outside diameter (in.) | Inside diameter (in.) | Cross-sectional area (sq.in.) |
|---|---|---|---|---|
| Steel | 30 | 0·31 | 0·21 | 0·0422 |
| Copper | 29 | 0·25 | 0·12 | 0·0384 |
| Aluminium | 21 | 0·50 | 0·19 | 0·1688 |
| Aluminium (single crystal) | 16 | 0·56 | 0·19 | 0·2151 |

from left to right at an angle $\tan^{-1} 2 \times 3{\cdot}59n$. For the steel specimen ($n=3$) this angle is $\tan^{-1} 21{\cdot}5$. Examination of the record will show that these lines do in fact slope to about this extent.

On increasing the load, for the next extension, it will be seen that with steel and aluminium (both plain and single-crystal) no extension occurs until the load has risen appreciably above that previously reached. There is then a rapid extension at almost exactly constant load, followed by a period in which both load and extension increase. The fact that the curve corresponding to the last part joins smoothly on to that representing the end of the previous extension suggested that the 'kink' in the curve was instrumental, but as the phenomenon is not observed with copper (see records) it is presumably characteristic of certain materials only, corresponding to something analogous to sticking.

With copper the absence of the 'kink' is so definite that it was sometimes found difficult to decide exactly where each separate extension ended, though the approximate point could be seen without difficulty, owing to the slightly deeper depression in the celluloid made by the scribing point resting in one position for some minutes. The marks seen in the copper records were therefore made deliberately, as mentioned above.

It will be seen that the 'kinks' in the curves for single crystals of aluminium decrease regularly as the load increases, and practically disappear when the breaking load is reached—see in particular no. 100, which was on the point of breaking at the end of the experiment.

The record for mild steel no. 94 shows the characteristic fall of stress at the yield-point. An actual *fall* of stress was not observed in all specimens, the variation being presumably due to slight differences in the rate of extension. The 'ripples' on the record are due to slight irregularity in turning the handle of the machine. They do not appear in the records for the other materials, presumably because these have a much smaller tendency to 'creep'.

All the specimens were thoroughly annealed before the experiments, but there is nevertheless an appreciable 'elastic range' for all except the plain aluminium. This material, however, hardens remarkably rapidly, having after 1 % extension an 'elastic range' of about 4000 lb./sq.in., and after 5 % extension a range of about 9000 lb./sq.in., these stresses being about one-third and three-fourths respectively of its ultimate breaking stress.

When plotted on true stress-strain co-ordinates the records for all the copper and all the plain aluminium specimens tested agree very closely with one another. The single-crystal specimens of aluminium differ considerably from one another, no. 99 having a lower yield point than no. 100, though it is ultimately stronger.

The work was carried out in the Cavendish Laboratory through the kindness of Sir Ernest Rutherford, to whom we wish to express our thanks.

We wish also to thank Professor Carpenter and Miss Elam for presenting us with single-crystal specimens of aluminium, and Mr W. W. Hackett, of Messrs Accles and Pollock Ltd., who presented us with specially annealed steel tubing from which the test-pieces were made. An analysis of this material (for which we are indebted to Mr W. E. Woodward, M.A.) showed that it contained 0·17 % carbon and 0·76 % manganese.

# 7

# THE PLASTIC EXTENSION AND FRACTURE OF ALUMINIUM CRYSTALS*

REPRINTED FROM

*Proceedings of the Royal Society*, A, vol. CVIII (1925), pp. 28–51

In the Bakerian Lecture, 1923,† an experiment was described in which a large single crystal of aluminium was stretched till it broke. The resulting distortions were measured at various stages of the test, and at the same time X-ray measurements were made to determine the orientation of the crystal axes. Various conclusions were drawn regarding the connection between the distortion and the crystal axes for the particular specimen with which the experiment was carried out.

The present paper describes similar experiments made with several more specimens, in order to find out whether the conclusions previously reached are general in their application, and to settle several points which the previous experiment left in doubt. The use of a new X-ray spectrometer, specially designed for the purpose, and other minor improvements have enabled us to obtain greater accuracy than was possible previously.

The chief result of the earlier work was that when a 'single crystal' bar of aluminium is stretched the whole distortion during a large part of the stretching is due to a simple shear parallel to an octahedral (111) plane, and in the direction of one of the three diad (110) axes lying in that plane. Of the twelve‡ crystallographically similar possible modes of shearing, the one for which the component of shear stress in the direction of shear was greatest was the one which actually occurred.

The first object of the present work was to determine the orientation of the crystal axes for several specimens. Assuming that the relationship outlined above holds in general, it is possible to predict the orientation of the plane on which slipping should occur. Three of the specimens were stretched and their distortions were analysed by the methods previously described. It was found that in each case the distortion was such as would be produced by shearing parallel to a single plane, and that this plane of slip was the one predicted. It was also found that the changes in orientation of the crystal axes relative to the axis of the specimen during the test were in good agreement with the prediction.

In the previous work it was not possible to make a detailed analysis of the deformation during the last stages of the stretching, but it was shown that the

* With C. F. ELAM.

† *Proc. Roy. Soc.* A, CII (1923), 643; paper 5 above, p. 63.

‡ Or 24 if shears in opposite senses parallel to the same line be regarded as distinct.

deformation ceased to be due to slipping on one crystal plane. Reasons were given for supposing that the deformation might be due to slipping on two crystal planes simultaneously, and it was shown that the crystal axis which formed the intersection of these two planes remained unstretched during the test, as it should if the distortion were, in fact, due to this cause. On the other hand, there are a number of other types of distortion which might also leave this particular axis unstretched. In the present work it is shown that, for a small distortion, the 'unstretched cone' passes through, or very close to, three other crystal axes, which, in fact, determine all the possible kinds of 'unstretched cone' which can be produced by double slipping of the type contemplated. This completes the proof that the slipping in the last stages of the test is of this type.

When the crystal begins to slip on a second plane, it seems likely that the rate of slipping on each plane would be the same. In this way they would remain inclined at equal angles to the axis of the specimen, but it was impossible to verify this suggestion from the previous measurements. This point has now been tested, and it is found that though the double slipping does, in fact, begin when the two planes get to the position in which they make equal angles with the axis, the rate of slipping on the original slip-plane is sometimes greater than it is on the new one. The two planes do not, therefore, necessarily remain at equal angles with the axis. The process cannot be followed very far, however, because the specimen usually breaks when only a comparatively small amount of double slipping has occurred.

One specimen, however (no. 24), was found whose longitudinal axis happened to be very close to a (112) axis. In this position double slipping should begin at once, and it was found that double slipping occurred during nearly the whole extension. In this case it was found that the amount of slipping on the two planes was practically equal, and the axis of the specimen remained very close to the (112) axis during the whole test.

Now that the slip-planes have been determined it is possible to calculate the tangential stress per square inch of the slip-plane acting in the direction of slip. This has been done for all the specimens on which measurements have been made, and it is found that this stress rapidly increases as the stretching of the specimen proceeds. A curve connecting tangential stress with extension is shown (fig. 10, p. 124), and it is found that all specimens give points which lie very close to the same curve. The stress goes on increasing right up to the breaking point, but in the last part of the test the increase is less rapid than in the first part. This leads to an explanation of the reason why the specimen stretches so uniformly and a discussion of the conditions which determine the extension at which the specimen will break.

## Method of Prediction of Planes and Directions of Slip

The orientation of the crystal axes of six single-crystal test-pieces were determined by means of X-rays. In order to predict how they would be distorted on being stretched, it was first necessary to devise a convenient method for determining which of the twelve possible types of slipping corresponded with the maximum component of tangential force tending to produce the slip. If $\theta$ represents the angle between the axis of the specimen and the normal to a possible slip-plane, and $\eta$ represents the angle between the projection of the axis of the specimen on the slip-plane and a possible direction of slip, the tangential component of stress tending to produce this kind of slipping is $(T/A)\cos\theta\sin\theta\cos\eta$, where $T$ is the total pull on the test-piece and $A$ is the area of its cross-section.

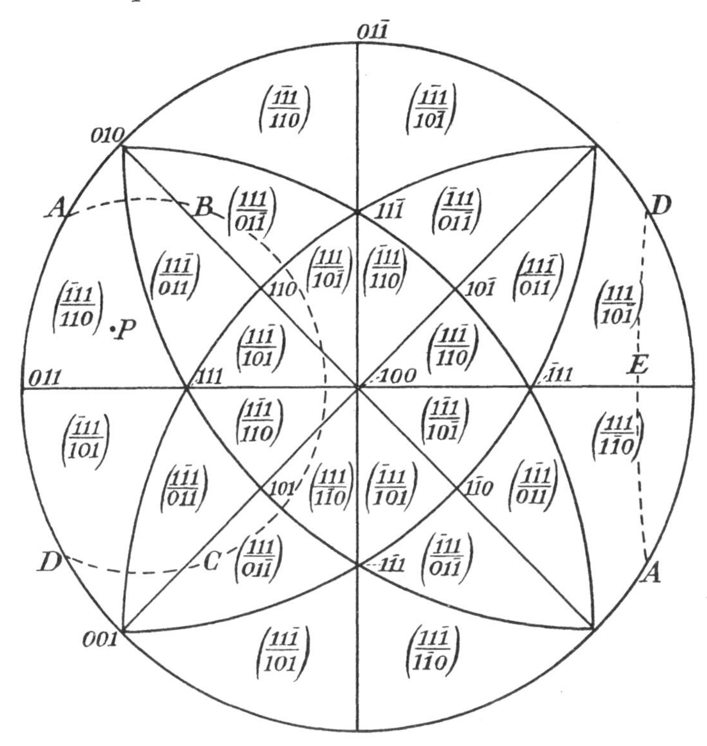


Fig. 1. Stereographic diagram for predicting slip-plane and direction of slip from X-ray measurements.

The problem was to find a convenient way to estimate which of the twelve possible values of $(T/A)\cos\theta\sin\theta\cos\eta$ was the greatest. For this purpose the stereographic projection of the crystal axes shown in fig. 1 was made. In this figure one of the cubic (100) axes is the centre and the four octahedral (111) axes are arranged symmetrically round it. To fit the axes determined by X-rays in any particular case to this figure it is only necessary to rotate them in such a way that any one of the cubic axes comes in to the centre of the figure. In each case the orientation of the crystal axes was found by determining the spherical polar co-ordinates $(\theta, \phi)$* of two crystal axes. Sometimes one (111) and one (100) axis were determined. Sometimes it happened that two (111) axes were found. In the latter case a (100) axis could be found by marking on the diagram the position of

* See Bakerian Lecture; paper 5 above, p. 72.

the external bisector of the angle (70° 32′) between the two (111) planes. This operation is easily performed graphically using a stereographic net.* It is then possible to superpose the crystal axes determined by X-rays on the diagram of fig. 1.

The rotation of the crystal axes into the position where the (100) axis is in the centre of the figure moves the axis of the specimen away from the centre to some point $P$ of the diagram (see fig. 1). With any given position of the point $P$ it is an easy matter to measure with the stereographic net the values of $\theta$ and $\eta$ corresponding with the twelve possible types of slipping, and thus to determine which corresponds with the maximum value of $\cos\theta\sin\theta\cos\eta$. This operation was performed for a very large number of points scattered all over the surface of fig. 1, and it was found that the whole sphere shown in the stereographic diagram could be divided up into areas such that all the points $P$ within any given area would correspond with a direction of the axis of the specimen for which one particular type of slipping might be expected to occur.

These areas are found by trial to be spherical triangles bounded by the six planes of type (110) and the three cubic planes of type (100). They are shown in fig. 1, and the following notation is adopted to show which type of slipping is applicable to each triangle. The symbol ($\bar{1}11/110$) means that the slipping is on the plane perpendicular to the axis ($\bar{1}11$) in the direction of the axis (110). It will be seen that the hemisphere represented in fig. 1 is divided into twenty-four triangles and that any given distinguishing symbol, such as ($\bar{1}11/110$), occurs twice. The two triangles so indicated correspond with slipping parallel to the same plane and to the same line in that plane, but in opposite directions.

To use the diagram (fig. 1) the point $P$ corresponding with the axis of the specimen is marked on it and the corresponding slip-plane and direction of slip are read off. The normal to each slip-plane and all the possible directions of slip are marked on the diagram, so that it is only necessary to rotate the crystal axes back to their original positions, as measured in the specimen, in order to predict the position of the slip-plane and direction of slip. This operation, which takes only a few minutes to perform, was carried out with six specimens (in addition to the original specimen described previously), but as the distortion measurements were carried out with only three of them the results are given in these three cases only (see table 3).

It will be noticed that the condition that $\cos\theta\sin\theta\cos\eta$ shall be as great as possible implies in most cases that the slip-plane is the octahedral plane which is most nearly at 45° to the axis of the specimen, but this is not always the case. In fig. 1 the cone which is the locus of all directions at 45° to the axis (111) is shown as a broken dotted line, *ABCDEA*. It will be noticed that it passes through the six triangles which correspond with slipping parallel to the six possible directions on the plane (111); it also passes through six other triangles (corresponding with slipping on the three other octahedral planes) near their corners. At points on the

* See Bakerian Lecture; paper 5 above, p. 72.

45° circle in these six corners the factor $\cos\theta\sin\theta$ is greater for the plane (111) than it is for any other octahedral plane, but the factor $\cos\eta$ reduces the product $\cos\theta\sin\theta\cos\eta$ to a less value than that corresponding with another octahedral plane. Unfortunately, we have not obtained any specimens corresponding with points in these corners, so it is not possible to distinguish between the hypothesis that $\cos\theta\sin\theta\cos\eta$ shall be as great as possible and the hypothesis that the slip-plane is the one which is nearest to 45° to the axis of the specimen.*

## Analysis of Distortion in Three Specimens during Stretching—Methods of Measurement

Three 'single-crystal' specimens were selected, two of them, nos. 68 and 59, were machined very carefully till they were accurately rectangular in section. The crystals are so soft that great care was necessary to ensure that they should not bend during this process, and special apparatus had to be designed to hold the specimen while it was being machined. They were then marked with the system of marking described previously,† but in this case special precautions were taken to ensure that the marks should be true and of uniform depth. For this purpose a special apparatus was designed for us by Dr G. F. C. Searle, F.R.S., in which the specimen was caused to slide along an optical bench under a small knife which pressed on it with a uniform pressure and thus made a scratch of uniform depth.

The third specimen, no. 72, was left round. It was marked by eight longitudinal scratches arranged at intervals of 45° round its cross-section, and by a number of circular scratches round the specimen in planes perpendicular to the axis, the system forming initially a rectangular network on the curved surface of the bar. The method used for determining the distortion was the same as that previously described, but in the case of the round specimen some slight modifications were necessary. The angles denoted by $\beta$ and $\gamma$ in the previous work could not be measured on a curved surface. The quantities $\lambda$ and $\cot\alpha$‡ required in the determination of the unstretched cone were determined more directly than they were in the case of the square section.

The specimen was mounted vertically in the spectrometer specially designed for the purpose of these experiments. A small microscope was mounted so that it would slide horizontally in a V groove which was directed towards the axis of rotation of the turn-table. The vertical plane containing vertical scratches at opposite ends of a diameter of the cross-section was then brought into coincidence with the axis of rotation of the spectrometer table. This was rendered easy by the use of a mechanical stage mounted on the turn-table and on to which the specimen was fixed. This mechanical stage was adjusted till the readings of the turn-table, when the two vertical scratches were seen on the cross-wires of the microscope,

* This or the equivalent hypothesis that $\cos\theta\sin\theta$ has the maximum possible value.

† Bakerian Lecture; paper 5 above, p. 64.

‡ Ibid. p. 70, equation (2).

8 BSP

differed by 180°. To find the angle corresponding with $\lambda$* between the vertical planes through any two pairs of opposite vertical scratches it was necessary to take two pairs of readings in the manner just described.

The angle between them was then the difference between the readings of the spectrometer table in the two cases. Except when the axis of rotation of the spectrometer table happened to coincide with the line of intersection of the two vertical planes whose inclination to one another was being measured, it was necessary to adjust the mechanical stage after completing the first pair of readings and before starting the second pair, but as the stage merely translated the specimen to a parallel position without rotating it no error was caused in that way.

In this way the $\phi$ co-ordinates of the four vertical planes through the opposite pairs of vertical scratches were measured. This method of measuring $\phi$ turned out to be so convenient that it was used with the square specimens as well as the round ones, replacing the method described previously.†

In the case of the round specimens the quantity $\tan \alpha$ was found by measuring directly $d$, the distance between a pair of opposite vertical scratches, and $d \tan \alpha$, the difference in height, when the specimen is vertical, between the points where these two vertical scratches intersect one of the cross scratches. This method of determining $\tan \alpha$ was found to be less accurate than the method applicable to sections originally rectangular, owing to the difficulty of measuring $d$ by means of a micrometer when the section had become elliptical in the course of the stretching.

## Verification of Predicted Position of Slip-plane

In table 1 are given the dimensions of the mean parallelepiped‡ for two specimens, nos. 59 and 68, for a series of extensions of the material. The positions of the axes which were determined by X-rays are given in table 2.§

Table 1. *Dimensions of mean parallelepiped for two test-pieces*

No. 59

| | | | | | | | |
|---|---|---|---|---|---|---|---|
| $\epsilon$ | | 1·000 | 1·1016 | 1·484 | 1·560 | 1·745 | 1·936 |
| $b$ | in. | 0·4003 | 0·3623 | 0·2725 | 0·2588 | 0·2314 | 0·215 |
| $c$ | in. | 0·3968 | 0·3977 | 0·3973 | 0·3968 | 0·3958 | 0·3921 |
| $\lambda$ | degrees | 90·0° | 90·9° | 96·2° | 97·3° | 100·1° | 102·5° |
| $\beta$ | degrees | 90° 0′ | 88° 51′ | 96° 30′ | 99° 23′ | 107° 44′ | 115° 27′ |
| $\gamma$ | degrees | 90° 1′ | 87° 37′ | 81° 37′ | 80° 28′ | 77° 56′ | 75° 29′ |

No. 68

| | | | | | | | | | |
|---|---|---|---|---|---|---|---|---|---|
| $\epsilon$ | | 1·000 | 1·0932 | 1·4115 | 1·516 | 1·5700 | 1·6212 | 1·659 | 1·6955 |
| $b$ | in. | 0·3460 | 0·3228 | 0·2672 | 0·2539 | 0·247 | 0·2412 | 0·2378 | 0·2340 |
| $c$ | in. | 0·3465 | 0·3401 | 0·3281 | 0·3259 | 0·3280 | 0·3232 | 0·3231 | 0·3226 |
| $\lambda$ | degrees | 90·0° | 85·9° | 75·3° | 72·6° | 71·8° | 70·5° | 69·5° | 68·9° |
| $\beta$ | degrees | 89° 57′ | 89° 37′ | 96° 21′ | 99° 41′ | 101° 44′ | 103° 27′ | 104° 38′ | 105° 38′ |
| $\gamma$ | degrees | 89° 58′ | 89° 16′ | 83° 48′ | 81° 5′ | 79° 57′ | 79° 25′ | 78° 31′ | 77° 46′ |

* Bakerian Lecture; paper 5 above, p. 65.

† Ibid. p. 69.

‡ Ibid. p. 66.

§ For explanation of co-ordinates $\theta$ and $\phi$, see ibid. p. 71.

The angles between two of the axes whose positions were determined by X-rays are shown in the last column of table 2. These angles should be 54° 44′ in the case when the two axes are (100) and (111) respectively, or 70° 32′ in the case when

Table 2. *Co-ordinates of crystal axes determined by X-ray measurements at different stages of test. The angles are measured in degrees*

No. 59

| | (111) | | $(\bar{1}11)$ | | (100) | | Measured angle between pairs of axes | |
|---|---|---|---|---|---|---|---|---|
| $\epsilon$ | $\theta$ | $\phi$ | $\theta$ | $\phi$ | $\theta$ | $\phi$ | | Degrees |
| 1·000 | 32·4 | 333·2 | 39·3 | 170·9 | — | — | (111), $(\bar{1}11)$ | 70·7 |
| 1·102 | 28·1 | 325·1 | 44·6 | 174·2 | — | — | (111), $(\bar{1}11)$ | 70·2 |
| 1·484 | — | — | 58·5 | 184·4 | 68·3 | 346·9 | $(\bar{1}11)$, (100) | 56·1 |
| 1·560 | — | — | 59·3 | 185·6 | 67·5 | 347·8 | $(\bar{1}11)$, (100) | 55·7 |
| 1·745 | — | — | 62·8 | 190·3 | 65·3 | 350·6 | $(\bar{1}11)$, (100) | 55·5 |
| 1·936 | — | — | 64·3 | 192·7 | 63·6 | 352·0 | $(\bar{1}11)$, (100) | 55·7 |

No. 68

| | $(1\bar{1}1)$ | | $(11\bar{1})$ | | (001) | | Measured angle between pairs of axes | |
|---|---|---|---|---|---|---|---|---|
| $\epsilon$ | $\theta$ | $\phi$ | $\theta$ | $\phi$ | $\theta$ | $\phi$ | | Degrees |
| 1·00 | 79·3 | 74·5 | 73·7 | 5·0 | 28·1 | 45·5 | $(1\bar{1}1)$, (001) | 55·7 |
| 1·09 | 80·2 | 74·5 | 68·8 | 2·4 | $(\bar{1}11)$ | | — | — |
| 1·41 | — | — | 60·9 | 359·0 | 56·9 | 143·0 | $(11\bar{1})$, $(\bar{1}11)$ | 70·9 |
| 1·52 | — | — | 59·7 | 358·7 | 59·7 | 141·7 | $(11\bar{1})$, $(\bar{1}11)$ | 71·4 |
| 1·57 | — | — | 59·3 | 359·6 | 60·5 | 140·9 | $(11\bar{1})$, $(\bar{1}11)$ | 70·9 |
| 1·62 | — | — | 59·3 | 357·6 | 59·6 | 137·7 | $(11\bar{1})$, $(\bar{1}11)$ | 71·2 |
| 1·66 | — | — | 59·1 | 356·6 | 60·6 | 137·9 | $(11\bar{1})$, $(\bar{1}11)$ | 71·3 |
| 1·70 | — | — | 58·1 | 357·4 | 61·6 | 138·1 | $(11\bar{1})$, $(\bar{1}11)$ | 69·7 |

No. 72

| | (100) | | $(\bar{1}11)$ | | (111) | | Measured angle between pairs of axes | |
|---|---|---|---|---|---|---|---|---|
| $\epsilon$ | $\theta$ | $\phi$ | $\theta$ | $\phi$ | $\theta$ | $\phi$ | | Degrees |
| 1·00 | 83·8 | 345·6 | — | — | 30·8 | 339·2 | (100), (111) | 53·3 |
| 1·02 | 84·5 | 345·6 | 43·5 | 174·6 | 30·0 | 338·2 | $(\bar{1}11)$, (111) | 72·7 |
| 1·05 | 81·2 | 346·5 | 45·2 | 176·0 | 28·9 | 336·3 | $(\bar{1}11)$, (100) | 54·8 |
| 1·10 | 79·7 | 347·4 | 49·4 | 177·4 | 26·2 | 333·0 | $(\bar{1}11)$, (100) | 54·2 |
| 1·20 | 74·8 | 348·9 | 52·8 | 181·7 | — | — | $(\bar{1}11)$, (100) | 54·3 |
| 1·40 | 69·4 | 352·2 | 57·6 | 184·2 | — | — | $(\bar{1}11)$, (100) | 54·9 |
| 1·51 | 67·2 | 353·8 | 60·0 | 190·8 | — | — | $(\bar{1}11)$, (100) | 54·9 |
| 1·60 | 64·5 | 355·8 | 62·3 | 192·8 | — | — | $(\bar{1}11)$, (100) | 55·5 |
| 1·70 | 63·8 | 358·0 | 62·5 | 196·3 | — | — | $(\bar{1}11)$, (100) | 56·7 |
| Broken | 61·5 | 358·5 | 65·8 | 197·0 | — | — | $(\bar{1}11)$, (100) | 55·6 |

they are two (111) axes. It will be noticed that the measured angles are very near to the true ones, and that there is a marked improvement in this respect over the measurements given in table 5 of the Bakerian Lecture.

The spherical polar co-ordinates of points on the unextended cone were next determined by the method previously described.* Before it was possible to make

* Using equations (1) and (3) in Bakerian Lecture.

full use of the accuracy of the measurements, it was necessary to obtain a more accurate stereographic net than had been available before. A very excellent net, 18 in. in diameter, was obtained from the United States Hydrographic Department. This was reduced to 12 in. and reproduced on cardboard by the lithographic department of the Cambridge University Press.

It seems unnecessary to give a table showing the complete results of these calculations, but it is of interest to exhibit one of them on a stereographic diagram as a specimen, to show the striking accuracy with which the calculated points lie

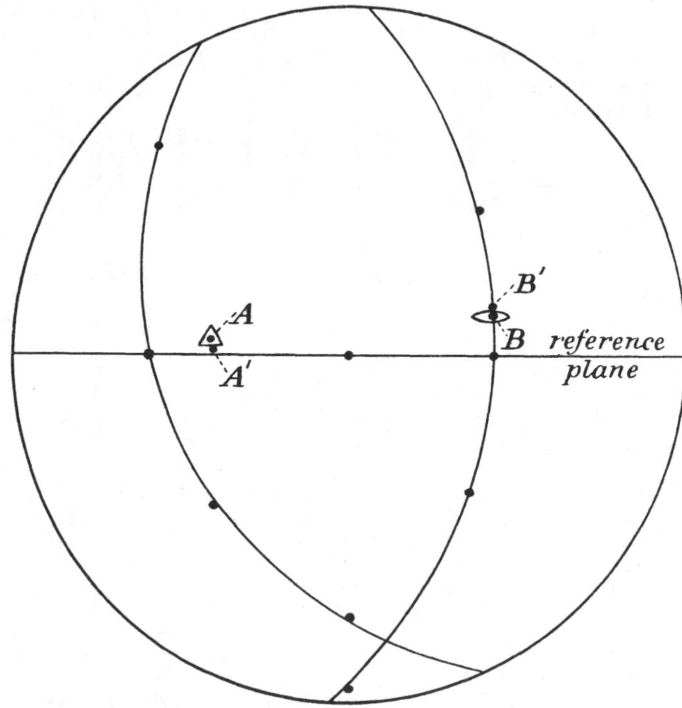


Fig. 2. Diagram showing accuracy with which unextended cone conforms to two planes. $A'$ is position of normal to slip-plane as measured by distortion; $A$ is position of normal to slip-plane as predicted by X-rays; $B'$ is direction of slip as measured by distortion; $B$ is direction of slip as predicted by X-rays.

on two great circles during the early stages of the test, when the slipping is on one plane only. Fig. 2 shows the unextended cone in the material at 10% extension for the extension from 10 to 40% in specimen no. 59. The two great circles which pass through the calculated points are shown. The position of the pole of one of them (the slip-plane) is shown at $A'$, and the point on the circle which is at 90° to the intersection of the two, and should therefore represent the direction of slipping, is shown at $B'$.

To compare the observed slip-plane with that predicted by X-ray measurements, the operations described above (p. 110) were performed, and the predicted slip-plane and direction of slip were found and marked on the diagram (fig. 2) at $A$ and $B$.

These operations were also carried out for two other specimens, nos. 68 and 72. The comparison between the predicted and observed slip-planes and directions of slip are given in table 3. It will be seen that in each case the agreement is good. It appears therefore that the small discrepancy found previously between the slip-plane and one of the octahedral planes was probably due to inaccuracies of measurement.

Table 3. *Comparison between observed and predicted positions of normals to slip-planes and directions of slipping for three specimens. $\theta$ and $\phi$ are spherical polar co-ordinates*

| Specimen no. | Extension (%) | Range of extension for distortion measurements (%) | Normal to slip-plane (degrees) $\theta$ | $\phi$ | Direction of slipping (degrees) $\theta$ | $\phi$ | |
|---|---|---|---|---|---|---|---|
| 72 | 5 | 5–40 | 43·0 | 178·0 | 49·0 | 22·0 | Distortion |
| | | | 45·2 | 176·0 | 49·0 | 21·5 | X-rays |
| 59 | 10 | 10–48 | 43·4 | 177·5 | 48·4 | 17·3 | Distortion |
| | | | 44·5 | 174·5 | 48·0 | 15·5 | X-rays |
| 68 | 10 | 10–40 | 49·0 | 146·7 | 43·2 | 337·4 | Distortion |
| | | | 47·0 | 147·0 | 41·6 | 334·5 | X-rays |

## Changes in Orientation of Crystal Axes during Test

The changes in orientation of the crystal axes during a test are best discussed by projecting the axis of the specimen on to a figure like that shown in fig. 1. If the distortion is due to slipping parallel to a given crystal plane along a given crystal axis, the longitudinal axis of the specimen will move relatively to the crystal axes towards the direction of slip. In figs. 3, 4 and 5 are shown two of the spherical triangles of fig. 1, and the successive positions of the axis of the specimens relative to the crystal axes for the three specimens, nos. 59, 68 and 72, are marked on them. It will be seen that in each case the predicted result is confirmed, the point representing the axis moving along a great circle (dotted lines in figs. 3, 4 and 5) towards the axis (110), which is the direction of slip. The numbers in each case refer to the percentage extension of the specimen. The numbers below the dotted arcs refer to the positions as measured by X-rays. The numbers above refer in each case to the position calculated on the assumption that slipping takes place only on the plane ($\bar{1}11$) and in the direction (110). It will be seen that in the part of the arc which lies in the left-hand triangle, $A$, the agreement between the observed and calculated positions is very good. On the other hand, as soon as the axis has crossed the line dividing $A$ from $B$ (i.e. the great circle containing directions which make equal angles with the two slip-planes ($\bar{1}11$) and ($11\bar{1}$)), the observed position of the axis lags behind the position it would occupy if distortion continued to be due to slipping on ($\bar{1}11$) alone. This change is proved later to be due to the fact that double slipping on the two planes ($\bar{1}11$) and ($11\bar{1}$) begins at this point. It will be noticed, however, that the axis does move over a little way into the triangle $B$, so that there is evidence that in the two cases, nos. 59 and 68, at any rate, the rate of slipping on the original slip-plane is greater than that on the new one. This again is confirmed by the distortion measurements (see p. 122).

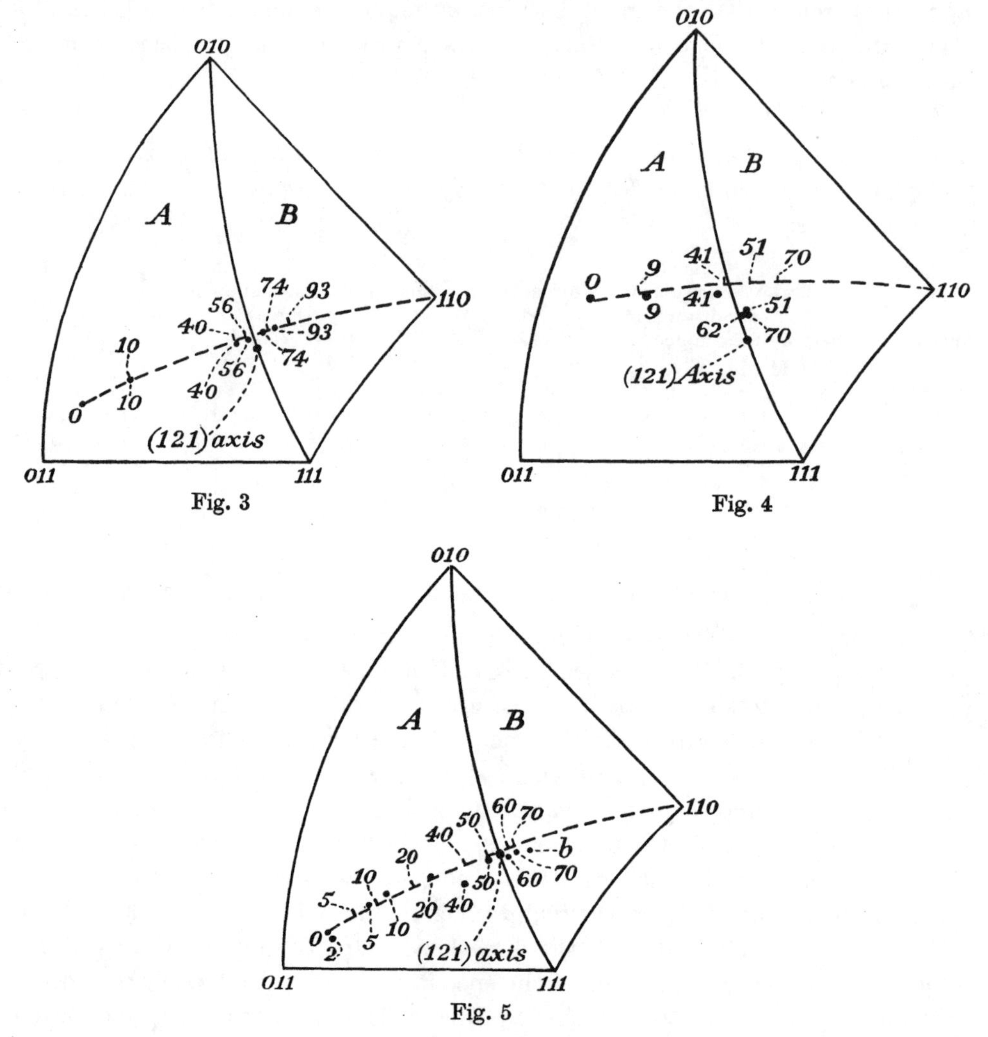


Fig. 3

Fig. 4

Fig. 5

**Figs. 3, 4, 5. Diagrams showing motion of axis of specimen relative to axes of crystal during a tensile test. Fig. 3 refers to specimen no. 59; fig. 4 refers to specimen no. 68; fig. 5 refers to specimen no. 72.**

## Deformation during the Last Stages of Stretching

It was shown previously that the deformation of the specimen described in the Bakerian Lecture during the extension from 40 to 73 % elongation was not due to slipping on one plane. It was pointed out that at some stage in the course of the extension from 40 to 73 % the axis of the specimen would get to a position in which it made equal angles with two possible slip-planes, and that if the law that slipping takes place in that direction for which the component of shear stress is greatest is true, then double slipping would begin at this point. If the law continued to hold, then the amount of the slipping on the two slip-planes would be equal, and the axis would continue to be symmetrically placed with regard to these two planes.

A consequence of this equal slipping would be that the 'unstretched cone' for any deformation which began and ended after double slipping had begun would be symmetrical with respect to the two slip-planes. It was not possible to test this before, because only one of the stages for which measurements were made was in the region in which double slipping was to be expected. On the other hand, two of the specimens (nos. 59 and 68) with which we are now concerned were carefully measured at several stages near the maximum extension, and it has been possible

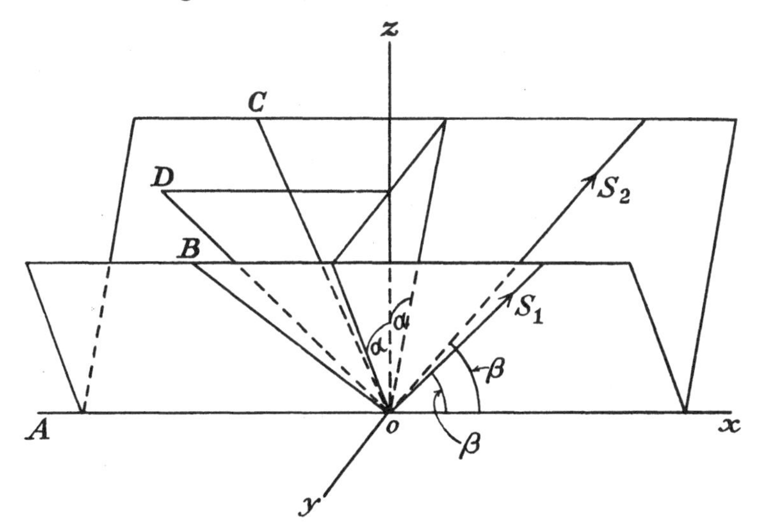


Fig. 6. Scheme of axes for analysis of distortion due to double slipping.

to determine the 'unextended cones' for ranges of extension which are wholly within the range in which double slipping was to be expected. Before giving the results of these tests, however, it is useful to calculate the equation to the 'unextended cone' which would result from two small unequal shearing strains parallel to two planes.

Let $2\alpha$ be the angle between the two planes, and let the material be strained so that it is sheared by a small amount $\delta S_1$ parallel to one plane in a direction making angle $\beta$ with their line of intersection, and by a small amount $\delta S_2$ parallel to the second plane in a direction also making an angle $\beta$ with the line of intersection. In the case of double slipping of the type we are going to consider $2\alpha$ is the exterior angle of the faces of an octahedron, namely, $70^\circ\ 32'$, and $\beta$ is $60^\circ$. Take axes $(x, y, z)$ so that $ox$ is the line of intersection of the two slip-planes whose equations are $y = z \tan\alpha$ and $y = -z \tan\alpha$. The scheme is shown in fig. 6.

If $\vartheta$ is the angle which any line through the origin makes with $ox$, and $\chi$ is the inclination of its projection on the plane $x=0$ to the axis of $z$, then the co-ordinates of a point on this line distant $r$ from the origin are

$$\left.\begin{aligned} x &= r\cos\vartheta, \\ y &= r\sin\vartheta\sin\chi, \\ z &= r\sin\vartheta\cos\chi. \end{aligned}\right\} \quad (1)$$

If the particle which was originally at the point $(xyz)$ shifts to

$$(x+\delta x,\ y+\delta y,\ z+\delta z),$$

owing to the strain,

$$\left.\begin{aligned} \delta x &= \delta S_1(z\sin\alpha - y\cos\alpha)\cos\beta + \delta S_2(z\sin\alpha + y\cos\alpha)\cos\beta, \\ \delta y &= \delta S_1(z\sin\alpha - y\cos\alpha)\sin\alpha\sin\beta - \delta S_2(z\sin\alpha + y\cos\alpha)\sin\alpha\sin\beta, \\ \delta z &= \delta S_1(z\sin\alpha - y\cos\alpha)\cos\alpha\sin\beta + \delta S_2(z\sin\alpha + y\cos\alpha)\cos\alpha\sin\beta. \end{aligned}\right\} \quad (2)$$

The condition that the line of particles between the origin and the point $(xyz)$ shall be unextended is

$$\delta(r^2) = \delta(x^2+y^2+z^2) = x\,\delta x + y\,\delta y + z\,\delta z = 0; \quad (3)$$

substituting from (1) and (2) in (3) it will be found that the equation to the 'unextended cone' is

$$\cot\vartheta = -\tfrac{1}{2}\tan\beta\,\frac{\delta S_1\sin 2(\alpha-\chi) + \delta S_2\sin 2(\alpha+\chi)}{\delta S_1\sin(\alpha-\chi) + \delta S_2\sin(\alpha+\chi)}, \quad (4)$$

or, if $\mu$ be written for

$$\frac{\delta S_1 - \delta S_2}{\delta S_1 + \delta S_2},$$

$$\cot\vartheta = -\tfrac{1}{2}\tan\beta\,\frac{\sin 2\alpha\cos 2\chi - \mu\cos 2\alpha\sin 2\chi}{\sin\alpha\cos\chi - \mu\cos\alpha\sin\chi}. \quad (5)$$

In the case we are considering,

$$\tan\beta = \tan\tfrac{1}{3}\pi = \sqrt{3},$$

$$\cos 2\alpha = \tfrac{1}{3},\quad \sin 2\alpha = 2\sqrt{2}/3,\quad \sin\alpha = 1/\sqrt{3},\quad \cos\alpha = \sqrt{2}/\sqrt{3},$$

so that

$$\cot\vartheta = -\frac{2\sqrt{2}\cos 2\chi - \mu\sin 2\chi}{2\cos\chi - 2\sqrt{2}\mu\sin\chi}. \quad (6)$$

Whatever value $\mu$ may have, this cone passes through four fixed points, namely, the points

$$\left.\begin{aligned} &A \quad (\vartheta = 0), \\ &B \quad (\vartheta = 120^\circ,\ \chi = +35^\circ\ 16'), \\ &C \quad (\vartheta = 120^\circ,\ \chi = -35^\circ\ 16'), \\ &D \quad (\vartheta = 144^\circ\ 44',\ \chi = 0). \end{aligned}\right\} \quad (7)$$

These four points are crystal axes; $A$, $B$ and $C$ are diad axes of type (110), while $D$ is a triad axis of type (111).

The relation between these four directions $OA$, $OB$, $OC$ and $OD$ and the directions of slip $S_1$ and $S_2$ are shown in fig. 6.

(6) is the equation to a quadric cone, and since a quadric cone can be made to pass through any five given directions in space, this cone can be made to pass through one other given direction besides those represented by $A$, $B$, $C$ and $D$.

If $(\vartheta, \chi)$ be regarded as the co-ordinates of this direction (6) may be regarded as a linear equation for determining $\mu$, and every possible quadric cone through the four points $A$, $B$, $C$, $D$ corresponds with some value of $\mu$. Hence, to prove that any small distortion is due to double slipping of the type contemplated it is only necessary to show that the unstretched cone passes through the four axes $A$, $B$, $C$, $D$.

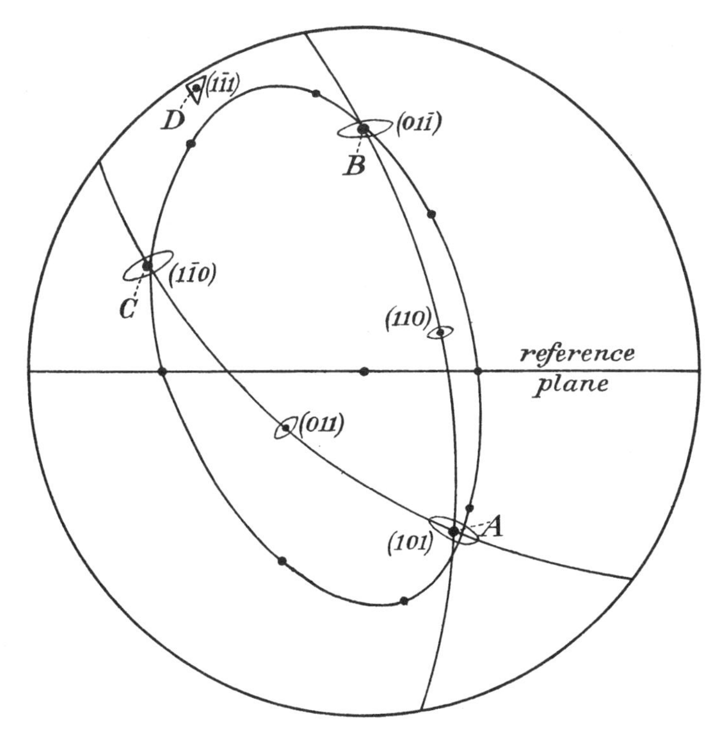


Fig. 7. Unstretched cone for extension from $\epsilon = 1{\cdot}74$ to $\epsilon = 1{\cdot}93$ in case of no. 59.

If the total extension in the interval for which the unstretched cone is determined is not small, then this cone will still pass through the axis $A$, but not through $B$, $C$ or $D$.

In fig. 7 is shown the unextended cone for the extension from 74 to 93 % in the case of no. 59, and in fig. 8 the cone for the extension of no. 68 from 51 to 69 %. The positions at 73 and 51 % extension are shown, and the axis of the specimen is in the centre of the figure. The two slip-planes perpendicular to the axes ($\bar{1}$11) and (11$\bar{1}$) are shown as great circles, and the directions of slip (110) and (011) are shown. The four axes $A$, $B$, $C$, $D$, through which it is predicted that the cone will pass if the distortion is due to double slipping on the two slip-planes (11$\bar{1}$) and

($\bar{1}11$), are shown on the figure. They are represented by the symbols (101), ($01\bar{1}$), ($1\bar{1}0$) and ($1\bar{1}1$) respectively. It will be seen that in each case the cone passes very nearly through all these axes.

It appears, therefore, that the distortion of the specimen during that part of the test when double slipping might be expected to occur is, in fact, such as could be produced by this double slipping. On the other hand, the asymmetry of the unstretched cones with respect to the slip-planes shows that in the case of both no. 59 and no. 68 there was greater slipping on the original slip-plane than on the second. This agrees in general with the deductions made on p. 115 from the X-ray

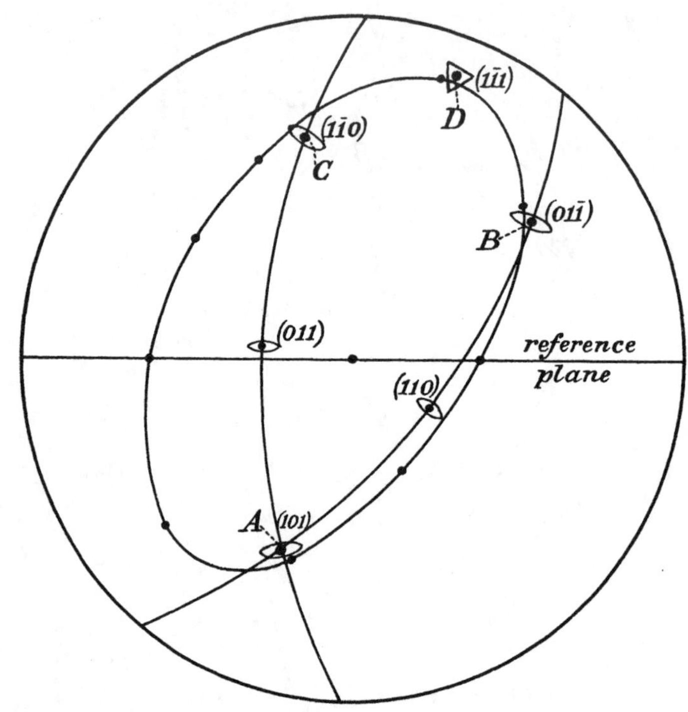


Fig. 8. Unstretched cone for extension from $\epsilon = 1{\cdot}51$ to $\epsilon = 1{\cdot}69$ in case of no. 68.

measurements. The accuracy of the X-ray measurements in the last stages of the test is not good enough to make any accurate determinations of the ratios of the amounts of slipping on the two slip-planes in the cases of specimens nos. 68 and 59; but by taking different values of $\mu$ in (6) and comparing the cones so obtained with the observed ones shown in figs. 7 and 8 it is found that for nos. 59 and 68 the amount of the slip on the original plane was nearly three times as great as that on the second slip-plane.

It is clear that such a large difference between the rates of slipping on the two slip-planes would quickly move the axis of the specimen into the triangle $B$ (figs. 4, 5), but in the cases of nos. 59 and 68 the specimen broke before much double slipping had taken place.

One specimen, no. 24, was found in which the initial position of the axis was quite close to the (121) axis. It was found that the axis of the specimen moved very little relative to the crystal axes during an extension of 42 %.* In this case, therefore, the amount of slipping on the two slip-planes must have been nearly equal. For the case of no. 24 the motion of the axis of the specimen relative to the crystal axes is shown in fig. 9. There is clearly very little motion of the specimen axis, but whatever motion there is appears to be in the direction of the (121) axis, though the X-ray measurements are not sufficiently good to enable any further deductions to be made.

## Stress Components Parallel to Plane of Slipping

During all the tests described in this and the earlier paper the tension was measured at every stage of the test.

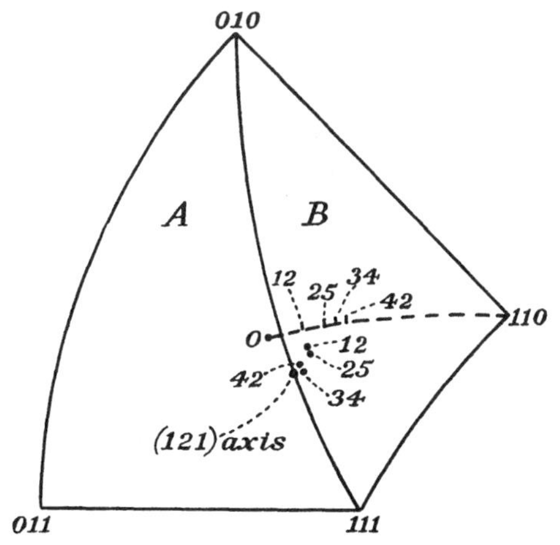


Fig. 9. Diagram showing motion of axis of specimen relative to axes of crystal during a tensile test. This figure refers to specimen no. 24.

The component of shearing stress parallel to the direction of slip, i.e.

$$T/A \cos\theta \sin\theta \cos\eta,\dagger$$

was calculated in each case. The results are given in table 4 and they are shown graphically on the diagram (fig. 10), in which the shearing stress in pounds per square inch is plotted against $\epsilon$. Two points about this diagram stand out clearly. In the first place, the points obtained with different specimens lie close together; and, secondly, the stress increases very rapidly at first but more slowly for larger extensions. The shear stress appears to be increasing up to the breaking point of the specimen. We are not able to put forward any quantitative explanation of this phenomenon—which is, of course, well known—but it appears to be connected

* The specimen broke at 42 % extension.

† See p. 111.

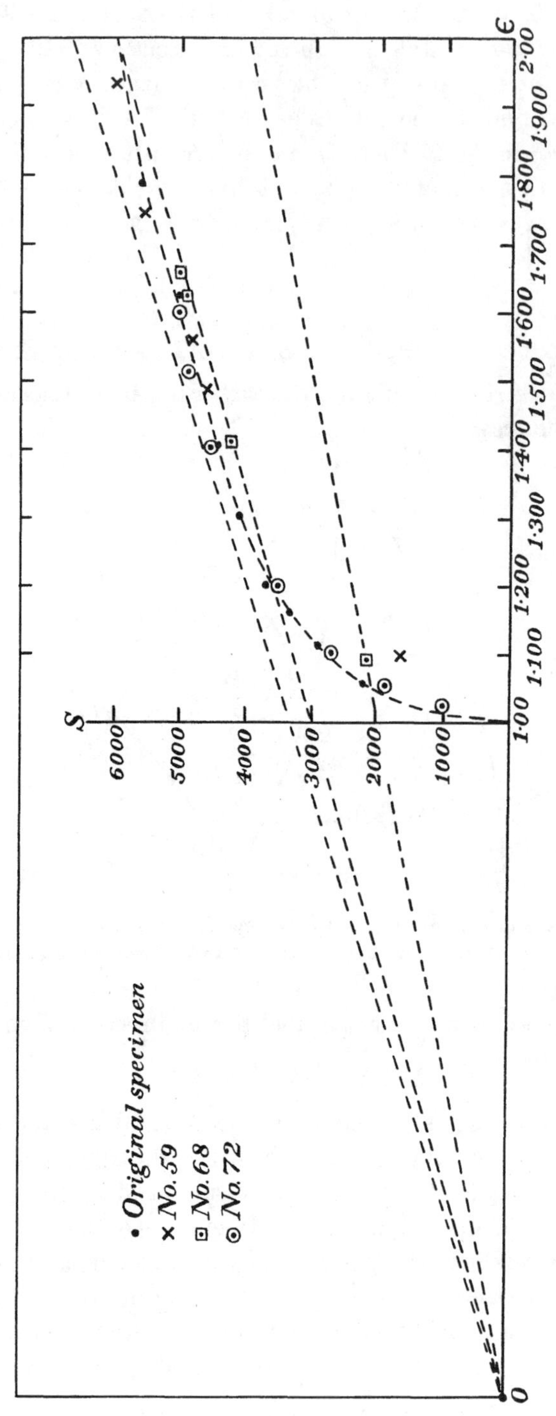


Fig. 10. $S$ is shearing stress measured in pounds per square inch parallel to the direction of slip.

with a certain amount of breaking up of the crystal, which is indicated by the broadening of the range of angles over which homogeneous X-rays are reflected from the crystal planes.

Table 4

| $\epsilon$ | $\theta°$ | $\phi°$ | $T$ (lb.) | $S$ (lb. per sq. in.) | $\tan^2\eta+\sec^2\eta\cot^2\theta$ | |
|---|---|---|---|---|---|---|
| 1·00 | 41 | 12 | — | — | 1·43 | |
| 1·053 | 43·5 | 11 | 690 | 2180 | 1·19 | |
| 1·110 | 46 | 10 | 770 | 2930 | 1·00 | |
| 1·161 | 48 | 9 | 950 | 3380 | 0·85 | Original specimen |
| 1·200 | 50 | 9 | 1010 | 3740 | 0·74 | cross-section |
| 1·304 | 53·5 | 8 | 1100 | 4140 | 0·58 | 0·1630 sq. in. |
| 1·404 | 56·5 | 8 | 1140 | 4480 | 0·47 | |
| 1·623 | 61 | 8 | 1210 | 5030 | 0·34 | |
| 1·785 | 61·5 | 8 | 1230 | 5570 | 0·32 | |
| 1·00 | 42·5 | 6 | — | — | 1·13 | |
| 1·093 | 47·5 | 5·5 | 490 | 2200 | 0·86 | |
| 1·411 | 58 | 4·5 | 810 | 4250 | 0·40 | |
| 1·516 | 59·7 | 7 | — | — | 0·36 | No. 68, cross- |
| 1·570 | 60·5 | 7 | — | — | 0·34 | section |
| 1·621 | 59·6 | 7 | 845 | 4946 | 0·36 | 0·1200 sq. in. |
| 1·659 | 60·6 | 7 | 848 | 5015 | 0·33 | |
| 1·695 | 61·6 | 7 | — | — | 0·31 | |
| 1·00 | 41·9 | 23 | — | — | 1·60 | |
| 1·02 | 43·5 | 20 | 548 | 1050 | 1·39 | |
| 1·050 | 45·3 | 18 | 963 | 1930 | 1·19 | |
| 1·103 | 49·4 | 17 | 1293 | 2760 | 0·90 | No. 72, cross- |
| 1·202 | 52·8 | 14·5 | 1593 | 3580 | 0·68 | section |
| 1·404 | 58·0 | 14 | 1844 | 4520 | 0·48 | 0·2493 sq. in. |
| 1·511 | 60·0 | 12 | 1903 | 4880 | 0·39 | |
| 1·602 | 62·3 | 10 | 1938 | 5040 | 0·32 | |
| 1·695 | 62·5 | 10 | — | — | 0·31 | |
| 1·00 | 40 | — | — | — | — | |
| 1·102 | 44 | 14 | 500 | 1650 | 1·20 | |
| 1·484 | 58·5 | 9 | 1135 | 4650 | 0·41 | No. 59, cross- |
| 1·560 | 59·3 | 8 | 1147 | 4860 | 0·38 | section |
| 1·745 | 62·8 | 8 | 1265 | 5550 | 0·29 | 0·1589 sq. in. |
| 1·936 | 64·3 | 7 | 1274 | 6000 | 0·26 | |

## Stability and Uniformity of Stretching due to Slipping on One Plane

One of the most striking features of the stretched specimens is their extraordinary uniformity. In fig. 11 is shown a photograph* of specimen no. 68 after stretching through an extension of 70 %. It will be seen that to the eye the specimen appears quite uniform. In order to obtain some numerical measure of the variations in the extension in different parts of the specimen, we may refer to table 1† of our previous paper. In the bottom row of that table the mean extension for twelve lengths (i.e. three lengths on each of the four faces of the specimen) is given. The variations in the extension may be estimated by taking the differences between

* Reproduced here as a line drawing.

† Bakerian Lecture; paper 5 above, p. 67.

the twelve individual values of $\epsilon$ and their mean value. Thus, in the first column of the table referred to, the mean value of $\epsilon$ is 1·053; the individual values at 1·047, 1·058, 1·052, 1·048, 1·053, 1·051, 1·051, 1·059, 1·054, 1·050, 1·056, 1·054.

Taking the differences between these values and the mean value 1·053, without regard to sign, the average variation in $\epsilon$ from the mean is found to be 0·0028.

The same process was repeated for each of the other subsequent stages of the test and the figures in the following table (5) were obtained. It will be seen that up to 20 % extension $\epsilon$ varies by only about 2 parts in 1000 in different parts of the specimen. It then begins to increase, till at the breaking point it rises to 4 parts in 100. Similar results were obtained with the other specimens.

It seems likely that the explanation of these phenomena must lie in the form of the stress-strain curve shown in fig. 10. If we imagine a very long specimen and suppose that during a tensile test slipping occurs in one part before it occurs in another part, the effect of slipping on one plane is to decrease the angle between the direction of slip and the axis of the specimen; and since the area of the plane of slip is unaltered by slipping on one plane, and the total pull at all sections of the specimen is the same, the tangential stress in the direction of the slip is increased by the slipping, even if the total pull on the specimen does not increase. Hence, if the resistance of the material to slip were not increased by the slipping, the material would go on slipping at the place where it first started and the specimen would break at that point without extending the rest of the material at all.

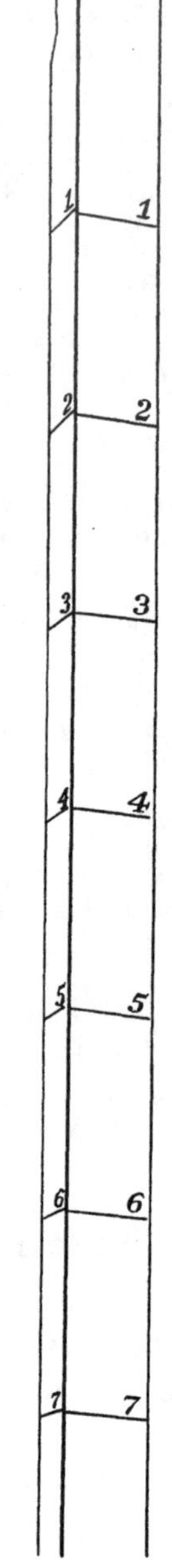

Fig. 11. Specimen no. 68, after extension of 70 %.

On the other hand, the curve (fig. 10) shows that the resistance to slipping increases as the extension proceeds. The necessary condition for uniform stretching is, therefore, fulfilled provided that the rate of increase of resistance to slipping is greater than the rate of increase of stress due to the decrease in angle between the axis of the specimen and the direction of slip. It will be noticed at once that the region 0–20 % extension, in which the stretching is uniform to the extent of 2 parts in 1000, corresponds to the region where the resistance to slipping is increasing at the greatest rate.

It remains to put these ideas into mathematical symbols. If $S$ is the resistance to slipping per sq.cm. of the slip-plane, and $B$ the area of the slip-plane, then

$$SB = T \sin\theta \cos\eta.^* \tag{8}$$

If the increase $dS$ in $S$ during an extension $d\epsilon$ were just sufficient to balance the increase in tangential stress due to change in angle between the direction of slip and the axis of the specimen, then

$$\frac{dS}{S} = \frac{d(\sin\theta\cos\eta)}{\sin\theta\cos\eta}.$$

Table 5

| $\epsilon$ | 1·053 | 1·110 | 1·161 | 1·200 | 1·304 | 1·404 | 1·623 | 1·785 |
|---|---|---|---|---|---|---|---|---|
| Extension (%) | 5·3 | 11·0 | 16·1 | 20·0 | 30·4 | 40·4 | 62·3 | 78·85 |
| Average variation in $\epsilon$ | 0·0028 | 0·0021 | 0·0026 | 0·0013 | 0·0053 | 0·0088 | 0·0207 | 0·0393 |

This is equivalent to putting $dT = 0$ when differentiating (8). Geometrical considerations give the connections between $\theta$ and $\eta$ and $\epsilon$. If $\theta_0$ and $\eta_0$ are the initial value of $\theta$ and $\eta$ when $\epsilon = 1$, then

$$\left.\begin{aligned} &\epsilon\cos\theta = \cos\theta_0, \\ \text{and}\quad &\epsilon\sin\theta\sin\eta = \sin\theta_0\sin\eta_0. \end{aligned}\right\} \tag{9}$$

Hence, differentiating and rearranging terms it will be found that

$$\frac{\epsilon}{S}\frac{dS}{d\epsilon} = \tan^2\eta + \sec^2\eta\cot^2\theta. \tag{10}$$

This gives us the condition that the specimen shall stretch uniformly while slipping on one plane instead of breaking, namely,

$$\frac{\epsilon}{S}\frac{dS}{d\epsilon} > \tan^2\eta + \sec^2\eta\cot^2\theta. \tag{11}$$

In this equation $\epsilon$, $S$ and $dS/d\epsilon$ can be taken from the curve (fig. 10), while $\eta$ and $\theta$ can be taken from stereographic diagrams.

## Conditions for Fracture

If $\frac{\epsilon}{S}\frac{dS}{d\epsilon}$ is less than $\tan^2\eta + \sec^2\eta\cot^2\theta$, the specimen will break, because if slipping occurs in any part of the specimen, the pull required to produce further slipping there is reduced. In this case the breakdown would be due to slipping on one plane only, and the appearance of the specimen would presumably indicate a fracture of this type.

The values of $\tan^2\eta + \sec^2\eta\cot^2\theta$ have been calculated for the three specimens,

* Compare p. 111.

nos. 59, 68 and 72, and set down in the last column of table 4. It will be seen that they decrease as the extension proceeds and that they lead to a limiting value in the neighbourhood of $\frac{1}{3}$.

The values of $\frac{\epsilon}{S}\frac{dS}{d\epsilon}$ may be taken from the diagram (fig. 10), but it is worth while to point out that it is easy to see from inspection of the curve whether $\frac{\epsilon}{S}\frac{dS}{d\epsilon}$ is greater or less than 1. If $dS/d\epsilon = S/\epsilon$, the tangent to the curve in fig. 10 will, if produced backwards, pass through the point $S=0$, $\epsilon=0$, which is marked 0 in the diagram (fig. 10), and $dS/d\epsilon$ will be $>$ or $<$ $S/\epsilon$, according as the tangent to the curve at any point is more steep or less steep than the line joining it to the point 0. In fig. 10 a few dotted lines have been drawn to pass through 0. It will be seen that in the earlier stages of the test, up to about $\epsilon = 1{\cdot}3$, the curve is steeper than the corresponding line through 0, but from about $\epsilon = 1{\cdot}35$ onwards the curve is practically parallel to the dotted lines, i.e. $\frac{\epsilon}{S}\frac{dS}{d\epsilon}$ is nearly equal to 1. From a comparison between the curve in fig. 10 and the values of $\tan^2\eta + \sec^2\eta\cot^2\theta$ given in table 4, it will be seen that fracture by slipping on a single plane is not possible for any of the four specimens.

As soon as slipping on two planes begins the conditions for stability are altered. To get some idea of the change in stability which will occur when double slipping begins, one may consider an idealized case. Suppose that the point representing the position of the axis in the stereographic figs. 3, 4, 5, approaches the position of the (112) axis owing to slipping on a single plane and that when it gets in that position double slipping occurs equally on both planes so that the direction of the axis of the specimen remains permanently along this crystal axis. After double slipping begins $\eta$ and $\theta$ remain constant, but the area of the slip-plane is no longer constant. Differentiating (8) under these conditions it is found that the condition for fracture is $dS/S + dB/B < 0$. The constancy of volume gives $dB/B + d\epsilon/\epsilon = 0$, so that the condition for fracture is that

$$\frac{dS}{S} < \frac{d\epsilon}{\epsilon} \quad \text{or} \quad \frac{\epsilon}{S}\frac{dS}{d\epsilon} < 1.$$

Before double slipping began the condition for stability was

$$\frac{\epsilon}{S}\frac{dS}{d\epsilon} < \tan^2\eta + \sec^2\eta\cot^2\theta,$$

and using the values $\eta = 10^\circ\ 53'$, $\theta = 61^\circ\ 52'$ which these angles have when the axis of the specimen lies along a (112) axis, it is found that

$$\tan^2\eta + \sec^2\eta\cot^2\theta = \tfrac{1}{3},$$

so that the condition for fracture is $\frac{\epsilon}{S}\frac{dS}{d\epsilon} < \frac{1}{3}$. It will be seen, therefore, that if rate of increase in resistance to slip remains unchanged the stability is much less for double slipping than for slipping on a single plane. In the ideal case considered above, for instance, if the material were such that $\frac{\epsilon}{S}\frac{dS}{d\epsilon}$ lay between $\frac{1}{3}$ and 1, the material would be stable and would stretch uniformly while the slipping took place on a single plane, but directly double slipping began it would be unstable and would break.

These conclusions are in accordance with observation. The slope of the curve at any point is never less than that of the dotted line through it at that point, so that $\frac{\epsilon}{S}\frac{dS}{d\epsilon}$ is never less than 1; and during the last part of the curve when the material is breaking $\frac{\epsilon}{S}\frac{dS}{d\epsilon}$ is practically equal to 1.

The chief conclusions to be drawn from the theory of fracture outlined above are:

(1) With a stress-strain curve of the type shown above fracture of aluminium single crystals cannot occur by slipping on a single plane.

(2) The geometrical conditions alone imply that fracture takes place more easily when double slipping occurs than when all the slip is confined to one plane.

(3) The specimen should break by double slipping when $\frac{\epsilon}{S}\frac{dS}{d\epsilon}$ is about equal to 1. For this reason, therefore, the parts of the stress-strain curve for which $dS/d\epsilon$ is less than $S/\epsilon$ do not appear on the diagram.

In conclusion, we wish to express our thanks to Sir Ernest Rutherford and to Professor H. C. H. Carpenter for the facilities which they afforded us for carrying out the work partly in the Cavendish Laboratory and partly in the Royal School of Mines.

# 8

# NOTES ON THE 'NAVIER EFFECT'

Paper written for the Aeronautical Research Committee (1925)

It has been observed that specimens of many metals elongate when a gradually increasing alternating stress is applied to them. This observation has been taken as evidence that the tangential stress necessary for slipping is greater when the stress perpendicular to the slip-planes is compressive than when it is tensile. Such an effect has been called the 'Navier effect'. The object of the present note is to show that the extension of a specimen under alternating load would be expected even if the 'Navier effect' does not exist, and to describe direct experiments with a single crystal of aluminium which show that if it exists in that case it is too small to be significant.

Consider first what result might be expected in an alternating stress experiment if the simplest possible conditions are assumed, consistent with existing evidence obtained in static tensile tests of single crystal aluminium bars.

In those tests it was found that the bar extends as the load is increased, but if at any stage the load is taken off and reapplied, the material only begins to extend again plastically when the load has reached the maximum load previously applied. Suppose an alternating stress of gradually increasing range and zero mean stress is applied and consider two cases:

*Case I*. There is a 'Navier' or other effect which prevents the direction of slip from being reversed. In this case the material will extend only at the time of maximum tension. During the rest of the cycle no hardening and no extension is taking place, so that the load-extension curve should be identical with the maximum load-extension curve during a static test.

*Case II*. There is no 'Navier' or other effect to prevent the material from slipping back when the maximum compression slightly exceeds the previous greatest tension. In this case reversed slipping would take place, and if the hardening effect of reversed slipping were identical with the effect of an equal amount of direct slipping then the maximum load-extension curve would lie above the load-extension diagram obtained in static tests. At first sight one might say that the material would simply harden without extending, because the small extensions which occur at the time of maximum tension would neutralize the small contractions occurring at the time of maximum compression; but this is not the case, for with the given conditions which are symmetrical with regard to reversed and direct slipping, the extension for a given small increase in tension would be greater than the compression for the same increase in compressive stress.

If $S$ is the resistance to slipping per sq.cm., $B$ the area of the slip-plane (which, in the case of single slipping, is constant), $\theta$ the angle between the axis of the test-piece and the normal to the slip-plane, $\eta$ the angle between the direction of slip and the line of greatest slope in the slip-plane, then

$$SB = T \sin\theta \cos\eta, \tag{1}$$

where $T$ is the tension.

From this, and the relations

$$\epsilon \cos\theta = \cos\theta_0, \tag{2}$$

$$\epsilon \sin\theta \sin\eta = \sin\theta_0 \sin\eta_0, \tag{3}$$

where $\epsilon$ is the ratio of the length of the specimen to its initial length and $\theta_0$ and $\eta_0$ are the initial values of $\theta$ and $\eta$, it can be shown that

$$\frac{dT}{T} = \left(\frac{\epsilon}{S}\frac{dS}{d\epsilon} - \tan^2\eta - \sec^2\eta \cot^2\theta\right)\frac{d\epsilon}{\epsilon}. \tag{4}$$

Equation (4) gives the increase in extension for small increase in $T$.

If now we consider the corresponding expression for the compression due to an equal small increase in compressive load, we must take this expression and reverse the signs of $T$, $\delta T$ and $\delta\epsilon$, but $\epsilon$ clearly does not reverse being necessarily positive.

It appears, therefore, that the ratio of the increase in extension for a given small increase in tension to the corresponding decrease in length for an equal increase in compressive stress is

$$R = \frac{\frac{\epsilon}{S}\frac{dS}{d\epsilon} + \tan^2\eta + \sec^2\eta\cot^2\theta}{\frac{\epsilon}{S}\frac{dS}{d\epsilon} - \tan^2\eta - \sec^2\eta\cot^2\theta}.$$

The values of $\frac{\epsilon}{S}\frac{dS}{d\epsilon}$ can be obtained from the stress-strain curve in a static tensile test. During the last part of the extension of aluminium crystals shortly before breaking occurs it is nearly equal to 1, whereas the value of $(\tan^2\eta + \sec^2\eta\cot^2\theta)$ is usually in the neighbourhood of 0·33. In that case, then, $R = 2$.

If then the 'Navier effect' is absent and reversed slipping occurs with the same shearing force as direct slipping, we should expect the material to extend during an alternating stress test but only about one-third as much as the extension in a static test carried to the same maximum load.

## Direct Experiments on 'Navier Effect'

It seems that alternating stress experiments are difficult to interpret from the point of view of the 'Navier effect'. On the other hand, static compression tests are difficult because if the specimen is long it buckles, and if it is short it becomes barrel-shaped when tested in the ordinary compression machine. In either case the distortion is non-uniform and the stress distribution is very complex. In order to

9-2

compare distortion under compression with that in a tensile test-piece it is essential to get uniform distortion, and after a large number of attempts I have succeeded in getting this, using disc-shaped test-pieces compressed between parallel steel plates which are highly polished and lubricated with grease. The compression is carried out in small steps, the specimen being taken out and re-greased between each successive small increase in load.

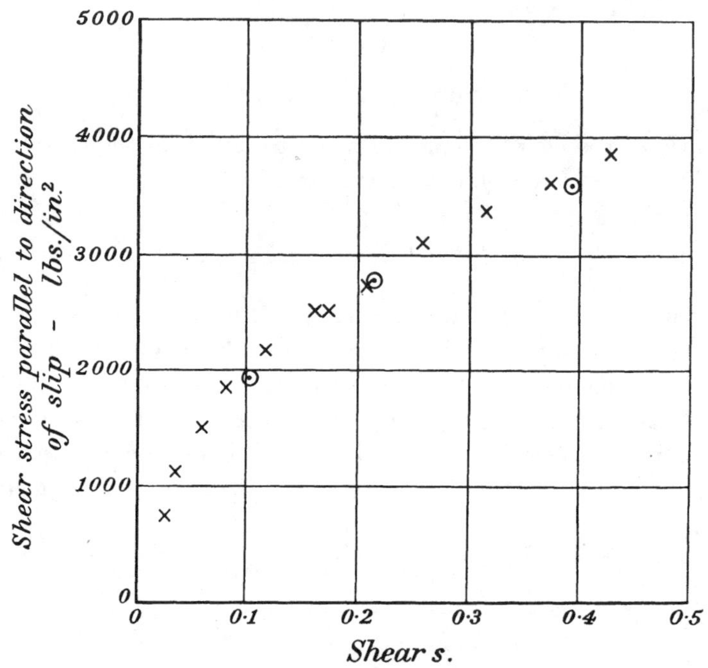


Fig. 1. Shear $S$ is amount of lateral movement of a plane parallel to basic slip-plane and unit distance from it, i.e. tan $\theta$. ⊙, no. 72 specimen in *tension*; ×, no. 69 specimen in *compression*.

Analysis of the distortion of these specimens shows that the distortion consists, in general, of slipping on one plane, and X-ray analysis shows that the plane is identical with the slip-plane in a tensile test-piece. Under these circumstances, if one assumes that the friction on the lubricated polished steel surface is negligible it is possible to find the connection between the amount of slipping and the resistance to slip. Having done this one can compare directly with the shear stress-strain curve obtained in tensile tests.

In this way an unambiguous test of the 'Navier effect' can be carried out. I have now worked out one case and the result is shown in fig. 1. The result is, of course, quite provisional and needs verification, but it is very remarkable that the curves for compression and tension are almost identical, indicating that the 'Navier effect' is small or else that it does not exist.

# 9

# THE DISTORTION OF CRYSTALS OF ALUMINIUM UNDER COMPRESSION*
# PART I

REPRINTED FROM

*Proceedings of the Royal Society*, A, vol. CXI (1926), pp. 529–51

The result of the work here described may be summed up in the statement that, so far as these experiments go, the distortion of a crystal of aluminium under compression is of the same nature as the distortion which occurs when a uniform single-crystal bar is stretched. The distortion is due to slipping parallel to a certain crystal plane and in a certain crystallographic direction, and the choice of which of twelve possible crystallographically similar types of slipping actually occurs depends only on the components of shear stress in the material and not at all on whether the stress normal to the slip-plane is a pressure or a tension.

In previous experiments on the distortion of aluminium crystals† a uniform bar was cut from a single crystal and subjected to a tension along its length. This form of test ensures a uniform stress in the central part of the bar, and it was found that the distortion was uniform, and that it conformed to very simple laws. These laws may be summarized as follows:

(*a*) The distortion is due to slipping, or shearing, parallel to one octahedral (111) plane in the direction of a diad axis (110).

(*b*) Of the twelve crystallographically similar possible types of slipping, in general only one occurs, namely, the one for which the component of shear stress in the direction of shear is the greatest.

It will be noticed that these laws take no account of any possible effect due to the component of pressure normal to the slip-plane. In the case of a tensile test there is always a tension perpendicular to the slip-plane, and the question naturally arises whether the distortion would follow the same laws if the component of stress perpendicular to the slip-plane were compressive. The experiments described below were designed partly to give information on this point, but chiefly to find out whether the simple laws found for the distortion due to stretching are applicable to other kinds of plastic strain. The chief difficulty lay in devising a kind of experiment in which the stress might be expected to be uniformly distributed through the material. So far as we are aware there is no form of test known to engineers which

* With W. S. FARREN.

† 'The distortion of an aluminium crystal during a tensile test', G. I. Taylor and C. F. Elam, Bakerian Lecture, *Proc. Roy. Soc.* A, CII (1923), 643 (paper 5 above, p. 63), referred to as B.L. in future, and 'The plastic extension and fracture of aluminium crystals', G. I. Taylor and C. F. Elam, *Proc. Roy. Soc.* A, CVIII (1925), 28 (paper 7 above, p. 109), referred to as P.E. in future.

gives a uniform plastic strain to a material except the extension of a uniform bar. In torsion tests, for instance, the stress is not uniform, even when applied to isotropic materials, and when a bar cut from a single crystal is twisted the stresses must be very complicated.

The known type of distortion which most nearly approaches the desired conditions occurs in the ordinary compression tests used by engineers. In these tests short cylindrical lengths of the material are compressed between parallel plates. The distortion is not uniform, for the compressed material usually assumes a barrel-shaped form, and in any case tests of this kind are unsuitable for single crystals because the distortion would necessarily be unsymmetrical and the distorted specimen would be of such a shape that the load could not be applied centrally. The distortion would therefore cease to be uniform as soon as any appreciable distortion had occurred.

This eccentricity of loading could be diminished by reducing the height of the specimen in comparison with its diameter, and if the specimen were cut in the form of a thin disc it could be made quite negligible. On the other hand, there is another factor which tends to give rise to non-uniformity of stress in the material—namely, the friction between the end-plates and the specimen. It is this friction which causes the specimens used in the ordinary engineer's compression tests to become barrel-shaped. The effect of this friction in making the stresses non-uniform must be greater for thin discs than for thick ones, so that there are two factors, each tending to give rise to non-uniformity of stress. The effect of one can be diminished by decreasing the height of the specimens, while the effect of the other can be diminished by increasing it. In these circumstances it became a matter for experiment to find out whether, when the friction of the specimen on the end-plates had been reduced as much as possible, a ratio of height to diameter could be chosen such that a cylindrical or disc-shaped specimen would be distorted uniformly under compression.

Specimens were prepared by cutting discs from single-crystal bars of circular cross-sections about 1·4 cm. in diameter. As it was necessary to find by means of X-ray reflections the directions of the crystal axes at various stages of the tests, special precautions were taken to ensure that the layer of material disturbed by the cutting tools should be as small as possible. The cylindrical piece of material from which the disc was to be cut was set inside a brass tube and concentric with it, the annular space between them being filled with sulphur. The brass tube was then mounted in a chuck in a lathe and the end faced off with a very fine cutting tool. The last 6 mm. of the brass tube was next turned off, leaving 6 mm. of the aluminium specimen projecting from it. The disc was then cut off by means of a very fine saw and the rough face trued up on a lathe with a fine tool so as to be parallel to the face already trued. The specimen was next mounted on a specially designed holder and its two faces were ground down on fine emery paper stuck to a piece of plate glass. The final thickness of the specimens was usually about 2·5 mm.

Preliminary experiments were made with specimens prepared in this way. They were pressed between parallel horizontal steel plates which had been faced in a lathe and polished with fine emery paper. It was found that the circular discs became elliptic in plan. In elevation or vertical section they ceased to be rectangular because the generators of the curved sides became inclined to the vertical. In general, a uniform distortion would change the shape of a circular disc in this way. In order to get some idea of how nearly the distortion was uniform, straight scratches were ruled on the plane faces of the specimen before compression, and these were examined afterwards to find out whether they had become bent or curved.

Fig. 1 A. Side view of specimen compressed without lubrication.

Fig. 1 B. Side view of specimen lubricated during compression.

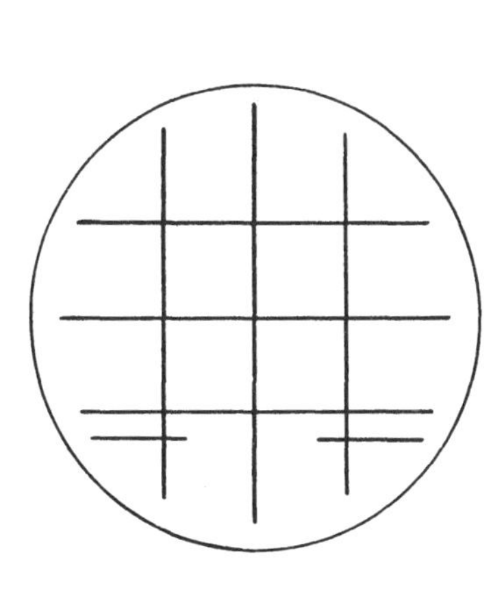

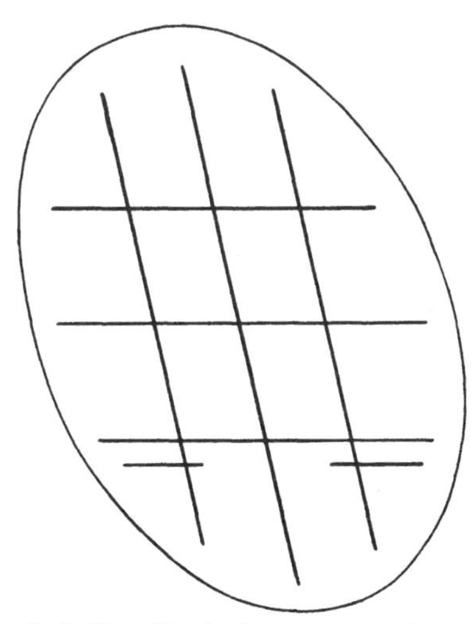

Fig. 2. Disc no. 59·10 after compression and similar disc before compression.

It was found that if the specimen was not lubricated with grease distortion was far from uniform. Straight scratches became curved and the specimen when seen sideways became barrel-shaped and skew, as shown in Fig. 1 A. When the specimen was lubricated with grease both these signs of non-uniformity diminished greatly, the compressed specimen appearing as in fig. 1 B. Finally, the friction was still further reduced, first, by hardening, grinding and polishing the steel plates with diamantine powder, and secondly, by carrying out the compressions in small steps, re-greasing the specimen before each small increase in compressive load.

When tests were carried out in this way the distortion seemed quite uniform, except occasionally in a small region close to the curved edge of the specimen. Photographs* of one of the specimens before and after compression are shown in fig. 2. The magnification of the photographs is approximately 4·5. The specimen is

* Reproduced here as line drawings.

marked in a manner to be described presently. It will be seen that straight lines* remain straight after compression, and that squares ruled on the surface become oblique parallelograms.

## MARKING THE SPECIMENS

In order to measure the distortion and to find its relationship with the crystal axes, it is necessary to rule fiducial marks on the original bar before cutting it into discs. (The orientation of the crystal axes was measured with reference to these marks before the specimen was cut up.) For this purpose the round, single-crystal bar was mounted on V-blocks on a surface plate and four generators of the cylinder were marked with a scribing block. These were spaced at intervals of 90° round the

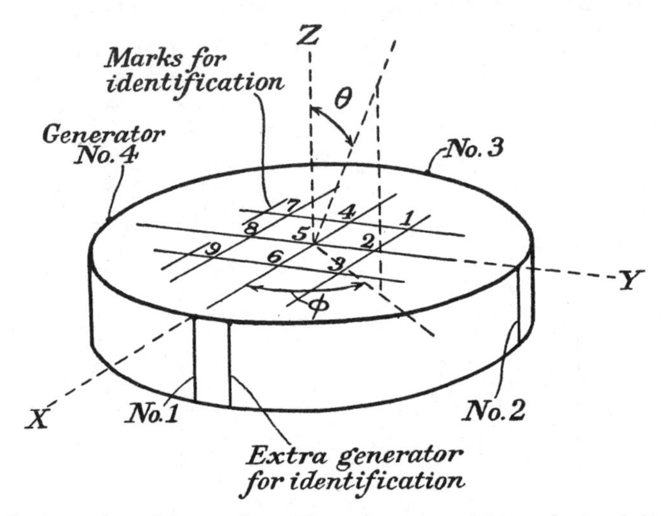


Fig. 3. System of marking and position of axes used in analysis of distortion.

specimens and numbered 1, 2, 3, 4. A fifth generator was also ruled in such a position that the marks could be distinguished after the specimen had been cut up. The bar was then cut into discs in the manner described above, and cross-marks were then made on the two plane faces so that the distortion could be measured.

The system of marking is shown in fig. 3. Two sets of three parallel lines were drawn intersecting in nine points numbered 1, 2, 3, 4, 5, 6, 7, 8, 9. The middle line of each series was ruled so that it passed approximately through a pair of opposite marked generators. Thus, the line through points 2, 5 and 8 marked approximately the axial plane in the original uncut specimen, which passed through the generators marked 2 and 4.

It was not found possible to mark the specimens satisfactorily by hand. The special marking machine which is shown diagrammatically in fig. 4 was therefore

* The particular specimen shown in the photograph was marked much more deeply than those used in the experiments described below, as it was found that the finer marks used in the latter specimens did not show up clearly in a photograph.

constructed. In that drawing $A$ is the specimen resting on a flat plate $B$ and pressed against a V-groove $C$ by a spring $D$. A V-block $E$ is mounted so that a cylindrical steel rod $F$ can slide along it, and a sharpened gramophone needle $G$ is rigidly attached to $F$ in such a position that it passes over the centre of the specimen, scratching a straight mark.

In order to compare different discs cut from the same single crystal, it was necessary to rule the scratches so that they were always in the same orientation with respect to the marked generators on the curved surface. To ensure this, a microscope $H$ (fig. 4) is mounted and focused on the curved surface of the specimen, which is then turned round in the V-groove $C$ till the generator marked 1 comes on to the

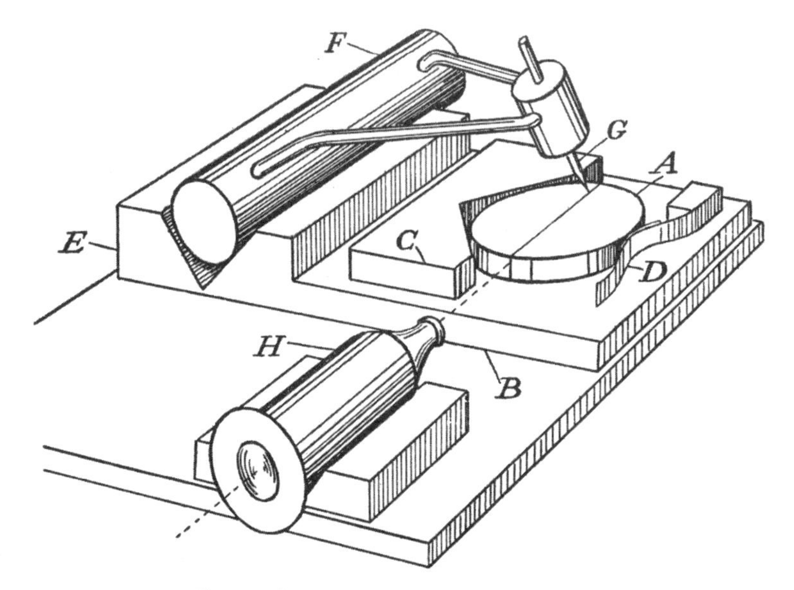


Fig. 4. Apparatus for ruling scratches.

cross-wires. The microscope is fixed in such a position that its axis is parallel to the V-groove $E$ and passes approximately through the centre of the specimen.

After ruling the middle line passing through the points 4, 5, 6 (fig. 3), the two parallel lines passing through points 1, 2, 3 and 7, 8, 9 were ruled. For this purpose two sliders were made similar to that shown at $F$ in fig. 4, but in one the scratching point was fixed nearer to the slider and in the other farther away. To rule the three lines 1, 4, 7; 2, 5, 8; and 3, 6, 9, the specimen was turned in the V-groove $C$ till the generator 2 came on the cross-wires of the microscope. Scratches were then made with the three sliders used before. Further short scratches were made to assist in identifying the ruled markings. These are shown in fig. 3, and they can also be seen in the photographs in fig. 2. After ruling the six lines on one side, the specimen was turned upside down and six similar lines were ruled on the bottom. The nine points so obtained were numbered 1′, 2′, 3′, 4′, 5′, 6′, 7′, 8′, 9′, the point 1′ being approximately under point 1, 2′ under 2, etc.

### Measurement of Specimens

Having marked the specimens, such measurements were made of the relative positions of the points as were necessary to enable the distortion to be calculated. For this purpose, a reading microscope was used which is capable of measuring rectangular co-ordinates on a plane.

All measurements both before and after compression were reduced to the rectangular system of co-ordinate axes $ox$, $oy$, $oz$, shown in fig. 3. The origin is the point 5 in the upper face; $ox$ lies along the line 4, 5, 6; $oy$ is perpendicular to it, so that before distortion $oy$ nearly coincides with the line 2, 5, 8. $z$ is measured upwards

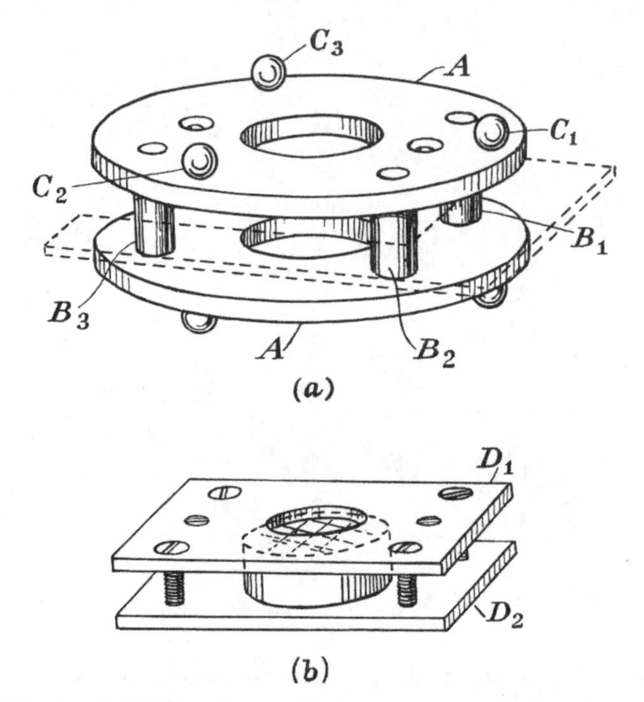


Fig. 5 *a*, *b*. Holders for use during measurement of specimen.

perpendicular to the face, so that if $t$ represents the thickness of the specimen, the $z$ co-ordinate of all points in the lower surface is $-t$.

Assuming the distortion to be uniform, it was necessary to measure the $x$ and $y$ co-ordinates of points 2, 4, 5, 6, 8 on the upper surface and also the $x$, $y$, and $z$ co-ordinates of the point 5′ on the lower surface, in order that a complete calculation of the distortion could be made. Measurements of the $x$ and $y$ co-ordinates of points on the lower surface gave rise to some trouble, and the following method was finally adopted for making them. The specimen was mounted in the holder, shown in fig. 5*a*. This consists of two brass discs $A$, $A$, rigidly fixed to one another by three steel pillars $B_1$, $B_2$, $B_3$. Three steel balls $C_1$, $C_2$, $C_3$, soldered to the top, and three similar balls to the bottom, form two alternative sets of legs which support the apparatus horizontally on the stage of the microscope either upright or upside down. The specimen was mounted horizontally in the middle of the holder, and in order that the ruled scratches might still be in focus when the holder was inverted to look at the lower surface of the specimen it was necessary to mount the specimen midway between the two sets of balls $C_1$, $C_2$, $C_3$. For this purpose a subsidiary holder was used, consisting of two steel plates, $D_1$, $D_2$ (fig. 5*b*), containing holes slightly smaller than the specimen, which was mounted between them so that the ruled scratches could be seen through the holes. This subsidiary holder was then mounted in the main holder, shown in fig. 5*a*, and packed up with brass and paper strips till the upper face of the specimen was in focus when the holder was upright, and the underside came into focus when it was reversed.

A small steel square in the form of a rectangular $L$ is fixed to the stage of the microscope at the level of the mid-points of the pillars $B_1$, $B_2$, $B_3$ when the holder rests on the stage. The pillars are so spaced in the holder that, when $B_2$ rests in the corner of the square touching both arms, $B_3$ and $B_1$ are each in contact with an arm of the square. The position of the square is indicated by the dotted lines in fig. 5*a*. The square is finally adjusted so that its arms are parallel to the axes of the co-ordinate system measured by the reading microscope.

Call these co-ordinates $\xi$ and $\eta$, and suppose the pillars $B_2$ and $B_3$ are originally in contact with the arm of the square which is parallel to the axis $\xi$. If the holder is reversed, and $B_2$ and $B_3$ again placed in contact with the arm parallel to $\xi$, the displacement of the holder is equivalent to a rotation through 180° about the axis $\eta$, together with a translation parallel to the axis $\xi$. Such a displacement leaves the $\eta$ co-ordinates of all points in the holder and the specimen unaltered. In this way the $\eta$ co-ordinates of points on the back of the specimen relative to points on the front can be obtained. Similarly, by reversing the specimen and bringing the pillars $B_1$, $B_2$ into contact with the arm of the steel square which is parallel to the axis $\eta$, the $\xi$ co-ordinates of points on the back of the specimen relative to points on the front can be measured.

It is seldom possible to mount the specimen in the holder so that the $x$, $y$ co-ordinate system of the specimen is accurately parallel to the $\xi$, $\eta$ system of the microscope. The angle between them can be calculated from the measured $\xi$, $\eta$ co-ordinates of the points 4 and 6. Thus, if this angle is $w$

$$\tan w = \frac{\eta_6 - \eta_4}{\xi_6 - \xi_4}. \tag{1}$$

In order to check the measurements this angle was also measured by means of a microscope with a rotating eye-piece which was kindly lent us by Dr A. Hutchinson, F.R.S. At the same time the angle $\chi$ between the two sets of scratches was measured.

## Calculation of Distortion

To find the nature of the distortion from the measurements made on the surface of the specimen the method previously adopted in the case of tensile test-pieces* was used. The unextended cone, or cone containing all the directions of lines of particles which are the same length after distortion that they were before, was found. To do this the extension of the material parallel to the two sets of scratches was found. If the measured $\xi$, $\eta$ co-ordinates of the point 4 are $\xi_4$, $\eta_4$ before compressing, and $\xi_4'$, $\eta_4'$ after compressing, the ratio of the final distance between points 4 and 6 to the initial distance is

$$\alpha=\frac{\{(\xi_6'-\xi_4')^2+(\eta_6'-\eta_4')^2\}^{\frac{1}{2}}}{\{(\xi_6-\xi_4)^2+(\eta_6-\eta_4)^2\}^{\frac{1}{2}}}. \tag{2}$$

Similarly, the ratio of the final to the initial length of lines parallel to the line joining the points 2 and 8 is

$$\beta=\frac{\{(\xi_2'-\xi_8')^2+(\eta_2'-\eta_8')^2\}^{\frac{1}{2}}}{\{(\xi_2-\xi_8)^2+(\eta_2-\eta_8)^2\}^{\frac{1}{2}}}. \tag{3}$$

If $\chi$ and $\chi'$ are the initial and final values of $\chi$, the angle between the two sets of scratches, then the co-ordinates $(x_1, y_1, 0)$ after compression of a point in the upper face of the specimen are related to the co-ordinates $(x_0, y_0, 0)$ of the same particle before compression by the equations

$$\left.\begin{aligned} x_1&=\alpha x_0+\left(\beta\frac{\cos\chi'}{\sin\chi}-\alpha\cot\chi\right)y_0,\\ y_1&=\beta\frac{\sin\chi'}{\sin\chi}y_0. \end{aligned}\right\} \tag{4}$$

If the particle whose displaced position is being discussed does not lie in the upper surface of the specimen, the transformation formulae equivalent to (4) are

$$\left.\begin{aligned} x_1&=\alpha x_0+ly_0+\mu z_0,\\ y_1&=my_0+\nu z_0,\\ z_1&=\gamma z_0, \end{aligned}\right\} \tag{5}$$

where

$$\left.\begin{aligned} l&=\beta\frac{\cos\chi'}{\sin\chi}-\alpha\cot\chi,\\ m&=\beta\frac{\sin\chi'}{\sin\chi}, \end{aligned}\right\} \tag{6}$$

and $\mu$, $\nu$, $\gamma$ are to be determined from the measurements of the specimen. Of these $\gamma$† evidently represents the ratio $t_1/t_0$, where $t_0$ and $t_1$ are the initial and final thick-

* B.L.; above, p. 71.

† In this work $\gamma$ is used to denote the ratio of the thickness after any given compression to the thickness before. The symbol $\epsilon$ is used to characterize the state of the material at any stage, and represents the ratio of the thickness at that stage to the initial thickness before any compressive stress had been applied. Thus, if the material were successively subjected to two compressions and the corresponding values of $\gamma$ were $\gamma_1$ and $\gamma_2$, the value of the characteristic $\epsilon$ would be 1 before starting the first test, $\gamma_1$ after the first test and before the second, and $\gamma_1\times\gamma_2$ after the second test.

nesses of the specimen. These were measured with a micrometer. To find $\mu$ and $\nu$ the co-ordinates of the point 5′ (i.e. the central point on the lower face of the specimen) are first found from the measurements. If these are $(X_0, Y_0, -t_0)$ before compression and $(X_1, Y_1, -t_1)$ after compression, equations (5) become

$$\left.\begin{aligned} X_1 &= \alpha X_0 + lY_0 - \mu t_0, \\ Y_1 &= mY_0 - \nu t_0, \end{aligned}\right\}$$

and solving these equations

$$\left.\begin{aligned} \mu &= \frac{X_1 - \alpha X_0 - lY_0}{-t_0}, \\ \nu &= \frac{Y_1 - mY_0}{-t_0}. \end{aligned}\right\} \qquad (7)$$

Hence, all the coefficients in the transformation formulae (5) can be found from the measurements.

## Calculation of the Unstretched Cone

To find the unstretched cone we must substitute from (5) in the equation

$$x_0^2 + y_0^2 + z_0^2 = x_1^2 + y_1^2 + z_1^2. \qquad (8)$$

The equation of the cone in the material before distortion is found by eliminating $x_1, y_1, z_1$, between (5) and (8). It is

$$x_0^2(\alpha^2 - 1) + y_0^2(m^2 + l^2 - 1) + z_0^2(\gamma^2 + \mu^2 + \nu^2 - 1) \\ + 2x_0 y_0 \alpha l + 2x_0 z_0 \alpha\mu + 2y_0 z_0(\mu l + m\nu) = 0. \qquad (9)$$

The equation of the cone in its distorted position could be found by eliminating $x_0, y_0, z_0$ from the same two equations. Expressing (9) in spherical polar co-ordinates chosen so that $\theta$ is the angle which the direction considered makes with the axis of $z$, and $\phi$ the angle between its projection on the plane $z = 0$ and the axis of $x$, equation (9) becomes

$$\{(\alpha^2 - 1)\cos^2\phi + (m^2 + l^2 - 1)\sin^2\phi + 2\alpha l \cos\phi \sin\phi\}\tan^2\theta \\ + \{2\alpha\mu\cos\phi + 2(l\mu + m\nu)\sin\phi\}\tan\theta + \gamma^2 + \mu^2 + \nu^2 - 1 = 0. \qquad (10)$$

Taking a series of value of $\phi$, the corresponding pairs of values of $\theta$ can be found from (10) and the cone can then be plotted in a stereographic diagram.

## Determination of Crystal Axes

In all cases the orientation of the crystal axes was determined* with reference to the ruled generators before the single crystal bar had been cut up into discs. The axes were then determined again by X-ray reflections taken at points on the plane faces of the disc-shaped specimens, and it was found that good agreement was obtained, the differences in orientation being never greater than 2°. This agreement showed that the method adopted for cutting up and marking the specimens was sufficiently

* The method used was that described by Dr A. Müller, *Proc. Roy. Soc.* A, cv (1924), 500.

accurate for our purpose, so that we could rely on knowing the orientation of the crystal axes of all the discs into which the original single-crystal bar had been cut.

After compression, the orientations of the crystal axes were redetermined, and to do this it was sometimes found necessary to etch away one of the faces in order to remove the layer of aluminium which had been in contact with the steel-compressing plate. In these circumstances, if the etching had to be carried to a depth which made the fiducial marks indistinguishable, it was still possible to rule a new mark on the etched face and to determine the orientation of the axes with reference to that. The specimen could then be mounted in the holder shown in fig. 5, and the angle between the new mark and the original fiducial marks on the unetched face determined by reversing the holder on the microscope stage.

A difficulty arose in setting the specimen up in the X-ray spectrometer. If the specimen was fixed to the plane face of a bar which was parallel to the axis of rotation of the spectrometer, only one reflection could usually be obtained with homogeneous X-rays from a copper or iron anticathode. To determine completely the orientation of the axes it is necessary to have two reflections, and in order to satisfy this requirement it was desirable to be able to rotate the specimen about an axis perpendicular to the axis of rotation of the spectrometer. For this purpose the holder shown in fig. 6 was designed.

Two pieces of brass angle $A$, $B$ (fig. 6 A), soldered together and to a piece of cylindrical brass rod $C$, form the frame of the holder, which is mounted on the universal table of the X-ray spectrometer, its lower end fitting in the attachment provided for the long cylindrical specimens for which the spectrometer was designed. Four pieces of half-round steel $D$ soldered to this frame, as shown, form two vees in which the cylindrical slider $E$ rests, touching at four points. The slider is prevented from rotating about its axis by a steel rod $F$ fixed to it, which bears against another steel rod $G$ fixed to the frame, making the fifth point of contact. The rod $G$ is adjusted to be parallel to the slider $E$ when the latter is resting on the vees. This can be done with all the accuracy necessary to ensure that there is no appreciable rotation of the slider as it is moved longitudinally.

The slider is kept firmly in position against all the five points of contact with the frame by the system of springing shown in figs. 6 A and B. A spring is attached to the end of the rod $F$ and to a pivoted triangular framework $H$. To the latter is fastened a rod $K$, which presses on the back of the slider $E$.

The slider is moved axially by the mechanism shown in fig. 6 A and C, which was copied from an instrument made by the Cambridge Instrument Company. A small grooved pulley $L$ is pressed against a piece of steel rod which is soldered to a cylindrical piece of brass $M$. When rotated by the knurled wheel shown it acts in the same way as a rack and pinion, but much more smoothly. The attachment to the slider is by means of the ball and wire shown in fig. 6 C; these are arranged so as to produce no constraint except the desired axial one.

To the upper end of the slider $E$ is attached a circular brass disc $N$, 2·5 cm. in diameter, shown in figs. 6 A and D. A second similar disc $P$ is held in frictional contact with $N$, and is capable of being rotated coaxially with it. The specimen $Q$ is stuck on to the face of the disc $P$ with shellac. The edges of the discs are graduated, $N$ with 40 and $P$ with 36 divisions. This forms a continuous vernier, and is a convenient arrangement when the available diameter is small.

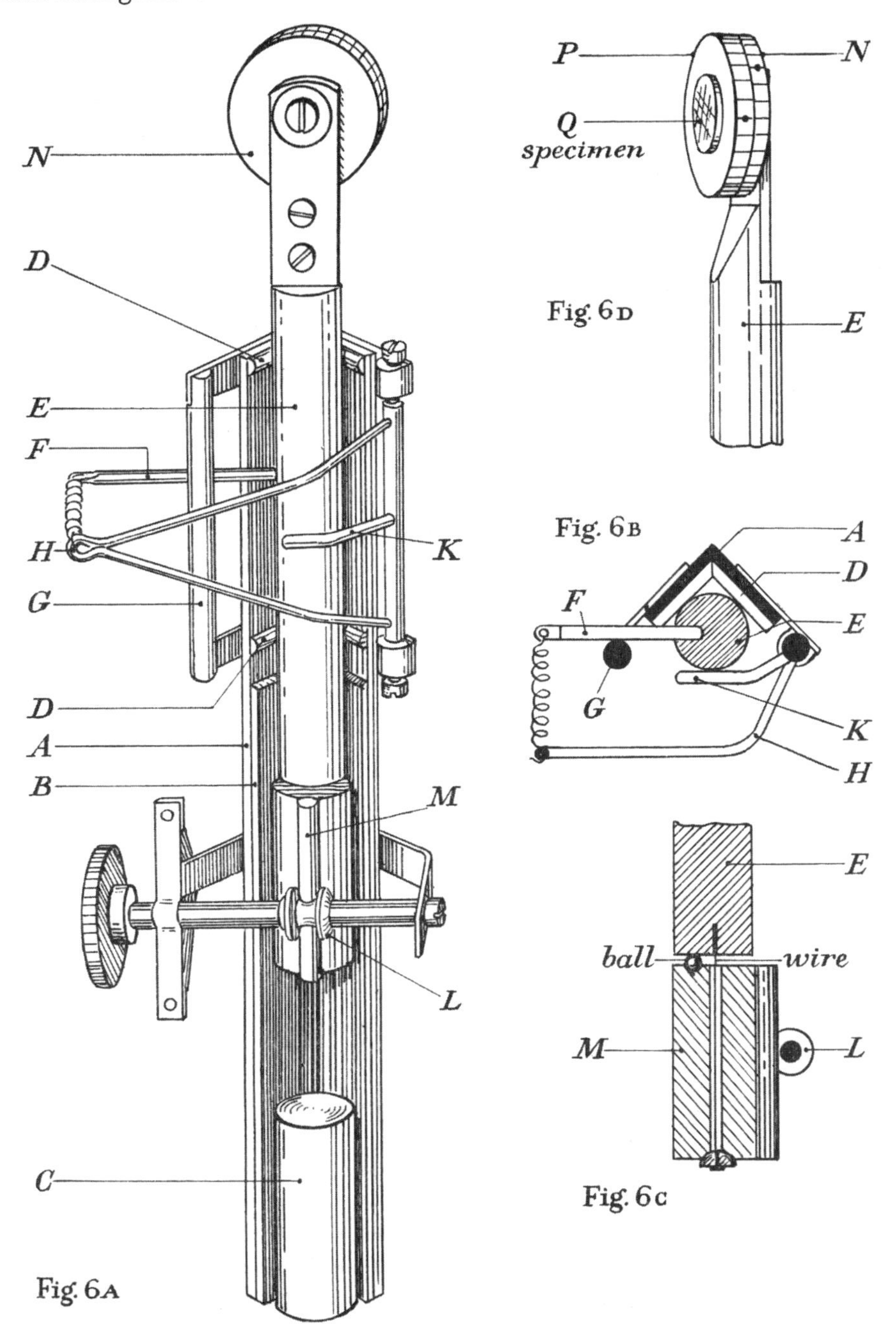


Fig. 6. Holder for mounting specimen in X-ray spectrometer.

In use the specimen is mounted as described, and any one of the lines ruled on its surface is focused in the microscope of the spectrometer. The disc $P$ is then rotated until this line remains on the cross-wire, when the slider $E$ is moved axially. By adjusting the universal table of the spectrometer, this line is then made to coincide with the axis of rotation of the table. Finally, the specimen is rotated in azimuth until its plane contains the direction of the incident beam of X-rays, as ascertained by sighting through the two apertures which define the beam. By reading the azimuth graduations on the universal table, and the vernier on the discs $N$ and $P$, the orientation of the specimen is completely determined, and that of the crystal planes can then be ascertained by noting the readings of both these scales when the appropriate X-ray reflections are obtained.

It will be seen that this holder serves two purposes. First, the up-and-down movement of the slider enables the specimen to be set so that the line selected (and not merely a point on it) is accurately on the axis of rotation of the spectrometer table. Secondly, the graduated discs enable the specimen to be rotated in its own plane through any desired angle. The reading of the vernier is to 1°, but ½° can be estimated with ease.

## Choice of Specimens

The orientation of the crystal axes with reference to the flat faces of the specimen can be represented on a stereographic diagram which is exactly equivalent to the figure used in the case of tensile tests, except that the centre of the projection now

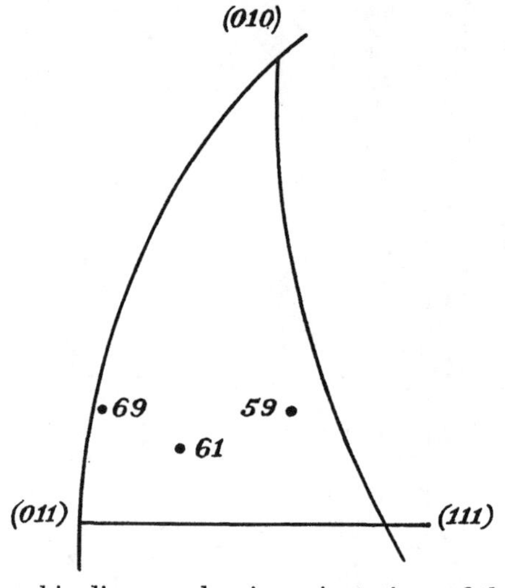


Fig. 7. Stereographic diagram showing orientations of three specimens.

represents the normal to the disc instead of the axis of the tensile test-piece. If the axes are now rotated till a cubic (100) axis is in the centre of the projection,* the position on this diagram of the point representing the normal to the plane of the flat

* This projection is shown in P.E., fig. 1 (p. 111, above).

faces of the specimen is sufficient to define completely the relationship between this normal and the crystal axes. By rotating the appropriate cubic axis into the centre it is always possible to make this point come into a given spherical triangle, the corners of which are a (100), a (110) and a (111) axis.*

For the experiments to be described three single-crystal bars were chosen. They were numbered 59, 61 and 69, and their representative points, which are shown in fig. 7, are well separated in the triangle. The disc-shaped specimens cut from no. 59 were numbered 59·1, 59·2, 59·3, ...; those from 61 were fig. 7, 61·1, 61·2, ..., etc. Many were spoiled in the preliminary experiments necessary to find out how to carry out the compression so as to get uniform distortion, and even when uniform distortion had apparently been obtained it was in some cases necessary to develop methods for ensuring that the friction between the aluminium and the steel plates was the minimum possible before we could be sure of being able to repeat measurements of compressive pressure with different specimens cut from the same bar.

## Experiments with No. 59

The first experiments were carried out with discs cut from no. 59. It was found that they became roughly elliptical when compressed. The distortion was uniform over the central part of the disc; but there were two regions, one at each end of the major axis, where the distortion was different from that in the middle. Ruled scratches

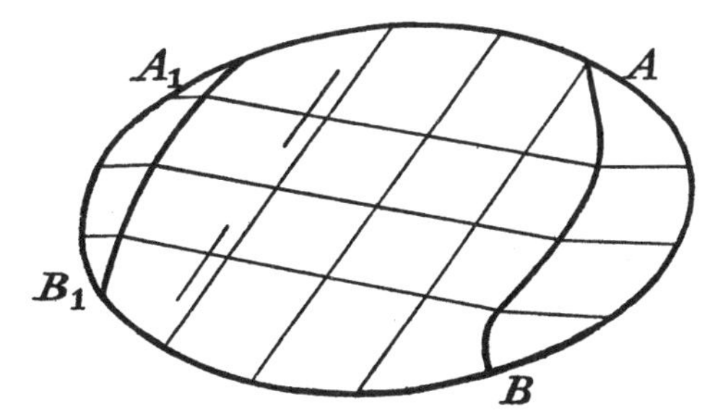


Fig. 8. Appearance of discs cut from no. 59 after compression.

which crossed into these regions were bent at their boundaries. The general appearance of the specimens after compression is that shown in fig. 8, where the two regions of extraordinary distortion are shown as $AB$ and $A_1B_1$. In the sketch the region $AB$ is larger than the region $A_1B_1$, but on the underside the region $A_1B_1$ would be identical in appearance with the region $AB$ on the upper face, and vice versa, the region of extraordinary distortion being thus symmetrical with respect to the centre of the specimen. The boundaries of these regions could be made visible in the compressed specimen by etching with caustic soda.

It will be seen later that the distortion in the main body of the specimen was due to slip on one crystal plane. In most parts of the specimen its section by a slip-plane consists of a four-sided figure with two straight parallel sides and two curved sides.

* See P.E., figs. 3, 4, 5 and 9.

In the neighbourhood of regions $AB$ and $A_1B_1$, these slip-planes cut the specimen in D-shaped figures with one straight side and one curved side. It seemed probable, therefore, that the regions of non-uniform distortion could in this case be reduced by reducing the thickness of the specimens. It was found that this was true, and it was found also that if the elliptic specimen was turned down into a circular disc, thereby removing the regions $AB$ and $A_1B_1$, the specimen would distort quite uniformly throughout its volume on further compression. The specimen shown in the photograph, fig. 2, was cut from no. 59, and was first compressed 8 % (i.e. to 92 % of its initial thickness). The regions of extraordinary distortion were then removed as described above, the specimen being turned down to the circular disc shown on the first photograph. It was then marked,* and finally compressed a further 33 % (i.e. to 62 % of its initial thickness), the result being shown in the second photograph. It will be seen that the marks remained straight throughout, the boundary being therefore elliptical.

These regions of extraordinary distortion occurred only in discs cut from no. 59. With nos. 61 and 69 it was found that the distortion was uniform throughout during the whole test.

59·6. The initial dimensions of disc no. 59·6 were 14·135 mm. diameter and 2·083 mm. thick. It was compressed by a load of 1 ton in one operation and its final thickness was 1·903 mm. The other measurements, when reduced in accordance with formulae (2) and (3) gave

$$\alpha=1{\cdot}00303, \quad X_0=-0{\cdot}217\text{ mm.},$$
$$\beta=1{\cdot}0937, \quad Y_0=-0{\cdot}144\text{ mm.},$$
$$\gamma=0{\cdot}9134, \quad X_1=-0{\cdot}212\text{ mm.},$$
$$\chi=90^\circ\,10', \quad Y_1=-0{\cdot}169\text{ mm.}$$
$$\chi'=91^\circ\,30',$$

Hence using (6) and (7) it is found that

$$l=-0{\cdot}0259, \quad m=1{\cdot}0934, \quad \mu=+0{\cdot}0345, \quad \nu=-0{\cdot}0326.$$

The equation of the unextended cone is

$$(0{\cdot}00607\cos^2\phi-0{\cdot}0516\cos\phi\sin\phi+0{\cdot}1960\sin^2\phi)\tan^2\theta + (0{\cdot}0692\cos\phi-0{\cdot}0731\sin\phi)\tan\theta-0{\cdot}1634=0. \quad (11)$$

This cone is represented by the broken line on the stereographic diagram shown in fig. 9. In this diagram the centre $O$ represents the normal to the face of the specimen, while the point $X$ represents the direction of the ruled line on the upper surface of the specimen which is taken as $\phi=0$.

It will be seen that the projection of the unstretched cone is a closed curve, but on laying the tracing paper on which it was drawn over a stereographic net it was found that this cone very nearly coincided with two planes. These two planes are

* The marks on this specimen were made specially deep—see footnote; above, p. 136.

shown as full lines in fig. 9. The figure is very similar to those found for the distortion in a tensile test.* It seems likely, therefore, that the distortion is due to slipping parallel to one crystal plane. It will be seen later† that the experiment can be carried out under conditions which give rise to an unextended cone which is accurately two planes. For this reason in the further analysis of specimen no. 59·6 the nearest two planes, namely, those shown in fig. 9, have been regarded as equivalent to the unextended cone.

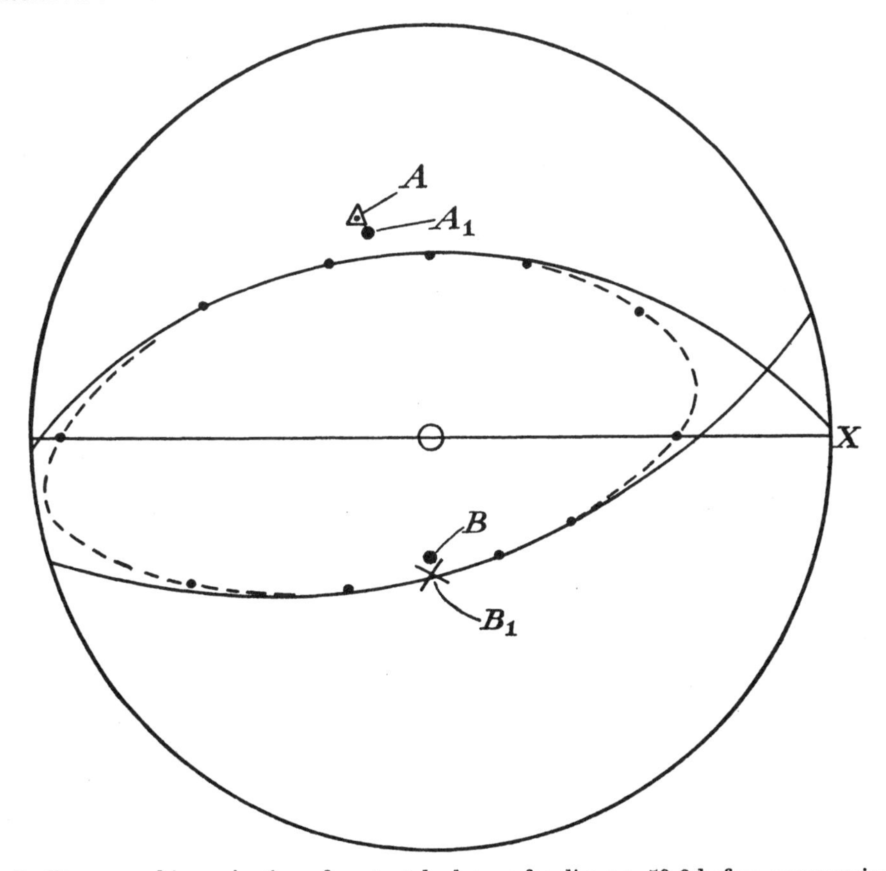


Fig. 9. Stereographic projection of unstretched cone for disc no. 59·6 before compression.

In the case of the tensile test-pieces used in our previous experiments, the stress must be uniformly distributed over the cross-section of the bar, and in these circumstances it was shown that, in general, slipping occurs only in one octahedral (111) plane and parallel to a (110) direction. It was shown further that, of the twelve possible crystallographically similar shears of this type, the one which occurs is that for which the component of shear stress in the direction of slip is the greatest. In the case of the compression tests here described, the stress cannot be quite

* P.E., fig. 2 (p. 116 above), and B.L., fig. 7 (p. 73 above).
† See p. 150 below.

IO-2

uniform at all points, because the friction between the steel plates and the specimens cannot be reduced to zero, but, on the other hand, it was found that the measures taken to reduce friction did have the effect of making the distortion uniform throughout the specimen. It appears likely, therefore, that the friction is not great enough to cause any considerable variation of stress in different parts of the specimen. Assuming this to be the case, it is possible, if the orientation of the crystal axes is known, to calculate the components of shear stress corresponding with each of the twelve kinds of shear which were shown to be possible in the case of bars in tension. Choosing that one of the twelve for which the component of shear stress in the direction of shear is greatest, we can predict the orientations of the plane and direction of shear. The method used for making this choice was identical with that previously described in the case of tensile test-pieces.*

If there is no friction between the specimen and the steel plates, the shear stress corresponding with any particular type of shear is equal in magnitude but opposite in sign to the shear stress which would exist in a tensile test-piece whose cross-section coincided with a flat face of the compression specimen, provided the tensile load in the one case was equal to the compressive load in the other. For the purpose of predicting the slip-plane and direction of slip, therefore, we may identify the normal to the surface of the compression specimen with the axis of the tensile test-piece and apply directly the method described previously.*

Applying this method to measurements of the orientation of the crystal axes made by Miss Elam, it was found that the co-ordinates of the pole of the predicted slip plane were ($\theta=59°$, $\phi=108°$), while those of the direction of slip were ($\theta=33°$, $\phi=269°$). These points are marked in fig. 9 as $A$ and $B$.

The pole of one member of the pair of planes determined by the distortion measurements is shown at $A_1$ in fig. 9. Its co-ordinates are $\theta=55°$, $\phi=107°$. The direction of slip is the point on the slip-plane which lies at 90° from the intersection of the two planes which constitute the unextended cone. The projection of this direction is shown at $B_1$ in fig. 9. Its co-ordinates are ($\theta=37°$, $\phi=270°$). It will be seen that the predicted positions of $A$ and $B$ are close to their observed positions $A_1$ and $B_1$.

59·7. Specimen no. 59·7 was first compressed by a pressure of 1·5 tons in one operation, so that its thickness changed from 1·806 to 1·472 mm. It was then measured and the orientation of its crystal axes determined by X-rays.

The measurements gave:

$$\left.\begin{array}{ll} \alpha=1{\cdot}0153, & X_0=-0{\cdot}254\text{ mm.,} \\ \beta=1{\cdot}1976, & Y_0=-0{\cdot}236\text{ mm.,} \\ \gamma=0{\cdot}817, & X_1=-0{\cdot}379\text{ mm.,} \\ \chi=89°\,45', & Y_1=-0{\cdot}173\text{ mm.,} \\ \chi'=94°\,10', & \end{array}\right\} \quad (12)$$

* P.E.; above, p. 111.

and the equation to the unextended cone in its second or compressed position is

$$(0{\cdot}0300\cos^2\phi - 0{\cdot}1480\cos\phi\sin\phi + 0{\cdot}292\sin^2\phi)\tan^2\theta + (0{\cdot}1766\cos\phi - 0{\cdot}0901\sin\phi)\tan\theta - 0{\cdot}511 = 0. \quad (13)$$

The calculated points of this cone are shown in the stereographic diagram (fig. 10) as crosses, and the cone is marked in as a broken line. As in the case of 59·6 the

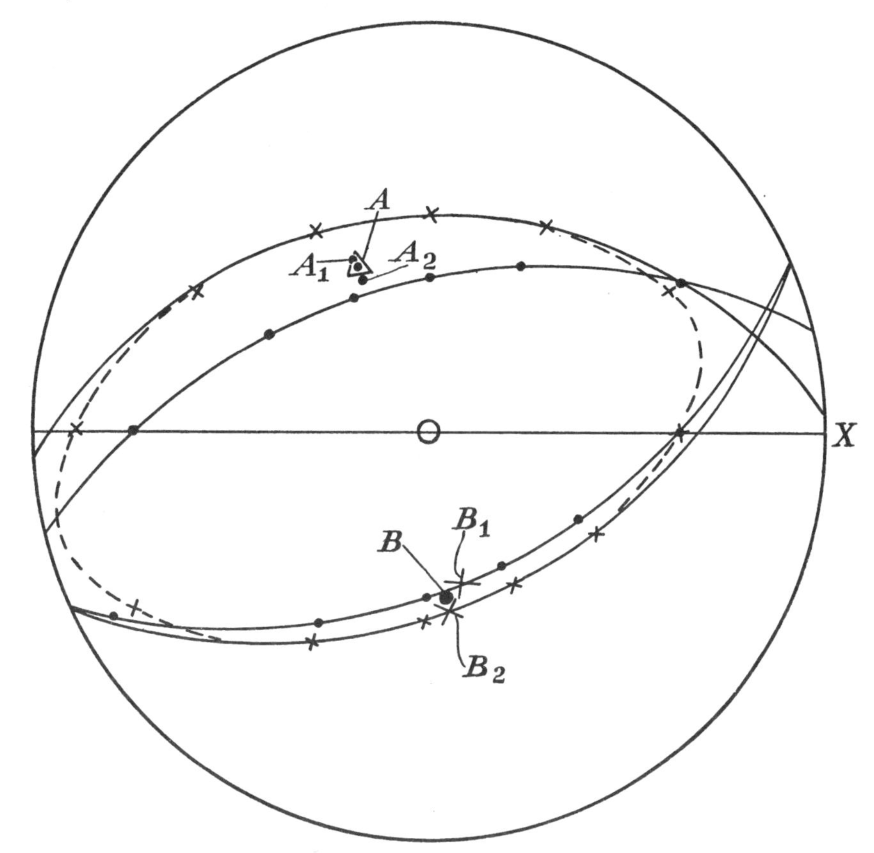


Fig. 10. Unstretched cones for 59·7 in material compressed to $\epsilon = 0{\cdot}817$. ×, calculated points for cone corresponding with compression from $\epsilon = 1{\cdot}0$ to $0{\cdot}817$. ●, calculated points for cone corresponding with compression from $\epsilon = 0{\cdot}817$ to $0{\cdot}817 \times 0{\cdot}8376 = 0{\cdot}684$.

unextended cone nearly, but not quite, coincides with two planes which are shown as full lines passing nearly through the crosses. Neglecting the difference between the actual unextended cone and these two planes, the pole of the slip-plane and the direction of slip are shown in fig. 10 at $A_2$ and $B_2$.

The predicted positions of these points derived from X-ray measurements by the methods described above are shown at $A$ and $B$. It will be seen that, except for the divergence of the unextended cone from two planes the agreement is good.

While these tests were being made, it was discovered that if the compression was carried out in two or three stages, the specimen being removed from the compression machine and re-greased after each operation, the diminution in thickness was greater than when the whole load was applied at once. By successively diminishing the increase in load at each stage, it was found that a limit was reached beyond which a further diminution of the increase in load corresponding with one stage in the operations produced no further increase in the amount of the compression for a given final compressive load. It was found, for instance, that with discs 14·3 mm. diameter, the curves connecting thickness of the specimen with compressive load were practically identical when the loads were increased by 0·05 and 0·1 ton at each stage, but that if the increase was 0·2 ton at each stage there was a measurable increase in the resistance of the specimen to compression. It seemed probable, therefore, that by the time the change in load had been reduced to 0·1 ton per stage, the friction had been reduced as much as it was possible to reduce it by this method. A reduction in friction increases the uniformity of the stress distribution in the specimen, and if the deviation of the unextended cone from two planes is due in any way to variations in stress in different parts of the specimen, one might expect the unextended cone to approximate more nearly to two planes when the compression is carried out in small stages.

To test this, specimen no. 59·7 was again compressed from 1·423 to 1·192 mm. thick in five stages. Before starting the second test, the specimen which had become elliptical during the first compression was cut down to a circular disc of diameter 12·34 mm., and re-marked with scratches ruled parallel to the original ones. The elements from which the first position of the unextended cone for the second test (i.e. the position in the material when $\epsilon = 0{\cdot}817$)* were calculated are

$$\left.\begin{array}{ll} \alpha = 1{\cdot}0224, & X_0 = -0{\cdot}363\,\text{mm.}, \\ \beta = 1{\cdot}1705, & Y_0 = +0{\cdot}031\,\text{mm.}, \\ \gamma = 0{\cdot}8376, & X_1 = -0{\cdot}412\,\text{mm.}, \\ \chi = 94^\circ\,30', & Y_1 = +0{\cdot}005\,\text{mm.} \\ \chi' = 100^\circ\,10', & \end{array}\right\} \qquad (14)$$

The equation to the cone is

$$(0{\cdot}0453\cos^2\phi - 0{\cdot}259\cos\phi\sin\phi + 0{\cdot}355\sin^2\phi)\tan^2\theta + (0{\cdot}053\cos\phi + 0{\cdot}044\sin\phi)\tan\theta - 0{\cdot}2972 = 0, \qquad (15)$$

and its stereographic projection is shown in fig. 10, where the points calculated from (15) are represented by round dots. It will be seen that in this case the cone is, to the limit of accuracy of our measurements, actually two planes. With a stereographic net 20 cm. diameter, it was not possible to detect any difference between the cone and two planes. In the light of the results obtained with tensile test-pieces, it appears, therefore, that the distortion is in this case a simple shear parallel to one plane. Slip on any other plane could have been detected even if it had given rise to only 1 or 2 % of the whole distortion.

The normal to the plane of slip in the second compression is shown at $A_1$ and the

* See footnote †; above, p. 140.

direction of slip at $B_1$ in fig. 10. It will be noticed that the distortion measurements determine two planes, one of which is the slip-plane. In each case one of the planes has nearly coincided with the position of the slip-plane predicted from knowledge of the orientation of the crystal axes, and that has been taken as the slip-plane. Fig. 10, however, enables us to choose between the two planes without making use of knowledge derived from X-rays. One member of the pair of planes determined

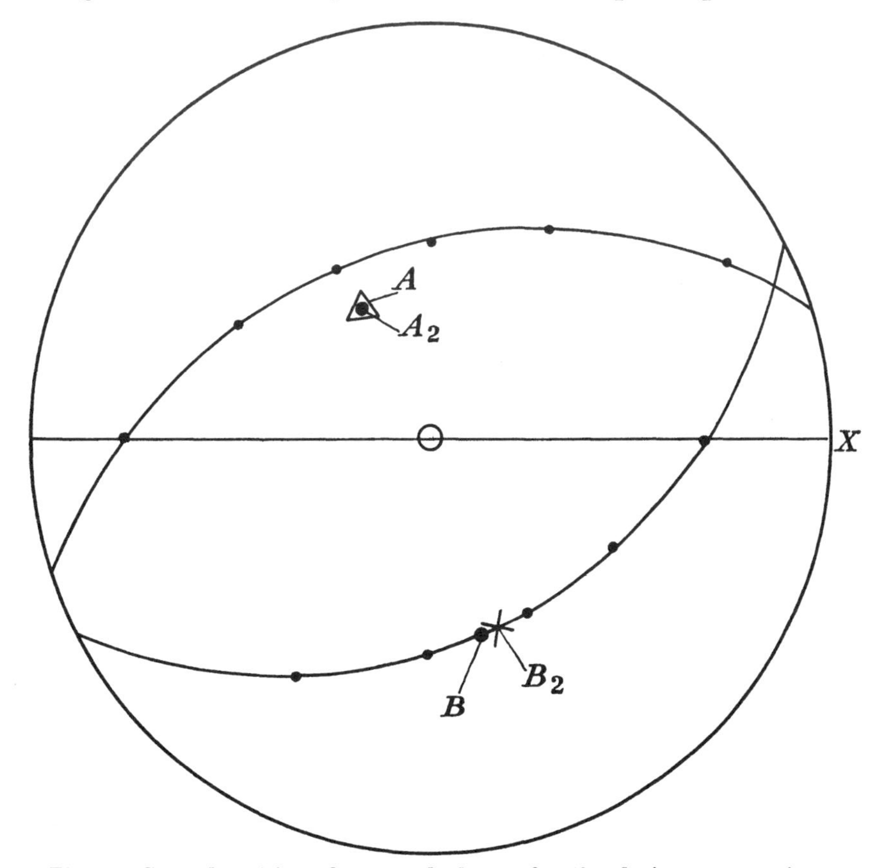


Fig. 11. Second position of unstretched cone for 59·7 during compression from $\epsilon = 0{\cdot}817$ to $0{\cdot}684$.

for the first compression nearly coincides with one of those derived from the second compression. This plane has remained unstretched and undistorted in both experiments. It is, therefore, the slip-plane. It will be noticed that this method of choosing between two possible slip-planes leads to the same result as the other method involving the use of X-rays.

*Second position of the unextended cone for the second compression test of* 59·7. The equation to this cone can be derived from the data (14). Its equation is

$$(0{\cdot}0433\cos^2\phi - 0{\cdot}210\cos\phi\sin\phi + 0{\cdot}241\sin^2\phi)\tan^2\theta + (0{\cdot}0576\cos\phi + 0{\cdot}0456\sin\phi)\tan\theta - 0{\cdot}426 = 0, \quad (16)$$

and its stereographic projection is shown in fig. 11, which corresponds with material compressed till its thickness is $\epsilon = 0{\cdot}817 \times 0{\cdot}9376 = 0{\cdot}684$* of its original thickness. As might be expected, the calculated points lie very accurately on two planes. The co-ordinates of the normal to one of these planes are ($\theta = 39°$; $\phi = 118{\cdot}5°$) and the co-ordinates of the corresponding direction of slip are ($\theta = 52°$; $\phi = 286°$). These directions are represented in fig. 11 by the points $A_2$ and $B_2$.

The corresponding directions predicted from X-ray measurements were: normal to slip-plane ($\theta = 39°$; $\phi = 118{\cdot}5°$); direction of slip ($\theta = 52°$; $\phi = 290°$). These are marked as $A$ and $B$ in the projection. It will be seen that the agreement is very good.

We desire to express our thanks to Miss C. F. Elam, who gave us the single-crystal bars used in this work, and carried out all the X-ray measurements. Without her help the work could not have been done. We wish also to acknowledge the assistance of Dr A. Hutchinson, F.R.S., who allowed us to use measuring apparatus in his laboratory, and of Messrs Armstrong Siddeley Motors, who kindly ground the hardened discs used in compressing the specimens.

Most of the work was carried out in the Cavendish Laboratory, through the kindness of Sir Ernest Rutherford. We are also indebted to Professor C. E. Inglis for allowing us to use the compression testing machines in the Cambridge Engineering Laboratory.

* See footnote †; above, p. 140.

# 10

# THE DISTORTION OF IRON CRYSTALS*

REPRINTED FROM

*Proceedings of the Royal Society*, A, vol. CXII (1926), pp. 337–61

## EXPERIMENTAL METHODS

In recent years several attempts have been made to determine what happens to iron crystals when they are strained, and various conflicting statements have been made as to the connection between the crystal axes and the nature of the strain. No reliable results have been obtained, however, partly because the largest crystals available were too small for accurate experiment, partly because workers have assumed that planes of slip coincide with crystal planes—an assumption which the experiments to be described prove to be erroneous—but chiefly because the analysis of strain has not been carried out in a systematic manner so as to obtain all information possible from external measurement of strained crystals.

The work of Professor Edwards and Mr Pfeil has now enabled us to obtain crystals sufficiently large for the purpose, and, in fact, all the material used in the experiments now to be described was cut from specimens very kindly supplied by them.

Two different methods were used for producing distortion. In the first the material was cut in the form of a uniform bar of rectangular cross-section, usually about 2 mm. square. The length was about 10 cm., of which some 4 cm. in the middle was occupied by a single crystal. Specimens of this type were marked with fine scratches and pulled in a tensile testing machine. The distortion was measured in the manner described in our Bakerian Lecture.†

In the second method circular discs about 6 mm. diameter × 1·4 mm. thick were cut from a crystal and compressed between polished steel plates. The distortion was measured by methods previously described.‡

Though the methods used were identical with those developed for dealing with aluminium crystals, the difference in the material necessitated small changes in the procedure. In the first place great care had to be taken to ensure that the depth of the surface layer to which the crystal lattice is distorted by grinding and polishing was as small as possible. The specimens were sawn with a fine saw or milled with a fine cutter. They were then ground down on emery paper of successive

* With C. F. ELAM.

† *Proc. Roy. Soc.* A, CII (1923) 643; paper 5 above, p. 63.

‡ Taylor and Farren, 'Distortion of crystals of aluminium under compression. Part I', *Proc. Roy. Soc.* A, CXI (1926), 529; paper 9 above, p. 133.

degrees of fineness till at least 0·2 mm. had been removed from each cut surface. Finally, all flat faces were polished till a mirror surface was obtained. The holder used for grinding and polishing the tension specimens was made, at the suggestion of Mr A. Woodward, from stainless steel kindly supplied to us by Dr W. H. Hatfield.

Owing to the small size of the specimens it was necessary to use special methods for marking them. The tension specimens were mounted on an adjustable table which could slide on a bed similar to a lathe bed. A cutter consisting of a safety razor blade broken across the middle was mounted so that it pressed, with a weight of 10 g., on the specimen as it passed under it. In this way a fine mark of very uniform width could be made. The specimens were marked with a longitudinal scratch down the middle of each face and cross-scratches on all four faces spaced at intervals of 5 mm.

The marks on the compression discs had to be slightly deeper and a weight of 50 g. on a sharpened gramophone needle was employed, the specimen being mounted on the traversing stage of our measuring microscope.

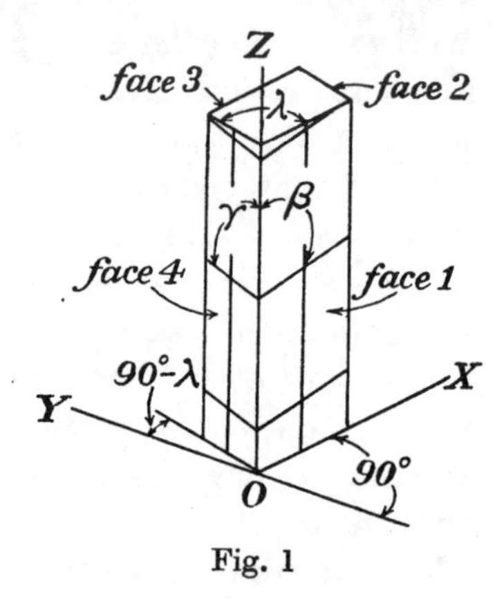


Fig. 1

*Measurements.* The most difficult measurement to make with the necessary accuracy was the angle between the faces of the tension specimens. For this purpose the lines down the middle of each face were used. The angle between the planes passing through pairs of marks on opposite faces was measured by a method previously described.* But the smallness of the specimen necessitated the use of a very good goniometer, which was kindly lent to us by Dr A. Hutchinson, F.R.S. With this instrument measurements of the angle between pairs of opposite faces could be relied on to 20 min. of arc.

*Methods of calculation.* The method adopted for representing the distortion was the same in the tension and the compression specimens. The cone containing all lines of particles which remained unstretched after the crystal had been distorted was determined. The difference in shape between the two types of specimen, however, necessitated some difference in the formulae used for deducing the equation of the unstretched cone from the external measurement.

## Equation to Unstretched Cone for Tension Specimens

In the case of tension specimens rectangular co-ordinates were chosen so that one edge was the axis $OZ$ and one face (called face 1) was the plane $y = 0$. The scheme is shown in fig. 1.

* Taylor and Elam, 'The plastic extension and fracture of aluminium crystals', *Proc. Roy. Soc.* A, CVIII (1925), 33; paper 7 above, p. 113.

The measured quantities were:

$\lambda$ the angle between the faces 1 and 4.

$b$ the width of faces 1 and 3.

$c$ the width of faces 2 and 4.

$\beta$ the angle between ruled scratches across face 1 and the axis $OZ$.

$\gamma$ the angle between ruled scratches across face 4 and the axis $OZ$.

$d$ the distance parallel to the axis of the specimen between successive cross-marks.

Using suffixes 0 and 1 to denote the conditions before and after stretching, the ratio of the final to the initial length is $d_1/d_0 = \epsilon$, and let $f = b_1/b_0$ and $g = c_1/c_0$. If $(x_1, y_1, z_1)$ are the co-ordinates in the strained material of a particle whose co-ordinates in the unstrained material were $(x_0, y_0, z_0)$, the formulae of transformation are

$$\left.\begin{aligned} x_1 &= fx_0 + ly_0, \\ y_1 &= my_0, \\ z_1 &= px_0 + qy_0 + \epsilon z_0, \end{aligned}\right\} \tag{1}$$

where

$$\left.\begin{aligned} l &= g\frac{\cos\lambda_1}{\sin\lambda_0} - f\cot\lambda_0, \\ m &= g\frac{\sin\lambda_1}{\sin\lambda_0}, \\ p &= -\epsilon\cot\beta_0 + f\cot\beta_1, \\ q &= \epsilon\cot\beta_0\cot\lambda_0 - f\cot\beta_1\cot\lambda_0 + g\frac{\cot\gamma_1}{\sin\lambda_0} - \epsilon\frac{\cot\gamma_0}{\sin\lambda_0}. \end{aligned}\right\} \tag{2}$$

The unstretched cone is given by

$$x_0^2 + y_0^2 + z_0^2 = x_1^2 + y_1^2 + z_1^2, \tag{3}$$

and eliminating $x_1, y_1, z_1$ from 1 and 3 the equation of the unstretched cone becomes

$$\begin{aligned} x^2(f^2+p^2-1) + y^2(l^2+m^2+q^2-1) + z^2(\epsilon^2-1) \\ + 2xy(fl+pq) + 2zx(\epsilon p) + 2yz(\epsilon q) = 0. \end{aligned} \tag{4}$$

If spherical polar co-ordinates are used, $\theta$ being the angle which the direction concerned makes with the axis of $z$ and $\phi$ the angle which its projection on the plane $z = 0$ makes with the axis of $x$, then

$$\left.\begin{aligned} x/z &= \tan\theta\cos\phi, \\ y/z &= \tan\theta\sin\phi, \end{aligned}\right\} \tag{5}$$

so that the equation of the unstretched cone in its first position before stretching the material is

$$\begin{aligned} \{(f^2+p^2-1)\cos^2\phi + 2(fl+pq)\cos\phi\sin\phi + (l^2+m^2+q^2-1)\sin^2\phi\}\tan^2\theta \\ + \{2\epsilon p\cos\phi + 2\epsilon q\sin\phi\}\tan\theta + \epsilon^2 - 1 = 0. \end{aligned} \tag{6}$$

To find the unstretched cone in its second position, in the stretched material, the simplest method is to reverse all the formulae, replacing measurements made before extension by corresponding ones in the stretched material and vice versa.

The formulae of transformation are then:

$$\left.\begin{aligned} x_0 &= f_1 x_1 + l_1 y_1, \\ y_0 &= m_1 y_1, \\ z_0 &= p_1 x_1 + q_1 y_1 + \epsilon_1 z_1, \end{aligned}\right\} \tag{7}$$

where $$l_1 = \frac{1}{g}\frac{\cos\lambda_0}{\sin\lambda_1} - \frac{1}{f}\cot\lambda_1, \text{ etc.,}$$

and the equation to the unstretched cone in its second position is identical with (6) except that each of the symbols inside the brackets has a suffix 1.

## Equation of Unstretched Cone for Compression Specimens

The scheme of marking specimens and the methods of measurement are described in a previous paper by one of the authors.* A photograph† of one of the specimens before compression is shown in fig. 2. Figs. 3 and 4 are photographs‡ taken after compression of specimens which were marked originally with six scratches in the square pattern shown in fig. 2. A rectangular system of co-ordinates was chosen

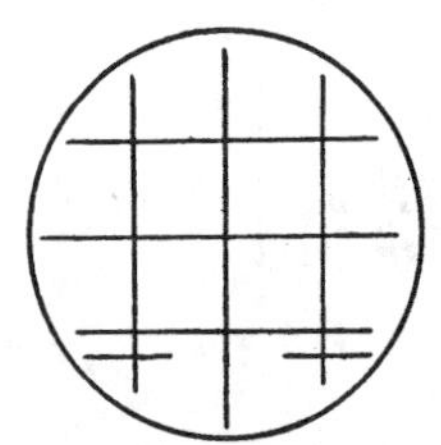

Fig. 2. Iron crystal Fe 8*c* before compression.

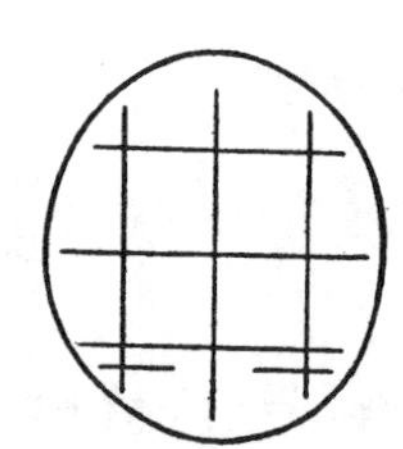

Fig. 3. Fe 7*c* after compression to $\epsilon = 0{\cdot}897$.

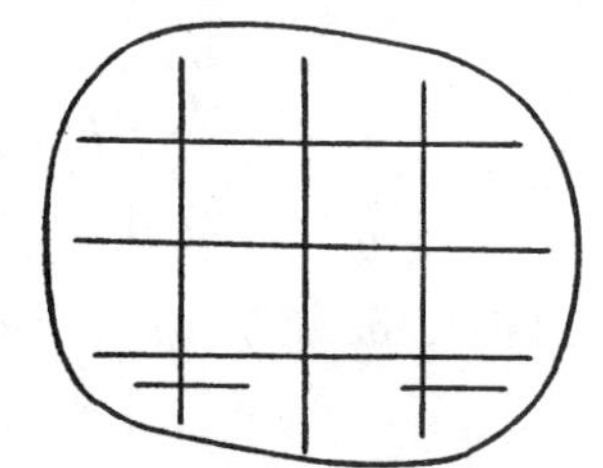

Fig. 4. Fe 3*c* after compression to $\epsilon = 0{\cdot}840$.

so that the origin was at the central point *o* of the nine points where the scratches intersect. The axis $OZ$ was vertical and perpendicular to the face of the specimen. $OX$ was along one of the central scratches and $OY$ was perpendicular to it. The directions of the axes are shown in fig. 5.

Measurements were made of the amount the material had stretched in the directions of the two sets of scratches. The ratios of the final to the initial length in these two directions are called $\alpha$ and $\beta$§ respectively. If the final and initial thickness of the specimen between its plane faces are $t_1$ and $t_0$, the compression is measured by $\gamma = t_1/t_0$. The angle between the two sets of scratches is $\chi_0$ before, and $\chi_1$ after, compression. The co-ordinates of the central mark on the under-side of

* Taylor and Farren, loc. cit.

† Reproduced here as a line drawing.

‡ Reproduced here as line drawings.

§ This notation is the same as that adopted in the previous work on compression tests. It uses some symbols which have already been used in other senses in this paper in connection with tensile specimens, but no confusion need arise.

the specimen are $(X_0, Y_0, -t_0)$ before and $(X_1, Y_1, -t_1)$ after compression. The formulae of transformation are:

$$\left.\begin{aligned} x_1 &= \alpha x_0 + l y_0 + \mu z_0, \\ y_1 &= m y_0 + \nu z_0, \\ z_1 &= \gamma z_0, \end{aligned}\right\} \qquad (8)$$

where

$$\left.\begin{aligned} l &= \beta \frac{\cos \chi_1}{\sin \chi_0} - \alpha \cot \chi_0, \\ m &= \beta \frac{\sin \chi_1}{\sin \chi_0}, \\ \mu &= \frac{X_1 - \alpha X_0 - l Y_0}{-t_0}, \\ \nu &= \frac{Y_1 - m Y_0}{-t_0}. \end{aligned}\right\} \qquad (9)$$

The equation of the unextended cone in its first, or unstrained, position is

$$\{(\alpha^2 - 1)\cos^2\phi + (m^2 + l^2 - 1)\sin^2\phi + 2\alpha l \cos\phi \sin\phi\}\tan^2\theta + \{2\alpha\mu\cos\phi + 2(l\mu + m\nu)\sin\phi\}\tan\theta + \gamma^2 + \mu^2 + \nu^2 - 1 = 0, \qquad (10)$$

where $\theta$ and $\phi$ are spherical polar co-ordinates which bear the same relation to the rectangular co-ordinates $x$, $y$, $z$ as they did in the case of the tension specimens (see equations (5)).

The most convenient method for finding the equation to the unextended cone in its second position in the compressed material is to replace in formulae (8), (9) and (10) measurements made before compression by those made after and vice versa. As in the case of tension specimens, suffixes are added to the coefficients in the transformation formulae. These then become

$$\left.\begin{aligned} x_0 &= d_1 x_1 + l_1 y_1 + \mu_1 z_1, \\ y_0 &= m_1 y_1 + \nu_1 z_1, \\ z_0 &= \gamma_1 z_1. \end{aligned}\right\} \qquad (11)$$

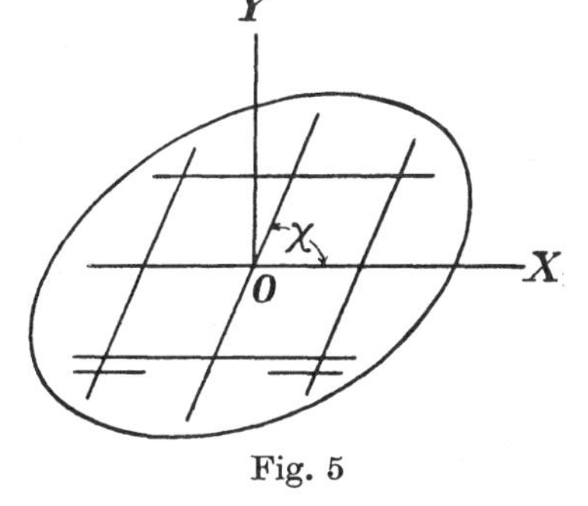

Fig. 5

In table 1 is given a list of the data used in calculating the unextended cones for three extension and four compression specimens. In the case of Fe1, the first of our tension specimens, the methods of calculation described in our Bakerian Lecture were used. In the case of all the other specimens we used the methods described in the present paper.

## Representation of Unstretched Cones

The unstretched cones are represented by means of a stereographic diagram of which the centre is the axis of $Z$ $(\theta = 0)$. The axis of $x$ $(\theta = 90°, \phi = 0)$ is represented by a radius marked in each of figs. 6–15.

The symbols common to all diagrams are explained in fig. 1. Points on the cone are found by taking values 0, $\pm 30$, $\pm 60$ and 90° for $\phi$ and calculating the two corresponding values of $\theta$ from the equation to the cone.

## Measurement of Orientation of Crystal Axes

The orientation of the crystal axes was determined by the methods used previously in the case of aluminium crystals.*

The $\alpha$ radiations from an iron anti-cathode were reflected from dodecahedral {110} planes in the crystal, the angle of reflection being 28·9°. In the case of aluminium, where reflections were obtained from {111} and {100} planes, two were sufficient to determine the orientation of all the axes, but in the case of iron two planes are not sufficient to determine the rest completely unless they are at right angles to one another. When two {110} planes making an angle of 60° have been found, they determine a {111} plane, but there are two alternative positions for the crystal lattice, and in this case it is necessary to find another {110} plane, not in the same {111} plane as the first two. In the case of the tension specimens it was not always possible to get reflections from three crystal planes owing to the limitations of the apparatus, but the ambiguity was resolved by cutting the specimen and polishing a plane perpendicular to the axis, after the test was finished. It was always possible to get a reflection from this new face and so to remove the ambiguities in the case of the distorted material. Since the motion of the {110} planes relative to the surface of the specimen during the distortion was not large and was, moreover, related to the distortion, no difficulty was encountered in identifying planes in the distorted specimen with those measured in the specimen before distortion. In this way the ambiguity was resolved in every case.

## Results of Extension Tests

Three specimens were stretched; Fe 1 was extended 15 %, Fe 3 and Fe 4 each about 9 %. It was found that they do not stretch very uniformly, a variation of 1 or even 2 % in a total of 9 % being found between the extension in successive 5 mm. lengths of the specimen. In order to make the best use of the measurements two or three sections 5 mm. long were taken in the part of the specimen where the stretching appeared to be most uniform. The measurements in these sections were averaged and used in the calculations. The figures given in table 1 are derived from these averages. In each case the unstretched cone was calculated for both positions, i.e. before and after stretching. The orientations of the crystal axes were likewise determined before and after stretching.

In figs. 6–11 the unstretched cones are represented on stereographic diagrams, each of the calculated points being shown by a round dot. On examining these diagrams it was found that the unstretched cones in every case coincided almost

* See Müller, *Proc. Roy. Soc.* A, CV (1924), 500; and Taylor and Farren, 'Distortion of crystals of aluminium under compression. Part I', *Proc. Roy. Soc.* A, CXI (1925), 529 (paper 9 above).

exactly with two planes. In each of the diagrams the pair of planes which passes most nearly through the calculated points is shown by means of two great circles.

This type of distortion was already familiar to us. It can be caused by uniform shearing or slipping parallel to either of the two unstretched planes in a direction

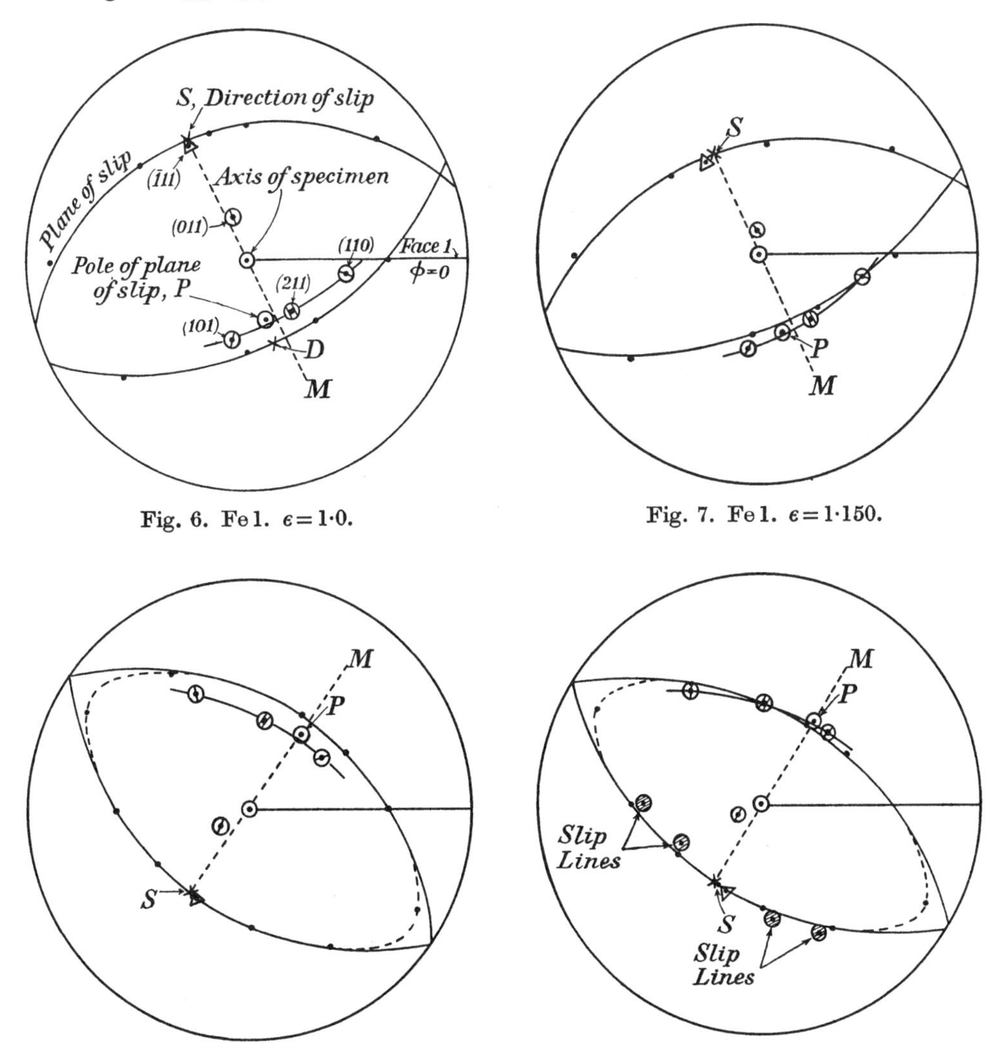


Fig. 6. Fe 1. $\epsilon=1{\cdot}0$.

Fig. 7. Fe 1. $\epsilon=1{\cdot}150$.

Fig. 8. Fe 3. $\epsilon=1{\cdot}0$.

Fig. 9. Fe 3. $\epsilon=1{\cdot}0915$.

at right angles to their line of intersection.* Accordingly, the point on each of the planes which corresponds with a possible direction of slip was marked on the diagram with a cross and the pole of each possible plane of slip was marked also.

The positions of the crystal axes were next determined from the X-ray measure-

* Bakerian Lecture; paper 5 above, p. 75.

ments and marked on the stereographic diagrams. From inspection of six diagrams similar to those shown in figs. 6–11 the following deductions were drawn:

(*a*) In each case one of the two alternative possible directions of slip was close to the pole of a {111} plane. The spherical polar co-ordinates of the pole of the {111} plane and the corresponding possible direction of slip are given in table 2, and are marked in each figure by means of the symbol △* and the direction of slip is represented by a cross ×.

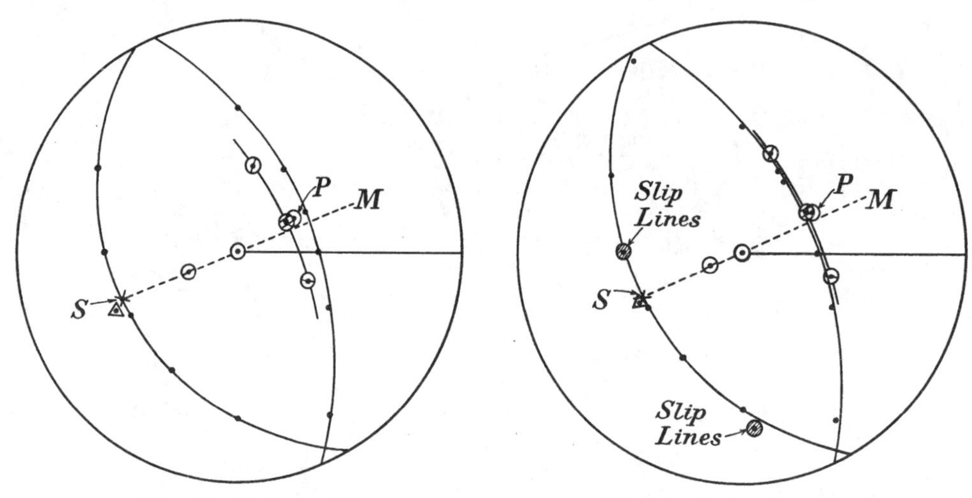


Fig. 10. Fe4. $\epsilon = 1{\cdot}0$.

Fig. 11. Fe4. $\epsilon = 1{\cdot}0879$.

The second possible direction of slip—that marked at *D* in fig. 6, but not marked in the rest of the diagrams—seemed to bear no relation to the crystal axes.

(*b*) The crystal axes move during the distortion so that they are nearly fixed relative to that one of the pair of unstretched planes which contains the pole of the {111} plane mentioned in (*a*). They move in the opposite direction to the other unstretched plane.

The material behaves therefore, both in regard to the motion of the crystal axes and in regard to the total distortion, as though this distortion were due to slipping on a crystal plane and in a direction parallel with the perpendicular of a {111} plane. It remains to discover how the plane of slip is related to the crystal axes. In an attempt to solve this question the pole of the plane determined by distortion measurements was marked on each diagram. These are shown at *P* in figs. 6–11, and an arc representing a portion of the ($\bar{1}11$) plane, which is nearly perpendicular to the direction of slip determined by distortion measurements, is also drawn. The plane contains the poles of three {110} planes and three {112} planes. In each diagram the poles of two {110} planes, (101) and (110) in fig. 6, one either side of *P*, are represented by the symbol ⊕, and the pole of the intermediate {112} plane,

* In fig. 6 the particular octahedral plane referred to is given the symbol ($\bar{1}11$) in order to conform to the conventions of crystallography.

(211) in fig. 6, is represented by the symbol ⓗ. It will be seen that in the case of Fe 4 the pole of the plane of slip coincides almost exactly with that of the {211} plane, so that the plane of slip coincides with this plane. On the other hand, in the case of Fe 1, the pole of the plane of slip is 14° away from the pole of the {211} plane, i.e. almost exactly half-way between the poles of the {101} and {211} planes. In the case of Fe 3, $P$ is nearer to the pole of the {101} plane than to the pole of the {211} plane.

It appears therefore that the distortion is such as can be produced by slipping parallel to a plane of particles in the material and that the direction of slip has a definite relation to the crystal axes, but that the plane of slip is not a definite crystal plane at all. On the other hand, the plane of slip does appear to be related to the distribution of stress. The pole of the slip plane lies close to the plane which contains the axis of the specimen and the direction of slip. In each of the diagrams (figs. 6–15) the straight line which represents this plane is shown as $SM$. If it were accurately true that $P$ in every case lies, as it does in fig. 8, on $SM$, then it would mean that of all possible planes through the given direction of slip, slipping occurs on the one for which the direction of slip lies along the line of greatest slope in the slip-plane, the axis of the specimen being supposed vertical.

Metallurgists are familiar with the conception of a plastic material which yields by slipping or shearing on a plane parallel to the plane of maximum shearing stress, and they have recently become familiar with the conception of a plastic crystal which yields by slipping on a crystal plane and in the direction of a crystal axis, the choice among crystallographically similar types of slipping being determined by the stress distribution. The conception now put forward is quite different from either of these. The direction of slipping is a crystal axis and the plane of slipping is determined chiefly by the stress distribution.

## Compression Experiments

Before going on to consider how this kind of plastic yielding could arise we shall describe further experiments in which the question is examined by means of a method which permits of much greater accuracy than was possible in the experiments we have so far described. The chief sources of error in the extension experiments were attributable (*a*) to the fact that the stretching of the specimens was not uniform, and (*b*) to the difficulty of determining the position of the specimen relative to the X-ray spectrometer in which it was fixed when measuring the orientation of its crystal axes. In regard to (*b*) the accuracy of the X-ray determinations relative to the spectrometer can be judged by the fact that the angle between two crystal axes never differed by more than 2° and seldom by more than 1° from the angle required by the cubic symmetry of the crystal. The accuracy of the distortion measurements can be gauged from the accuracy with which the points of the unstretched cone lie on two planes and from a comparison of the volume of the specimen before and after stretching. It is improbable that the calculated direction of slip is in error by more than 3°. On the other hand, there seems to be a possibility that an error of 5 or 6° may occur in determining the position of a tension specimen in our X-ray spectrometer. This position is determined either by rotating the specimen till the line of sight lies along one face when looking through the holes which confine the beam of X-rays, or else by adjusting the X-ray spectrometer till the axis of rotation of the table is in the plane passing through the vertical scratches on opposite faces. The accuracy of both these methods depends on the smallest dimension of the cross-section of the specimen, which, for various reasons, could not be greater than about 1·5 mm. The first method is also liable to inaccuracy owing to the rounding of the faces due to polishing and to the unevenness of the surface when the material has been stretched. The second method depends on a high degree of accuracy in the axis of the turn-table of the spectrometer. With our X-ray spectrometer this source of error was appreciable, but the measurements of $\lambda$ used in our distortion calculations were made with a first-class goniometer the axis of which was quite good enough for our purpose.

The use of compression specimens reduces these sources of error till they are less than the other errors to which both types of specimen are liable. In the first place, the diameter of the flat faces on which the measurements are made are three times as great as the thickness of the largest tension specimen we are able to use. In the second place, the conditions of the experiment kept the faces flat and parallel during compression. This last point is important because it was sometimes found possible to obtain extraordinary uniformity of distortion through the whole volume of a compression specimen. The photographs shown in figs. 2, 3 and 4 illustrate this. Fig. 2 shows a specimen before distortion marked in squares. Fig. 3 shows Fe 7*c* after compression till its thickness was *$\epsilon = 0{\cdot}897$ of its initial thickness, i.e. a compression of about 10 %. It will be seen that after compression the ruled lines remained straight. Fig. 4 shows a specimen Fe 3*c* in which the ruled scratches were so straight and so nearly parallel after compression to 84 % of its initial thickness that we were unable to detect any want of uniformity.†

In some of the compression specimens the distortion was not quite uniform, the ruled scratches becoming slightly bent or curved. In some of these cases it was possible to say that the want of uniformity arose from imperfections in the crystal which only came to light when the specimen was compressed. In all cases where there was an obvious lack of uniformity in distortion (three out of the nine

* The symbol $\epsilon$ is here used to denote a state of the material, i.e. the total amount of compression the crystal has undergone since it was formed. The symbol $\gamma$ is used to denote the ratio of the thickness at the end of any experiment to that at the beginning, so that if $t_0$ is the unstrained thickness, $t_1$ that after one compression and $t_2$ that after a second compression, $t_2$ is the value of $\gamma$ used in calculating the unstretched cone for $t_1$ the second compression. After the first compression $\epsilon = t_1/t_0$ and after second compression $\epsilon = t_2/t_0$.

† This specimen was originally circular, but three flats were ground on its curved edge for a reason which will be explained later, so that the plan form is not elliptical, as is always the case when a circular disc is compressed.

Table 1. *Measurements and data from which unstretched cones were calculated*

*Extension tests*

$\epsilon, f, g, l, m, p, q$ are non-dimensional. $\beta_0, \beta_1, \gamma_0, \gamma_1, \lambda_0, \lambda_1$ are angles, and are expressed in degrees.

Fe 1. $\epsilon=1\cdot150, f=0\cdot9894, g=0\cdot9000, \beta_0=90\cdot0°, \beta_1=92\cdot9°, \gamma_0=89\cdot9°, \gamma_1=85\cdot5°, \lambda_0=90\cdot0°, \lambda_1=84\cdot3°$.

Fe 3. $\epsilon=1\cdot0915, f=0\cdot9748, g=0\cdot9481, \beta_0=90\cdot0°, \beta_1=89\cdot8°, \gamma_0=90\cdot0°, \gamma_1=89\cdot8°, \lambda_0=89\cdot9°, \lambda_1=94\cdot0°$.
$l=-0\cdot0679, m=0\cdot9460, p=0\cdot0034, q=0\cdot0033$.

Unstretched cone in first position is

$$(0\cdot0498 \cos^2\phi+0\cdot1325 \cos\phi \sin\phi+0\cdot0992 \sin^2\phi)\tan^2\theta +(-0\cdot0074 \cos\phi-0\cdot0072 \sin\phi)\tan\theta-0\cdot1914=0.$$

In second position it is

$$(0\cdot0525 \cos^2\phi+0\cdot1508 \cos\phi \sin\phi+0\cdot1229 \sin^2\phi)\tan^2\theta +(-0\cdot0059 \cos\phi-0\cdot0062 \sin\phi)\tan\theta-0\cdot1606=0.$$

Fe 4. $\epsilon=1\cdot0879, f=0\cdot9379, g=0\cdot9856, \beta_0=90\cdot0°, \beta_1=93\cdot4°, \gamma_0=90\cdot0°, \gamma_1=90\cdot9°, \lambda_0=90\cdot0°, \lambda_1=93\cdot6°$.
$l=-0\cdot0618, m=0\cdot9836, p=-0\cdot0552, q=-0\cdot0147$.

Unstretched cone first position is

$$(0\cdot1173 \cos^2\phi+0\cdot1144 \cos\phi \sin\phi+0\cdot0275 \sin^2\phi)\tan^2\theta +(0\cdot1202 \cos\phi+0\cdot0320 \sin\phi)\tan\theta-0\cdot1835=0.$$

In second position it is

$$(0\cdot1397 \cos^2\phi+0\cdot1446 \cos\phi \sin\phi+0\cdot0381 \sin^2\phi)\tan^2\theta +(0\cdot0992 \cos\phi+0\cdot0314 \sin\phi)\tan\theta-0\cdot1551=0.$$

*Compression tests*

$X_0, Y_0, X_1, Y_1, t_0, t_1$ are expressed in millimetres. $\alpha, \beta, \gamma, l, m, \mu, \nu$ are non-dimensional.

Fe 3c. First compression from $\epsilon=1\cdot0$ to $\epsilon=0\cdot908$.
$\alpha_1=0\cdot9117, \beta_1=0\cdot9988, \gamma_1=1\cdot1022, X_0=0\cdot118, Y_0=+0\cdot111, X_1=-0\cdot069, Y_1=+0\cdot114,$
$l_1=+0\cdot0207, m_1=+0\cdot9994, \mu_1=+0\cdot0385, \nu_1=+0\cdot002, t_0=1\cdot630, t_1=1\cdot479$.

Second position of unstretched cone is

$$(0\cdot1690 \cos^2\phi-0\cdot0378 \cos\phi \sin\phi+0\cdot0008 \sin^2\phi)\tan^2\theta +(-0\cdot0702 \cos\phi-0\cdot0056 \sin\phi)\tan\theta-0\cdot2162=0.$$

Total compression from $\epsilon=1\cdot0$ to $\epsilon=0\cdot840$.

Fe 3c. $\alpha_1=0\cdot8460, \beta_1=0\cdot9954, \gamma_1=1\cdot1915, X_0=-0\cdot118, Y_0=+0\cdot111, X_1=-0\cdot078, Y_1=0\cdot116,$
$l_1=+0\cdot0472, m_1=+0\cdot9974, \mu_1=+0\cdot8417, \nu_1=0\cdot004, t_0=1\cdot630, t_1=1\cdot368$.

Second position of unstretched cone is

$$(0\cdot2843 \cos^2\phi-0\cdot0798 \cos\phi \sin\phi+0\cdot0030 \sin^2\phi)\tan^2\theta +(-0\cdot0704 \cos\phi-0\cdot0119 \sin\phi)\tan\theta-0\cdot4214=0.$$

Fe 5c. Compression from $\epsilon=1\cdot0$ to $\epsilon=0\cdot888$.
$\alpha_1=0\cdot9966, \beta_1=0\cdot8898, \gamma_1=1\cdot1268, X_0=-0\cdot042, Y_0=+0\cdot002, X_1=-0\cdot051, Y_1=-0\cdot119,$
$l_1=-0\cdot0346, m_1=+0\cdot8903, \mu_1=-0\cdot0038, \nu_1=-0\cdot0825, t_0=1\cdot475, t_1=1\cdot309$.

Second position of unstretched cone is

$$(0\cdot00677 \cos^2\phi+0\cdot0689 \cos\phi \sin\phi+0\cdot2064 \sin^2\phi)\tan^2\theta +(0\cdot0076 \cos\phi+0\cdot1466 \sin\phi)\tan\theta-0\cdot2665=0.$$

Fe 6c. Compression from $\epsilon=1\cdot0$ to $\epsilon=0\cdot902$.
$\alpha_1=0\cdot9951, \beta_1=0\cdot9078, \gamma_1=1\cdot1083, X_0=-0\cdot036, Y_0=-0\cdot171, X_1=-0\cdot045, Y_1=-0\cdot226,$
$l_1=+0\cdot0275, m_1=+0\cdot9082, \mu_1=-0\cdot0106, \nu_1=-0\cdot0243, t_0=1\cdot575, t_1-1\cdot421$.

Second position of unstretched cone is

$$(0\cdot0098 \cos^2\phi-0\cdot0548 \cos\phi \sin\phi+0\cdot1744 \sin^2\phi)\tan^2\theta +(0\cdot0211 \cos\phi+0\cdot0477 \sin\phi)\tan\theta-0\cdot2290=0.$$

Fe 7c. Compression from $\epsilon=1\cdot0$ to $\epsilon=0\cdot897$.
$\alpha_1=0\cdot997, \beta_1=0\cdot894, \gamma_1=1\cdot115, X_0=-0\cdot016, Y_0=-0\cdot027, X_1=-0\cdot025, Y_1=-0\cdot113,$
$l_1=+0\cdot0204, m_1=+0\cdot894, \mu_1=-0\cdot009, \nu_1=0\cdot072, t_0=1\cdot340, t_1=1\cdot202$.

Second position of unstretched cone is

$$(0\cdot006 \cos^2\phi-0\cdot041 \cos\phi \sin\phi+0\cdot201 \sin^2\phi)\tan^2\theta +(0\cdot018 \cos\phi+0\cdot129 \sin\phi)\tan\theta-0\cdot248=0.$$

II-2

specimens tried) the specimen was rejected after compression. In order to save the great waste of time which the determination of the crystal axes of the rejected specimens would have entailed, the orientation of the crystal axes was not measured before compression. The crystal axes of the four specimens Fe 3*c*, Fe 5*c*, Fe 6*c* and Fe 7*c* were determined after compression, and the second (distorted) position of the unextended cone was calculated in each case. The equations to the cones and also the data used in calculating them are given in table 1.

## Results of Compression Tests

Stereographic diagrams for the compression tests similar to those used in the case of extension specimens are shown in figs. 12–15. It will be seen that they give the same result as the tension tests. In each case the point $S$, which represents the

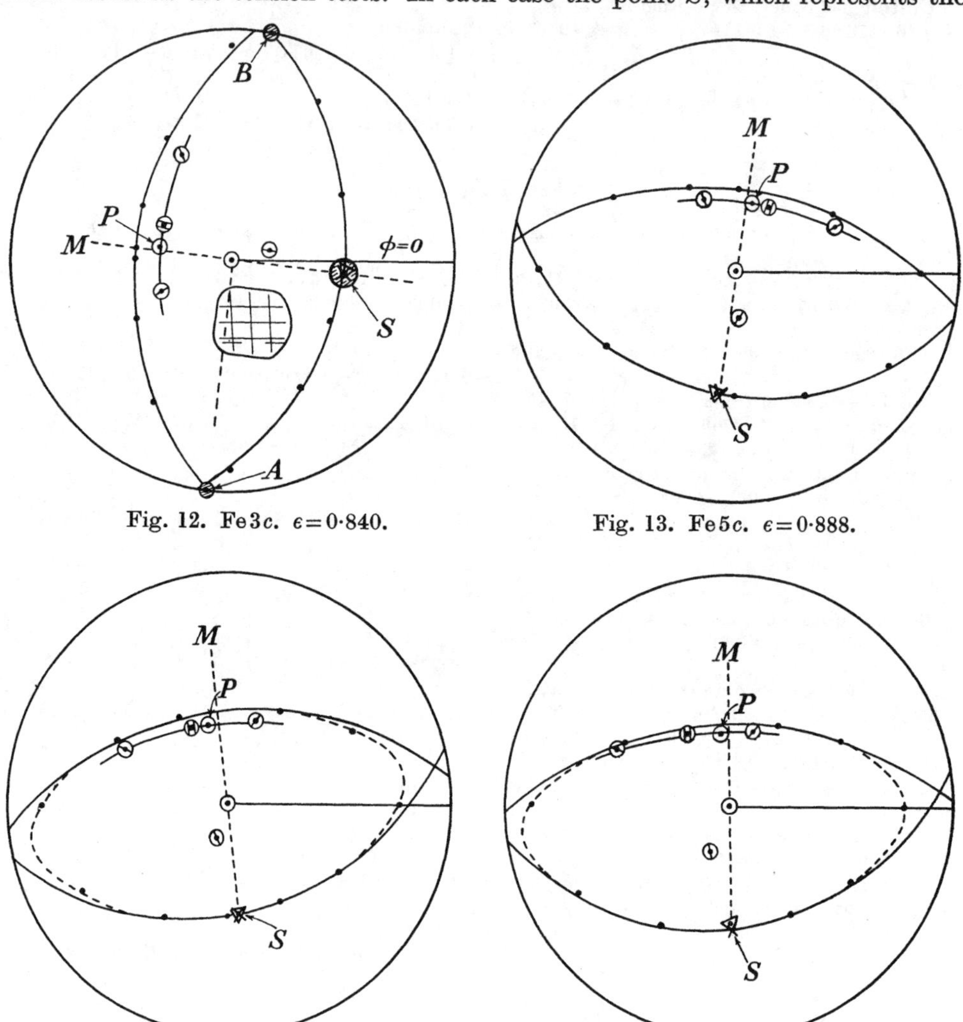


Fig. 12. Fe 3*c*. $\epsilon=0{\cdot}840$.

Fig. 13. Fe 5*c*. $\epsilon=0{\cdot}888$.

Fig. 14. Fe 6*c*. $\epsilon=0{\cdot}902$.

Fig. 15. Fe 7*c*. $\epsilon=0{\cdot}897$.

direction of slip deduced from the distortion measurements, lies very close to the triangle which marks the position of a pole of a {111} plane determined by X-rays. The improvement in the accuracy of the measurements has produced a corresponding improvement in the agreement between these two directions. This is shown in the second half of table 2, where it will be seen that the maximum angle

Table 2. *Co-ordinates of direction of slip and normal to* {111} *plane*

*Extension specimens*

| | Fe 1 | | Fe 3 | | Fe 4 | |
|---|---|---|---|---|---|---|
| | $\theta$ | $\phi$ | $\theta$ | $\phi$ | $\theta$ | $\phi$ |
| Before stretching: | | | | | | |
| Direction of slip | 61 | 116 | 48 | 233½ | 58½ | 202 |
| Normal to {111} plane | 58 | 117 | 49½ | 235½ | 62 | 205½ |
| After stretching: | | | | | | |
| Direction of slip | 52 | 114½ | 43 | 238 | 52 | 204 |
| Normal to {111} plane | 50½ | 120 | 44½ | 246 | 53 | 206 |

*Compression specimens*

| | Fe 3c | | Fe 5c | | Fe 6c | | Fe 7c | |
|---|---|---|---|---|---|---|---|---|
| | $\theta$ | $\phi$ | $\theta$ | $\phi$ | $\theta$ | $\phi$ | $\theta$ | $\phi$ |
| After compression: | | | | | | | | |
| Direction of slip | 53 | 356 | 56 | 262 | 51½ | 276 | 55 | 272 |
| Normal to {111} plane | 53½ | 355 | 56½ | 261 | 51½ | 276 | 54 | 271 |

Table 3

| | (degrees) | Angle between pole of {112} plane and pole $P$ of slip-plane (degrees) | Angle between $P$ and axial plane $SM$ through direction of slip (degrees) |
|---|---|---|---|
| Fe 3c | 53 | 9½ | 2 |
| Fe 5c | 56 | 8 | 3 |
| Fe 6c | 51½ | 8 | 4½ |
| Fe 7c | 55 | 14 | 4 |

between them is only 1½°. Figs. 12–15 and the third column of table 3 show also that the plane of slip is not a definite crystal plane, but that, as in the case of the tension specimens, it is nearly perpendicular to the vertical plane which passes through the direction of slip and the axis of the specimen. In the diagrams this plane is represented by a dotted line $SM$. It will be seen that for each of the four compression specimens the pole of the plane of slip is inclined through a small angle away from the plane $SM$ towards the pole of the nearest {112} plane and therefore away from the pole of the nearest {110} plane. The measured values of this small angle are given in column 4 of table 3. Its maximum value is only 4½°, but, on the other hand, an error of this magnitude could not have occurred in our measurements, so that the effect represented by it must be a real one.

The conclusions to be drawn from these results are that slip can occur on any plane, not necessarily a crystal plane, which passes through the normal of a {111} plane. The fact that the pole of the slip-plane is so close to the plane *SM* seems to show that the resistance to shear does not vary much, as the plane of slip takes up different positions round the pole of the {111} plane. The fact that the pole of the plane of slip is inclined through a small but measurable angle away from the plane *SM* and towards the pole of the nearest {112} plane seems to indicate that the resistance to slipping is rather less on planes near {112} planes than it is on those in the neighbourhood of {110} planes.

## Mechanism of Slip in Iron Crystals

The question now arises how it can happen that a material can slip parallel to a crystal axis but on a plane which is related to the direction of stress rather than to the orientation of the crystal axes. Where slipping occurs on a crystal plane, as it does in all metals which have been examined in the past, the distortion may be represented by a model consisting of a pack of cards capable of sliding over one another, the cards being ribbed or grooved so that they can only slide on one another in one direction. A corresponding model for representing the distortion of iron crystals might consist of a bundle of rods or pencils, and in order that there may be three-fold symmetry about their axes they might be hexagonal in section.*

Another way in which we could conceive the distortion of iron taking place is by slipping on two crystal planes which both pass through the normal of the {111} plane, the direction of slip on each of them being parallel to their line of intersection. Such planes might be two {110} planes or two {112} planes. By adjusting the ratio of the amount of slip on one plane to that on the other the total distortion can be equivalent to slip on any given plane through their line of intersection.

It is impossible to distinguish between these two hypotheses by distortion or X-ray measurements. On the other hand, an examination of the slip lines which appear on a polished surface of the crystal when it is strained seems to furnish the clue required. In each of the tensile specimens these slip lines appeared. They were not straight, but, on the other hand, they appeared to run in a definite direction, and measurements of this direction could frequently be repeated in different parts of the same face with a probable error of about 2°. Plate 1*a*, *b* and *c* show typical examples of the lines in parts of the specimens where they are free from imperfections in the crystal. In many parts these imperfections make measurement of slip lines impossible, but some reasonably good measurements were obtained in the case of crystals Fe 3 and Fe 4. The orientation of the slip lines on two faces of Fe 4 and on two faces and two extra flats ground on the corners in Fe 3 are represented by the symbol ⊛ in figs. 9 and 11.† It will be seen that they lie close to the plane of slip in each case. In table 4 are given the measured values of the angle between

* Such a model cannot without modification fully represent the crystal because a cubic crystal like iron has no hexagonal axis of symmetry, but only trigonal, i.e. axis through diagonal of cube which is threefold. For the present purpose, however, this point is immaterial.

† These two flats are not those referred to later, p. 170.

the axis of the specimen and (*a*) the slip lines, (*b*) the trace of the plane of slip determined by distortion measurements on the face of the specimen, (*c*) the trace of the nearest {110} plane, (*d*) the trace of the nearest {112} plane, (*e*) the trace of the second unstretched plane determined by distortion measurements. It will be seen that the maximum angle between the slip lines and the trace of the plane of slip is 4°. This is less than the errors of measurement in the case of slip lines on the face of tension specimens. On the other hand, the maximum angle between slip lines and the traces of crystal planes are 31° in the case of the nearest {110} plane and 26° in the case of the nearest {112} plane.

Table 4

| | Fe 3 | | Fe 4 | |
|---|---|---|---|---|
| | Face 1 (degrees) | Face 2 (degrees) | Face 1 (degrees) | Face 2 (degrees) |
| Inclination to axis of specimen of: | | | | |
| (*a*) Slip lines | 125 | 127 | 124 | 105 |
| (*b*) Trace of plane of slip* | 121 | 131 | 124½ | 109 |
| (*c*) Trace of nearest {110} plane | 126 | 115 | 133 | 74 |
| (*d*) Trace of nearest {112} plane | 99 | 132 | 123 | 109 |
| (*e*) Trace of second unstretched plane* | 61 | 50 | 37 | 62 |

* From distortion measurements.

It appears, therefore, that the slip lines mark the intersection of the plane of slip determined by distortion measurements with the surface of the specimen. They have no relation to the crystal planes except, as in the case of Fe 4, where the plane of slip happens to coincide with a crystal plane owing to an accidental element of symmetry in the original orientation of the crystal axes in the specimen. It will be seen later that this result is confirmed with considerable accuracy in the compression experiments.

It is worth noticing that the slip lines confirm our identification of one of the two unstretched planes determined by distortion measurement as the plane of slip. This is shown clearly by the figures in rows *a*, *b* and *e* in table 4, where row *e* shows the position of the trace of the second unstretched plane on the surface of the specimen. It seems, therefore, that if we had no X-ray measurements, we should still be able to make our choice by observing slip lines. In fact, the information furnished by rough indications of slip lines on the curved surfaces of the compression discs were used in our later measurements to guess which of the two planes was the slip plane and so to limit the range of setting angles employed during our search for positions of the specimen in which X-ray reflections could be obtained. Much labour was saved in this way, the search for these positions being very laborious.

## Interpretation of Results

We are now in a position to interpret our results. In the first place, we can reject the hypothesis that the distortion is due to slipping on two or more crystal planes passing through the normal of a {111} plane. This hypothesis would explain the nature of the distortion and the position of the crystal axes in relation to the slip phenomena, but it is contrary to previous experience with other metals. In the cases of aluminium, copper, silver and gold,* planes of slipping which are crystal planes have been proved to exist when a bar of the material is stretched. The material slips on one crystal plane only till the change in orientation of the crystal axes, as the distortion increases, bring another crystallographically similar plane into such a position that the two planes are symmetrically placed with respect to the direction of stress. Slipping then occurs on both planes. In the case of iron, the two planes of slip required in the hypothesis we are discussing would not, except in special accidental circumstances, be similarly orientated in relation to the axis of the specimen, so that the shearing stress on one plane would be different from that on the other. We should, therefore, expect slipping on one plane only, namely, that for which the shear stress was the greatest.

On occasions when slipping on two planes has been proved to exist, i.e. when the two planes are symmetrically placed, we have sometimes been able to observe slip lines, and in each case there have been two sets of lines crossing one another. Observations of this kind have also been made in the case of single crystals under alternating stresses.†

Among our iron crystals we had one example in which the axis of the specimen lay very nearly in a $\{100\}$ plane so that the normals of two $\{111\}$ planes, i.e. two possible directions of slip, were equally inclined to the axis of the specimen. On stretching this crystal two sets of slip lines appeared crossing one another. It appears therefore that double slipping occurs in iron when the shear stress is equal on two possible planes of slip, and that when it does occur two sets of slip lines appear on the surface. In all cases where the crystal axes were not symmetrically placed with respect to the axis of the specimen, only one set of slip lines was observed, and that set did not coincide with any crystal plane.

The weight of evidence is therefore strongly against the hypothesis that the distortion is due to slip on two crystal planes.

There remains the hypothesis that the crystal does not divide itself into sheets when a shearing stress is applied, but into rods or pencils. It is clear that any uniform distortion of such a system due to slipping of the rods on their neighbours must be a uniform shear with a plane of slip which contains the direction of the length of the rods, but may be in any orientation round this direction. If these rods were of molecular dimensions and if the shear were uniform right down to molecular dimensions, so that every rod bore exactly the same relation to its neighbours as every other rod, any marks or slip lines which could appear on the surface of a strained crystal would depend on the shape of the rods and their relative positions, not on the direction of the plane of slip. This can be easily under-

* The case of aluminium is treated in our Bakerian Lecture, loc. cit. (paper 5 above). The cases of copper, silver and gold which have lattices similar to aluminium will be treated in a paper by one of us which will be published shortly.

† Gough, Hanson and Wright, *Phil. Trans.* (1925).

PLATE 1

(a) × 100

(b) × 45

(c) × 350

PLATE 2

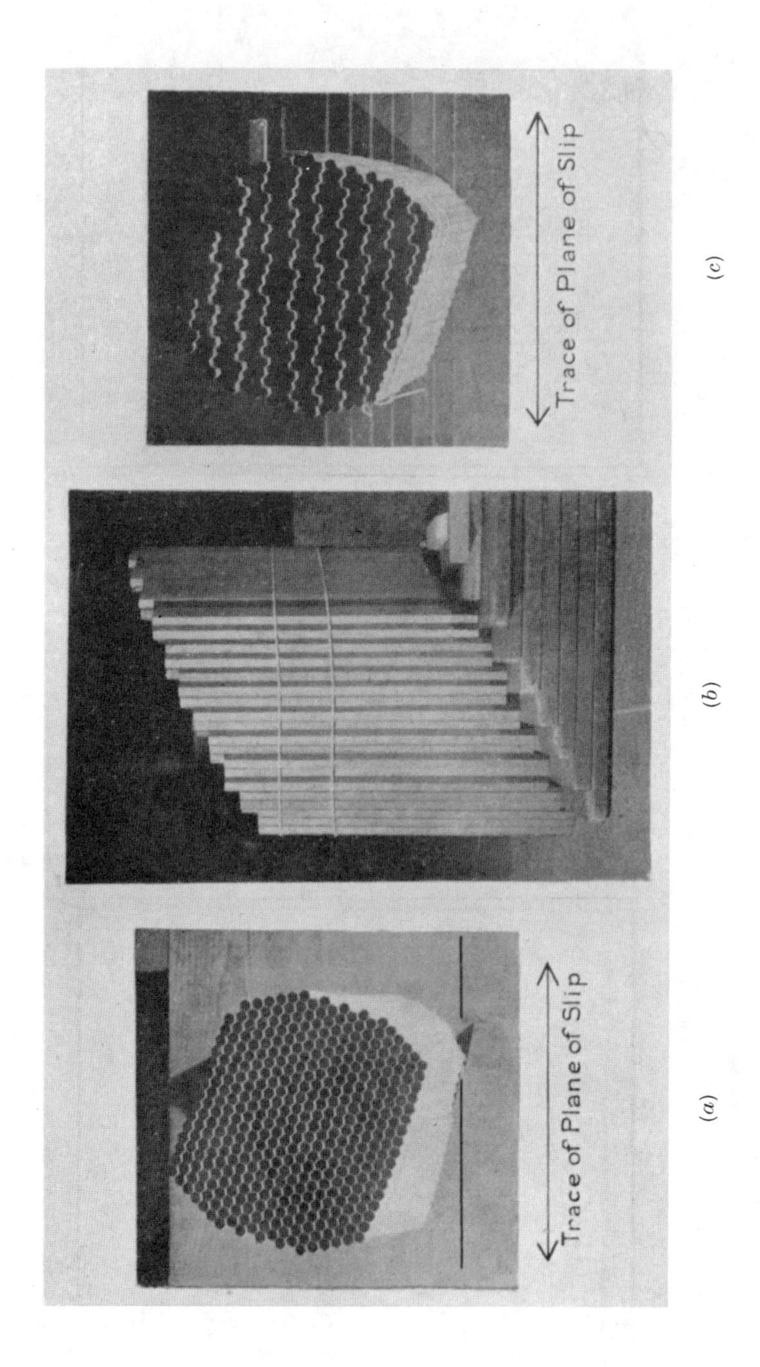


(a) (b) (c)

PLATE 3

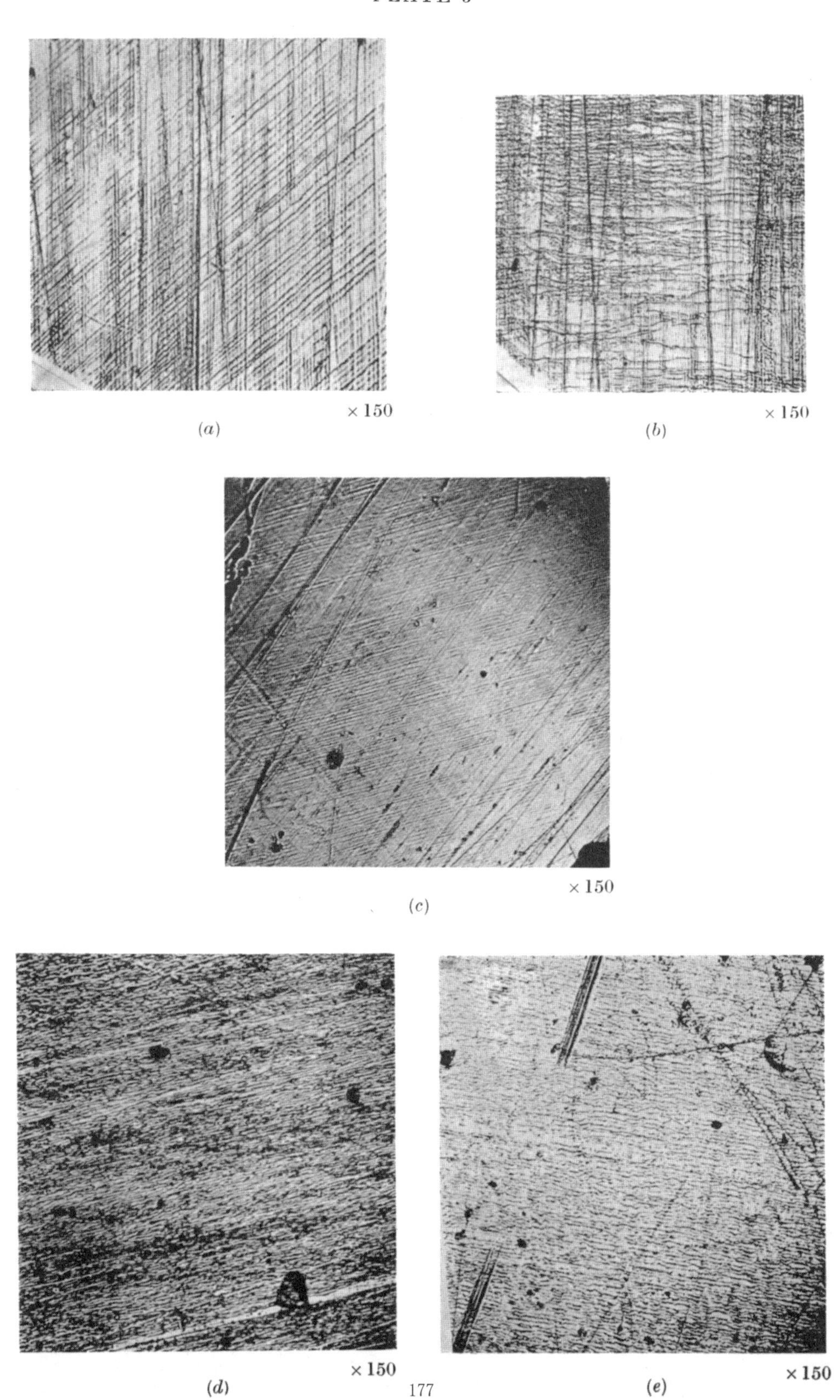

stood by thinking of a model. Plate 2*a* is a photograph, taken from above, of a bundle of hexagonal rods standing with their axes vertical. These rods pack together so as to fill a volume without interstices. The ends of the rods are all cut square (i.e. at right angles to their axes), and before distortion they were all at the same level so that the end of the bundle was a plane perpendicular to the axes of the rods. Owing to the fact that the rods were accurately made, the outline of the hexagonal ends could not be seen before distortion. The bundle was given a uniform shear by sloping the board on which it was standing. The direction of the trace of the slip plane on the plane of the ends of the rods is shown as a line below the photograph. The whole bundle was lighted obliquely so as to show up the projecting parts of the surfaces of slip. It will be seen that the 'model slip lines' which appear on the surface are traces of crystal planes on the plane end of the bundle. The orientation of the plane of slip which would be determined by external distortion measurement may affect the relative brightness of these 'model slip lines', but not their direction.

As a matter of fact, the distortion of which plate 2*a* is a model cannot occur because it would involve an alteration in the spacing of the atoms in the crystal lattice, which X-ray analysis shows does not occur. In any distortion large bundles of rods must stick together. When they slip they must slip at least one atomic distance along the direction of slip—they may slip much more.

In the model one atomic distance is represented by a given fraction of the diameter of each rod. In order, therefore, to see what the model looks like when it is forced to shear in such a way that the slip of each rod on its neighbour is one atomic distance at every point where there is any slip, the sloping board which produced the effect shown in plate 2*a* was replaced by a pile of drawing-boards the thickness of each of which represented one atomic distance. These were arranged in steps so that the slope of the plane which touched the edges of all the steps was the same as that used in the first experiment. The model was then placed on the steps with the rods vertical and they were pressed down as far as they would go. Plate 2*b* shows a side view of the model in this position.

Plate 2*c* is a top view of the bundle of rods in the same position as they occupied when the photograph for plate 2*b* was taken. The point of view, the method of illumination, the amount of shear and the orientation of the plane of slip are identical with those applicable to plate 2*a*. It will be seen that the 'model slip lines' are very different from those shown in plate 2*a*. In general they run parallel to the trace of the plane of slip which is marked below the photograph. Neglecting the details of their structure in general direction, they are straight lines which have no connection with crystal planes, but when examined in detail they have a jagged appearance. It has been known for a long time that slip lines on the surface of iron crystals are not straight in detail. We have found that when single crystals are strained so that the shear is uniform throughout a finite volume, the general direction of the slip lines can be found with fair accuracy, and this direction is constant over the whole of a flat surface ground and polished on the outside of a crystal. We have already shown that this direction coincides with the trace of the plane of slip derived from distortion measurements. It appears, therefore, that our model does in fact reproduce with remarkable accuracy all the facts connected with the distortion of iron crystals which we have brought forward.

### Prediction of Variation in Character of Slip Lines with Orientation of Plane of Slip to Surface of Specimen

The success of our model in fitting together the known facts about slipping in iron crystals naturally led us to consider whether it would enable us to predict any hitherto unknown properties of iron. The prediction which seemed most suitable for immediate verification was a possible variation in the character of the slip lines with changes in the orientation of the surface of the specimen in relation to the direction of slip. It will be seen that whatever the form of the bounding surface of the individual rods of the model may be (in our model they were hexagonal, but that is not essential) the jagged elements of the lines of slip will flatten out and approximate to the general direction of the slip lines, which will therefore become almost straight, when the surface of the specimen is nearly parallel to the direction of slip. In the limit when the direction of slip is in the surface of the specimen, the slip lines should appear quite straight provided that the surface is flat and that the lines can be made visible. Conversely, the most jagged slip lines may be expected when the surface of the specimen is most nearly at right angles to the slip lines.

In order to test this prediction one of the extension specimens, Fe 3, for which the orientation of the direction of slip had already been worked out from distortion measurements, was taken, and two of its edges were ground down in such a way that two new faces were formed. Both were parallel to the axis of the specimen and one of them was also parallel to the direction of slip. The second was perpendicular to the first, so that it made the greatest possible angle with the direction of slip. The specimen became therefore a prism whose section was an irregular hexagon. Referring to fig. 9, it will be seen that the dotted line $SM$ represents the orientation of the first of these new faces. This line makes an angle of 57° with face 1, which corresponds with the axis $\phi = 0$. Accordingly the new faces were ground at angles of 57° and 147° with face 1. The faces were then polished and the specimen again extended. Plate 3*a* shows the appearance of the slip lines on the new face cut parallel to the direction of slip, while plate 3*b* shows the slip lines in the face cut at 147° to face 1. It will be seen that our prediction is verified in a most striking manner. The slip lines in plate 3*a* are remarkably continuous; they are also straighter than any others we had obtained before, as can be seen by comparing plate 3*a* with plate 1*a*, *b*, *c*. The vertical lines in plate 3*a* and *b* are due to irregularities in the surface of the specimen which we attribute to somewhat inexpert polishing. It will be seen that the small wobbles in the slip lines shown in plate 3*a* are associated with these grinding and polishing marks. If the surface had been flatter, the slip lines would have been even straighter.

The slip lines shown in plate 3*b* are, as we predicted they should be, more jagged or curved than any of the others. In order to confirm this result we ground new faces on another specimen—this time a compression disc, Fe 3*c*. The specimen was first compressed to $\epsilon = 0{\cdot}908$. The distortion at this stage was calculated and the direction of slip in the distorted specimen found. Its co-ordinates were $\theta = 52°$, $\phi = 354°$. Three flat faces were then ground on the curved surface of the specimen so that they were all parallel to the normal to its upper and lower faces. Two of them were parallel to the direction of slip, i.e. they were parallel to the plane $\phi = 354°$. The third was cut at right angles to these so that its equation was $\phi = 84°$. These extra flat faces are seen edge-on in fig. 4, and the relation between them and the co-ordinate axes is shown in fig. 12, where an outline copy of fig. 4 is shown in position on the stereographic diagram representing the distortion of Fe 3*c*.

The specimen was next compressed to $\epsilon = 0{\cdot}840$. Plate 3*c* shows the appearance of the surface ground parallel to the line of slip. Some difficulty was experienced in showing any slip lines at all on this surface and careful arrangement of the top illumination was necessary. It will be seen that the slip lines are extraordinarily straight. Their direction can be measured to 0·5°. The inclination of the slip lines to the trace of the flat top of the specimen was found to be 37°, so that their inclination to the normal to the flat top was 53°. This direction is shown in fig. 12 as the shaded circle *S*. The inclination of the direction of slip is given in table 2. It is 53°, so that there is perfect agreement between the slip lines and the trace of the slip plane.

Plate 3*d* shows the slip lines on the flat face $\phi = 84°$. It will be seen that they are jagged or curved as our theory led us to predict, but that it is possible to measure their general direction with a probable error of about 4°. This direction happened to be parallel to the trace of the flat top of the specimen. Its position is represented by the point *A* in fig. 12.

Plate 3*e* shows the slip lines which appear on the top surface of the specimen (the surface which had been in contact with the sheet plates during compression). It will be seen that here again they are very much bent, but that their general direction can be determined with a probable error of about 3°. This direction, which was measured as $\phi = 80°$, is marked on the stereographic projection of fig. 12 at *B*. Again, the plane of slip coincides to the limit of accuracy of our measurements with the general direction of the slip lines.

## Conclusion

The final result of this work is to show that the mechanism of distortion in iron crystals is subject to laws which are quite different from those which govern the slip phenomena in any metal hitherto investigated. The particles of the metal stick together along a certain crystallographic direction and the resulting distortion may be likened to that of a large bundle of rods which slide on one another. The rods stick together in groups, or smaller bundles of irregular cross-section; and the slip lines which appear on a polished surface are the traces of these bundles on that surface. When the distortion of the crystal in bulk is a uniform shear these bundles stick together to form plates of irregular thickness, but lying in general with their planes parallel to the plane of slip determined by external measurements of the surface. The plane of these plates is determined by the direction of the principal stress. It has no direct relationship with the crystal axes.

This conception was arrived at entirely as a result of external measurements of the specimens and measurements of the orientation of their crystal axes. The fact that it appears to explain the nature of the slip lines is therefore remarkable. The slip lines, which are curved in detail, preserve a general direction which can be measured in cases where the distortion is uniform, and this direction coincides with the trace on the polished surface of the specimen of the plane of slip determined by external measurements. The slip lines appear to have no direct relationship with any of the principal crystal planes.

Perhaps the most telling point in favour of the theory is that it has enabled us to predict the hitherto unknown fact that if the crystal is cut with a polished surface parallel to the direction of slip the slip lines are all straight. When there is an appreciable angle between the polished surface and the direction of slip the slip lines are jagged or curved, and the greater this angle the more jagged they become; but even so they preserve a general direction which is easily measured and is in agreement with the distortion measurements.

### Bearing of these Conclusions on Previous Work

This completes the work up to a definite stage. Before concluding, it may be of interest to bring out the connection between our results and those of previous workers. Nearly all our predecessors* have assumed that the crystal has a plane of slipping which is a crystal plane, and in most cases they have attempted to correlate the slip lines with traces of crystal planes. If the work described in this paper is accepted, it is clear that that method was foredoomed to failure. One notable exception, however, is to be found in the work of Osmond and Cartaud,† who point out that among the slip lines which occur when an iron crystal is strained, curved ones predominate. They were unable to find any relationship between these curved lines either as a whole or in their detail and the crystallographic planes. In order to produce slip lines which are straight as a whole, though curved in their details, it is essential in the light of our work here described to subject a single crystal to a *uniform* distribution of stress. It was only after a careful study of the methods by which a uniform strain can be produced that we were able to discover the relationship between slip lines and distortion.

In conclusion, we should like to express our thanks to Sir Ernest Rutherford for

* Rosenhain and Ewing, *Phil. Trans.* A, XCIII (1900), 353; Howe, *Metallography of Steel*; Polanyi, *Z. Kristallog.* LXI (1925), 49; Weissenberg, ibid. p. 58; Mark, ibid. p. 71.

† Osmond and Cartaud, *J. Iron Steel Inst.* no. 111 (1906).

allowing the work to be carried out in the Cavendish Laboratory, to Professor Carpenter for the use of his laboratory in which all the micro-photography and the X-ray measurements connected with the tension specimens were done, to Dr A. Hutchinson for allowing us to use several of his crystallographic measuring instruments, and to Professor Inglis for allowing us the continuous use of a Buckton compression machine in his laboratory. We also wish to express our gratitude to Professor Edwards and Mr Pfeil for supplying the large crystals of iron with which the work was carried out, and to Mr W. S. Farren for much help in designing special apparatus for marking, holding and measuring our specimens.

## DESCRIPTION OF PLATES

### Plate 1

(*a*) Slip lines on polished surface of iron crystal. Magnification 100. The lines parallel to the broad black line which is a scratch made by a razor blade are marks made in polishing.
(*b*) Slip lines. Magnification 45. Polish marks vertical.
(*c*) Slip lines (horizontal in photograph). Polish marks at angle of about 60° to slip lines. Magnification 350.

### Plate 2

(*a*) Top view of model, showing appearance where no rods stuck together during distortion.
(*b*) Side view of model, showing hexagonal rods standing vertically on slope of steps.
(*c*) Top view of model, showing appearance when rods slip a definite amount or not at all.

### Plate 3

(*a*) Slip lines on face of Fe 3 cut parallel to direction of slip. Magnification 150.
(*b*) Slip lines on face of Fe 3 cut at maximum possible angle with direction of slip. Magnification 150.
(*c*) Slip lines on vertical face of Fe 3*c* cut parallel to direction of slip. Magnification 150.
(*d*) Slip lines on vertical face of Fe 3*c* cut at maximum angle (53°) to direction of slip. Magnification 150.
(*e*) Slip lines on flat top of iron crystal. These were not actually taken on Fe 3*c*, because the specimen got scratched too much, but they were taken on another almost identical specimen. Angle to direction of slip, 37°. Magnification 150. Slip lines parallel to bottom of photograph.

# 11

# THE DISTORTION OF SINGLE CRYSTALS OF METALS

REPRINTED FROM

*Proceedings of the 2nd International Congress for Applied Mechanics (Zürich, 1926).* Zürich: Orell Füssli (1927), pp. 46–52

Our knowledge of the distortion of metallic crystals began, I think, with the work of Rosenhain and Ewing, who showed that when certain metals are strained and the polished surface examined under a microscope a number of straight lines appear. These lines appear to consist of steps, and the authors came to the conclusion that they represent the traces of crystal planes on the surface of the metal and that they are due to slipping on those planes. Very little advance beyond this was possible, however, till the discovery of methods for finding, by means of X-rays, the orientation of crystal axes in grains of metal which have no crystal faces but are crystals as far as their internal structure is concerned.

Simultaneously with the development of systems of analysis by X-rays there has been a great development during the last few years of methods for producing large metallic crystals. Methods of slow crystallization from the molten state and of crystallization on to a nucleus have been developed in Germany, America, Russia and other countries, while methods of growing large crystals from small ones in the solid state have been developed chiefly in England. As a result of these researches we are now in a position to develop the theory of distortion of metals to an almost unlimited extent. For this reason it seems opportune at the present time to discuss the methods that may most usefully be employed in the solution of problems of this type. This is specially so because the problems are essentially mathematical in character and most of the workers in this field are experimental physicists.

The simplest elements of the problem are the following: A piece of metal which consists of a single crystal throughout its volume is subjected to a given distortion. Its shape and size can be measured before and after distortion and the orientation of its crystal axes determined in each case. In some cases lines due to the distortion appear on the surface. It is required to find the relationship between the distortion, the crystal axes and these slip lines.

In all cases except, I believe, in the work of Miss Elam and myself on aluminium crystals,* the principle of the methods used has been to begin by finding a kind of test which gives uniform distortion and also one set of slip lines on the surface. In cases like that of tin, zinc and bismuth these conditions are satisfied by a simple

* G. I. Taylor and C. F. Elam, 'The distortion of an aluminium crystal during a tensile test', *Proc. Roy. Soc.* A, CII (1923), 643; and 'The plastic extension and fracture of aluminium crystals', ibid. CVII (1925), 28; papers 5 and 7 above.

tensile test of a uniform bar. It is then assumed that the meaning of the slip lines is known (in many cases it is indeed obvious), and that they are the traces of planes of slipping on the surface of the specimen. The next procedure is to determine the orientation of the crystal axes and to show that one of the crystal planes coincides with the plane containing the slip lines. If this is done before and after the stretching it can be proved in most cases that the crystal axes remain fixed relative to the plane containing the slip lines. Then from the external measurements of the specimen (i.e. in the case of a round bar the major and minor axes of the strained elliptic section, and the orientation of lines originally marked on the specimen in a plane perpendicular to the axis) one can show (*a*) that the plane containing the slip lines is undistorted by the strain and (*b*) that the motion of the axis of the specimen relative to the plane of particles referred to in (*a*) during the strain is such as can be accounted for by slipping in a certain direction on that plane. This direction is then shown to coincide with one of the crystal axes.

By these and methods which are essentially though not in detail the same, Polanyi and his fellow-workers have obtained some very interesting results. They have shown that certain metals, notably tin, are distorted under pure tensile stress by slipping parallel to a crystal plane, in the direction of a crystal axis. These methods, however, depend on the assumption that the distortion is always due to slipping parallel to a crystal plane and that the slip lines mark out that plane on the surface of the specimen. They depend in fact on knowing the form which the answer will take before starting to solve the problem, and the work consists in verifying the correctness of the original assumptions. There are, however, many cases to which these methods do not apply. In the case of aluminium for instance, Miss Elam and I have shown that under certain definable conditions the distortion is due to simultaneous slipping on two crystal planes and she has extended that result to the cases of silver, copper and gold. Again, the case of iron has proved an insoluble problem to metallurgists who have arrived at an enormous number of mutually contradictory results because, as I shall show later, the lines of slip, though definitely related to the distortion, are related only indirectly to the crystal axes. Then again one may want to experiment with metals under conditions where slip lines are not observable, as, for instance, when aluminium crystals are tested under compression* between parallel steel plates, a type of experiment to which I shall refer later.

To-day I propose to describe the general method which I have developed for dealing with all problems of this kind. The advantage of this method lies in the fact that it gives all information which can possibly be obtained from external measurements of a specimen uniformly distorted, and that it makes no assumptions in regard to the nature of slip lines. The method consists in determining all the directions which remain unchanged in length during a uniform distortion. I will not go into the details of how this is done. It is quite simple in idea though sometimes laborious in practice. I will content myself with a general description of the method.

In order to get the greatest possible accuracy special precautions are necessary

* G. I. Taylor and W. S. Farren, *Proc. Roy. Soc.* A, CXI (1926), 529; paper 9 above.

in the experiment to ensure uniformity in the distortion. I need hardly go into the details of the experimental methods, but in order to illustrate the degree of uniformity which can be obtained, photographs* of a circular disc cut from an aluminium crystal before and after compression between parallel steel plates are shown in figs. 1 and 2. It will be seen that the squares ruled on the surface before compression become oblique parallelograms after compression. The uniformity of the distortion may be judged by the remarkable straightness of the ruled lines in the distorted specimen.

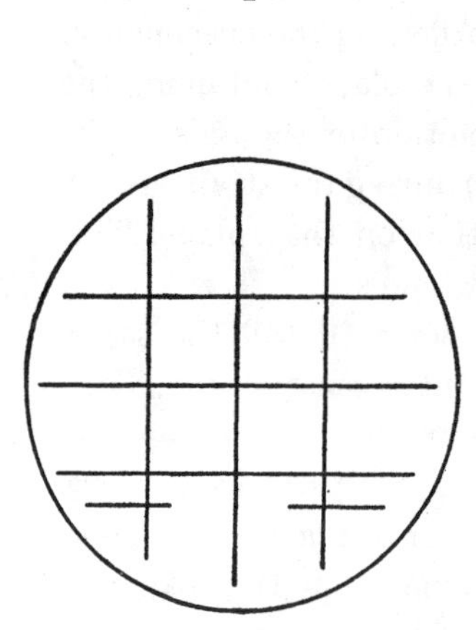

Fig. 1. Disc cut from an aluminium crystal marked in squares ready for compression.

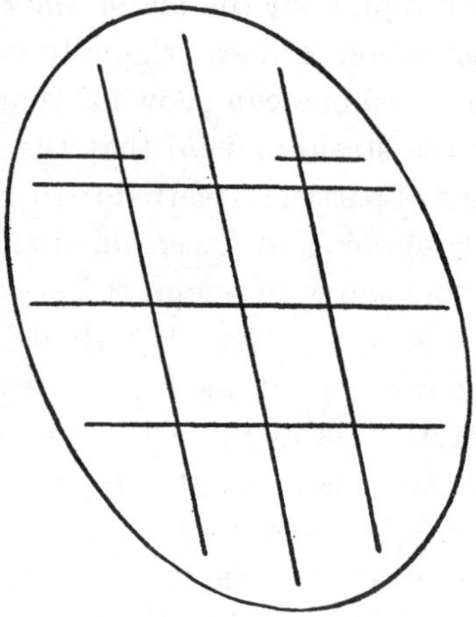

Fig. 2. Similar disc similarly marked, photographed after compression.

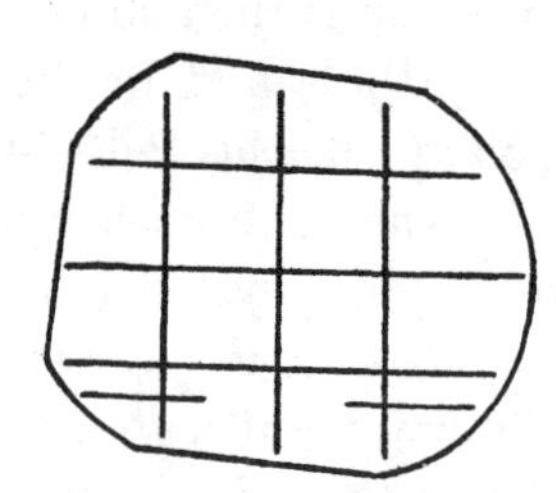

Fig. 3. Disc cut from iron crystal originally marked in squares, appearance after compression.

Fig. 3 is a photograph* after distortion of a flat piece cut from an iron crystal and compressed between parallel plates. This piece was originally marked in squares with the same pattern as the aluminium specimen shown in fig. 1. It will be seen that the uniformity of distortion is remarkable.

From measurements of the shape and size of similar elements in the material before and after distortion it is a simple matter to obtain formulae connecting the positions of a particle before and after straining. For this purpose rectangular co-ordinate axes are chosen in some convenient manner; thus in the case of compression specimens marked as shown in fig. 1, the axis of $z$ may be taken perpendicular to the surface of the specimen. The axis of $x$ may be parallel to one set of ruled scratches and the axis of $y$ also in the surface of the specimen but perpendicular to the axis of $x$ so that before distortion it is parallel to the second set of scratches. With this choice of axes the formulae connecting the co-ordinates $(x_1, y_1, z_1)$ of the particle in the distorted material with $(x_0, y_0, z_0)$ the co-ordinates of the same particle before distortion are:

$$\left.\begin{aligned} x_1 &= ax_0 + ly_0 + \mu z_0, \\ y_1 &= my_0 + \nu z_0, \\ z_1 &= \gamma z_0. \end{aligned}\right\} \qquad (1)$$

* Reproduced here as line drawings.

PLATE 1

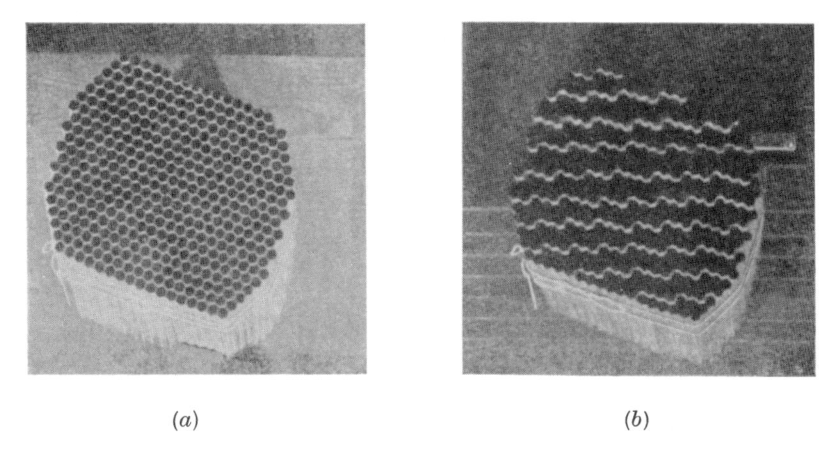

(a) (b)

Photographs of model to represent slipping in iron crystals.

In these formulae $\alpha$, $l$, $\mu$, $m$, $\nu$, $\gamma$ are found by simple formulae from measurements of the specimen.

It is now a simple matter to calculate the equation to the conical surface on which all unstretched directions lie. The condition that the distance of the particle $x$, $y$, $z$ from the origin, is unchanged by distortion is

$$x_0^2+y_0^2+z_0^2=x_1^2+y_1^2+z_1^2. \tag{2}$$

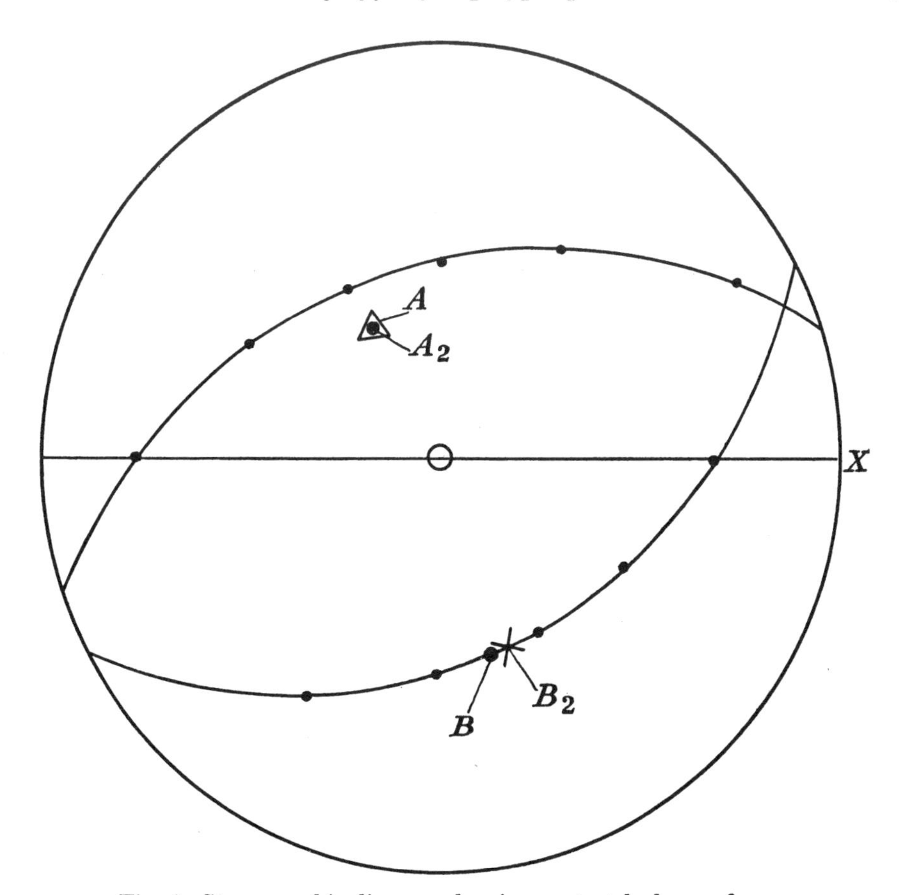


Fig. 4. Stereographic diagram showing unstretched cone for aluminium under compression.

Substituting for $x_1$, $y_1$, $z_1$ from (1) and (2) we get as the equation to the unstretched cone in its first position, i.e. its position before distortion,

$$(\alpha^2-1)x^2+(m^2+t^2-1)y^2+(\gamma^2+\mu^2+\nu^2-1)z^2+2xy(\alpha l)+2zx(\alpha\mu)+2yz(l\mu+m\nu)=0.$$

It will be seen that the unstretched cone is necessarily a cone of the second degree. In order to connect the position of this cone with that of the crystal axes (the orientation of which was measured in our experiments by X-rays), a stereographic diagram may be used. Fig. 4 is a diagram of this type. It shows the projection of the unstretched cone in the case of a compression experiment with a specimen cut from the same aluminium crystal as that shown in figs. 1 and 2. The calculated points are marked by round dots. This figure is characteristic of all such figures derived either from compression or extension experiments, except when the crystal axes happen to be in certain symmetrical positions with regard to the principal direction of stress.

On examining figures of this kind with a stereographic net it is found that the points of the unstretched cones lie almost exactly on two planes, shown in fig. 4 as two circular arcs. Simple considerations show that a uniform shearing or slipping of the material parallel to either of these planes and in a direction at right angles to their line of intersection would give rise to an unstretched cone of this kind.

At first sight it may seem a defect in the method that two alternative positions for the plane of slip are determined, but a moment's consideration will show that every conceivable method which attempts to find the plane of slip from external measurements of two states of the material only must suffer from this disadvantage.

There are now three ways in which we can find out which of these two planes is the plane of slip.

(*a*) We can determine the crystal axes before and after straining and at the same time calculate the two corresponding positions of the unstretched cone. If it is found that the crystal axes remain fixed relative to one of the planes and not relative to the other, the slip-plane is clearly the one with respect to which they remain fixed. In addition to this the plane with respect to which the crystal axes remain fixed is usually, as in the case of aluminium, a crystal plane. In fig. 4 the pole of the slip-plane is marked at $A_2$, a (111) plane determined by X-rays at $A$ the direction of slip at $B_2$ and the nearest (110) direction at $B$. It will be seen that this method of representation brings out very clearly the accuracy with which, in the case of aluminium under compression, the slip-plane and direction of slip agree with a crystal plane and a crystal axis.

(*b*) We can examine the slip lines, if there are any. In many cases they are the traces of the slip-plane, and even if they are not marked clearly enough to permit of measurement they may indicate which of the two alternative planes to choose.

(*c*) We may carry out the distortion in two stages, so as to obtain measurements for three states of the material: (1) unstretched, (2) after first distortion, (3) after second distortion. We can then work out the second position $A$, of the unstretched cone for the transition from states (1) to (2), and the first position $B$, of the cone for the transition from states (2) to (3). Both these cones exist in the material in state (2). It is found that one of the planes in $A$ coincides in position with one of the planes in $B$. This is the slip-plane because it remains unstretched during both distortions. The other plane in $A$ does not coincide with the corresponding plane in $B$, that is to say, the second plane of particles which is unstretched by the first distortion does not remain unstretched in the second distortion. Hence from the external measurements alone we can determine which of the two parts of the unstretched cone is the slip plane.

### Examples of the Use of the Method

Our first use of the method was to determine the plane and direction of slip for tension specimens of aluminium. In this work we made no use of slip lines which in our experiments sometimes did not appear and were very indistinct when they did. The problem, however, may also be solved by the method first described in which the answer to the problem is first assumed and then proved to be correct. The result is similar to that of other workers with other materials, the distortion

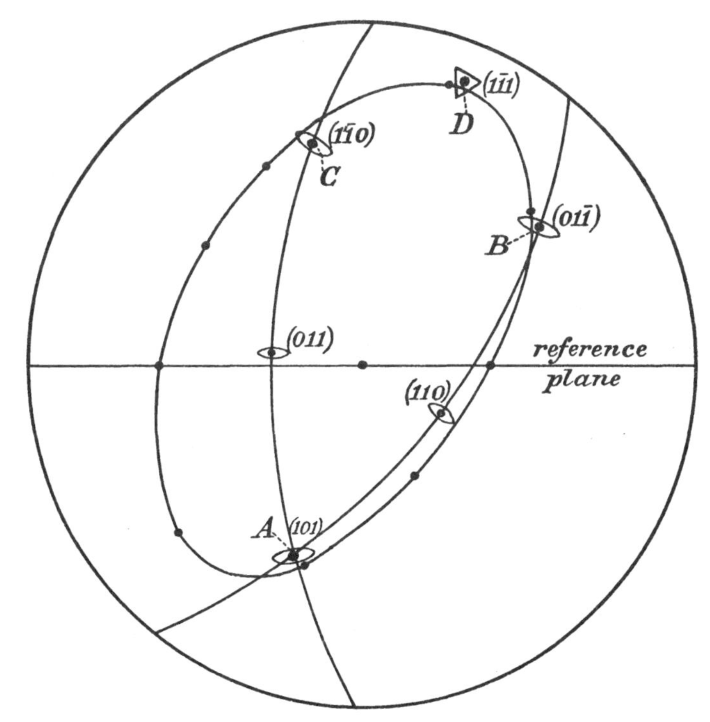


Fig. 5. Stereographic diagram showing unstretched cone for aluminium when slipping is on two planes.

being entirely due, except in special cases, to slipping on one crystal plane in the direction of a crystal axis. Experiments with compression specimens give identical results.

Since most of us are familiar with this type of distortion I will pass on to the consideration of the cases in which the crystal axes have moved, as a result of distortion, into symmetrical positions with respect to the principal direction of stress. Here there is an *a priori* reason to suppose that the material might slip on two planes at once, and an examination of the unstretched cones has enabled us to prove this to be the case.

The oval curve in fig. 5 is a stereographic projection of the unstretched cone for a small extension of a specimen when two possible slip-planes were nearly sym-

12-2

metrically placed with respect to the axis. The two slip-planes are shown as two arcs of circles $AC$ and $AB$. The unstretched cone nearly passes, as it should if the hypothetical double slipping actually occurs, through the point of intersection of the two planes. From a consideration of the nature of the distortion due to slipping through two small, but not necessarily equal shears on the two planes $AB$ and $AC$ it can be proved that the unstretched cone must pass also through three other crystal axes, making four in all. These are marked in fig. 5 as $A$ (101), $B$ $(01\bar{1})$, $C$ $(1\bar{1}0)$ and $D$ $(1\bar{1}1)$. It will be seen that the cone passes very close to all these points. By choosing different ratios of the amount of slip on the two slip-planes we can find a distortion of this type which will give rise to every possible second-degree cone which passes through the four points $A$, $B$, $C$, $D$. Hence we have a complete mathematical proof that the particular cone shown in fig. 5 can be due to simultaneous slipping on the two planes $(11\bar{1})$ and $(1\bar{1}1)$ in the directions (011) and (110) respectively.

A similar result has been obtained from compression tests, but in that case the four crystal axes are not the same as those which apply to tension tests because the crystal axes during compression move till the axis of the specimen (normal to flat faces) gets into a (100) plane whereas in extension it moves to a (110) plane.

## Distortion of Iron Crystals*

The other application of the method to which I wish to direct your attention concerns iron crystals. The problem presented by the slip phenomena in iron crystals is one which has completely baffled everyone in spite of the immense amount of work which has been done on the subject. The slip bands in iron crystals are in most cases wildly irregular, being curved, jagged and branched. On the other hand, X-ray analysis shows that the grains themselves have a quite perfect crystal structure. The photograph† (fig. 3) shows that the distortion of such a crystal may be extremely uniform. We have experimented with tension specimens 1·5 mm. square in cross-section and containing 4 cm. in the middle cut from a single crystal, and we have used the compression method employing discs 6 or 7 mm. diameter and 1·4 mm. thick. The compression specimens are susceptible of greater accuracy of measurement than the tension specimens, but both give the same result.

Fig. 6 shows the unextended cone for a tension specimen in the first position while fig. 7 shows the second, or distorted position of the same cone. Fig. 8 shows the second or distorted position of the cone for the compression specimen shown in fig. 3. It will be seen that in each case the points of the unextended cone lie almost exactly on the two planes represented by the two circular arcs in the projection. The distortion then is due to slip or shear parallel to one or other of these two planes.

In each of these cases and in seven others the crystal axes were determined

* Taylor and Elam, *Proc. Roy. Soc.* A, CXII (1926), 337; paper 10, above.

† Reproduced here as a line drawing.

by X-rays and their positions marked on the diagram. It was found that in every case one of the two possible directions of slip ($S$ in fig. 6) concided almost exactly with a (111) axis* which is marked by a triangle in figs. 6–8. The other ($D$ in fig. 6) had no connection with the crystal axes.

On examining figs. (6) and (7) it will be seen that during the distortion the crystal axes move relative to the reference axes so that they remain fixed relative to that one of the two possible slip-planes which contains the (111) axis. These two facts taken together prove that this plane can be regarded as a plane of slip.

The direction of slip is therefore a crystal axis. To find how the plane of slip is related to the crystal axes the pole $P$ of the plane of slip was marked on each diagram.

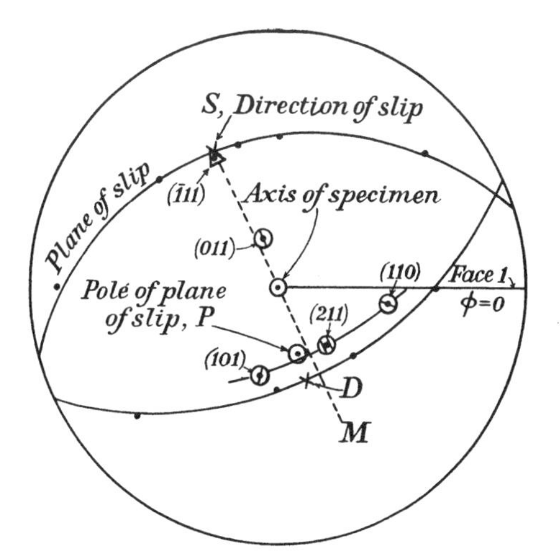


Fig. 6. First position of an unstretched cone for iron crystal, stretched.

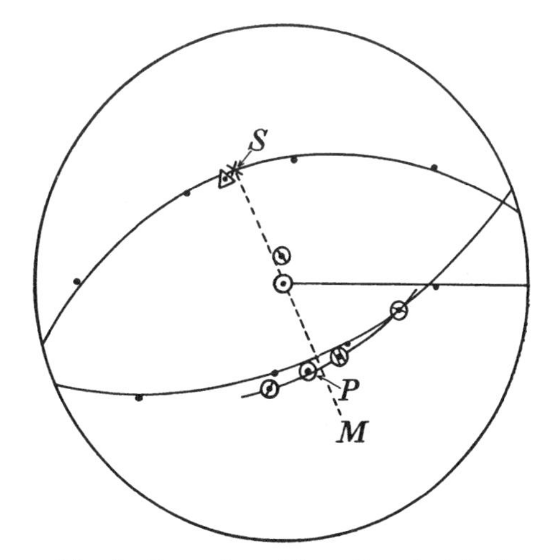


Fig. 7. Second position of unstretched cone for iron crystal, stretched.

It lies necessarily on or very close to the (111) plane because the direction of slip coincides with or is very close to the (111) axis. A portion of the (111) plane is drawn as an arc of a circle in each diagram (figs. 6–8). This plane contains three (110) and three (112) axes. The positions of the two nearest (110) axes are marked by the symbol ⦸, while the position of the nearest (112) axis is shown as ⊛ on the diagrams. It was found that $P$ could occupy any position on the (111) plane. In different specimens with different orientations of the crystal axes in relation to the principal direction of stress, the position of $P$ varied from coincidence with the (112) axis to 18° away from it. The point $P$ was in fact always close to the direction of maximum slope in the slip-plane and seemed therefore to be related chiefly to the direction of stress rather than to the orientation of the crystal axes.

The question now arises how it can happen that a material can slip parallel to a crystal axis, but on a plane which is related to the direction of stress rather than to the orientation of the crystal axes.

* The term 'axis' is used here to indicate the normal to a crystal plane.

Two theories suggest themselves immediately: (*a*) The material may slip simultaneously on two planes parallel to their line of intersection. It might for instance slip on two (110) planes or two (112) planes. (*b*) The particles of the material may stick together along the (111) directions and may not slip on crystal planes at all. The resulting distortion may be likened to that of a large bundle of rods which slide on one another. There is a strong inherent improbability that hypothesis (*a*) is true because the two slip-planes would in general have different orientations to the principal direction of stress. Previous experience with metals of cubic symmetry has shown that under these conditions slip confines itself to that plane among crystallographically similar planes, for which the shear stress is the greatest,

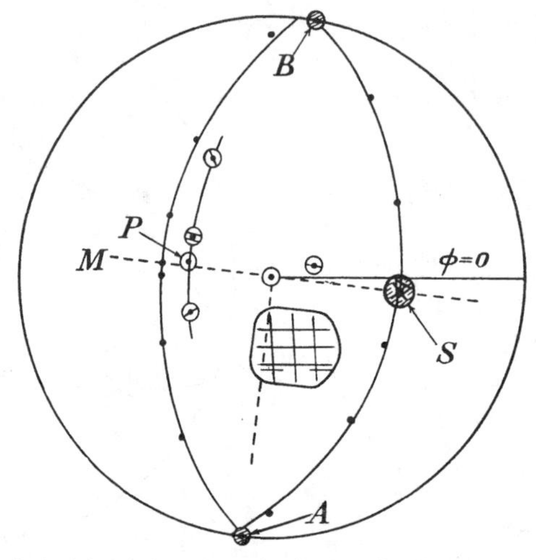


Fig. 8. Second position of unstretched cone for iron crystal, compressed.

so that if there is a tendency for the material to slip on crystal planes we should expect it to slip on one plane only. On the other hand, it is not possible to distinguish between (*a*) and (*b*) from distortion and X-ray measurements alone. Both are capable of accounting fully for all the facts so far brought forward.

Observation of slip lines, however, furnishes us with the clue required. In our experiments it was found that the slip lines were much curved or jagged, but that when the distortion is *uniform* (i.e. when straight scratches on the surface of a specimen remain straight after distortion) they preserve a general direction which is constant all over a flat surface, and can be measured in most cases with a probable error of 3°. This direction was found to correspond with the trace of the slip-plane found from distortion measurements. The error was never greater than 4° for the tension specimens and 3° for the compression specimens. These are both within the limits of our experimental error. The directions of the slip lines on the top face and on two flat faces ground on the curved edge of the compression disc shown in fig. 8 are shown in fig. 8 by means of the symbol ⊘. It will be seen that they are very close to the plane of slip. It appears therefore that the general direction of the slip lines marks out the trace of the plane of slip found in distortion measurements and has therefore no direct connection with the crystal axes.

These observations appear to offer conclusive evidence against hypothesis (*a*) because whenever we have observed slip lines on occasions when double slipping has been proved to occur, two sets have been visible crossing one another. In the case of our tests there was never more than one set in cases where distortion measurements showed a single direction of slip. Hypothesis (*b*), however, offers a complete explanation of the slip lines. Plate 1*a* shows a bundle of hexagonal rods standing vertically and photographed from above. Initially they were all standing on a horizontal board, and as they were all the same length the top surface was a horizontal plane and the traces of the faces of the rods could not be seen. In fig. 9 they have been given a shear by tilting the board on which they are standing. The general direction of the lines which appear are parallel to 'crystal planes', i.e. parallel or perpendicular to the faces of the rods. The distortion represented by this, however, is an impossible one, because if each rod is representing a molecular cell the slip of every rod on its neighbour represents a fraction of a molecular distance and the crystal lattice distances are changed, an effect which X-ray analysis proves to be unknown. Plate 1*b* represents the same model distorted in the same way as the model shown in (*a*), but the rods have been caused to stick together in such a way that they slip at least one 'molecular distance' on their neighbours if they slip at all. The trace on the horizontal plane of the top of the rods of the slip-plane which would be determined by external measurements is represented by the lines parallel to the bottom of the photograph, which will be seen at each side of the model. It will be seen that the 'model slip lines' which appear are bent or jagged, and that the jagged elements of which they are composed are directly related to the crystal axes, but that their general direction is parallel to the trace of the plane of slip determined by external measurements.

This conception of the nature of slip in iron crystals leads to a remarkable prediction, namely, that if a face is cut and polished so that it contains the direction of slip, the slip lines, if they can be made visible, will all be straight, whereas they will be more and more jagged as the surface of the specimen is more and more inclined to the direction of slip.

The compression specimen shown in fig. 3 was first compressed by a small amount and the direction of slip calculated. Three flat faces were then cut parallel to the normal to the face. These can be seen in fig. 3 and their position relative to the slip direction is shown in fig. 8, where an outline drawing of the specimen is placed in position on the diagram.

A microphotograph of the slip lines on one of the faces cut parallel to the direction of slip is shown in plate 1*c*. It will be seen that the slip lines (horizontal in the figure, the diagonal marks are due to polishing) are remarkably straight. Plate 1*d* shows a photograph of the slip lines taken on the face cut at the maximum possible angle (53°) with the direction of slip. It will be seen that they are jagged or curved, but that they preserve a general direction which can be measured with a probable error of about 3°. Plate 1 *e* shows the slip lines on the top surface of the specimen which made an angle of 37° with the direction of slip.

It is remarkable that a conception based entirely on distortion and X-ray measurements should lead to a prediction about the character of slip lines which is capable of so striking a verification.

In conclusion, I may say that I hope I have succeeded in making clear the advantages of using a general method in dealing with distortion of crystals instead of verifying particular hypotheses by special measurements.

# 12

# THE DISTORTION OF CRYSTALS OF ALUMINIUM UNDER COMPRESSION

## PART II. DISTORTION BY DOUBLE SLIPPING AND CHANGES IN ORIENTATION OF CRYSTAL AXES DURING COMPRESSION

REPRINTED FROM

*Proceedings of the Royal Society*, A, vol. CXVI (1927), pp. 16–38

Changes in the orientation of crystal axes during compression of a disc cut from a single crystal of aluminium are discussed. They are in accordance with the prediction made on assumption that the crystal slips as determined by distortion measurements. As in case of tensile test-pieces, crystal axes always take a position where two possible planes of slip are symmetrically disposed in relation to the stress, but in this case, the orientation of crystal axes relative to normal to flat surfaces of specimen is quite different from their orientation relative to axis of tensile test-piece.

After the axes have taken the symmetrical position, Laue photographs show that they remain there, even when distortion is very great. Tests were continued till thickness of specimen was only 0·28 its original thickness. Distortion during period when crystal axes remain in symmetrical position is due to slipping on two symmetrically disposed planes of slip.

The bearing of the results on structure of rolled metals is discussed and distortions of cubical blocks of material, due to various types of double slipping, are compared, to find out which is most likely to occur.

It has been shown in Part I* of this paper that the distortion of an aluminium crystal during compression between parallel plates is identical in kind with the distortion of a tensile test-piece cut from a single crystal. It has been shown also that the method previously devised for predicting the slip-plane and direction of slip in a tensile test from a knowledge of the orientation of the crystal axes applies also in the case of compression, provided that the normal to the two parallel steel plates in the one case is identified with the axis of the tensile test-piece in the other. The direction of slip in one case is, of course, exactly opposite to that in the other. One might be inclined to suppose, therefore, that the motion of the normal to the flat surface of the specimen relative to the crystal axes during compression would be exactly opposite to that of the motion of the axis of the tensile test-piece during extension, but this is not the case. The normal to the disc must move, relative to the crystal axes, in a great circle towards the pole of the slip-plane, because the trace of the slip-plane on the surface of the specimen remains fixed relative to the crystal

* Taylor and Farren, *Proc. Roy. Soc.* A, CXI (1926), 529; paper 9 above.

axes. The normal to the specimen is necessarily perpendicular to this trace; it can, therefore, only move on a great circle relative to the crystal axes, and this great circle passes through the pole of the slip-plane because that also is perpendicular to the trace.

To verify this, discs cut from specimen no. 59 were used. The orientations of the crystal axes were determined by reflection of X-rays at four stages of the compression, namely, at $\epsilon=1\cdot0$ (i.e. material uncompressed), $\epsilon=0\cdot9134$, $\epsilon=0\cdot817$ and $\epsilon=0\cdot684$.* In each case the orientation was determined by finding reflections of homogeneous X-rays from two different planes, usually two of type {111}.† The co-ordinates of the three cubic axes (100),† (010) and (001), relative to the system described in Part I (Paper 9), were then calculated. The results are given in columns 2–7 of table 1.

Table 1. *Co-ordinates of cubic axes of discs cut from crystal no.* 59 *at various stages of compression. Co-ordinates measured in degrees*

| | (010) | | (001) | | (100) | |
|---|---|---|---|---|---|---|
| $\epsilon$ | $\theta$ (degrees) | $\phi$ (degrees) | $\theta$ (degrees) | $\phi$ (degrees) | $\theta$ (degrees) | $\phi$ (degrees) |
| 1·00 | 36 | 180 | 64 | 47 | 67 | 305 |
| 0·9134 | 36 | 197 | 57 | 43 | 77 | 305 |
| 0·817 | 35 | 204 | 57 | 45 | 80 | 309 |
| 0·684 | 35 | 219 | 55 | 43 | 87 | 311 |

To find the motion of the normal to the specimen relative to the crystal axes the co-ordinates of the three cubic axes given in table 1 were used. These were set out in a stereographic diagram, the centre of which represented the normal to the surface of the specimen. The whole system was then rotated by means of a stereographic net till the axis (100), whose co-ordinates are given in columns 6 and 7 of table 1, was at the point opposite to the centre of the stereographic figure, i.e. at infinity in the complete projection. The remaining two cubic axes (010) and (001) were then on the bounding circle of the stereographic diagram. Each diagram was then rotated about its centre till the crystal axes coincided in all four cases. The points representing the successive positions of the normal to the specimen were then found to lie in the spherical triangle whose vertices are represented by (010), (011), and (111). Their positions are shown by means of round dots at $P_0$, $P_1$, $P_2$, $P_3$ in fig. 1, which covers only a portion of the complete stereographic diagram.

Fig. 1 may be compared with similar diagrams used in the case of tensile tests.‡ In the case of tensile test-pieces it was shown that when slipping on one crystal plane occurs the axis of the specimen moves relative to the crystal axes in a great circle towards the direction of slip. In fig. 1 an arc of this great circle is shown by the

* $\epsilon$ is the ratio of the final to the initial thickness of the disc, see Part I, paper 9 above.

† In Parts II and III of this paper the symbol {111} is used to represent the plane {111}, while the symbol (111) is used to represent a direction, the normal to the plane {111}.

‡ 'Plastic extension and fracture of aluminium crystals', *Proc. Roy. Soc.* A, CVIII (1925), 28 (paper 7 above), figs. 3, 4, 5 and 9.

broken line $TP_0S$, where $S$ is the normal to the plane {110} which is the direction of slip, and $P_0$ is the point representing the initial condition of the specimen. If the motion of the normal to the specimen during compression were exactly opposite to that of the axis of a tensile test-piece, points $P_1$, $P_2$, $P_3$ would lie on the arc $P_0T$. It will be seen that they do not do so.

On the other hand, the considerations given above lead to the prediction that $P_0$, $P_1$, $P_2$, $P_3$ should lie on the great circle which passes through $P_0$ and the normal to the plane {111} which is the slip-plane. An arc of this circle is shown as the line $P_0M$. It will be seen that the observed points do lie close to this arc.

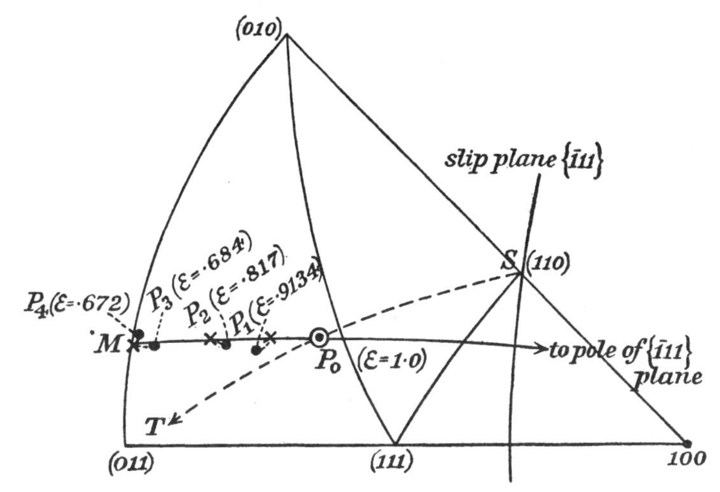


Fig. 1. Stereographic diagram showing motion of normal to compression disc no. 59 relative to crystal axes.

As the specimen is compressed the point $P$ representing the normal to the flat surface of the compression disc moves from $P_0$ towards $M$, and there is a simple relationship between the amount of compression and the distance of $P$ from $P_0$ along the arc. If $\theta$ is the angle between the normal to the slip-plane and the normal to the disc, $\epsilon$ the ratio of the thickness of the specimen to its initial thickness, and $\theta_0$ the initial value of $\theta$ when $\epsilon = 1$, then it can be shown that

$$\sin\theta = \epsilon\sin\theta_0. \tag{1}$$

In the case of specimen no. 59 the values of $\theta$ were found by laying the complete stereographic figure, of which fig. 1 is a part, over the stereographic net. Their values were 59·3°, 49·9°, 46·6°, 38·5°, corresponding with $\epsilon = 1{\cdot}0$, 0·9134, 0·817 and 0·684 respectively. Using the value $\theta_0 = 59{\cdot}3°$ the values of $\theta$ calculated from equation (1) are 59·3°, 51·7°, 44·6°, 36·0°. The predicted positions of the normal to the surface of the specimen corresponding with these values are shown in fig. 1 by means of crosses, and short lines have been drawn connecting predicted positions and the positions determined by X-ray analysis. It will be seen that the agreement is fairly good.

## Further Experiments with No. 59

As the discs are compressed the point $P$ (fig. 1) moves along the arc towards the position $M$, where the arc cuts the $\{100\}$ plane. At this stage the crystal axes come into a special position in relation to the plane face of the disc. The axes are symmetrical with respect to a plane through the normal to the specimen. To calculate the compression at which this should occur the angle between the point $M$ and the direction $(\bar{1}11)$ was measured and found to be $36{\cdot}5^{\circ}$. Using formula (1) it was found that the corresponding value of $\epsilon$ is 0·69. Accordingly, it is to be expected that if the orientation of the crystal axes were measured at this stage it would be found to possess this type of symmetry.

The point $P_3$ (fig. 1) which corresponds to $\epsilon = 0{\cdot}684$ should, in fact, be almost exactly in the cubic plane $\{100\}$. It will be seen that this symmetrical position has almost but not quite been reached.

In the course of some experiments with disc 59·9 an X-ray photograph was taken by the Laue method, using heterogeneous X-rays, at the stage when $\epsilon = 0{\cdot}672$. Since this value of $\epsilon$ differs little from the critical value $\epsilon = 0{\cdot}69$, and since the incident beam of X-rays was perpendicular to the plane of the disc, the spots of the Laue photograph should be very nearly symmetrically placed with respect to a line through the central spot. The photograph is shown in plate 1*a*. It will be seen that this prediction is verified with considerable accuracy.

By measuring the plate and constructing a stereographic diagram showing the position of the normal to the plane corresponding with each spot, no difficulty was found in identifying them. The two most intense ones close below the central spot correspond with the normals $(1\bar{1}1)$ and $(11\bar{1})$. The plane of symmetry through the normal to the specimen was found to be of type $\{100\}$, and it was found that the nearest (110) direction in this plane made an angle of $8^{\circ}$ with the normal to the specimen. This enables us to place the point $P_4$ corresponding with $\epsilon = 0{\cdot}627$ on the diagram of fig. 1. It will be seen that $P_4$ is very close to its predicted position.

On looking at the Laue photograph (plate 1*a*) it will be seen that the spots are not diffuse; in fact, they are very little larger than the central spot. It appears, therefore, that from the point of view of distortion measurements, at any rate, the material may still be regarded as consisting of a single crystal with axes possessing a definite orientation, when it has been compressed to a fraction 0·67 of its original thickness.

## Experiments with Specimen No. 61

In order to find out whether the results obtained with no. 59 apply to other specimens with axes in a different orientation, discs were cut from specimen no. 61, the characteristic point for which is shown in fig. 7 of Part I of this paper (above, p. 144). It was found that in all cases, including that of no. 59, the distortion during the first 2 % of compression was not due entirely to slipping on one plane, but that after this initial stage had been passed the regular régime described above set in. Accordingly disc no. 61·9 was compressed till its thickness was reduced by 3 %. It was then measured and compressed through another 15 % and re-measured. These measurements gave the following elements* necessary for calculating the equation to the unstretched cone:

$$\alpha = 0{\cdot}9990, \qquad X_0 = +0{\cdot}152,$$
$$\beta = 1{\cdot}1240, \qquad Y_0 = -0{\cdot}176,$$
$$\gamma = 0{\cdot}8890, \qquad X_1 = +0{\cdot}055,$$
$$\chi = 90^\circ\,15', \qquad Y_1 = -0{\cdot}123,$$
$$\chi' = 89^\circ\,15',$$

and the equation to the first position of the unstretched cone is

$$(-0{\cdot}0020\cos^2\phi + 0{\cdot}038\cos\phi\sin\phi + 0{\cdot}264\sin^2\phi)\tan^2\theta + (0{\cdot}0723\cos\phi - 0{\cdot}0650\sin\phi)\tan\theta - 0{\cdot}2076 = 0. \quad (2)$$

The stereographic projection of this cone is shown in fig. 2. As in the case of no. 59, the points lie very nearly on two planes. These planes are shown as arcs of circles.

The difference between the cone represented by equation (2) and the two planes is only appreciable in the neighbourhood of the intersection of the two planes. In this region very small changes in the measurements of the specimen make comparatively large changes in the position of the unextended cone. All directions in this region are, in fact, changed very little in length by the compression. The exact form of that part of the cone is therefore not very significant. The actual cone represented by (2) is shown by the broken curve in fig. 2. It will be seen that the unextended cone divides the whole system into two regions, one which contains all directions which are extended during compression and the other containing all those which are reduced in length. The fact that the cone cuts the bounding circle of the projection means that there is a certain range of directions on the flat surface of the disc which are decreased in length when the specimen is compressed. In no. 61·9 the ruled line ($\phi = 0, \theta = 90^\circ$) happens to be in this range. The actual measured lengths between points 4 and 6† were 5·986 mm. before and 5·980 mm. after compression, the contraction therefore being 1 part in 1000. The contraction which would be predicted on the assumption that the distortion is entirely due to slipping on one plane is about two parts in 1000. The fact that a contraction is actually observed is remarkable. I find that if the amount of slip on any other crystallographically similar plane had been as great as 2 % of that on the slip-plane whose pole is at $A_1$ (fig. 2), then the contraction along the ruled line ($\phi = 0, \theta = 90^\circ$) would not have been observed. The existence of this contraction amounts therefore almost to a proof that at any rate 98 % of the distortion is due to slipping on one plane.

The crystal axes were determined by Müller's method before the disc had been compressed. Two octahedral planes were determined and the co-ordinates of their

* For the meaning of these symbols see Part I, p. 140.

† See Part I, p. 136.

poles were ($\theta=46°$, $\phi=255°$), ($\theta=37°$, $\phi=50°$). The former of these two is the slip-plane and the predicted direction of slip was found for these two to be ($\theta=46°$, $\phi=100°$). These are marked as $A$ and $B$ on fig. 2, and the observed pole of the slip-plane and direction of slip were found from the unextended cone. They are marked as $A_1$, $B_1$ in fig. 2, and their co-ordinates are $A_1$ ($\theta=45°$, $\phi=257\frac{1}{2}°$), $B_1$ ($\theta=46\frac{1}{2}°$, $\phi=98°$). The fact that $A_1$ is close to $A$, and $B_1$ close to $B$ shows that the relationship

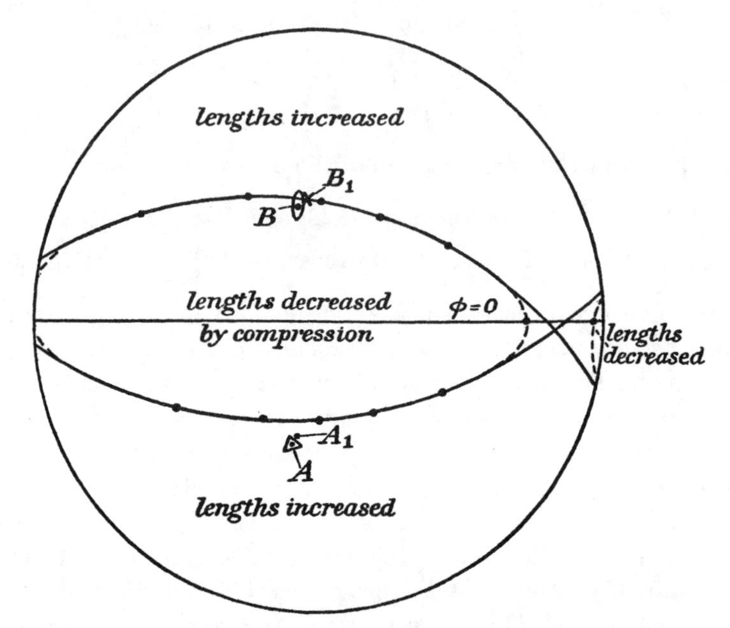


Fig. 2. Unstretched cone for compression of disc no. 61·9 during régime of slipping on one plane.

between the distortion and the crystal axes is quite regular, but it must be remembered that the distortion figure gives the position of the slip-plane when the material had already been compressed nearly 3 %, whereas the axes were measured before compression began, so that exact agreement is not to be expected.

## SLIPPING ON TWO CRYSTAL PLANES

In the case of tensile test-pieces* it has been shown that the axis of the specimen moves, relatively to the crystal axes, till it gets to a position in which it is equally inclined to two octahedral planes of type {111}. Under these conditions double slipping occurs and the axis continues to be equally inclined to the two slip planes gradually moving to the position of the normal (112).

In the case of compression tests an analogous result holds. For specimen no. 59 the point $M$ (fig. 1) represents the position of the normal to the surface of the specimen when double slipping begins. In the case of no. 61, the point $A$ (fig. 3)

* *Proc. Roy. Soc.* A, CVIII (1925), 38; paper 7 above, p. 117.

represents the initial position of the normal to the slip-plane. The line $AB$ is an arc of the great circle through the pole of the slip-plane and $B$ is the point where this arc cuts the {100} plane which is represented by the edge of the stereographic diagram. The angle between $A$ and the pole of the slip-plane is 46°, and the angle between $B$ and this pole is found from fig. 3 to be 36°. Double slipping should occur therefore when

$$\epsilon = \sin 36°/\sin 46° = 0{\cdot}817.$$

If it is true that slipping occurs on the plane and in the direction where the shear stress is a maximum after double slipping has begun, the rate of shear on the two slip-planes should be equal. The effect of equal double slipping would be to keep the normal to the specimen in the {100} plane, but it would very gradually approach the direction (011).

The orientation of three specimens numbered 61·2, 61·7 and 61·8 were measured by X-rays, after compression to $\epsilon = 0{\cdot}786$, $0{\cdot}769$ and $0{\cdot}781$ respectively. In each case the normal to the specimen was found to lie exactly on the {100} plane and at 7° from the direction (011).

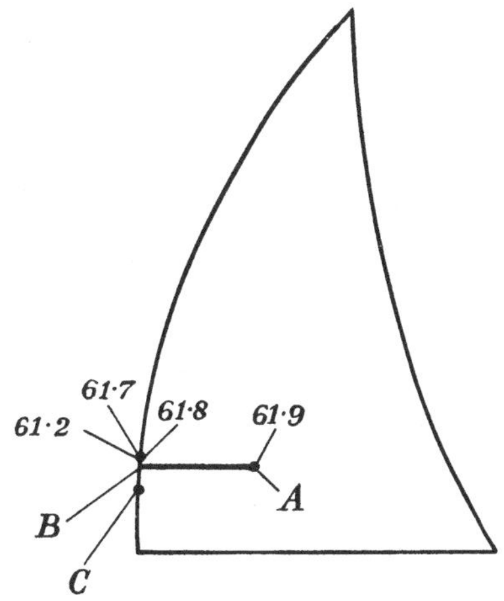


Fig. 3. Stereographic diagram showing positions measured by X-rays of normal to specimens cut from no. 61 before and after compression.

The theory of double slipping outlined above implies that the representative point would move slowly towards the normal (011) after double slipping has begun. During the compression from $\epsilon = 0{\cdot}817$ to $\epsilon = 0{\cdot}769$, the point would move towards the normal (011) by an amount which is considerably less than 1°, so that to the limit of accuracy of the measurements the representative points of 61·2, 61·7 and 61·8 should coincide with $B$ (fig. 3). It will be seen that they do so.

In order to find out whether in a large distortion the amount of slip on the two planes continues to be equal, one specimen, no. 61·19, was compressed in small stages till its thickness was only 0·298 of its initial thickness. At this stage the orientation of its axes was measured by Müller's method. Reflections were obtained from the two slip-planes. Using a certain arbitrary line on the surface of the specimen as $\phi = 0$ their co-ordinates are found to be ($\theta = 35\frac{1}{2}°$, $\phi = 354\frac{1}{2}°$) and ($\theta = 37°$, $\phi = 187°$). The representative point found from these axes is very nearly on the {100} plane and is at $4\frac{1}{2}°$ from the direction (011). It is shown at $C$ in fig. 3. It appears, therefore, that during the very large distortion which accompanies compression from $\epsilon = 0{\cdot}817$ to $\epsilon = 0{\cdot}289$, the amount of slip on the two slip-planes is equal and the {100} plane remains, to within 1°, perpendicular to the flat faces of the specimen.

The spots on the plates due to the reflected X-rays were rather diffuse so that considerable breaking up of the crystal had evidently occurred, but the orientations of the planes of maximum intensity of reflection were still measurable to an accuracy of about 1° and the angle between the two measured normals to the octahedral planes was found to be 72°, which is within 2° of the angle between the faces of an octahedron.

## Analysis of Distortion Measurements when Material Slips on Two Planes

Several sets of measurements were made in order to determine the nature of the distortion during the régime of double slipping. Measurements of specimens nos. 59 and 61 seemed to show that the distortion can be accounted for by supposing that double slipping of the type predicted does take place, but that during the earlier stages at any rate, the amount of slip is greater on the original slip-plane than it is on the second slip-plane. This result is similar to that found in the case of tensile tests.* When the amounts of slip on the two planes are not equal it is difficult to calculate an equation for the unextended cone except in the case when the amounts of slip are small. On the other hand, the equation to the unextended cone can be calculated for any amount of compression, provided the amount of slipping on the two planes is equal. In order to compare calculated with observed distortions specimen no. 69 was chosen. The crystal axes of this specimen were orientated in such a way that double slipping was to be anticipated even in the earlier stages of compression. The initial position of the axis of the specimen was in fact very nearly in a cubic plane, as can be seen by inspection of fig. 7 in Part I of this paper (above, p. 144).

A disc, 69·10, cut from this crystal was compressed in small stages from $\epsilon=1\cdot0$ to $\epsilon=0\cdot812$. It was then cut circular and compressed in twelve stages to $\epsilon=0\cdot667$, the surface being covered with a layer of grease before each increase in compression. It was then re-marked and again cut circular and then compressed to $\epsilon=0\cdot5665$. The unextended cones were worked out for the compression from $\epsilon=0\cdot812$ to $\epsilon=0\cdot667$ and from $\epsilon=0\cdot667$ to $\epsilon=0\cdot5665$. These cones are shown in fig. 4. Their planes of symmetry are shown by means of broken lines. It will be seen that in each case the axis of the specimen lies very close to the plane of symmetry. This proves that if the distortion is due to double slipping the amount of slipping on the two planes concerned is nearly equal.

It remains to show that the cones are actually ones which could be due to double slipping. For this purpose it is necessary to calculate the unstretched cone for equal double slipping and to compare this with one of those shown in fig. 4. An expression for displacements of particles due to small shears $\delta S_1$ and $\delta S_2$ parallel to two planes making an angle $2\alpha$ with one another is given on p. 41 of 'Plastic extension and fracture of aluminium crystals'.† In the case where the two shears are equal so that $\delta S_1=\delta S_2$ the equations (2) of that paper become (after correcting an obvious misprint where $\cos\beta$ is written for $\sin\beta$ in the expression for $\delta y$)

$$\left.\begin{aligned}\delta x&=-z\delta s\sin\alpha\cos\beta,\\ \delta y&=y\delta s\cos\alpha\sin\alpha\sin\beta,\\ \delta z&=-z\delta s\sin\alpha\cos\alpha\sin\beta,\end{aligned}\right\}\qquad(3)$$

* *Proc. Roy. Soc.* A, CVIII (1925), 39; paper 7 above, p. 117.

† Paper 7 above, p. 120.

PLATE 1

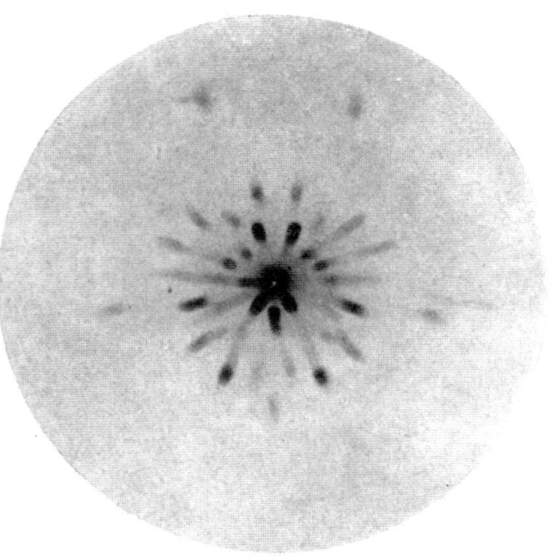

*a*, Laue photograph of specimen no. 59·9 taken when $\epsilon = 0{\cdot}672$.

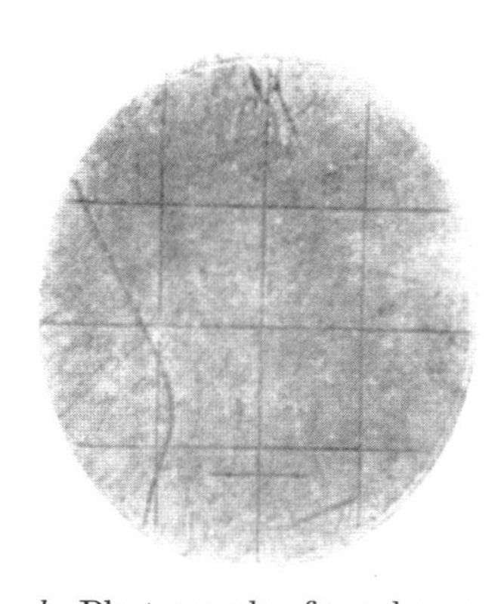

*b*, Photograph of marks on no. 69·9 after compression to $\epsilon = 0{\cdot}2776$, showing uniformity of distortion even when the distortion is very large.

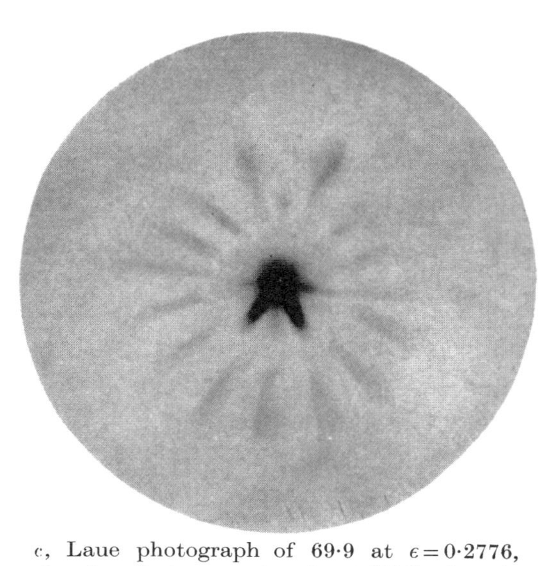

*c*, Laue photograph of 69·9 at $\epsilon = 0{\cdot}2776$, showing that symmetry about {100} plane is preserved when the distortion is very great.

where $-\frac{1}{2}\delta s$ has been written for $\delta S_1$ or $\delta S_2$. The negative sign is taken in order that $s$ may have a positive value during a compression test, and the factor $\frac{1}{2}$ has been inserted in order to make the amount of shear for a given value of $s$ comparable with shear by slipping on one plane. Since $\alpha$ and $\beta$ are constant these equations can be integrated, the resultant transformation equations being

$$\left.\begin{aligned} x &= x_0 + z_0 \cot\beta \sec\alpha(1/A - 1), \\ y &= y_0 A, \\ z &= z_0/A, \end{aligned}\right\} \tag{4}$$

where $A = e^{s\cos\alpha\sin\alpha\sin\beta}$ and $s = \int ds$.

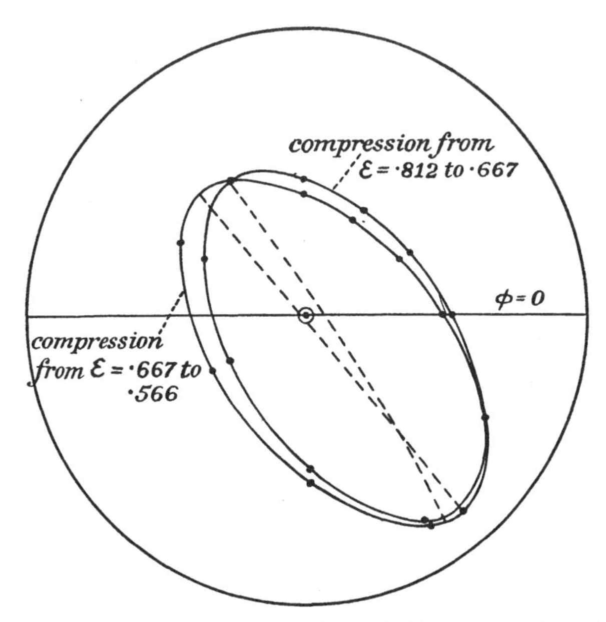


Fig. 4. Unstretched cones for compression of no. 69·10 during régime of double slipping.

These equations give the motion of particles of the material relative to the system of co-ordinate axes shown in fig. 6 of 'Plastic extension and fracture of aluminium crystals'.*

In the case when the symmetrical position of the crystal axes is arrived at by compression instead of by extension, the value of $\alpha$ is not the same as it was in the case previously discussed, $2\alpha$ is now the internal instead of the external angle between the faces of an octahedron so that $\alpha = 54° 44'$; $\beta$ is 60° as before, so that $\cos\alpha = 1/\sqrt{3}$, $\sin\alpha = \sqrt{2}/\sqrt{3}$, $\sin\beta = \frac{1}{2}\sqrt{3}$, $\cos\beta = \frac{1}{2}$. Hence

$$\cos\alpha\sin\alpha\sin\beta = 1/\sqrt{6} = 0{\cdot}408$$

* Paper 7 above.

and $\cot\beta\sec\alpha=1$. So that formulae (4) become

$$\left.\begin{aligned}x&=x_0+z_0(1/A-1),\\ y&=y_0A,\\ z&=z_0/A.\end{aligned}\right\}\tag{5}$$

And $$A=e^{0\cdot 408s}\quad\text{or}\quad s=\sqrt{6}\log_e A=5\cdot 640\log_{10}A.$$

The equation to the unstretched cone may now be written down directly, it is: $x^2+y^2+z^2=x_0^2+y_0^2+z_0^2$, and using (5) this gives for the cone in its first, or unstrained position,

$$A^2(A+1)y_0^2-2z_0^2-2Ax_0z_0=0.\tag{6}$$

Before we can express the unextended cone in terms of $\gamma$, the ratio of the thickness of the disc after compression to that before compression, it is necessary to find the relationship between $\gamma$ and $A$. Let $i_0$ be the angle between the normal to the surface of the specimen and the bisector of the two slip-planes which is the axis of $z$ in equations (5). To find $\gamma$ in terms of $A$ and $i_0$ consider the area of a rectangle in the flat surface of the specimen before and after compression. Let the sides of this rectangle be initially of length $c$ and $d$, the side of length $c$ being along the axis of $y$ (which is parallel to the surface when the specimen is in the symmetrical position assumed for double slipping) and the side of length $d$ being in the plane of symmetry $y=0$.

If the origin of co-ordinates be taken at a corner of this rectangle, then before compression the co-ordinates of one corner are $(0, c, 0)$ while those of another are $(d\cos i_0, 0, -d\sin i_0)$.

Applying the transformation formulae (5) it will be seen that the co-ordinates of these two points after distortion are

$$(0, Ac, 0)\quad\text{and}\quad(d\cos i_0-d\sin i_0(1/A-1), 0, -d\sin i_0/A).\tag{7}$$

The area of the rectangle before distortion is $cd$, the area after distortion is $cdA\{[\cos i_0-\sin i_0(1/A-1)]^2+\sin^2 i_0/A^2\}^{\frac{1}{2}}$, and since the volume of the material is unchanged by compression this area multiplied by $\gamma$ must be equal to $cd$. Hence

$$A^2(1+2\sin i_0\cos i_0)-2A(\sin i_0\cos i_0+\sin^2 i_0)+2\sin^2 i_0-1/\gamma^2=0.\tag{8}$$

This equation gives $A$ and hence $s$ when $i_0$ is known from X-ray measurements and $\gamma$ has been measured.

To find the position of the normal to the specimen relative to the crystal axes after compression it is only necessary to note that it is in the plane $y=0$ and at right angles to one of the sides of the rectangle $cd$, thus if $i$ represents the angle which the normal to the specimen makes with the axis $z$ after compression, we find from (7)

$$\tan i=-\left\{\frac{-d\sin i_0/A}{d\cos i_0-d\sin i_0(1/A-1)}\right\}$$

or $$\cot i=A(\cot i_0+1)-1.\tag{9}$$

To apply these formulae to specimen no. 69 the value of $i_0$ was found by X-ray reflections to be 12°. Inserting this value in (8) and (9) the equations for $A$ and $i$ are

$$1{\cdot}4067A^2 - 0{\cdot}493A + 0{\cdot}0862 - 1/\epsilon^2 = 0, \qquad (10)$$

and

$$\cot i = 5{\cdot}705A - 1, \qquad (11)$$

and inserting observed values of $\epsilon$ table 2 shows the values obtained for $A$ and $i$.

Table 2

| $\epsilon$ | $A$ | $i$ (degrees) |
|---|---|---|
| 1·00 | 1·000 | 12 |
| 0·812 | 1·199 | 9·7 |
| 0·667 | 1·427 | 8·0 |
| 0·566 | 1·653 | 6·8 |

In the case of specimen no. 69·10 complete distortion measurements were made for the compression between $\epsilon = 0{\cdot}667$ and $\epsilon = 0{\cdot}566$. In order to calculate the corresponding unextended cone due to double slipping the value

$$A = 1{\cdot}653/1{\cdot}427 = 1{\cdot}158$$

must be used in (6). Inserting this value in (6) the equation to the unextended cone in the material at $\epsilon = 0{\cdot}667$ is found to be

$$2{\cdot}900y_0^2 - 2{\cdot}316x_0z_0 - 2z_0^2 = 0. \qquad (12)$$

In order to compare this with the unextended cone derived from external measurements it is necessary to transform the axes so that the axis of the specimen is the axis of $z$. The transformation formulae necessary are

$$\left.\begin{aligned} x_0 &= x'\cos i + z'\sin i, \\ y_0 &= y', \\ z_0 &= -x'\sin i + z'\cos i, \end{aligned}\right\} \quad \text{where} \quad x'/z' = \tan\theta\cos\phi,\ y'/z' = \tan\theta\sin\phi. \qquad (13)$$

And $\theta$ and $\phi$ are the spherical polar co-ordinates with respect to axes for which $\theta = 0$ is the normal to the specimen and ($\theta = 90°$, $\phi = 0$) is the line in the surface of the specimen which is symmetrical with respect to the distortion. Taking from table 2 the appropriate value $i = 8{\cdot}0°$ and inserting it in equations (13) and inserting the values of $x_0$, $y_0$ so found in (12), the equation to the unextended cone is found to be

$$(0{\cdot}281\cos^2\phi + 2{\cdot}900\sin^2\phi)\tan^2\theta - 1{\cdot}675\cos\phi\tan\theta - 2{\cdot}279 = 0. \qquad (14)$$

A stereographic projection of this cone is shown in fig. 5, and in order to compare it with the unextended cone calculated from the measured distortion, the cone for the compression from $\epsilon = 0{\cdot}667$ to $\epsilon = 0{\cdot}566$ was taken from fig. 4 and rotated about the normal to the specimen till the cubic crystal plane which should be symmetrically disposed with respect to the distortion coincides with the co-ordinate plane $\phi = 0$ of the calculated cone. This plane is represented in fig. 5 by a vertical diameter of the bounding circle of the projection. It will be seen that the calculated cone derived from the assumption of double slipping is very close to the cone deduced from the distortion measurements, but that the latter is rather longer and narrower than the former.

This characteristic appears in all cases that I have examined, and the fact that the cones always coincide at one end, namely, the end which contains the intersection of the two slip-planes marked $A$ in fig. 5, suggests that there may be a certain amount of slipping on the same two slip-planes but in one of the other possible directions of slip.

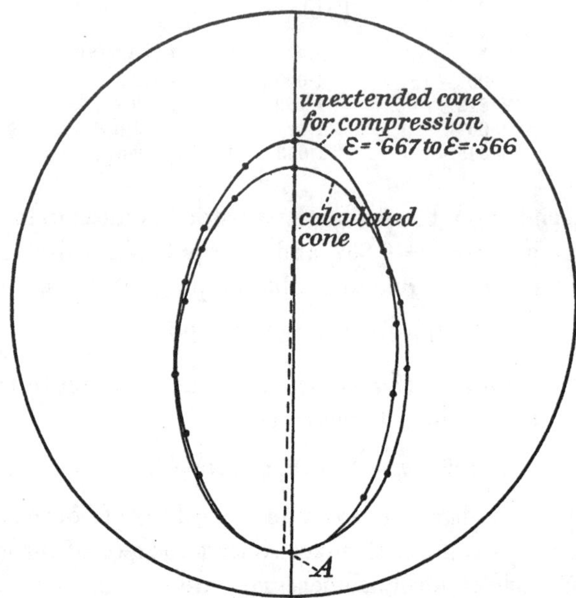


Fig. 5. Comparison between measured form of unstretched cone for 69·10 and that deduced from assumption of equal amounts of slip on two symmetrical planes of slip.

## Condition of the Specimens when the Distortion is very Great

In the cases of specimens nos. 61 and 69 the tests were continued till the distortion was very large. The details of the manner in which the tests were carried out will be described in Part III (paper 13), where the stress measurements are discussed, but for the present purpose it is sufficient to describe the last stage of the test in the case of one specimen, no. 69·9. This specimen had been compressed in four stages, the disc being cut circular three times. The distortion was found to be uniform all the time, and the thickness at the end of the fourth stage was 0·349 of the original thickness. It was then cut circular, the thickness being left unaltered, and marked in squares thus .

It was then compressed till its thickness was only 0·2776 of its original thickness.

A photograph of the final condition of the specimen is shown in plate 1*b*. It will be seen that the distortion is still quite uniform, the straight lines remaining straight.

The three lines which run vertically, down the page, i.e. parallel to the major axis of the elliptic upper surface of the specimen, had been scratched so as to be parallel to the direction (100) in the crystal. The fact that the disc is distorted into an ellipse with its major axis parallel to the direction (100) shows that the symmetry is preserved, and the general features of the distortion are in accordance with the prediction based on equal slipping on two planes; but we can apply a far more searching test as to how far the assumption is true. The extensions of the specimen in the flat upper surface parallel to the major and minor axes of the ellipse were measured. The distance between the intersections of two outer vertical lines and the middle horizontal line (i.e. between points 4 and 6 in the notation employed in Part I, p. 136) was 3·185 mm. before compression and 3·226 mm. after compression. The extension ratio of the material parallel to the minor axis may be defined as the ratio of the two lengths $3{\cdot}226/3{\cdot}185 = 1{\cdot}0185$. In the same way the extension ratio parallel to the major axis was found to be $4{\cdot}024/3{\cdot}246 = 1{\cdot}240$.

The extension ratios parallel to these two directions can be calculated on the assumption of equal double slipping. The value of $A$ corresponding with the compression from $\epsilon = 0{\cdot}349$ to $\epsilon = 0{\cdot}2776$ was found to be $A = 1{\cdot}250$. Referring to equation (7) it will be seen that the rectangle there discussed whose sides were $c$ and $d$ is in fact a rectangle whose sides are parallel to the scratches shown in the photograph (plate 1*b*). Formula (7) shows therefore that the extension ratio in the direction of the major axis should be equal to $A$ because the initial length one side was $c$ and the final length $Ac$. The predicted extension ratio in this direction was therefore $A = 1{\cdot}250$. Comparing this with the observed extension ratio, 1·240, it will be seen that the agreement is very good.

The extension ratio in the direction of the minor axis is, from (7)

$$\{[\cos i_0 - \sin i_0(1/A - 1)]^2 + \sin^2 i_0/A^2\}^{\frac{1}{2}}. \tag{15}$$

At the beginning of the last stage of compression of 69·9 the value of $\epsilon$ was 0·349. At this stage in the compression the value of $i$ was found by calculation to be 4·2°. Using this value for $i_0$ in (15), and putting $A = 1{\cdot}250$, the predicted extension ratio is found to be 1·014. Comparing this with the observed value 1·0185 it will be seen that again the agreement is very good.

In order to throw into as clear a light as possible the relationship between the distortion and the crystal axes the diagram fig. 6 has been prepared. This shows the specimen no. 69·9 at the end of the test. The two planes of slip $\{111\}$ and $\{\bar{1}11\}$ and also the plane of symmetry $\{100\}$ are shown as material sheets, and the directions of slip (110) and $(1\bar{1}0)$ are marked on them. The figure may be of some assistance in understanding the manner in which the normal to the specimen approaches the direction (011) with gradual reduction of the angle $i$ as the compression proceeds.

Top and side views of the specimen are also shown in fig. 6. These may be compared with the photograph (plate 1*b*).

The accuracy with which the predicted distortion is realized in practice is very surprising when the amount of breaking up which the crystal has suffered is taken into consideration. Plate 1*c* shows a Laue photograph of 69·9 taken with the X-ray beam perpendicular to the plane faces of the specimen. This photograph was taken at the stage of maximum compression, i.e. at $\epsilon = 0{\cdot}2776$. It may be compared with plate 1*a*, but in doing so it must be remembered that the distance of the photographic plate from the specimen was 1·97 cm. in the photograph (plate 1*c*), whereas it was only 1 cm. in the photograph of plate 1*b*. Allowing for this difference it will be seen that the two photographs are essentially of the same nature so far as type of symmetry is concerned. (The value of $i$ was less in the specimen photographed in

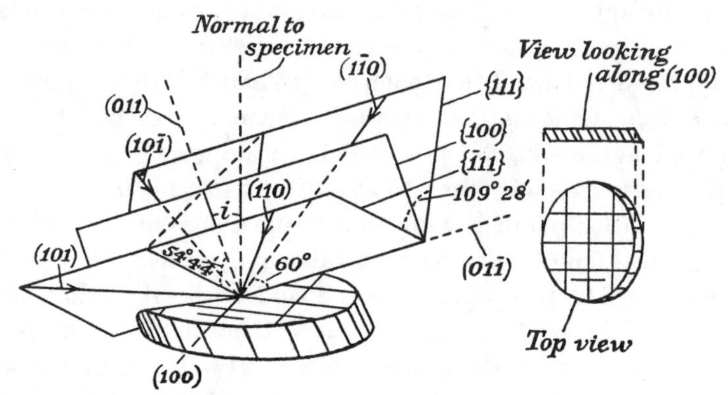


Fig. 6. Diagram showing positions of planes and directions of slip in relation to the distortion of a compressed disc during régime of double slipping.

plate 1*c* than in the other.) The diffuseness of the spots in (*c*), and the smallness of their number as compared with those of (*a*), show that the material can no longer be regarded, from the point of view of X-rays, as a single crystal, even approximately. On the other hand, the external measurements of the specimen show that the distortion is almost exactly what one would expect from a single crystal placed with its axes in the symmetrical position indicated by the spots in the Laue photograph (plate 1*c*). It may be noted that the value of $s$ appropriate to the distortion of 69·9 at $\epsilon = 0{\cdot}2776$ is $s = 2{\cdot}8$. That is to say, the distortion is equivalent to a distortion by single slipping, in which any two planes of particles parallel to the slip plane move relatively to one another through a distance 2·8 times as great as their distance apart.

## Application of Foregoing Results to the Structure of Rolled Sheets of Metal

The nature of the distortion which occurs when metals are rolled is not well understood. Regarding the sheet as a whole, one can say that there is an extension in the direction of rolling, a contraction in thickness, and that there is little or no extension or contraction in the line perpendicular to the direction of rolling. If the distortion is uniform through the thickness of the sheet then the principal axes of strain are along and perpendicular to the direction of rolling and perpendicular to the surface of the sheet. Lines ruled on the surface of the sheet perpendicular to the direction of rolling remain perpendicular to that direction, and lines of particles originally perpendicular to the surfaces of the sheet remain so.

In rolled metals, X-ray analysis shows that the crystal axes tend to take up certain preferred orientations, and as the rolling proceeds these orientations become 'stable' and remain fixed relative to the surface of the sheet. One of the fundamental problems in considering the distortion of metals during rolling is to find how these preferred orientations can remain fixed in direction as the rolling proceeds. The first successful attempt at solving a problem of this type appears to be that of Miss Elam and myself,* when we showed that, if a single crystal of aluminium could be extended indefinitely, the orientation of the axes would come into a definite position such that the axis of the specimen was normal to a plane of type {112}.

This result was used by Polanyi and Weissenberg,† who pointed out that when the axes were in this orientation the reduction in the cross-section of the bar was confined to one direction, that of a crystal direction of type (110), the dimension of the bar in the direction of type (111) which is perpendicular to (110) and (112) remaining constant. For this reason they predicted that the 'stable' orientations in a rolled bar would be such that the crystal direction (112) was in the direction of rolling, (111) in the surface of the sheet but perpendicular to the direction of rolling, and (110) perpendicular to the sheet.

I had noticed this peculiarity of the distortion due to double slipping in a tensile test-piece when the crystal axes were in this 'stable' position, but had hesitated to apply it to the distortion of a rolled sheet because (*a*) it seems unlikely that the distortion in rolling is really uniform through the thickness of the sheet, and (*b*) the distortion of the crystal cannot be that of a rolled sheet as a whole, even if uniformity of distortion is assumed. It is true that Weissenberg's type of slipping makes the breadth of the sheet constant, the extension along the direction of rolling being equal to the contraction in thickness of the sheet. It is true also that lines of particles perpendicular to the surface remain perpendicular to it; but lines of particles initially in the surface of the sheet and perpendicular to the direction of rolling do not remain perpendicular to the direction of rolling.

To me this seems a fatal objection to Polanyi and Weissenberg's theory. It is possible that adherents of the theory might postulate an equal number of crystals with orientations in the two alternative positions consistent with the three given directions. The distortion of the material, as a whole, might be such that lines ruled across the material would, on the whole, remain perpendicular to the direction of rolling, though each element of the line became inclined to this direction; but the conditions under which the separate crystals could adjust themselves so as to fit in with the distorted positions of their neighbours are very difficult to visualize. One could, for instance, assume that they were very long in the direction of rolling, thus

* 'Distortion of an aluminium crystal during a tensile test', *Proc. Roy. Soc.* A, CII (1923), 643; paper 5 above.

† See G. Masing and M. Polanyi, *Naturwissenschaften*, vol. II.

getting rid of the necessity for considering the adjustments at their ends. If the two orientations occurred alternately in grains very much elongated in the direction of rolling, they could adjust themselves, so far as movement relative to their nearest neighbours situated in the same plane of rolling as themselves was concerned, but it is geometrically impossible for them to adjust themselves to the movements of neighbours in the direction perpendicular to the plane of the sheet. For this reason Weissenberg's type of slipping could only exist if the grains extended right through the sheet, their boundaries being planes parallel to the direction of rolling and perpendicular to the sheet. This distribution of grains seems highly artificial.

### Application of Distortion by Compression between Parallel Plates

Referring to fig. 6 and to equation (9) it will be seen that the 'stable' position of axes to which the compressed discs tend is one for which the direction (011) is perpendicular to the disc, and the plane {100} is the plane of symmetry of the distorted specimen. In this position the distortion of the specimen is such that there is no extension in the direction $(01\bar{1})$ (which is in the face of the specimen when $i=0$). There is a reduction in thickness in the direction (011) and a corresponding extension in the direction (100). Lines ruled on the surface in the directions (100) and $(01\bar{1})$ remain at right angles to one another after compression.

On the other hand, lines of particles originally at right angles to the surface of the specimen do not remain so, as will be seen by looking at the side view of the specimen sketched in fig. 6. It appears therefore that this type of double slipping, like that of Polanyi and Weissenberg, does not give rise to a distortion which is identical with that which occurs in rolling when the assumption is made that the distortion is uniform through the thickness of the sheet.

It will be noticed that when $i=0$ two more possible directions of slip, $(10\bar{1})$ and (101), come into positions which are symmetrical with respect to the two directions of slip $(1\bar{1}0)$ and (110) and the surface of the specimen. Their positions are shown in fig. 6. Under these circumstances it is to be expected that slipping might occur in these two new directions as well as in the old ones. As has already been pointed out, the elongation of the unextended cone beyond its predicted position for the extension of 69·10 from $\epsilon=0{\cdot}667$ to $\epsilon=0{\cdot}566$ (see fig. 5) suggests that a certain amount of slipping in the two new directions $(10\bar{1})$ and (101) had already occurred. An examination of the unextended cone for 69·9 from $\epsilon=0{\cdot}349$ to $\epsilon=0{\cdot}2776$ showed that this effect was getting more pronounced as the compression increased and $i$ decreased, the cone being more elongated than that shown in fig. 5.

If the amount of slip in both the two directions on each slip-plane is equal, then the distortion of the compressed disc is identical with that assumed for rolling. If, therefore, the crystal axes are placed in accordance with the scheme shown in the right-hand bottom corner of fig. 7, the distortion due to slipping in two directions on each of the two planes of slip is identical with that assumed for rolling. Cubes in

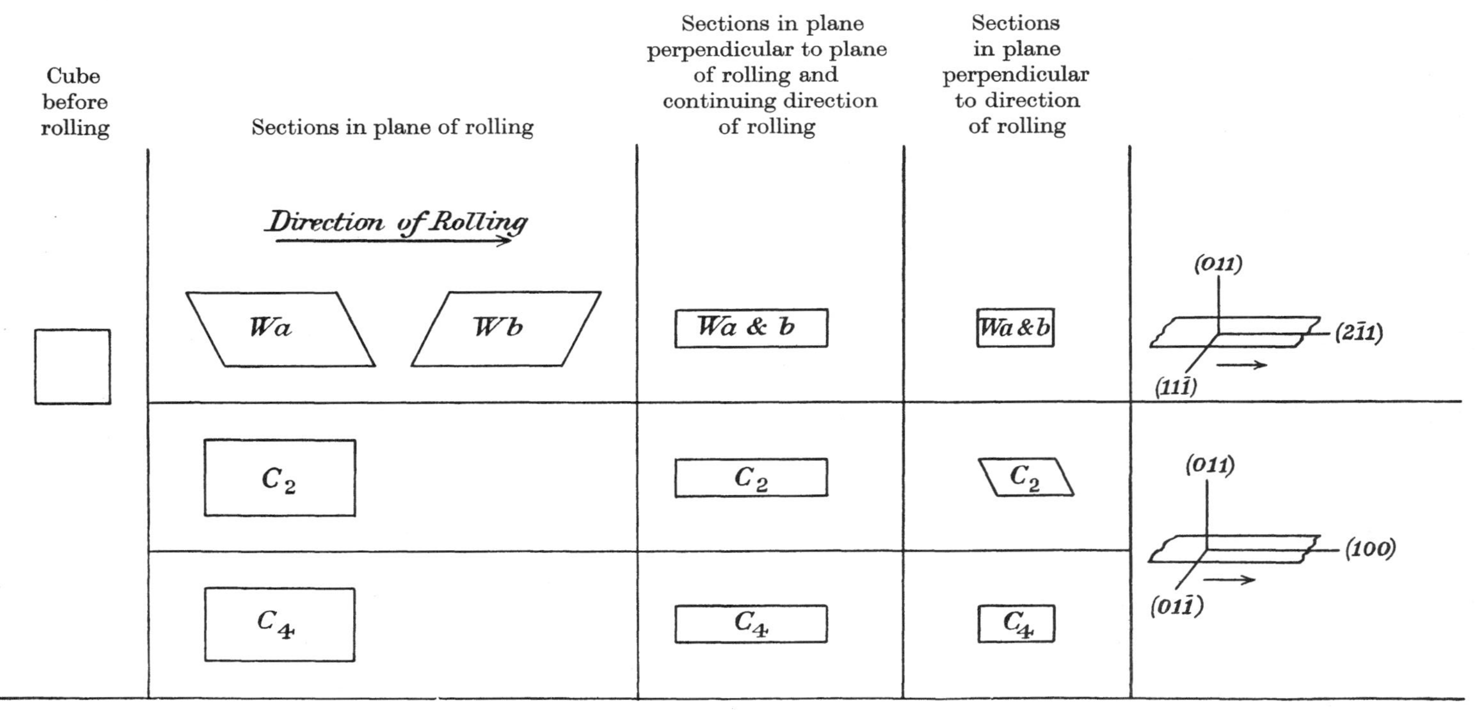


Fig. 7. Diagram showing sections by three perpendicular planes, of shapes into which a cube is distorted (*Wa* and *Wb*) by Polanyi and Weissenberg's type of double slipping which occurs in tensile test-pieces; ($C_2$) by double slipping of type which occurs in compression specimens; ($C_4$) by double slipping on two planes in two directions on each plane. This type might be expected to occur in compression discs as the normal to the specimen gets close to its ultimate 'stable' position.

the material with their edges perpendicular to the surface of the sheet, and parallel to the direction of rolling, are distorted into rectangular blocks with one dimension unchanged.

In order to illustrate the differences in the kinds of distortion which arise from these various kinds of double slipping, the diagram shown in fig. 7 has been prepared. This shows the distortion of a cube whose faces and edges were originally parallel and perpendicular to the plane of rolling and the direction of rolling. $Wa$ and $Wb$ represent the distorted shape of the cube when the distortion is due to double slipping of Polanyi and Weissenberg's type. There are two possible shapes because there are two possible positions of the crystal axes, and in Weissenberg's theory both these two types of orientation occur. $C_2$ represents the distorted shape when the distortion is due to slipping of the type found in compression specimens, on two planes and in one direction on each plane. $C_4$ represents the distorted shape of the cube when the distortion is due to slipping in two directions in each of two planes. Inspection of fig. 7 shows that $C_4$ is the only one of these types which satisfies the assumed condition that the distortion shall be uniform through the thickness of the rolled sheet. It seems possible that if the distortion in rolling were uniform through the thickness of the sheet the slipping would, in fact, be of this type, and the 'stable' position of the crystal axes would be that shown in the sketch at the right-hand bottom corner of fig. 7.

## Comparison with X-ray Results

Perhaps the most complete experiments on the orientation of crystal axes in rolled foil are those of Owen and Preston*. These authors, using both photographic and ionization methods, got results nearly in agreement with previous work by Wever.† They conclude that there are two sets of crystals in the material whose orientations are mirror images of one another in the plane of rolling. The individual members of each set are not all exactly in one orientation, but are rotated through a range of about 10° round the transverse direction, i.e. the line in the plane of rolling at right angles to the direction of rolling. This direction, which is of type (110), is common to all crystals of both sets. This feature comes out very clearly in a photograph taken with homogeneous X-rays while the specimen was rotated about the transverse direction. In this case evidently the photograph should, if the spread of crystal orientation is entirely due to rotation about this direction, be similar to a photograph of a single crystal rotated about the same axis. This photograph (Owen and Preston's fig. 12) is, in fact, the only one of their photographs which in the least resembles a rotation photograph of a single crystal.

The crystals arrange themselves so that they all have a crystal direction of type (110) in the transverse direction. Directions of type (111) lie in the direction of rolling and in a range of about 10° on either side of it.

* 'The effect of rolling on the crystal structure of aluminium', *Proc. Phys. Soc.* XXXVIII, part 2 (1926).

† *Z. Phys.* XXVIII (1924), 69.

Comparing this result with that deduced from compression experiments combined with the assumption of uniform distortion through the thickness of the sheet, it will be seen that the crystal direction (110) sets itself in the sheet and across the direction of rolling in both cases, but in the disposition predicted from the assumption of uniform distortion the direction (100) should set itself along the direction of rolling, whereas in the observed foils (100) directions lie in ranges of 10° from 55° to 65° on either side of the plane of rolling.*

The results of Owen and Preston could be explained by double slipping in two directions on two planes, if the distortion during rolling is not uniform through the thickness of the material. Suppose, for instance, that the distortion is such that there is no extension or shear in the transverse direction, but lines of particles originally normal to the surface of the sheet are curved in the distorted material

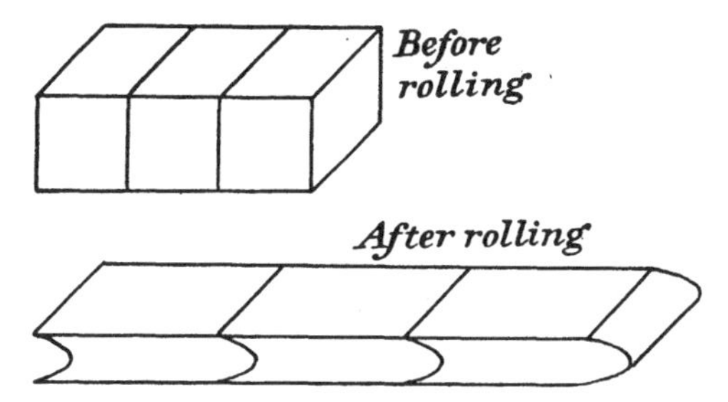


Fig. 8. Diagram illustrating possible changes of shape in rectangular bar when rolled under conditions where distortion was not uniform through thickness of material.

(the curves, of course, being in planes perpendicular to the transverse direction). The sketch in fig. 8 shows the effect of distorting in this way a rectangular bar marked with lines parallel to its edges. In this case the distortion of each element could be accounted for by slipping parallel to two planes only. These two {111} type planes intersecting in the transverse direction, which would therefore be of type (110), might be in any orientation round the transverse direction. The exact orientation at any given distance from the middle of the sheet would depend on the relationship between the total extension of the material in the direction of rolling, and the amount of shear of the outer layers of material relative to the inner ones.

In order that there may be no lateral motion of layers parallel to the plane of the sheet the amount of slip in the two directions on each plane of slip must be equal to one another, but the amount of slipping on one plane may be different from that on the other, and for any given orientation of the principal axes of strain a ratio of the amounts of slip on the two planes could be found which would bring the crystal axes into a 'stable' position so that further extension would keep them in that position.

It must be remembered, however, that the distortion due to rolling is probably even more complicated than has been considered here. Though the resultant strain is necessarily of this type, the directions of the shear stresses in the material on entering the rollers are probably quite different from those in the metal when it

* See Owen and Preston, loc. cit. fig. 8.

leaves them. This question has been studied in a very interesting paper by von Kármán,* who bases his work on the fact that the friction between the roller and the metal is a necessary part of the action of a rolling machine. He points out that the direction of friction may change sign as the material passes from the point where contact with the roller is first made to the point where the metal leaves the rollers. The total hydrostatic† pressure in the material also varies in a complicated manner.

For these reasons it is impossible in the present state of knowledge to do more than give a possible reason why the crystals set themselves with a (110) direction in the plane of rolling, and at right angles to the direction of rolling. The explanation of why the orientation of the crystal axes round the transverse direction is what it is, must be left till the distortion of metals, and particularly single crystals of metals in the process of rolling, has been studied in greater detail.

In conclusion, I wish to express my thanks to Sir Ernest Rutherford for allowing the work to be carried out in the Cavendish Laboratory, and to Miss C. F. Elam for giving me the single crystals used in the work, and also for carrying out some of the earlier determinations of orientation of crystal axes by X-rays. Most of the X-ray work was done with a Hadding tube presented to me by Professor P. Debye. The Laue photographs shown in plate 1*a* and *c* were very kindly made for me by Dr A. Müller in the Davy Faraday Laboratory. I wish also to express my thanks to Mr W. S. Farren, with whom I began this work, for his continued interest and help.

* *Z. Angew. Math. Mech.* v (1925), 139.

† Sum of the principal stresses.

# 13

# THE DISTORTION OF CRYSTALS OF ALUMINIUM UNDER COMPRESSION

## PART III. MEASUREMENTS OF STRESS

REPRINTED FROM

*Proceedings of the Royal Society*, A, vol. CXVI (1927), pp. 39–60

Several experiments were devised to find out whether it is possible to measure the internal shearing stresses in a compressed disc, and the conditions under which it is possible to do so. Equations are developed for analysing shear stresses parallel to planes of slip and distortion due to double and single slipping. The relationship between shear stress and amount of shear is found for tensile and for compression specimens, when slipping is confined to one plane. The experimental results in the two cases are identical. The fact that the component of force normal to plane of slip is a pressure in one case and a tension in the other makes no measurable difference to resistance to slipping for given amount of slip.

During double slipping resistance to shear increases more rapidly for a given total amount of slipping than when all slip is confined to one plane. The experiments cover a large range and show that resistance to shear goes on increasing up to greatest amounts of distortion used.

When a bar cut from a single crystal of a soft metal is extended, the relationship which exists between the extension at any stage of the test and the force necessary to produce that extension may depend on a number of circumstances: the orientation of the crystal axes, the manner in which the load is applied, the time during which the load is on, and so forth. When a specimen of aluminium at room temperature is extended the extension does not seem to depend on the manner of loading provided the whole load is not applied very suddenly. Thus, if a given extension $\epsilon$ is obtained with a given load $T$ and the specimen is then unloaded and afterwards gradually loaded again, then except for the small elastic extension and sometimes a very small plastic extension the specimen will not begin to extend again till the load attains the value $T$. It will then extend as though the load had been increasing since the beginning of the test. In the case of aluminium, therefore, at a given temperature the load necessary to produce a given extension seems to depend chiefly on the orientation of the crystal axes. In seeking to find out how the load depends on the orientation one is naturally guided by the observed type of deformation. In the case where the deformation is due entirely to slipping on a given crystal plane and in a given crystallographic direction, the force which acts across this crystal plane can conveniently be resolved into three components: a normal force perpendicular to the plane, a shearing force parallel to the direction of slip and a shearing force perpendicular to the direction of slip. Since the strengthening of the material depends on the amount of distortion it is essential in comparing specimens whose crystal axes have different orientations in relation to the direction of stress to decide on some system for measuring the distortion. Clearly the correct method in the case of slipping on one plane is to calculate the amount of shear parallel to the plane of slip. Referring to fig. 1, if $ABCD$ is the undistorted position of a rectangular block of material, and if $AD$ is parallel to a slip plane, the distorted position of the block is $AB'C'D$. The amount of shear will be defined as $BB'/BA$, that is, $\tan(B\hat{A}B')$. This will be denoted by the symbol $s$. The shearing stress parallel to the direction of slip will be represented by the symbol $S$ and the normal stress is $N$. One of the chief purposes of the work here described was to find out whether $S$ can be regarded simply as a function of $s$, or whether the component of force normal to the plane of slip has any effect on the resistance to shear as it would have if the phenomenon was analogous to that of solid friction. This question has already been studied by E. Schmid,* who found in the case of bismuth and some other metals that the normal force does not appear to affect the resistance to shear. In his case he found $S$ and $N$ for a number of tension specimens, the normal force was therefore always a tension. In the present work the values of $S$ found in extension experiments are compared with those found in compression experiments, so that the stress normal to the plane of slip is a pressure in one case and a tension in the other.

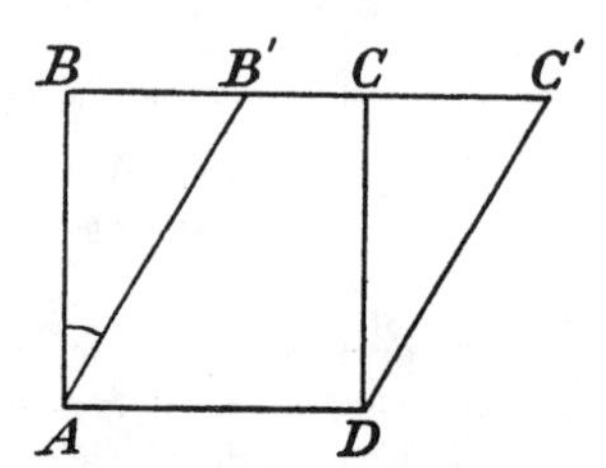

Fig. 1. Diagram illustrating definition of amount of slip, $s$.

## Experimental Details

The determination of the distribution of stress in the interior of a plastic body is a matter of great difficulty unless it is possible to ensure that the stress is uniform. In the case of soft metals like aluminium in which the resistance to distortion increases steadily with the amount of distortion, it seems that a non-uniform stress would produce a non-uniform distortion. Conversely, therefore, the existence of a uniform distortion implies a uniform distribution of stress. It is only when the distribution of stress is uniform that one can calculate the stress at any point in a material from a knowledge of the total external forces applied to the specimen. For this reason, therefore, in determining the strength of materials it is essential to arrange the tests so that the distortion is uniform.

In the case of tests in which uniform bars of soft metals are extended no special precautions to ensure uniformity of distortion are necessary. The inherent stability, conferred by the rapid increase of stress with increasing extension makes the distortion uniform automatically. In the case of discs cut from single crystals of aluminium and compressed between steel plates, the conditions under which uniform

* 'Ueber die Festigkeit und Plasticität von Wismutkristallen', *Z. Phys.* XXXIX (1926), 359; and 'Ueber die Schubverfestigung von Eisenkristallen bei Plastischer Deformation', *Z. Phys.* XL (1926), 54.

distortion can be obtained were described in Part I.* The considerations put forward above indicate that when the distortion is uniform the stress must be uniformly distributed. The existence of tangential friction at the flat surface when the specimen is squeezed into a larger and thinner disc would necessarily tend to give rise to a non-uniform distribution of stress and therefore presumably to non-uniform distortion. One is therefore driven to the conclusion that the precautions taken to ensure uniform distortion have so reduced the friction of the specimen on the steel plates that its effect on the stress in the material is negligible compared with the stress due to the normal load.

This conclusion is surprising, but unless it is true it is not possible to deduce the values of the shear stress $S$ from measurements of the total compressive force on the specimen. For this reason several experiments were undertaken to verify the fact that the friction is really negligible. Some of these will now be described.

## Compressive Force on Discs of Different Thicknesses

As was pointed out in Part I the effect of friction must be greater when the disc is thin than when it is thick. Two discs of thickness 1·5 mm. were cut from specimen no. 59 and two of thickness 3 mm. The initial diameter of all of them was 14 mm. One specimen of each thickness was compressed in a few stages, the load increasing by 1000 lb. each time. The remaining two were compressed in many stages, the load increasing each time by 100 lb. The discs were covered with a layer of grease before each increase in load.

It was found that in the first case when the compression was carried out in a few stages the load necessary to produce a given compression was greater for the thin disc than for the thick one. On the other hand, when the load was increased in 100 lb. stages the loads necessary to produce a given compression were identical, and lower than those necessary to produce the compression in a few stages.

If the loads were increased in 200 lb. stages the amount of compression in the thick and thin discs was still nearly though not quite identical, but when the increase in load for each stage was 500 lb. the load necessary to produce a given compression was appreciably increased. It appears, therefore, that for aluminium discs 14 mm. diameter one may increase the load in stages of 200 lb. at a time, before the friction becomes appreciable compared with the normal load.

## Experiment to Find an Upper Limit to the Effect of Friction on Measurements of Compressive Force

As a final and apparently conclusive test as to whether the effect of the friction between the specimen and the steel plates has a measurable effect on the stresses in the specimen, an experiment was devised in which the friction on two identical discs could be applied so as to act in opposite directions in the two cases. To understand

* 'The distortion of crystals of aluminium under compression. Part I', *Proc. Roy. Soc.* A, CXI (1926), 529–31; paper 9 above, pp. 133–5.

how this end can be attained it is necessary to consider the direction of relative motion of the surfaces of the specimen and the steel plates. If a uniform circular disc of non-crystalline material or of soft metal not cut from a single crystal be compressed between parallel plates without lubrication, the friction on both upper and lower surfaces of the specimen is radial and directed inwards. When a piece cut from a single crystal is compressed it becomes elliptical, and if there were no motion of the upper surface as a whole relative to the lower surface, i.e. if the line joining the centres of the upper and lower elliptic faces remained perpendicular to the plane of the surfaces, the resultants of the friction forces on the upper and on the lower surfaces would each be zero. Actually, however, there is a shearing motion of the upper surface relative to the lower one.* If the upper surface of the specimen moves to the right and the lower one to the left one might expect a resultant friction acting towards the left at the upper surface and to the right at the lower surface. It appears that the effect of these friction forces would be to decrease the shearing force on the plane of slip when a given normal load is applied.

On the other hand, if the upper steel plate could be moved laterally with respect to the lower one as the load was applied the resultant friction would be made to act in any assigned direction.

In the experiment now to be described two identical discs, 69·6 and 69·10, were cut from the same crystal, no. 69. These were compressed between parallel polished steel plates which had a lateral as well as a normal movement relative to one another. In the case of 69·6 the lateral movement of the plates was in the opposite direction to the relative motion of the upper and lower surfaces of the specimen. In the case of no. 69·10 it was in the same direction. The way in which this end was attained is illustrated diagrammatically in fig. 3. In that drawing $A_1$ and $A_2$ are the polished steel plates. These were mounted on two solid steel sliders $B_1$ and $B_2$ each 1 in. square. The ends of these sliders were cut at 45° to take the steel plates which were carefully fitted to them. The sliders moved without any detectable play between two parallel steel blocks $C$ and $D$ and were carefully greased before every experiment. The apparatus was made so carefully that although no play could be detected in any position, the block $B_1$ could slide under its own weight very slowly from top to bottom of the slide.

It will be seen that when the plates $A_1$ and $A_2$ approach one another so as to decrease the distance between them by a given amount they also move an equal amount laterally because the total movement is at 45° to their polished surfaces.

The distortion of specimens cut from no. 69 has already been examined.† The direction of relative lateral motion of the upper and lower surfaces of the specimen during compression was, therefore, known. This was marked on the upper surface of each of the specimens by means of an arrow. During the tests the specimen was always placed on the lower steel plate with its marked surface up. In the case of 69·6 the specimen was always set so that the arrow pointed upwards along the line

* See Part I, vol. XCI (1926), p. 531; paper 9 above, p. 135.

† See Part II; paper 12 above, pp. 192 and 196.

of greatest slope on the 45° plate, while in 61·10 the arrow pointed downwards. A side view of a specimen in the machine before compression is shown in fig. 2A. In figs. 2B and C are shown the two specimens 69·10 and 69·6 after compression.

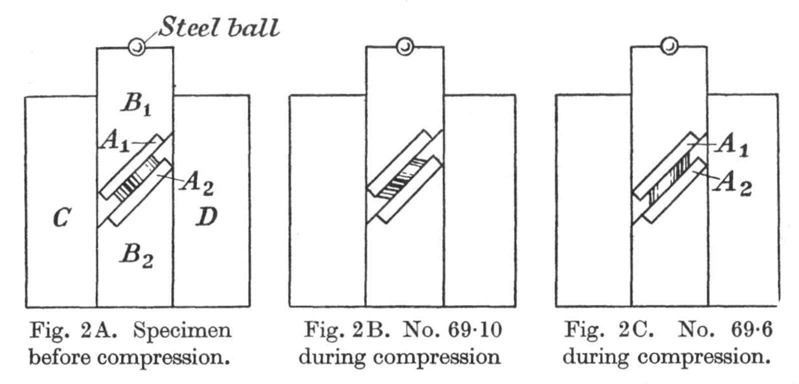


Fig. 2A. Specimen before compression.

Fig. 2B. No. 69·10 during compression

Fig. 2C. No. 69·6 during compression.

Fig. 2. Sketch of apparatus for compression of discs between plates set at 45° to their direction of motion.

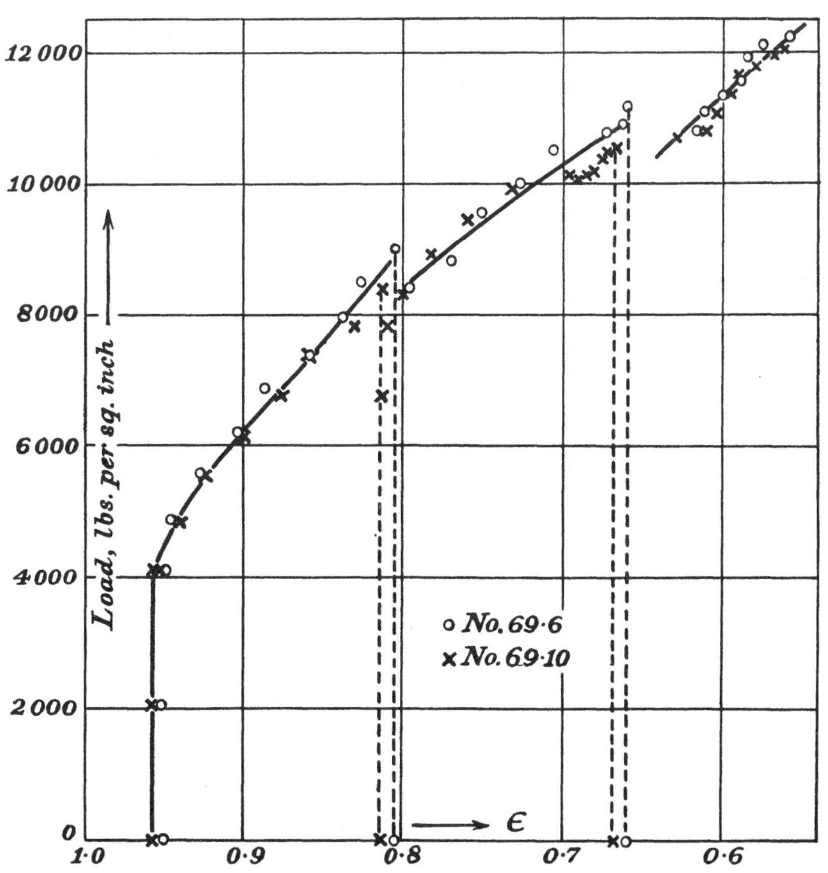


Fig. 3. Comparison between loads when specimens are compressed with different directions of lateral motion of steel plates.

These sketches illustrate, qualitatively at any rate, why the friction should tend to help the normal force in producing deformation in the case of 69·10 and to prevent it in the case of 69·6.

It seems unnecessary to give the results of these experiments in tabular form, but they are shown graphically in fig. 3, where the crosses refer to experiments with disc no. 69·10 and the circles to those with 69·6. The ordinates represent the load applied to the slider $B$, in pounds per square inch of the area of the flat surfaces of the specimens. The abscissae are the values of $\epsilon$, the ratio of the thickness of the disc to its initial thickness. It will be seen that in spite of the difference in the direction of the friction in the two cases the two sets of points fall almost exactly on the same curve. We may, therefore, deduce that the effect of friction on the load required to produce a given compression is negligible.

It will be noticed that the curves are divided into three parts. During the course of the experiment the shearing of the top surface relative to the bottom gave rise to eccentricity of loading and to non-uniformity of distortion towards the edges of the specimens. At two stages of the test, namely, $\epsilon = 0{\cdot}804$ and $\epsilon = 0{\cdot}660$ in the case of 69·6, and at $\epsilon = 0{\cdot}813$ and $\epsilon = 0{\cdot}667$ in the case of 69·10, the specimen was reduced in area and turned circular again in order to keep the distortion from becoming non-uniform, and the breaks in the curve are due to these operations.

Another point which will be noticed is that the readings begin at $\epsilon = 0{\cdot}950$ in one case and $\epsilon = 0{\cdot}958$ in the other. This is because the specimens were first compressed about 5 % between parallel plates without lateral motion in order to verify that the direction of lateral movement of the top surface of the specimen relative to the bottom was in the direction it was expected to be.

## Methods of Calculation

It has been seen that when compression experiments are carried out in the manner described above the friction is so small that no appreciable error is likely to occur if its effects are neglected. Under these circumstances one can calculate the component of shear stress parallel to a slip-plane in a compression specimen by the method that was previously used in the case of tension specimens. If $\theta$ is the angle between a plane of slip and the flat surface of the specimen, $\eta$ the angle between the direction of slip and the line of greatest slope in the slip-plane (the specimen being placed with its flat surfaces horizontal), then the component of shear stress parallel to the direction of slip is $S = (P/a)\sin\theta\cos\theta\cos\eta$, where $a$ is the area of cross-section of the specimen and $P$ is the total load.

To find $s$, the amount of shear corresponding with any degree of compression, one must know the nature of the distortion. Two sets of formulae will be developed, the first for dealing with cases like that of no. 59 discussed in Part I (paper 9) where the distortion is due to slip on one plane, and the second for dealing with cases like that of 69 discussed in Part II (paper 12), where there is equal slipping on two slip-planes situated symmetrically with respect to the axis of the specimen. In cases where the orientation of the axes changes from an unsymmetrical to a symmetrical position during the course of the test both sets of formulae will be used, the first for the stage of single slipping and the second for the stage of double slipping.

## FORMULAE FOR SINGLE SLIPPING

*Compression tests.* Take rectangular co-ordinates $x$, $y$, $z$, so that the plane $z=0$ is parallel to the plane of slip, $Ox$ is the direction of slip and 0 is in the upper surface of the specimen. The spherical polar co-ordinates of the normal to the surface of the specimen are $(\theta, \eta)$ where $\theta$, $\eta$ have the meaning assigned them before. The relation between $\theta$, $\eta$ and $x, y, z$ is

$$\left.\begin{aligned} x&=r\sin\theta\cos\eta,\\ y&=r\sin\theta\sin\eta,\\ z&=r\cos\theta,\end{aligned}\right\} \tag{1}$$

where $r$ is the distance of the point considered from the origin.

Let $(\theta_0, \eta_0)$ be the values of $(\theta, \eta)$ before distortion. Consider two particles $A$ and $B$, both in a surface of the specimen. $A$ is situated at distance $a$ from the origin in the line of intersection of the surface of the disc and the slip-plane. Its co-ordinates before distortion are evidently

$$(x=-a\sin\eta_0, \quad y=a\cos\eta_0, \quad z=0).$$

$B$ is situated at distance $b$ from the origin along a line perpendicular to $OA$, and its co-ordinates before distortion are

$$(x=-b\cos\theta_0\cos\eta_0, \quad y=-b\cos\theta_0\sin\eta_0, \quad z=b\sin\theta_0).$$

The area of the rectangle contained between $0A$ and $0B$ is $ab$.

After distortion due to a shear of amount $s$ defined in accordance with the scheme of fig. 1, the co-ordinates of $A$ remain unaltered because it is in the slip-plane, which is fixed with respect to the co-ordinates that we are considering. The co-ordinates of $B$ become

$$(x=-b\cos\theta_0\cos\eta_0-sb\sin\theta_0, \quad y=-b\cos\theta_0\sin\eta_0, \quad z=b\sin\theta_0).$$

The perpendicular $p$ from the new position of $B$ on to the line $OA$ is given by

$$\begin{aligned}(p^2/b^2)&=\sin^2\theta_0+\{\cos\eta_0(-\cos\theta_0\cos\eta_0-s\sin\theta_0)+\sin\eta_0(-\cos\theta_0\sin\eta_0)\}^2\\ &=1+2s\sin\theta_0\cos\theta_0\cos\eta_0+s^2\sin^2\theta_0\cos^2\eta_0.\end{aligned}$$

The area of the parallelogram included between $0A$ and $0B$ is therefore

$$ab(1+2s\sin\theta_0\cos\theta_0\cos\eta_0+s^2\sin^2\theta_0\cos^2\eta_0)^{\frac{1}{2}}.$$

Since there is no change in volume of the material the thickness multiplied by the area of this parallelogram must be constant. Hence the relationship between $s$ and $\epsilon$ is

$$\epsilon^{-2}=1+2s\sin\theta_0\cos\theta_0\cos\eta_0+s^2\sin^2\theta_0\cos^2\eta_0. \tag{2}$$

The shear stress $S$ is

$$(P/a)\sin\theta\cos\theta\cos\eta. \tag{3}$$

14-2

If $a_0$ is the area of the original specimen then $a = a_0/\epsilon$. It has already been pointed out that during a compression test in which the distortion is due to slipping in one plane the axis of the specimen moves relative to the crystal axes in a great circle towards the normal to the slip-plane. Hence $\eta$ is constant and equal to $\eta_0$. For compression experiments the connection between $\theta$ and $\theta_0$ is

$$\sin\eta = \epsilon \sin\theta_0.$$

Hence
$$S = (P\epsilon^2/a_0)\sin\theta_0\cos\eta_0(1-\epsilon^2\sin^2\theta_0)^{\frac{1}{2}}. \tag{4}$$

*Extension tests.* The connection between $\epsilon$ and $s$ for tensile test-pieces can be found in the same way as for compression tests.

Using rectangular co-ordinates $x, y, z$ which are related to the slip-plane and slip direction in the same way as before (see equation (1)), the initial co-ordinates of a point on the axis of the specimen at distance $r$ from the origin are

$$(r\sin\theta_0\cos\eta_0,\quad r\sin\theta_0\sin\eta_0,\quad r\cos\theta_0).$$

After the material has been subjected to a shear $s$ parallel to the plane $z = 0$, and in the direction of the axis of $x$, the co-ordinates are

$$(r\sin\theta_0\cos\eta_0 + sr\cos\theta_0,\quad r\sin\theta_0\sin\eta_0,\quad r\cos\theta_0),$$

$\epsilon$ the ratio of the final to the initial length* is given by

$$\epsilon^2 = (\sin\theta_0\cos\eta_0 + s\cos\theta_0)^2 + \sin^2\theta_0\sin^2\eta_0 + \cos^2\theta_0$$

or
$$\epsilon^2 = 1 + 2s\cos\theta_0\sin\theta_0\cos\eta_0 + s^2\cos^2\theta_0. \tag{5}$$

The shearing stress $S$ is $S = (T/a)\sin\theta\cos\theta\cos\eta$, where $T$ is the total force applied and $a$ is the area of cross-section of the test-piece. Using the formulae† connecting $\eta$ with $\theta_0$, $\eta_0$

$$\left.\begin{aligned} \epsilon\cos\theta &= \cos\theta_0, \\ \epsilon\sin\theta\sin\eta &= \sin\theta_0\sin\eta_0, \end{aligned}\right\}$$

$$S = (T/a_0)\cos\theta_0\{1-\epsilon^{-2}(\cos^2\theta_0 + \sin^2\theta_0\sin^2\eta_0)\}^{\frac{1}{2}}. \tag{6}$$

## EXPERIMENTAL RESULTS FOR SLIPPING ON ONE PLANE

*Extension experiments.* In the course of the experiments in which single-crystal bars were extended, the force was measured at each stage of the test and the results are given on p. 45 of 'Plastic extension and fracture of aluminium crystals'.‡ Unfortunately, when these tests were made the force was only measured at a few stages in the extension, since it was not at that time realized that it would be of interest to have more points on the curves.

* $\epsilon$ has slightly different meanings in tensile and compression experiments. The same symbol is used in the two cases to make the formulae consistent with those given in previous papers but no confusion need arise.

† 'Plastic extension and fracture of aluminium crystals', *Proc. Roy. Soc.* A, CVIII (1925), 49; paper 7 above, p. 127.

‡ Paper 7 above, p. 125.

It was found that all the specimens had about the same strength for any given value of the extension. On the other hand, all the specimens except no. 72 had been cut square so that the surface at any rate had already been slightly distorted before the test began. For this reason only the measurements on no. 72, which was not cut after being formed into a single crystal, are here analysed. The other specimens gave almost identical results, but, as will be seen in the curve (fig. 10, P.E.), they were less regular.

Referring to P.E., fig. 5, it will be seen that no. 72 slips by single slipping up to 50 % extension, but that between 50 and 60 % extension the crystal axes get into the position for double slipping. Accordingly the analysis of no. 72 has only been carried out up to the stage represented by $\epsilon = 1{\cdot}511$.

For no. 72 the X-ray examination gave $\theta_0 = 41{\cdot}9°$, $\eta_0 = 23°$. Using these values in (5) and (6) and the values of $T$ given in table 4, P.E. p. 45, the values of $S$ and $s$ given in table 1 were found.

Table 1

| $\epsilon$ | $s$ | $S$ (lb. per sq. in.) |
|---|---|---|
| 1·000 | 0 | 0 |
| 1·050 | 0·105 | 1930 |
| 1·103 | 0·214 | 2760 |
| 1·202 | 0·392 | 3580 |
| 1·404 | 0·734 | 4590 |
| 1·511 | 0·902 | 4880 |

The relationship between $S$ and $s$ is shown in fig. 4, where the points are marked thus $\odot$.

*Compression experiments.* Of the three single-crystal bars from which compression discs were cut, only two were capable of yielding results about slipping on one crystal plane, because one of them, no. 69, happened to have its crystal axes so disposed that double slipping began almost at the beginning of the test. It has been shown in Parts I and II that slipping on one crystal plane occurred in no. 59 from $\epsilon = 1{\cdot}0$ to $\epsilon = 0{\cdot}67$, while in no. 61 it occurred from $\epsilon = 1{\cdot}0$ to $\epsilon = 0{\cdot}815$. Accordingly, the measurements have been analysed in these regions by means of formulae (2) and (4). The results are given in table 2. In the case of no. 61 the distortion was uniform from the beginning of the test, whereas in the case of no. 59 the distortion at the edges of the specimen was not uniform in the early stages of the test.* The test on disc no. 59·9 was carried out in two series of stages. The initial diameter of the disc was 14·135 mm. and its thickness was 2·220 mm. It was first compressed by small steps to a thickness of 1·938 mm., i.e. to $\epsilon = 0{\cdot}874$. It was then cut down to a disc 8·005 mm. diameter and compressed again till its thickness was 1·496 mm., so that $\epsilon = 0{\cdot}674$. During the compression from $\epsilon = 1{\cdot}0$ to $\epsilon = 0{\cdot}874$ the distortion at the edge of the specimen was not uniform so that the value of $S$ calculated from (4) is not quite the true value, but the values obtained from $\epsilon = 0{\cdot}874$ to $\epsilon = 0{\cdot}674$, where the distortion was uniform, should be true values of the resistance to slipping. On

* See Part I; paper 9 above, p. 145.

Table 2. *Values of S and s for discs 59·9 and 61·17 with data from which they have been calculated*

| Thickness (mm.) | Load (lb.) | $e$ | $S$ (lb. per sq. in.) | $s$ |
|---|---|---|---|---|
| Disc 59·9, $\theta_0=57°$, $\eta_0=13°$. Initial diameter 14·135 mm. | | | | |
| 2·220 | 0 | 1·00 | 0 | 0·000 |
| 2·220 | 112 | 1·00 | 199 | 0·000 |
| 2·220 | 224 | 1·00 | 388 | 0·000 |
| 2·218 | 336 | 0·999 | 596 | 0·002 |
| 2·215 | 448 | 0·998 | 796 | 0·004 |
| 2·210 | 560 | 0·995 | 994 | 0·011 |
| 2·208 | 672 | 0·994 | 1190 | 0·013 |
| 2·189 | 784 | 0·987 | 1400 | 0·027 |
| 2·176 | 896 | 0·980 | 1600 | 0·043 |
| 2·168 | 1010 | 0·977 | 1805 | 0·049 |
| 2·143 | 1200 | 0·966 | 2135 | 0·074 |
| 2·105 | 1400 | 0·950 | 2490 | 0·114 |
| 2·046 | 1600 | 0·922 | 2810 | 0·171 |
| 2·007 | 1800 | 0·904 | 3120 | 0·212 |
| 1·938 | 2000 | 0·874 | 3390 | 0·281 |
| Disc 59·9 cut down to 8·005 mm. diameter. | | | | |
| 1·938 | 0 | 0·874 | 0 | 0·281 |
| 1·849 | 500 | 0·833 | 2890 | 0·378 |
| 1·845 | 550 | 0·831 | 3160 | 0·383 |
| 1·843 | 600 | 0·830 | 3450 | 0·385 |
| 1·808 | 650 | 0·814 | 3670 | 0·426 |
| 1·766 | 700 | 0·796 | 3840 | 0·477 |
| 1·720 | 750 | 0·774 | 3970 | 0·530 |
| 1·681 | 800 | 0·757 | 4110 | 0·577 |
| 1·643 | 850 | 0·740 | 4250 | 0·624 |
| 1·615 | 900 | 0·728 | 4400 | 0·679 |
| 1·595 | 950 | 0·718 | 4560 | 0·684 |
| 1·554 | 1000 | 0·700 | 4630 | 0·738 |
| 1·530 | 1050 | 0·689 | 4740 | 0·776 |
| 1·496 | 1100 | 0·674 | 4810 | 0·823 |
| Disc 61·17, $\theta_0=46°$, $\eta_0=17°$. Initial diameter 14·16 mm. | | | | |
| 3·114 | 0 | 1·000 | 0 | 0 |
| 3·102 | 100 | 0·9962 | 194 | 0·007 |
| 3·090 | 200 | 0·9923 | 389 | 0·016 |
| 3·086 | 300 | 0·9910 | 581 | 0·019 |
| 3·084 | 400 | 0·9904 | 774 | 0·021 |
| 3·070 | 500 | 0·986 | 967 | 0·030 |
| 3·064 | 600 | 0·984 | 1157 | 0·033 |
| 3·048 | 700 | 0·979 | 1340 | 0·044 |
| 3·033 | 800 | 0·974 | 1530 | 0·056 |
| 3·010 | 900 | 0·967 | 1715 | 0·073 |
| 2·986 | 1000 | 0·960 | 1910 | 0·089 |
| 2·950 | 1100 | 0·950 | 2040 | 0·108 |
| 2·928 | 1200 | 0·940 | 2200 | 0·120 |
| 2·892 | 1300 | 0·929 | 2350 | 0·153 |
| 2·864 | 1400 | 0·920 | 2500 | 0·176 |
| 2·831 | 1500 | 0·909 | 2640 | 0·196 |
| 2·795 | 1600 | 0·899 | 2780 | 0·225 |
| 2·763 | 1700 | 0·887 | 2900 | 0·253 |
| 2·737 | 1800 | 0·879 | 3120 | 0·274 |
| 2·713 | 1900 | 0·871 | 3160 | 0·291 |
| 2·668 | 2000 | 0·857 | 3250 | 0·327 |
| 2·631 | 2100 | 0·845 | 3350 | 0·356 |
| 2·601 | 2200 | 0·835 | 3460 | 0·383 |
| 2·560 | 2300 | 0·822 | 3530 | 0·417 |
| 2·530 | 2325 | 0·812 | | |
| 2·510 | 2330 | 0·806 | | |

the other hand, the distortion of no. 61 was uniform over the whole disc during compression from $\epsilon = 1{\cdot}0$ to $\epsilon = 0{\cdot}815$, so that values obtained from formulae (2) and (4) may be expected to be accurate.

The results together with the experimental data are given in table 2. They are also shown graphically in fig. 4, which also shows the results of extension tests.

Inspection of that figure reveals two facts. First, the points corresponding with the compression of 59·9 from $\epsilon = 0{\cdot}874$ to $\epsilon = 0{\cdot}674$ lie in a continuous curve with the points representing the compression of 61·17 from $\epsilon = 1{\cdot}0$ to $\epsilon = 0{\cdot}815$. The points representing the compression of 59·9 from $\epsilon = 1{\cdot}0$ to $\epsilon = 0{\cdot}874$ are above the curve representing the compression for 61·17. This is evidently the effect of want of uniformity in the distortion of 59·9 in the early stages of compression, for the

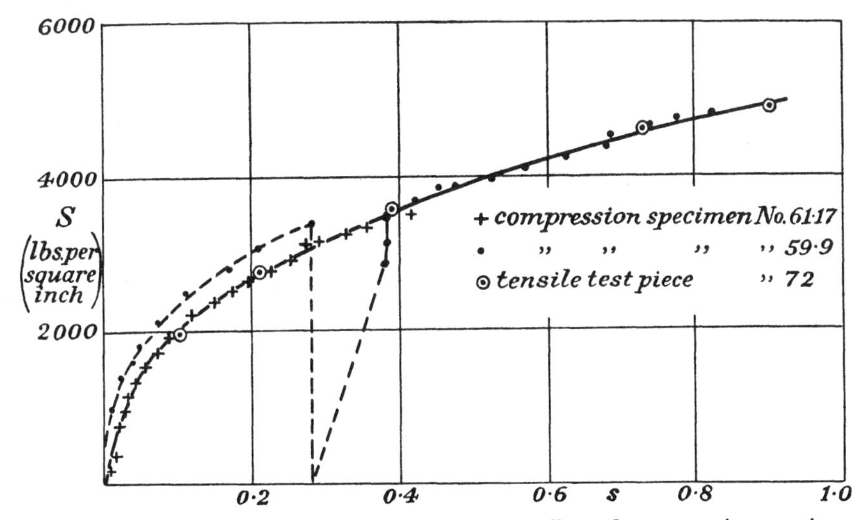


Fig. 4. Relationship between $S$ and $s$ for tensile and compression specimens when slipping on one plane.

difference between the two specimens disappears as soon as the non-uniform part of the specimen has been removed. The points corresponding with the early stages of compression for 59·9 have been connected by a dotted curve in fig. 4. The portion between $\epsilon = 0{\cdot}874$ and $\epsilon = 0{\cdot}833$ is missing because the first load applied after cutting down the specimen was greater than it should have been.

The second point which should be noticed is that the points deduced from extension experiments lie on the same curve as the points deduced from compression experiments. This is interesting because it means that it is immaterial whether the component of force normal to the slip-plane is a tension or a compression; the resistance to slipping for a given amount of distortion is identical in the two cases.

It has been shown by Dr E. Schmid* that in the case of stretched crystals the resistance to shear parallel to the slip-plane is independent of the component of tension normal to the plane of slip. The present measurements extend this result

* Loc. cit.

to a case where the force normal to the slip-plane changes from a tensile to a compressive force.

*Double slipping.* In the case of a material which slips equally on two planes it is necessary to define a quantity to represent the amount of slipping. Such a definition has been given in Part II where the total amount of slip $s$ defining the state of the material at any stage is the sum of all the elements of shear on the two planes separately. Thus a small amount $\delta s$ of double slipping is conceived to consist of two small shears each equal to $\frac{1}{2}\delta s$ on each of the two planes of slip concerned. These two elements of single slipping, which go to make up the element of double slipping, may be conceived so far as the geometry of the distortion is concerned to occur consecutively—not simultaneously.

With slipping on one crystal plane we have seen that each increase in the amount of shear implies a corresponding increase in the resistance to shear. If each of the two elements of shear $\frac{1}{2}\delta s$ which combine to make an element $\delta s$ of double slipping contribute the same increase of resistance to shear on all the crystal planes that they would if the material were undergoing distortion by slipping on one plane, then it will be seen that the values of $S$ for any given value of $s$ would be the same whether the distortion had proceeded by single or by double slipping. If, therefore, the values of $S=(P/a)\cos\theta\sin\theta\cos\eta$ and $s$, as defined, are calculated, and the results plotted on a diagram similar to that of fig. 4, any difference between the curve for double slipping and the curve for single slipping will be due to the effect of simultaneous slipping, on two planes as distinct from the sum of the effects of slipping on each of the planes considered separately.

*Formulae for double slipping.* The connection between $\epsilon$ and $s$ has been given in equation (8) of Part II.

It is

$$A^2(1+2\sin i_0\cos i_0)-2A(\sin i_0\cos i_0+\sin^2 i_0)+2\sin^2 i_0-\epsilon_2^{-2}=0, \tag{7}$$

where

$$s_2=5\cdot640\log_{10}A. \tag{7A}$$

$\epsilon_2$ is the ratio of the thickness at the stage concerned to the thickness when double slipping began. $s_2$ is the amount of shear which has occurred since the crystal axes got into the symmetrical position for double slipping. If $s_1$ is the amount of shear which occurred by single slipping before the double slipping began, then $s$, the total amount of shear, may be defined as $s=s_1+s_2$. If $\epsilon_1$ is the value of $\epsilon$ when double slipping began, then evidently $\epsilon$, defined as the ratio of the thickness at any stage to the initial thickness at the beginning of the test, is equal to $\epsilon_1\epsilon_2$. The symbol $i_0$ represents the angle between the line of intersection of the two planes of slip and the flat face of the specimen at the time when double slipping began (see fig. 8, Part II).

The component of shear stress $S$ is

$$(P/a)\cos\theta\sin\theta\cos\eta, \tag{8A}$$

where $a$ is the area of the disc and $P$ the total load. In the symmetrical position both

$\theta$ and $\eta$ are evidently functions of $i$ the angle between the line of intersection of the slip-planes and the plane of the disc. The slip-planes each make an angle of $54° 44'$ with the symmetrical $\{100\}$ plane which contains the normal to the plane of the disc. The direction of slip makes an angle of $60°$ with the intersection of the two slip-planes. Using these relations it can be shown that

$$\cos\theta \sin\theta \cos\eta = \frac{1}{\sqrt{6}}\left(\frac{1+\tan i}{1+\tan^2 i}\right). \tag{8B}$$

The relationship between $i$ and $i_0$ is given in equation (9), Part II. It is

$$\cot i = A(\cot i_0 + 1) - 1. \tag{9}$$

### Application to Nos. 69·9 and 61·17

In the case of no. 69 where the crystal axes were in the position for double slipping at the beginning of the test, $i_0 = 12°$ and $s_1 = 0$.

The equation for $A$ is

$$1{\cdot}4067A^2 - 0{\cdot}4928A + 0{\cdot}0862 - \epsilon^{-2} = 0. \tag{10}$$

Using this equation a series of values of $\epsilon$ were taken and the corresponding values of $A$ were found. These are given in columns 1 and 2 of table 3. Using the formula (7 A) the values of $s$ given in column 3 were found. The relationship between $s$ and $\epsilon$ was then plotted in the curve shown in fig. 5.

To find the corresponding values of $\cos\theta \sin\theta \cos\eta$ the values of $\cot i$ were first calculated by means of equation (9). The corresponding values of $i$ in degrees are given in column 4 of table 3. These were then inserted in equation (8 B) and the values of $\cos\theta \sin\theta \cos\eta$ given in column 5 were calculated. These were plotted on the same diagram (fig. 5), as the values of $s$ and the two curves were used with the measured values of $P$ and $\epsilon$ to deduce the values of $S$ and $s$ given in columns 4 and 5 of table 4.

It will be noticed that in (8 A) the area of the disc occurs. This was not measured in each case. The compression was carried out in a series of stages, at the beginning of each of which the specimen was cut circular. Since there is no change in volume the value of $a\epsilon$ is constant during each of these stages, and the value of $a$ was calculated from this relationship.

In the case of 61·17 the specimen was first compressed to $\epsilon = 0{\cdot}815$ and the formulae applicable to single slipping were used. The value of $s_1$ at the end of the stage of slipping on one plane was $s_1 = 0{\cdot}435$. The value of $i_0$ at this stage was $7{\cdot}2°$. The values of $S$ and $s$, i.e. $s_1 + s_2$, corresponding with each value of $\epsilon$ during the period of double slipping are given in columns 4 and 5 of table 5.

In both cases the test was carried out in several stages. Specimen no. 69·9 was compressed in five stages. At the beginning of the first stage the thickness was 2·904 mm. and the diameter 14·2 mm. It was then compressed to a thickness of 2·189 mm., i.e. to $\epsilon = 0{\cdot}754$ in 23 steps, the specimen being removed from between the steel plates and covered with a layer of grease before each of these 23 increases in load.

Table 3. *Showing the calculated values of $A$, $i$ and $\cos\theta \sin\theta \cos\eta$ assuming equal double slipping and $i_0 = 12°$*

| $\epsilon$ | $A$ | $s$ | $i$ (degrees) | $\cos\theta \sin\theta \cos\eta$ |
|---|---|---|---|---|
| 1·00 | 1·000 | 0·000 | 12 | 0·473 |
| 0·95 | 1·045 | 0·108 | 11·4 | 0·471 |
| 0·90 | 1·096 | 0·224 | 10·8 | 0·468 |
| 0·85 | 1·150 | 0·346 | 10·2 | 0·466 |
| 0·80 | 1·215 | 0·477 | 9·6 | 0·463 |
| 0·75 | 1·285 | 0·614 | 9·0 | 0·461 |
| 0·70 | 1·368 | 0·768 | 8·4 | 0·458 |
| 0·65 | 1·461 | 0·929 | 7·7 | 0·455 |
| 0·60 | 1·571 | 1·107 | 7·1 | 0·452 |
| 0·55 | 1·698 | 1·298 | 6·6 | 0·450 |
| 0·50 | 1·352 | 1·510 | 6·0 | 0·447 |
| 0·45 | 2·040 | 1·747 | 5·4 | 0·443 |
| 0·40 | 2·275 | 2·014 | 4·8 | 0·439 |
| 0·35 | 2·577 | 2·320 | 4·2 | 0·435 |
| 0·30 | 2·978 | 2·674 | 3·6 | 0·431 |
| 0·25 | 3·542 | 3·097 | 3·0 | 0·428 |
| 0·20 | 4·390 | 3·624 | 2·3 | 0·424 |
| 0·10 | 8·603 | 5·265 | 1·2 | 0·416 |
| 0 | ∞ | ∞ | 0 | 0·408 |

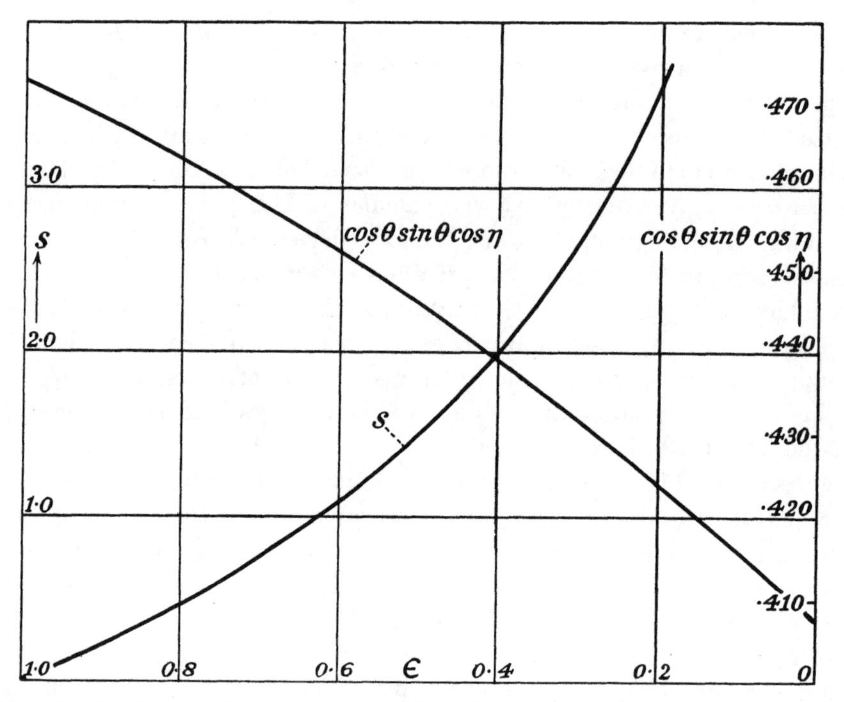


Fig. 5. Calculated values of $s$ and $\cos\theta \sin\theta \cos\eta$ during compression of discs cut from specimen no. 69.

Table 4. *Double slipping from beginning of test*

No. 69·9. First stage. $i_0 = 12°$. Initial diameter 14·42 mm.

| Thickness (mm.) | Load (lb.) | *e* | *S* (lb. per sq. in.) | *s* |
|---|---|---|---|---|
| 2·904 | 0 | 1·00 | 0 | 0·000 |
| 2·895 | 100 | 0·997 | 192 | 0·006 |
| 2·892 | 200 | 0·996 | 385 | 0·008 |
| 2·885 | 400 | 0·994 | 767 | 0·012 |
| 2·875 | 600 | 0·990 | 1145 | 0·020 |
| 2·859 | 800 | 0·984 | 1520 | 0·032 |
| 2·833 | 1000 | 0·976 | 1880 | 0·050 |
| 2·795 | 1200 | 0·963 | 2220 | 0·080 |
| 2·753 | 1400 | 0·948 | 2550 | 0·111 |
| 2·717 | 1600 | 0·935 | 2860 | 0·142 |
| 2·674 | 1800 | 0·921 | 3160 | 0·174 |
| 2·617 | 2000 | 0·901 | 3430 | 0·220 |
| 2·562 | 2200 | 0·882 | 3790 | 0·260 |
| 2·510 | 2400 | 0·864 | 3930 | 0·304 |
| 2·490 | 2400 | 0·857 | 3910 | 0·320 |
| 2·462 | 2500 | 0·848 | 4020 | 0·342 |
| 2·424 | 2600 | 0·834 | 4100 | 0·380 |
| 2·388 | 2700 | 0·822 | 4190 | 0·415 |
| 2·366 | 2800 | 0·814 | 4300 | 0·432 |
| 2·340 | 2900 | 0·806 | 4400 | 0·453 |
| 2·305 | 3000 | 0·794 | 4500 | 0·485 |
| 2·277 | 3100 | 0·784 | 4570 | 0·510 |
| 2·246 | 3200 | 0·773 | 4660 | 0·545 |
| 2·212 | 3300 | 0·762 | 4720 | 0·570 |
| 2·189 | 3400 | 0·754 | 4800 | 0·595 |

No. 69·9. Second stage. Cut down to disc 10·36 mm. diameter.

| Thickness (mm.) | Load (lb.) | *e* | *S* (lb. per sq. in.) | *s* |
|---|---|---|---|---|
| 2·189 | 0 | 0·754 | 0 | 0·595 |
| 2·186 | 1000 | 0·753 | 3790 | 0·600 |
| 2·180 | 1100 | 0·750 | 4150 | 0·606 |
| 2·169 | 1200 | 0·746 | 4500 | 0·620 |
| 2·158 | 1250 | 0·743 | 4670 | 0·629 |
| 2·154 | 1300 | 0·742 | 4840 | 0·630 |
| 2·138 | 1350 | 0·736 | 4980 | 0·650 |
| 2·135 | 1400 | 0·735 | 5160 | 0·656 |
| 2·064 | 1450 | 0·710 | 5180 | 0·731 |
| 2·064 | 1500 | 0·710 | 5350 | 0·731 |
| 2·034 | 1550 | 0·700 | 5430 | 0·768 |
| 2·000 | 1600 | 0·689 | 5500 | 0·803 |
| 1·995 | 1650 | 0·687 | 5650 | 0·809 |
| 1·986 | 1700 | 0·684 | 5800 | 0·820 |
| 1·950 | 1750 | 0·671 | 5850 | 0·860 |
| 1·938 | 1800 | 0·667 | 5980 | 0·873 |
| 1·928 | 1850 | 0·664 | 6110 | 0·885 |
| 1·878 | 1900 | 0·646 | 6110 | 0·943 |
| 1·872 | 1950 | 0·644 | 6220 | 0·950 |
| 1·840 | 2000 | 0·634 | 6280 | 0·984 |
| 1·830 | 2050 | 0·630 | 6400 | 1·000 |
| 1·830 | 2050 | 0·630 | 6400 | 1·000 |
| 1·790 | 2100 | 0·616 | 6410 | 1·050 |

Table 4. (*cont.*)

| Thickness (mm.) | Load (lb.) | $\epsilon$ | $S$ (lb. per sq. in.) | $s$ |
|---|---|---|---|---|
| No. 69·9. Third stage. Cut down to 8·560 mm. diameter. | | | | |
| 1·790 | 0 | 0·616 | 0 | 1·050 |
| 1·760 | 1150 | 0·606 | 5740 | 1·090 |
| 1·750 | 1200 | 0·602 | 5940 | 1·105 |
| 1·720 | 1250 | 0·592 | 6090 | 1·149 |
| 1·670 | 1300 | 0·575 | 6140 | 1·213 |
| 1·612 | 1350 | 0·554 | 6110 | 1·305 |
| 1·606 | 1400 | 0·553 | 6350 | 1·310 |
| 1·582 | 1450 | 0·544 | 6440 | 1·345 |
| 1·526 | 1500 | 0·525 | 6430 | 1·423 |
| 1·480 | 1550 | 0·509 | 6430 | 1·490 |
| 1·400 | 1700 | 0·482 | 6640 | 1·615 |
| 1·398 | 1750 | 0·481 | 6830 | 1·620 |
| 1·397 | 1800 | 0·481 | 7000 | 1·620 |
| 1·387 | 1850 | 0·477 | 7140 | 1·635 |
| 1·321 | 1900 | 0·455 | 6950 | 1·735 |
| No. 69·9. Fourth stage. Cut down to 7·047 mm. diameter. | | | | |
| 1·321 | 0 | 0·4550 | 0 | 1·735 |
| 1·316 | 800 | 0·4530 | 5840 | 1·740 |
| 1·314 | 850 | 0·4525 | 6200 | 1·740 |
| 1·310 | 900 | 0·4510 | 6550 | 1·750 |
| 1·308 | 950 | 0·4505 | 6920 | 1·750 |
| 1·304 | 1000 | 0·4490 | 7240 | 1·760 |
| 1·278 | 1050 | 0·4400 | 7400 | 1·805 |
| 1·224 | 1100 | 0·4213 | 7460 | 1·900 |
| 1·178 | 1150 | 0·4055 | 7460 | 1·992 |
| 1·121 | 1200 | 0·3860 | 7400 | 2·093 |
| 1·104 | 1250 | 0·3800 | 7580 | 2·130 |
| 1·085 | 1300 | 0·3735 | 7710 | 2·173 |
| 1·032 | 1350 | 0·3553 | 7610 | 2·285 |
| 1·013 | 1400 | 0·3490 | 7710 | 2·330 |
| No. 69·9. Fifth stage. Cut down to 6·950 mm. diameter. | | | | |
| 1·013 | 0 | 0·3490 | 0 | 2·330 |
| 1·010 | 900 | 0·3480 | 6650 | 2·335 |
| 0·991 | 950 | 0·3413 | 6990 | 2·385 |
| 0·982 | 1000 | 0·3382 | 7170 | 2·400 |
| 0·951 | 1050 | 0·3273 | 7270 | 2·480 |
| 0·927 | 1100 | 0·3190 | 7420 | 2·537 |
| 0·905 | 1150 | 0·3116 | 7560 | 2·595 |
| 0·875 | 1200 | 0·3013 | 7600 | 2·670 |
| 0·859 | 1250 | 0·2958 | 7780 | 2·710 |
| 0·837 | 1300 | 0·2880 | 7860 | 2·770 |
| 0·818 | 1350 | 0·2817 | 7980 | 2·820 |
| 0·806 | 1400 | 0·2776 | 8140 | 2·855 |

Table 5

No. 61·17. Second stage of compression. Double slipping begins at $\epsilon = 0{\cdot}815$ and then $i_0 = 7{\cdot}2°$ and $s_1 = 0{\cdot}435$. Disc cut down to 13·863 mm. diameter.

| Thickness (mm.) | Load (lb.) | $\epsilon$ | $S$ (lb. per sq. in.) | $s_1 + s_2$ |
|---|---|---|---|---|
| 2·508 | 1800 | 0·805 | 3480 | 0·455 |
| 2·496 | 1900 | 0·802 | 3650 | 0·460 |
| 2·483 | 2000 | 0·798 | 3830 | 0·473 |
| 2·474 | 2100 | 0·795 | 4000 | 0·480 |
| 2·456 | 2200 | 0·789 | 4160 | 0·500 |
| 2·438 | 2250 | 0·783 | 4230 | 0·512 |
| 2·420 | 2300 | 0·777 | 4260 | 0·535 |
| 2·403 | 2350 | 0·772 | 4340 | 0·545 |
| 2·397 | 2400 | 0·770 | 4430 | 0·550 |
| 2·394 | 2450 | 0·769 | 4520 | 0·552 |
| 2·366 | 2500 | 0·760 | 4540 | 0·578 |
| 2·353 | 2550 | 0·756 | 4610 | 0·590 |
| 2·344 | 2600 | 0·753 | 4680 | 0·600 |
| 2·328 | 2650 | 0·748 | 4730 | 0·615 |
| 2·306 | 2700 | 0·740 | 4760 | 0·638 |
| 2·302 | 2750 | 0·739 | 4850 | 0·642 |
| 2·275 | 2800 | 0·731 | 4875 | 0·670 |
| 2·254 | 2850 | 0·724 | 4920 | 0·694 |
| 2·238 | 2900 | 0·719 | 4960 | 0·708 |
| 2·226 | 2950 | 0·715 | 5010 | 0·720 |
| 2·218 | 3000 | 0·712 | 5080 | 0·730 |
| 2·203 | 3050 | 0·708 | 5130 | 0·740 |
| 2·197 | 3100 | 0·706 | 5200 | 0·750 |
| No. 61·17. Third stage. Cut down to disc 10·028 mm. diameter. | | | | |
| 2·198 | 0 | 0·706 | 0 | 0·750 |
| 2·122 | 1500 | 0·682 | 5280 | 0·825 |
| 2·091 | 1550 | 0·671 | 5370 | 0·860 |
| 2·060 | 1600 | 0·662 | 5450 | 0·890 |
| 2·039 | 1650 | 0·654 | 5550 | 0·916 |
| 2·009 | 1700 | 0·645 | 5660 | 0·945 |
| 1·992 | 1750 | 0·640 | 5740 | 0·965 |
| 1·970 | 1800 | 0·632 | 5850 | 0·990 |
| 1·942 | 1850 | 0·623 | 5910 | 1·022 |
| 1·914 | 1900 | 0·614 | 5990 | 1·052 |
| 1·899 | 1950 | 0·610 | 6100 | 1·075 |
| 1·859 | 2000 | 0·597 | 6100 | 1·125 |
| 1·849 | 2050 | 0·593 | 6220 | 1·138 |
| No. 61·17. Fourth stage. Cut down to disc 8·522 mm. diameter. | | | | |
| 1·849 | 0 | 0·593 | 0 | 1·140 |
| 1·848 | 1100 | 0·593 | 5490 | 1·140 |
| 1·838 | 1150 | 0·590 | 5710 | 1·155 |
| 1·823 | 1200 | 0·585 | 5900 | 1·170 |
| 1·821 | 1250 | 0·584 | 6140 | 1·175 |
| 1·814 | 1300 | 0·582 | 6370 | 1·185 |
| 1·798 | 1350 | 0·577 | 6540 | 1·203 |
| 1·745 | 1400 | 0·560 | 6570 | 1·278 |
| 1·715 | 1450 | 0·559 | 6680 | 1·315 |
| 1·706 | 1500 | 0·548 | 6860 | 1·325 |
| 1·652 | 1550 | 0·530 | 6870 | 1·400 |
| 1·640 | 1600 | 0·526 | 6910 | 1·417 |
| 1·605 | 1650 | 0·515 | 6970 | 1·460 |
| 1·540 | 1700 | 0·495 | 6900 | 1·555 |
| 1·502 | 1750 | 0·482 | 6900 | 1·615 |

Table 5. (*cont.*)

No. 61·17. Fifth stage. Cut down to disc 7·531 mm. diameter.

| Thickness (mm.) | Load (lb.) | $\epsilon$ | $S$ (lb. per sq.in.) | $s_1+s_2$ |
|---|---|---|---|---|
| 1·502 | 0 | 0·482 | 0 | 1·615 |
| 1·495 | 900 | 0·480 | 5670 | 1·625 |
| 1·493 | 1000 | 0·479 | 6280 | 1·628 |
| 1·492 | 1050 | 0·479 | 6600 | 1·630 |
| 1·488 | 1100 | 0·478 | 6890 | 1·635 |
| 1·485 | 1150 | 0·477 | 7160 | 1·640 |
| 1·446 | 1200 | 0·464 | 7280 | 1·705 |
| 1·348 | 1250 | 0·433 | 7050 | 1·863 |
| 1·342 | 1300 | 0·432 | 7280 | 1·870 |
| 1·309 | 1350 | 0·420 | 7360 | 1·940 |
| 1·275 | 1400 | 0·409 | 7420 | 1·998 |
| 1·248 | 1450 | 0·401 | 7530 | 2·040 |
| 1·208 | 1500 | 0·388 | 7540 | 2·110 |
| 1·190 | 1550 | 0·382 | 7650 | 2·140 |
| 1·175 | 1600 | 0·377 | 7800 | 2·172 |

The successive stages were:

Stage 2 cut down to diameter 10·36 mm. and compressed from $\epsilon = 0{\cdot}754$ to 0·616 in 22 steps.

Stage 3 cut down to diameter 8·56 mm. and compressed from $\epsilon = 0{\cdot}616$ to 0·455 in 23 steps.

Stage 4 cut down to diameter 7·047 mm. and compressed from $\epsilon = 0{\cdot}455$ to 0·349 in 13 steps.

Stage 5 cut down to diameter 6·950 mm. and compressed from $\epsilon = 0{\cdot}349$ to 0·2776 in 11 steps.

In the case of no. 61·17 the stages were:

Stage 1 diameter 14·15 mm. compressed from $\epsilon = 1{\cdot}0$ to $\epsilon = 0{\cdot}806$ in 26 steps.
Stage 2 diameter 13·86 mm. compressed from $\epsilon = 0{\cdot}806$ to $\epsilon = 0{\cdot}706$ in 22 steps.
Stage 3 diameter 10·028 mm. compressed from $\epsilon = 0{\cdot}706$ to $\epsilon = 0{\cdot}593$ in 12 steps.
Stage 4 diameter 8·522 mm. compressed from $\epsilon = 0{\cdot}593$ to $\epsilon = 0{\cdot}482$ in 14 steps.
Stage 5 diameter 7·531 mm. compressed from $\epsilon = 0{\cdot}482$ to $\epsilon = 0{\cdot}3775$ in 15 steps.

The results given in tables 4 and 5 are represented in fig. 6. In that diagram the dots refer to experiments with 69·9 and the crosses to experiments with 61·17. Dotted lines have been drawn corresponding with the values of $\epsilon$ at which the specimens were cut down. For comparison the curve for single slipping has been transferred from fig. 4 and marked on fig. 6.

It will be seen that the curve for 69·9 lies above the single slipping curve, thus proving that the material hardens more rapidly when a given total amount of slipping is divided equally between two planes than when it is all on one plane. The effect is not very large, amounting roughly to about 20 %.

Another point which is noticeable in the curve is that the resistance to shear goes on increasing steadily up to the end of the test when the distortion is very great.

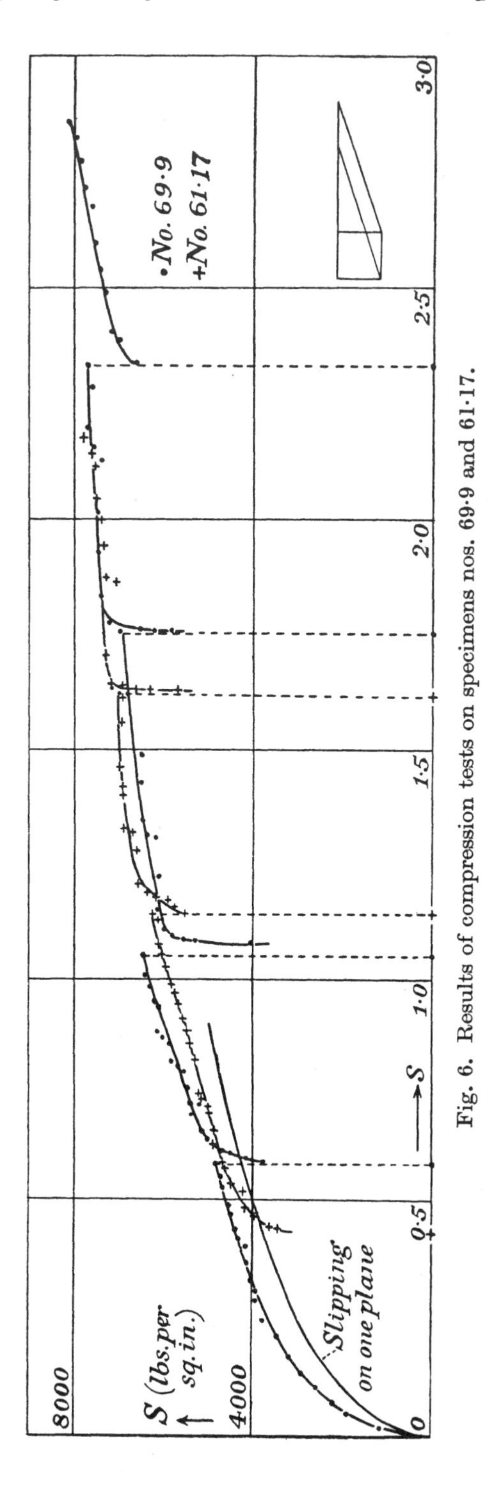


Fig. 6. Results of compression tests on specimens nos. 69·9 and 61·17.

A small diagram is inset in fig. 6 to show the amount of shear corresponding with the last point on the curve. This represents the distorted position of a cube sheared by single slipping by an amount equal to the maximum value of $s$.

It will be noticed that the curve corresponding with the last stage in the compression of 69·9 is not quite continuous with the curve for the first four stages of the compression. This is probably due to the fact that the material was disturbed by scratching six lines on the surface before the last compression in order to measure the distortion in the last stage.*

In conclusion, I wish to express my thanks to Sir Ernest Rutherford for allowing the work to be carried out in the Cavendish Laboratory, to Miss Elam for supplying the single crystals with which the experiments were carried out, and to Mr Farren for his help in designing some of the apparatus used, particularly that shown in fig. 2. The compression measurements were made with a Buckton compression machine in the Cambridge Engineering Laboratory, by kind permission of Professor Inglis.

* This was described in Part II, paper 12 above.

# 14

# THE DEFORMATION OF CRYSTALS OF β-BRASS

REPRINTED FROM

*Proceedings of the Royal Society*, A, vol. CXVIII (1928), pp. 1–24

It is shown that $\beta$-brass, which has a crystal structure similar to that of $\alpha$-iron, behaves in a similar, though not identical, manner when distorted.

The peculiar feature of the distortion of iron crystals, namely, the fact that slip does not occur on a definite crystallographic plane, is repeated in $\beta$-brass within a certain range of orientations of the crystal axes in the specimen. On the other hand, in another range of orientation slip occurs on a definite crystal plane of type {110}. The conditions which determine which of these types of distortion will occur in any given case are investigated, and it is shown that the determining cause is the variation in resistance to shear which occurs as the plane of slip rotates about the direction of slip. This variation is calculated from the experimental results within the range to which they apply, and it is shown that resistance to shear is least when the plane of slip coincides with a crystal plane of type {110}. On either side of this position shear stress increases linearly, there being a discontinuity in the rate of change of shear strength with orientation of plane of slip. This peculiar property is also possessed by the model consisting of hexagonal rods proposed by the author and Miss Elam, and by Mr Hume Rothery's model, which, from this point of view, is identical with the model of hexagonal rods.

The assumptions made in a recent paper by Mr Hume Rothery are discussed, and it is shown that in a body-centred cubic structure like $\alpha$-iron they lead, when analysed, to the conclusion that slipping should always take place parallel to a crystal plane of type {110}, and not, as he stated, to the conclusion that slip should occur on the non-crystallographic planes which were actually found.

It is shown also that in $\beta$-brass resistance to slipping in one direction on a given plane of slip is not the same as resistance offered to slipping in the opposite direction. Such a difference is to be expected from crystallographic symmetry, but was not observed in $\alpha$-iron.

In a recent paper* it has been shown that the behaviour of single crystals of iron when deformed by external pressures or tension is different from that of single crystals of any other metal which had been examined up to that time. In all other metals the material distorts in such a way that planes of particles parallel to definite crystal planes remain undistorted while they slip over one another parallel to some definite crystallographic direction in that plane.

In the case of iron there is no such plane. The particles of the material appear to cling together in lines or rods instead of in planes. These rods, which are in the direction in which the atoms of the crystal are closest together, are capable of sliding over one another. In any uniform distortion due to slipping of this type it is a geometrical necessity that there must be one set of parallel planes, containing the

* 'The distortion of iron crystals', *Proc. Roy. Soc.* A, CXII (1926), 337; paper 10 above.

direction of the axes of the rods, which remains unstretched and undistorted, and in this sense the material has a plane of slipping, but it is not a definite crystal plane, being differently orientated with respect to the crystal axes in different specimens.

It seemed probable that this peculiarity in the nature of the distortion of iron is connected with the crystal structure. Metals which crystallize in face-centred cubes such as aluminium, copper, silver and gold all behave in the same way when distorted. The question therefore naturally arises as to whether other metals besides iron which crystallize in body-centred cubes would behave like iron in distortion, or whether the peculiar behaviour of iron is connected with some peculiarity in the atom of iron itself. The metals which crystallize in body-centred cubes are, however, mostly unsuitable for experiments of the kind previously carried out with face-centred cubic metallic crystals. Tungsten, for instance, seems at present to be unobtainable in the form of large ductile crystals, and its hardness would in any case make compression experiments difficult to carry out.

On the other hand, it was pointed out to me by Miss Elam that the alloy $\beta$-brass which crystallizes in a body-centred lattice, has recently been obtained in the form of large crystals by Dr Tamura working with Professor Carpenter. This alloy consists of almost equal numbers of copper and zinc atoms. Its structure was determined recently by Owen and Preston and independently by Phragmen and Westgren, who agreed in showing that the lattice of $\beta$-brass is a body-centred cube in which atoms of zinc are situated on one simple cube lattice while the atoms of copper are situated on an identical lattice which interpenetrates the first. The structure is therefore identical with that of iron except that in the one case the two interpenetrating lattices contain atoms of the same metal while in the other they contain different kinds of atoms.

It seemed desirable therefore to carry out distortion measurements with $\beta$-brass, and I was able to do so through the kindness of Miss Elam who procured for me two strips of $\beta$-brass containing about eight crystals large enough to cut into specimens suitable for carrying out distortion measurements.

These strips were prepared by Dr Tamura who treated them by the method of Carpenter and Elam so that some of the crystals grew large. The dimensions of the strips were 10 cm. long × 11 mm. wide × 3 mm. thick, and some of the larger crystals occupied as much as 5 cm. of their length. An analysis, for which I have to thank Miss Elam, showed that the brass used contained 51·3 % copper atoms and 48·7 % zinc atoms in some specimens and 53 % copper in others. The method adopted was in most cases first to determine by means of X-rays the orientation of the crystal axes of one of the crystals relative to the surface and edge of the strip. Both compression and tensile tests were carried out. The compression specimens were prepared by cutting out circular discs. In some cases their flat surfaces were parallel to the two faces of the original strip, but in others the flat faces were ground in definite orientations with respect to the crystal axes by means of a crystallographer's tripod grinding machine kindly lent to me by Professor A. Hutchinson, F.R.S. The tensile specimens were rectangular, approximately square in section, but they had enlarged ends which were bored to take a steel pin which fitted a holder arranged to give a central load. They were arranged in such a way that the central portion was cut from a single crystal. The relationship between the tensile specimens and the original strip is represented in fig. 1 where the boundaries of the crystals are represented by dotted lines. Marks were made on the specimen before cutting and grinding in order to preserve a knowledge of the orientation of the crystal axes; but in every case this orientation was redetermined with reference to the marks used in making the distortion measurement.

In preparing both types of specimen a layer about 0·7 mm. thick was ground off each face of the original strip, and before ruling the scratches necessary for observing the distortion they were lightly etched to make sure that they contained no small crystals inserted in the main crystal.

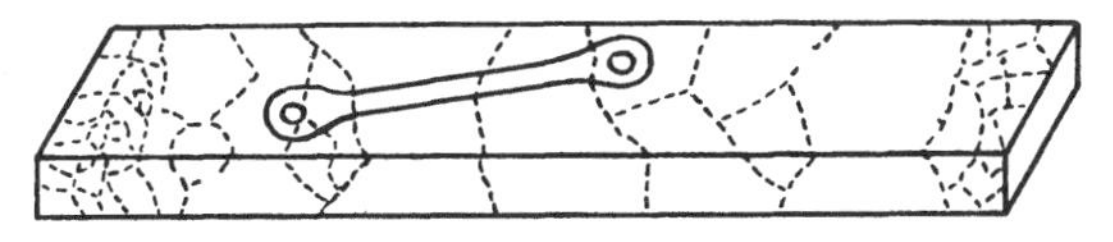

Fig. 1

The specimens were then marked with systems of marks previously described.* The compression specimens were compressed between polished and greased steel plates† till the thickness was reduced by 5–15 %, and the tension specimens were stretched a corresponding amount. It was found as in the case of aluminium and iron crystals that in general the distortion was uniform, ruled lines on the surface remaining straight after compression. The specimen was then measured and the equation to the unstretched cone calculated.

Ten cases in all were analysed, namely, seven compression and three tensile tests. In three of them, numbered $\beta 4{\cdot}1$, $\beta 2{\cdot}1$ and $T\beta 1$, the orientation of the crystal axes was determined only after distortion. In those cases the second or distorted position of the unstretched cone‡ was determined, while in the remaining cases the orientation of the axes was determined before distortion and the first position of the cone was calculated. The unstretched cones were marked on a stereographic figure in a manner previously described.§ Two typical specimens of these diagrams are shown in figs. 2 and 3. It will be seen that in each case the cone very nearly coincides with the two great circles which are represented by two arcs of circles. The distortion might be due to slipping parallel to either of the two planes represented by these two great circles. The two possible directions of slip are represented by the points $S$ and $S'$ (fig. 2) at right angles to the line of intersection of the two planes.

* For compression specimens see Taylor and Farren, 'The distortion of crystals of aluminium under compression. Part I, *Proc. Roy. Soc.* A, CXI (1926), 533; paper 9 above, p. 136. For tensile specimens see Taylor and Elam, 'The distortion of iron crystals', ibid. CXII (1926), 338; paper 10 above, p. 154.

† See above, p. 135.

‡ For definition of unstretched cone see 'The distortion of an aluminium crystal during a tensile test', *Proc. Roy. Soc.* A, CII (1923), 650; paper 5 above, p. 69.

§ Paper 9 above, p. 140, and paper 10 above, p. 155.

15-2

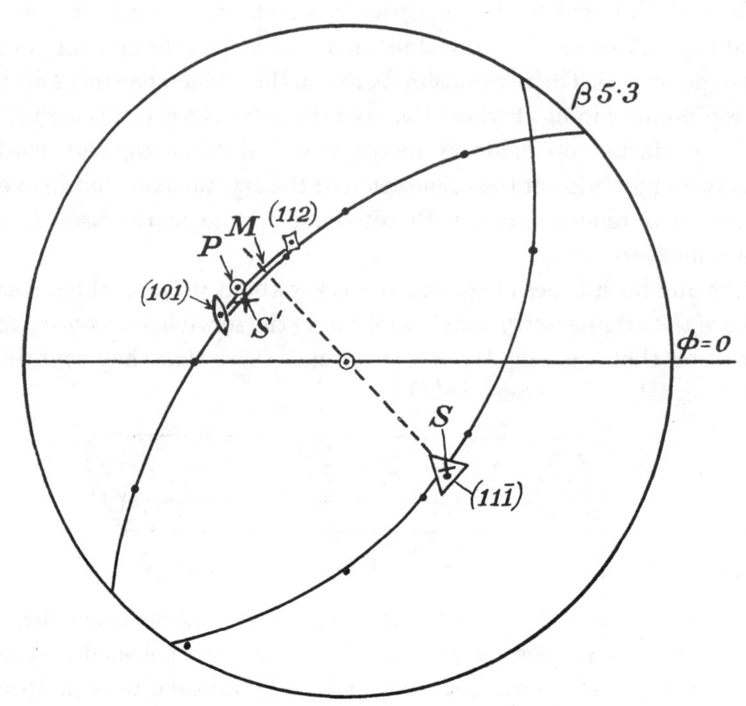


Fig. 2. Unstretched cone and crystal axes for compression specimen $\beta$ 5·3.

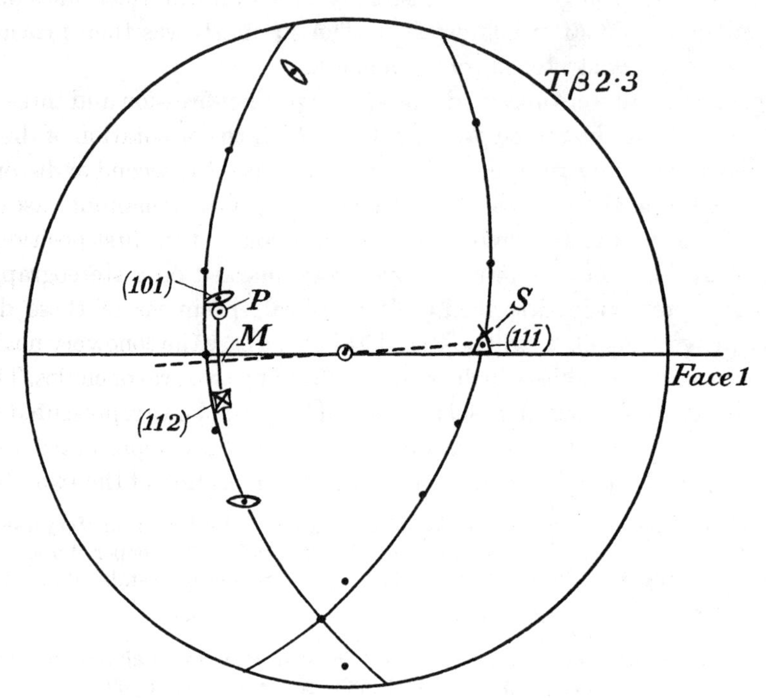


Fig. 3. Unstretched cone and crystal axes for tensile test-piece $T\beta$ 2·3.

## Orientation of Crystal Axes

In each case the orientation of the crystal axes with reference to the ruled scratches on the surface of the specimen was determined by Müller's method using a spectroscope previously described.* The $K_\alpha$ and $K_\beta$ reflections from an iron anticathode were used. Using the value of the lattice constant given by Owen and Preston and also by Westgren and Phragmen† the reflecting angle for the $K_\alpha$ radiation from planes of type {110} is 27° 40′. For the $K_\beta$ radiation it is 24° 50′.

To determine the orientation of the crystal axes reflections were obtained from two planes of type {110} which were perpendicular to one another, or from three planes mutually at 60° and not intersecting in one line. In each case the positions of the directions actually determined by X-rays were marked in a stereographic diagram on tracing paper and certain other directions deduced from them, such as those of type (111), were also marked on the diagrams.

Table 1. *Results of distortion measurements*

| | Direction of slip | | Normal to {111} plane | |
|---|---|---|---|---|
| | $\theta$ (degrees) | $\phi$ (degrees) | $\theta$ (degrees) | $\phi$ (degrees) |
| $T\beta 1$ | 36 | 97 | 35 | 96 |
| $T\beta 2{\cdot}3$ | 47½ | 6 | 47½ | 4 |
| $T\beta 2{\cdot}4$ | 50 | 224 | 49 | 224 |
| $\beta 2{\cdot}1$ | 59 | 342½ | 58 | 338 |
| $\beta 4{\cdot}1$ | 64 | 78½ | 65½ | 79½ |
| $\beta 5{\cdot}1$ | 61 | 343 | 60 | 347 |
| $\beta 5{\cdot}2$ | 58½ | 1 | 59 | 4½ |
| $\beta 5{\cdot}3$ | 48½ | 313½ | 49 | 313 |
| $\beta 3{\cdot}2$ | 48 | 208 | 50 | 208½ |
| $\beta 6{\cdot}2$ | 50 | 4 | 52 | 6½ |

The tracing paper containing the projection of the crystal axes was then laid over the projection of the unextended cone. It was found in each case that one of the two possible directions of slip (which are either of them capable of producing the measured change in shape of the specimen) very nearly coincided with a direction of type (111) (i.e. the normal to the plane {111}).

These directions are marked in figs. 2 and 3 by the symbol △ and the direction of slip determined by distortion measurements by a cross ×. Table 1 gives the co-ordinates of the direction of slip determined from external measurements and the nearest (111) direction. It will be seen that the direction of slip is, as in the case of iron, parallel to the normal to a plane of type {111}.

The pole of the plane of slip determined by external measurements lies therefore in a crystal plane of type {111}. This plane contains three directions of type (110) and three of type (211). It was found in some cases the pole of the plane of slip

* Paper 9 above, p. 143.

† Owen and Preston using β-brass containing 48·3 % Zn give the side of unit cube as 2·946 Å. Westgren and Phragmen using β-brass containing 46·9 % Zn give it 2·945 Å.

coincided with a direction of type (110), while in others it did not coincide with either of these crystal directions, but lay at some intermediate position which differed in different specimens. In no case did it coincide with a direction of type (211).

## Comparison with Iron Crystals

It will be seen that the results are similar to those obtained with iron crystals in that the direction of slip is a crystal direction of type (111) while the plane of slip varies according to the orientation of the crystal axes relative to the direction of the principal stress. On the other hand, the relationship between the orientation of the crystal axes relative to the direction of principal stress, and the orientation of the plane of slip among all the possible planes passing through the given direction of slip is quite different from that found for iron. This difference was first noticed owing to the occurrence of specimens of $\beta$-brass in which the plane of slip coincided with a plane of type {110}. The simplest way to represent the difference seems to be to mark on each stereographic diagram the point $M$ (figs. 2, 3), where the pole of the plane of slip would be if the resistance to shear on all possible planes passing through the given direction of slip were the same. To do this it is only necessary to draw a straight line (dotted line $SM$, figs. 2, 3) through $S$ and the centre of the diagram. The point $M$ is where this line cuts the circle whose pole is $S$. In the case of iron crystals it was shown* that $P$, the pole of the observed plane of slip, was close to $M$, but that in compression specimens, at any rate, there was a small though distinct tendency to deviate towards the nearest direction of type (211). In the case of tension specimens there seemed to be a slight tendency the other way.

In the case of $\beta$-brass it was found that $P$ deviated more from $M$ than it did in the case of iron crystals, but in the opposite direction, that is, towards the nearest pole of type (110). This will be seen in figs. 2 and 3 where the positions of the poles (101) and (112) are marked. These are the nearest poles of the required type to $M$, the position where the pole of the slip-plane would be if the resistance to shear on all planes passing through the direction of slip $(11\bar{1})$ were the same.

It was found that if the orientation of the crystal axes is such that $M$ falls near the pole (101) then the plane of slip coincides with the plane {101}, but if it falls more than, say, 15° away from (101), then $P$ lies at some point intermediate between $M$ and (101). It may safely be deduced therefore that among all possible planes through a given direction of slip of type (111), the resistance to shear is least when the plane of slip coincides with one of type {101}, but this question will now be examined systematically.

## Choice of Direction of Slip

In order to make a complete exploration of all possible orientations of the direction of principal stress in relation to the crystal axes the point representing the position of the normal to the flat surfaces of the compression disc was marked in each case

* 'The distortion of iron crystals', paper 10 above, p. 166.

on a diagram representing the cubic axes of the crystal. It has already been shown* that by a proper choice of one of the three cubic axes as the centre of the projection this point can always be made to lie in one particular spherical triangle. The points representing the axes of all the specimens measured are shown in fig. 4. It will be seen that they are fairly well distributed over a greater part of the area of the triangle whose points are (111), (110), (100). One specimen, however, $\beta 5{\cdot}1$, is represented just outside this triangle. Using another choice of axes it could have been made practically to coincide with $\beta 5{\cdot}2$, but for reasons to be given later it is projected outside the triangle In every case where the point representing the axis was in the triangle whose points are (111), (110) and (100) it was found that the direction of slip was the line $(11\bar{1})$ which is represented by the point $S$ in fig. 4.

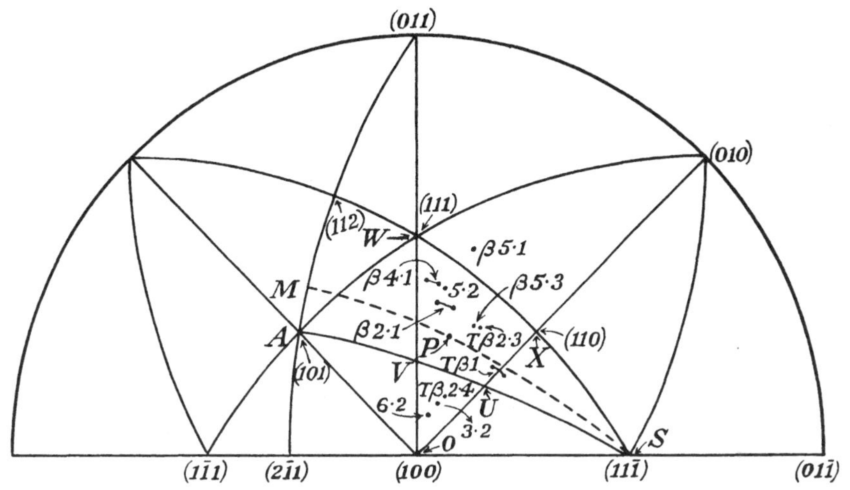


Fig. 4. Positions of axes of specimens relative to crystal axes.

## Representation of Results

It seems unnecessary to give the details of the X-ray measurements from which these points were found, but sufficient data are given in table 2 to enable a reader to reconstruct the fig. 4. In column 1 the number of the specimen is given, column 2 contains the angle $\xi$ between the axis of specimen† and the direction of slip $(11\bar{1})$.

As a second co-ordinate the azimuth $\chi$ of the plane containing the axis of the specimen and the direction of slip is given in column 3. $\chi$ is measured from the direction (101) (marked as $A$ in fig. 4) so that if $M$ is the point of intersection of the azimuth circle $SM$ with the plane {111}, $\chi$ is the arc $AM$. $\chi$ is reckoned as positive when $M$ lies between (101) and (011) and negative when it lies between (101) and $(2\bar{1}1)$. The values of $\xi$ and $\chi$ given in table 2 are sufficient for the construction of fig. 4. The points $M$ and $S$ correspond with those similarly marked in figs. 2 and 3.

* Taylor and Elam, 'The plastic extension and fracture of aluminium crystals'. *Proc. Roy. Soc.* A, CVIII (1935), 28; paper 7 above, p. 109.

† Direction of principal stress, i.e. normal to plane face in case of compression discs or longitudinal axis in the case of tension specimens.

The results of all the distortion measurements are given in column 4 of table 2 which contains the angle $\psi$ between the direction (101) and the pole of the plane of slip (which lies in the plane $\{11\bar{1}\}$).

In previous work on aluminium for which the slipping is confined to a given crystal plane the orientation of the plane and direction of slip were defined by co-ordinates $\theta$ and $\eta$. If the axis of the specimen is supposed vertical, $\theta$ was the slope of the slip-plane to the horizontal and $\eta$ was the angle between the direction of slip and the line of greatest slope on the slip-plane. $\theta$ and $\eta$ are connected with $\xi$ and $\chi-\psi$ by the equations

$$\left.\begin{aligned} \sin\theta\cos\eta &= \cos\xi \\ \text{and}\qquad \cos\theta &= \sin\xi\cos(\chi-\psi). \end{aligned}\right\} \tag{1}$$

Table 2

| | $\xi$ | $\chi$ | $\psi°$ |
|---|---|---|---|
| $\beta 2{\cdot}1$ | 59 | +18 | +11 |
| $\beta 4{\cdot}1$ | 65 | +22 | +17 |
| $\beta 5{\cdot}1$ | $58\frac{1}{2}$ | +36 | +43 |
| $\beta 5{\cdot}2$ | $59\frac{1}{2}$ | +23 | +17 |
| $\beta 5{\cdot}3$ | $48\frac{1}{2}$ | $+17\frac{1}{2}$ | + 9 |
| $\beta 3{\cdot}2$ | 50 | −11 | 0 |
| $\beta 6{\cdot}2$ | $52\frac{1}{2}$ | −16 | − 2 |
| $T\beta 2{\cdot}3$ | $47\frac{1}{2}$ | +18 | + 3 |
| $T\beta 1$ | 36 | + 7 | 0 |
| $T\beta 2{\cdot}4$ | 50 | − 8 | $-\frac{1}{2}$ |

$\xi$ is inclination of axis of specimen to direction of slip.

$\chi$ is the angle between the normal to the crystal plane {101} and the plane containing the direction of slip and the axis of the specimen.

$\psi$ is the angle between plane of slip and crystal plane {101}.

If $P$ is the total pressure on a compression specimen of area $a$ or the total pull in a tension specimen of cross-section $a$, then the component of shearing force parallel to the direction of slip is $F=(P/a)\cos\theta\sin\theta\cos\eta$. From (1) it will be seen that

$$F=(P/a)\cos\xi\sin\xi\cos(\chi-\psi). \tag{2}$$

## Analysis of Results

The results of all the distortion measurements and X-ray analysis which are summed up in table 2 are illustrated graphically in fig. 5. In that diagram the abscissae represent directions on the plane $\{11\bar{1}\}$, the point $M$ being represented by a circle (⊙). and the pole of the slip-plane by a cross (×). The angle $\chi-\psi$ is represented by the line connecting the circle and the cross. The vertical lines represent 10° intervals, and the principal crystal directions (011), (112), (101), $(21\bar{1})$ which occur at intervals of 30° round the plane $\{11\bar{1}\}$ are marked at the bottom of the figure.

It will be seen that in all ten cases, namely, seven compression specimens, $\beta 2{\cdot}1$, $\beta 4{\cdot}1$, $\beta 5{\cdot}1$, $\beta 5{\cdot}2$, $\beta 5{\cdot}3$, $\beta 3{\cdot}2$, $\beta 6{\cdot}2$, and three tension specimens $T\beta 1$, $T\beta 2{\cdot}3$, $T\beta 2{\cdot}4$, the pole of the slip-plane lies away from the point $M$ in the direction of the nearest pole of type (101). From equation (2) it will be seen that if $F$, the resistance to shear, were equal on all planes through the given direction of slip, then $P$ would be least when $\chi = \psi$ so that the pole of the slip-plane would coincide with $M$. The fact that the pole of the plane of slip lies away from $M$ towards the nearest pole of type (101) must therefore mean that of all possible planes through the direction of slip $(11\bar{1})$, the resistance to shear is least when it coincides with a crystal plane of type (110). For purposes of comparison the corresponding points for the iron crystals previously examined are put at the top of the diagram, fig. 5.

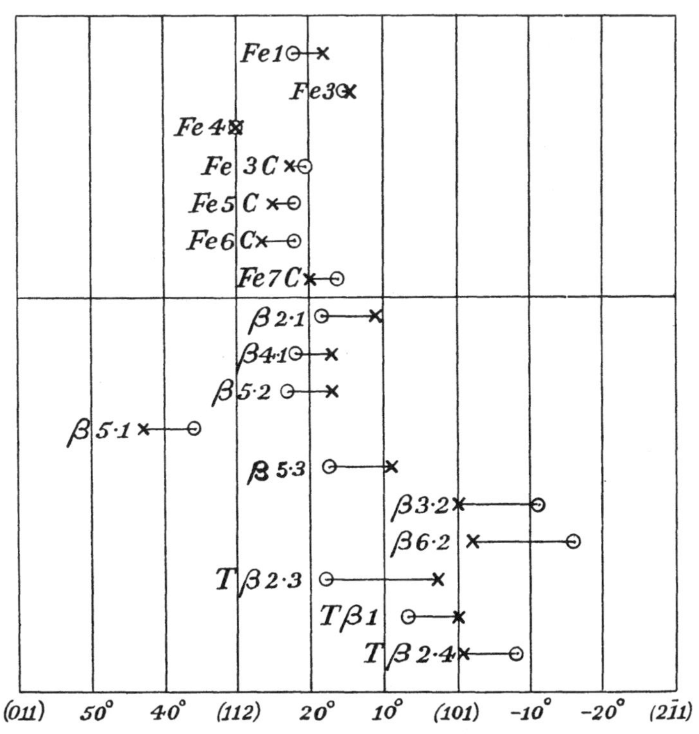


Fig. 5. Relative positions of $M$ (⊙) and pole of slip-plane (×)

## Experiments with Compression Discs β5·1 and β5·2

Since the resistance to shear is a minimum when the plane of slip coincides with a crystal plane of type {110} it seems probable that the resistance would be a maximum when it coincides with one of the crystal planes of type {112} which lies midway between the planes of type {110}. To test this, a crystal, β5, was chosen for which the normal to its flat surface was close to a plane of type {110}. Two compression discs β5·1 and β5·2 were ground from it in such a way that when the point representing the normal to the flat face of β5·2 was projected on to the diagram of fig. 4, so as to be just inside the triangle whose vertices are (111), (110) and (100), then the point representing β5·1 fell just outside this triangle when the same cubic axis was turned into the centre of the figure. By turning another cubic axis into the centre the point representing $\beta 5{\cdot}1$ could have been made to lie inside the triangle, but for the purpose of the present experiment it was more convenient to project it in the manner stated above.

Referring to fig. 5 it will be seen that for $\beta 5{\cdot}2$, $M$ was 7° to the right of the pole (112) while the pole of the plane of slip was 13° to the right; for $\beta 5{\cdot}1$, $M$ was 6° to the left of (112) while the pole of the plane of slip was 13° to the left of (112). This seems to indicate a maximum of resistance to shear in the neighbourhood of planes of type {112}.

## Calculation of Resistance to Shear

It has been shown in the case of aluminium and some other metals that at any given temperature the resistance to shear depends only on the amount of distortion, and not on the component of force normal to the plane of slip, or on the component of shear parallel to the plane of slip and transverse to the direction of slip. If it is

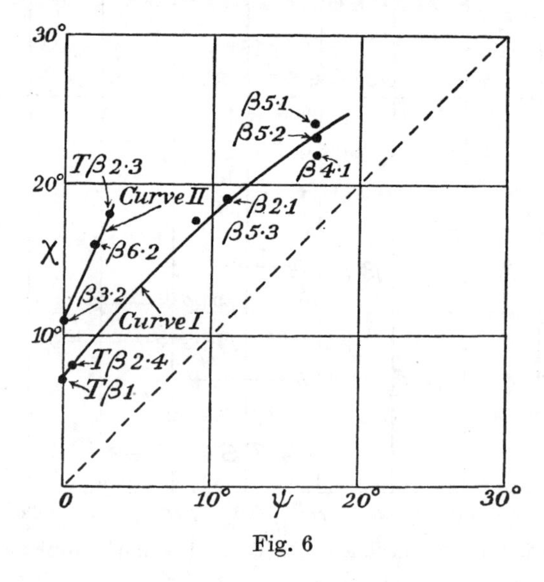


Fig. 6

assumed that this is a general law governing distortion of metallic crystals, then $F$, the resistance to shear, would be a function of the amount of distortion and of the angle $\psi$. It will be seen later that experimental evidence on this point is furnished by the results of these experiments. The actual value $\psi$ for any given value of $\xi$ and $\chi$ would be determined by the condition that $P$ is a minimum as $\psi$ varies, thus from (2), $\dfrac{F}{\cos(\chi-\psi)\cos\xi\sin\xi}$ is a minimum, so that if $F$ is regarded as a known function of $\psi$ and independent of $\xi$, the equation for determining $\chi$ may be written

$$\frac{1}{F}\frac{dF}{d\psi} = +\tan(\chi-\psi). \tag{3}$$

Conversely, if the relationship between $\chi$ and $\psi$ is known (3) may be used for determining the relationship between $F$ and $\psi$. Accordingly, we proceed to examine how far these experiments give us the relationship between $\chi$ and $\psi$.

In fig. 6 the abscissae represent values of $\psi$ and the ordinates the corresponding values of $\chi$. The numerical magnitudes only are set out, the signs being for the moment disregarded, and the results of both tension and compression experiments are shown. It will be seen that the points in fig. 6 lie very nearly on the two smooth curves which are distinguished by the numbers I and II. Both these curves cut the axis of $\chi$ so that for a range of $\chi=0$ to $\chi=7°$ in one case and from $\chi=0$ to $\chi=11°$ in the other the slip-plane is actually the crystal plane $\{101\}$.

## Influence of the Sense of the Direction of Slip

The question naturally arises, what is the difference between the two curves? It will be seen that the upper curve, II, contains the points corresponding to the tension specimen $T\beta 2{\cdot}3$, and the compression specimens $\beta 3{\cdot}2$ and $\beta 6{\cdot}2$. The lower curve, I, contains the tension specimen $T\beta 2{\cdot}4$ and the compression specimens $\beta 5{\cdot}1$, $\beta 5{\cdot}2$, $\beta 5{\cdot}3$, $\beta 4{\cdot}1$, $\beta 2{\cdot}1$. The point corresponding to $T\beta 1$ might belong to either curve because it is on the axis $\psi=0$ and below both the points where these curves cut this axis.

An inspection of fig. 4 shows that the triangle $WXO$ may be divided into two parts $XUVW$ and $UOV$ by the great circle joining the direction of slip, $(11\bar{1})$, with $(101)$. $XUVW$ contains the points representing compression specimens which give points on curve I and tension specimens which give points on curve II. $UOV$ contains all the points representing tension specimens which give points on curve 1 and compression specimens which give points on curve II. This is exactly what we should expect if the resistance to shear on a given plane differs according to the *sense* of the direction of slip. Consider for instance two compression specimens $C_1$ and $C_2$ and suppose that the representative point for $C_1$ lies in $XUVW$ and that for $C_2$ lies in $UOV$. Suppose further that the angle $\psi$ for $C_1$ is $\Psi$ and that the value of $\psi$ for $C_2$ is $-\Psi$. From the point of view of crystallographic symmetry the *distortion* in the case of $C_1$ is identical with that of $C_2$ except that the *senses* of the direction of slip are opposite in the two cases. On the other hand consider two *tension* specimens $T_1$ and $T_2$, the representative points for which lie in $XUVW$ and $UOV$ respectively; the values of $\psi$ are $\Psi$ and $-\Psi$. The distortion for $T_1$ is, from the point of view of crystallographic symmetry, identical with that for $C_2$ while that for $T_2$ is identical with that for $C_1$.

In order to illustrate the difference the diagrams shown in fig. 7 have been prepared to represent a unit cube of the crystal structure viewed perpendicular to the direction of slip and with the eye looking along the plane of slip. The arrows pointing in opposite directions above and below the plane of slip indicate the direction and sense of the slipping motion. The three cubes on the left represent distortions corresponding with points on curve I, fig. 6, while the three cubes on the right represent distortions corresponding with points on curve II.

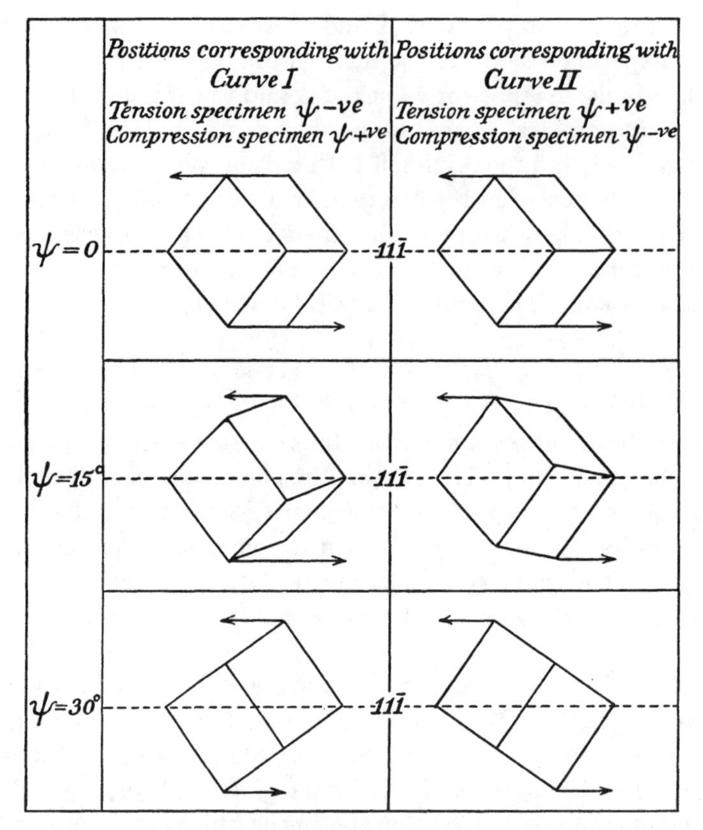


Fig. 7. Diagram illustrating crystallographic difference between shearing in opposite directions.

## Variations of Resistance to Shear with Orientation of Plane of Slip

One may infer from the fact that for a given value of $\psi$ curve II lies above curve I in fig. 6, that the resistance to shear is greater for specimens corresponding with points on curve II than for points on curve I. To calculate the resistance to shear for various orientations of the plane of slip round the given direction of slip, equation (3) may be used; and here it may be remarked that the curves in fig. 6 confirm the assumption made in deducing that equation, that the resistance to shear is independent of the component of force normal to the plane of slip. The ratio of the component normal to the plane of slip to the component of shearing force in the direction of slip is $\dfrac{\cos\theta}{\sin\theta\cos\theta\cos\eta}$ and from equation (2) this is equal numerically to $\sec\xi$, but it is positive for compression specimens and negative for tension specimens.

The value of $\xi$ varies from 36° to 65°, but in spite of this variation the curves of fig. 6 show that the value of $\chi$ depends only on $\psi$ and not on $\xi$, so that the assumption used in equation (3) seems to be justified. If $F_0$ represents the value of the resistance to shear when the slipping is on the plane (101) then on integrating (3) it will be found that

$$\log_\epsilon (F/F_0) = \int_0^\psi \tan(\chi - \psi)\, d\psi. \tag{4}$$

Taking the values of $\chi$ from curve I or II one can determine graphically the values of $\int_0^\psi \tan(\chi-\psi)\,d\psi$ for the range of values of $\psi$ covered by the experiments. The results of this operation are shown in fig. 8 where the ratio $F/F_0$ is given for a series of values of $\psi$. It will be seen that as the plane of slip passes through the crystal

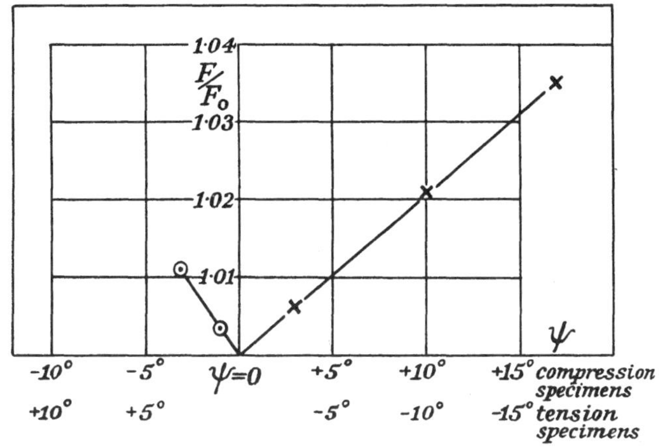


Fig. 8. Curve showing variations in resistance to shear in β-brass as $\psi$ varies.

plane $\psi = 0$, i.e. $\{101\}$, the resistance to slipping passes through a minimum and $dF/d\psi$ changes abruptly both its sign and its magnitude. The resistance increases linearly at a rate of 0·2 % for 1° change in $\psi$ over the range $\psi = 0$ to $\psi = 17°$. In the range $\psi = 0$ to $\psi = -3°$ the resistance increases linearly by 0·37 % per degree increase in $-\psi$.

## Comparison with Results of Direct Measurements of Force

The consistency of the experimental results affords good evidence that we are really measuring changes in resistance to shear with changes in the orientation of the plane of slip round the given crystallographic direction of slip. Unfortunately it is not possible to verify the results by direct measurement of the shearing force in several specimens, the crystal axes of which are in different orientations. Measurements of shearing force at different stages of the tests were in fact made, and they are discussed below. The shearing force depends to a small extent, as we have seen, on $\psi$, but it depends very much more on the value of $s$, the amount of shear. Each compression specimen gives us a curve connecting $F$ the resistance to shear, and $s$ for a given value of $\psi$. At first sight one might suppose that if we obtain a series of such curves for different values of $\psi$, the variation of $F$ with $s$ might be found by comparing the values of $F$ for a given value of $s$. This is not the case, however, for there is no reason to suppose when a series of different crystals are compared after having sheared through the same amount, $s$, but parallel to planes having different values of $\psi$, that they will have the same properties. It will not be possible to discuss the meaning of direct observations of shearing force until some adequate theory can be advanced to explain quantitatively the 'hardening' or strengthening of a metal with increase in distortion. On the other hand the consistency of the observations giving a 'one-to-one' connection between $\psi$ and $\chi$, among specimens covering a large range of values of $s$ and $\xi$ seems to show that the variations of $F$ with $\psi$ is the same whether a large or a small amount of slip has taken place. For this reason there seems some hope that it may be possible to account for the connection between $F$ and $\psi$ from a knowledge of the properties of the crystal lattice itself, apart from any knowledge of the way in which $F$ depends on $s$.

## Direct Measurement of Shearing Force

Each of the seven compression discs here described was compressed in several stages. The total compressive load was measured at each stage and methods described in a previous paper were used to find $S$ the component of shear stress parallel to the plane of slip, and $s$ the amount of shear. These results are given in table 3 and are

Table 3. *Results of measurements of compressive force. $S$ is shear strength parallel to slip-plane in lb. per square inch*

| | | | | | | | | | | | |
|---|---|---|---|---|---|---|---|---|---|---|---|
| $\beta$5·1 | $S$ | 9100 | 13700 | 17800 | 22000 | 26000 | | | | | |
| | $s$ | 0·027 | 0·058 | 0·093 | 0·136 | 0·189 | | | | | |
| $\beta$5·2 | $S$ | 7100 | 13900 | 20400 | 26700 | | | | | | |
| | $s$ | 0·027 | 0·075 | 0·131 | 0·199 | | | | | | |
| $\beta$5·3 | $S$ | 3250 | 6350 | 9200 | 10100 | | | | | | |
| | $s$ | 0·014 | 0·060 | 0·133 | 0·149 | | | | | | |
| $\beta$2·1 | $S$ | 2300 | 4580 | 6800 | 9000 | 11100 | 13200 | 15200 | 17300 | 19200 | 20800 |
| | $s$ | 0·0104 | 0·0182 | 0·0270 | 0·0465 | 0·0688 | 0·0933 | 0·122 | 0·149 | 0·164 | 0·208 |
| $\beta$4·1 | $S$ | 3500 | 6900 | 10300 | 13500 | 16600 | 19600 | 22400 | 25340 | | |
| | $s$ | 0·007 | 0·031 | 0·049 | 0·077 | 0·119 | 0·164 | 0·196 | 0·232 | | |
| $\beta$3·2 | $S$ | 3100 | 6300 | 9100 | 11800 | 14600 | | | | | |
| | $s$ | 0·028 | 0·056 | 0·096 | 0·139 | 0·812 | | | | | |
| $\beta$6·2 | $S$ | 3100 | 6050 | 8800 | | | | | | | |
| | $s$ | 0·010 | 0·070 | 0·131 | | | | | | | |

exhibited graphically in fig. 9. Inspection of that figure shows that the curves for different specimens are not the same. The difference may be due to slight differences in the material. There might, for instance, be small included crystals in some specimens and not in others. This possibility, though always present, is unlikely to be the cause of the difference between specimens $\beta$5·1, $\beta$5·2 and $\beta$5·3, because they were all cut from the same crystal. $\beta$5·1 and $\beta$5·2 were cut so that their planes of slip were crystallographically similar though their shapes and sizes were not the same. It will be seen that the corresponding curves in fig. 9 are nearly coincident. Specimen $\beta$5·3, on the other hand, was cut in a different orientation so that $\psi = 8\frac{1}{2}°$. It will be seen that its resistance to shear is very much less than that of $\beta$5·1 or $\beta$5·2. It will be seen that specimens $\beta$5·1, $\beta$5·2, $\beta$4·1 and $\beta$2·1 for which the planes of slip have the higher values of $\psi$ have also the higher resistance to shear. Specimens $\beta 3{\cdot}2$, $\beta 6{\cdot}2$ and $\beta 5{\cdot}3$ which have slipped on planes near the crystal plane of type $\{110\}$ have a low resistance to shear. The values of $\psi$ are given in brackets in fig. 9.

These direct force measurements are given here because they appear fairly consistent, but it is necessary to state that the shearing force was deduced from the total pressure on the specimen making the assumption that the friction of the specimen on the greased surfaces of the polished steel plates produced only a negligible effect on the shear stress. In the case of aluminium specimens somewhat elaborate

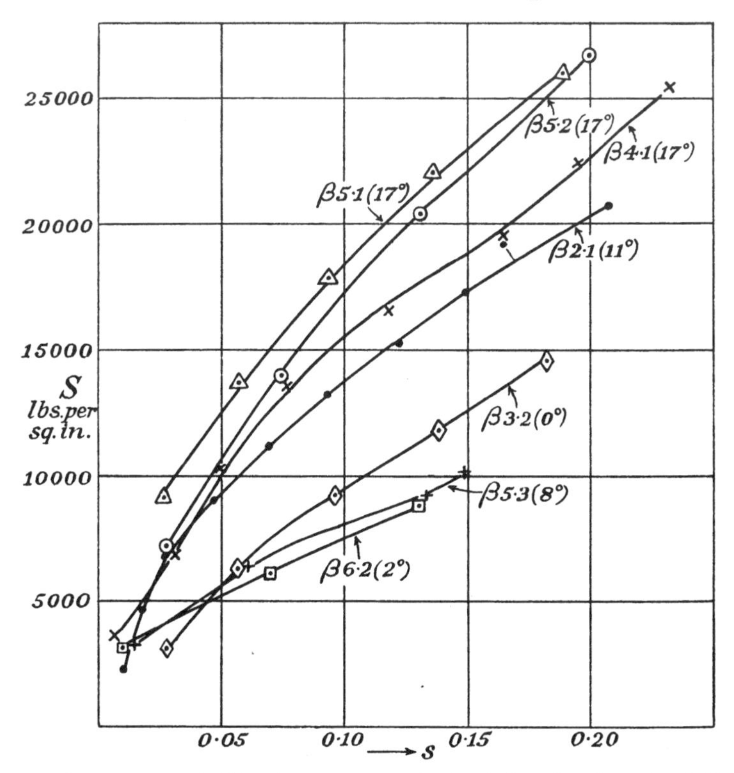


Fig. 9. Shear strength $S$ in lb. per square inch of the slip-plane for different values of the amount of shear.

experiments were made to find out whether this assumption is true and it was found to be so. The method used in the case of specimens of $\beta$-brass was the same as that used in the case of aluminium, but it must be remembered that pressures in the former case were considerably greater than they were in the latter. On the other hand the observed uniformity of the distortion is strong evidence that the friction has no appreciable effect.

## Slip Lines

Though I made repeated efforts I was unable to see slip lines, so I cannot say whether they are of the same type as those of $\alpha$-iron or not. Attempts by metallurgists skilled in polishing also failed to reveal them.

### Connection between Choice of Plane of Slip and Crystal Structure

At the present time one of the models for representing crystal structure in pure metals, which is most in favour among metallurgists, seems to be that of a pile of similar ionized atoms mutually repelling one another but held together by a kind of 'glue' of electrons moving about in the spaces between them. This is the type of model used for instance by Frenkel in discussing the properties of metals. Recently Mr Hume Rothery* has proposed alternatively that the electrons instead of moving about occupy definite positions between the atoms, though he does not explain how they can maintain those positions.

In seeking to predict from such a model which of the crystal planes would afford least resistance to slipping, the principle which has commonly been adopted is to imagine a close packed pile of spheres whose centres are at the lattice points of the crystal structure. Slipping parallel to various crystal planes is then contemplated. In order that the spheres may move over one another it is necessary for the two portions into which the material is divided by the plane of slip to move apart, thus doing work against the attractions of the electrons moving between the atoms, and it is supposed that the condition which determines which among the crystallographic planes shall be a plane of slip is that this distance of separation shall be a minimum.

Another method which has been used gives a similar result. The spheres are imagined to contract in radius, their centres remaining fixed. At first it is not possible to place any plane in such a position that it misses all the spheres. At a certain stage of the contraction it becomes possible to place a sheet parallel to one of the crystallographic planes in such a position that it does not cut any of the spheres. As the contraction proceeds it becomes possible to place sheets parallel to other crystallographic planes, but the plane of slip is the first of those 'ways through' to appear. It is obvious that purely geometrical conceptions of this kind are not capable of explaining distortion of metals, but, as has frequently been pointed out, they give the right result in the case of face-centred cubic metals like aluminium.

In the case of a body-centred structure like iron the first 'ways through' which appear as the spheres are contracted towards their centres are not a series of parallel planes as they are in the case of face-centred cubic structures, but they are the faces of a pile of hexagonal rods. The view looking down the lines of atoms which are the centres of these hexagonal rods at the stage when the 'ways through' first appear is shown in fig. 10. The 'ways through' are shown dotted. In our paper on the 'distortion of iron crystals', Miss Elam and I described the deformation in terms of these hexagonal rods, but for reasons which will shortly be explained we made no reference to the piles of atoms lying along them. Other metallurgists, however, placed the ions in position along our hexagonal rods, and concluded from data which were purely geometrical that the type of distortion which we observed for iron could

* *Phil. Mag.* IV (1927), 1017.

be explained without any detailed reference to the forces between the atoms of the structure. Shortly after our paper appeared, for instance, a distinguished metallurgist wrote to me that he had been inclined to doubt our experimental results till it had occurred to him that in a body-centred pile of spheres the first 'ways through' seen as the spheres contract towards their centres consist of a network of hexagonal prisms instead of the set of parallel planes which first appear in the case of face-centred piles. More recently a similar idea has been put forward by Mr Hume Rothery.*

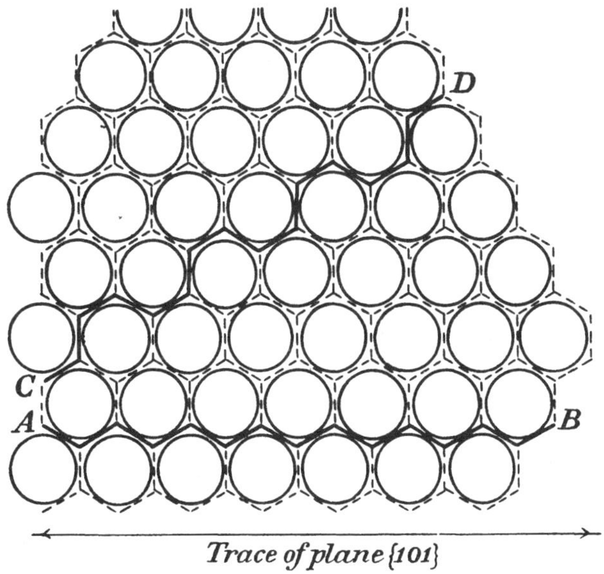


Fig. 10. Diagram illustrating crystal structure of body-centre cubic crystals, when viewed along a cube diagonal.

The reason why we did not put forward ideas of this kind, though they were in our minds when we put forward the 'hexagonal rod' description of distortion, was that if pursued to their logical conclusion they led to the wrong result. Assuming for the moment that the direction of slip is along the lines where the atoms are closest together, namely, one of the directions of type (111), the choice of the plane of slip must depend on the relative amounts of the resistances to shear parallel to the various planes which pass through the given direction of slip. In order to predict therefore which of these various planes will be the plane of slip we must first predict the way in which resistance to shear varies as the plane of slip rotates round the direction of slip.

To do this it is necessary to make some assumption as to the cause of the resistance to shear, or, in other words, the cause of the dissipation of energy which occurs when a metal crystal is distorted. The simplest assumption that can be made is that the energy is lost when one pile of atoms moves relative to its immediate neighbours.

* *Phil. Mag.* IV (1927), 1035.

If it is supposed that a certain definite amount of energy is dissipated between each pair of piles when the two move through a given distance relatively to one another, then the resistance to shear will be the same as that of a pile of hexagonal rods when they are placed so that their sides are parallel to planes of type {112} provided that the friction at every interface is the same. This assumption of equal friction between any neighbouring pair of piles of ions is necessarily implied, though it is not specifically stated, in Mr Hume Rothery's work, for in order to distinguish between different types of slipping he takes as his criterion that the resistance to slipping will be least when the depth to which the ions of one pile penetrate during a slipping motion into the spheres of repulsion of those in the neighbouring pile is least. This necessarily implies that when the depths of penetration in two cases are the same, the resistance to slipping between the piles is the same.

Making this assumption we can now work out the resistance to shear of any plane of slip which passes through the direction of slip. Let $\psi$ be the angle between any plane of slip and the plane {101}.

Let $F$ be the magnitude of the shear stress which will just cause slip on the plane $\psi$ and let $f$ be the resistance to slipping experienced by unit length of each pile of ions as it slips past its neighbour. Let $a\sqrt{3}$ be the distance between the central lines of two neighbouring piles of ions, so that $a$ is the breadth of a face of one of the hexagonal rods. It is required to find $F$ in terms of $f$, $a$ and $\psi$. Consider a length $l$ of the trace of the slip-plane on a plane perpendicular to the direction of slip. This trace, though a straight line from the point of view of an observer who cannot see the individual atoms, may be regarded as consisting of microscopic elements which are the sides of the hexagons. In fig. 10 the section of a slip-plane parallel to {101} is shown at $AB$. The section of a slip-plane parallel to {112} is shown at $CD$. There are three types of these elements inclined at 30°, $-30°$ and 90° respectively to the plane {101}. Suppose that $\psi$ lies in the range 0–30° and that in the breadth $l$ of the slip-plane there are $n_1$ elements of the first type, $n_2$ of the second, and $n_3$ of the third.

We then have the following relationships which can easily be verified by looking at fig. 10:

$$\left.\begin{aligned} n_1 - n_2 &= n_3, \\ (n_1 + n_2)\, a \cos 30° &= l \cos \psi, \\ (n_1 - n_2)\, a \sin 30° + a n_3 &= l \sin \psi, \\ (n_1 + n_2 + n_3) f &= Fl. \end{aligned}\right\} \qquad (5)$$

Eliminating $n_1$, $n_2$ and $n_3$ from these equations it is found that

$$F = \frac{4f}{3a} \cos (\psi - 30°).$$

If $F_0$ is the value of $F$ when $\psi = 0$

$$F/F_0 = \tfrac{1}{2}\sqrt{3} \cos (\psi - 30°), \qquad (6)$$

and when the angle $\psi$ is between 0 and $-30°$

$$F/F_0 = \tfrac{1}{2}\sqrt{3}\cos(\psi + 30°).$$

Now let us apply the condition given in equation (3) which determines the value of $\psi$ when $\chi$ is given and the relationship between $F/F_0$ and $\psi$ is known.

It is

$$\frac{1}{F}\frac{dF}{d\psi} = \tan(\chi - \psi);$$

differentiating (6) it will be found that

$$\frac{1}{F}\frac{dF}{d\psi} = -\tan(\psi - 30°).$$

When $\chi$ has any positive value less than 30° there is no possible positive value of $\psi$ less than 30° for which $\tan(\chi - \psi) = -\tan(30° - \psi)$. In fact, for all orientations of the crystal axes except special symmetrical ones for which $\chi = 30°$ (i.e. ones for which the axis of the specimen lies in a crystal plane of type {110}) the plane of slip would be a crystal plane of type {110}. This conclusion follows directly from an analysis of the effect of Mr Hume Rothery's hypothesis, no other hypothesis than that necessarily implied in his discussion being employed. Since in the case of iron crystals the plane of slip is not in general a plane of type {110}, it follows that the hypothesis that the conditions of slipping are determined by the geometrical considerations put forward by Mr Hume Rothery is unsound.

## Comparison between Slipping in β-Brass and Model of Hexagonal Rods

Though we cannot hope to explain the choice of slip-planes in metal crystals in the manner of Mr Hume Rothery, using only simple geometrical data, yet there is one very remarkable feature of the slip phenomena in crystals of $\beta$-brass which is reproduced in our model in which we represented piles of ions by hexagonal rods (and from this point of view our model complies with the requirements of Mr Hume Rothery's hypothesis). The rate of variation of resistance to shear with the angle $\psi$ is discontinuous at the point $\psi = 0$ both in the model and in $\beta$-brass. This can be seen by comparing fig. 11 which represents $F/F_0$ for the model, with fig. 8 which represents a limited range of values of the same quantity deduced from distortion and X-ray measurements of $\beta$-brass.

In view of this point of similarity it is of interest to enquire why the model shows this peculiar feature. Referring to fig. 10 it will be seen that when the slip-plane coincides with the crystal plane {101}, i.e. when $\psi = 0$, all the elements composing the plane of slip are at angles of $\pm 30°$ to its general direction. A small change $\delta\psi$ in $\psi$ introduces elements of the slip-plane at right angles to its general direction, the number of them being proportional to the absolute value of $\delta\psi$ irrespective of whether $\delta\psi$ is positive or negative.

On the other hand, when the slip-plane is at $\psi = 30°$ so that it coincides with the crystal plane {112} as in $CD$ (fig. 10), the three types of element composing the whole slip-plane make angles of 0 and $\pm 60°$ with the general direction. A small change $\delta\psi$ in $\psi$ does not change the number of elements at angle 0°. If $\delta\psi$ is positive there is an increase proportional to $\delta\psi$ in the number of elements at $+60°$ together with an equal decrease in the number at $-60°$. Small changes in $\delta\psi$ therefore leave the total number of elements unaltered so that the rate of change in resistance to shear is zero when the plane of slip coincides with a crystal plane of type {112}.

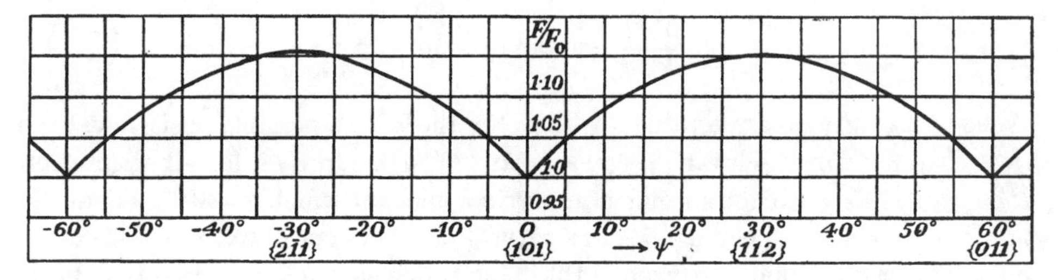


Fig. 11. Theoretical curve showing variation in shear strength, as plane of slip rotates about the direction of slip.

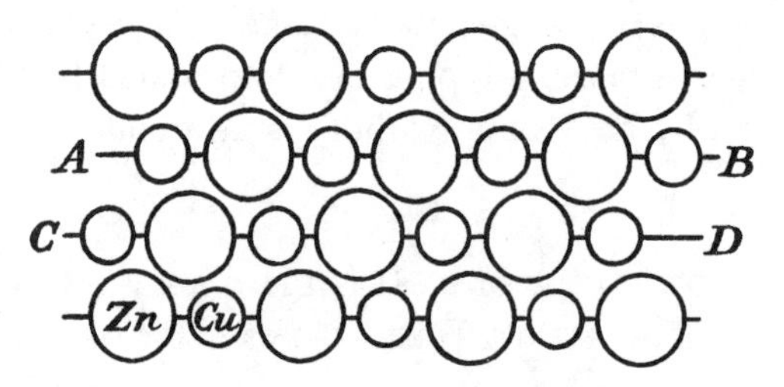


Fig. 12. Atomic spacing in neighbouring lines of atoms parallel to the cube diagonal.

It will be seen that in the case of $\beta$-brass there is not only a sudden change of sign in the rate of increase in $F/F_0$ with $\psi$ at $\psi = 0$, but there is also a sudden change in its absolute magnitude. This corresponds presumably with a difference in the resistance to slipping between neighbouring hexagonal prisms or piles of atoms, according to the direction of motion. That such a difference is to be expected from the point of view of crystallographic symmetry has already been pointed out. Fig. 12 shows a pair of neighbouring piles of ions $AB$ and $CD$. Regarding $CD$ as fixed it will be seen that the motion of $AB$ from $A$ to $B$ is crystallographically different from that from $B$ to $A$. It should be noticed, however, that the hypothesis that the resistance to relative motion between two piles of ions depends only on the direction of motion does not account for the asymmetry of the curve of resistance as shown in fig. 8, for it is a geometrical necessity that among the elements which make up the slip-plane the direction of slip across half of them must be the type illustrated in fig. 12 when the pile $AB$ moves from $A$ to $B$ while the other half is of the opposite kind typified by a motion from $B$ to $A$.

It seems to me that it is useless to put forward hypotheses about the manner in which slip will occur in a metal crystal until some theory is put forward which will account, in a quantitative manner, for the way in which energy is lost during the process of distortion. So far no progress whatever has been made in this direction.

In conclusion, I should like to express my thanks to Sir Ernest Rutherford for permission to carry out the work in the Cavendish Laboratory, and to Miss Elam, Professor Carpenter and Dr Tamura for the material with which these experiments were carried out.

# 15

# RESISTANCE TO SHEAR IN METAL CRYSTALS

REPRINTED FROM

*Transactions of the Faraday Society*, vol. XXIV (1928), pp. 121–5

When a single crystal is strained so that it shears parallel to a crystal plane, the resistance to shear rapidly increases with the amount of shear and at the same time X-ray examination of the material shows that it has broken into fragments which remain crystals but are rotated through various small angles from the original orientation of the single crystal. It seems certain that the two phenomena are associated, but it is not at all clear how the breaking up of the crystals is responsible for the increased resistance to shear.

The theory most commonly put forward as an explanation is that in a perfect crystal a very small shearing force parallel to a crystal plane will cause all the atoms in one plane to slide over the atoms in the neighbouring parallel plane. The increased resistance to shear in an imperfect crystal is due to the fact that rows of atoms are held up at any imperfection in the crystal structure. This theory has many forms; the displaced material may for instance form keys in the crystal plane or, alternatively, the slipping may be held up at the boundaries between crystal fragments whose crystal axes are orientated at slightly different angles.

All such variations of the theory are identical in principle, and it seems to me that they all fail to do what is required of them. Suppose, for instance, that the crystal has only one set of parallel planes of slip, and suppose that it offers a very large resistance to cleavage across the planes of slip and a large resistance to compression. Under these circumstances a 'key' between two neighbouring planes of atoms would lock them together completely over their whole extent whatever their area might be. When a 'key' had formed between every neighbouring pair of planes the material would seize up completely. If the number of keys per unit volume depends on the amount of distortion and not on the size of the specimen, then an obvious consequence of the theory would be that large specimens would seize up more quickly than small ones. This is certainly not the case.

In the form of the theory which assumes the slipping to be held up at the boundaries of crystal fragments into which the single crystal is divided, the difficulty is more pronounced still, for the material would in that case be completely locked or seized up as soon as any breaking at all had occurred.

One must therefore assume that if the distortion is due to slipping parallel to crystal planes, more than one crystal plane is involved. In crystals of cubic symmetry like aluminium the symmetry alone shows that there must be many planes crystallographically similar to any plane of slipping, and these must also be planes of slip.

As soon as the possibility of slipping parallel to a number of crystal planes is admitted, the whole theory falls to the ground. Taking the case of aluminium for instance, there are twelve possibilities of slipping, namely, four planes of slip and three directions on each plane. By combining a limited number of these it is possible for a crystal in any orientation to suffer a distortion identical with that of the main body of the original single crystal when sheared parallel to one crystal plane of slip. The shearing force necessary to produce this distortion is of course least when the crystal axes of the fragment are parallel to those of the main body of the crystal, but mathematical analysis shows that however the fragment is orientated, the resistance to shear parallel to the plane of slip of the main body of the crystal cannot be greater than about twice the minimum value.

It appears therefore that however favourably the fragments are orientated for increasing the resistance to shear, they cannot increase it to more than twice the resistance of the original single crystals, and if the actual distribution of crystal axes found from X-ray measurements is taken into account it is found that the increase in resistance to shear cannot amount to more than 2 or 3 % of the original resistance of the unbroken crystal.

For these reasons I do not think that the breaking up of the crystal can be used to explain the increase in resistance to shear which occurs when a single-crystal specimen of a metal like aluminium is distorted if the force necessary to move one plane of atoms over its neighbour is assumed to be small. The only alternative seems to be to imagine that this force is not small, but that there are regions in the material where a high concentration of stress can occur so that very high values of the shear stress can be reached locally although the mean shear stress is small.

The idea that stress may be concentrated at the end of a crack is an old one. It has been used with great success by Dr A. A. Griffith in discussing the strength of glass, and earlier by Professor Inglis in another connection. It has, I think, been recognized among workers with single crystals of metals that there is a possibility that concentrations of stress may occur, but I do not remember seeing any attempt to explore the possibilities which this concentration afford for explaining the phenomena observed when single-crystal specimens are distorted.

There are three things which require explanation:

(1) The form of the load-extension curve, i.e. the rapid increase in resistance to shear with increasing distortion.

(2) The fact that a certain definite proportion of the energy used in producing the distortion is absorbed in the material, the remainder being dissipated in the form of heat. As the shear strength of the material increases the amount of energy

necessary to produce a given distortion increases. The amount of energy absorbed in the material also increases and in the same ratio*.

(3) The distribution of the crystal fragments.

I will take number (3) first. Up to the present not much information is available. Some years ago I believe Professor Joffé made some experiments with rock-salt which he bent. He came to the conclusion that fragments of the material broke away from the rest and acted like rollers, rotating about axes in the plane of slipping and transverse to the direction of slip. I do not know whether experiments have been carried out to see whether a similar type of breaking up occurs in metal crystals. Dr Goucher in the course of X-ray examination of strained tungsten crystals observed that the spreading of the fragments was due to rotation about a single line transverse to the length of his specimen. The distortion of his specimen was, however, very non-uniform, and he attributed, correctly I imagine, the spreading he observed to a curvature of the whole plane of slip which would make the orientation of the crystal fragments different on the two sides of his specimen.

No experiments seem to have been made in which precautions have been taken to ensure that the distortion of the specimen as a whole is uniform.

In the course of some experiments on the behaviour of aluminium crystals under compression the distortion of a certain specimen was examined completely and proved to be uniform and to be due to slipping parallel to a crystal plane of type (111).† This specimen was cut parallel to the plane of slip, as calculated from measurements of marks on the surface of the specimen. It was ground and etched till the effect of the saw cuts had disappeared. This was verified by applying the same treatment to an undistorted specimen. The X-ray reflections in this case were perfect ones.

The specimen was then mounted on an X-ray spectrometer and exposed to a narrow beam of homogeneous X-rays from an iron anti-cathode. The beam was limited by passing through two small circular holes. Two sets of photographs were taken (*a*) with the axis of rotation of the spectrometer table in the plane of slip and transverse to the direction of slip, and (*b*) with the axis of the spectrometer table lying along the direction of slip. The exposures were each of 8 min. duration, and the intensity of the incident beam was kept constant while each set was being taken. Before each exposure the photographic plate was lowered in a vertical slide through a few millimetres, and the table was turned till the plane surface of the specimen was parallel to the axis of the beam of X-rays. Under these circumstances the direct beam produced a semicircular spot which marked the position of the incident beam.

A series of exposures was made, the setting angle of the spectrometer table

* 'The heat developed during plastic extension of metals', W. S. Farren and G. I. Taylor, *Proc. Roy. Soc.* A, CVII (1925), 436; paper 6 above, p. 97.

† 'The distortion of crystals of aluminium under compression. Part I', G. I. Taylor and W. S. Farren, *Proc. Roy. Soc.* A, CXI (1926), 529; paper 9 above, p. 133.

PLATE 1

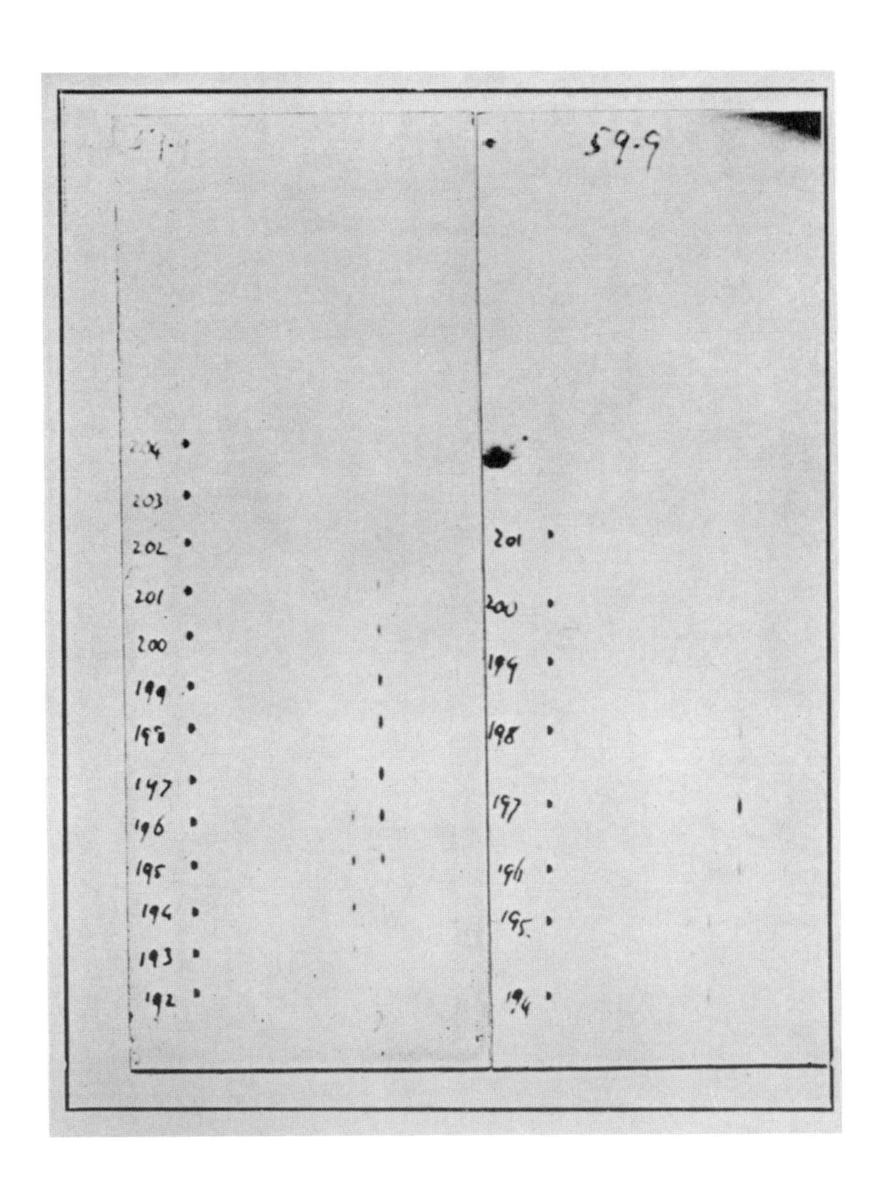

59.9
203
202
201
200
199
198
197
196
195
194
193
192
201
200
199
198
197
196
195
194

being increased by 1° from one exposure to the next. In this way the photographs shown in plate 1 were made. The left-hand one was taken with the specimen mounted on the spectrometer in position (*a*) and the right-hand one in position (*b*). The black semicircular dots on the left of each photograph are the centre spots and the figures written beside them are the setting angles. The setting angle when the specimen was parallel to the incident beam was 172·5° in each set. The reflected spots are seen on the right; the darker series are due to $K_\alpha$ radiation for which the reflecting angle is 24·3°, and the fainter series to $K_\beta$ radiation.

It will be seen that in position (*a*) the spots are small but the range of angles over which reflection takes place extends over 10 or 12°. On the other hand, in position (*b*) the spots are elongated into appreciable arcs of Debye rings, and the range of setting angles at which reflection occurs is limited to 2° from 196 to 198°. It will be noticed also that the spot of maximum intensity is at 197°. On one side of this maximum the reflections only extend through a range of 2–195°. On the other side, however, there was still a visible spot after 8 min. exposure at 206°.

These photographs prove that all the parts of the material which have rotated through angles greater than 2° have rotated about the transverse direction in the plane of slip. Some of them have rotated as much as 9° about this direction, and, moreover, in all cases the direction of rotation is the same. This direction of rotation is in fact the same as that which might be expected if the detached portions of the crystal acted as rollers between the slip-planes, but it is very much less than would be necessary if they actually played the part of rollers.

## Rotation of Material in Neighbourhood of Regions of Concentration of Stress

If the slipping of one crystal plane on another does not take place simultaneously over the whole of the section of the specimen by a plane of slip, then at the ends of a region of slipping there will be a concentration of stress. The displacements of the material above and below this region relieve the stress in the middle and concentrate it at the ends. If the material is considered as elastic except in so far as it has slipped over a limited area of the slip-plane, the displacements and stresses are those round a limited crack in a material when a shearing stress is applied parallel to its plane. Considering a crack as the limiting case of a very long ellipse, these can be calculated using the theory of elasticity.

I have worked this out with the following results:

(1) The stress concentrates at the ends of the crack, and the direction of maximum shear stress in the region where it is greatest is that of the crack itself—so that it would tend to extend in the same direction.

(2) Throughout the whole of the two regions of stress concentration, i.e. at the two ends of the crack, the material is rotated about the transverse direction and the direction of rotation is everywhere the same, namely, that in which rollers in

the crack would rotate if they rubbed on the top and bottom of the crack. There is a concentration of rotation in the regions of concentration of stress.

The rotation of an elastic material at the ends of a limited crack or area of slipping is therefore exactly what is observed by means of X-rays in a strained single crystal.

Returning now to the load-extension curve. The load at which slipping takes place will depend, according to this theory, on the ratio in which the maximum stress is increased above the applied stress. This depends, as was pointed out by Inglis and by Griffith, on the curvature of the end of the crack and on its length. The curvature at the end of the crack may be taken as depending on the atomatic structure and therefore independent of its length. At the point of greatest concentration the stress is increased, as in Griffith's case, in a ratio which is proportional to the square root of the length of the crack. Considered as a two-dimensional system the stress is, as it were, drained out of an area proportional to the square of the length of the crack, or region of slipping. In the early stages when slip has occurred in a few places only, the ends of a crack can concentrate stress from a large region; consequently the ratio of the maximum stress to the applied stress is large, and a small stress applied to the specimen will raise the stress at the point where it is greatest to the value necessary for slipping.

As the slipping proceeds on more and more planes the undisturbed parts from which stress can be drained into regions of stress concentration become smaller and smaller, so that effectively (i.e. from the point of view of stress concentration) the newly formed cracks are smaller and smaller. The ratio in which stress is increased therefore decreases, and the stress which must be applied to the specimen in order that slipping may occur at a point of stress concentration therefore rises.

It appears therefore that a theory of this kind can be made, qualitatively at any rate, to account for the rapid rise in the load-extension curve at small values of the extension.

Turning now to the energy absorbed in the material, according to this theory it would be in the form of strain energy in the regions of stress concentration. Using the value appropriate to a crack in an elastic medium under shear stress parallel to the direction of the crack, I find that if, during the formation of a small strain $\delta s$, $n\delta s$ cracks are formed, each of length $2c$, then the ratio of energy retained in the material to energy dissipated in the form of heat is constant if $nc^2$ is constant. If one makes the assumption that the length of a new crack formed in a piece of previously uncracked material is proportional to the linear dimensions of this piece, then $nc^2$ would be constant, and the observed constancy of the ratio of energy absorbed to heat dissipated would be explained.

## Conclusion

It seems that a consistent theory might be formed on the assumption that the resistance to shear parallel to a crystal plane of slip is not small, and that much greater concentrations of stress can occur in an unbroken crystal than in one which is already filled with regions of stress concentration. The alternative theory that the resistance to shear is weak in a single crystal and that it is strengthened by the formation of imperfections which prevent rows of atoms from slipping seems untenable in the case of metals like aluminium, though it may be applicable to metals like zinc, for which there are fewer possible crystallographically similar modes of slipping.

# 16

# THE PLASTIC DISTORTION OF METALS*

REPRINTED FROM

*Philosophical Transactions of the Royal Society*, A, vol. CCXXX (1931), pp. 323–62

The plasticity of metals has been the subject of many recent papers, but, owing to the complexity of the subject, there is but little agreement between different researches. Attempts to extract simple generalizations from the very complex phenomena have been made chiefly in two directions: (1) Engineers have used test bars of certain specially simple form, such as uniform round bars which they have subjected to twisting or tension, and they have found the effect on their test of varying physical conditions. (2) Mathematicians have assumed an ideal plastic material and have given it properties which may or may not be possessed by some real material. They have then analysed the distributions of stress and strain when this ideal material is subjected to given external forces or distortions.

The main problem in plasticity is to determine the internal stresses and strains in a plastic body when given external loads and strains are applied to its outer surfaces. The common engineering tests are incapable of supplying information in any conditions of stressing except those under which the particular test concerned was carried out, thus a pure tensile test of an annealed copper bar yields a load-extension curve in which the true elastic limit is extremely low, the load rapidly increases, with plastic extension. A twisted bar yields a torque-angle curve of the same type, and the relationship between these two curves is a specially simple example of the type of problem which must be investigated before it will be possible to analyse the internal stresses and strains in any of the more complex problems of plastic distortion, such for instance as that of analysing the internal stresses during the drawing of a wire through a draw-plate. To connect tests made with different types of loading it is necessary to develop some theory of plasticity which takes account of the most essential observed properties of plastic materials but leaves out of consideration those which appear to be of less importance in connection with the particular set of phenomena under discussion. Thus it happens that a theory which affords a useful means of representing plastic phenomena in a steel which has a high elastic limit may concern itself with the conditions at the elastic limit, defined as the stress at which the first deviation from perfect elasticity is observable. For copper, however, the elastic limit is very low indeed, even after it has been hardened by cold working. On the other hand, if a gradually increasing load of any type is applied to annealed copper the distortion increases, and if the load stops increasing the distortion stops increasing, except for a very small increase which occurs at the

* With H. QUINNEY.

highest loads which copper can maintain. This increase is, however, too small to affect any of the results given in this paper, so that at any stage of the process the metal may be said to have a definite strength with regard to the particular type of stress which is being applied. If the load be removed and gradually applied again, plastic distortion begins at a load which is considerably lower than the maximum load applied during the first loading, but the distortion remains very small till the highest load applied in the first loading is reached. It then begins to increase rapidly, and after quite small distortion during the earliest stages of the second loading it has a strength which is identical with that which it would have had if it had been loaded in one operation without the intermediate unloading.

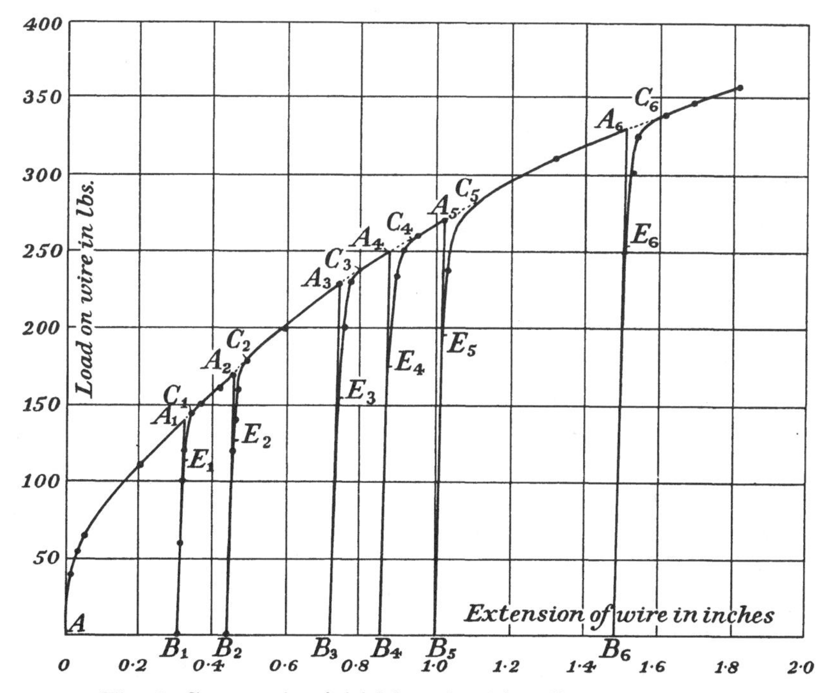


Fig. 1. Copper wire: initial length 9·9 in., diameter 0·127 in. Repeated tensile test (load increasing).

This is represented in the curve, fig. 1, which shows the load-extension curve for an annealed copper wire. At the points $A_1, A_2, \ldots$ the loads were removed and gradually replaced. The stress-strain curves during the removal of the load are shown as $A_1B_1, A_2B_2, \ldots$. They correspond with an elastic contraction. The reloading curves are shown as $B_1E_1C_1A_2$, $B_2E_2C_2A_3, \ldots$, while the result of testing another piece of the same wire without unloading at $B_1B_2\ldots$ is shown as the continuous curve $AA_1C_1A_2C_2\ldots$. The parts $A_1C_1$, $A_2C_2$, where the curve for a continuously increasing load does not coincide with that for the unloaded and reloaded wire, are shown dotted in fig. 1. It will be seen that on reloading a small but finite plastic distortion represented by the horizontal projection of the dotted portion $AC$ must be given to the metal before it attains the same strength on the test piece which was never unloaded.

We thus distinguish two points on each reloading curve: (i) the elastic limit represented by the points $E_1E_2$ in fig. 1 and (ii) the points $C_1C_2$... where the reloading curve joins or becomes parallel to the load-extension curve for continuously increasing load. The points $E_1E_2$ do not seem to be very well defined, but we may define the strength of the material under direct tension at the beginning of the second loading as the ordinate at which the curve $A_2C_1$ would, if produced backwards, cut the vertical corresponding with the length of the specimen at the beginning of the second loading. Since the curve $A_2C_1$ when produced backwards is found as shown in fig. 1 to pass through the point $A_1$, this definition ensures that the strength of the material at the beginning of the second loading shall be the same as that at the end of the first loading.

This interpretation of the strength of a plastic material makes it possible to compare the strengths of the material when subjected to different distributions of stress. To compare the strength of a circular tube in torsion with the strength under a direct load one may first load the tube till the extension has some definite value, then remove the load and start twisting. The torque-angle curve will be found to be similar to the part $B_1E_1C_1A_2$ of the curve in fig. 1 which corresponds to the second loading. It has a very rapid rise corresponding with $B_1E_1C_1$ (fig. 1) followed by a gradual rise corresponding with $C_1A_2$. The torque to be compared with the direct load at the end of the first loading of the tube is that found by producing the second part of the torque-angle curve backwards till it cuts the axis of zero angle. This process sounds rather artificial but in actual operation it is very simple. The curves, figs. 6*a*, *b*, *c*, are actual examples, the produced portions of the curves being shown dotted. It will be seen that in every case the early part of the curve is so steep compared with the succeeding portion which corresponds with $C_1A_2$ of fig. 1 that there is little latitude for possible variation in the extrapolated torque corresponding with zero angle of twist.

The method just described for defining the strength of plastic materials is essentially the same as that used by Lode in his work on the plasticity of soft metals. It applies well to copper and annealed iron and to mild steel, but in the case of aluminium there is in many cases an appreciable creep or gradually increasing strain with constant stress. Under these conditions the stress-strain curve is not unique, but it is still possible to discuss the relationship between the distortions produced by various distributions of stress because on reloading a specimen after removing the initial load in the way previously described for copper (see fig. 1) the load at which a rapid increase in strain occurs is definite even though a slow creeping begins at a lower load.

After defining the strength of the material in this way we are in a position to compare the behaviour of real materials with an ideal isotropic plastic body which has the property that the resistance to distortion bears a definite relationship to the total amount of distortion from the initial annealed condition. A complete specification of the properties of such a material must include: (1) the particular function of the stress components which must rise to a certain value (which defines the strength of the material) before plastic flow begins; (2) the relationship between the plastic distortion produced and the type of applied stress; (3) the relationship between the amount of distortion (cold work) and the resistance to further distortion. Of these (1) has been the subject of many researches which have resulted in the announcing of several mutually incompatible laws, the best known of which are (*a*) the hypothesis known as Guest's law or Mohr's hypothesis that the maximum shear stress determines the beginning of plastic flow independently of the other components of shear stress, and (*b*) von Mises's hypothesis that the sum of the squares of the principal shear stresses must rise to a given value before plastic distortion begins. This we will call von Mises's first hypothesis to distinguish it from another which will be discussed later.

On the other hand (2), a knowledge of which is quite as necessary as (1) in many problems of plasticity, has received but little attention. The reason for this neglect is, no doubt, that in the case of the simplest tests which can be applied to an isotropic or plastic material, the density of which is unaltered by strain, the type of distortion which occurs is simply a matter of symmetry; thus the only possible distortion of a uniform tube under pure tension is a uniform extension parallel to its axis, together with a uniform contraction of the material equal to half the extension in all directions at right angles to the axis. In the case of a twisted tube the distortion must be a pure shear which transforms generators of the tube into spirals without altering its external shape or dimensions. If, however, these two types of stressing are combined, a tube being subjected to direct tensile load and to torsion simultaneously, symmetry alone is not sufficient to determine the resulting distortion, and it is by the use of tests of this kind that we have investigated the relationship between distortion and stress distribution.

## General Theory of Relationship between Stress and Strain in a Plastic Solid

*Specification of stress.* The mathematical specification of stress in a solid is discussed in all books on elasticity. It can most simply be represented by means of a stress quadric* which is so placed that its principal axes are in the direction of the principal stresses, so that the components of shear stress parallel to the principal planes are zero. The principal shear stresses $\tau_1 \tau_2 \tau_3$ are defined as the shear stresses parallel to the planes which pass through one of the principal stress axes and are at 45° to the other two. The stress acts in the direction at right angles to the principal stress axis which lies in the plane concerned. If $\sigma_1, \sigma_2, \sigma_3$ are the principal stresses then

$$\tau_1 = \tfrac{1}{2}(\sigma_2 - \sigma_3), \quad \tau_2 = \tfrac{1}{2}(\sigma_3 - \sigma_1), \quad \tau_3 = \tfrac{1}{2}(\sigma_1 - \sigma_2).$$

The importance of the principal shear stresses lies in the fact that it is found experimentally that the addition of a total hydrostatic pressure to any stress system does not produce any plastic strain. This depends only on the shear stresses. The addition

* Love's *Elasticity*, 4th ed. § 50.

of a constant pressure to $\sigma_1, \sigma_2$ and $\sigma_3$ leaves $\tau_1, \tau_2$ and $\tau_3$ unaffected, so that $\tau_1, \tau_2, \tau_3$ are sufficient to define the stress system so far as plastic distortion is concerned; and since $\tau_1+\tau_2+\tau_3=0$, the direction and magnitude of two of the principal shear stresses are sufficient to define the stress at any point. The equation to the stress quadric when referred to the principal axes of stress is

$$\sigma_1 x^2+\sigma_2 y^2+\sigma_3 z^2=\text{constant}. \tag{1}$$

*Specification of strain.* Any small strain may be represented by a strain quadric* which is a surface possessing the property that the reciprocal of the square of its central radius vector in any direction is proportional to the extension of a line in that direction. The principal planes of this quadric are so orientated that there is no component of shearing strain parallel to them. The equation of the strain quadric referred to its principal axes is

$$e_1 x^2+e_2 y^2+e_3 z^2=\text{constant}, \tag{2}$$

where $e_1, e_2, e_3$ are the (small) extensions in the directions of the principal axes. In the case of materials which do not change their density during distortion

$$e_1+e_2+e_3=0.$$

*Directional relationship between stress and strain.* All experiments on the plasticity of metals which we have been able to find recorded have been carried out under conditions in which symmetry alone ensures that the directions of the principal axes of stress shall be identical with those of strain. Thus in the case of a uniform bar extended by a direct load the stress and strain quadrics are both spheroids with the principal axis of revolution in the direction of the axis of the bar. In all theories of plastic flow in isotropic solids such as that of Levy and von Mises it has been assumed without experimental test and sometimes without any explicit statement that the directions of the principal stresses are identical with those of principal strains. It is difficult to imagine any other relationship between these directions, but it is worth while noticing that any departure from isotropy in the material might be expected to falsify this relationship unless, as in the case of a drawn metal rod or tube under direct load, the directions of principal stress are identical with the principal directions of the non-isotropic properties. In the work which will be described later, experiments are carried out in which the directions of the principal stresses are variable at will, and it is proved experimentally that the directions of the axes of the stress quadric do in fact coincide with those of the strain quadric.

*Other relationships between stress and strain.* When the directions of the principal axes of stress are known the complete specification of stress contains two variables, which may be regarded as (1) the absolute magnitude of any one of the principal shear stresses or of any homogeneous function of them, and (2) any one ratio between the principal shear stresses or between linear combinations of them. The same remarks apply to the specification of strain. When there is no change of density and

* Love's *Elasticity*, 4th ed. §11.

the directions of the principal axes are known, two variables only are needed for the complete specification of any small strain. These variables will now be considered separately.

1. (*a*) *Absolute magnitude of stress.* With a plastic body of the type here considered a definite stress is required to produce a small plastic strain, and various empirical laws have been put forward to represent the strength of the material to resist different types of stress; thus according to Mohr's hypothesis of Guest's law, plastic strain takes place when the absolute value of either $\tau_1$ or $\tau_2$ or $\tau_3$ rises to a certain value $\kappa$. According to the first hypothesis of von Mises, plastic strain begins when $\tau_1^2+\tau_2^2+\tau_3^2$ rises to the value $2\kappa^2$. In either case $\kappa$ is defined as half the lowest direct stress at which a tensile specimen can be extended plastically. On applying these empirical laws to the case of a tube twisted by application of a torque after the value of $\kappa$ has been determined by a direct tensile test it will be seen that according to Mohr's hypothesis the greatest of the principal shear stresses in torsion should be $\kappa$, while according to that of von Mises it should be $2\kappa/\sqrt{3}$. The comparison between the strengths of tubes under direct load and under torsion therefore affords a means of comparing these theories. The experiments to be described later verify the substantial accuracy of von Mises's hypothesis in the case of copper and aluminium, not only in the case of the particular stress distribution of a twisted tube, but for all ratios of the principal shear stresses.

(*b*) *Absolute magnitude of strain.* The absolute magnitude of the strain when any given stress is applied depends on the rate of increase in strength of the material with increasing strain. This varies very widely with the material, and for the present at any rate we may without loss of interest regard it as indeterminate.

2. For any given material the ratio of any pair of principal shear stresses may be regarded as a definite function of the ratio of the corresponding shear strains. Von Mises has proposed the law that these ratios are always equal to one another as can be proved to be the case for a viscous fluid. This hypothesis will be called von Mises's second hypothesis. It is worth noticing that, if the stress system be reduced by addition or subtraction of a uniform hydrostatic pressure to one in which $\sigma_1+\sigma_2+\sigma_3=0$, von Mises's second hypothesis may be completely expressed by the statement that the stress quadric is similar and similarly orientated to the strain quadric.

## Work of Lode

The only work in which these theories have been put to the test of experiment seems to be that of Lode,* who tested thin-walled metal tubes under direct load when subjected simultaneously to internal pressure. By varying the ratio of the internal pressure to the direct load he was able to obtain all possible ratios of the principal shear stresses, and by measuring the extension and also change in the diameter of the tube he was able to measure the ratios of the corresponding principal shear

* 'Versuche über den Einfluss der mittleren Hauptspannung auf das Fliessen der Metalle, Eisen, Kupfer und Nickel', *Z. Phys.* XXXVI (1926), 913.

stresses. It has already been pointed out that the ratio of any two independent linear combinations of shear stresses can be used to define the stress. The particular combination used by Lode was

$$\mu = 2\left(\frac{\sigma_2 - \sigma_3}{\sigma_1 - \sigma_3}\right) - 1,$$

and the corresponding variable for defining the strain was

$$\nu = 2\left(\frac{e_2 - e_3}{e_1 - e_3}\right) - 1,$$

where $e_1$ was the extension of the material in the direction of the length of the tube, $e_3$ was that in the direction of the radius and $e_2$ was the tangential extension obtained by measuring the change in mean radius of the tube. According to von Mises's second theory the relationship $\mu = \nu$ should hold.

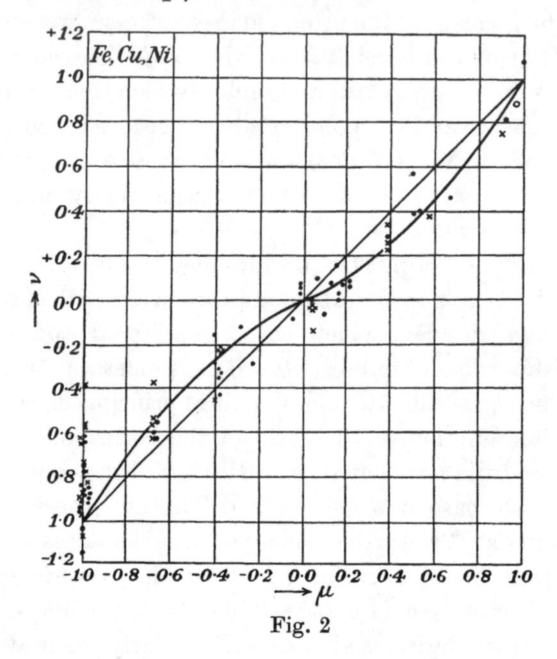


Fig. 2

Lode's results are shown in fig. 2 where the ordinates represent $\nu$ while the abscissae are $\mu$. Since Lode's tubes were thin-walled the stress $\sigma_2$ may be considered as great compared with $\sigma_3$ and Lode took $\sigma_3 = 0$, so that $\mu = 2\sigma_2/\sigma_1 - 1$. When there was no internal pressure $\sigma_2 = 0$, so that $\mu = -1$, and as $\sigma_2$ increased $\mu$ increased from $-1$ through zero till when $\sigma_2 = \sigma_1$, $\mu = +1$. Further increases in $\sigma_2$ were not considered because in an isotropic material it is sufficient to limit the range of experiment to cases for which $\sigma_1 > \sigma_2 > \sigma_3$.

For an isotropic material the stress quadric degenerates into a spheroid when

$$\mu = -1, \ 0 \text{ or } +1,$$

so that symmetry alone necessitates that $\mu = \nu$ at those three points.

On examining Lode's diagram, fig. 2, it will be seen that some of the experimental points, particularly for $\mu = -1$, which correspond with direct tension in the absence of internal pressure, do not satisfy this condition, so that either there is a considerable possible error in the measurements or else the tubes were not isotropic. In spite of the fact that the probable error appears to be of about the same order as the effect measured, Lode's results seem to show a distinct tendency for the absolute value of $\nu$ to be less than that of $\mu$.

In view of the uncertainty so clearly shown in Lode's diagram, fig. 2, it cannot be definitely stated whether von Mises's hypothesis that $\mu = \nu$ is true or untrue, it was largely to find out whether another method could be devised to give more reliable results that the work here described was undertaken.

## Method of the Present Work

The method adopted for obtaining all possible ratios of the principal shear stresses was to subject a tube simultaneously to a direct load and to torsion. By adjusting the ratio of direct load to torque any desired value of $\mu$ between $-1$ for no torque and 0 for no direct load could be obtained. This covers the whole range of ratios of principal shear stresses because for an isotropic material symmetry shows that the $\mu$, $\nu$ curve must be symmetrical about the point $\mu = 0$, $\nu = 0$. In searching for the most probable sources of error in Lode's experiments attention was concentrated first on testing tubes for want of isotropy so that the results with non-isotropic tubes could be rejected, and secondly on improving on Lode's method of measuring changes in diameter of the tubes. Both these objects were attained by the same device, namely, measuring the change in internal volume of the tube during the strain by filling it with water and measuring the movement of the water column in a capillary tube which was directly connected with the tube under test.

Each tube under test was first extended by means of a total load $W$, applied directly, and the change in internal volume measured. It is clear that when a tube of isotropic material, which does not change its density under strain, is stretched there should be no change in internal volume, because each element of the wall of the tube contracts equally in all directions at right angles to its length, so that the ratio of the internal to external diameter will remain unchanged. Hence the ratio of the volume of the bore (i.e. the space inside the inner surface of the tube), to that of the material of the tube (i.e. the space between the walls of the tube) remains unchanged, and consequently since the change in density is assumed zero the internal volume must remain unchanged by the stretching. After any end-effects have been eliminated any change in internal volume on stretching by a direct load indicates either that the material is not isotropic or that there is an appreciable change in density on stretching. The change in density can be measured independently, and in the case of copper and aluminium it is found to be very small. The change in internal volume which results from the small measured change in density can be calculated, and if for any particular tube the observed change in

17-2

internal volume was found to be much greater than this (as sometimes occurred under special heat treatment) the tube was rejected.

The load was next partially removed till a fraction $mW(0 < m < 1)$ remained and a gradually increasing torque applied, the angle of twist and the extension being measured. The torque-angle curve and the twist-extension curve were thus obtained; specimens of these are given in fig. 7 *a, b, c*. At the same time the change in internal volume was measured and expressed in terms of the extension. Specimens of the resulting curves are shown in fig. 8.

By varying $m$ from 0 to 1 these measurements make possible: (i) a comparison between the strength of the material under direct load and that under any other ratios of shear stresses, so that von Mises's first hypothesis can be compared with that of Mohr and Guest, (ii) a verification that the orientations of the principal axes of stress coincide with those of strain, (iii) measurements of $\nu$ in terms of $\mu$, and hence a test of von Mises's second hypothesis that $\mu=\nu$. The method is particularly suitable for testing whether $\mu=\nu$, because if $\mu=\nu$ the internal volume remains unchanged whatever combination of direct load and torque is applied. Hence the measured changes in internal volume afford a means of making a direct measurement of deviations from the expected relationship.

## Analysis of Stress and Strain in a Thin-walled Tube subjected to Combined Direct Load and Torque

To represent the stress and strain quadrics axes through any point in the material of the tube are taken, $x$ parallel to the length of the tube, $y$ tangential, $z$ radial. The scheme is shown in fig. 3. If $mW$ is the total load, $r_m$ the mean radius of cross-section of the tube, $t$ its thickness, $G$ the applied torque, the stress system is represented in the usual notation by

$$\left.\begin{aligned} X_x&=mP_0=P=\frac{mW}{2\pi r_m t}, \quad Y_y=0, \quad Z_z=0,\\ X_y&=S=\frac{G}{2\pi r_m^2 t}, \quad Y_z=0, \quad Z_x=0, \end{aligned}\right\} \tag{3}$$

where $P_0$ is the stress in the material at the end of the first loading, $P$ is the stress in the direction of the length of the tube during the combined stress, and $S$ is the shear stress.

We may suppose that a length $l_0$ of the material is first stretched to length

$$l=l_0(1+e_0)$$

by direct load. Next, owing to the action of the combined direct load $mW$ and torque $G$ the tube stretches to length $l+\delta l$ and twists through an angle $\chi$. This strain may be represented by the components

$$\left.\begin{aligned} e_{xx}&=e, \quad e_{yy}=-\tfrac{1}{2}e-f, \quad e_{zz}=-\tfrac{1}{2}e+f,\\ e_{xy}&=s, \quad e_{yz}=0, \quad e_{zx}=0. \end{aligned}\right\} \tag{4}$$

The extension $$e=\delta l/l=\delta l/l_0(1+e_0).$$

The shear $s = r_m \chi / l$, and since the material does not change in density during the extension from $l_0$ to $l$, $r_m = r_{m_0}(1+e_0)^{-\frac{1}{2}}$, where $r_{m_0}$ is the mean radius of the tube before the first stretching. Hence

$$s = r_{m_0} l_0^{-1}(1+e_0)^{-\frac{3}{2}} \chi. \tag{5}$$

As has already been pointed out, if the density and internal volume of the tube do not change $e_{yy} = e_{zz} = -\frac{1}{2}e$, so that the strain $f$ in (4) is proportional to the change in internal volume, in fact

$$\frac{\text{change in internal volume}}{\text{internal volume of tube}} = 2\,\frac{\text{change in radius}}{\text{mean radius}} + \frac{\text{change in length}}{\text{length}} = 2e_{yy} + e = -2f. \tag{6}$$

The strain quadric is

$$ex^2 - (\tfrac{1}{2}e + f)\,y^2 - (\tfrac{1}{2}e - f)\,z^2 + sxy = \text{constant}. \tag{7}$$

To find the direction of its principal axes put

$$x = x'\cos\phi - y'\sin\phi, \quad y = x'\sin\phi + y'\cos\phi. \tag{8}$$

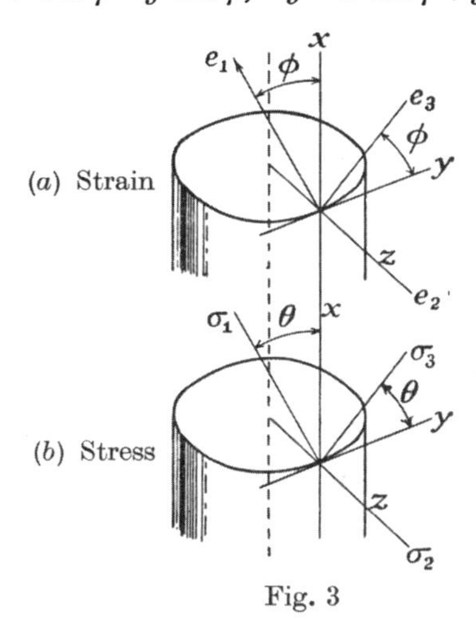


Fig. 3

The strain quadric then becomes

$$\begin{aligned} &x'^2\{e\cos^2\phi - (\tfrac{1}{2}e+f)\sin^2\phi + s\sin\phi\cos\phi\} \\ &\quad + y'^2\{e\sin^2\phi - (\tfrac{1}{2}e+f)\cos^2\phi - s\cos\phi\sin\phi\} \\ &\quad + x'y'\{-(\tfrac{3}{2}e+f)\sin 2\phi + s\cos 2\phi\} - (\tfrac{1}{2}e - f)\,z^2 = 0. \end{aligned} \tag{9}$$

The strain quadric is now referred to its principal axes if the coefficient of $x'y'$ is zero, i.e. if

$$\tan 2\phi = s/(\tfrac{3}{2}e + f), \tag{10}$$

and substituting this value of $\phi$ in the coefficients of $x'^2$ and $y'^2$ in (9) the principal extensions are found to be

$$\left.\begin{aligned} e_1 &= \tfrac{1}{4}e - \tfrac{1}{2}f + (\tfrac{3}{4}e + \tfrac{1}{2}f)\sec 2\phi, \\ e_3 &= \tfrac{1}{4}e - \tfrac{1}{2}f - (\tfrac{3}{4}e + \tfrac{1}{2}f)\sec 2\phi, \\ e_2 &= -\tfrac{1}{2}e + f. \end{aligned}\right\} \tag{11}$$

It will be noticed that $e_1 > e_2 > e_3$. The directional relationships between these quantities are shown in fig. 3*a*.

Using Lode's variable

$$\nu = 2\left(\frac{e_2 - e_3}{e_1 - e_3}\right) - 1,$$

it will be found that

$$\nu = \left(\frac{-1 + 2f/e}{1 + \frac{2}{3}f/e}\right)\cos 2\phi. \tag{12}$$

The stress quadric is

$$Px^2 + 2Sxy = \text{constant}, \tag{13}$$

and substituting $x = x'\cos\theta - y'\sin\theta, \quad y = x'\sin\theta + y'\cos\theta,$
this becomes

$$x'^2(P\cos^2\theta + 2S\cos\theta\sin\theta) + y'^2(P\sin^2\theta - 2S\cos\theta\sin\theta) + 2x'y'\{-P\cos\theta\sin\theta + S(\cos^2\theta - \sin^2\theta)\} = \text{constant}. \tag{14}$$

If $\theta$ is now chosen so that $x'y'$ are principal axes of stress, the coefficient of $x'y'$ in (14) is zero so that

$$\tan 2\theta = 2S/P. \tag{15}$$

The principal stresses are then

$$\sigma_1 = \tfrac{1}{2}P(1 + \sec 2\theta), \quad \sigma_2 = 0, \quad \sigma_3 = \tfrac{1}{2}P(1 - \sec 2\theta). \tag{16}$$

The directions of these stresses are shown in fig. 3*b*. Lode's variable

$$\mu = 2\left(\frac{\sigma_2 - \sigma_3}{\sigma_1 - \sigma_3}\right) - 1$$

is found from (16) to be

$$\mu = -\cos 2\theta. \tag{17}$$

Comparing (6), (12) and (17) it will be seen that if there is no change in internal volume of the tube during distortion so that $f = 0$, $\nu = -\cos\phi$, and if in addition the directions of the principal axes of stress and strain are the same so that $\theta = \phi$, then $\mu = \nu$. Thus after verifying experimentally that $\theta = \phi$, as is done later in table 2, the observed change in internal volume is a measure of the deviation from von Mises's relationship in accordance with the equation

$$\frac{\mu}{\nu} = \frac{1 - \dfrac{2f}{e}}{1 + \dfrac{2f}{3e}}.$$

## Relationship between $m$ and $S/P_0$ according to Mohr's and von Mises's Hypotheses

It seems clear that when $m$ is nearly equal to 1, so that the direct load is only slightly reduced after the first stretching, a small torque will suffice to cause plastic distortion. When $m = 0$, so that there is no direct load, as we have already seen $S/P_0 = 0{\cdot}5$ according to Mohr's hypothesis that the maximum principal shear stress alone determines whether plastic distortion shall occur, and $S/P_0 = 1/\sqrt{3}$ according to von Mises's hypothesis. To calculate the intermediate values we may use the expressions for $\sigma_1, \sigma_2, \sigma_3$ given in (16). According to Mohr's hypothesis $\sigma_1 - \sigma_3 = P_0$, so that from (16) $\sigma_1 - \sigma_3 = P \sec 2\theta = P_0$ and $P/P_0 = m$, so that

$$m = \cos 2\theta, \tag{18a}$$

and since $2S/P = \tan 2\theta$,

$$S/P_0 = \tfrac{1}{2} m \tan 2\theta = \tfrac{1}{2} m \left(\frac{1}{m^2} - 1\right)^{\frac{1}{2}} = \left(\frac{1-m^2}{4}\right)^{\frac{1}{2}}. \tag{18b}$$

According to von Mises's first hypothesis

$$(\sigma_1 - \sigma_2)^2 + (\sigma_2 - \sigma_3)^2 + (\sigma_3 - \sigma_1)^2 = 2P_0^2,$$

so that from (16)

$$\tfrac{1}{4} P^2 (1 + \sec 2\theta)^2 + \tfrac{1}{4} P^2 (1 - \sec 2\theta)^2 + P^2 \sec^2 2\theta = 2P_0^2,$$

or since

$$P = mP_0, \quad m^2 (\tfrac{1}{2} + \tfrac{3}{2} \sec^2 2\theta) = 2. \tag{19a}$$

Hence

$$S/P_0 = \tfrac{1}{2} m \tan 2\theta = \left(\frac{1-m^2}{3}\right)^{\frac{1}{2}}. \tag{19b}$$

## Experimental Details

A sketch of the apparatus used is shown in fig. 4. The tube under test, which is about $\frac{1}{4}$ in. external diameter, is shown as $A$. $BB$ are the upper and lower fixings for gripping the ends of the tube. The upper fixing $B$ is supported by a steel ball $C$ on a fixed support by means of a connecting carriage $D$ which is prevented from rotating by fixed stops $EE'$. A cylindrical drum $F$, 8 in. diameter, rests on the lower end-fixing, $B'$. The direct load $W$ is attached to this through a steel ball pivot similar to that at the top but not shown in the sketch, and the torque is applied by means of threads passing over ball-bearing pulleys $GG'$ supporting equal weights $pp$. The angle of twist is measured by a fixed pointer $H$ reading on a protractor $K$ fixed to the upper surface of the drum $F$. The change in internal volume is measured by reading the position of the meniscus $J$ of the water in the capillary tube $L$ on the scale $M$. The capillary tube is connected with the upper end-fixing $B$ and the whole apparatus filled with water through the glass stopcock $O$ at the lowest point of the lower fixing $B'$. The extensions are read by telescopes (not shown in fig. 4) mounted on a vertical steel rod. These can be focused either on fine ring marks $QQ$ on the specimen or on the upper carriage $D$ and on a corresponding steel point $R$ attached to the axis of the lower fixing $B'$ in such a way that it can be centralized so that it remains in focus during the rotation of the drum $F$.

The chief difficulty in the design of the apparatus lay in making end-fixings which would remain watertight when the tube was stretched and at the same time limit to a small area the end-effects or variation of the distortion at the end from the uniform state of the middle of the tube. This was accomplished by expanding the ends of the specimens into conical flanges by means of a specially designed press. The ends, so formed, were then gripped in the end-fixings $BB'$ of fig. 4. These are shown in detail in fig. 5. In this diagram $A$ is the specimen with its conical ends.

$C$ is a steel sleeve (split for convenience) with conical upper and lower faces. $J$ is a conical plug with a central hole to let the water through and a plane upper face which makes a watertight seating in the upper nut $L$ of the fixing. The lower nut $F$ which rests on the carriage is split and threaded with a deep thread so that by screwing the two parts $L$ and $F$ hard together a great pressure can be brought to act on the flanged upper portion of the specimen. The large distortions so produced in the material of the conical flange harden copper, iron and aluminium to

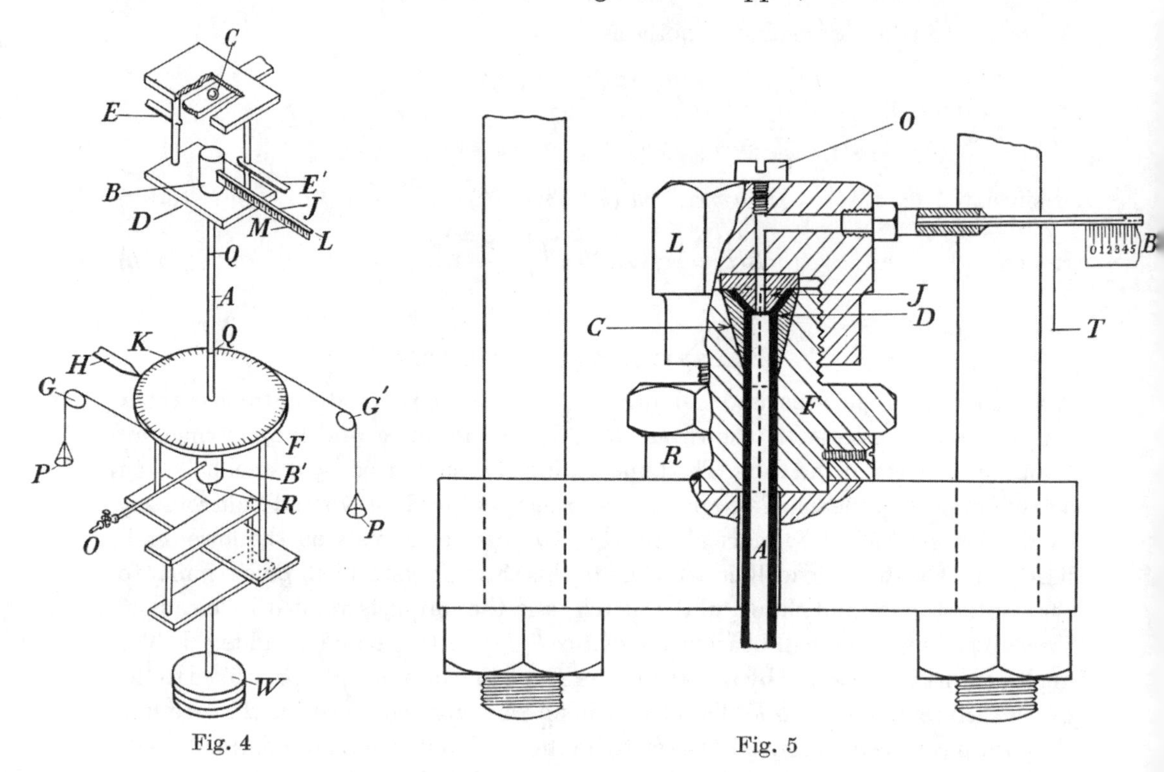


Fig. 4

Fig. 5

such an extent that when the tube is afterwards stretched no observable plastic flow takes place in the material of the flange. The stretching, in fact, begins at the corner $D$ where the flange connects with the parallel part of the tube. The ring $R$ is merely used for holding the two portions of the split nut $F$ in place.

It will be noticed that in this description the sleeve $C$ plays no part; it might, in fact, be part of $F$, when the apparatus is used in the manner described. When it was used for soft metals like lead and cadmium which do not harden appreciably by cold work it was found that the method just described does not ensure that there shall be no flow inside the material of the flange. In these cases, therefore, a cylindrical steel plug with a small central hole was made to fit the bore of the specimen. The flat faces of the split sleeve $C$ were then filed so that when the nut $L$ was screwed home, the soft metal was gripped between the split sleeve $C$ and the cylindrical plug. In this way it was found possible to grip the lead tube.

To fill the apparatus with water the plug $O$ and the capillary tube $T$ were unscrewed, alcohol was then passed through from the stopcock in the lower fixing (see fig. 4) in order to remove grease. Distilled water from which air had been removed by boiling was then passed through till it overflowed at $O$ (fig. 5). The plug $O$ was then replaced and afterwards the capillary tube $T$, the water flowing all the time. To regulate the position of the water meniscus in $T$ the water was allowed to flow back through the apparatus by opening the lower stopcock. As a final test whether any bubbles were caught in the apparatus air pressure was applied to the capillary tube. If the meniscus moved an appreciable distance down the tube, bubbles were indicated and no tests were carried out till they had been removed.

In planning the work it had been intended to do all tests with two different lengths of tube so that by subtracting the results of the two sets of measurements from one another end-effects could be eliminated, but on carrying out this plan in a few cases, it was found that the end-effects were too small to be appreciable when the tube was $11\frac{1}{2}$ in. long by $\frac{1}{4}$ in. diameter. Accordingly tubes of this length were used throughout the experiments and no allowance was made for end-effects.

## Heat Treatment

The copper tubes were supplied in a hardened condition and they were annealed at 650° C. for about 36 hr. This produced grains the linear dimensions of which were small compared with the wall thickness. For comparison we also annealed some tubes after a slight straining at considerably higher temperatures. The large grains thus produced in many cases stretched from the inner to the outer wall, and in these cases we found that the material behaves in regard to distortion as though it were not isotropic. This question is discussed later in connection with fig. 8.

The copper tubes were composed of what the makers described as high conductivity copper, containing not more than 0·2 % impurity. A comparison between the mechanical properties of these tubes and those of electrolytic copper wire of the highest purity obtainable, commercially, showed us that the small amount of impurity present in our tubes was not sufficient to make them differ appreciably from what might be expected if we had been able to use electrolytic copper throughout.

The aluminium tubes were supplied in a small-grained annealed condition and no heat treatment was required. The aluminium was described by the makers as containing 99·7–99·8 % aluminium.

The mild steel was too hard in the condition in which it was supplied. Its carbon content was from 0·12 to 0·18 %. It was, therefore, annealed *in vacuo* at about 920° C. In some cases the same mild steel was decarburized in a stream of hydrogen at 920° C. This gave large crystals, and the results of our experiments again indicated that the material so treated did not behave in an isotropic manner. To obtain smaller crystals some of the tubes were decarburized at 650° C. for 5 or 6 days. These tubes were found to behave more like small-grained copper tubes though they still appeared to be less isotropic in regard to their distortion.

## NUMERICAL DATA

Diameter of torque drum 8·22 in.

*Copper tubes.* $l_0 = 11{\cdot}5$ in., external diameter 0·248 in., $t = 0{\cdot}036$ in., $r_{m_0} = 0{\cdot}106$ in.,* initial area of section $= 2\pi r_{m_0} t = 0{\cdot}02395$ sq.in.

$$P = \frac{mW(1+e_0)}{0{\cdot}02395}, \tag{20a}$$

$$S = \frac{8{\cdot}22p(1+e_0)^{\frac{3}{2}}}{(0{\cdot}02395)(0{\cdot}106)} = 3230\,(1+e_0)^{\frac{3}{2}} p \text{ lb. per sq.in.}, \tag{21a}$$

where $p$ is the weight in lb. in the scale pans of the torque system

$$\tan 2\phi = \frac{2s}{3e(1+\frac{2}{3}f/e)} = \tfrac{2}{3} r_{m_0} (1+e_0)^{-\frac{1}{2}} (1+\tfrac{2}{3}f/e)^{-1} \left(\frac{\chi}{\delta l}\right),$$

or if $\chi$ is measured in degrees instead of angular measure

$$\tan 2\phi = 0{\cdot}001234(1+e_0)^{-\frac{1}{2}} (1+\tfrac{2}{3}f/e)^{-1} \left(\frac{\chi}{\delta l}\right). \tag{22a}$$

To connect the reading of the meniscus in the capillary tube with $f/e$, the area of cross-section of the capillary is 0·0002607 sq.in., so that if $d$ is the increase in the reading of the meniscus in the capillary tube expressed in inches when the length of the specimen is increased by $\delta l$, then from (6)

$$2f = \frac{0{\cdot}0002607}{(11{\cdot}5)(\pi)(0{\cdot}106)^2} = 0{\cdot}00065d,$$

and remembering that
$$e = \frac{\delta l}{l_0(1+e_0)}$$

this gives
$$\frac{f}{e} = 0{\cdot}0037(1+e_0)\frac{d}{\delta l}. \tag{23a}$$

*Aluminium tubes.* $l_0 = 11{\cdot}5$ in., external diameter 0·2516 in., $t = 0{\cdot}0352$ in., $r_{m_0} = 0{\cdot}1082$ in., initial area of section 0·0240 sq.in.

$$P = \frac{mW(1+e_0)}{0{\cdot}0240}, \tag{20b}$$

$$S = 3190(1+e_0)^{\frac{3}{2}} p, \tag{21b}$$

$$\tan 2\phi = 0{\cdot}00126(1+e_0)^{-\frac{1}{2}} (1+\tfrac{2}{3}f/e)^{-1} \left(\frac{\chi}{\delta l}\right), \tag{22b}$$

$$\frac{f}{e} = 0{\cdot}00355(1+e_0)\left(\frac{d}{\delta l}\right). \tag{23b}$$

* It will be seen that $t$ is not very small compared with $r_{m_0}$, calculations were made to find the values of the errors introduced by neglecting $t$ in comparison with $r_{m_0}$, and it was found that in all the formulae the error depended on neglecting $(\frac{1}{2}t/r_{m_0})^2$ in comparison with unity. In our tubes $((\frac{1}{2}t/r_{m_0})^2$ was about 0·01, and we estimate that the maximum error in any of our results which can arise from the neglect of $(\frac{1}{2}t/r_{m_0})^2$ in comparison with unity is 3 %, the average being about 2 %.

*Iron and steel tubes.* $l_0 = 11{\cdot}5$ in., external diameter 0·2487 in., $r_{m_0} = 0{\cdot}1054$ in., initial area of cross-section 0·02507 sq. in.

$$P = \frac{mW(1+e_0)}{0{\cdot}02507}, \tag{20c}$$

$$S = 3112(1+e_0)^{\frac{3}{2}}p, \tag{21c}$$

$$\tan 2\phi = 0{\cdot}001226(1+e_0)^{-\frac{1}{2}}(1+\tfrac{2}{3}f/e)^{-1}\left(\frac{\chi}{\delta l}\right) \tag{22c}$$

$$= 0{\cdot}00373(1+e_0)\left(\frac{d}{\delta l}\right). \tag{23c}$$

## Representation of Results

The results of the measurements can best be presented in the form of curves showing the relationship between various pairs of simultaneous measurements.

*Torque extension.* The torque-extension curves for copper are given for various values of $m$ in fig. 6*a*, for aluminium in fig. 6*b*, for mild steel and decarburized iron in fig. 6*c*. It will be seen that in all cases except two the second or slowly rising portion has been produced backwards in a broken line to the axis to obtain the virtual value of $p$ corresponding with the beginning of plastic distortion in the idealized plastic body. It will be seen that there is little scope for ambiguity in the values of $p$ so obtained. These values are given in the sixth line of table 1. The values of $p$ for copper tube $E$ for which $m = 0{\cdot}025$ and the mild steel tube for which $m = 0{\cdot}1$ were obtained in a similar manner, but the torque-angle curve was found to be more satisfactory than the torque-extension curve in these cases because the extension

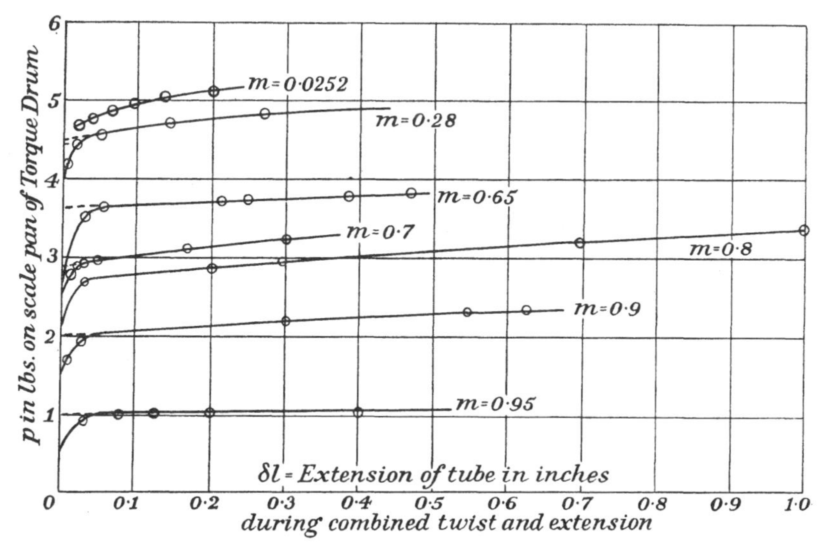


Fig. 6*a*. Copper tubes: length 1·5 in., external diameter 0·248 in., internal diameter 0·175 in.

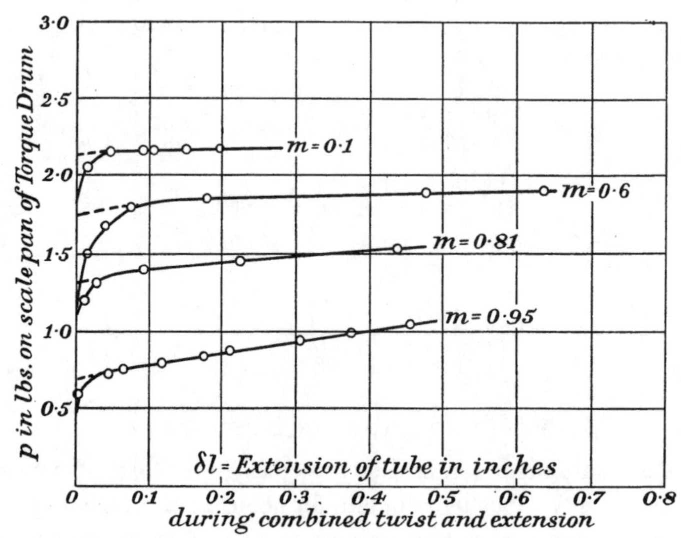


Fig. 6*b*. Aluminium tubes: length 11·5 in., external diameter 0·2516 in., internal diameter 0·1812 in.

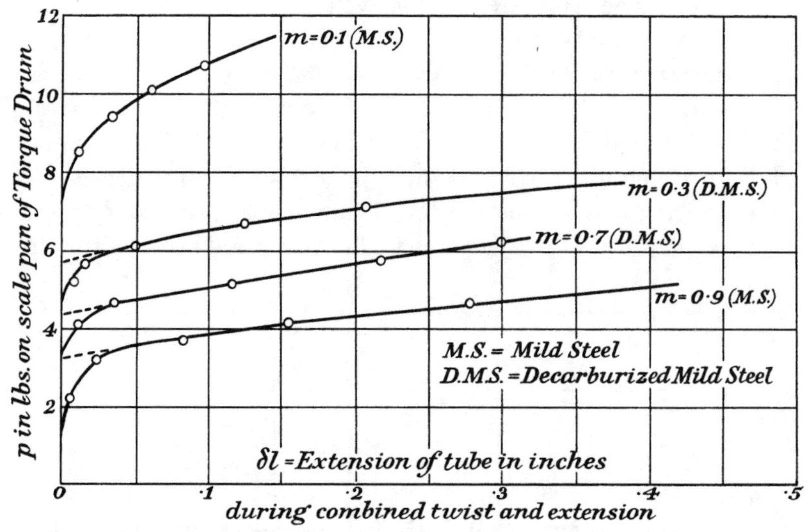


Fig. 6*c*. Mild steel and decarburized iron: length 11·5 in.

for small values of $m$ is too small, most of the distortion being due to twisting. The values of $p$ given in table 1 were inserted in formulae (21 $a, b, c$) to obtain the values of $S$ given in line 7 of table 1. The complete data concerning the initial stretching including $P_0$, the direct stress at the end of the first direct loading and $1+e_0$ are given in lines 4 and 5 of table 1. The values of $\tan 2\theta = 2S/mP_0$ are given in line 8.

*Twist extension.* The twist-extension curves are given in figs. 7 *a*, *b* and *c*. It will be seen that in all cases the observed points fall remarkably well on straight lines. From these lines can be found directly the ratio $\chi/\delta l$ needed in formulae (22 $a, b, c$)

for calculating $\phi$, the angle of inclination of the axes of the strain ellipsoid to the centre lines of the specimen. The values of $\chi/\delta l$ so found are given in line 9 of table 1.

## Change in Internal Volume Extension

In every test the position of the water meniscus in the capillary tube was recorded both during the initial direct loading of the specimen and during the subsequent combined direct load and torque. Two typical specimens of the resulting curves are

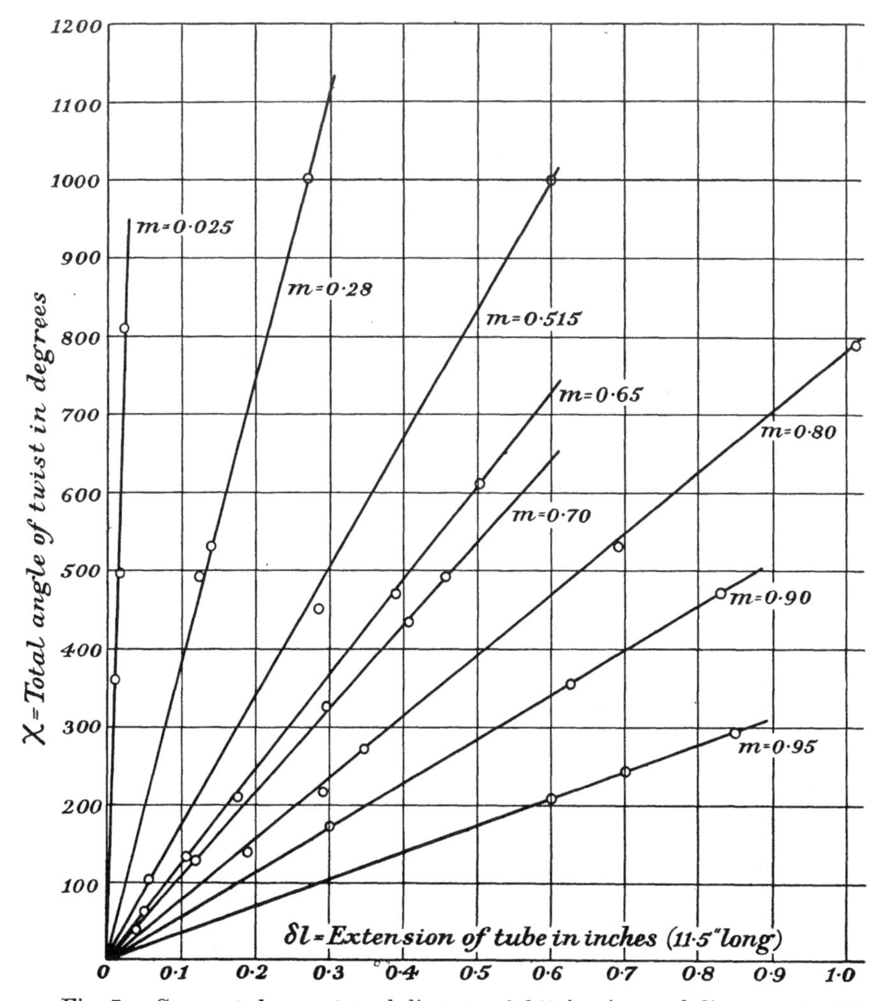


Fig. 7*a*. Copper tubes: external diameter 0·248 in., internal diameter 0·176 in.

shown in fig. 8 with the actual observations from which the curves are drawn. The initial state is represented by *A* in each case. The changes in volume during the initial direct loading are shown in the portion *AB* of each curve, while the effect of applying combined torque and direct load is shown in the portion *BC* in each case.

It will be seen that in the case of aluminium there is a small increase in internal volume. Since we were unable to account for this increase either as an elastic effect, an effect of the small change in density, or as an end-effect, we came to the conclusion that it must be the result of a small residual want of isotropy in the material which had not been removed by heat treatment. The same remarks apply to the copper, but in this case there is a decrease in volume instead of an increase. On starting the compound twist and direct load with copper, iron or aluminium tubes there was in every case a comparatively large decrease in internal volume represented by the

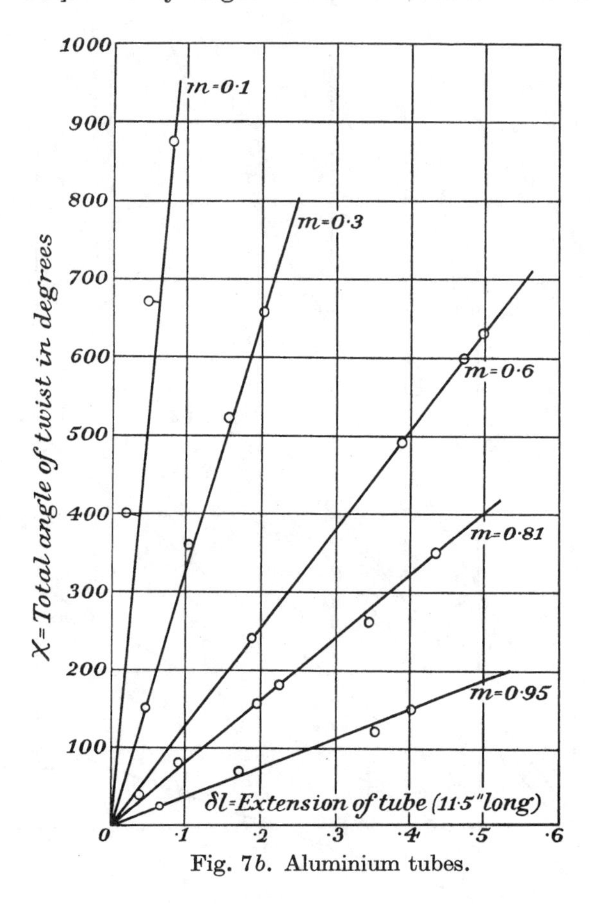


Fig. 7*b*. Aluminium tubes.

portions *BC* of the curves. And it was the slope of *BC* which was used in every case to determine the ratio $d/\delta l$ used in connection with formulae (23 $a, b, c$), for determining $f/e$.

That a large change in internal volume during a direct extension can occur owing to want of isotropy is shown very clearly by the broken curve in fig. 8, which was obtained with a copper tube which had been annealed at such a temperature that the crystal grains were very large; in fact, many of them stretched through the whole thickness of the walls from the inner to the outer surface of the tube. The existence of crystals comparable in linear dimensions with the thickness of the wall of the tube might be expected to make the tube behave as though the material were not isotropic, for in that case the resistance to a type of distortion which involves principally contraction of the material in the radial direction might be very different from a distortion which involves principally contraction in the tangential direction. All results obtained from tubes which showed increases or decreases in volume during

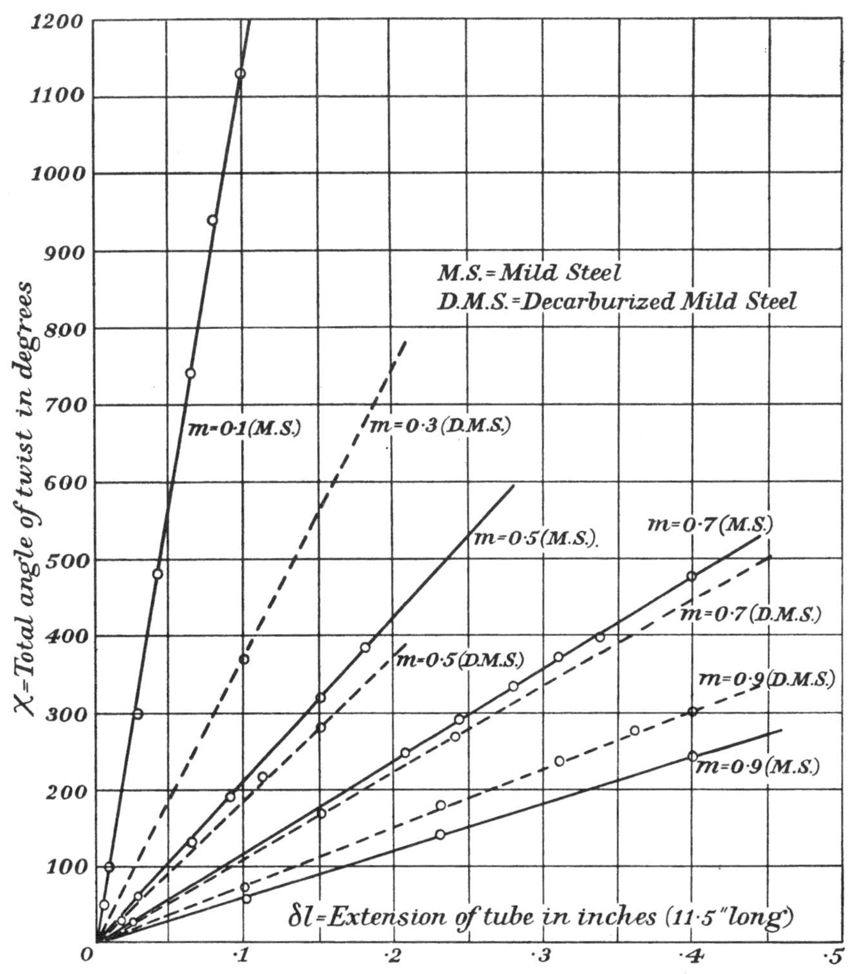


Fig. 7*c*. Mild steel and decarburized mild steel tubes.

the initial direct extension comparable with those obtained during the combined torsion and extension were rejected on the ground that the material of which they are composed is not isotropic or that it behaves as an anisotropic material.

In determining the ratio $d/\delta l$ for various values of $m$ the slopes of the curves of volume change are required only during the action of the combined stress system. In order to facilitate comparison between results for different values of $m$ the portions $BC$ of all curves similar to those shown in fig. 8 are collected together in the three figures, 9*a*, which refers to copper tubes, 9*b* to aluminium and 9*c* to steel. It will be seen that all these curves are, within the limits of accuracy of our measurements, straight lines. The change in internal volume is therefore proportional to the extension, and the values of the ratio $d/\delta l$ taken from figs. 9*a*, *b* and *c* are given in line 10 of table 1. The values of $f/e$ obtained from formulae 23*a*, *b* and *c* are given in line 11 of table 1. The values of tan $2\phi$ calculated by means of formulae 22*a*, *b*, *c* from the figures for $\chi/\delta l$ and $f/e$ given in lines 9 and 11 are given in line 12 of table 1.

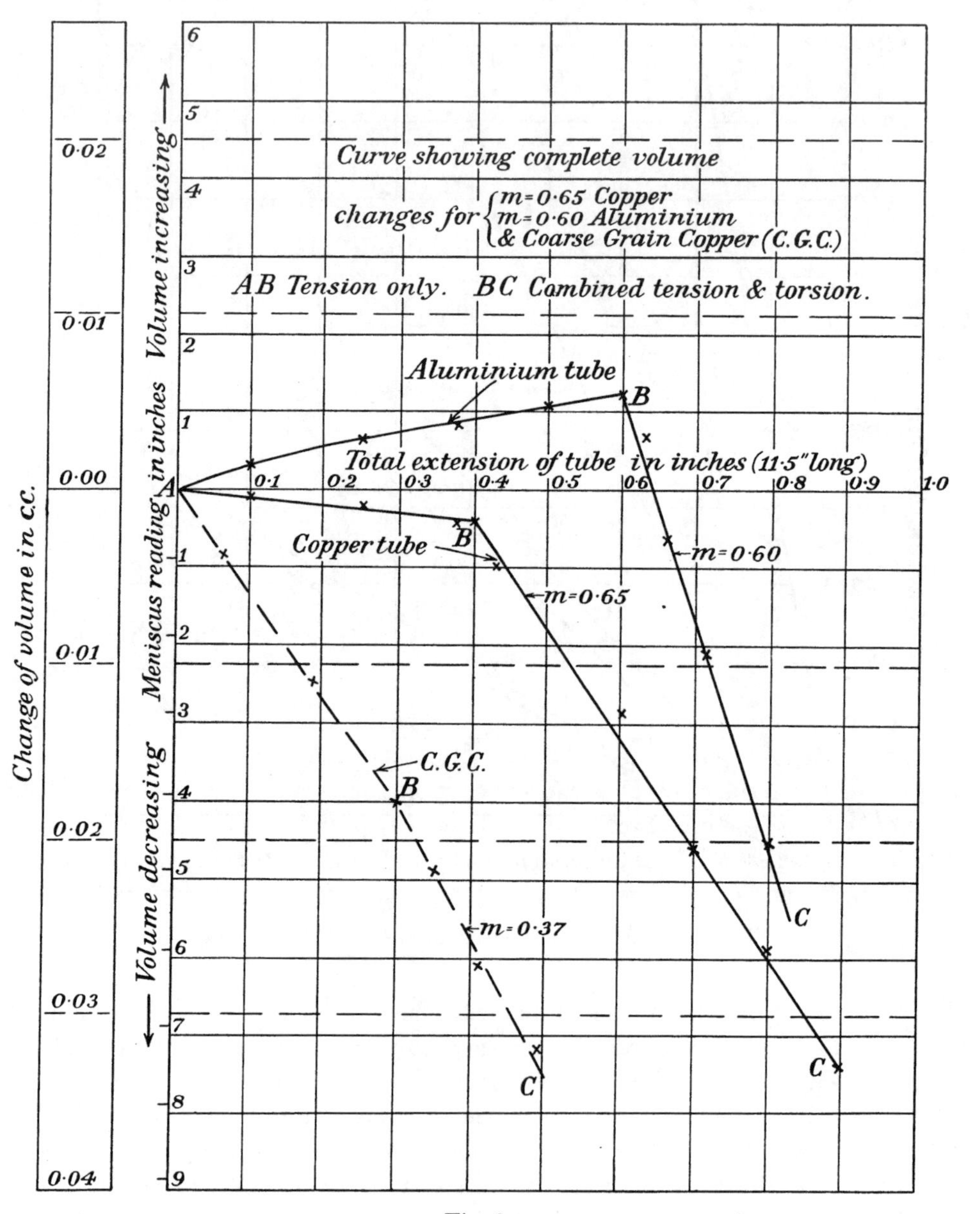


Fig. 8

## INTERPRETATION OF RESULTS

*Orientation of the principal axes of stress and strain.* The inclinations $\theta$ and $\phi$ of the principal axes of stress and strain respectively to the axis of the tube are collected in table 2. It will be seen that the agreement is very good in all cases except that of copper tube $A$. The maximum angle between the measured orientation of the principal stress axes and the principal strain axes, i.e. $\theta-\phi$ is 1·9°, the average

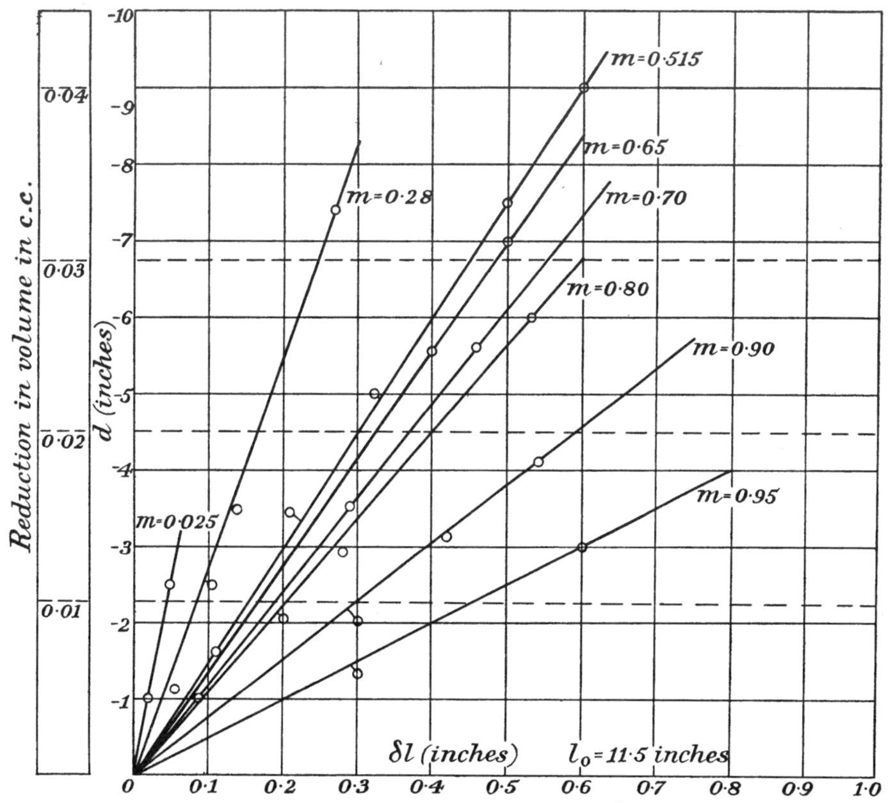


Fig. 9$a$. Copper tubes: external diameter 0·248 in., internal diameter 0·176 in.

difference taken without regard to sign being 0·64°. If it is assumed that the tubes are isotropic on account of the heat treatment they have received, this table, showing that $\theta=\phi$, may be taken as proving that the principal axes of stress and strain in an isotropic plastic material are coincident. If, on the other hand, this is taken as axiomatic in the definition of isotropic material, then the figures in table 2 may be taken as indicating that the material of the tubes is isotropic.

*Relationship between ratios of principal shear stresses and ratios of corresponding principal shear strains, i.e. between $\mu$ and $\nu$.* Taking the values of $\theta$ and $\phi$ from table 2 and the value of $f/e$ from table 1, the values of $\mu=-\cos 2\theta$ and

$$\nu=\frac{-1+2f/e}{1+\frac{2}{3}f/e}\cos 2\phi$$

shown in table 3 were calculated. These are plotted for copper, iron and aluminium separately in the curves of fig. 10. It will be seen that the deviation from von Mises's assumed relationship $\mu=\nu$ which was suspected by Lode is definitely established. Comparing fig. 10, which covers the range $\mu=-1$ to $\mu=0$, with Lode's diagram (fig. 2), which covers the range $\mu=-1$ to $\mu=+1$, it will be seen that fig. 10 corresponds with the lower left-hand quarter of Lode's diagram, but the condition of

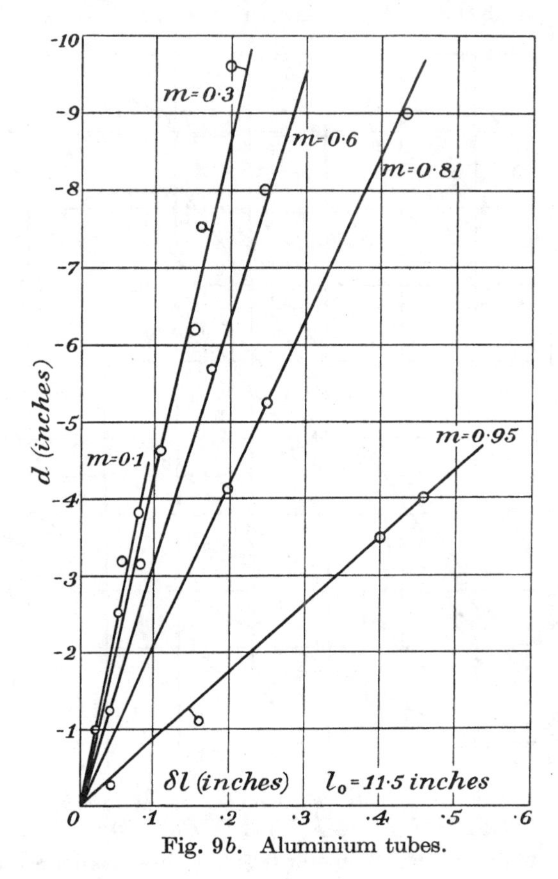


Fig. 9*b*. Aluminium tubes.

symmetry in an isotropic material ensures that the $(\mu\nu)$ curve shall be symmetrical about the point $\mu=0$, $\nu=0$, so that fig. 10 covers the whole range of all possible ratios of the principal shear stresses and strains. It will be seen that the scattering which is so notable a feature of the points in Lode's diagram has disappeared, the points in fig. 10 lying on a curve which is very similar to that shown in fig. 2 which Lode marked on his diagram to represent roughly the most probable variation of $\nu$ with $\mu$.

## Comparison with Plasticity of Glass at High Temperatures

It has already been pointed out that von Mises's second assumption that $\mu=\nu$ would be true if the equations of plastic flow were identical with those of a very viscous fluid. It is known that when glass is heated to such a temperature that it flows slowly under a direct load or a bending load, it behaves like a very viscous fluid and, in fact, slow plastic distortion of this type is used in order to measure the viscosity of glass at temperatures at which it is beginning to soften. As has been mentioned in connection with plastic flow in metals, most of these tests have been carried out under conditions like that of extension under a direct load, where the type of distortion is determined by the condition of symmetry alone.

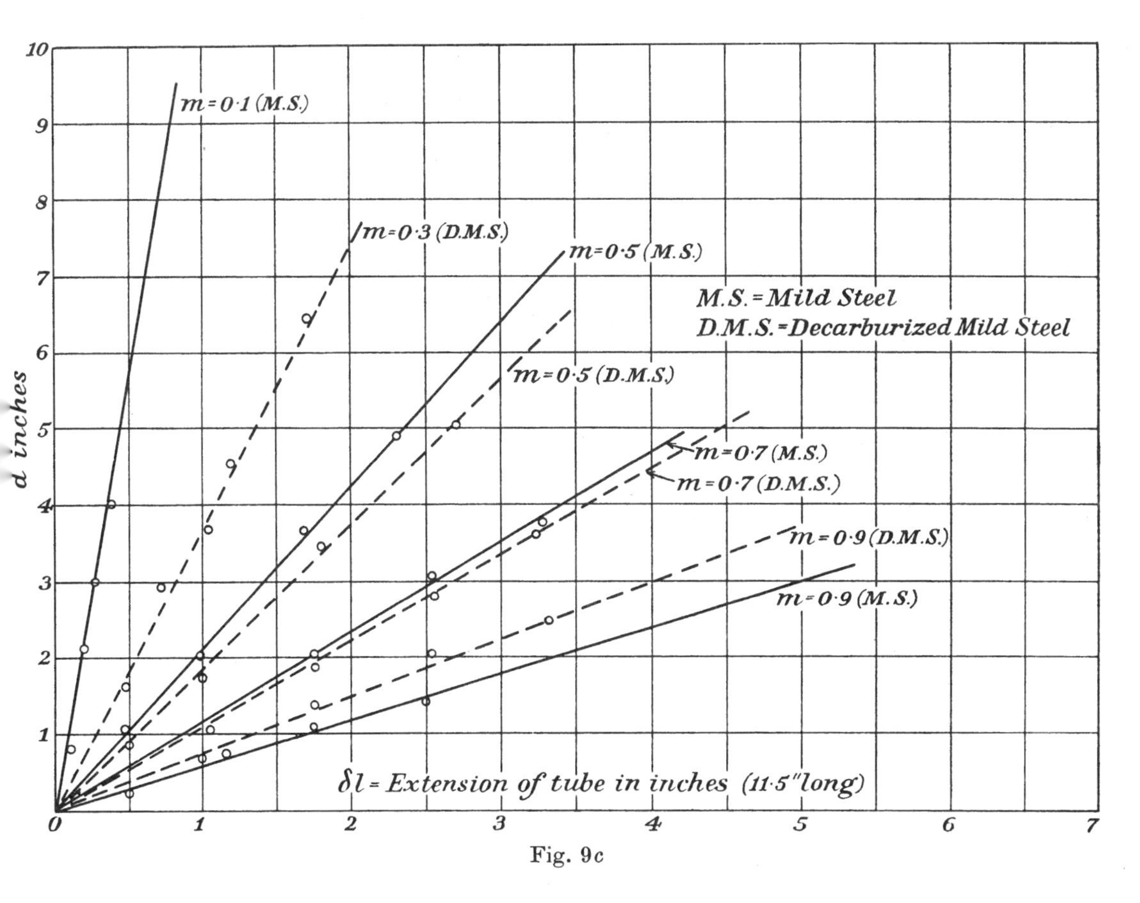


Fig. 9c

Experiments on the plasticity of glass have in fact been limited to determining the rate of distortion for different loads under varying conditions of temperature, etc. It seems that there would be no inconsistency in any existing experiments if the distortion of glass under load were not that of a very viscous fluid but were similar to that of metals for which $\nu$ is not equal to $\mu$. It seemed worth while therefore to carry out tests with glass tubes similar to those already described with metal tubes. These are described in the Appendix. It was found that the change in volume of the bore of a heated glass tube subjected to combined load and extension was so small as to be hardly measurable. The small volume changes actually measured were used to calculate $\mu$ and $\nu$, and the points so obtained are marked in fig. 10. It will be seen that in the case of glass, von Mises's second hypothesis $\mu = \nu$ is fulfilled, the

18-2

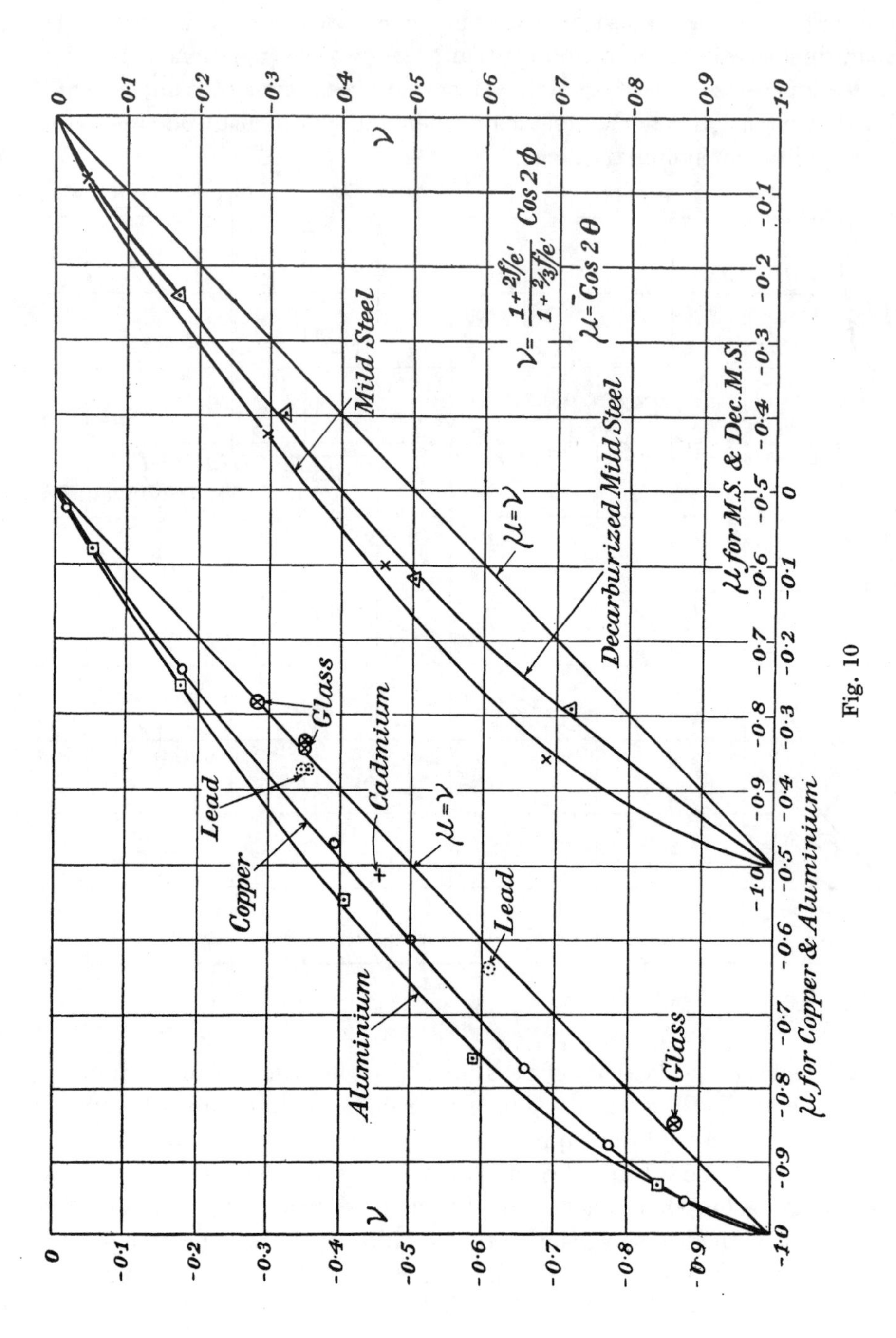
Mild Steel
Decarburized Mild Steel
$\nu = \frac{1+2f/e'}{1+2/3 f/e'} \cos 2\phi$
$\mu = \cos 2\theta$
$\mu = \nu$
$\mu$ for M.S. & Dec. M.S.
$\mu$ for Copper & Aluminium
Glass
Cadmium
Lead
Copper
Aluminium
$\nu$

Fig. 10

small deviation of the observed points from the straight line $\mu = \nu$ being due partly to experimental error and partly to the fact that the thickness of the wall is not quite negligible. This indicates not only that heated glass does behave like a viscous fluid in regard to its distortion, but also that the method of measuring $\mu$ and $\nu$ by observing changes in internal volume of a tube under combined extension and twisting is susceptible of considerable accuracy.

## Experiments with Lead and Cadmium

It was found that lead tubes and cadmium tubes change their internal volumes much less than copper, iron or aluminium tubes when subjected to combined extension and twisting. In both cases we did not find that the change from pure

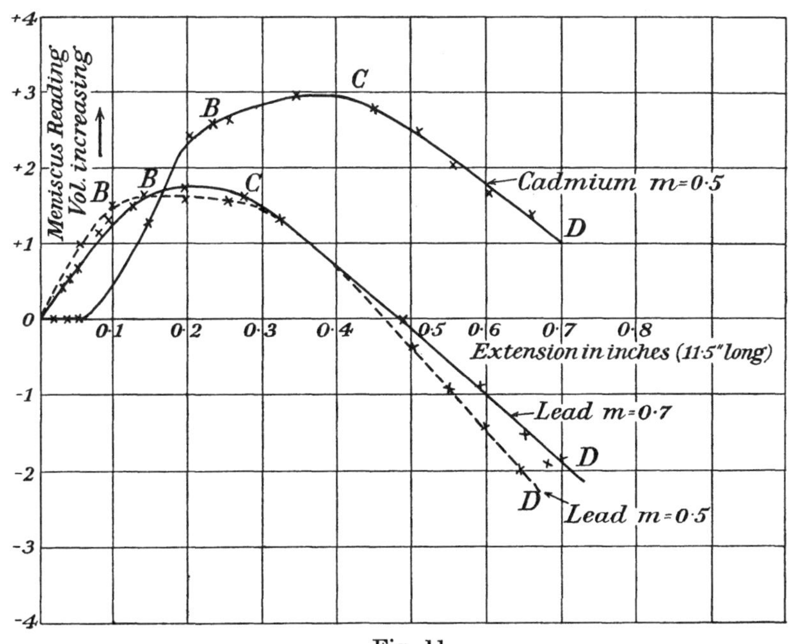


Fig. 11

extension to extension combined with torsion produced a very definite bend in the curve connecting internal volume with extension. This is shown in fig. 11. The part of the curve corresponding with pure extension is from $O$ to $B$. At the condition represented by $B$ the load was reduced and torque applied. The subsequent internal volume-extension curve is shown as $BCD$. It will be seen that there is no very marked sudden bend in the curve at $B$, such as always occurred with copper, aluminium and iron.

It was at first thought that the change in volume during the extension without torsion was due to some end-effect, but a large number of different ways of fixing the ends such as casting enlarged ends, gripping the ends round a steel peg to keep the tube from collapsing inwards, and soldering on brass end-pieces, all gave the same result, so that we were forced to the conclusion that the effect is a real one and that the tubes therefore behave as though they are not isotropic.

For the purpose of comparison with copper and aluminium we calculated the values of $\mu$ and $\nu$ from the observed slopes of the parts $CD$ of the curves of fig. 11 and the results are marked on fig. 10, but in view of the fact that the early parts of the curve were not horizontal we are not inclined to place much confidence in the accuracy of the result. Subject to these remarks, however, it will be seen that lead and cadmium behave in a manner which is intermediate between that of iron and copper and that of glass.

### Comparison between von Mises's First Hypothesis for the Stress at which Plastic Distortion begins, namely, $\tau_1^2+\tau_2^2+\tau_3^2=2\kappa^2$, and Mohr's Hypothesis $|\tau_1|=\kappa$

The consequences of these two hypotheses in terms of the quantities measured in combined twist-extension experiments have been expressed in equations (18) and (19). It will be seen that there are two separate methods by which the two hypotheses can be compared based on (1) equations (18*b*) and (19*b*), and (2) equations (18*a*) and (19*a*).

(1) The observed values of $S/P_0$ can be plotted against values of $m$. The resulting curve on either hypothesis should be an ellipse, but von Mises's hypothesis leads to an ellipse $S/P_0=\left(\frac{1-m^2}{3}\right)^{\frac{1}{2}}$ which has ordinates $2/\sqrt{3}$ times as great as the corresponding ellipse $S/P_0=\left(\frac{1-m^2}{4}\right)^{\frac{1}{2}}$ appropriate to Mohr's hypothesis. The comparison between the observed values of $S/P_0$ and the values predicted on Mohr's and von Mises's hypotheses are given in table 4 and are represented in figs. 12*a*, *b*, *c*, where the two ellipses are also marked. It will be seen that in the cases of copper and aluminium, von Mises's hypothesis fits the observations within the limits of experimental error.

(2) A more interesting comparison can be made by means of equations (18*a*) and (19*a*). These equations give the relationship between $m$ and $\theta$ according to the two hypotheses, and if used directly depend on the same experimental data as those given in table 4. On the other hand, if it is assumed, as has been done by all previous writers on plasticity of isotropic materials, that $\theta=\phi$, then equations (18*a*) and (19*a*) may be used to give the relationship between $m$ and $\phi$; thus according to Mohr's hypothesis

$$m\sec 2\phi = m\sqrt{(1+\tan^2 2\phi)}=1,$$

and according to von Mises's hypothesis $m\sqrt{(1+0{\cdot}75\tan^2 2\phi)}=1$. Now $m$ is the ratio of the direct load applied during the combined stress to the maximum direct load during the preliminary direct stress. This is a straightforward measurement to which no ambiguity can attach. $\phi$ is derived from the linear relationship between the observed extension and the observed twist during the application of the combined load. The assumption that $\theta = \phi$ therefore makes possible a comparison between von Mises's and Mohr's hypotheses which is quite independent of the measurements $p$ given in line 6 of table 1, which were found by producing backwards the flat parts of the torque-extension curves in the dotted lines of figs. 6 *a*, *b* and *c*.

The comparison between $m\sqrt{(1+\tan^2 2\phi)}$ and $m\sqrt{(1+0{\cdot}75\tan^2 2\phi)}$ with 1·00 therefore provides a method of comparing Mohr's and von Mises's hypotheses without the necessity for the theoretically ambiguous proceeding used to obtain $p$ from figs. 6 *a*, *b* and *c*. This comparison is given in table 5, and it will be seen that in the case of copper and aluminium it provides a striking confirmation of von Mises's hypothesis. For copper the average value of $m\sqrt{(1+0{\cdot}75\tan^2 2\phi)}$ is 1·006, while the average value of $m\sqrt{(1+\tan^2 2\phi)}$ is 1·082. The greatest deviation of $m\sqrt{(1+0{\cdot}75\tan^2 2\phi)}$ from 1·00 is 2·0 % and the average deviation is only 0·9 %. The greatest value of $m\sqrt{(1+\tan^2 2\phi)}$ is 1·138.

The case of aluminium provides confirmation of von Mises's hypothesis which is nearly as good as that of copper, for the mean value of $m\sqrt{(1+0{\cdot}75\tan^2 2\phi)}$ is 1·000, while the greatest deviation from 1·00 is 5 %. The mean value of $m\sqrt{(1+\tan^2 2\phi)}$ is 1·093, while the greatest deviation from 1·00 is 18·3 %.

*Mild Steel and Iron.* For mild steel and decarburized mild steel the mean values of $m\sqrt{(1+0{\cdot}75\tan^2 2\phi)}$ were 1·059 and 1·078 respectively, while the mean values of $m\sqrt{(1+\tan^2 2\phi)}$ were 1·144 and 1·187. Thus it appears that von Mises's hypothesis is nearer the truth than Mohr's.

It will be seen also that the experimental points in fig. 12*c* lie well above both the theoretical ellipses, thus indicating that the strength for combined extension and torsion is greater than that indicated in either of von Mises's or Mohr's theories. It seems probable, however, that this is an effect due to want of isotropy for, as is shown in fig. 12*c*, it is more marked with specimens which had been heat-treated so that the crystal grains were larger than in those which had a less coarse structure. The change in internal volume of steel tubes during the preliminary stretching indicated the same thing. It is well known that the load-extension curve for mild steel has the peculiar property that after a certain definite extension (which depends on the heat treatment and the carbon content) it ceases rising, becomes flat or even descending with increasing extension. This flat part of the curve is succeeded by a further steady rise. This is shown for one particular specimen in the upper curve in fig. 13. The corresponding changes in internal volume of the tube in terms of the reading of the meniscus in the capillary tube of our apparatus is shown on the same diagram. It will be seen that in the elastic range which extends to about 500 lb. load the internal volume increases. This portion is shown as *OA* in fig. 13. This increase in volume is well known in the theory of elasticity. It depends on Poisson's ratio, being zero for truly incompressible substances with Poisson's ratio 0·5. When the material of the tube begins to yield plastically the internal volume begins to decrease till at the point *B*, where the load has risen to 790 lb., it begins to rise again. This change begins suddenly at the same instant that the load ceases to increase with increasing extension, and during the period of constant or decreasing load the increase in internal volume is very rapid. At the point $C$ where the extension is 0·26 in., i.e. 22 %, the load begins to increase and simultaneously the internal volume

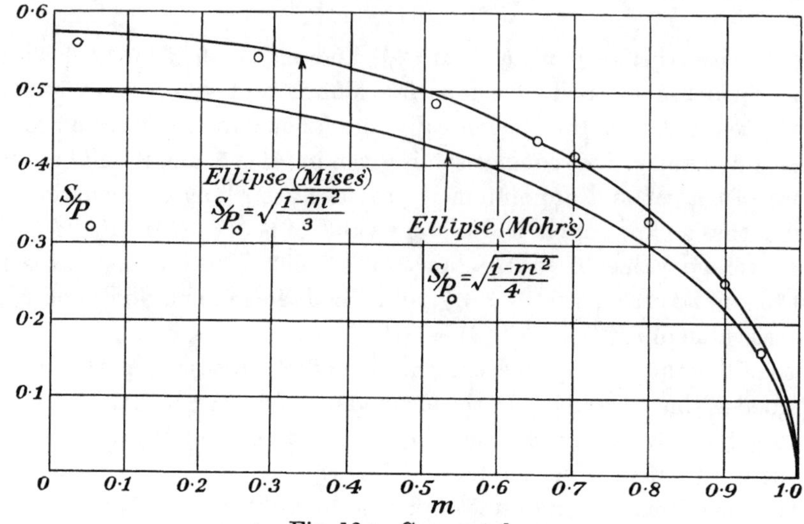


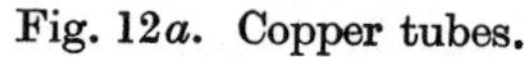
Fig. 12$a$. Copper tubes.

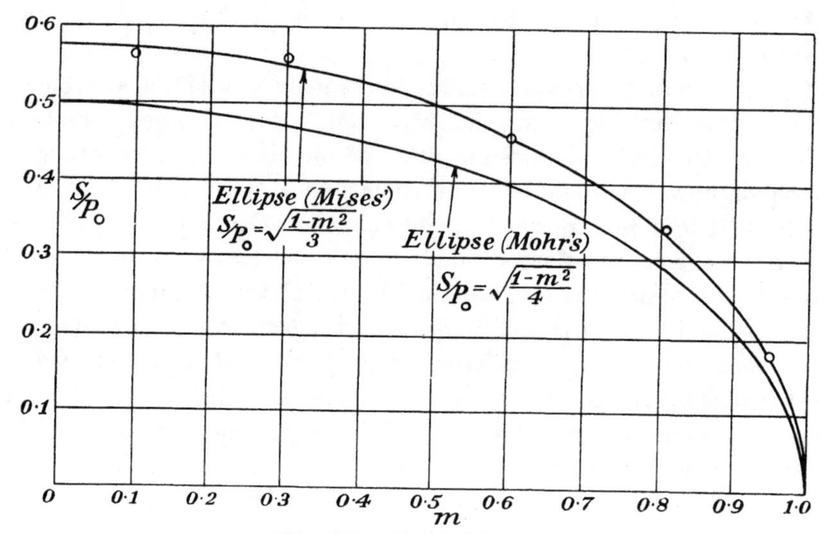


Fig. 12$b$. Aluminium tubes.

decreases. These changes in internal volume are far greater than can be accounted for by the small changes in density which are known to occur when mild steel is stretched plastically. They show therefore that the material is not behaving in an isotropic manner in the tube.

In the case of decarburized mild steel the volume changes during the initial stretching of the material are usually, but not always, greater than they are in the case of fine-grained copper and aluminium. The load-extension and volume change curves for a specimen of fine-grained decarburized mild steel are shown in fig. 13. It will be seen that the removal of the carbon from the steel has removed the flat part of the load-extension curve and also the discontinuities in the curve of change in internal volume.

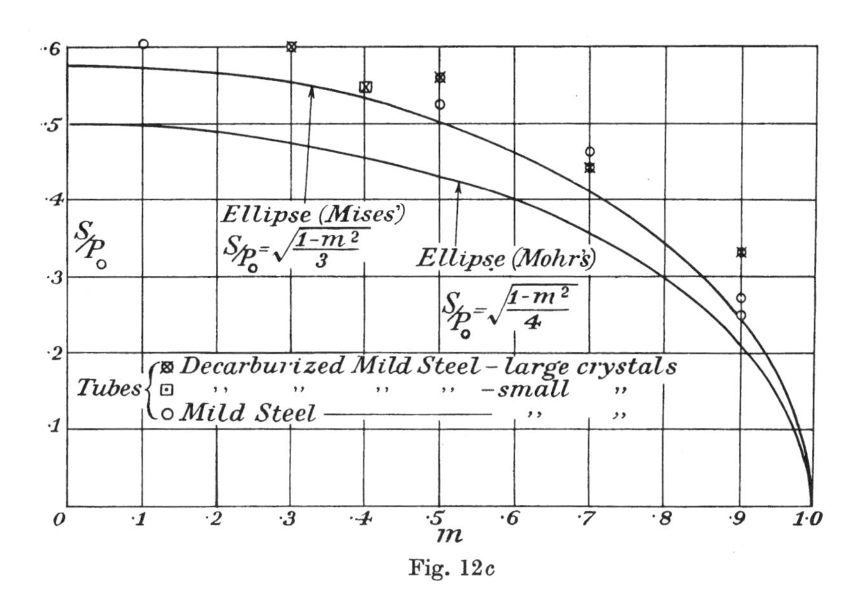


Fig. 12c

## Comparison between Load-extension Curve and Complete Shear, Shear-stress Curve

It was pointed out above that for a complete solution of any problem in plasticity it is necessary to know the resistance to further distortion after any given total strain. Many observations have been made of the load-extension curve over a large range of strains, and some have been made of the torque-angle curve for round rods. The connection between the torque-angle curve for a rod and that for a tube is known,* in fact one may be deduced from the other by purely mathematical processes. From the torque-angle curve for a tube, the shear stress–shear curve can be obtained, but for a complete analysis of plasticity the whole effect of the past strain history of an element of metal would have to be known. This leads to such complexity that further progress is impossible unless some simple generalization can be made to give results of sufficient accuracy for practical purposes. The general similarity of the $(P, e)$ and the $(S, s)$ curves is well known, and it might be suggested that the resistance to further distortion depends only on the amount of work which has been done on the material since it was in its initial annealed state.

* See Nadai, 'On the mechanics of the plastic state in metals', *Trans. Amer. Soc. Mech. Engrs* (1929).

If this were true the quantity $\kappa$ in the expression for Mohr's and von Mises's hypotheses would be a function of the work done; thus if

$$e=\frac{l-l_0}{l_0},$$

where $l_0$ is the initial length in the annealed state of a specimen subjected to tensile load, and $P$ is the stress, the total work done per unit volume is

$$Q=\int_0^e \frac{P de}{1+e}, \quad \text{so that} \quad \frac{dQ}{d(\log e)}=P.$$

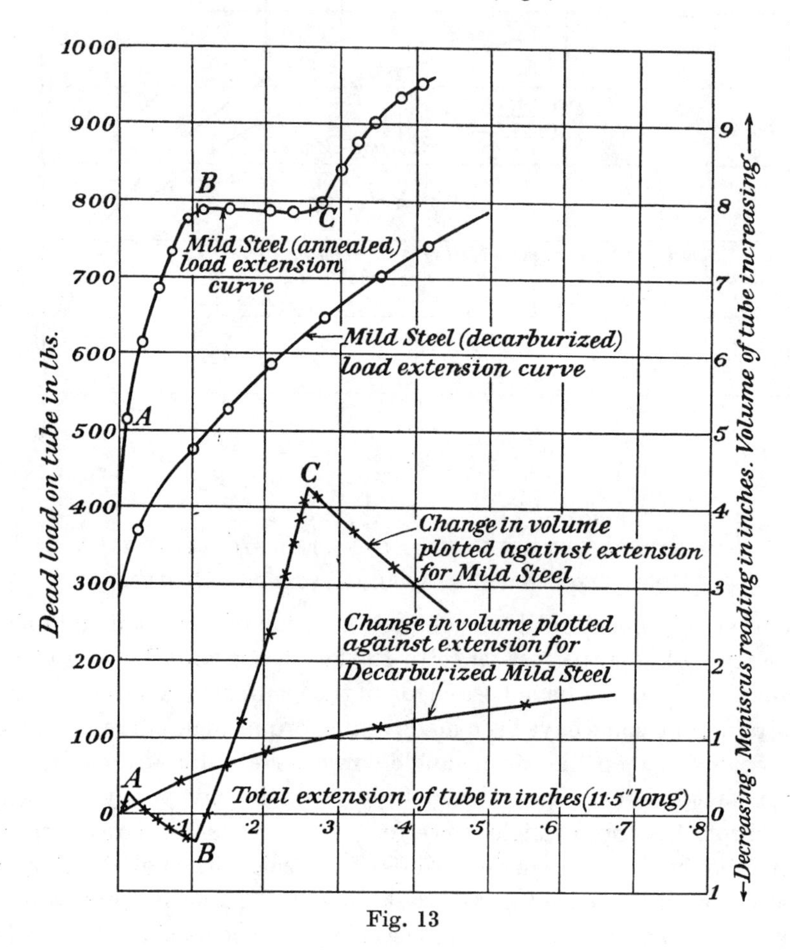


Fig. 13

If $s$ is the shear during a torsion test and $S$ the shear stress,

$$Q=\int S ds, \quad \text{so that} \quad S=\frac{dQ}{ds}.$$

According to Mohr's hypothesis $S=\kappa$, and according to von Mises's hypothesis $2\kappa/\sqrt{3}$, so that if $\kappa$ be a function of $Q$ only, then the $(S, s)$ and $(P, e)$ curves would be identical according to Mohr's theory if $\log(1+e)=\frac{1}{2}s$ and $P=2S$, or according to von Mises's theory if $\log(1+e)=s/\sqrt{3}$ and $P=S\sqrt{3}$.

This hypothesis that $\kappa$ is a function of $Q$ only is compared with observation in fig. 14. The $(S, s)$ curve was found by twisting annealed copper tubes. To find the $(P, \log(1+e))$ curve to be expected according to Mohr's hypothesis the observed values of $S$ were multiplied by 2 and the observed values of $s$ were divided by 2. For

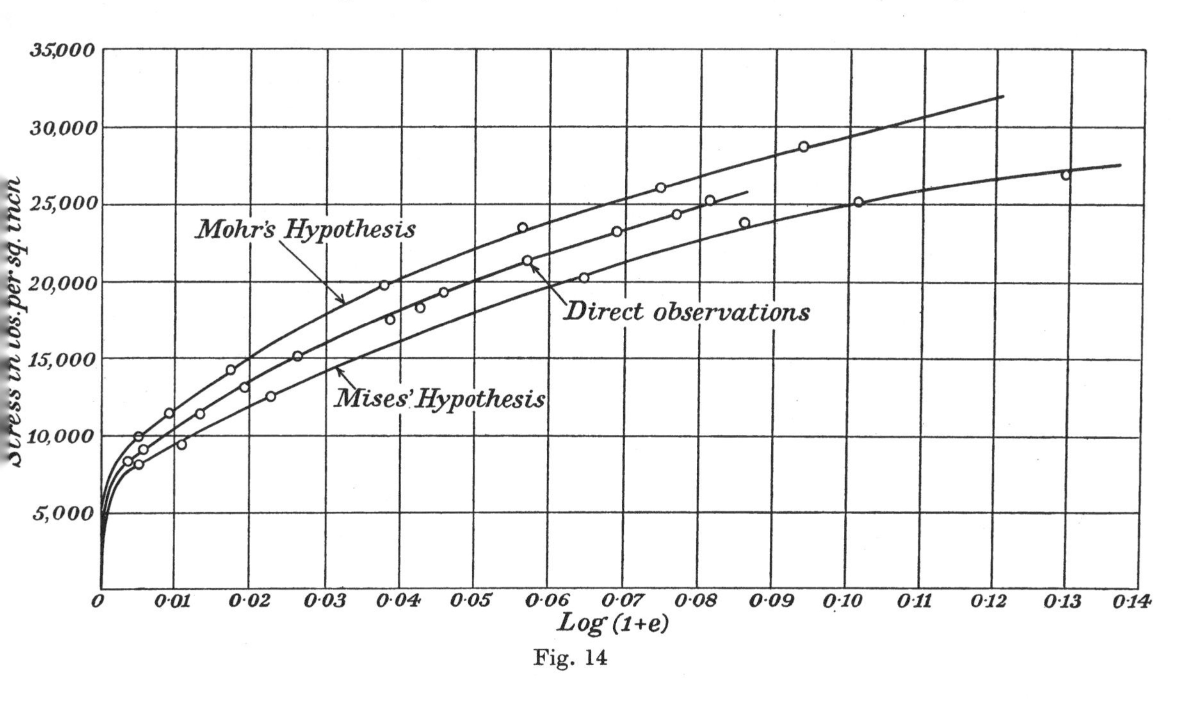


Fig. 14

the $(P, \log(1+e))$ curve to be expected according to von Mises's hypothesis the observed values of $S$ were multiplied by $\sqrt{3}$ and for the corresponding observed values of $s$ were divided by $\sqrt{3}$. The results of these operations are shown in fig. 14 together with the values observed during a direct loading of a similar tube. It will be seen that the observed curve falls between those predicted by von Mises's and Mohr's hypotheses. For most practical purposes therefore it seems probable that it will be sufficiently accurate to assume that $\kappa$ is a function of $Q$ only.

Another assumption for predicting the $(P, e)$ curve from the $(S, s)$ curve has been made by Nadai,* who takes $e=\frac{2}{3}s$, $P=S\sqrt{3}$; this leads in the case of copper to a rather closer agreement with the observations, but as it appears to be purely empirical there seems to be no way in which this result could be extended to the more general case of distortion where the principal stresses may have any assigned ratios.

* Loc. cit.

## APPENDIX

### *Experiments with Glass Tubes*

In order to compare the distortion of heated glass with that of metals subjected to the same type of stress distribution, glass tubes about 40 in. long, 0·264 in. external and 0·192 in. internal diameter were hung vertically from one end. An electric furnace was placed round the middle portion of the tube, and a thick copper tube between the furnace and the glass served to give an even temperature distribution. The lower end of the tube was gripped in a fixing at the centre of a torque drum, and a direct load was hung centrally below the middle of the drum. The ends of the glass tube were ground flat and the internal volume was measured before and after the test by filling it with water which was retained by a glass cover-slip stuck to the tube with Chatterton's Compound.

To carry out a test the load and torque were applied while the glass tube was cold. The temperature of the furnace was then gradually raised till plastic flow began. The furnace was then maintained at such a temperature that very slow plastic flow occurred. The total increase in length, $\delta l$, and an angle of twist, $\chi$, were observed, and when $\delta l$ was about 0·35 in., the furnace was cut off and the glass allowed to cool.

It was found, as in the case of metal tubes, that the extension was proportional to the twist. From the applied load and torque and the internal and external diameters of the tube the values of $P$ and $S$ can be found. These are given in columns 2 and 3 of table 6. The value of $\tan 2\theta = 2S/P$ are given in column 4. The total angle of twist, $\chi$, in degrees, the total extension, $\delta l$ in inches and the increase in internal volume $\delta v$ in cubic inches are given in columns 5, 6 and 7 of table 6.

To find $f/e$ from $\delta v$ formula (6) must be used. This formula applies to a thin-walled tube. In a tube the thickness of the walls of which are not negligible compared with the radius, the strain $f$ varies with the radius. The value with which we are concerned is the value of $f$ at the mean radius $r_{m_0}$ so that the 'internal volume' which comes into equation (6) must be the volume enclosed by the cylinder which lies half-way between the inner and the outer wall; thus if $l_h$ is the length of the heated portion of the tube (6) becomes

$$2f = \frac{-\delta v}{\pi r_{m_0}^2 l_h},$$

and since $e = \delta l/l_h$

$$\frac{f}{e} = -\frac{\delta v}{2\pi r_{m_0}^2 \delta l}. \tag{24}$$

It will be noticed that $l_h$ does not enter into this formula, so that it is not necessary to know how much of the tube is actually undergoing distortion. Using the observed mean value $r_{m_0} = 0·114$ in. the values of $f/e$ found from (24) are given in column 8 of table 6. It will be seen that they are always very small, the greatest value, namely, 0·01, is within the limits of experimental error.

To find $\phi$, the inclination of the principal axes of strain to the axis of the tube, the values of $\chi$, $\delta l$ and $f/e$ given in columns 5, 6 and 8, table 6, are substituted in the

$$\tan 2\phi = \tfrac{2}{3} r_{m_0} (1 + \tfrac{2}{3} f/e)^{-1} \left(\frac{\chi}{\delta l}\right)$$

formula; the values of $\tan 2\phi$ so found are given in column 9, table 6.

For comparison the values of $2\theta$ and $2\phi$ are given in columns 10 and 11. It will be seen that they are equal to one another, the greatest observed difference between them being only 1 degree.

*Relationship between $\mu$ and $\nu$.* Referring to formulae (12) and (17) it will be seen that a value of $f/e = 0{\cdot}01$ makes the factor

$$\frac{-1 + 2f/e}{1 + \tfrac{2}{3} f/e}$$

equal to $-0{\cdot}974$, so that the relationship between $\mu$ and $\nu$ would in that case be $\nu/\mu = 0{\cdot}974 (\cos 2\phi / \cos 2\theta)$. The difference of the factor $\cos 2\phi/\cos 2\theta$ from $1{\cdot}00$ owing to the fact that the measured values of $\theta$ and $\phi$ are not exactly equal to one another is as great as the greatest deviation of the factor $(-1 + 2f/e)/(1 + \tfrac{2}{3} f/e)$ from $1{\cdot}00$. We have already seen that values of $f/e$ are so small as to be within the limits of experimental error, so that the variations of the measured values of $\mu/\nu$ from unity are due mainly to the deviation of the measured value of $\theta$ from exact equality with the measured value of $\phi$. The values of $\mu = -\cos 2\theta$ and $\nu = \dfrac{-1 + 2f/e}{1 + \tfrac{2}{3} f/e} \cos 2\phi$ are given in columns 12 and 13 of table 6, and the results are marked in fig. 10 for comparison with the results of experiments with metal tubes.

In conclusion, we wish to express our thanks to Professor Inglis for allowing the work to be done in the Engineering Laboratory at Cambridge, to Messrs Thomas Bolton and Sons for the trouble they took to produce suitable copper tubes for us, to Messrs Accles and Pollock for presenting us with steel tubes and to the British Aluminium Company for aluminium tubes.

## Table 1

### *Annealed copper tubes*

| Name of tube ... | E | $J_1$ | III | C | G | $J_2$ | A | F |
|---|---|---|---|---|---|---|---|---|
| Maximum load $W$ lb. | 674 | 674 | 674 | 674 | 674 | 674 | 576 | 575 |
| $m$ | 0·0252 | 0·28 | 0·515 | 0·65 | 0·80 | 0·90 | 0·95 | 0·7 |
| $1+e_0$ | 1·120 | 1·150 | 1·117 | 1·117 | 1·125 | 1·150 | 1·185 | 1·09 |
| $P_0$ lb./sq. in. | 31,580 | 32,200 | 31,450 | 31,450 | 31,600 | 32,200 | 28,000 | 26,200 |
| $p$ lb. | 4·65 | 4·52 | 3·98 | 3·62 | 2·70 | 2·0 | 1·00 | 2·9 |
| $S$ lb./sq. in. $=3210(1+e)^{\frac{3}{2}}p$ | 17,700 | 17,440 | 15,100 | 13,720 | 10,380 | 7,940 | 4,170 | 10,600 |
| $\tan 2\theta=\frac{2S}{mP_0}$ | 44·5 | 4·02 | 1·87 | 1·344 | 0·820 | 0·557 | 0·314 | 1·160 |
| $\chi$ degrees/$\delta l$ in. | 3,700 | 3,700 | 1,680 | 1,220 | 782 | 561 | 350 | 1,068 |
| $d$ in./$\delta l$ in. | 52 | 27·4 | 15·0 | 14·0 | 11·3 | 7·72 | 5·0 | 12·2 |
| $f/e=0{\cdot}0037(1+e_0)\frac{d}{\delta l}$ | 0·215 | 0·117 | 0·062 | 0·0578 | 0·047 | 0·033 | 0·0218 | 0·0492 |
| $\tan 2\phi=\frac{0{\cdot}00123(1+e_0)^{-\frac{1}{2}}}{1+\frac{2}{3}f/e}\times\frac{\chi}{\delta l}$ | 37·7 | 3·94 | 1·883 | 1·368 | 0·880 | 0·616 | 0·389 | 1·216 |

### *Annealed aluminium tubes*

| Tube reference ... | VI | III | II | IV | V |
|---|---|---|---|---|---|
| Maximum load $W$ lb. | 305 | 305 | 305 | 305 | 305 |
| $m$ | 0·1 | 0·3 | 0·6 | 0·81 | 0·95 |
| $1+e_0$ | 1·07 | 1·065 | 1·08 | 1·08 | 1·07 |
| $P_0$ lb./sq. in. | 13,600 | 13,600 | 13,700 | 13,700 | 13,650 |
| $p$ lb. | 2·17 | 2·15 | 1·75 | 1·32 | 0·72 |
| $S$ lb./sq. in. $=3190(1+e_0)^{\frac{3}{2}}p$ | 7,670 | 7,550 | 6,270 | 4,750 | 2,540 |
| $\tan 2\theta=\frac{2S}{mP_0}$ | 11·30 | 3·690 | 1·525 | 0·855 | 0·393 |
| $\chi$ degrees/$\delta l$ in. | 1,093 | 3,300 | 1,265 | 800 | 342·5 |
| $d$ inches/$\delta l$ in. | 48·8 | 48·0 | 32·0 | 21 | 8·75 |
| $f/e=0{\cdot}00355(1+e_0)\frac{d}{\delta l}$ | 0·1855 | 0·1815 | 0·123 | 0·0804 | 0·0332 |
| $\tan 2\phi=\frac{0{\cdot}00126(1+e_0)^{-\frac{1}{2}}}{1+\frac{2}{3}f/e}\times\frac{\chi}{\delta l}$ | 11·83 | 3·60 | 1·416 | 0·915 | 0·407 |

### *Mild steel and decarburized mild steel tubes*

| | Annealed mild steel | | | | Decarburized mild steel | | | |
|---|---|---|---|---|---|---|---|---|
| Name of tube ... | I | II | III | IV | V | VI | VII | IX |
| Maximum load $W$ lb. | 931 | 942 | 927 | 952 | 744 | 741 | 744 | 744 |
| $m$ | 0·5 | 0·9 | 0·1 | 0·7 | 0·5 | 0·9 | 0·3 | 0·7 |
| $1+e_0$ | 1·035 | 1·038 | 1·035 | 1·04 | 1·04 | 1·04 | 1·04 | 1·04 |
| $P_0$ lb./sq. in. | 38,500 | 39,100 | 38,400 | 39,600 | 30,900 | 30,800 | 30,900 | 30,900 |
| $p$ lb. | 6·30 | 3·30 | 7·30 | 5·70 | 5·4 | 3·25 | 5·8 | 4·4 |
| $S$ lb./sq. in. $=3112\,(1+e_0)^{\frac{3}{2}}p$ | 20,600 | 10,800 | 23,500 | 18,800 | 17,900 | 10,650 | 18,800 | 14,500 |
| $\tan 2\theta=\frac{2S}{mP_0}$ | 2·14 | 0·614 | 12·2 | 1·35 | 2·30 | 0·77 | 4·06 | 1·33 |
| $\chi$ degrees/$\delta l$ in. | 2,100 | 600 | 11,300 | 1,175 | 1,860 | 750 | 3,700 | 1,125 |
| $d$ in. /$\delta l$ in. | 24·7 | 18·5 | 45 | 22·5 | 24·7 | 19·2 | 30 | 19·25 |
| $f/e=0{\cdot}00373(1+e_0)\frac{d}{\delta l}$ | 0·0955 | 0·0717 | 0·17 | 0·0874 | 0·0956 | 0·074 | 0·117 | 0·0746 |
| $\tan 2\phi=\frac{0{\cdot}00122(1+e_0)^{-\frac{1}{2}}}{1+\frac{2}{3}f/e}\times\frac{\chi}{\delta l}$ | 2·37 | 0·686 | 12·0 | 1·325 | 2·10 | 0·85 | 4·10 | 1·2850 |

Table 1 (*cont.*)

*Lead tubes and cadmium tube*

| | Lead | | | Cadmium |
|---|---|---|---|---|
| Name of tube ... | I | II | III | I |
| Maximum load $W$ lb. | 52 | 52 | 52 | 140 |
| $m$ | 0·5 | 0·5 | 0·7 | 0·5 |
| $(1+e_0)$ | 1·01 | 1·01 | 1·02 | 1·015 |
| $P_0$ lb./sq. in. | 2,650 | 2,650 | 2,650 | 5,880 |
| $p$ lb. | 0·431 | 0·430 | 0·268 | 0·75 |
| $S$ lb./sq. in. $=3830(1+e_0)^{\frac{3}{2}}p$ | 1,662 | 1,660 | 1,062 | 2,940 |
| $\tan 2\theta=\dfrac{2S}{mP_0}$ | 2·51 | 2·50 | 1·21 | 1·670 |
| $\chi$ degrees/$\delta l$ in. | 2,000 | 2,000 | 945 | 1,525 |
| $d$ in./$\delta l$ in. | 10 | 10 | 8 | 7·5 |
| $f/e=0{\cdot}000307\,(1+e_0)\dfrac{d}{\delta l}$ | 0·0356 | 0·0356 | 0·0288 | 0·0270 |
| $\tan 2\phi=0{\cdot}00124\,\dfrac{(1+e_0)^{-\frac{1}{2}}}{1+\frac{2}{3}f/e}\times\dfrac{\chi}{\delta l}$ | 2·40 | 2·35 | 1·14 | 1·800 |

Table 2. *Comparison of values of $\theta$ and $\phi$*

*Annealed A.C. copper tubes*

| Name of tube... | E | $J_1$ | III | C | G | $J_2$ | A | F |
|---|---|---|---|---|---|---|---|---|
| $2\theta$ (degrees) | 88·8 | 76·0 | 61·9 | 53·4 | 39·4 | 29·1 | 17·4 | 49·3 |
| $2\phi$ (degrees) | 88·5 | 75·8 | 62·0 | 53·8 | 41·4 | 31·6 | 21·3 | 50·6 |

*Annealed aluminium tubes*

| Name of tube ... | VI | III | II | IV | V |
|---|---|---|---|---|---|
| $2\theta$ | 85·0 | 74·8 | 56·7 | 40·6 | 21·5 |
| $2\phi$ | 85·2 | 74·5 | 54·6 | 42·5 | 22·1 |

*Mild steel and annealed mild steel*

| Name of tube ... | I | II | III | IV | V | VI | VII | VIII |
|---|---|---|---|---|---|---|---|---|
| $2\theta$ | 65·0 | 31·6 | 85·3 | 53·5 | 66·5 | 37·6 | 76·2 | 53·2 |
| $2\phi$ | 67·1 | 34·2 | 85·2 | 53·0 | 64·6 | 40·4 | 76·3 | 52·2 |

*Lead tubes and cadmium tube*

| | Prepared from pure lead in suitable die | | | Cadmium |
|---|---|---|---|---|
| Name of tube ... | I | II | III | I |
| $2\theta$ | 68° 12′ | 68° 12′ | 50° 24′ | 59° 6′ |
| $2\phi$ | 67° 14′ | 66° 54′ | 48° 42′ | 61° 0′ |

## Table 3

### *Annealed copper tubes*

| Reference ... | E | $J_1$ | III | C | G | $J_2$ | A | F |
|---|---|---|---|---|---|---|---|---|
| $\mu = -\cos 2\theta$ | −0·021 | −0·242 | −0·471 | −0·597 | −0·773 | −0·874 | −0·954 | −0·652 |
| $\nu = \dfrac{-1+2f/e}{1+\frac{2}{3}f/e}\cos 2\phi$ | −0·011 | −0·171 | −0·393 | −0·502 | −0·659 | −0·775 | −0·876 | −0·554 |

### *Annealed aluminium tubes*

| Name of tube ... | VI | III | II | IV | V |
|---|---|---|---|---|---|
| $\mu = -\cos 2\theta$ | −0·087 | −0·262 | −0·549 | −0·759 | −0·930 |
| $\nu = \dfrac{-1+2f/e}{1+\frac{2}{3}f/e}\cos 2\phi$ | −0·046 | −0·168 | −0·405 | −0·589 | −0·845 |

### *Mild steel and decarburized mild steel tubes*

| Name of tube ... | I | II | III | IV | V | VI | VII | VIII |
|---|---|---|---|---|---|---|---|---|
| $\mu = -\cos 2\theta$ | −0·423 | −0·852 | −0·082 | −0·595 | −0·399 | −0·792 | −0·239 | −0·599 |
| $\nu = \dfrac{-1+2f/e}{1+\frac{2}{3}f/e}\cos 2\phi$ | −0·296 | −0·677 | −0·0495 | −0·470 | −0·325 | −0·618 | −0·170 | −0·498 |

### *Lead tubes and cadmium tube*

| | Lead | | | Cadmium |
|---|---|---|---|---|
| Reference ... | I | II | III | I |
| $\mu = -\cos 2\theta$ | 0·372 | 0·371 | 0·637 | 0·5135 |
| $\nu = \dfrac{-1+2f/e}{1+\frac{2}{3}f/e}\cos 2\phi$ | 0·348 | 0·356 | 0·61 | 0·450 |

## Table 4

### *Copper tubes*

| Reference ... | E | $J_1$ | III | C | G | $J_2$ | A | F |
|---|---|---|---|---|---|---|---|---|
| $m$ | 0·025 | 0·28 | 0·515 | 0·65 | 0·80 | 0·90 | 0·95 | 0·70 |
| $S/P_0$ | 0·560 | 0·541 | 0·480 | 0·436 | 0·329 | 0·247 | 0·149 | 0·405 |
| $\sqrt{\dfrac{1-m^2}{3}}$ | 0·577 | 0·554 | 0·495 | 0·438 | 0·346 | 0·251 | 0·180 | 0·412 |
| $\sqrt{\dfrac{1-m^2}{4}}$ | 0·500 | 0·480 | 0·428 | 0·380 | 0·300 | 0·218 | 0·155 | 0·357 |

### *Aluminium tubes*

| Reference ... | VI | III | II | IV | V |
|---|---|---|---|---|---|
| $m$ | 0·10 | 0·30 | 0·60 | 0·81 | 0·95 |
| $S/P_0$ | 0·564 | 0·555 | 0·458 | 0·347 | 0·186 |
| $\sqrt{\dfrac{1-m^2}{3}}$ | 0·574 | 0·550 | 0·461 | 0·337 | 0·180 |
| $\sqrt{\dfrac{1-m^2}{4}}$ | 0·498 | 0·477 | 0·400 | 0·292 | 0·155 |

Table 4 (*cont.*)

*Mild steel and decarburized iron*

| Reference ... | I | II | III | IV | V | VI | VII | VIII | IX |
|---|---|---|---|---|---|---|---|---|---|
| $m$ | 0·5 | 0·9 | 0·1 | 0·7 | 0·5 | 0·9 | 0·3 | 0·7 | 0·4 |
| $S/P_0$ | 0·535 | 0·276 | 0·612 | 0·474 | 0·579 | 0·345 | 0·608 | 0·469 | {0·55, 0·56, 0·56} |
| $\sqrt{\frac{1-m^2}{3}}$ | 0·500 | 0·251 | 0·574 | 0·412 | 0·500 | 0·251 | 0·550 | 0·412 | 0·53 |
| $\sqrt{\frac{1-m^2}{4}}$ | 0·433 | 0·218 | 0·498 | 0·357 | 0·433 | 0·218 | 0·477 | 0·357 | 0·46 |

*Lead tubes and cadmium tube*

| | Lead | | | Cadmium |
|---|---|---|---|---|
| Reference ... | 1 | 2 | 3 | 1 |
| $m$ | 0·5 | 0·5 | 0·7 | 0·5 |
| $S/P_0$ | 0·622 | 0·622 | 0·421 | 0·417 |
| $\sqrt{\frac{1-m^2}{3}}$ | 0·50 | 0·50 | 0·412 | 0·50 |
| $\sqrt{\frac{1-m^2}{4}}$ | 0·435 | 0·435 | 0·375 | 0·435 |

Table 5

| $m$ | $\tan 2\phi$ | $m\sqrt{(1+\tan^2 2\phi)}$ (=1·00 Mohr's hypothesis) | $m\sqrt{(1+0{\cdot}75\tan^2 2\phi)}$ (=1·00 von Mises's hypothesis) | |
|---|---|---|---|---|
| 0·28 | 3·94 | 1·138 | 0·995 | Copper |
| 0·515 | 1·883 | 1·097 | 0·985 | Copper |
| 0·65 | 1·368 | 1·101 | 1·008 | Copper |
| 0·70 | 1·216 | 1·102 | 1·016 | Copper |
| 0·80 | 0·880 | 1·060 | 1·005 | Copper |
| 0·90 | 0·616 | 1·057 | 1·020 | Copper |
| 0·95 | 0·389 | 1·020 | 1·001 | Copper |
| | | Mean 1·082 | Mean 1·006 | |
| 0·10 | 11·83 | 1·183 | 1·029 | Aluminium |
| 0·30 | 3·60 | 1·121 | 0·983 | Aluminium |
| 0·60 | 1·410 | 1·040 | 0·950 | Aluminium |
| 0·81 | 0·915 | 1·097 | 1·031 | Aluminium |
| 0·95 | 0·407 | 1·024 | 1·007 | Aluminium |
| | | Mean 1·093 | Mean 1·000 | |
| 0·1 | 12·0 | 1·205 | 1·039 | Annealed mild steel |
| 0·5 | 2·37 | 1·286 | 1·141 | Annealed mild steel |
| 0·7 | 1·325 | 1·141 | 1·067 | Annealed mild steel |
| 0·9 | 0·686 | 1·090 | 1·046 | Annealed mild steel |
| | | Mean 1·144 | Mean 1·059 | |
| 0·3 | 4·10 | 1·266 | 1·106 | Decarburized mild steel |
| 0·5 | 2·10 | 1·163 | 1·038 | Decarburized mild steel |
| 0·7 | 1·285 | 1·138 | 1·047 | Decarburized mild steel |
| 0·9 | 0·85 | 1·180 | 1·115 | Decarburized mild steel |
| | | Mean 1·187 | Mean 1·078 | |

Table 6. *Glass tubes*

| Tube no. | $P$ (lb. sq. in.) | $S$ (lb. sq. in.) | $\chi$ $\tan 2\theta = 2S/P$ | $\chi$ (°) | $\delta l$ (in.) | $\delta v$ (cu. in.) |
|---|---|---|---|---|---|---|
| 1 | 127 | 214 | 3·35 | 1035 | 0·418 | +0·00006 |
| 2 | 127 | 179 | 2·80 | 1095 | 0·550 | +0·00006 |
| 3 | 127 | 170 | 2·70 | 638 | 0·330 | −0·00030 |
| 4 | 162 | 51·2 | 0·632 | 137 | 0·296 | +0·00024 |

| Tube no. | $\frac{f}{e} = -\frac{\delta v}{0 \cdot 0817 \delta l}$ | $\tan 2\phi$ | $2\theta$ (°) | $2\phi$ (°) | $\mu$ | $\nu$ |
|---|---|---|---|---|---|---|
| 1 | −0·0017 | 3·30 | 73·4 | 73·1 | −0·286 | −0·290 |
| 2 | −0·0013 | 2·65 | 70·4 | 69·4 | −0·336 | −0·353 |
| 3 | +0·0111 | 2·57 | 69·7 | 68·8 | −0·347 | −0·351 |
| 4 | −0·0099 | 0·615 | 32·3 | 31·6 | −0·845 | −0·872 |

# 17

# THE DISTORTION OF WIRES ON PASSING THROUGH A DRAW-PLATE*

REPRINTED FROM

*Journal of the Institute of Metals*, vol. XLIX (1932), pp. 187–99

Composite copper wires $\frac{1}{8}$ in. in diameter, each consisting of two wires of semicircular section, were pulled through various draw-plates. Photographs are reproduced showing the distortion of the cross-sections. These are treated in a quantitative manner, measurements of the ratio of distortion to increase in length being found for various reductions in area and angle of taper. The principal results are: (1) When an annealed wire is drawn through successive holes in a draw-plate, each of 3° taper, so that it suffers equal proportional reductions in area at each draught, the distortion of the cross-section rapidly diminishes until (in the example of figs. 1–6) it has ceased to be measurable after four draughts. (2) For any given reduction in area the distortion increases as the angle of taper increases. (3) When the reduction in area is small and the angle of taper large so that the length of wire in contact with the draw-hole is small compared with its diameter, the distortion of the central part is independent of the angle of taper, but the distortion in the outer part increases as the angle of taper increases. (4) For a given reduction in area and angle of taper the distortion in the outer part of the wire is greater when the drawing is done in two stages than when the whole operation is done in one draught. The distortion in the central portion is the same in the two cases. (5) The fact that the distortion of the outer part of the wire varies so much with the angle of taper of the draw-hole, whilst the inner part is little affected, seems to explain why different X-ray analysts obtain consistent results for the structure of the inner parts of drawn wire, but widely divergent results for the outer layers.

In some recent papers† and discussions on the mechanics of wire-drawing, attention was directed to the distortion suffered by planes of particles initially at right angles to the axis of a wire as they pass through the die. It has been suggested to the authors by Mr W. E. Alkins that members of the Institute of Metals might be interested in some photographs which they took rather more than a year ago and which illustrate this distortion. The authors' experiments were made with annealed copper wire of semicircular cross-section and $\frac{1}{8}$ in. in diameter. The flat face was polished and engraved with fine scratches running parallel and perpendicular to the axis of the wire over a length of some 20 or 30 diameters. The wire was then bent double in the middle so that the flat polished faces were in contact. It was then filed sufficiently conical to enable the bend to be pushed far enough through the die for a three-jaw grip to hold it. The die was then very carefully adjusted

* With H. QUINNEY.

† (*a*) Alkins and Cartwright, *J. Inst. Metals*, XLVI (1931), 293–303. (*b*) Francis and Thomson, *J. Inst. Metals*, XLVI (1931), 313–37.

19-2

until the axis of the conical die-hole was vertical. The chuck, which was constrained to move vertically, was then moved downwards at a constant speed. These precautions in regard to centring and setting up were found to be necessary in order to ensure that the two halves pull through the die in an exactly similar manner. Small errors in setting up the die and chuck resulted in unsymmetrical distortion as revealed by the scratches or by a residual curvature of the wire after drawing.

It is evident that the division of the wire into two halves* should not affect the stresses or strains which occur during the process of wire-drawing, because, owing to symmetry, the stress exerted across any axial plane must be entirely normal to the surface, so that the stresses in a split wire should be identical with those in a solid wire. The same considerations do not, of course, apply to rods or wires built up of concentric tubes.

In the first series of experiments the wire was fixed to the slide rest of a lathe, and transverse scratches were made by means of a weighted razor blade of such a depth that they were just visible after passing through five successive draw-holes, so that the length increased 3·7 times. The diameter of the wire was initially 0·1235 in., and the diameters after passing through the five holes were 0·1140, 0·0978, 0·0842, 0·0720 and 0·0640 in. The dies used were of the type used in commerce, and were kindly presented to the authors by Mr W. E. Alkins; their angle of taper was approximately 3° (i.e. the angle between the surface of the die and the axis of symmetry), but they were not exactly conical.

The successive appearances of the scratches are shown in pl. 1 *a*—*f*. It will be seen that they resemble parabolas. The actual forms of the transverse scratches were obtained by measurement of enlarged photographs. They are shown in fig. 1, which contains five curves drawn to scale. No. 1 is the form after 1 draught, no. 2 after 2 draughts, etc.

The curves show that, in addition to the longitudinal extension and lateral compression, there is a shearing strain parallel to the axis which increases to a maximum value at the surface. Pl. 1 *b*–*f* enable the amount of shearing strain to

* This method has been used by Siebel, and it was the appearance of his paper just after the present authors had completed the work here described that made it seem unnecessary for them to publish their results. The fact, however, that their results were obtained with ⅛ in. wires drawn through ordinary die-holes, whilst Siebel used large rods, and the fact that they have analysed their photographs and given quantitative interpretations of them, have made them reconsider their original decision not to publish them. Some of the photographs are not technically so good as Siebel's. This is due chiefly to the small size of the wire as compared with Siebel's, and to the fact that the present authors' method of annealing produced grains which were too large a fraction of the diameter of the wire. E. Siebel, 'Die Formänderung bei technischen Formgebungsverfahren', *Naturwissenschaften*, XIX (1931), 515.

*Note added 26 May 1932.* The photographs referred to above are reproduced in Professor Körber's lecture to the Institute on 11 May 1932 (*J. Inst. Metals*, XLVIII (1932), 317). Readers may therefore compare the two sets, and will see that there is good qualitative agreement between them. The more quantitative interpretation of the present authors' photographs which are given in the following pages may be regarded as an extension of the work described by Professor Körber.

be compared with the extension at each stage. The shearing strain is the small angle through which a transverse line initially straight is bent during the passage through the die.

If $x_1, x_2, \ldots, x_5$ are the heights of the centres of the curves of figs. 2, 3, …, 6 above their bases, and $d_0, d_1, d_2, \ldots, d_5$ the initial diameter and the diameters of the wire after the first, second, …, fifth draughts, the measured values of $x_n$ and $d_n$ (i.e. the values of $x$ and $d$ after the $n$th draught) are given in the second and third columns of table 1. If $\epsilon_n$ is the ratio of the length of wire after the $n$th draught to its initial length, then, if there is no appreciable change in density, $\epsilon_n = \left(\frac{0\cdot1235}{d_n}\right)^2$, and the values of $\epsilon_n$ are given in column 4.

Table 1

| 1 | 2 | 3 | 4 | 5 | 6 | 7 | 8 | |
|---|---|---|---|---|---|---|---|---|
| $n$ | $d_n$ | $x_n$ | $\epsilon_n$ | $\epsilon_n/\epsilon_{n-1}$ | $x_n - x_{n-1}\left(\frac{\epsilon_n}{\epsilon_{n-1}}\right)$ | $ds_n$ | $\dfrac{ds_n}{\left(\frac{\epsilon_n}{\epsilon_{n-1}}-1\right)}$ | Draught |
| 0 | 0·1235 | 0 | 1·00 | — | — | — | — | — |
| 1 | 0·1140 | 0·0046 | 1·17 | 1·17 | 0·0046 | 0·16 | 0·9 | First |
| 2 | 0·0978 | 0·0111 | 1·59 | 1·36 | 0·0049 | 0·20 | 0·5 | Second |
| 3 | 0·0842 | 0·0174 | 2·15 | 1·35 | 0·0024 | 0·11 | 0·3 | Third |
| 4 | 0·0720 | 0·024 | 2·94 | 1·37 | 0·00 | 0 | 0 | Fourth |
| 5 | 0·0640 | 0·029 | 3·72 | 1·27 | — | 0 | 0 | Fifth |

Assuming as a rough approximation that the curves of pl. 1*b*—*f* are parabolas, the maximum shearing strain occurs at the surface of the wire, and in the case of the first draught it is

$$ds_1 = 4x_1/d_1. \tag{1}$$

To obtain $ds_n$, the amount of shearing strain which occurs during the $n$th draught, it is necessary to remember that after the first draught the scratched transverse line is no longer straight at the beginning of the operation. Even if no further shearing strain occurred during subsequent draughts, the parabolas would become higher and narrower, owing to the lateral contractions and longitudinal extension of the wire. To allow for this effect, the geometry of the system shows that one must use the expression

$$ds_n = 4\left\{\frac{x_n - x_{n-1}(\epsilon_n/\epsilon_{n-1})}{d_n}\right\}. \tag{2}$$

This is, of course, identical with expression (1) for the first draught, because when $n=1$, $x_{n-1}=0$. The values of these shearing strains $ds_n$ have been calculated from the figures of columns 2, 3 and 4 of table 1. They are given in column 6 of the same table.

It is clear that at any given stage of the proceedings the shearing strain might be expected to increase with the amount of extension. The percentage

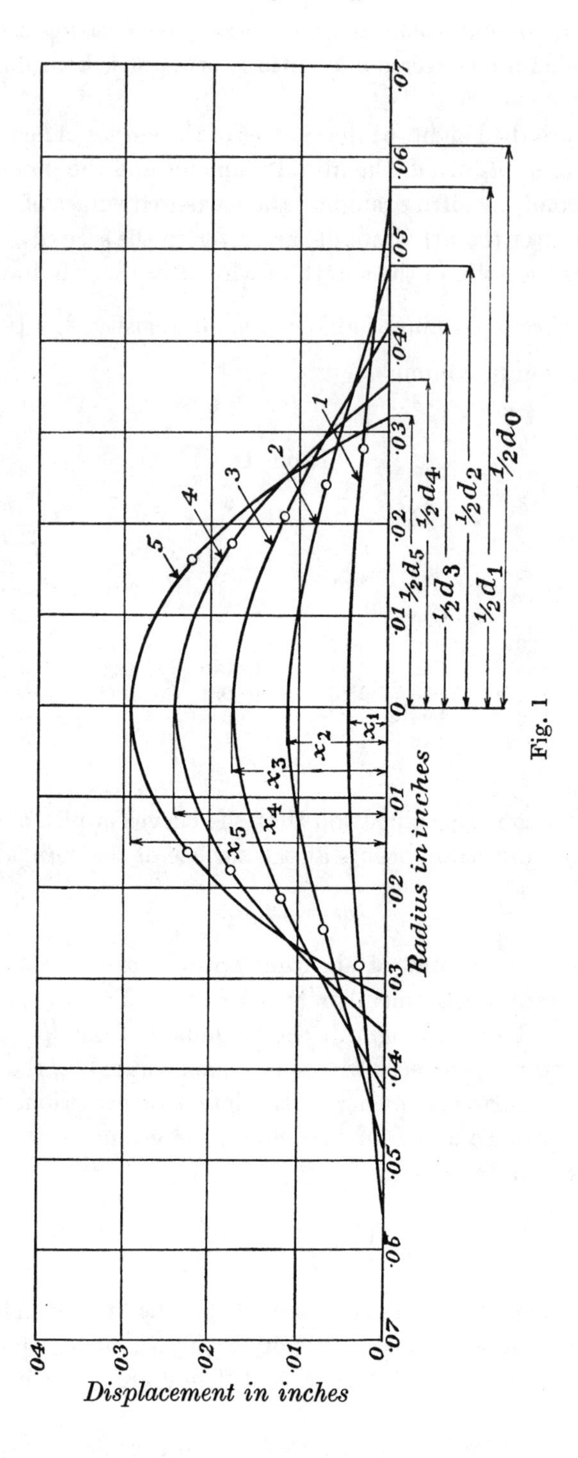
Displacement in inches
Radius in inches
·04
·03
·02
·01
0
·07
·06
·05
·04
·03
·02
·01
0
·01
·02
·03
·04
·05
·06
·07
1
2
3
4
5
$x_1$
$x_2$
$x_3$
$x_4$
$x_5$
$\frac{1}{2}d_0$
$\frac{1}{2}d_1$
$\frac{1}{2}d_2$
$\frac{1}{2}d_3$
$\frac{1}{2}d_4$
$\frac{1}{2}d_5$

Fig. 1

extension during the $n$th draught is $100\left(\frac{\epsilon_n}{\epsilon_{n-1}}-1\right)$. The quantity $\left(\frac{\epsilon_n}{\epsilon_{n-1}}-1\right)$, which is the ratio of the extension of the wire during the $n$th draught to the length of the wire before the $n$th draught, may be called the proportional extension, occurring during the $n$th draught.

The quantity which expresses the ratio of the maximum shearing strain to this extension is evidently $ds_n\Big/\left(\frac{\epsilon_n}{\epsilon_{n-1}}-1\right)$. The values of this have been calculated, and they are given in column 8, table 1.

It will be seen that in the first draught $ds_n\Big/\left(\frac{\epsilon_n}{\epsilon_{n-1}}-1\right)=0{\cdot}9$, in the second draught it is 0·5, in the third it is 0·3; in the fourth and fifth it is too small to be measurable. These experiments show, therefore, that in drawing copper wire through these dies it was only in the first three draughts, whilst the metal was rapidly hardening, that the distortion was appreciable. During the fourth and fifth draughts no further measurable* distortion occurred, the longitudinal extension and lateral contraction being uniform across the section.

A first glance at plate 1*e* and *f* would suggest that the distortion is increasing, because the fourth parabola is higher and narrower than the third, and the fifth is higher and narrower than the fourth; but the figures in columns 2 and 3 of table 1 show that these changes in shape are merely due to the lateral contraction and longitudinal extension, for in the absence of shearing strain the height of a parabola would increase in the ratio $\frac{\epsilon_n}{\epsilon_{n-1}}$. It will be seen from column 3 that $\frac{x_4}{x_3}=\frac{0{\cdot}024}{0{\cdot}0174}=1{\cdot}38$, whilst from column 5, $\frac{\epsilon_4}{\epsilon_3}=1{\cdot}35$. For the fifth draught $\frac{x_5}{x_4}=\frac{0{\cdot}29}{0{\cdot}24}=1{\cdot}21$, whilst $\frac{\epsilon_5}{\epsilon_4}=1{\cdot}27$. Thus in these two draughts $\frac{x_n}{x_{n-1}}=\frac{\epsilon_n}{\epsilon_{n-1}}$, so that the whole measurable distortion was an extension without additional shear.

### Observations with more Rapidly Tapering Die-holes

The dies with which the series of observations already described were made had only a small taper of about 3°. With a view to finding out whether a more rapid contraction of the die-hole would increase the distortion of cross-sections, three dies were prepared, all designed to draw down from 0·127 to 0·1075 in. The first was one of Mr Alkins's dies with taper 3°, the second had a taper of 15°, and the third had a taper of 30°, and in order to observe the actual distortions occurring everywhere during the process of drawing, the specimens were ruled with four longitudinal and a large number of transverse scratches. The split wire was then drawn through the plate until the middle of the draw-plate was approximately in the middle of the engraved area. The draught was then stopped and the

* I.e. measurable by the somewhat insensitive methods here described.

specimen was drawn out backwards, special precautions again being taken to ensure that this operation was carried out symmetrically.

With this technique the distortions occurring within the draw-plate could be observed, but before any reliable results could be obtained it was necessary to improve on the methods of engraving used in the first series of experiments. To this end an engraving machine was made with an accurate feed screw of 0·5 mm. pitch, and of the quality used in first-class optical instruments. The engraving tool was a safety-razor blade which had been broken across the middle. It was mounted on a pivoted holder so that its cutting edge was at 45° to the horizontal plane of the polished surface of the specimen. The weight of the holder was held by a long and weak spiral spring, and the depth of the cut was controlled by weighting this holder until the knife pressed with a known force on the specimen. The pressure used on the blade varied from 5 to 50 g., and the scratch made was invisible to the naked eye.

The engraving machine which held the specimen had two perpendicular movements like a slide rest. The transverse marks were made by sliding the specimen by hand under the tool, and, after engraving each line, the feed screw was turned through a given angle so that the lines were equidistant. Although the marks could not be seen by the naked eye in ordinary lights, they could be illuminated by special types of lighting, and enlarged photographs could be taken. Some of these are shown in pls. 2–4.

Plate 2*a* shows the effect of drawing through a die-hole with 3° taper from 0·127 to 0·107 in., the reduction in area being therefore 29 %. The length of wire on contact with the draw-plate was $L=\frac{1}{2}\left(\frac{0\cdot127-0\cdot107}{0\cdot127}\right)\frac{d_0}{\tan 3^\circ}=1\cdot5d_0$, where $d_0$ is the diameter of the wire before drawing.

Plate 2*b* shows the distortion when the angle of taper is 15°, the reduction in area being the same as before—namely, 29 %. The length of wire in contact with the plate was $0\cdot29d_0$.

Plate 3 shows the distortion when the angle of taper is 30°, the reduction in area being 33 %. In this case the length of wire in contact with the plate was $L=0\cdot14d_0$.

Since the reduction in area is nearly the same in all cases, plate 2*a*, *b* and plate 3 are comparable.

It will be seen that the distortion of the cross-section increases rapidly as the taper increases, and the length of contact consequently decreases. In the case of the 30° taper the precautions taken to ensure symmetry do not seem to have been very successful; the chord joining the ends of the curved transverse line is not perpendicular to the axis. In order to compare the distortion of plane sections for various angles of taper, the transverse lines in each photograph were measured by means of a reading microscope fitted with a stage and measuring screw for moving the photograph under the microscope in a direction at right angles to that of the motion of the microscope. The co-ordinates thus measured of points on any transverse scratch were then set out on a diagram as shown in curve I, fig. 2.

PLATE 1

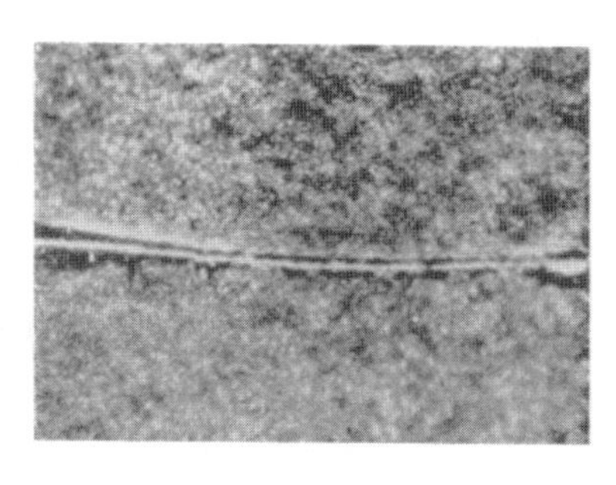

(*a*) (*b*)

*a*, Original wire as marked. Diameter 0·1235 inch. Magnification 17·7 diameters.

*b*, First draught: load required to draw through die, 115 pounds. Diameter after drawing, 0·114 inch. Magnification 17·7 diameters.

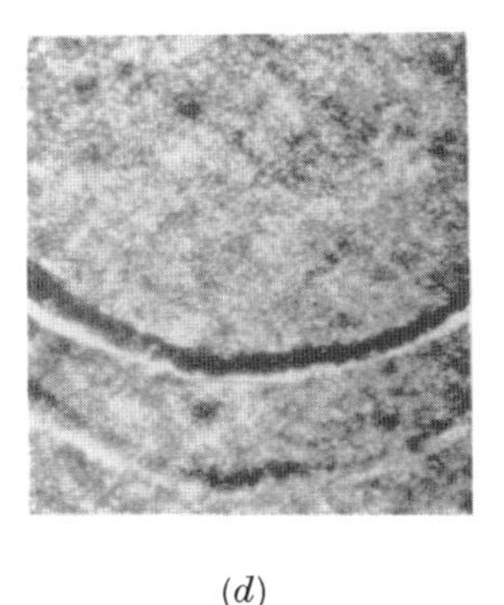

(*c*) (*d*)

*c*, Second draught: load required to draw through die, 230 pounds. Diameter after drawing, 0·0978 inch. Magnification 17·7 diameters.

*d*, Third draught: load required to draw through die, 180 pounds. Diameter after drawing, 0·0842 inch. Magnification 17·7 diameters.

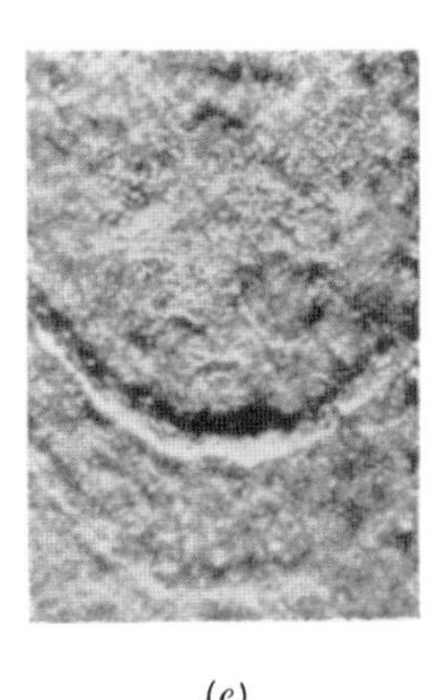

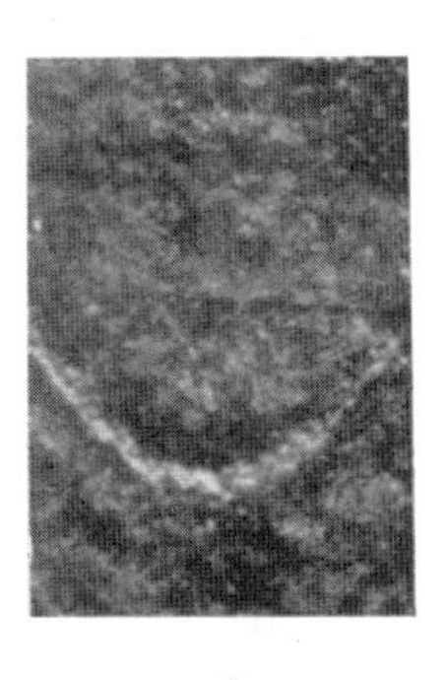

(*e*) (*f*)

*e*, Fourth draught: load required to draw through die, 110 pounds. Diameter after drawing, 0·0720 inch. Magnification 17·7 diameters.

*f*, Fifth draught: load required to draw through die, 58 pounds. Diameter after drawing, 0·0640 inch. Magnification 17·7 diameters.

×12·1

*a*, 3° taper.

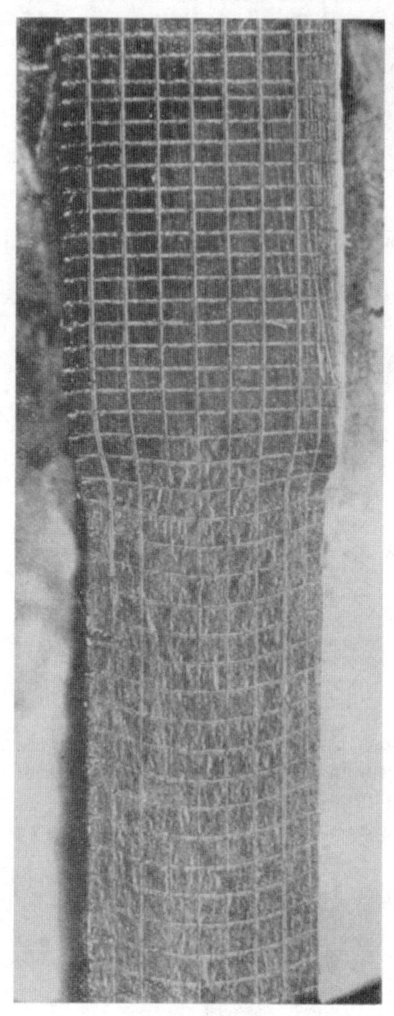

×12·1

*b*, 15° taper.

PLATE 3

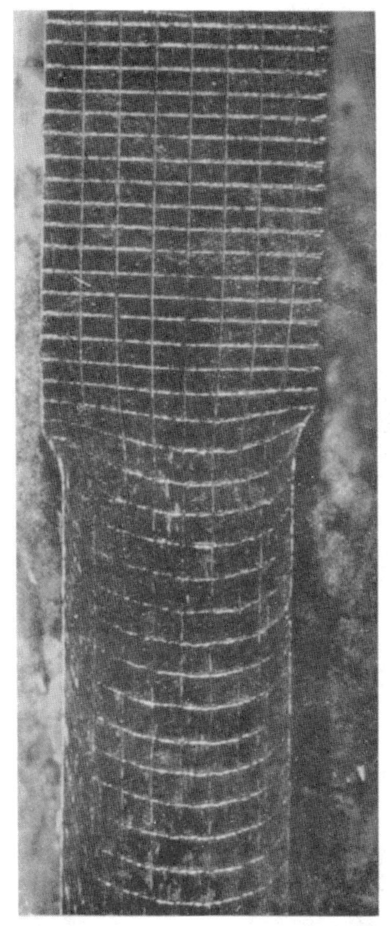

×12·1

30° taper.

PLATE 4

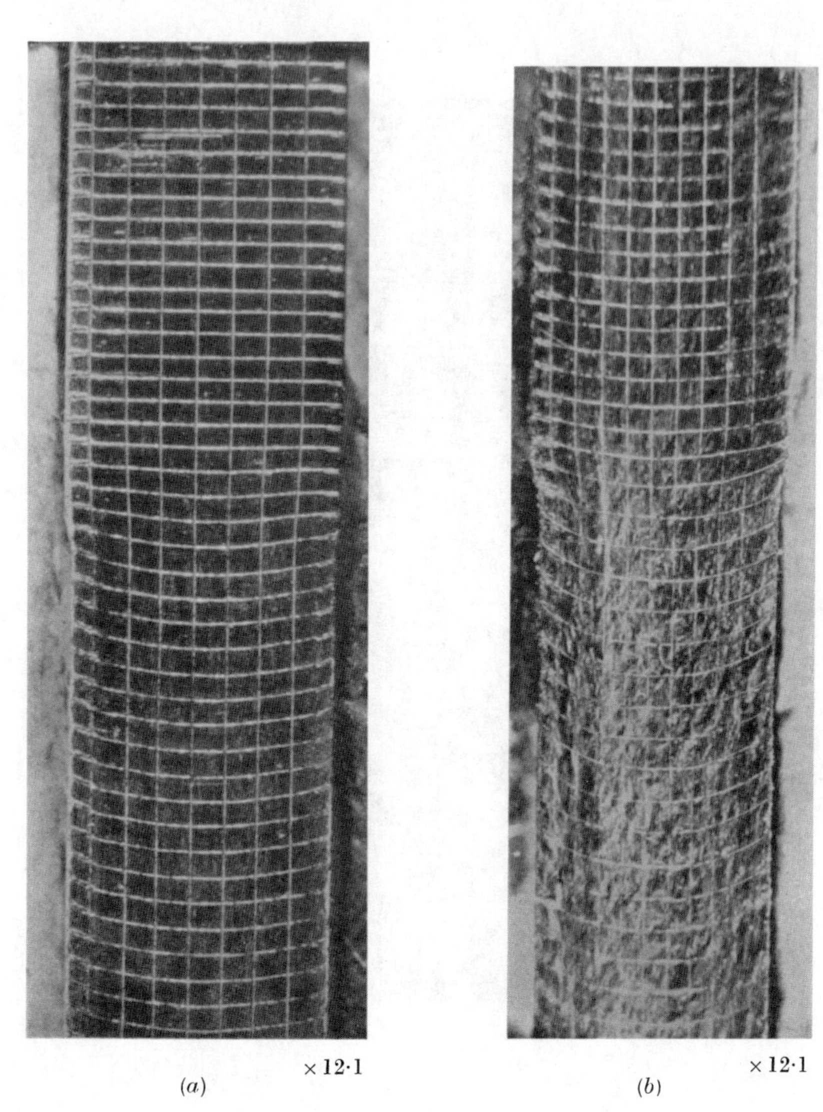


*a*, 15° taper. Two-stage draught at the first stage.
*b*, 15° taper. Two-stage draught at the second stage.

In this diagram the ordinates represent the displacement of the transverse lines parallel to the axis of the wire, whilst the abscissae represent distances from the central axis of the wire. The scale of the ordinates is increased in the ratio 3:1, in order to show up more clearly the principal characteristics of the distortion.

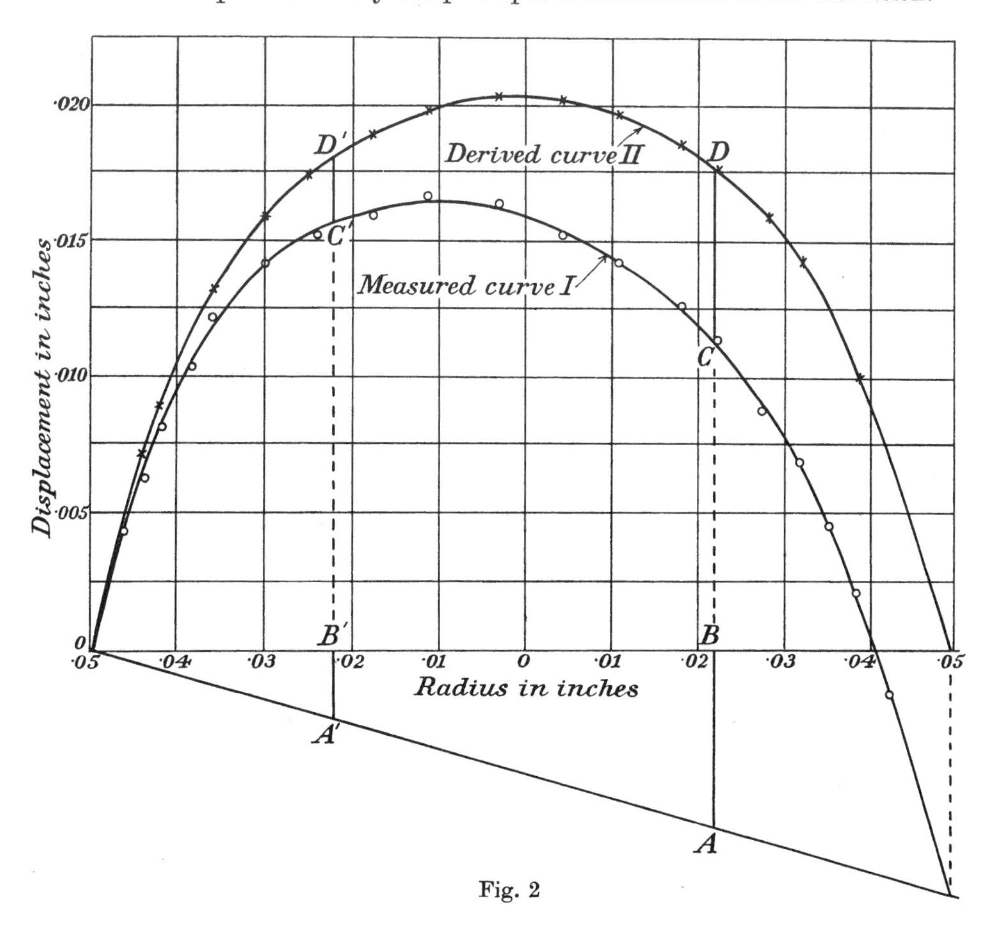


Fig. 2

It will be seen that curve I, which represents the distortion of the cross-section after drawing through the die with 30° taper, is noticeably unsymmetrical. In the other cases the asymmetry was scarcely appreciable, but in this case there must have been some accidental want of symmetry in the drawing of the wire. In order to eliminate as far as possible errors due to this cause, and thus make it possible to compare the distortions produced by various draw-holes, the curves were shifted in the manner represented in fig. 2, so that they represent, as in curve II, the displacement of the transverse section relative to the plane through its outer edge.

The curves derived in this way from plate $2a$, $b$ and plate 3 are shown as $A$, $B$ and $C$ in fig. 3. It will be seen that with the 3° taper the longitudinal displacement of the central portion of the wire relative to its outer surface is 0·004 in., with 15° taper it is 0·0115 in., whilst with 30° taper it is 0·0205.

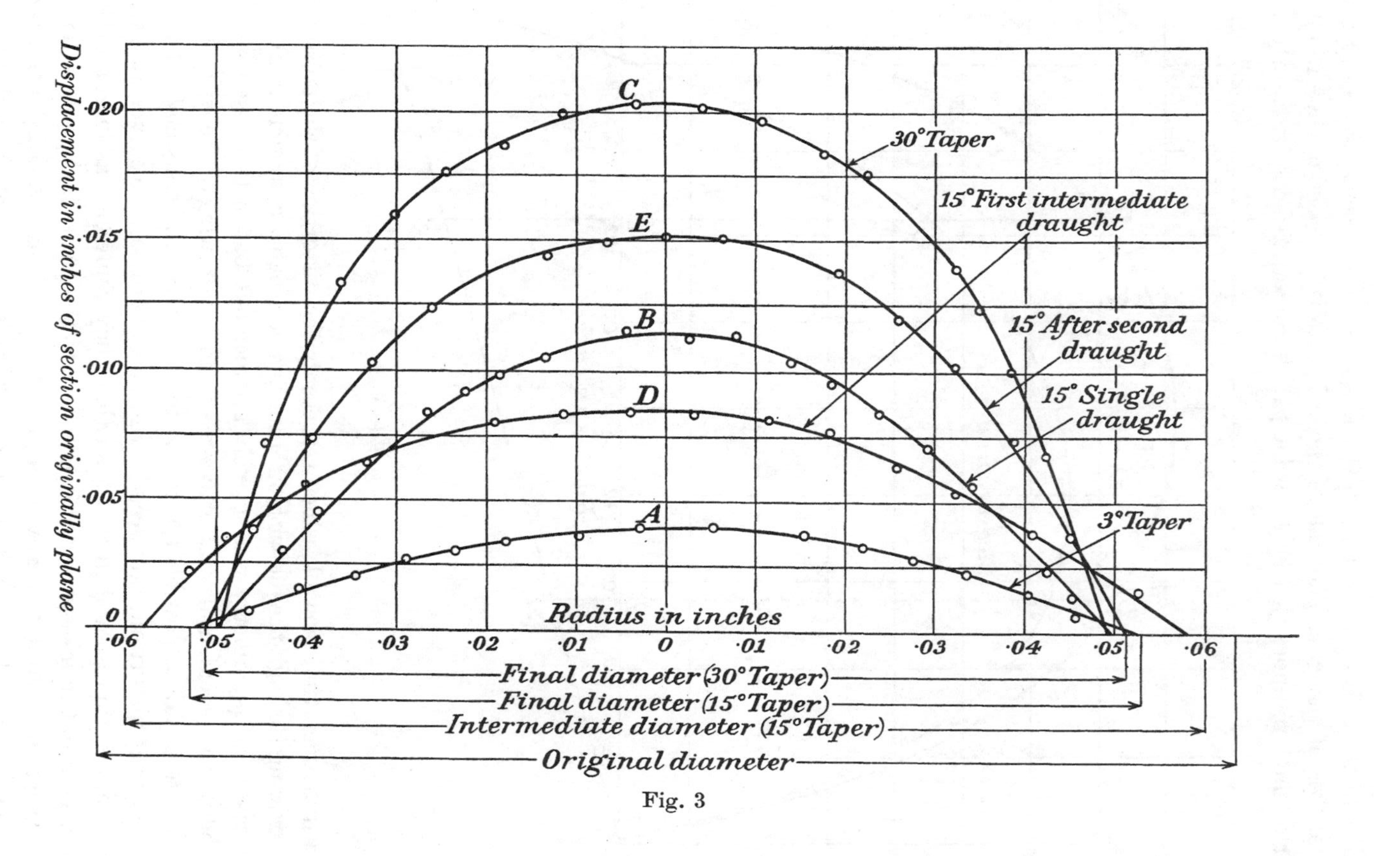

C
30° Taper
15° First intermediate draught
E
B
15° After second draught
D
15° Single draught
A
3° Taper
·020
·015
·010
·005
0
·06 ·05 ·04 ·03 ·02 ·01 0 ·01 ·02 ·03 ·04 ·05 ·06
Radius in inches
Final diameter (30° Taper)
Final diameter (15° Taper)
Intermediate diameter (15° Taper)
Original diameter
Displacement in inches of section originally plane


Fig. 3

## Effect of Length of Contact between Wire and Draw-hole

For any given percentage decrease in area of cross-section, the length of wire in contact with the draw-hole decreases as the angle of taper increases. The effect of decreasing the length of contact can be investigated independently of the effect of increasing the angle of taper, by comparing the distortion produced by drawing a wire through a hole of given taper in one single stage with the distortion produced by drawing in two stages to the same final diameter using the same angle of taper. The second stage would, of course, be a draught through the same hole that was used for the single-stage draught. A plate was accordingly made with a draw-hole of 15° taper and minimum diameter 0·120 in. This is intermediate between the initial diameter of the wire 0·127 in. and the final diameter 0·107 in.

If the wire remains unchanged in diameter until it first touches the draw-plate, and then ceases to decrease as it leaves the region of contact, the length of the wire in contact with the plate is

$$L=\frac{d_1-d_2}{2\tan\alpha} \quad \text{or} \quad L=d_1\left[\frac{d_1-d_2}{2d_1\tan\alpha}\right], \tag{3}$$

where $\alpha$ is the angle of taper and $d_1$, $d_2$ are the diameters before and after passing through the draw-plate. Using this formula with $\alpha=15°$, a decrease in diameter from 0·127 to 0·120 in. corresponds with length of contact $0{\cdot}10d_1$, whilst a decrease from 0·120 to 0·107 corresponds with a length of contact $0{\cdot}19d$. The single-stage reduction, however, corresponds with $L=0{\cdot}10d_1+0{\cdot}19d_1=0{\cdot}29d_1$.

The distortion during the reduction from 0·127 to 0·120 in. is shown in plate 4*a*, whilst the distortion during the second stage from 0·120 to 0·107 in. is shown in plate 4*b*. The measured co-ordinates of the transverse lines are shown in curves *D* and *E* in fig. 3. Comparing curve *B*, which represents the distortion when the wire is reduced in area by 29 % in one draught, with curve *E*, which represents the distortion occurring when the same reduction in area is carried out in two stages, it will be seen that the total distortion is considerably less in the former case than it is in the latter. Thus reducing the length of contact increases the amount of distortion for a given reduction in area and the same angle of taper. If, for instance, a draw-plate were constructed containing a very large number of holes, each of slightly less diameter than its predecessor, but all having the same angle of taper, this result would lead one to expect that the distortion of the cross-section would be greater if the wire were drawn through every hole than if it were drawn through every second hole.

## Variation in Distortion from the Centre to the Surface of the Wire

In comparing the three curves *B*, *C*, *E*, fig. 3, it will be seen that, although they stand up from their bases to very different heights, they are nearly parallel curves in the middle part of the wire. The differences between them are confined to the outer part of the wire. Curve $A$, however, is flatter than $B$, $C$ or $E$ throughout its length. This can be seen more clearly by displacing the curves parallel to the axis of the wire until their vertices coincide. This has been done in fig. 4, and it will be seen that the curves $B$ and $E$ are very nearly coincident over three-fifths of their diameter. The curve $C$ is slightly steeper than $B$ or $E$ in the central part, but it must be remembered that the final diameter was slightly smaller in the case of the 30° taper than in the case of the 15° taper. Making due allowance for this, it will be seen that when $L$, the length of wire in contact with the die, is a fraction (less than one-third) of its diameter, the distortion of the central three-fifths of the wire depends only on the total reduction in area, being independent of the actual value of $L$ or of the angle of taper of the die. On the other hand, when $L = 1{\cdot}5d_0$ as in the case of curve $A$, the distortion in the central part is considerably less than for curves $B$, $C$ and $E$, where $L = 0{\cdot}29d_0$, $0{\cdot}14d_0$, $0{\cdot}19d_0$ and $0{\cdot}10d_0$.

Comparing curves $B$ and $E$, a decrease in $L$ from $0{\cdot}29d_0$ for $B$ to $0{\cdot}10d_0$ and $0{\cdot}19d_0$ for $E$ corresponds with an increase in the amount of distortion in the outer part of the wire. This might be expected, for the regions of greatest stress must be concentrated near the part of the wire where there is contact with the die-hole, so that a decrease in $L$ may be expected to correspond with an increase in concentration of shearing stress in the outer parts of the wire and a consequent increase in distortion there. On the other hand, the stress distribution in the central part of the wire would scarcely be expected to depend much on the exact length of the region of contact, so that the fact that the distortion in the central part of the wire is the same in the three cases $B$, $C$ and $E$ is also quite understandable.

## The Structure of Drawn Wire

The structure of drawn wire has been examined by several investigators by means of X-rays, but the results are not in good agreement. Schmid* found that there is a considerable degree of preferred orientation throughout the wire, but in the central part the orientation is more pronounced than in the outer part. W. A. Wood,† on the other hand, found preferred orientation only in the central part of the wire in a region extending out to less than half the radius. In the rest of the wire, comprising four-fifths of its total area of cross-section, he found practically no preferred orientation of the crystal axes. Unfortunately, none of the workers with X-rays has given precise descriptions of the exact conditions of drawing. Comparison of the four curves of fig. 4 which represent the distortions of plane cross-sections under varying conditions of drawing for one given reduction in area, namely 29 %, shows that enormous variations in the distortion of the outer layers are likely to occur as the conditions of drawing are varied. On the other hand, the distortion in the central portion of the wire, out to half its radius, is not nearly so

* *Z. Metallkunde*, XX (1928), 375.
† *Phil. Mag.* [vii], XI (1931), 611.

sensitive to the conditions of drawing, and in any case the distortion there is small, so that the agreement between various X-ray workers as to the structure of the central parts of the wire and their disagreement as to the structure of the outer part may be due entirely to differences in the manner in which their wires were

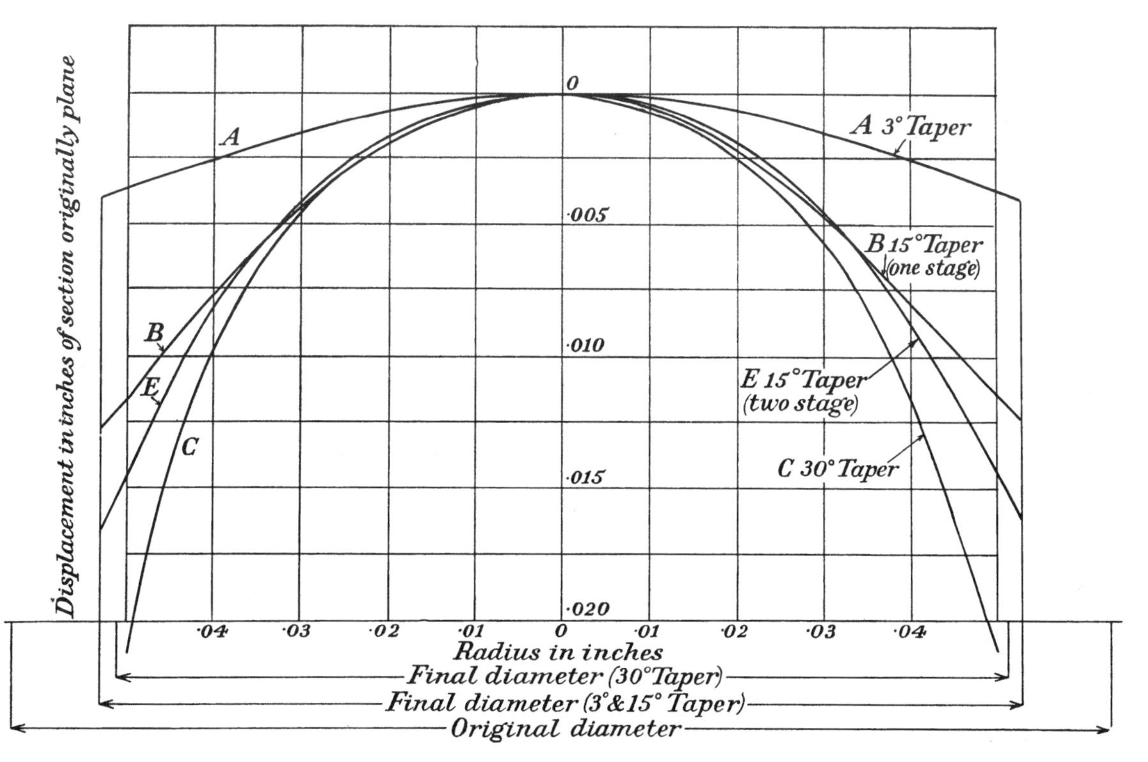


Fig. 4

drawn. It is perhaps significant that the part of the wires $B$, $C$, $E$ for which the distortion was unaffected by varying the value of $L$, extended from the middle to a radius of 0·03 in., i.e. to 0·6 of the total radius of the wire. The area over which W. A. Wood found preferred orientation stretched out from the centre to a distance of 0·4 time the radius.

In conclusion, the authors wish to express their thanks to Lord Rutherford and Professor Inglis for permission to carry out this work in their laboratories in Cambridge.

# 18

# THE BUCKLING LOAD FOR A RECTANGULAR PLATE WITH FOUR CLAMPED EDGES

REPRINTED FROM

*Zeitschrift für angewandte Mathematik und Mechanik*, vol. XIII (1933), pp. 147–52

## Introduction and Summary

The stability of a rectangular plate subjected to thrusts in its plane and perpendicular to its edges has been completely discussed (*a*) when the edges are simply supported,* (*b*) when two opposite edges are simply supported and two others clamped.† The stability problem for an infinite strip subjected to equal and opposite shearing stresses along its edges has also been solved.‡

In each of these cases the solution is relatively simple because a mode of disturbance is possible in which the displacement is a simple harmonic function of one of the co-ordinates, and in such cases the displacement can be represented by, at most, four terms of simple mathematical form. When all the edges are clamped no simple displacement of this type is possible and no analysis of the stability of a rectangular plate subject to this edge condition has yet been given. Failing a complete solution, Timoshenko and others have assumed an arbitrary form of disturbance which satisfies the edge conditions but does not satisfy the proper differential equation. By calculating the strain energy of this arbitrary displacement they found an approximate value for the magnitude of the edge thrust which causes instability. Recently Sezawa § has rediscussed the subject assuming a simple form of disturbance which satisfies the differential equation but does not satisfy the edge conditions except at the corners and the mid-points of the sides.

In the present paper a formal solution of the problem is given when all four edges are clamped. The forms of displacement which satisfy both the clamped edge condition and the differential equation of neutral equilibrium are found, and the magnitudes of the thrusts which are necessary in order to maintain these displacements are expressed as values which cause an infinite determinant to vanish.

The numerical work necessary for completing the solution in any given case is very heavy, but in one case, that of a square plate subjected to equal thrusts perpendicular to all its edges, some simplification is possible owing to symmetry. In this case the lowest critical edge thrust is calculated by finding the lowest roots of

* G. Bryan, *Proc. Lond. Math. Soc.* XXII (1891), 54.

† E. Reissner, *Zbl. Bauverw.* (1909), 93.

‡ R. W. Southwell and S. Skan, *Proc. Roy. Soc.* A (1924), 582.

§ K. Sezawa. 'On the buckling under edge thrusts of a rectangular plate clamped at four edges', *Rep. Aero. Res. Inst. Tokyo*, no. 69.

the determinants formed by taking the first 1, 2, 3, 4 and 5 rows and columns of the infinite determinant. It is found that the convergence is extremely rapid, the successive values of $X=4Pa^2/D\pi^2$ being 5·00, 5·30, 5·308, 5·307, 5·304, where $P$ is the thrust and $2a$ is the length of the side of the square. $D=\dfrac{Eh^2}{12(1-\sigma^2)}$ and $h$ is the thickness of the plate, $E$ Young's modulus and $\sigma$ Poisson's ratio.

The value obtained for $X$ by Sezawa in this case was 5·61, while that obtained by the Ritz strain-energy method is 5·33. The extremely close agreement between this and the true value 5·30 is probably accidental.

The critical value of $X$ for the square plate with simply supported edges is 2·0, while if two opposite edges are clamped and the others simply supported it is 3·86.

## Formal Solution of the Problem

The differential equation for elastic displacement $w$ of a plane sheet subjected to stresses $P_1$ and $P_2$ parallel to rectangular axes $\xi$, $\eta$ is

$$\frac{\partial^4 w}{\partial \xi^4}+2\frac{\partial^4 w}{\partial \xi^2 \partial \eta^2}+\frac{\partial^4 w}{\partial \eta^4}+\frac{P_1}{D}\frac{\partial^2 w}{\partial \xi^2}+\frac{P_2}{D}\frac{\partial^2 w}{\partial \eta^2}=0, \tag{1}$$

where $D=\frac{1}{12}Eh^2(1-\sigma^2)^{-1}$. The sketch (fig. 1) shows the rectangular sheet loaded along its edges. Writing $\xi=2ax/\pi$, $\eta=2bx/\pi$ the rectangle whose sides are $\xi=\pm a$, $\eta=\pm b$ is transformed into a square whose sides are $x=\pm\frac{1}{2}\pi$, $y=\pm\frac{1}{2}\pi$ and (1) becomes

$$\frac{1}{a^4}\frac{\partial^4 w}{\partial x^4}+\frac{2}{a^2b^2}\frac{\partial^4 w}{\partial x^2\partial y^2}+\frac{1}{b^4}\frac{\partial^4 w}{\partial y^4}+\frac{4P_1a^2}{D\pi^2}\left(\frac{1}{a^4}\frac{\partial^2 w}{\partial x^2}\right)+\frac{4P_2b^2}{D\pi^2}\left(\frac{1}{b^4}\frac{\partial^2 w}{\partial y^2}\right)=0. \tag{2}$$

The problem is to find a solution of (2) other than $w=0$ which satisfies the condition $w=0$ at $x=\pm\frac{1}{2}\pi$ and $y=\pm\frac{1}{2}\pi$ and also the conditions $\partial w/\partial x=0$ at $x=\pm\frac{1}{2}\pi$ and $\partial w/\partial y=0$ at $y=\pm\frac{1}{2}\pi$.

The method adopted is first to find solutions of the differential equation (2) which satisfy $w=0$ at the edges. It is then shown that an infinite number of these can be combined together in such a way that the remaining 'clamped edge' condition $\partial w/\partial x=\partial w/\partial y=0$ is also satisfied provided a certain relationship exists between the size and shape of the plate and the thrusts $P_1$ and $P_2$.

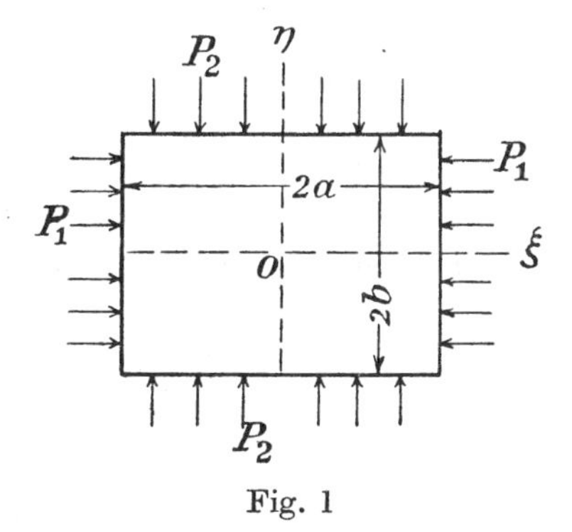

Fig. 1

In the work which follows we shall only consider displacements which are symmetrical with respect to both axes, so that $x$ and $y$ occur in $w$ only as even functions.

The function $e^{\alpha y}\cos nx$ satisfies the condition $w=0$ at $x=\frac{1}{2}\pi$ provided $n$ is an odd integer and it also satisfies (2) provided

$$\frac{a^4}{b^4}\alpha^4-2n^2\alpha^2\frac{a^2}{b^2}+n^4-\frac{4P_1a^2}{D\pi^2}n^2+\frac{4P_2b^2}{D\pi^2}\frac{a^4\alpha^2}{b^4}=0. \tag{3}$$

By a suitable combination of the four terms of type $e^{\alpha y}\cos nx$ an even function is obtained which satisfies $w=0$ at $y=\pm\frac{1}{2}\pi$. These combinations will take different forms according as the roots of (3) are real, imaginary or complex.

Type I. If the two roots of (3), regarded as a quadratic in $a^2$, are positive, calling them $\alpha_n^2$, $\beta_n^2$, the appropriate form is

$$w=(\cosh\alpha_n y\cosh\tfrac{1}{2}\beta_n\pi-\cosh\beta_n y\cosh\tfrac{1}{2}\alpha_n\pi)\cos nx.$$

Type II. If the two roots of (3) regarded as quadratic in $\alpha^2$ are positive and negative let them be $-\alpha_n^2, \beta_n^2$. The required term is

$$w=(\cos\alpha_n y\cosh\tfrac{1}{2}\beta_n\pi-\cosh\beta_n y\cos\tfrac{1}{2}\alpha_n\pi)\cos nx.$$

Type III. If the four roots are pure imaginaries $\pm i\alpha$, $\pm i\beta$ the form is

$$w=(\cos\alpha_n y\cos\tfrac{1}{2}\beta_n\pi-\cos\beta_n y\cos\tfrac{1}{2}\alpha_n\pi)\cos nx.$$

Type IV. If the roots of (3) are complex they must be of the form $\pm\alpha\pm i\beta$ and the appropriate form for $w$ is

$$w=(\cosh\alpha_n y\cos\beta_n y\sinh\tfrac{1}{2}\alpha_n\pi\sin\tfrac{1}{2}\beta_n\pi$$
$$-\sinh\alpha_n y\sin\beta_n y\cosh\tfrac{1}{2}\alpha_n\pi\cos\tfrac{1}{2}\beta_n\pi)\cos nx.$$

In the work which follows only terms of type I will be referred to, assuming that in applying the results terms of other types are substituted where necessary.

If $w$ is a function satisfying (1) then when $w$ and $dw/dn$ are given at all points of a closed boundary, $w$ is in general determined at all points within it, $dw/dn$ being the rate of change in $w$ along a normal to the boundary. It appears therefore that in order that a series of terms of type I may be capable of representing all possible values of $w$ inside a square consistent with $w=0$ at its edges, it must be capable of representing arbitrary symmetrical distributions of $\partial w/\partial y$ along $y=\pm\frac{1}{2}\pi$ and of $\partial w/\partial x$ along $x\pm=\frac{1}{2}\pi$.

The single series

$$w=\sum_{n\text{ odd}}{}_n A(\cosh\alpha_n y\cosh\tfrac{1}{2}\beta_n\pi-\cosh\beta_n y\cosh\tfrac{1}{2}\alpha_n\pi)\cos nx$$

is capable of representing any assigned distribution of $\partial w/\partial y$ along $y=\pm\frac{1}{2}\pi$ because any even function of $x$ which vanishes at $\pm\frac{1}{2}\pi$ can be expanded between the limits $\pm\frac{1}{2}\pi$ in a cosine series of odd multiples of $x$.

Similarly the series

$$w=\sum_{n\text{ odd}} B_n(\cosh\gamma_n x\cosh\tfrac{1}{2}\delta_n\pi-\cosh\delta_n y\cosh\tfrac{1}{2}\gamma_n\pi)\cos ny$$

can represent any assigned symmetrical distribution of $\partial w/\partial x$ along the edges $x=\pm\frac{1}{2}\pi$, and if $\pm\gamma_n\pm\delta_n$ are the roots of the biquadratic

$$\frac{b^4}{a^4}\gamma^4-2n^2\gamma^2\frac{b^2}{a^2}+n^4-\frac{4P_2b^2}{D\pi^2}n^2+\frac{4P_1a^2}{D\pi^2}\frac{b^4}{a^4}\gamma^2=0, \tag{4}$$

each term of the series satisfies (2).

Next consider the double series

$$w = \sum_{n\,\text{odd}} A_n(\cosh\alpha_n y\cosh\tfrac{1}{2}\beta_n\pi - \cosh\beta_n y\cosh\tfrac{1}{2}\alpha_n\pi)\cos nx$$
$$+ B_n(\cosh\gamma_n x\cosh\tfrac{1}{2}\delta_n\pi - \cosh\delta_n x\cosh\tfrac{1}{2}\gamma_n\pi)\cos ny, \quad (5)$$

at $y = \pm\frac{1}{2}\pi$,

$$\frac{\partial w}{\partial y} = \sum_{n\,\text{odd}} A_n(\alpha_n\sinh\tfrac{1}{2}\alpha_n\pi\cosh\tfrac{1}{2}\beta_n\pi - \beta_n\sinh\tfrac{1}{2}\beta_n\pi\cosh\tfrac{1}{2}\alpha_n\pi)\cos nx$$
$$+ (-1)^{\frac{1}{2}(n+1)} nB_n(\cosh\gamma_n x\cosh\tfrac{1}{2}\delta_n\pi - \cosh\delta_n x\cosh\tfrac{1}{2}\gamma_n\pi). \quad (6)$$

If all the $B$'s were known the second series in (6) would be a known function, $F(x)$, which could be expanded in a cosine series of odd multiples of $x$ between the limits $x = \pm\frac{1}{2}\pi$. The given arbitrary distribution of $\partial w/\partial y$ could also be expressed in such a series, and the values of $A_n$ would then be determined by equating coefficients of $\cos nx$. Similarly, if the $A$'s were known the values of $B_n$ could be determined so that $\partial w/\partial x$ could have any arbitrary distribution along $x = \pm\frac{1}{2}\pi$.

The double series (5) appears therefore to be capable of representing a function for which $\partial w/\partial y$ along $y = \pm\frac{1}{2}\pi$ and $\partial w/\partial x$ along $x = \pm\frac{1}{2}\pi$ have any given symmetrical distributions. The determination of the actual values of the $A$'s and $B$'s in any given case would necessitate the solution of an infinite series of linear equations.

If $\partial w/\partial x = 0$ along $x = \pm\frac{1}{2}\pi$ and $\partial w/\partial y = 0$ along $y = \pm\frac{1}{2}\pi$ the solution of this series of equations would in general yield the result that all the $A$'s and all the $B$'s are zero; but if a certain special relationship exists between the dimensions of the sheet and $P_1$ and $P_2$, namely that obtained by eliminating all the $A$'s and $B$'s from the system of linear equations, then a displacement can exist in neutral equilibrium. The sheet is then in a condition of neutral stability, and all the $A$'s and $B$'s can be determined in terms of any one of their number.

To carry out the operations indicated above it is convenient after putting $x = \frac{1}{2}\pi$ in (5) to expand each term of the $A$ and the $B$ series in (5) in a cosine series of even multiples of $y$. The coefficient of $\cos sy$ in the series so obtained is then equated to zero for each value (even) of $s$ in order that the condition $\partial w/\partial x = 0$ may be satisfied at all points of the edges $x = \pm\frac{1}{2}\pi$.

The necessary cosine series valid between $y = \pm\frac{1}{2}\pi$ are

$$\cosh\alpha y = \frac{4\alpha}{\pi}\sinh\tfrac{1}{2}\alpha\pi\left(\frac{1}{2\alpha^2} - \frac{\cos 2y}{\alpha^2+2^2} + \frac{\cos 4y}{\alpha^2+4^2} - \dots\right), \quad (7)$$

$$\cos\alpha y = \frac{4\alpha}{\pi}\sin\tfrac{1}{2}\alpha\pi\left(\frac{1}{2\alpha^2} - \frac{\cos 2y}{\alpha^2-2^2} + \frac{\cos 4y}{\alpha^2-4^2} - \dots\right), \quad (8)$$

and if terms of type IV occur the expansions of $\cos\alpha y\cosh\beta y$ and $\sin\alpha y\sinh\beta y$ are also needed.

Inserting expansions of this type for $\cos ny$, $\cosh\alpha_n y$, $\cosh\beta_n y$ in the right-hand side of (5) it is found that the condition $\partial w/\partial x = 0$ at $x = \pm\frac{1}{2}\pi$ is satisfied if for every (even) value of $s$

$$\frac{4}{\pi}(-1)^{\frac{1}{2}s}\sum_{n\,\text{odd}}(-1)^{\frac{1}{2}(n+1)} n(a_{ns}A_n - b_{ns}B_n) = 0, \quad (9)$$

where $a_{ns}$, $b_{ns}$ take different forms according to whether the corresponding term is of type I, II, III or IV.

If $\alpha_n$, $\beta_n$ occur in a term of type I

$$a_{ns}=\frac{\alpha_n \cosh\frac{1}{2}\beta_n\pi \sinh\frac{1}{2}\alpha_n\pi}{\alpha_n^2+s^2}-\frac{\beta_n \cosh\frac{1}{2}\alpha_n\pi \sinh\frac{1}{2}\beta_n\pi}{\beta_n^2+s^2}. \tag{10}$$

If they occur in a term of type II

$$a_{ns}=\frac{\alpha_n \cosh\frac{1}{2}\beta_n\pi \sin\frac{1}{2}\alpha_n\pi}{\alpha_n^2-s^2}-\frac{\beta_n \cos\frac{1}{2}\alpha_n\pi \sinh\frac{1}{2}\beta_n\pi}{\beta_n^2+s^2}; \tag{11}$$

similarly if $\gamma_n$ and $\delta_n$ occur in a term of type I

$$b_{ns}=(\gamma_n \sinh\tfrac{1}{2}\gamma_n\pi \cosh\tfrac{1}{2}\delta_n\pi-\delta_n \sinh\tfrac{1}{2}\delta_n\pi \cosh\tfrac{1}{2}\gamma_n\pi)(n^2-s^2)^{-1}; \tag{12}$$

or if they occur in a term of type II

$$b_{ns}=(-\gamma_n \sin\tfrac{1}{2}\gamma_n\pi \cosh\tfrac{1}{2}\delta_n\pi-\delta_n \sinh\tfrac{1}{2}\delta_n\pi \cos\tfrac{1}{2}\gamma_n\pi)(n^2-s^2)^{-1}. \tag{13}$$

Permuting $x$ and $y$ the condition $\partial w/\partial y=0$ at $y=\pm\frac{1}{2}\pi$ leads to

$$\frac{4}{\pi}(-1)^{\frac{1}{2}s}\Sigma(-1)^{\frac{1}{2}(n+1)}n(c_{ns}B_n-d_{ns}A_n)=0, \tag{14}$$

where again $c_{ns}$, $d_{ns}$ take various forms according as they are derived from terms of types I, II, III or IV. For instance, if $\alpha_n\beta_n$ and also $\gamma_n\delta_n$ are each derived from terms of type I

$$c_{ns}=\frac{\gamma_n \cosh\frac{1}{2}\delta_n\pi \sinh\frac{1}{2}\gamma_n\pi}{\gamma_n^2+s^2}-\frac{\delta_n \cosh\frac{1}{2}\gamma_n\pi \sinh\frac{1}{2}\delta_n\pi}{\delta_n^2+s^2}, \tag{15}$$

$$d_{ns}=(\alpha_n \sinh\tfrac{1}{2}\alpha_n\pi \cosh\tfrac{1}{2}\beta_n\pi-\beta_n \sinh\tfrac{1}{2}\beta_n\pi \cosh\tfrac{1}{2}\alpha_n\pi)(n^2-s^2)^{-1}. \tag{16}$$

Eliminating the $A$'s and $B$'s between (9) and (14) we obtain the infinite determinant equation:

$$\Delta=\begin{vmatrix} a_{10} & b_{10} & a_{30} & b_{30} & a_{50} & \cdots \\ d_{10} & c_{10} & d_{30} & c_{30} & d_{50} & \cdots \\ a_{12} & b_{12} & a_{32} & b_{32} & a_{52} & \cdots \\ d_{12} & c_{12} & d_{32} & c_{32} & d_{52} & \cdots \\ \cdots & \cdots & \cdots & \cdots & \cdots & \cdots \end{vmatrix}=0. \qquad (17),(18)$$

When $a$ and $b$ are fixed the only variables remaining in $\Delta$ are $P_1$ and $P_2$, so that $\Delta=0$ is a critical equation determining the special values of $P_1$ and $P_2$ at which elastic displacements can exist in neutral equilibrium when all four edges of the plate are clamped.

Though the analysis here given takes account only of symmetrical modes the same method could be applied to find antisymmetrical modes.

The solution expressed by (17) is merely formal. To find out whether it can be used to determine actual values of the buckling loads for a rectangular plate we must examine its convergence. For this purpose we may form a series of finite determinants $\Delta_1, \Delta_2, \Delta_3, \ldots, \Delta_N$ by taking $2, 4, 6, \ldots, 2N$ rows and columns starting at the left-hand top corner of $\Delta$, so that $\Delta_1 = \begin{vmatrix} a_{10} & b_{10} \\ d_{10} & c_{10} \end{vmatrix}$, etc. If it is found that any root of $\Delta_N = 0$ converges to a definite limit as $N$ increases this root represents a possible condition for which an elastic displacement can exist in neutral equilibrium.

It is difficult to give the matter any general treatment, accordingly I have taken a particular case, namely that of a square plate under equal thrusts perpendicular to all its four edges, and have calculated the smallest roots of $\Delta_1, \Delta_2, \Delta_3, \Delta_4$ and $\Delta_5$ with a view to finding out whether they converge to a definite limit.

## Stability of a Square Plate with Clamped Edges under Uniform Thrust in all Directions in its Plane

Taking $2a$ as the length of the side and $P$ as the load per unit length applied along the four sides we shall represent $4Pa^2/\pi^2 D$, which is a pure number, by $X$.

We may express the problem in non-dimensional form as follows:

'It is required to find the lowest value of $X$ for which $\Delta^4 w + X\Delta^2 w = 0$ can be satisfied over the square contained between $x = \pm\frac{1}{2}\pi$, $y = \pm\frac{1}{2}\pi$, subject to the edge conditions $w=0$, $\partial w/\partial x = 0$ along $x = \pm\frac{1}{2}\pi$ and $w = 0$, $\partial w/\partial y = 0$ along $y = \pm\frac{1}{2}\pi$.'

In the present case (3) becomes

$$(\alpha^2 - n^2)^2 + X(\alpha^2 - n^2) = 0 \quad \text{or} \quad (\alpha^2 - n^2 + X)(\alpha^2 - n^2) = 0, \tag{19}$$

while (4) has the same form, namely,

$$(\gamma^2 - n^2 + X)(\gamma^2 - n^2) = 0. \tag{20}$$

It appears therefore that if $n^2 > X$ the corresponding terms in both series of (5) are of type I and

$$\alpha_n = \gamma_n = \sqrt{(n^2 - X)}, \quad \beta_n = \delta_n = n. \tag{21}$$

If $n^2 < X$ the corresponding terms of both series are of type II and

$$\alpha_n = \gamma_n = \sqrt{(X - n^2)}, \quad \beta_n = \delta_n = n. \tag{22}$$

No terms of types III or IV occur.

It seems likely that the lowest value of $X$ consistent with instability will correspond with a displacement which is symmetrical with respect to $x$ and $y$; accordingly, we may search for a solution in which $A_n = B_n$. We may therefore assume for $w$ the series

$$\begin{aligned} w = \sum_{n<X^{\frac{1}{2}}} A_n[&(\cosh\tfrac{1}{2}n\pi \cos\sqrt{(x-n^2)}\,y - \cos\tfrac{1}{2}\pi\sqrt{(X-n^2)}\cosh ny)\cos nx \\ &+ (\cosh\tfrac{1}{2}n\pi\cos\sqrt{(X-n^2)}\,x - \cos\tfrac{1}{2}\pi\sqrt{(X-n^2)}\cosh nx)\cos ny] \\ + \sum_{n>X^{\frac{1}{2}}} A_n[&(\cosh\tfrac{1}{2}n\pi\cosh\sqrt{(n^2-X)}\,y - \cosh\tfrac{1}{2}\pi\sqrt{(n^2-X)}\cosh ny)\cos nx \\ &+ (\cosh\tfrac{1}{2}n\pi\cosh\sqrt{(n^2-X)}\,x - \cosh\tfrac{1}{2}\pi\sqrt{(n^2-X)}\cosh nx)\cos ny], \end{aligned} \tag{23}$$

where $n$ is an odd integer.

It seems clear that no value of $X$ less than 1 can lead to a solution of the problem because even with simply supported edges (Bryan's problem) the smallest value of

$X$ giving neutral equilibrium is 2·0. The effect of clamping the edges must necessarily be to raise the load at which instability occurs. On the other hand, it seems likely that there will be a solution in the range $1 < X < 9$. In this range of values of $X$ the only terms of type II which occur are those for which $n = 1$. The next step is to calculate $a_{ns}$, $b_{ns}$. Since $\alpha_n = \gamma_n$ and $\beta_n = \delta_n$ it will be seen that $a_{ns} = c_{ns}$ and $b_{ns} = d_{ns}$.

Hence both (9) and (14) reduce to

$$\Sigma(-1)^{\frac{1}{2}(n+1)} nA_n(a_{ns} - b_{ns}) = 0. \tag{24}$$

In the special case of symmetrical displacements of a square plate under uniform edge load the number of rows and columns in $\Delta$ is therefore halved. The finite determinant $\Delta_N$ which in general contains $2N$ rows and columns is therefore reduced in this special case to one of $N$ rows and columns, and the term in the $n$th column and $s$th row is $a_{ns} - b_{ns}$.

The formulae used in calculating these terms are

$n = 1$:

$$a_{1s} - b_{1s} = \cos\tfrac{1}{2}\pi\sqrt{(X-1)}\cosh\tfrac{1}{2}\pi[\sqrt{(X-1)}\tan\tfrac{1}{2}\pi\sqrt{(X-1)}\{(X-1-s^2)^{-1} + (1-s^2)^{-1}\} + \tanh\tfrac{1}{2}n\pi\{(1-s^2)^{-1} - (1+s^2)^{-1}\}], \tag{25)*$$

$n > 1$:

$$a_{ns} - b_{ns} = \cosh\tfrac{1}{2}\pi\sqrt{(n^2-X)}\cosh\tfrac{1}{2}n\pi[\sqrt{(n^2-X)}\tanh\tfrac{1}{2}\pi\sqrt{(n^2-X)} \times \{(n^2-X+s^2)^{-1} - (n^2-s^2)^{-1}\} + n\tanh\tfrac{1}{2}n\pi\{(n^2-s^2)^{-1} - (n^2+s^2)^{-1}\}]. \tag{26)*$$

Writing $c_{ns}$ for the terms in the square brackets in (25) and (26) the equation for the single variable $X$ is now

$$\Delta = \begin{vmatrix} c_{10} & c_{30} & c_{50} & \cdots \\ c_{12} & c_{32} & c_{52} & \cdots \\ c_{14} & c_{34} & c_{54} & \cdots \\ \cdots & \cdots & \cdots & \cdots \end{vmatrix} = 0. \tag{27}$$

## Search for Roots of $\Delta = 0$

A preliminary search made it seem improbable that (27) has any root less than 4. Accordingly the 25 values of $c_{ns}$ were calculated for $n = 1, 3, 5, 7, 9$ and $s = 0, 2, 4, 6, 8$, for each of four values of $X$, namely, $X = 4·0$, 5·0, 5·2, 5·4. The full determinant up to 5 rows and columns is set out in detail below in table 1 for the case $X = 5·2$ in order that the general character of the determinant may be seen. The values of $\Delta_1$, $\Delta_2$, $\Delta_3$, $\Delta_4$ and $\Delta_5$ were then calculated by taking 1, 2, 3, 4 and 5 rows and columns of $\Delta$. Their values are given in table 2. Inspection of the figures in that table shows that a root of $\Delta_1 = 0$ occurs at $X = 5·0$ (i.e. where $\tan\sqrt{(X-1)}\,\tfrac{1}{2}\pi = 0$) but that the corresponding roots of $\Delta_2, \Delta_3, \Delta_4$ and $\Delta_5$ all lie between $X = 5·2$ and $X = 5·4$. Using a quadratic interpolation formula applied to the values of the $\Delta$'s at $X = 5·0$, 5·2 and 5·4 the roots given at the foot of table 2 are obtained. It will be seen that the

* When $s = 0$ these values must be multiplied by $\tfrac{1}{2}$ (see (7) and (8)).

convergence is extremely rapid; the root of $\Delta_1$ is $X=5{\cdot}0$, that of $\Delta_2$ is $X=5{\cdot}310$ while those of $\Delta_3$, $\Delta_4$ and $\Delta_5$ are 5·308, 5·307, 5·304 respectively. The calculations were made with the help of a slide rule so that the last figure can hardly be relied on, particularly in the case of $\Delta_5$, but it seems certain that the root of $\Delta_4$ differs from that of $\Delta_2$ by only 1 part in 1000. We may therefore take $X=5{\cdot}30$ as the final result of our calculations. A square plate with clamped edges therefore becomes unstable when subjected to thrusts $P$ per unit length perpendicular to its edges if $4Pa^2/D\pi^2>5{\cdot}30$.

Table 1. *Construction of* $\Delta_5$ *for* $X=5{\cdot}2$

| $n=$ | 1 | 2 | 3 | 4 | 5 | $s$ |
|---|---|---|---|---|---|---|
| | 0·1974 | 0·2964 | 0·0467 | 0·0157 | 0·0074 | 0 |
| | 0·2545 | 0·2290 | 0·0407 | 0·0149 | 0·0070 | 2 |
| $\Delta_5=$ | −0·1403 | −0·2175 | +0·0640 | 0·0145 | 0·0066 | 4 |
| | −0·0611 | −0·0561 | −0·0520 | +0·0298 | 0·0097 | 6 |
| | −0·0339 | −0·0314 | −0·0171 | −0·0263 | +0·0180 | 8 |

Table 2

| $X$ | $\Delta_1$ | $10\Delta_2$ | $10^2\Delta_3$ | $10^4A_4$ | $10^6\Delta_5$ |
|---|---|---|---|---|---|
| 4·0 | −1·032 | −2·85 | −2·00 | −6·8 | −14·4 |
| 5·0 | 0 | −0·822 | −0·733 | −3·4 | − 9·4 |
| 5·2 | +0·197 | −0·302 | −0·296 | −1·31 | − 3·7 |
| 5·4 | +0·403 | +0·262 | +0·286 | +1·34 | + 3·9 |
| Root $X=$ | 5·00 | 5·310 | 5·308 | 5·307 | 5·304 |

## Comparison with Approximate Methods

By his approximate method Sezawa obtained for this case $4Pa^2/D\pi^2=5{\cdot}61$. That his result would be a little too high might have been anticipated. The value obtained by the Ritz method assuming the arbitrary displacement $w=\cos^2 x\cos^2 y$ which satisfies the boundary conditions but not the differential equation, is $4Pa^2/D\pi^2=5{\cdot}33$.* That this should be so close to the true value 5·30 is rather surprising. It seems probable that the displacement $w=\cos^2 x\cos^2 y$ happens by accident to be rather near to the true form of the displacement which can exist in neutral equilibrium.

* For this calculation I am indebted to Dr H. L. Cox.

# 19

# THE LATENT ENERGY REMAINING IN A METAL AFTER COLD WORKING*

REPRINTED FROM

*Proceedings of the Royal Society*, A, vol. CXLIII (1934), pp. 307–26

Measurements of the latent energy remaining in metal rods after severe twisting are described. Very much more cold work can be done on a metal in torsion than in direct tension. It is found that as the total amount of cold work which has been done on a specimen increases the proportion which is absorbed decreases. Though saturation was not fully reached even with twisted rods, curves representing the experimental results for copper indicate that it would have been reached at a plastic strain very little greater than the strain at fracture. The amount of cold work necessary to saturate copper with latent energy at 15° C. is thus found to be slightly greater than 14 cal. per g.

By using compression instead of torsion, it was found possible to do much more cold work on copper than this, and compression tests revealed the fact that the compressive stress increases with increasing strain till the total applied cold work was equivalent to 15 cal. per g. No further rise in compressive stress occurred with further compression even though the specimen was compressed till its height was only 1/53rd of its original height.

The fact that the absorption of latent energy and the increase in strength with increasing strain both cease when the same amount of cold work has been applied suggests that the strength of pure metals may depend only on the amount of cold work which is latent in them.

When a metal is subjected to plastic distortion (cold working) most of the work done reappears in the form of heat, but a certain proportion remains latent and is no doubt associated with the changes to which cold working give rise in the physical properties of the metal. When the metal is heated all this latent heat must be released before the melting-point is reached, and when it is dissolved the latent heat must appear as a heat of solution. It is therefore possible to measure the latent heat of cold working either when energy is put into the metal of when it is released. In the former case, the work done and the heat evolved during plastic deformation are measured. The difference is the latent energy of cold working. This method has been adopted by Farren and Taylor† and by Hort,‡ who found that within the range of their experiments $5\frac{1}{2}$–$13\frac{1}{2}$ % of the work done remains latent in the metal.

Some attempts have been made to measure the latent energy of cold working when it is released, but the results are not always comparable with those of Farren and Taylor, and when they are the two kinds of measurements do not seem to be

* With H. QUINNEY.

† *Proc. Roy. Soc.* A, CVII (1925), 422; paper 6 above.

‡ *Mitt. Forsch. Arb. dtsch. Ingenieur*, XLI (1907).

in agreement. In the experiments of Farren and Taylor the cold work was done by stretching a rod of metal by a direct load. With such a system of loading only a very limited amount of work can be done on the specimen before it breaks. With copper, for instance, the maximum latent energy which they measured was equivalent to a rise in temperature of only 0·83° C. In order to release the latent energy of copper it is probably necessary to raise its temperature to 500° C. or more. The latent energy would appear as the difference between the amounts of heat necessary to raise the temperature of two equal specimens to 500° C. when one of them had been subjected to cold work and the other had not. To measure it would be equivalent to measuring the difference between the specific heat of two metals when that difference is only 1 part in 600.

It is difficult to attain such accuracy in heat measurements, but by using metals which have been subjected to much more severe cold working than is attainable under a direct load, greater latent heat can be obtained and the accuracy of the measurements of the energy released on heating correspondingly increased.

It is well known that much more cold work can be done on a metal rod by twisting it than by stretching it. The amount of work which can be done in direct extension depends on the nature of the load-extension curve. The condition for fracture by instability owing to the formation of a local 'neck' is

$$\frac{l_0}{T}\frac{dT}{dl} < 1 \quad \text{or} \quad \frac{d \log T}{d \log (l/l_0)} < 1,$$

where $T$ is the tensile stress, $l$ is the length, and $l_0$ the initial length of the test specimen. The criterion for fracture can therefore be found by plotting the load-extension curve on a logarithmic scale. Fracture will then occur when the tangent to the curve makes an angle 45° with the axes. When a round bar is twisted it will not fail owing to the formation of a local neck until the shear stress has ceased to increase with increasing strain. (In fact, twisted specimens usually fail for reasons other than this kind of instability.)

If the tensile stress rises very rapidly for a very small extension and then continues to rise much more slowly as the strain increases to large values, a tensile specimen will break at a very small extension, but a twisted rod may suffer very great distortion before breaking. The results of tensile and torsion tests on two identical nickel rods are shown in fig. 1. The nickel was used in the hardened condition as supplied by Messrs Henry Wiggin and Co. In the tensile test the rod suffered no measurable plastic distortion till a stress of 76,800 lb. per sq.in. was reached. It fractured when the stress was 84,800 lb. per sq.in., and the extension was then $0{\cdot}012 l_0$.

In the torsion test the first measurable plastic strain occurred when $q$, the mean shear stress, was 46,000 lb. per sq.in.; $q$ then gradually increased with increasing twist to 68,000 lb. per sq.in. when fracture occurred. The amount of twist is expressed in fig. 1 by the non-dimensional quantity $ND/l$, where $N$ is the number of turns, $l$ the length of the specimen, and $D$ its diameter. It will be shown later that this is $\pi$ times the shear strain in the surface layers of the rod. The maximum value of $ND/l$ attained before fracture was 0·82.

In order to represent these two tests on the same diagram the theory of von Mises may be used. According to this theory if $T$ is the stress at which plastic flow begins in a rod subjected to direct tension and $q$ the shear stress for plastic flow under the action of a pure shear, $q = T/\sqrt{3}$. Similarly, in order to find the small increment in $ND/l$ which is equivalent, so far as work done on the specimen is concerned, to a small direct extension $(l - l_0)/l_0$ it is necessary to multiply* by

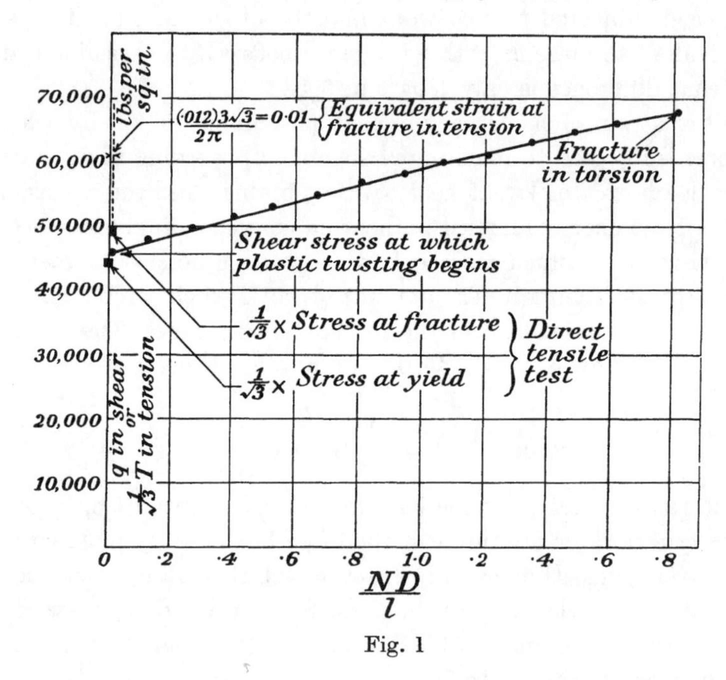


Fig. 1

$3\sqrt{3}/2\pi$. The results of the tensile test are represented in fig. 1 by points whose ordinates are $76{,}800/\sqrt{3} = 44{,}300$ lb. per sq.in. for the plastic yield point and $84{,}800/\sqrt{3} = 49{,}000$ lb. per sq.in. at fracture. The abscissa representing the strain at fracture is $(0{\cdot}012)\ 3\sqrt{3}/2\pi = 0{\cdot}01$.

It will be seen that the total work done on the specimen during the test, which is proportional to the area under the stress-strain curve in each case, is about 80 times as great for the twisted rod as it was for the rod broken in direct tension.

In the work to be described later cold work was done by twisting instead of direct extension. The maximum latent energy left in copper after severe twisting was equivalent to a rise in temperature of about 15° C., i.e. 18 times as much as that used in Farren and Taylor's experiments. The release of this amount of energy can be measured and in a later communication we hope to be able to describe the apparatus with which such measurements have been made.

* For $\rho\delta W = T(l - l_0)/l_0 = \dfrac{2\pi}{3}\, q\delta\,(ND/l)$, see equation (2), p. 316.

In the experiments of Farren and Taylor it was found that the energy left in a metal after distortion is a definite fraction of the work done on it. This fraction varied with the nature of the metal, but appeared to be constant for various amounts of distortion in spite of the fact that the resistance to extension rapidly increased as the extension increased. It seems unlikely that it would be possible to increase the amount of latent energy indefinitely by doing cold work; accordingly, one of our objectives in measuring the latent energy due to cold work in twisted bars was to find out whether it goes on increasing proportionately to the work done when the cold working is very severe. Work recently published by Rosenhain and Stott* shows that when copper or aluminium wire is drawn through a die a rather smaller proportion of energy is absorbed than was found by Farren and Taylor. In their apparatus the measured work done included the work done against friction between the wire and the die so that the proportion absorbed was necessarily smaller than that measured with apparatus in which all the work is expended in straining the material, but after making due allowance for the friction they still found that the proportion of energy absorbed was rather smaller than that found by Farren and Taylor. The present experiments confirm this result.

### Measurement of Latent Energy Produced by Cold Work

To measure the latent energy it is necessary to measure simultaneously the work done and the heat evolved, and in order to avoid loss of heat it is necessary to perform the whole experiment rapidly.

### Measurement of Work Done

To measure the work done during a rapid twisting of a bar, a self-recording machine was designed which produced diagrams in which the ordinates represent torque and the abscissae angle of twist between two sections of the twisted bar.

The specimens were round in section, $\frac{3}{8}$ in. diameter, and had square ends $\frac{3}{4} \times \frac{3}{4}$ in. Their length was 15 in., but the round part only occupied 11 in. of this. One square end fitted into the headstock of a lathe and could be twisted at any desired speed by a geared electric motor. The torque was applied at the other end by the lever arrangement and spring balance shown in fig. 2. This was designed to apply a pure couple in such a way that it would be recorded directly as a movement of the spring balance. The diagram is self-explanatory, except for the arrangement of the part where the torque is transmitted to the specimen. The arm, $ABC$, carries a ball-race at $B$ which is housed on the loose headstock of the lathe. It also carries two pins, $D$, $E$, which engage with a carrier fitting over the square end of the specimen. This arrangement permits the specimen to extend longitudinally† and ensures that no bending moments are applied. The torque was recorded by means

* *Proc. Roy. Soc.* A, CXL (1933), 9.

† When a bar is twisted plastically it usually grows in length.

of a steel tape (fig. 2), one end of which was attached to the torque arm, while the other passed round a recording drum which carried the paper on which the record was made.

The twist was recorded by means of two discs which were attached to the specimen at points 10 in. apart. These carried steel tapes, one of which drove a spindle and the other a nut. This nut carried a slide rest to which the recording pencil was attached so that the position of the pencil on the recording drum depended only on the relative rotations of the nut and spindle.

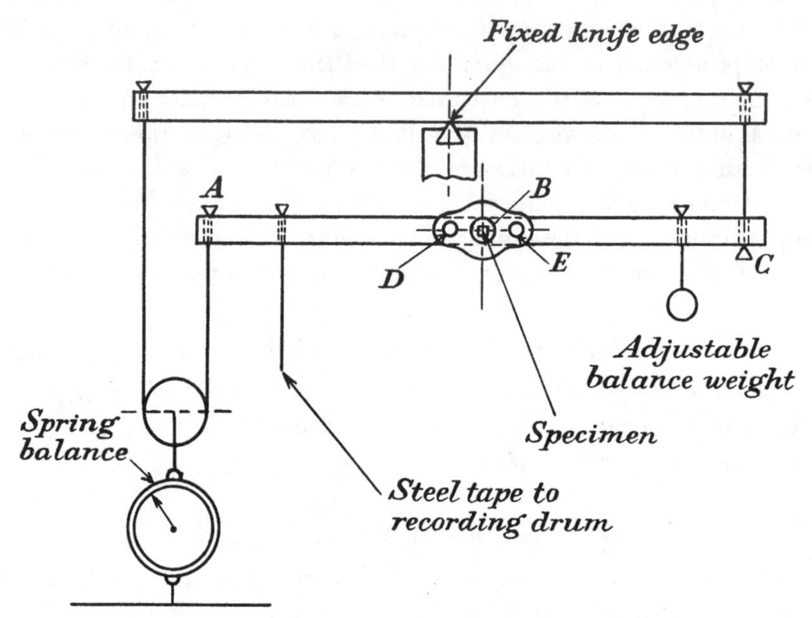


Fig. 2. Arrangement of torque recorder.

## Measurement of Heat Evolved

Two methods were used. The first was to measure the rise in the temperature of the surface of a specimen after rapidly twisting it through about one turn. The second was to remove it so rapidly from the lathe after finishing the twisting that the heat generated in the central part of the specimen had not penetrated through the square ends, and drop it into a calorimeter. The two methods proved to give results in good agreement with one another.

The first method is the same as that used by Farren and Taylor, but the fact that the specimen is now twisted instead of being pulled introduces a new difficulty. More cold work is done in the outer layers than near the middle of the specimen, so that if the experiment is done very rapidly the outer layers are heated more than the inner ones. A certain time must elapse before the temperature is equalized over the cross-section, but the reading must be made before the wave of cooling penetrates from the ends of the specimen to the central part where the temperature measurements are made. The temperature was measured by means of an iron-constantan thermocouple connected directly to a galvanometer.

The movement of a spot of light reflected from the galvanometer mirror was recorded photographically on a drum rotating at a uniform speed.

In the experiments of Farren and Taylor the thermocouple was inside the specimen and its leads passed down the middle of the hollow specimen so that the junction itself and its leads certainly took up the temperature of the specimen. In the case of solid specimens this method of ensuring good thermal contact is not possible. The method adopted in the present experiment was to solder the thermo-junction to the middle of a small square of sheet silver. This was tightly bound by silk threads to the outer surface of the specimen so that the junction itself was on the underside of the silver. The thermocouple wires were insulated from the silver and the specimen by enamel except at the point where they were soldered. This method ensured that the junction was not cooled by conduction along the thermo-couple wires.

In order to remove any further uncertainties that may exist in the use of a thermocouple applied to the outside of a specimen and to allow for cooling of the specimen during the time taken to twist it and for possible cooling of the thermo-couple by the silk threads which bound it to the specimen, the heat input corresponding with the observed temperature-time curve was determined independently after each experiment. For this purpose the temperature-time record and the torque-angle curve were taken for, say, one turn of the lathe head. The specimen and its thermocouple were then removed from the torque apparatus and after cooling a heavy current up to 100 amp. was passed through the specimen and maintained for the same length of time as that for which the torque had been applied. By adjusting this heating current it was found possible to reproduce exactly the temperature-time curve obtained in the torque test provided the twisting had not been carried out so quickly that the equalization of temperature between the outside and inside of the specimen could not take place. The current and the potential drop between the square ends of the specimen were then measured and the energy input calculated.

The cooling is due chiefly to conduction from the ends of the specimen. It was shown by one of the present writers that during a time equal to $T = 0{\cdot}014\rho\sigma l^2/\kappa$ after the generation of heat the cooling effect of the ends causes a drop in temperature in the middle of the specimen which is less than 0·6% of the total rise. In this expression $l$ is the length of the specimen which in the present case is 30 cm., for iron $\rho = 7{\cdot}8$, $\kappa$ the conductivity is 0·14 and $\sigma$ the specific heat is 0·106, so that $T = 80$ sec.

In the temperature-time record, fig. 3, $A$ represents the time at which the twisting began, $B$ represents the time it was complete. The time occupied by the operation, namely, 45 sec., is considerably less than the 80 sec. necessary for the cooling effect of the ends to penetrate to the middle, yet it is sufficiently long to ensure that the temperature is uniform across the section. It will be seen in fig. 3

that no measurable drop in temperature occurred till the time represented by $C$. The time interval between the beginning of the experiment and the first observable drop in temperature was found to be about 95 sec. This agrees well with the theoretical value $T=80$ sec.

For copper specimens, owing chiefly to their greater conductivity, $T$ was only 11·5 sec.; the twisting being carried out in 6 or 7 sec., leaving 4 or 5 sec. available for the temperature measurement, before the waves of cooling from the ends produced an appreciable effect in the middle.

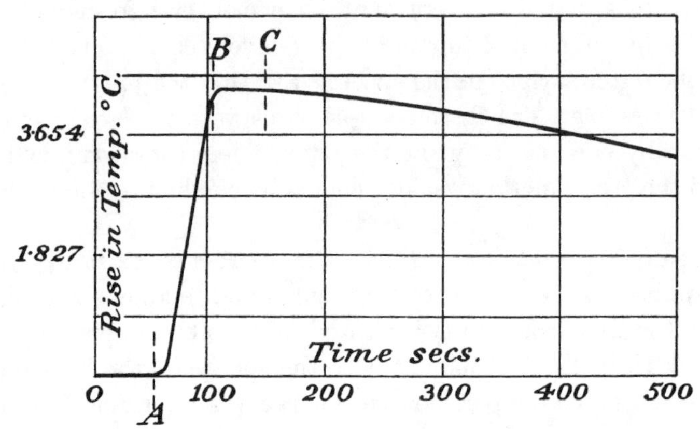


Fig. 3. Temperature-time record during plastic twisting.

### Expression of Results in Non-dimensional Form

The shear strain at any point in the twisted rod is $2\pi Nr/l$, where $N$ is the number of turns in length $l$. A non-dimensional expression for representing twist is therefore $ND/l$. If the shear stress depends only on the shear strain, it may be represented by an expression of the form $q=12G/\pi D^3$, where $D$ is the diameter of the specimen, $G$ is the applied torque and $q$ is the uniform shear stress which would give rise to the torque $G$. $q$ is approximately the average shear stress over the cross-section.

*Results.* The results of tests on an annealed mild steel bar and on a decarburized mild steel bar are given in tables 1 and 2. Each bar was twisted through successive small amounts, usually about 1 turn, and the value of $ND/l$ given in column 1 corresponds with the total strain from the initial annealed condition of the bar. The stress given in column 2 is expressed in lb. per sq.in. and is the mean value of $q$ between the beginning and end of one experiment; thus, referring to table 1, the mean value of $q$ during the fifth stage of twisting was 43,700 lb. per sq.ft. and during this stage the strain increased from $ND/l=0{\cdot}1126$ to $ND/l=0{\cdot}1452$. The work done during a twist of $\delta N$ turns is $2\pi G\delta N$, so that the work done on unit mass is

$$\delta W=8G\,\delta N/D^2 l\rho. \tag{1}$$

Expressed in non-dimensional form,

$$\delta W=\frac{2\pi q}{3\rho}\,\delta\left(\frac{ND}{l}\right), \tag{2}$$

where $\delta(ND/l)$ is the change in $ND/l$ during the experiment under consideration. If $q$ is expressed in lb. per sq.in. and $\delta W$ is expressed in calories per gram of metal,

$$\delta W = \frac{2\pi q}{3\rho J}\frac{(453\cdot 6)\,(981)}{(2\cdot 54)^2}\,\delta\left(\frac{ND}{l}\right). \qquad (3)$$

Table 1. *Mild steel annealed* in vacuo. *Test using thermojunction*

| 1 | 2 | 3 | 4 | 5 | 6 |
|---|---|---|---|---|---|
| $ND/L$ | $q$ (lb. per sq.in.) | $\delta W$ (cal./g.) | $\delta T$ (° C) | $\delta H$ (cal./g.) | $\frac{\delta W - \delta H}{\delta W}\times 100$ |
| 0·01655 | 14,730 | 0·106 | 0·975 | 0·1033 | 2·55 |
| 0·0455 | 26,200 | 0·331 | 2·76 | 0·292 | 11·76 |
| 0·0824 | 35,400 | 0·569 | 4·7 | 0·498 | 12·6 |
| 0·1126 | 39,250 | 0·515 | 4·3 | 0·456 | 11·4 |
| 0·1452 | 43,700 | 0·621 | 5·4 | 0·545 | 12·2 |
| 0·1768 | 44,250 | 0·606 | 4·98 | 0·528 | 12·85 |
| 0·2068 | 45,750 | 0·600 | 5·05 | 0·535 | 10·85 |
| 0·2447 | 47,400 | 0·781 | 6·60 | 0·700 | 10·37 |
| 0·2757 | 48,800 | 0·658 | 5·45 | 0·577 | 12·3 |
| 0·3052 | 49,100 | 0·632 | 5·25 | 0·556 | 12·0 |
| 0·3355 | 49,700 | 0·656 | 5·50 | 0·583 | 11·1 |
| 0·3660 | 50,800 | 0·676 | 5·52 | 0·615 | 9·02 |

$W = 6\cdot 75$ cal. per g. $H = 5\cdot 99$ cal. per g. $W - H = 0\cdot 76$ cal. per g.

Table 2. *Decarburized mild steel. Test using thermojunction*

| 1 | 2 | 3 | 4 | 5 | 6 |
|---|---|---|---|---|---|
| $ND/L$ | $q$ (lb. per sq.in.) | $\delta W$ (cal./g.) | $\delta T$ (°C.) | $\delta H$ (cal./g.) | $\frac{\delta W - \delta H}{\delta W}\times 100$ |
| 0·0429 | 16,650 | 0·311 | 2·9 | 0·3075 | — |
| 0·0808 | 26,050 | 0·430 | 3·61 | 0·383 | 13·5 |
| 0·1175 | 31,400 | 0·502 | 4·16 | 0·441 | 11·7 |
| 0·1537 | 33,850 | 0·534 | 4·42 | 0·469 | 12·1 |
| 0·1896 | 35,250 | 0·552 | 4·65 | 0·493 | 10·7 |
| 0·2255 | 36,800 | 0·576 | 5·04 | 0·534 | 7·3 |
| 0·2599 | 39,050 | 0·586 | 4·92 | 0·522 | 10·9 |
| 0·2940 | 39,700 | 0·590 | 4·90 | 0·520 | 11·9 |
| 0·3296 | 40,400 | 0·627 | 5·24 | 0·556 | 11·7 |
| 0·3655 | 41,500 | 0·650 | 5·46 | 0·579 | 12·45 |
| 0·4014 | 42,200 | 0·660 | 5·60 | 0·594 | 10·0 |
| 0·4369 | 42,200 | 0·654 | 5·70 | 0·605 | 7·5 |
| 0·4724 | 43,050 | 0·667 | 5·80 | 0·615 | 7·8 |
| 0·5084 | 43,400 | 0·681 | 5·98 | 0·634 | 6·9 |

$W = 7\cdot 92$ cal. per g. $H = 7\cdot 25$ cal. per g. $W - H = 0\cdot 67$ cal. per g.

For steel $\rho = 7\cdot 85$, and since $J = 4\cdot 18 \times 10^7$,

$$\delta W = 0\cdot 000436 q \delta(ND/l). \qquad (4)$$

Thus in the fifth stage of twisting of the mild steel specimens of table 1, $q = 43{,}700$ lb. per sq.in., $\delta(ND/l) = 0\cdot 1452 - 0\cdot 1126 = 0\cdot 0326$, so that $\delta W = 0\cdot 621$ cal. per g. The values of $\delta W$ calculated in this way are given in column 3.

The observed rise in temperature $\delta T$ is given in column 4 and the quantity of heat $\delta H$ necessary to raise the metal through $\delta T$° C., namely, $\sigma\delta T$, where $\sigma$ is the specific heat, is given in column 5. For steel $\sigma=0{\cdot}106$ so that the figures in column 5, table 1, are derived from those of column 4 by multiplying by 0·106.

The difference between the work done on the specimen and heat given out is $\delta W-\delta H$, so that the proportion of the work done on the metal which remains latent in it is $(\delta W-\delta H)/\delta W$. This is expressed as a percentage in column 6. It will be seen that during successive stages the proportion of heat remaining latent is very nearly constant and equal to 11 % of the work done on the steel. There is, however, some evidence of a slight falling off during the last stage of twisting, when the proportion of latent energy falls to 9 %. In this case the total work done $W$ is found by adding all the figures in column 3. Thus $W=6{\cdot}75$ g.cal. This is equivalent to a rise in temperature of 64° C. $H$, the total heat emitted, is found by adding the figures in column 5. $H$ is 5·99 cal., so that the total latent energy in the specimen at the end of the experiment is $W-H=0{\cdot}76$ cal. per g., which is equivalent to a rise in temperature of 7·1° C.

The corresponding results for decarburized mild steel (nearly pure iron) are given in table 2. It will be seen, referring to the last column of the table, that there is a very definite falling off in the proportion of energy which remains latent. From 12 % in the early stages it falls to about 7·5 % in the last stage of the test. The energy latent in the metal at the end of the experiment was 0·67 cal. per g. which is equivalent to a rise in temperature of 6·3° C.

## Heat Measurement with a Calorimeter

The method just described has two definite defects. First, it assumes that the specific heat of the specimen remains constant during the test and, secondly, there is always a certain element of uncertainty in measurements made with a thermocouple applied to the outside of a specimen. For these reasons it was decided to repeat the experiments using a calorimeter to measure $\delta H$ directly. For this purpose it was necessary to reduce the loss of heat from the ends of the specimen as much as possible. This was accomplished (*a*) by making the large square section ends long so that the heat which was generated in the small-diameter central part of the specimen would remain in it till the wave of temperature had penetrated down the square ends as far as the grips; (*b*) by inserting a heat-insulating material between the grips and the specimen; (*c*) by carrying out each stage of the twisting as rapidly as possible; (*d*) by redesigning the grips so that the specimen could be removed and dropped in the calorimeter within 1 sec. of the end of the twisting test.

By varying the time during which the specimen remained on the machine after twisting and before dropping it into the calorimeter, we were able to estimate the loss of heat and we found that when the experiment was carried out as rapidly as possible, the loss was considerably less than 1 % of the heat generated. This is so small that we did not attempt to use such corrections. The rise in temperature of the water in the calorimeter was measured by means of a Beckmann thermometer.

*Results.* Measurements using a calorimeter were carried out with bars made of pure copper, mild steel, and decarburized mild steel (i.e. nearly pure iron). The results are given in tables 3, 4 and 5. In these tables the columns 1, 2 and 3 give $ND/l$, $q$ and $\delta W$. Column 4 gives $\delta H$ which is now measured directly (in tables 1 and 2 it was found by multiplying the observed temperature rise $\delta T$ by a value for the specific heat of the metal taken from physical tables). Column 5 gives the percentage of the energy used during each stage of twisting which remains latent in the specimen. Column 6 gives the total work done on the specimen expressed in calories per gram of metal. Column 7 gives the total latent energy due to twisting, expressed in calories per gram, which remains in the metal at the end of each stage of the test.

Table 3. *Annealed pure copper. Test using calorimeter*

| 1<br>$ND/L$ | 2<br>$q$<br>(lb. per sq.in.) | 3<br>$\delta W$<br>(cal./g.) | 4<br>$\delta H$<br>(cal./g.) | 5<br>$\frac{\delta W-\delta H}{\delta W}\times 100$ | 6<br>$W$<br>(cal./g.) | 7<br>$W-H$<br>(cal./g.) |
|---|---|---|---|---|---|---|
| 0·0594 | 8,680 | 0·1974 | 0·1785 | 9·58 | 0·1974 | 0·0189 |
| 0·1230 | 14,320 | 0·3488 | 0·3235 | 7·25 | 0·5462 | 0·0439 |
| 0·1870 | 17,000 | 0·4168 | 0·3885 | 7·64 | 0·9630 | 0·0722 |
| 0·2670 | 19,550 | 0·5995 | 0·554 | 7·60 | 1·5625 | 0·1177 |
| 0·3598 | 21,690 | 0·7710 | 0·718 | 8·17 | 2·3335 | 0·1707 |
| 0·4195 | 23,250 | 0·5310 | 0·4805 | 9·52 | 2·8645 | 0·2212 |
| 0·4875 | 24,290 | 0·6328 | 0·5815 | 8·10 | 3·4973 | 0·2725 |
| 0·577 | 25,330 | 0·8690 | 0·791 | 8·58 | 4·3663 | 0·3505 |
| 0·642 | 26,380 | 0·6570 | 0·590 | 10·20 | 5·0233 | 0·4175 |
| 0·7035 | 26,900 | 0·6340 | 0·576 | 9·15 | 5·6573 | 0·4755 |
| 0·756 | 27,320 | 0·5495 | 0·490 | 8·58 | 6·2068 | 0·5350 |
| 0·817 | 27,750 | 0·648 | 0·587 | 9·42 | 6·8548 | 0·5960 |
| 0·876 | 28,200 | 0·637 | 0·584 | 8·32 | 7·4918 | 0·6490 |
| 0·972 | 28,800 | 1·058 | 0·945 | 10·67 | 8·5498 | 0·7620 |
| 1·067 | 29,160 | 1·062 | 0·965 | 9·08 | 9·6118 | 0·8590 |
| 1·153 | 30,200 | 0·995 | 0·911 | 8·45 | 10·6068 | 0·9430 |
| 1·253 | 30,350 | 1·162 | 1·066 | 8·25 | 11·7688 | 1·0390 |
| 1·278 | 30,550 | 0·2924 | 0·268 | 8·34 | 12·0612 | 1·0634 |
| 1·340 | 30,700 | 0·730 | 0·696 | 4·66 | 12·7912 | 1·0974 |
| 1·395 | 31,220 | 0·6578 | 0·634 | 4·19 | 13·4490 | 1·1210 |
| 1·426 | 31,400 | 0·3729 | 0·355 | 4·80 | 13·8219 | 1·1391 |
| 1·448 | 31,580 | 0·266 | 0·256 | 3·72 | 14·0879 | 1·1490 |

Table 4. *Annealed mild steel. Test using calorimeter*

| 1<br>$ND/L$ | 2<br>$q$<br>(lb. per sq.in.) | 3<br>$\delta W$<br>(cal./g.) | 4<br>$\delta H$<br>(cal./g.) | 5<br>$\frac{\delta W-\delta H}{\delta W}\times 100$ | 6<br>$W$<br>(cal./g.) | 7<br>$W-H$<br>(cal./g.) |
|---|---|---|---|---|---|---|
| 0·067 | 30,000 | 0·870 | 0·769 | 11·62 | 0·870 | 0·161 |
| 0·1353 | 39,200 | 1·158 | 1·070 | 8·23 | 2·028 | 0·189 |
| 0·201 | 43,800 | 1·240 | 1·117 | 9·92 | 3·268 | 0·312 |
| 0·267 | 46,650 | 1·336 | 1·209 | 9·51 | 4·604 | 0·439 |
| 0·3335 | 49,000 | 1·420 | 1·293 | 8·95 | 6·024 | 0·566 |
| 0·4322 | 51,700 | 2·20 | 2·020 | 8·19 | 8·224 | 0·746 |
| 0·499 | 52,580 | 1·529 | 1·398 | 8·57 | 9·753 | 0·877 |
| 0·566 | 53,800 | 1·558 | 1·427 | 8·42 | 11·311 | 1·008 |
| 0·635 | 54,680 | 1·605 | 1·507 | 6·10 | 12·916 | 1·106 |
| 0·702 | 55,500 | 1·620 | 1·519 | 6·72 | 14·536 | 1·267 |

In the previous work of Farren and Taylor, it was pointed out that the proportion $(\delta W-\delta H)/\delta W$ of the work done during cold working which remains latent in the metal is nearly constant over the range covered by these experiments, though the change in the strength of the material in the same range is very great.

It has been suggested that the latent energy due to cold work which can be retained in a metal may not increase indefinitely as the amount of cold work increases, but that a stage may ultimately be reached in which cold work can still be done on the metal, but no further latent heat can be absorbed. To show how far the present results support this view the values of $100(\delta W-\delta H)/\delta W$ given in tables 1–5 are plotted against $ND/l$ in figs. 4, 5 and 6. It will be seen from those diagrams that for annealed mild steel and decarburized mild steel the proportion of applied cold work which remains latent decreases as $ND/l$ increases. For decarburized mild steel (fig. 5), the absorption has decreased to 2·9 % of the applied work when $ND/l=0{\cdot}59$, and the curve seems to suggest that at the maximum observed value of $ND/l$, namely, 0·59, the metal has reached a state in which it is nearly saturated with latent energy. From table 2 it will be seen that the total latent energy is then 0·66 cal. per g.

Table 5. *Decarburized mild steel. Test using calorimeter*

| 1 | 2 | 3 | 4 | 5 | 6 | 7 |
|---|---|---|---|---|---|---|
| $ND/L$ | $q$ (lb. per sq.in.) | $\delta W$ (cal./g.) | $\delta H$ (cal./g.) | $\frac{\delta W-\delta H}{\delta W}\times 100$ | $W$ (cal./g.) | $W-H$ (cal./g.) |
| 0·0625 | 21,300 | 0·5790 | 0·5322 | 8·08 | 0·5790 | 0·0468 |
| 0·1469 | 32,600 | 1·1980 | 1·0890 | 8·10 | 1·7770 | 0·1558 |
| 0·2408 | 36,400 | 1·487 | 1·3480 | 9·35 | 3·264 | 0·295 |
| 0·3220 | 39,550 | 1·398 | 1·280 | 8·44 | 4·662 | 0·413 |
| 0·4092 | 41,650 | 1·580 | 1·4690 | 7·03 | 6·242 | 0·524 |
| 0·506 | 42,500 | 1·790 | 1·700 | 5·03 | 8·032 | 0·614 |
| 0·5900 | 43,400 | 1·586 | 1·540 | 2·90 | 9·618 | 0·660 |

For annealed mild steel (fig. 4) the diagram indicates that at the maximum observed value of $ND/l$, namely, 0·70, the material is not yet saturated with latent energy, though the proportion of the applied cold work which is absorbed and remains latent is only about half what it was in the initial annealed state. When $ND/l=0{\cdot}7$, $W-H=1{\cdot}27$ cal. per g. For copper, fig. 6 shows that the proportion of the applied cold work which remains latent is nearly constant up to $ND/l=1{\cdot}0$, and that after this stage of the twisting has been reached the absorption of latent energy rapidly decreases till the metal becomes saturated at about $ND/l=1{\cdot}6$. The maximum measured value of $W-H$ is 1·15 cal. per g. at $ND/l=1{\cdot}45$, and fig. 6 suggests that this may be nearly the maximum possible latent energy which the metal can retain at the temperature (15° C.) at which the measurements were made.

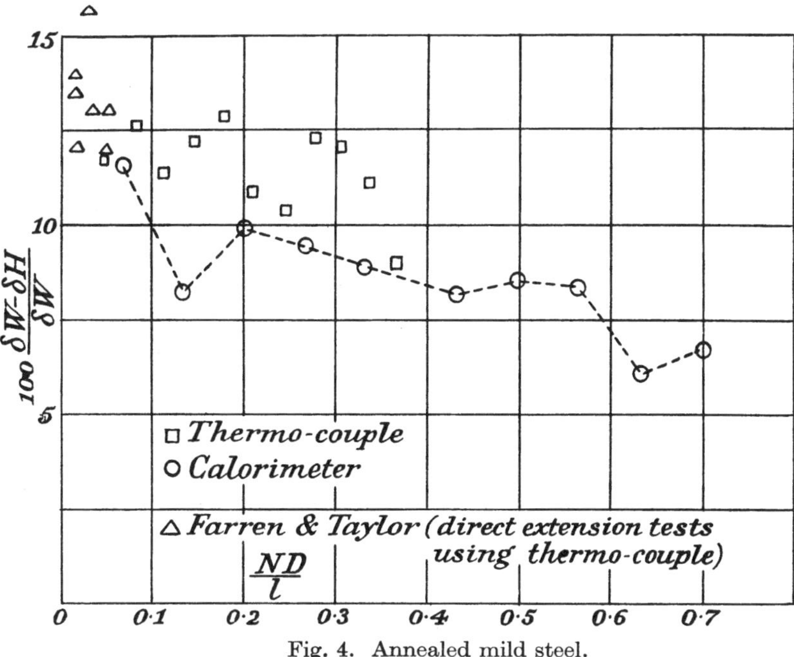


Fig. 4. Annealed mild steel.

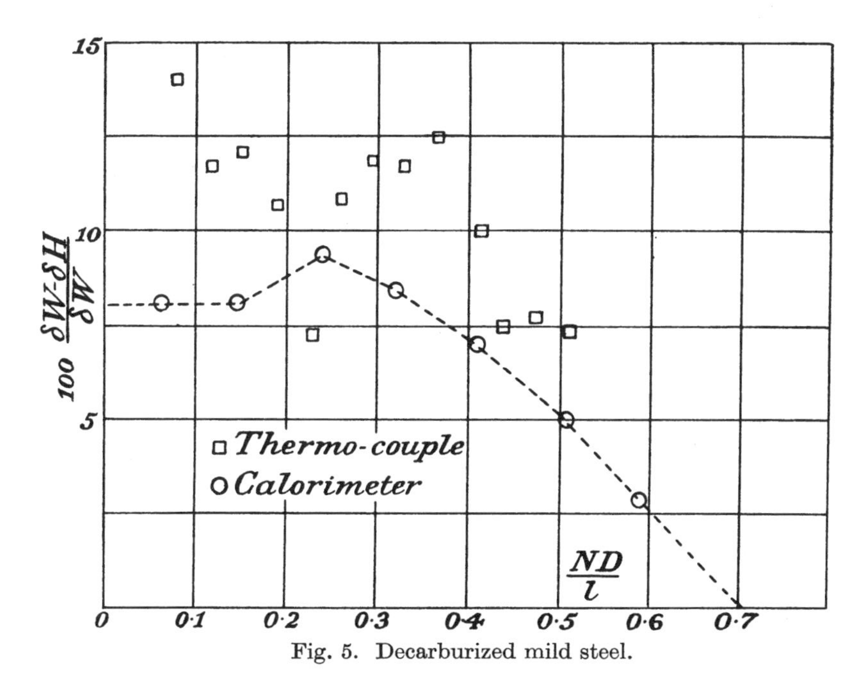


Fig. 5. Decarburized mild steel.

Figs. 4, 5. Percentage of work done which remains latent in specimen subjected to plastic twisting.

### Comparison with Measurements of Farren and Taylor

It is not possible to compare these results directly with those of Farren and Taylor because the type of distortion was different in the two experiments, but if it be assumed that the condition of the metal depends only on the amount of applied cold work retained latent in it irrespective of the distribution of the applied stresses, then a virtual value of $ND/l$ can be calculated* for which a twisted rod would have received the same amount of cold work as that given by the direct load in Farren and Taylor's experiments. The values of $100(\delta W - \delta H)/\delta W$ given by Farren and Taylor have been plotted in figs. 4 and 6 at the appropriate virtual values of

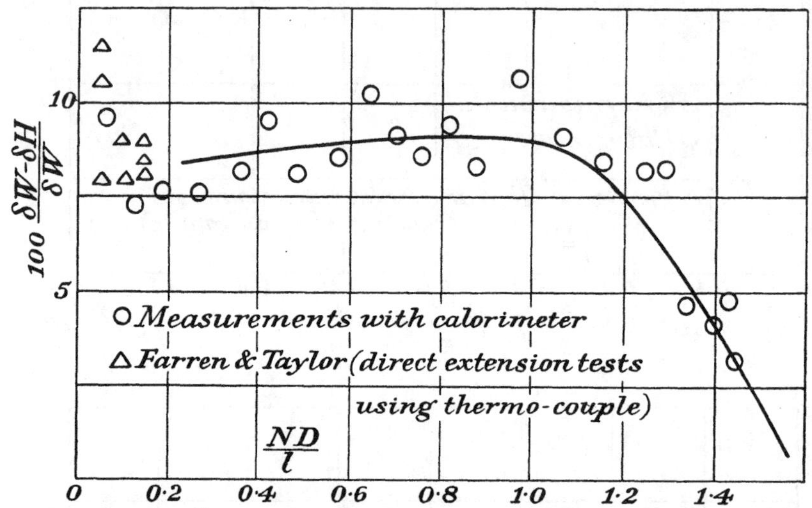


Fig. 6. Percentage of work done which remains latent in copper specimen subjected to plastic twisting.

$ND/l$. It will be seen that the agreement with the present results is good, but that a direct load is capable of giving to these metals only a very small fraction of the latent energy which they can contain.

### Connection between Strength and Latent Energy

It appears that both the strength and the latent energy of the material increase with increasing amounts of cold work. For both these, however, there seems to be a limit beyond which there is no further increase with further application of cold work. The question may naturally be asked, are these two limits identical? Does the attainment of maximum strength in a metal occur when the absorption of latent energy reaches its greatest possible value?

The values of $q$ given in columns 2 of tables 1–5 are mean values of the shear stress over the section. The value of the shear stress $q_s$ at the surface of the

* See p. 316 above.

specimen may be found from the measured values of $q$ by means of the following formula*

$$q_s = q + \frac{1}{3}\frac{ND}{l}\frac{dq}{d(ND/l)}. \tag{5}$$

The relationship between $q$ and $ND/l$ for annealed mild steel and decarburized mild steel is shown in figs. 7 and 8 and the values of $q_s$ found by applying (5) to

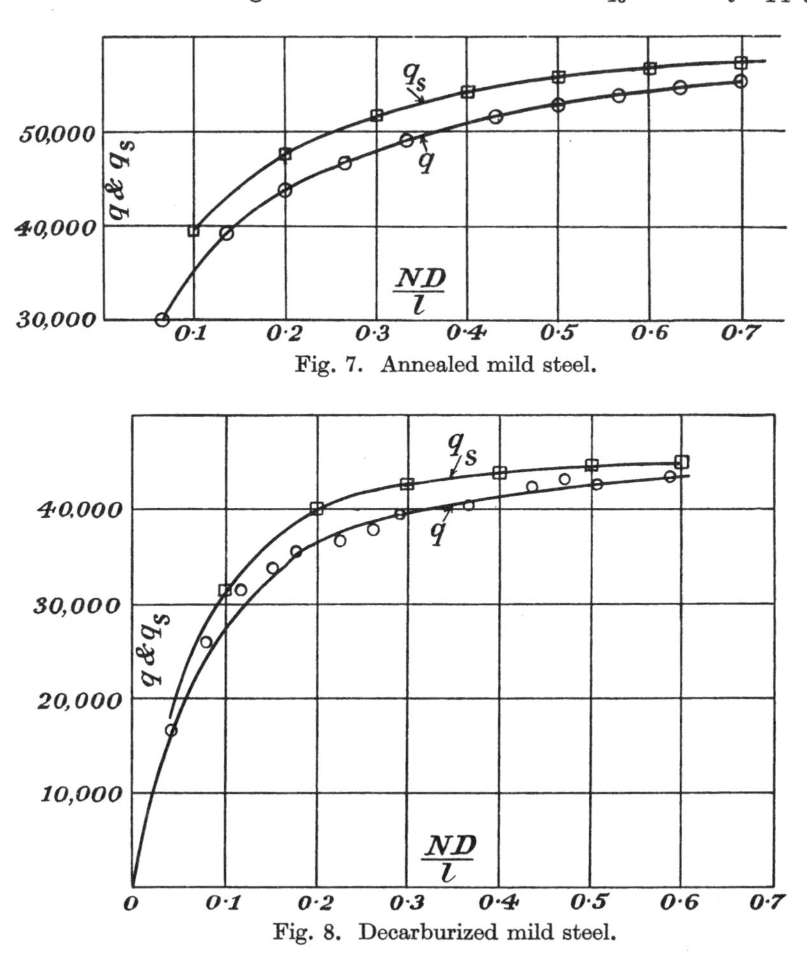


Fig. 7. Annealed mild steel.



Fig. 8. Decarburized mild steel.

Figs. 7, 8. Stress-strain curves during plastic twisting.

these curves are also given in the same figure. It will be seen that for decarburized mild steel (fig. 8), $q_s$ increases rapidly up to $ND/l = 0{\cdot}4$, but that at this point the increase practically ceases. Turning to fig. 5 it will be seen that the proportion of applied cold work which becomes latent in the metal does not appear to decrease till about $ND/l = 0{\cdot}4$, but that it decreases rapidly when $ND/l$ rises above 0·4. For annealed mild steel, fig. 7, $q_s$ increases up to $ND/l = 0{\cdot}7$ and it will be noticed in

* This relation is given by Nadai in a different form in his *Plasticity*, p. 128.

21-2

fig. 4 that $(\delta W - \delta H)/\delta W$ does not suffer any rapid decrease in this range. There is a gradual decrease, but saturation with latent energy is only reached outside the range of our experiments.

The values of $q$ given in table 3 for copper are the figures which must be used in calculating $q_s$, but they were obtained during the rapid twisting of the specimen. Copper is capable of withstanding a greater stress when the rate of deformation is large than when it is small. This effect seems to be considerably greater in twisted specimens than it is when the distortion is uniform. It is not possible, therefore, to deduce from table 3 the point in the test at which the strength of the material ceases to increase with increasing cold work. For this reason independent experiments were made under conditions ensuring uniform distortion to find out how much cold work must be done on copper before it attains its maximum strength.

### The Load-extension Curve for Pure Copper

The two most convenient methods for producing uniform distortion in soft metals are to extend a long bar of uniform section or to compress a short cylinder or disc between parallel plates, the ends being lubricated with grease. The conditions under which the distortion is uniform in the latter method were studied by one of the present writers* in connection with the distortion of single crystals of aluminium. It was then found that if the load was increased only slightly between successive stages of the experiment, uniform distortion was obtained if the specimen was greased before each application of the load. The effect of the friction between the flat faces of the specimen and the parallel steel plates was found to be inappreciable.

It has been pointed out that a uniform bar extended by a direct load necessarily breaks long before the material reaches its maximum strength. A compressed disc, however, can be subjected to far greater amounts of distortion than a bar under direct load or even one subjected to torsion. In carrying out our measurements, therefore, it was necessary first to compare the curve representing $T$ as a function of $\log (l/l_0)$ in an extension experiment with that representing $P$ as a function of $\log (h_0/h)$ in a compression experiment. Here $T$ and $P$ are the stresses (expressed in lb. per sq.in.), $l$ is the length and $l_0$ the initial length of the extended bar, $h$ is the thickness and $h_0$ the initial thickness of the compressed disc. If these curves are identical, it seems that the effect of the friction of the ends of the compressed disc on the steel plates is negligible and the $(P, \log h_0/h)$ curve truly represents the relationship between strength and the amount of distortion. The compression experiments can then be continued far beyond the stage at which the material reaches its maximum strength.

The results of such tests are shown in figs. 9 and 10. In the extension experiment the bar was loaded till the extension was 20 %. This corresponds with $\log l/l_0 = 0{\cdot}18$

* *Proc. Roy. Soc.* A, cxi (1926), 531; paper 9 above, p. 135.

and is represented by a dotted line in figs. 9 and 10. In fig. 9 the observations in the extension experiments are represented by crosses, while those in the compression experiments are shown as round dots. It will be seen that the compression and extension curves nearly coincide so that the compression results may be used with confidence outside the range in which they can be directly compared with those obtained by tensile loading.

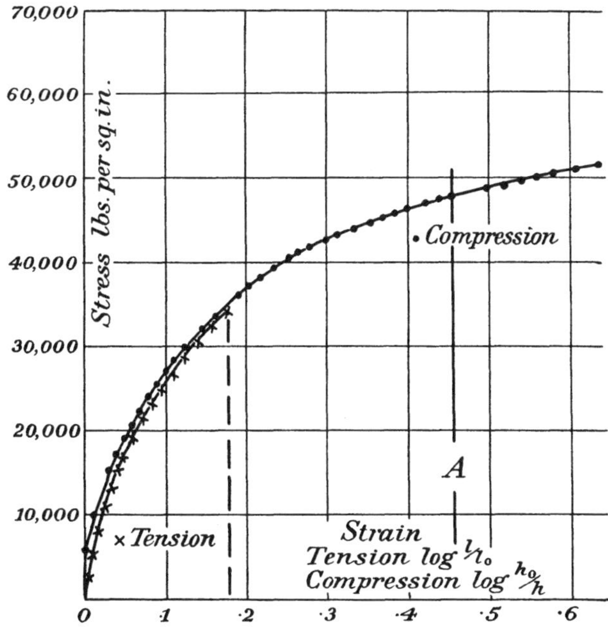


Fig. 9. Comparison between stress-strain curves for copper in direct extension and direct compression.

The complete stress-strain curve in compression is shown in fig. 10, but for clearness only a few of the points representing the observations are marked.

A short cylinder or disc of annealed copper 0·4770 in. high × 0·4390 in. diameter was first compressed in 31 stages till its thickness was 0·61 of its original thickness. The results of this test are shown by means of 31 dots in fig. 9 and 11 dots in fig. 10. The line $A$ at $\log h_0/h = 0{\cdot}46$ marks the end of this stage of the experiment.

At this stage the specimen was 0·3007 in. thick × 0·55 in. diameter, and since the effect of the friction on the ends increases as the ratio of the diameter to the thickness increases it was thought better to reduce the radius; accordingly the specimen was cut down to 0·2795 in., the thickness remaining unaltered.

The reduced specimen was then further compressed till its thickness was 0·1178 in. The corresponding value of $\log h_0/h$ was then 1·40. The end of this stage of the test is marked in fig. 10 by the line $B$. The specimen was then cut down to 0·1973 in. diameter and compressed by frequent small increments in load till its thickness was 0·0260. At the end of this third stage $\log h_0/h = 2{\cdot}91$. This is indicated by the line $C$ in fig. 10.

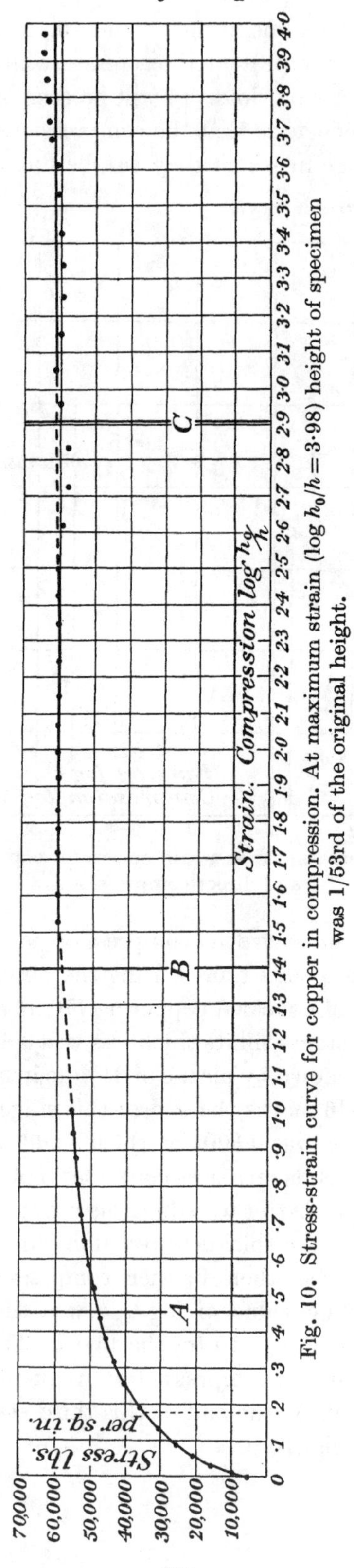


Fig. 10. Stress-strain curve for copper in compression. At maximum strain ($\log h_0/h = 3{\cdot}98$) height of specimen was 1/53rd of the original height.

The remainder of the points marked in fig. 10, namely those corresponding with strains from $\log (h_0/h) = 2{\cdot}91$ to $3{\cdot}98$, were obtained with another specimen cut from the same sample of copper.

Fig. 10 shows that the compressive stress rises with increase in strain till at

$$\log (h_0/h) = 1{\cdot}5, \text{ i.e. } h/h_0 = 0{\cdot}22,$$

the maximum value of 60,000 lb. per sq.in. is attained. No further increase was observed, although the specimen was compressed till at $\log (h_0/h) = 3{\cdot}98$ its thickness was only 1/53rd of its original thickness. It is worth noticing that 60,000 lb. per sq.in. is about equal to the tensile strength of hard-drawn copper wires.

## Cold Work necessary to Raise Strength to Maximum

The work done on unit volume of material during compression from thickness $h_0$ to thickness $h$ is

$$\int_h^{h_0} P\,d\,(\log h_0/h).$$

This can be found by means of a planimeter from the curve of fig. 10. During the course of the compression to $\log h_0/h = 1{\cdot}5$ the work done is found in this way to be $5{\cdot}78 \times 10^9$ ergs per c.c. Since the density of copper is 8·93 this is equivalent to 15·5 cal. per g. of copper.

## Comparison between Cold Work necessary to Saturate Metal with Latent Energy and that necessary to Give Maximum Strength

It has already been pointed out in connection with fig. 6 that the absorption of latent energy has nearly ceased at $ND/l = 1{\cdot}45$, and from table 3 it will be seen that at this stage $W = 14{\cdot}1$ cal. per g. The cold work necessary to saturate copper with latent energy at room temperature (about 15° C.) is therefore roughly the same as that necessary to raise the metal to its maximum strength, namely, 15·5 cal. per g.

Another way in which this question might be treated is to make use of von Mises's hypothesis concerning the criterion for plastic deformation and to assume also that the state of the material depends only on the amount of cold work done on it irrespective of whether it has been distorted by pure extension or by pure shear. Using these hypotheses $s$, the amount of shear equivalent to compression from thickness $h_0$ to thickness $h$ is $\sqrt{3} \log (h_0/h)$, so that the maximum strength of a twisted tube would be attained when $s = \sqrt{3}(1{\cdot}5) = 2{\cdot}6$. Since $s = \pi ND/l$ the maximum strength of the copper might be expected to be attained in the outer layers of a twisted copper rod when $ND/l = 2{\cdot}6/\pi = 0{\cdot}83$. Using Mohr's hypothesis instead of von Mises's, the result would have been $ND/l = 2(1{\cdot}5)/\pi = 0{\cdot}95$. It has been pointed out in discussing fig. 6 that the latent energy absorbed in twisting a copper bar bears an almost constant ratio to the applied cold work up to $ND/l = 1{\cdot}0$, and that at that point it suddenly begins to decrease. It seems significant that this decrease occurs at a stage of twisting which so nearly coincides with that at which the maximum strength is reached in the outer layers of the specimen.

In conclusion, we wish to express our thanks to Professor Inglis for allowing us to carry out this work in the Engineering Laboratory at Cambridge, and to Mr Parkes and Mr Jacobsohn for assistance in carrying out the work.

# 20

# FAULTS IN A MATERIAL WHICH YIELDS TO SHEAR STRESS WHILE RETAINING ITS VOLUME ELASTICITY

REPRINTED FROM

*Proceedings of the Royal Society*, A, vol. CXLV (1934), pp. 1–18

The theory of von Mises that plastic flow begins when a certain quadratic function of the principal stresses reaches a certain value has been found to hold with materials like copper, aluminium, pure iron, nickel and certain types of mild steel.* According to this theory the ratio

$$R = \frac{\text{shear yield stress in torsion}}{\text{tensile yield stress}}$$

should be $1/\sqrt{3}$ or $R = 0{\cdot}577$, and tests with tubes made of the material above mentioned do in fact give this result. The experiments of Cook† and others on steel tubes show that for some steels the observed value of $R$ is very close to 0·50, which is the value it would have according to Mohr's maximum stress difference hypothesis.

Two reasons have been suggested for this very marked discrepancy between the results of experiments with different materials. In the first place with soft metals it is sometimes difficult to observe any definite yield point at which elastic failure occurs so that the most reliable results are then obtained by extrapolation from measurements in which a considerable amount of plastic flow had already occurred. Von Mises's law might therefore be a law of plastic flow rather than of elastic breakdown. The steels used by Cook had a very definite and easily observed yield point both in tension and torsion so that so far as these experiments are concerned, Mohr's maximum stress difference hypothesis may be a general law governing elastic breakdown in complex stress distributions.

The second possible explanation is that the steels for which the ratio $R$ was found experimentally to be 0·50 were all steels which suffered a marked drop in stress at the yield point, while the materials for which $R$ was found to be nearly 0·577 did not possess this characteristic. It may therefore be true in general that the law of elastic breakdown is that of Mohr for materials for which the stress drops at the yield point and that of von Mises for materials which do not possess this property. One of the objects of this paper is to inquire whether there is any reason which would lead one to suspect that this may be so.

* Taylor and Quinney, *Phil. Trans.* A, CCXXX (1932), 323; paper 16 above.

† *Proc. Roy. Soc.* A, CXXXVII (1932), 559.

First consider the mechanism of breakdown in a material the stress of which does not drop at the yield point. The grains are initially self-stressed, the average stress being zero. When any stress is applied and gradually increased it might be expected that the greatest stress would occur in those grains which had initially an internal stress of the same type as the applied stress. These grains would be the first to yield and after yielding the stress in them would still increase with increasing average stress, but not so rapidly as that in the grains which were still in elastic strain. This process would rapidly bring about a condition in which the stress in each grain is the same as that of the whole mass so that the law of plastic yielding of each grain would be identical with that of the whole mass. It seems, therefore, that when von Mises's law of plastic yielding is observed in metals which have no drop in stress at the yield point we are really observing in the mass a condition which applies separately to each grain and is a property of the metal itself as distinct from its granular structure.

Next let us consider the mechanism of elastic breakdown in materials the stresses in which suddenly decrease at the yield point.* In this case, as in the previous one, the most stressed grains must be the first to yield, but when they yield the release of the stress which they were withstanding throws a heavier stress on the surrounding grains, some of which will yield in their turn. The material will yield as a whole when on the average each grain which yields throws such strain on the surrounding grains that at least one further grain yields without increase in mean stress. The effect of yielding of successive grains is therefore to make the stress distribution increasingly heterogeneous. This is the converse of the effect of yielding in a material which does not suffer a drop in stress at the yield point.

The condition of yielding of the material in bulk depends partly on the condition of yielding of the grains themselves and partly on the manner in which the yielding of one grain concentrates stress on its neighbours. From observations on the condition of yielding of material in bulk each grain of which has a drop in stress at the yield point, we can therefore make no direct deduction about the yielding of the individual grains. If we knew the shape of the grains and were able to carry out the necessary analysis we might be able to determine by means of the theory of elasticity the connection between the yielding of the grains and that of the material in bulk. Though this ideal is quite unattainable it seems possible to gain some insight into the problem by considering the effect on stress distribution in the surrounding metal of the yielding of grains of special forms for which the necessary calculations can be carried out.

In order to put the problem in a mathematically definable form it is necessary to make some simple assumption about the change which occurs in the stress inside a grain when it yields. The assumption which will be made is that at the instant a grain yields all shear stresses are released but the compressibility remains unaltered, so that any change in the volume of the grain which may result from its plastic deformation is accompanied by a corresponding change in the hydrostatic pressure

* This includes all metals for which the stress in each grain drops as it yields, even though no drop in stress can be observed in the material as a whole.

of the material which has deformed. Thus from the point of view of stress distribution the grain becomes virtually a liquid possessing the same compressibility as the elastic material surrounding it.

The yield point of the material as a whole will be assumed to correspond to the value of the average stress which is just capable of raising the stress, at the point of maximum stress concentration, to such a value that local yielding begins there. We thus distinguish between the conditions of yielding of each element of the material and that of the whole mass. It will be noticed that the model which has been chosen possesses the property that if the yield condition of each element is independent of the hydrostatic pressure the yield condition of the whole mass is also independent of hydrostatic pressure. This condition would obviously not have been satisfied if we had used as our model an elastic material containing an empty hollow to represent the region where a grain had yielded.

We now proceed to consider two special cases:

(1) The material which yields fills a sphere.

(2) It fills an elliptic cylinder.

In order to see the effect of granular structure on the yield condition we shall find the value in each case of the ratio

$$R=\frac{S}{P}=\frac{\text{shear yield stress in torsion}}{\text{tensile yield stress}},$$

making the alternative assumptions that the yield condition of each element of the mass is ($a$) that of von Mises and ($b$) the maximum stress difference condition of Mohr. If the yield condition of the granular mass is identical with that of each elementary grain we should find $R=1/\sqrt{3}=0{\cdot}577$ for ($a$) and $R=0{\cdot}500$ for ($b$).

## Yielding through a Spherical Volume

The equations for the elastic displacement due to a spherical hole in a material which is subject to a simple shearing strain have been given by Love.* If the system be referred to axes parallel to the principal axes of stress, the displacements due to an extension $\gamma$ per unit length parallel to $ox$ and a contraction $\gamma$ per unit length parallel to $oy$ are

$$\left.\begin{aligned}
u&=\gamma\left[x+\frac{2x}{9\lambda+14\mu}\left\{3(\lambda+\mu)\frac{a^5}{r^5}+5\mu\frac{a^3}{r^3}\right\}-15\frac{a^2-r^2}{r^7}\left(\frac{\lambda+\mu}{9\lambda+14\mu}\right)a^3x(x^2-y^2)\right],\\
v&=-\gamma\left[y+\frac{2y}{9\lambda+14\mu}\left\{3(\lambda+\mu)\frac{a^5}{r^5}+5\mu\frac{a^3}{r^3}\right\}+15\frac{a^2-r^2}{r^7}\left(\frac{\lambda+\mu}{9\lambda+14\mu}\right)a^3y(x^2-y^2),\right.\\
w&=-\gamma\left[15\frac{a^2-r^2}{r^7}\left(\frac{\lambda+\mu}{9\lambda+14\mu}\right)a^3z(x^2-y^2)\right],
\end{aligned}\right\}\quad(1)$$

where $a$ is the radius of the sphere $r^2=x^2+y^2+z^2$ and $\lambda$ and $\mu$ are the two elastic constants. The volume of the hollow remains unchanged by this strain so that if it

* *Mathematical Theory of Elasticity*, 4th ed. p. 252.

were filled with liquid of the same compressibility as the elastic material the pressure inside would remain zero At the surface of the sphere $r=a$ the strain components are given by

$$\left.\begin{aligned}
\frac{e_{xx}}{K\gamma}&=(\lambda+2\mu)\left(1-\frac{2x^2}{a^2}\right)+2(\lambda+\mu)\frac{x^2(x^2-y^2)}{a^4},\\
\frac{e_{yy}}{K\gamma}&=-(\lambda+2\mu)\left(1-\frac{2y^2}{a^2}\right)+2(\lambda+\mu)\frac{y^2(x^2-y^2)}{a^4},\\
\frac{e_{zz}}{K\gamma}&=2(\lambda+\mu)z^2\frac{(x^2-y^2)}{a^4},\\
\frac{e_{yz}}{K\gamma}&=2(\lambda+2\mu)\frac{yz}{a^2}+4(\lambda+\mu)yz\frac{(x^2-y^2)}{a^4},\\
\frac{e_{zx}}{K\gamma}&=-2(\lambda+2\mu)\frac{zx}{a^2}+4(\lambda+\mu)zx\frac{(x^2-y^2)}{a^4},\\
\frac{e_{xy}}{K\gamma}&=4\frac{xy(x^2-y^2)}{a^4}(\lambda+\mu),
\end{aligned}\right\}\tag{2}$$

where $$K=\frac{15}{9\lambda+14\mu}.$$

From the first three of these $$\frac{\Delta}{K\gamma}=-2\mu\left(\frac{x^2-y^2}{a^2}\right),$$

where $$\Delta=\frac{\partial u}{\partial x}+\frac{\partial v}{\partial y}+\frac{\partial w}{\partial z}.$$

We are now in a position to find the mean stress at which the yield point is first attained at some point in the field.

(*a*) Take first case (*a*) where the material yields when von Mises's function

$$F=(\sigma_1-\sigma_2)^2+(\sigma_2-\sigma_3)^2+(\sigma_3-\sigma_1)^2, \tag{3}$$

first attains a given value $F$. In this expression $\sigma_1, \sigma_2, \sigma_3$ are the principal stresses. Expressed in terms of the strains and elastic constants

$$F=4\mu^2[(e_{xx}-e_{yy})^2+(e_{yy}-e_{zz})^2+(e_{zz}-e_{xx})^2]+6\mu^2[e_{yz}^2+e_{zx}^2+e_{xy}^2]. \tag{4}$$

Consider the values of $F$ over the surface of the sphere. The form of the expressions for the strain components shows that $F$ has a maximum or minimum value at the end of three diameters parallel to the principal stresses. At the point $(a, o, o)$

$$e_{xx}=K\gamma\lambda, \quad e_{yy}=-K\gamma(\lambda+2\mu), \quad e_{zz}=e_{yz}=e_{zx}=e_{xy}=0.$$

Substituting these values in (4) the value of $F$ is

$$4\mu^2K^2\gamma^2(6\lambda^2+12\lambda\mu+8\mu^2).$$

At the point $(o, o, a)$

$$e_{xx}=K\gamma(\lambda+2\mu), \quad e_{yy}=-K\gamma(\lambda+2\mu), \quad e_{zz}=0,$$

so that $$F=4\mu^2K^2\gamma^2(6\lambda^2+24\lambda\mu+24\mu^2). \tag{5}$$

At the point $(o, a, o)$ $F$ is the same as at $(a, o, o)$. The greatest value of $F$, therefore, occurs at the point $(o, o, a)$ where the axis of $z$ cuts the sphere. The position of this point is shown graphically in the diagrams, figs. 1 (*a*) and (*b*).

To relate $\gamma$ with $S$ the shear stress in the material at great distances from the hole, notice that $\Delta = 0$ so that $S = 2\mu\gamma$. Hence from (5) the value of $S$ which raises the value of von Mises's function to the critical value $F$ at the point where it is greatest is given by

$$F = S^2 K^2 (6\lambda^2 + 24\lambda\mu + 24\mu^2). \tag{6}$$

When the material is subjected to a pure tensile stress $P$ parallel to the axis of $x$ the hole would increase in volume on applying the load but by adding a hydrostatic pressure equal to $\frac{1}{3}P$ the sum of the three principal stresses is zero. The applied stress is then the sum of two pure shear stresses. The strain at great distances from the hole is also the sum of two pure shearing strains parallel to the planes $xy$, $xz$ respectively and the volume of the strained hole is equal to that of the hole before straining.

If the three principal strains at great distances from the hole are

$$e_{xx} = 2\gamma, \quad e_{yy} = -\gamma, \quad e_{zz} = -\gamma,$$

the stress components at $(a, o, o)$ are

$$e_{xx} = 2\lambda K\gamma, \quad e_{yy} = e_{zz} = -K\gamma(\lambda + 2\mu), \quad e_{xy} = e_{zx} = e_{yz} = 0,$$

so that

$$F = 4\mu^2 K^2 \gamma^2 (18\lambda^2 + 24\lambda\mu + 8\mu^2).$$

At $(o, a, o)$

$$e_{xx} = 2K\gamma(\lambda + 2\mu), \quad e_{yy} = -K\gamma\lambda, \quad e_{zz} = -K\gamma(\lambda + 2\mu),$$

so that

$$F = 4\mu^2 K^2 \gamma^2 (18\lambda^2 + 60\lambda\mu + 56\mu^2). \tag{7}$$

The greatest value of $F$ occurs, therefore, on the equator, that is on the diametral plane perpendicular to the direction of greatest tension. The position of this equator is shown graphically in fig. 1 (*c*). From the elastic equation

$$X_x = 2P/3 = \lambda\Delta + 2\mu e_{xx} = 4\mu\gamma,$$

it will be seen that

$$P = 6\mu\gamma,$$

so that the tensile stress $P$ which will just cause the material to yield at the point of greatest stress concentration is given by

$$F = \tfrac{4}{36} P^2 K^2 \mu^2 (18\lambda^2 + 60\lambda\mu + 56\mu^2). \tag{8}$$

The shear yield stress may be compared with the tensile yield stress by dividing (8) by (6) thus

$$\frac{S^2}{P^2} = \frac{18\lambda^2 + 60\lambda\mu + 56\mu^2}{9(6\lambda^2 + 24\lambda\mu + 24\mu^2)}. \tag{9}$$

Remembering that Poisson's ratio $\sigma$ is equal to $\frac{1}{2}\lambda/(\lambda + \mu)$ it will be found that (9) gives $S/P = 0{\cdot}525$ when $\sigma = \frac{1}{4}$, $S/P = 0{\cdot}536$ when $\sigma = \frac{1}{3}$, and $S/P = 1/\sqrt{3}$ when $\sigma = \frac{1}{2}$. Comparing these with the value $S/P = 1/\sqrt{3}$ applicable to the homogeneous material it will be seen that if the yielding of each element of the material is determined by von Mises's law the yielding of the granular medium containing spherical holes will also be determined by von Mises's law provided $\sigma = \frac{1}{2}$. For any other value of $\sigma$ this is no longer true. For materials for which $\sigma = \frac{1}{4}$, for instance, the value 0·525 is nearer to 0·500 than to 0·577, so that the condition which determines yielding is nearer to that of Mohr than that of von Mises.

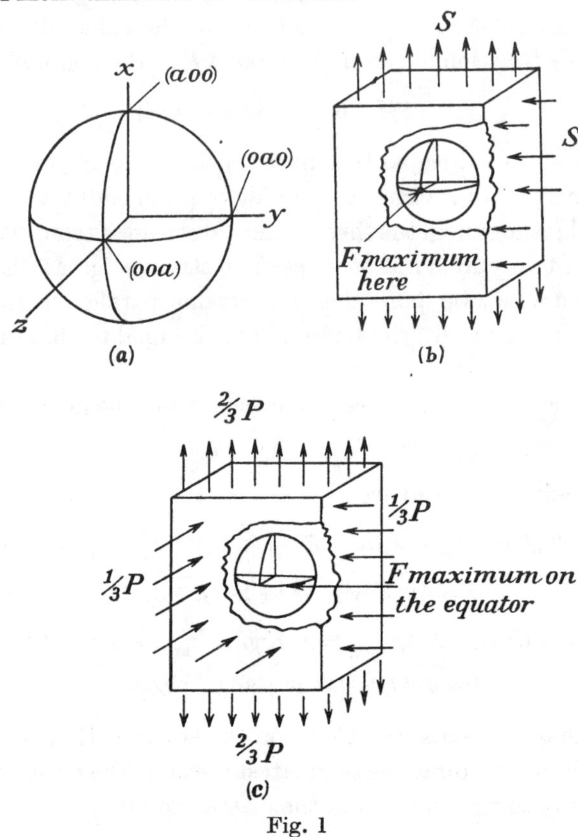


Fig. 1

(*b*) The maximum stress difference at the point of greatest stress concentration is found to be $2SK(\lambda + 2\mu)$ for a pure shear stress $S$, and $PK(\lambda + 2\mu)$ for a pure tensile stress $P$, so that if each element of the material is subject to Mohr's law that yielding begins when the maximum stress difference rises to a certain value, then $S/P = \frac{1}{2}$. Thus it appears that when each element obeys Mohr's law the granular mass also obeys Mohr's law.

### Grains in the Form of Elliptic Cylinders

*Empty elliptic cavity.* The distribution of stress round an elliptic cylindrical cavity in a material in a state of plane stress or plane strain was first found by Inglis.* It was afterwards rediscovered by Pöschl† who gave an expression for the stress function due to a hole whose major axis is at any given angle $\frac{1}{2}\pi + \alpha$ to the

* *Proc. Inst. Nav. Arch.* 14 March 1913.

† *Math. Z.* XI (1921), 89.

direction of a simple tensile stress $p$. Using elliptic co-ordinates derived from the transformation

so that
$$\left.\begin{aligned} z=x+iy&=c\cosh(\xi+i\eta)=\cosh\zeta,\\ x&=c\cosh\xi\cos\eta,\\ y&=c\sinh\xi\sin\eta. \end{aligned}\right\} \tag{10}$$

Pöschl's expression for the stress function is

$$\psi=\frac{pc^2}{8}\{\sinh 2\xi-\cos 2\alpha e^{-2(\xi-\xi_0)}-2(\cos 2\xi_0+\cos 2\alpha)\xi$$
$$+[\cosh 2(\xi-\xi_0)-1]e^{2\xi_0}\cos 2(\eta-\eta_0)\}, \tag{11}$$

where $\xi=\xi_0$ is the equation to the elliptic hole. From this Pöschl deduces the stress $\widehat{\eta\eta}$ round the surface of the hole. His expression* is

$$\widehat{\eta\eta}=p\frac{\sinh 2\xi_0-\cos 2\alpha+e^{2\xi_0}\cos 2(\eta-\alpha)}{\cosh 2\xi_0-\cos 2\eta}. \tag{12}$$

Consider now the stress distribution when the material is subjected to a simple shear so that at great distances from the hole $X_x=p$, $Y_y=-p$. At the surface of the hole

$$\widehat{\eta\eta}=2p\left[\frac{-\cos 2\alpha+e^{2\xi_0}\cos 2(\eta-\alpha)}{\cosh 2\xi_0-\cos 2\eta}\right]. \tag{13}$$

The stresses which must be applied to give the distributions represented by (12) and (13) are shown graphically in figs. 2 (*a*) and (*b*).

For any given value of $\alpha$ it is possible to find the maximum value of $\widehat{\eta\eta}$, but in order that our model may represent a granular structure which is isotropic we must imagine that it contains elliptic grains placed at every possible angle to the principal axes of stress. Thus to find the maximum possible value of $\widehat{\eta\eta}$ which can occur anywhere in the field we must find the maximum value of $\widehat{\eta\eta}$ when both $\eta$ and $\alpha$ are allowed to vary. The corresponding values of $\eta$ and $\alpha$ are then given by the simultaneous equations

$$\frac{\partial}{\partial\eta}(\widehat{\eta\eta})=0 \quad\text{and}\quad \frac{\partial}{\partial\alpha}(\widehat{\eta\eta})=0. \tag{14}$$

Differentiating (13) it will be found that these reduce to

and
$$\left.\begin{aligned} \tan 2\alpha[\cosh 2\xi_0\cos 2\eta-1]+\sinh 2\xi_0\sin 2\eta=0\\ \tan 2\alpha[1-e^{2\xi_0}\cos 2\eta]+e^{2\xi_0}\sin 2\eta=0. \end{aligned}\right\} \tag{15}$$

interpreted to mean that if the material is full of cracks the cracks will extend in a direction perpendicular to the tensile stress, i.e. at 45° to the direction of the shearing stress. The orientations of the holes for maximum stress are shown in fig. 2 (*c*).

*Yielding over elliptic area.* It has already been pointed out that the concentration of stress due to an empty hole cannot in general be similar to that resulting from the yielding of the material to shear stress while still retaining its elastic resistance to hydrostatic compression or expansion. In the particular case of a spherical hole in a material subject to shear stresses only, the volume of the hole is unaltered so that

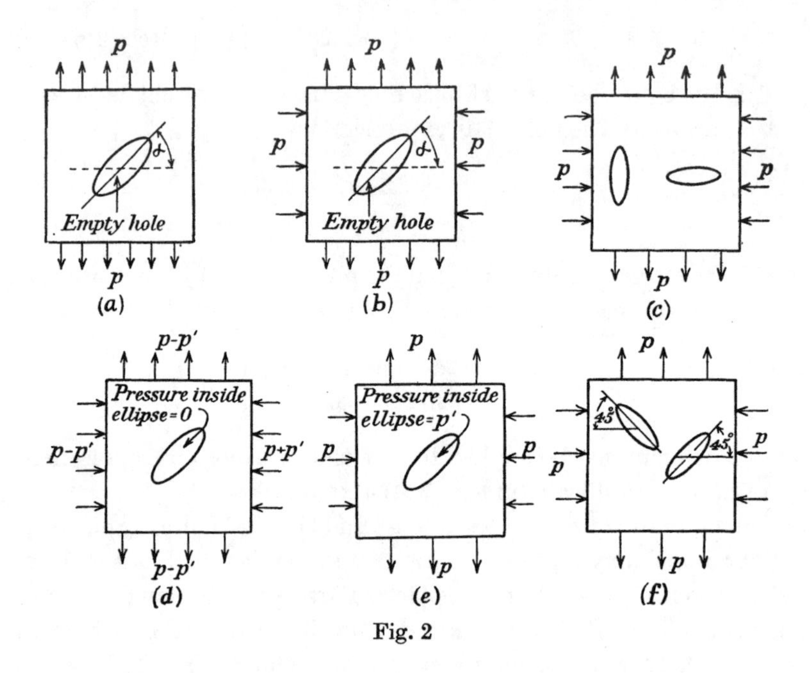


Fig. 2

the pressure of the material within it is zero and it is immaterial whether we imagine the hole to be filled or empty. This does not obtain for the elliptic hole so that Pöschl's expression (13) cannot be applied directly to find the stress concentrations due to local yielding. On the other hand we can calculate the change in the volume of the hole in Pöschl's case due to the applied stress, and can then apply a pressure uniform in all directions till the hole regains its original volume. The hole can then be regarded as being full of the material which has yielded and is still at zero pressure. Since further changes in hydrostatic pressure of the whole system do not then affect the stress differences, or shear stresses, we can calculate in this way as a problem in plane strain the distribution of shear stress round an elliptic grain which has yielded.

## Calculation of Change in Volume

We will first take Pöschl's case of a simple stress in one direction and calculate the change in volume of the elliptic hole. For this purpose it is necessary to find the displacements. A general method for doing this when the stress function is known is given by Love.* To apply this method we first find

$$\Delta = \frac{1}{2(\lambda+\mu)}\left(\frac{\partial^2\psi}{\partial x^2}+\frac{\partial^2\psi}{\partial y^2}\right).$$

The co-ordinates $\xi$, $\eta$ are related to $x$, $y$ by the relation

$$x+iy = c\cosh(\xi+i\eta) = c\cosh\zeta, \tag{16}$$

and the modulus of transformation is

$$h = \frac{2}{\sqrt{\{c^2(\cosh 2\xi - \cos 2\eta)\}}},$$

so that
$$\frac{\partial^2\psi}{\partial x^2}+\frac{\partial^2\psi}{\partial y^2} = h^2\left(\frac{\partial^2\psi}{\partial \xi^2}+\frac{\partial^2\psi}{\partial \eta^2}\right).$$

Applying this to Pöschl's expression (11)

$$\Delta = \frac{p}{2(\lambda+\mu)(\cosh 2\xi-\cos 2\eta)}[\sinh 2\xi(1+e^{2\xi_0}\cos 2\alpha)$$
$$+e^{2\xi_0}\sin 2\alpha\sin 2\eta + e^{2\xi_0}\cos 2\alpha\{-\cosh 2\xi+\cos 2(\eta-\alpha)\}]. \tag{17}$$

Next find $2i\overline{w}$ (where $\overline{w}$ is the rotation) from the equations for equilibrium and dilatation, namely,

$$(\lambda+2\mu)\frac{\partial\Delta}{\partial x} - 2\mu\frac{\partial\overline{w}}{\partial y} = 0, \quad (\lambda+2\mu)\frac{\partial\Delta}{\partial y} + 2\mu\frac{\partial\overline{w}}{\partial x} = 0,$$

and express $(\lambda+2\mu)\Delta + 2i\mu\overline{w}$ as a function of $\zeta$. It will be found

$$(\lambda+2\mu)\Delta + 2i\mu\overline{w} = \frac{p(\lambda+2\mu)}{2(\lambda+\mu)}[(1+e^{2\xi_0}\cos 2\alpha + ie^{2\xi_0}\sin 2\alpha)\coth\zeta - e^{2\xi_0}\cos 2\alpha]. \tag{18}$$

The function $\Xi + i\mathrm{H}$ is then defined by the equation

$$\Xi + i\mathrm{H} = \int\{(\lambda+2\mu)\Delta + 2i\mu\overline{w}\}\frac{dz}{d\zeta}d\zeta$$
$$= \frac{p(\lambda+2\mu)}{2(\lambda+\mu)}c\{(1+e^{2\xi_0}\cos 2\alpha + ie^{2\xi_0}\sin 2\alpha)\sinh\zeta - e^{2\xi_0}\cos 2\alpha\cosh\zeta\}. \tag{19}$$

The displacements $u$, $v$ parallel to the axes $x$ and $y$ are then given by

$$2\mu u = -\frac{\partial\psi}{\partial x} + \Xi, \quad 2\mu v = -\frac{\partial\psi}{\partial y} + \mathrm{H}.$$

* *Mathematical Theory of Elasticity*, 4th ed. p. 204.

The displacements $u_\xi, u_\eta$ parallel to $\xi$ and $\eta$ are

$$2\mu u_\xi = -h\frac{\partial\psi}{\partial\xi} + R\left[\frac{\Xi + i\mathbf{H}}{hc\sinh\zeta}\right]$$
$$2\mu u_\eta = -h\frac{\partial\psi}{\partial\eta} + I\left[\frac{\Xi + i\mathbf{H}}{hc\sinh\zeta}\right],$$

where $R$ and $iI$ represent the real and imaginary parts.

Taking the real part of $\dfrac{\Xi + i\mathbf{H}}{\sinh\zeta}$ and differentiating Pöschl's expression for $\psi$, it is found that

$$\left.\begin{aligned} u_\xi = &-\frac{pc^2h}{8\mu}[\cosh 2\xi + \cos 2\alpha\, e^{-2(\xi-\xi_0)} - (\cosh 2\xi_0 + \cos 2\alpha) \\ &\qquad + \sinh 2(\xi - \xi_0)\, e^{2\xi_0} \cos 2(\eta - \alpha)] \\ &\qquad + \frac{p(\lambda+2\mu)}{4\mu h(\lambda+\mu)}\left[1 + e^{2\xi_0}\cos 2\alpha - \frac{c^2h^2}{2} e^{2\xi_0}\cos 2\alpha \sinh 2\xi\right], \\ u_\eta = &-\frac{pc^2h}{8\mu}\cos 2\,[(\xi-\xi_0) - 1]\, e^{2\xi_0} \sin 2(\eta - \eta_0) \\ &\qquad + \frac{p(\lambda+2\mu)}{4\mu h(\lambda+\mu)}\left[e^{2\xi_0}\sin 2\alpha + \frac{c^2h^2}{2} e^{2\xi_0}\cos 2\alpha \sin 2\eta\right]. \end{aligned}\right\} \quad (20)$$

The change in volume of the hole is $\int_0^{2\pi} \frac{u_{\xi_0}}{h}\, d\eta$, where $u_{\xi_0}$ is the value of $u_\xi$ at $\xi = \xi_0$. From (20) it will be found that

$$\frac{1}{h} u_{\xi_0} = \frac{p(\lambda+2\mu)\, c^2}{8\mu(\lambda+\mu)}[\cos 2\alpha + \cosh 2\xi_0 - (1 + e^{2\xi_0}\cos 2\alpha)\cos 2\eta], \quad (21)$$

so that the change in volume due to a stress $p$ acting at angle $\frac{1}{2}\pi + \alpha$ to the major axis of the ellipse is

$$\frac{c^2p}{8\mu}\left(\frac{\lambda+2\mu}{\lambda+\mu}\right) 2\pi(\cos 2\alpha + \cosh 2\xi_0). \quad (22)$$

Replacing $\alpha$ by $\alpha - \frac{1}{2}\pi$ and $p$ by $-p$ in (22) the increase in volume of the hole due to a stress $-p$ acting at angle $\alpha$ is

$$\frac{c^2p\pi}{4\mu}\left(\frac{\lambda+2\mu}{\lambda+\mu}\right)(\cos 2\alpha - \cosh 2\xi_0), \quad (23)$$

and adding (22) and (23) it will be seen that the change in volume due to a pure shear stress formed by combining the two principal stresses $p$ at angle $\frac{1}{2}\pi + \alpha$ and $-p$ at angle $\alpha$ is

$$\frac{\pi c^2 p}{2\mu}\left(\frac{\lambda+2\mu}{\lambda+\mu}\right)\cos 2\alpha. \quad (24)$$

The change in volume due to uniform pressure $p'$ acting in all directions in the plane $xy$ can be found by subtracting (22) from (23). It is

$$-\frac{\pi c^2 p'}{2\mu}\left(\frac{\lambda+2\mu}{\lambda+\mu}\right)\cosh 2\xi_0. \quad (25)$$

In order that the change in volume due to the combination of the shear stress $(p, -p)$ with the uniform pressure $(p', p')$ may be zero it is therefore necessary that

$$p \cos 2\alpha - p' \cosh 2\xi_0 = 0,$$

so that

$$\frac{p'}{p} = \frac{\cos 2\alpha}{\cosh 2\xi_0}. \tag{26}$$

## Calculation of Stress Distribution due to Yielding inside the Ellipse

We are now in a position to calculate the stress distribution round an elliptic portion of material which has yielded in shear but not in uniform compression. The greatest stresses are likely to occur on the ellipse itself. Pöschl's expression for the stress $\widehat{\eta\eta}$ at the surface due to a single stress $p$ at angle $\frac{1}{2}\pi + \alpha$ is given in (12). For the shearing stress ($p$ at angle $\frac{1}{2}\pi + \alpha$, $-p$ at angle $\alpha$)

$$\widehat{\eta\eta} = 2p\left[\frac{-\cos 2\alpha + e^{2\xi_0}\cos 2(\eta - \alpha)}{\cosh 2\xi_0 - \cos 2\eta}\right]. \tag{27}$$

For a uniform tension $-p'$ applied equally in all directions in the plane

$$\widehat{\eta\eta} = -2p'\frac{\sinh 2\xi_0}{\cosh 2\xi_0 - \cos 2\eta}. \tag{28}$$

Hence for the combination which does not change the volume of the hole it is necessary to add (27) and (28) at the same time replacing $p'$ by

$$p\frac{\cos 2\alpha}{\cosh 2\xi_0}.$$

The resulting expression

$$\widehat{\eta\eta} = 2p\left[\frac{-\cos 2\alpha(1 + \tanh 2\xi_0) + e^{2\xi_0}\cos 2(\eta - \alpha)}{\cosh 2\xi_0 - \cos 2\eta}\right]. \tag{29}$$

The external stresses which give this value for the stress at the surface of the ellipse are shown graphically in fig. 2 (*d*).

When the externally applied stress is a pure shear stress $(p, -p)$ the pressure in the cavity is $-p'$ but the shear stresses are the same as in the case represented in fig. 2 (*d*). This condition is shown in fig. 2 (*e*).

To find the maximum possible value of $\widehat{\eta\eta}$ when $\alpha$ is allowed to assume any value we can proceed as for the empty elliptic hole, thus

$$\frac{\partial}{\partial \eta}\widehat{\eta\eta} = 0, \tag{30}$$

and

$$\frac{\partial}{\partial \alpha}\widehat{\eta\eta} = 0. \tag{31}$$

22-2

Differentiating (29) the equations for $\eta$ and $\alpha$ reduce to

$$\sin 2\alpha(\cosh 2\xi_0 \cos 2\eta - 1) = \cos 2\alpha \sin 2\eta \frac{\sinh^2 2\xi_0}{\cosh 2\xi_0}, \tag{32}$$

$$\sin 2\alpha(1 - \cosh 2\xi_0 \cos 2\eta) + \cosh 2\xi_0 \cos 2\alpha \sin 2\eta = 0. \tag{33}$$

The solution of these equations is

$$\left.\begin{aligned} \cos 2\alpha &= 0, \\ \cos 2\eta &= \frac{1}{\cosh 2\xi_0}. \end{aligned}\right\} \tag{34}$$

It appears, therefore, that for any given shape of ellipse, i.e. for any given value of $\xi_0$, the maximum value of $\widehat{\eta\eta}$ is attained when $\alpha = \frac{1}{4}\pi$, i.e. when the axes of the ellipse are at 45° to the directions of the principal stresses. Such faults are shown graphically in fig. 2 ($f$). The maximum value of $\widehat{\eta\eta}$ is then

$$\frac{4p}{1 - e^{-2\xi_0}},$$

or if the applied stress is expressed in terms of a shearing stress $S$ the maximum value of $\widehat{\eta\eta}$ is

$$\frac{4S}{1 - e^{-2\xi_0}}. \tag{35}$$

In an elongated ellipse $\xi_0$ is small and the maximum value of $\widehat{\eta\eta}$ is approximately $2S/\xi_0$. In this case it is of interest to see how the maximum value of $\widehat{\eta\eta}$ over the surface of an ellipse of given orientation $\alpha$ varies with $\alpha$. Using only the equation $\partial(\widehat{\eta\eta})/\partial\eta = 0$ it will be seen from (32) that when $\xi_0$ is small $\cos 2\eta = 1/\cosh 2\xi_0$ as before, and inserting this in the expression for $\widehat{\eta\eta}$ it is found that the maximum value of $\widehat{\eta\eta}$ is approximately $2S \sin 2\alpha/\xi_0$. Thus when the elongated ellipse lies with its axes parallel to the principal axes of stress so that $\alpha = 0$ then $\widehat{\eta\eta} = 0$ and there is no concentration of stress due to the yielding. Faults of this kind are shown in fig. 2 ($c$).

## Propagation of Faults and Luder's Lines

The result just obtained is a remarkable one. It brings out very clearly the profound difference which exists between the concentration of stress produced by an elongated empty hole and that produced by failure of shear stress in an elongated volume without change in resistance to compression or expansion. In the former case failure is propagated as a crack running perpendicular to the principal stress and in the latter as a fault running in a direction inclined at 45° to the directions of the principal stresses.

It seems that the hypothetical substance which forms the subject of the preceding analysis has a property which is possessed by all materials in which Luder's lines can be produced. Faults can propagate themselves at 45° to the principal axes of stress when the load is less than that necessary for yielding throughout the mass.

These faults can start from any hole or groove where there is an initial concentration of stress and the release of shear stress without release of compressive stress ensures that the fault will propagate itself at 45° to the direction of the principal stresses.

## YIELD CRITERION AS CONDITION UNDER WHICH FAULTS CAN BE PROPAGATED

It has been seen that the presence in a material of spherical volumes within which the shear stresses have disappeared causes the whole mass to yield according to a law which is nearer to Mohr's than von Mises's hypothesis, even though each element of the material obeys von Mises's law.

It will now be shown that the effect of replacing these spherical faults by faults occupying elongated elliptic cylinders is to accentuate this effect so that in the limit narrow faults will be propagated only when the maximum stress difference rises to a certain value irrespective of whether each element of the material itself obeys Mohr's or von Mises's law.

Consider a fault in the form of an elliptic cylinder $\xi = \xi_0$ with its axes at 45° to the principal stresses; a shear stress $S$ produces a maximum value of $\widehat{\eta\eta}$ equal to $4S/(1-e^{-2\xi_0})$, the other principal stresses in plane strain are

$$\widehat{\xi\xi} = 0 \quad \text{and} \quad \widehat{zz} = \frac{4\sigma S}{1-e^{2\xi_0}},$$

where $\sigma$ is Poisson's ratio. Hence von Mises's function is

$$F = \frac{32S^2}{(1-e^{-2\xi_0})^2}(1-\sigma+\sigma^2).$$

In the limit $\xi_0 \to 0$ this becomes

$$F = 8S^2\xi_0^{-2}(1-\sigma+\sigma^2). \tag{36}$$

We can obtain the distribution of stress round the ellipse when a simple stress $P$ is applied by starting with a shearing stress system $(\frac{1}{2}P, -\frac{1}{2}P, 0)$. With this system of applied stress the values of the three principal stresses at the point of maximum stress concentration are

$$\widehat{\eta\eta} = \frac{2P}{1-e^{-2\xi_0}}, \quad \widehat{zz} = \frac{2\sigma P}{1-e^{-2\xi_0}}, \quad \widehat{\xi\xi} = 0.$$

Next apply a uniform stress $\frac{1}{2}P$ in all directions, and finally apply an additional stress $-\frac{1}{2}P$ perpendicular to the plane of the shear stress. The stresses at the maximum concentration are then

$$\widehat{\xi\xi} = \tfrac{1}{2}P, \quad \widehat{\eta\eta} = \frac{2P}{1-e^{-2\xi_0}} + \tfrac{1}{2}P, \quad \widehat{zz} = \frac{2\sigma P}{1-e^{2\xi_0}},$$

and at some distance from the crack they are

$$X_x = P, \quad Y_y = 0, \quad Z_z = 0.$$

Von Mises's function is then

$$F=4P^2\left[\left(\frac{1-\sigma}{1-e^{-2\xi_0}}+\frac{1}{4}\right)^2+\frac{1}{(1-e^{-2\xi_0})^2}+\left(\frac{\sigma}{1-e^{-2\xi_0}}-\frac{1}{4}\right)^2\right].$$

In the limit when $\xi_0 \to 0$ this is

$$F=\frac{8P^2}{4\xi_0^2}(1-\sigma+\sigma^2). \tag{37}$$

Comparing (36) and (37) it will be seen that when $\xi_0 \to 0$, i.e. when the ellipses are very elongated

$$S^2/P^2=\tfrac{1}{4} \quad \text{or} \quad S/P=\tfrac{1}{2}.$$

Thus when each element of the material fails according to von Mises's law the material as a whole fails according to the maximum difference law of Mohr. It is obvious that if each element of the material fails according to Mohr's law the material as a whole does so also. It seems that when failure is by propagation of narrow faults the law of failure of the granular mass will be that of Mohr whatever the law of failure of individual grains may be.

Perhaps this result might have been anticipated without analysis, for the kind of fault which can give rise to large concentrations of stress is essentially two dimensional in character. The greatest stresses occur near its edge and the large accompanying strains are due to displacements in the plane perpendicular to that edge. The stress component parallel to the edge is only increased in proportion to the stresses in the plane perpendicular to the edge by an amount proportional to Poisson's ratio. There is no concentration of a stress externally applied to the direction parallel to the edge. The resulting concentration in the value of von Mises's function depends therefore almost entirely on the components of stress in the plane perpendicular to the edge of the fault, thus the intermediate principal stress in the granular mass as a whole does not affect the maximum value of von Mises's function at the point of greatest stress concentration in a fault.

## Summary and Conclusions

Plastic substances may be divided into two classes according to whether the stress increases or decreases after yielding. In the former the stress in the grains which have yielded increases more slowly with increasing strain than in those which are still elastic. The effect of successive yielding of grains is therefore to reduce the internal stresses to a state of uniformity. The yield condition of the whole granular structure is therefore identical with that of each grain. In materials which lose their power to withstand shear stresses when they yield, the yielding of successive grains might be expected to produce an increasingly heterogeneous distribution of internal stresses.

To trace the effect of this heterogeneity on the yield properties of the whole granular mass a hypothetical material is imagined which has the property that when any portion yields all shear stresses vanish but the compressibility remains unaltered.

The distributions of stress in the neighbourhood of grains which have yielded when their boundaries are spheres and elliptic cylinders are discussed. It is shown that when each element of the material fails according to the law of von Mises the effect of spherical faults is to make the whole mass fail according to a law which is intermediate between that of von Mises and Mohr. Narrow elliptic faults on the other hand cause the whole mass to fail according to Mohr's law of maximum stress difference even though each element fails according to von Mises's law. When the law of failure of each element is that of Mohr the law of failure of the whole mass is also that of Mohr.

The hypothetical material has the property that the greatest stress concentrations occur when elongated faults lie at 45° to the directions of the principal stresses, so that faults once started would propagate themselves along lines at 45° to the principal stresses. This property seems to have its counterpart in real materials which can exhibit Luder's lines. The fact that the greatest concentration of stress occurs when a fault lies at 45° to the direction of the principal stresses is in striking contrast to the case of a crack or fault in which compressive or tensile stress is released as well as shear stress. In that case the greatest concentration of stresses occurs when the fault is perpendicular to the direction of greatest tension so that faults of that type would be propagated in the direction of one of the principal stresses.

# 21

# THE MECHANISM OF PLASTIC DEFORMATION OF CRYSTALS

## PART I. THEORETICAL

REPRINTED FROM

*Proceedings of the Royal Society*, A, vol. CXLV (1934), pp. 362–87

The fact that the macroscopic distortion of metallic crystals is a shear parallel to a crystal plane and in a crystal direction and the fact that this remains true even when the distortion is large shows that the plastic strain must be chiefly due to the sliding of one plane of atoms over its immediate neighbour in such a way that the perfect crystal structure is re-formed after each atomic jump. It is supposed that slipping occurs over limited lengths $L$ of the slip-plane, and it is shown that this type of plastic strain necessarily gives rise to elastic stresses near the two dislocations which occur at the two ends of each of these lengths $L$.

It is then shown that the assumption that such dislocations will migrate through the crystal, owing perhaps to temperature agitation, under the influence of even the smallest shear stress leads to a definite picture of the mechanics of plastic distortion. This theory of strain hardening is expressible in quantitative form and gives a parabolic relationship between stress and plastic strain, namely, $S/\mu\sqrt{s}=\kappa\sqrt{(\lambda/L)}$. This expression is in good agreement with the results of experiment with metals which crystallize in the cubic system.

The observed parabolic relationship is then used in connection with the formula to determine $L$ which is found, at room temperatures, to be of order of magnitude $10^{-4}$ cm. This is of the same order of magnitude as the observed spacings of faults in metals and rock salt. According to this theory the part played by the system of faulting or mosaic structure is to limit the free motion of centres of dislocation. The actual strain takes place inside the 'blocks' of the mosaic structure and the crystallographic nature of the faults, i.e. whether they are boundaries of dendrites, a super-structure or merely 'pores' is immaterial from the point of view of the theory.

Experiments on the plastic deformation of single crystals, of metals and of rock salt have given results which differ in detail but possess certain common characteristics.

In general the deformation of a single crystal in tension or compression consists of a shear strain in which sheets of the crystal parallel to a crystal plane slip over one another, the direction of motion being some simple crystallographic axis. The measure of this strain, which will be represented by $s$, is the ratio of the relative lateral movement of two parallel planes of slip to the distance between them. Thus it is defined in the same way as the shear strain considered in the theory of elasticity.

The resistance to shear, which will be denoted by $S$, is defined as the component of shear stress in the direction of slip which must act parallel to the slip-plane in order that plastic deformation may occur.

It has been found that when the results of tests on single crystals of a metal are analysed the stress-strain curve which represents $S$ as a function of $s$ is independent of the stress normal to the slip-plane and of the components of shear stress perpendicular to the direction of slip. Thus the $(S, s)$ curve is a unique curve which defines the strength of the single crystal at any stage of distortion.

When the $(S, s)$ curves for single crystals of different metals are compared they are found to differ considerably in detail, but they all possess one general characteristic—a very small stress will produce a small plastic deformation, but as the

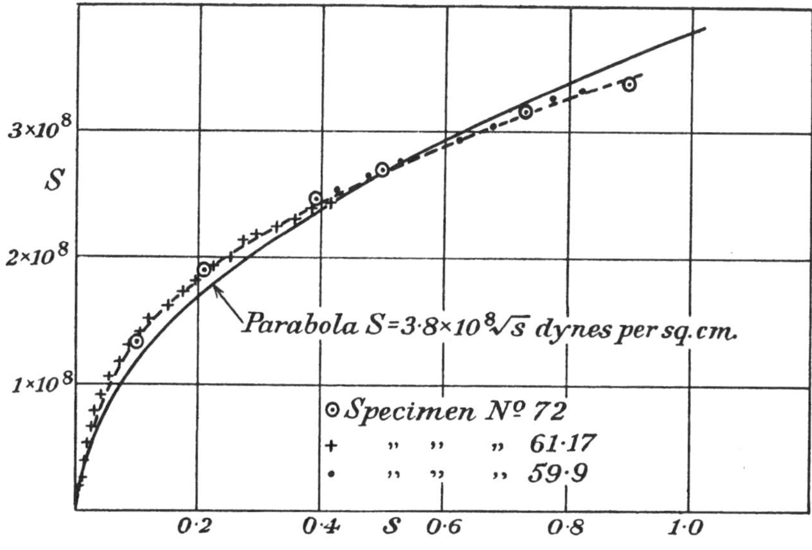


Fig. 1. Aluminium crystals.

deformation increases the stress necessary to increase it also increases. With some crystals* it has been found difficult to assign a definite stress at which plastic distortion begins; with others, such as rock salt and zinc, experimenters have found such a limit, but it is very small compared with the strength ultimately attained by the material. In some such crystals it has been found that the observed lower limit of strength depends very much on the degree of purity of the material, thus Schmid,† working with single crystals of zinc, found that a decrease in the total amount of metallic impurity from 0·03 to 0·002 % causes the stress at which the plastic deformation begins to decrease from 94 to 49 g. per sq.mm.

Except for these differences in the early stages of distortion the $(S, s)$ curves for many metallic single crystals are very similar. Some of them are shown in figs. 1 2 and 3. Fig. 1 refers to aluminium. The data from which this curve has been constructed were given in a previous paper.‡ In fig. 1, however, the unit of stress has

* E.g. aluminium.

† *Z. Phys.* LXXV (1932), 538.

‡ Taylor, *Proc. Roy. Soc.* A, CXVI (1927), 51; paper 13 above, p. 215.

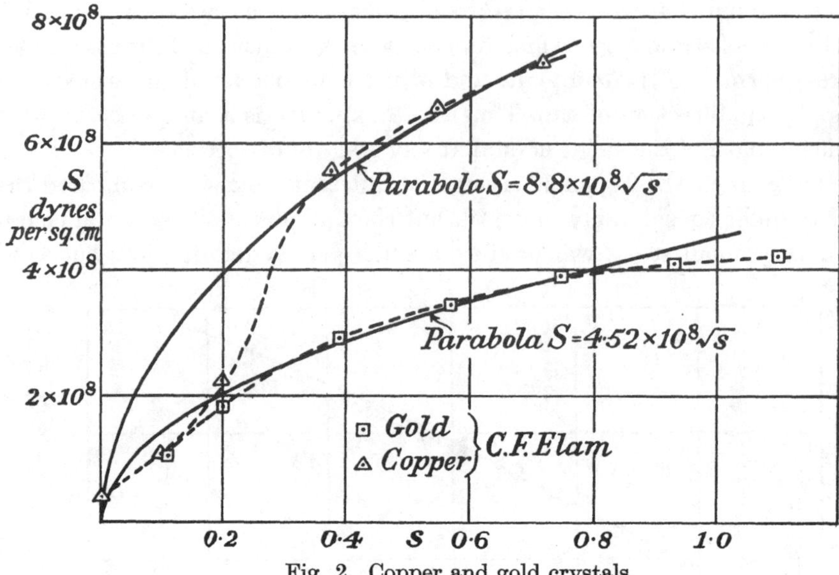

Fig. 2. Copper and gold crystals.

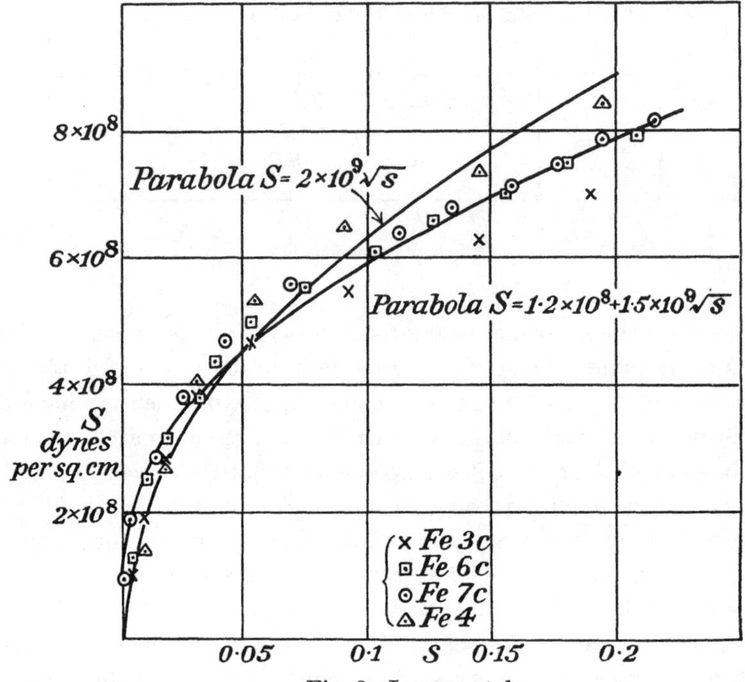

Fig. 3. Iron crystals.

been changed from lbs. per square inch to dynes per square centimetre. Fig. 2 shows the result of tests on single crystals of copper and gold.* These tests were made by Dr Elam and published in the form of curves giving $S$ for various extensions of the

* Elam, *Proc. Roy. Soc.* A, CXII (1926), 289.

specimens. The additional data necessary for calculating $s$ were kindly furnished by Dr Elam, and the results set forth in fig. 2 were calculated by means of the formulae previously given.*

Fig. 3 shows the results of tests on four single crystals of iron. The crystals in question were compressed and the geometry of their distortion analysed and published in a previous paper.† The stress data required for the $(S, s)$ curve had been recorded, but were not included in the published results.

It will be seen that for iron and aluminium at any rate the curves resemble the parabolas.

Such parabolas have been drawn in figs. 1, 2 and 3 in order that they may be compared with the results of observation. For copper (fig. 2) the first three observed points do not lie on the parabola, while for iron (fig. 3), it seems that the experiments indicate a finite value (about equal $1{\cdot}2 \times 10^8$ dynes per sq.cm.) for the stress at which plastic distortion first occurs. The parabolas $S = 1{\cdot}2 \times 10^8 + 1{\cdot}5 + 10^9 \sqrt{s}$ and $S = 2{\cdot}0 \times 10^9 \sqrt{s}$ have been marked on the figure for comparison with the measured stresses.

## Theories of Strain Hardening

Several attempts have been made to explain why plastic deformation increases the strength of metals (and rock salt), but they have mostly been of a qualitative nature. Such explanations are principally of three types. One type uses the observed fact that, when a single crystal breaks down under load, small portions of it are rotated into an orientation differing from that of the main body of the crystal. It is supposed that the perfect crystal can slip under the action of a very small stress, but that the portions of the crystal which are rotated relatively to the main body act as locking keys and hold up the slipping of the surrounding crystal, enabling the whole system to withstand a much greater load than the perfect crystal can support.

When the crystal contains *only* one set of planes of easy glide parallel to one crystallographic plane this theory might be capable, after some modification, of explaining a very rapid strengthening with increasing distortion. Where, however, the portions inside and outside a surface of misfit are supposed to be perfect crystals, but with crystallographic axes oriented in two different directions, easy gliding in the outer crystal would produce a definite strain (i.e. alteration of shape) in the surface of misfit while easy gliding in the inner crystal would produce a different strain. Thus the only types of strain which are possible as a result of easy gliding in both the inner and the outer crystal are geometrically inconsistent at the boundary between them. If no other type of plastic strain is possible, then the inner crystal completely locks the portion of the outer crystal within which the planes of easy glide cut any portion of the inner crystal.

* Paper 13 above, equations (2), (4), (5) and (6).

† Taylor, *Proc. Roy. Soc.* A, CXII (1926), 338; paper 10 above, p. 154.

When the crystals are capable of slipping equally easily on several crystallographically similar planes, as, for instance, when the crystals have cubic symmetry, this explanation of hardening due to cold work is no longer applicable, because, by an appropriate combination of easy glides, the inner crystal can be given exactly the same strain as the outer one. Under these conditions, therefore, there would be no geometrical inconsistency at the surface between them.

It is true that the stress necessary to cause any given kind of strain is likely to vary slightly according to whether the crystallographic axes are arranged in the most favourable or the most unfavourable orientation for the particular type of strain considered, but this effect is necessarily very small compared with the very large effect which the theory was designed to explain.

We may now turn to a second type of explanation. A perfect crystal is again supposed to be capable of withstanding only a very small stress, but every real crystal contains surfaces of misfit which form a mosaic structure or, as some crystallographers have described them, a system of lineages. These surfaces of misfit are supposed to hold up slipping to an increasing extent as the amount of distortion proceeds. This explanation, though at first sight promising, entails very great difficulty as soon as it is examined in detail.

The idea that the boundary can hold up slip in the interior of a perfect portion of the crystal might be interpreted in two ways. In one the surfaces of misfit may be regarded as being capable by themselves of carrying the whole applied stress, the intervening portions of perfect crystal structure being in a stress-free state. The strength would then be like that of a honeycomb, depending entirely on the boundaries which receive no assistance from the contents of each cell. Even if it were possible to explain the existence of any given strength on this principle it would be extremely difficult to explain why this increases with the amount of plastic strain.

In the other interpretation any thin sheet bounded by two parallel slip-planes must move as a whole. Each perfect portion of the crystal can support any stress system for which there is no component of shear stress parallel to the slip-planes, and the varying orientations of neighbouring portions which are crystallographically perfect might enable the whole system to support a large total stress in spite of the fact that each portion was excessively weak to one type of shear stress. This interpretation would make theory identical with the first type already discussed. It breaks down if the crystal has several crystallographically similar possible slip-planes.

The third type of explanation is the inverse of the first two. A perfect crystal is supposed to be capable of withstanding a very large stress. The observed weakness of metal crystals is attributed to concentrations of stress due to internal surfaces of misfits or cracks, and the increasing strength with increasing plastic strain is attributed to an increase in the number of faults or cracks.* As the number of such faults increases the ratio of maximum stress in a region of stress concentration to

* Taylor, *Trans. Faraday Soc.* xxiv (1928), 121; paper 15 above.

mean stress in the material would be expected to decrease. The mean stress necessary to cause a given maximum stress in a region of stress concentration must, therefore, increase as the number of faults per unit volume of the material increases.

In the present paper an attempt will be made to present a theory of this kind in a quantitative form. It will be found that the idea that a system of faulting exists in the structure of an apparently perfect crystal arises naturally in the course of the work.

## Significance of Direction of Slip in Crystals

Perhaps the most remarkable feature of the distortion phenomena of single crystals is that during the whole process of slipping parallel to one crystal plane they slip in a definite crystallographic direction. Regarding the matter from the atomic point of view the finite rigidity of the deformed crystal implies that each atom is situated in a position of stable equilibrium in the field of force due to its neighbours, and to intervening free electrons if there are any. Plastic strain must be due to stability interchanges in which atoms jump, owing to thermal or other agitation and to the applied stress, from one position of equilibrium to another. If these jumps are due to thermal agitation, they must be regarded as random occurrences though, since each jump alters the field of force in its neighbourhood, the probability that any given jump will occur may change very greatly when a jump occurs in a neighbouring position. From this point of view the whole macroscopic phenomenon of gliding must be regarded as the integrated effect of individual jumps, and since the direction of the resultant glide is parallel to a line of atoms in the crystal, it is difficult to avoid the conclusion that a great majority at any rate of the individual jumps are in this direction too.

If this is so, the length of the jump is likely to be equal to the spacing of atoms along this line, and the mechanism of slipping may be more like the simple shift shown in fig. 4 ($c$), in which the whole of the material on one side of a definite plane shifts through the length of one lattice cell, than has often been supposed. It is proposed, therefore, to examine how this ideal condition differs from what is observed in real materials. In the first place this ideal slipping would leave the material in the form of a perfect crystal and the strength would be unaltered by the distortion. In the second place to shift the whole of the upper row of atoms simultaneously over the lower row would necessitate the application of a stress comparable with, though no doubt less than, the elastic moduli of the material. Further, it would leave no room for an explanation of the large observed effect of temperature on plastic distortion.

## Slipping over a Portion of the Slip-plane

It seems that the whole situation is completely changed when the slipping is considered to occur not simultaneously over all atoms in the slip-plane but over a limited region, which is propagated from side to side of the crystal in a finite time. It has been observed by Joffé* that the first sign of breakdown when a stressed crystal of rock salt is examined between nicol prisms is the appearance of a bright line, indicating distorted material, which is propagated along a slip-line from side to side of the crystal. The stresses produced in the material by slipping over a portion of a plane are necessarily such as to give rise to increased stresses in the part of the plane near the edge of the region where slipping has already occurred, so that the propagation of slip is readily understandable and is analogous to the propagation of a crack.

If, in order to retain the explanation of the observed constancy of the direction of slip, we assume that the propagation of a line of slipping along a slip-plane leaves a perfectly well-ordered crystal arrangement in its wake, we obtain a system represented pictorially in fig. 4. In the diagram (*a*) represents the atoms in the lattice of a crystal block; (*b*) the condition of this block when a slip of one atomic spacing has been propagated from left to right into the middle; and (*c*) the block after the unit slip, or 'dislocation' as we may call it, has passed through from left to right.

## Positive and Negative Dislocations

The configurations *a*, *b* and *c* of fig. 4 represent the passage of a dislocation in which the atoms above the operative slip-planes are compressed in the direction of slip while those below are expanded. It is clear that a similar dislocation can exist which is the mirror image in the slip-planes of that shown in *b*, fig. 4. This configuration which is shown in fig. 4 (*e*) will be called a negative dislocation to distinguish it from fig. 4 (*b*) which will be called a positive dislocation. The passage of a positive dislocation across a crystal block from left to right produces the same effect as the passage of a negative one from right to left, namely, the shift of the upper part of the block through one atomic spacing to the right relative to the lower part. This is illustrated in fig. 4 by the sequence of configurations *d*, *e*, *f*.

It seems likely that the motion of a 'unit dislocation' may be determined by the stress and the temperature. It is known that in the case of metals there is a recrystallization temperature, below the melting point, at which large irregularities in structure such as those which must exist at the boundaries of crystal grains with different crystal orientations can be propagated through the material. It may well happen therefore that there is a definite temperature probably lower than the recrystallization temperature, at which a unit dislocation of the type described above may be capable of free movement in either direction along the crystal plane. Let us call this hypothetical temperature $T_D$.

This picture seems adequate to account for the fact observed at high temperatures that slipping may occur at very low stresses in single crystals and that, after an indefinite amount of distortion, the crystal is still nearly in its original condition, except for change in shape; it is still a single crystal and still has very little strength.

* *The Physics of Crystals*, p. 47.

The distortion would then be accounted for by the propagation of centres of dislocation along the slip-planes. These would enter the specimen at one side and leave it at the other as shown in fig. 4.

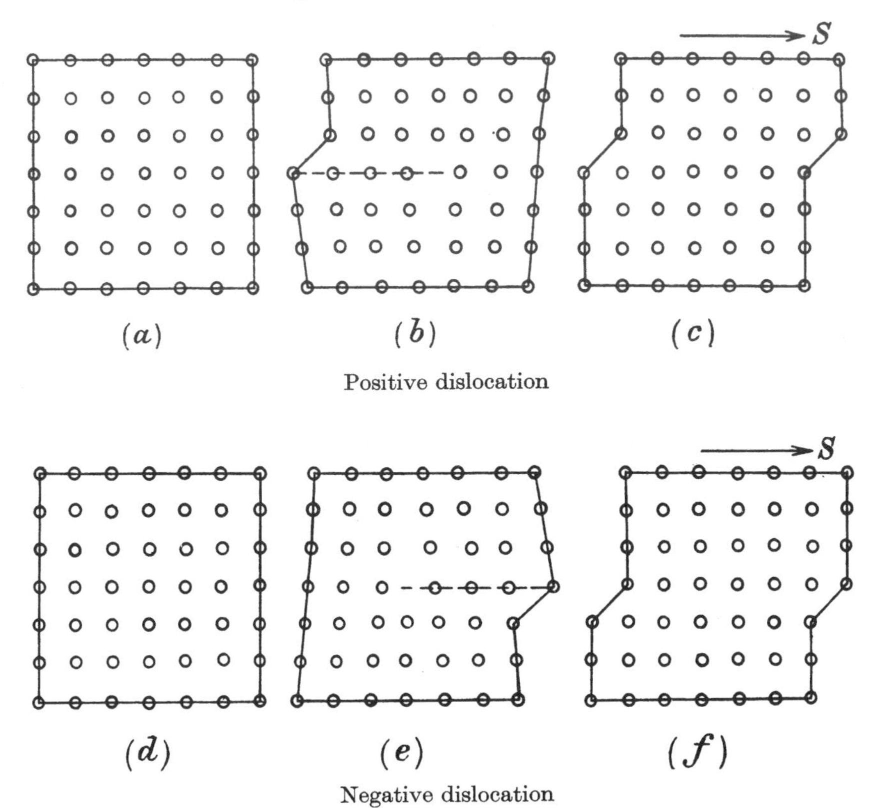


Fig. 4. Positions of atoms during the passage of a dislocation.

## Atomic Model of a Dislocation

Some insight into the manner in which a dislocation might be propagated along a slip-plane may be gained by introducing a conception due to Dehlinger* to explain recrystallization and described by him as 'Verhakung' or 'hooking'. In theories of the equilibrium of crystal lattices an atom is supposed to place itself in a position of minimum potential energy in relation to its neighbours. Heat motions will agitate the atom so that it moves about in the neighbourhood of this position of minimum potential energy, but, until a certain temperature is reached, the chance that the atom will escape across the 'potential barriers' which surround it is extremely small. In a perfect crystal structure one might, for instance, consider the potential along a line $CD$ spaced midway between two regularly spaced lines of atoms $A_0, A_1, A_2$ and $B_0, B_1, B_2$ (fig. 5).

* *Ann. Phys.* II (1929), 749.

If the positions of minimum potential due to each of the lines considered separately occur at the points $C_0, C_1, C_2$, where $C_0$ is midway between $A_0$ and $B_0$ atoms placed at these points would remain in equilibrium there and the equilibrium of a cubic structure might be partially explained in this way.

For the present purpose we may assume that the potential along $CD$ due to either row can be represented approximately by

$$-A\cos(2\pi x/\lambda), \tag{1}$$

where $\lambda$ is the lattice spacing, and $x$ is measured from the position $C_0$ midway between $A_0$ and $B_0$.

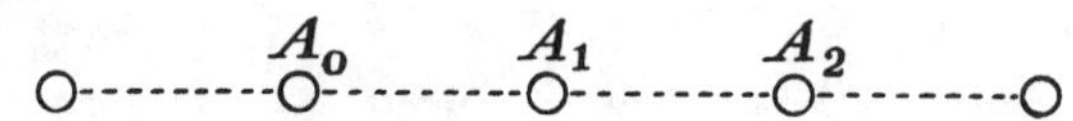


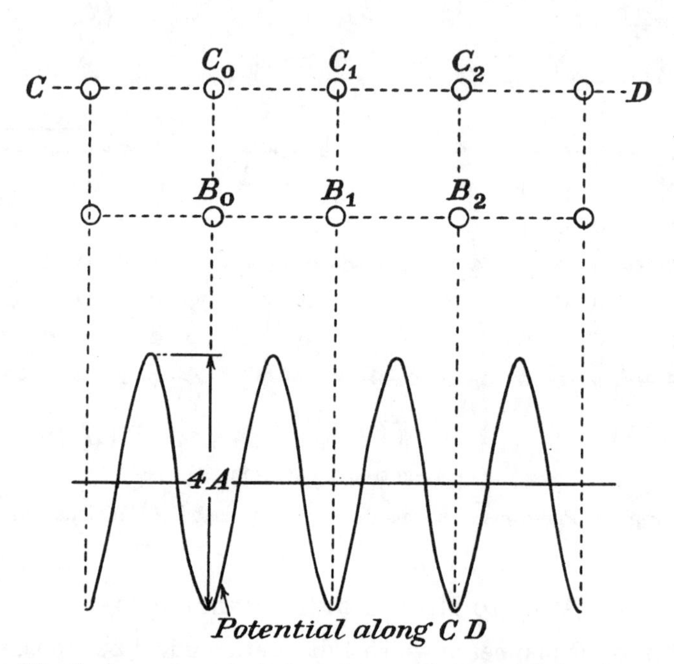


Fig. 5. Arrangement of atoms in a perfect crystal structure.

The total potential due to both rows, namely, $A_0, A_1, A_2, \ldots$ and $B_0, B_1, B_2, \ldots$ is then

$$\phi = -2A\cos(2\pi x/\lambda). \tag{2}$$

The height of the potential hill or barrier which must be climbed by any atom in order to escape altogether from its equilibrium position is then $4A$. This is represented graphically by the sine curve of amplitude $2A$ which is shown in fig. 5.

In the neighbourhood of a positive dislocation the atoms in the slip-plane above the centre of dislocation are compressed in the direction of slip while those in the plane below it are extended. At some distance from the centre the spacings along lines of atoms above and below the path of the centre of dislocation become regular and equal to the normal spacing for a perfect crystal, but if the number of atoms in a length $L$ of a line of atoms which passes below the centre of a unit dislocation is $N$, then the number in the same length $L$ on a line which passes above the centre is $N+1$.

To calculate the distribution of potential on the line of atoms along which the actual centre of dislocation passes it would be necessary to solve the very difficult problem of finding the equilibrium positions of all the surrounding atoms. Though this is not yet possible the general nature of the potential distribution on the line along which the centre of the dislocation passes may perhaps be inferred from the results of calculation of potential due to an arbitrary distribution of atoms which has the essential characteristic that $N+1$ atoms in a line above the centre correspond with $N$ atoms below it.* The simplest arrangement of this kind is when $N+2$ atoms in the upper line are evenly spaced over a length $(N+\frac{1}{2})\lambda$, while the $N+1$ atoms in the lower line are evenly spaced over the same length. This arrangement is shown in fig. 6 for $N=3$. Outside this range we may suppose the two rows to have the spacing $\lambda$ of the unstrained crystal. Making the same approximation as that for a regular lattice, the potential along the line $CD$ midway between these two lines is

$$P=-A\cos 2\pi\frac{x}{\lambda}\left(\frac{N+1}{N+\frac{1}{2}}\right)-A\cos 2\pi\frac{x}{\lambda}\left(\frac{N}{N+\frac{1}{2}}\right). \tag{3}$$

The nature of this function is well known. It is shown graphically in fig. 6 for the case $N=3$. In the length $(N+\frac{1}{2})\lambda$ it has $N+2$ minima including those at the two ends of the range. The heights of the 'potential hills' and the depths of the 'potential valleys' decrease from each end to the centre of the range. If $N$ is an odd number, the shallowest depression on the potential curve is in the centre of the range. If we now imagine that each of the potential depressions except the central one is occupied by an atom we have a rough picture of a possible equilibrium distribution of atoms in the neighbourhood of a centre of dislocation. The actual centre of the dislocation may be regarded as being situated at the point $O$ (fig. 6), in the middle of the central vacant potential hollow.

It has been pointed out that in a perfect crystal the atoms may be expected to remain within their potential barriers till a certain temperature is reached at which they can jump these barriers. At this temperature the atom $C_0$ (fig. 6) in the disturbed lattice might be able to jump into the space occupied by $C_1$. At lower temperatures an atom would not be capable of surmounting potential hills between $C_0$ and $C_1$, but might still be able to jump the lower hills nearer the centre of the disturbance. The lowest temperature at which a jump can be made is that which will just permit either of the two central atoms to jump one of the lowest hills into the vacant central hollow $O$ (fig. 6). This temperature we may call $T_D$. It may be noticed that even at the absolute zero of temperature some energy of agitation still persists in the crystal. If this is large enough to permit jumps to take place, then $T_D=0°$ abs.

* These numbers include the atoms at each end of the line so that $N+1$ of the compressed atomic spacings in the upper row correspond with $N$ spacings in the extended lower row.

When one of the two central atoms has jumped into the central hollow, the potential hollow which is left vacant now becomes the centre of the dislocation and we must imagine that the neighbouring atoms readjust themselves by a continuous process (i.e. without jumps) into new positions of equilibrium so that the potential hollow newly vacant becomes the shallowest one. This idea has already been expressed, without attempting to give it a definite atomic interpretation, in the statement made on p. 350 that a temperature $T_D$ exists above which the centres of dislocation can move freely in either direction along the slip-plane.

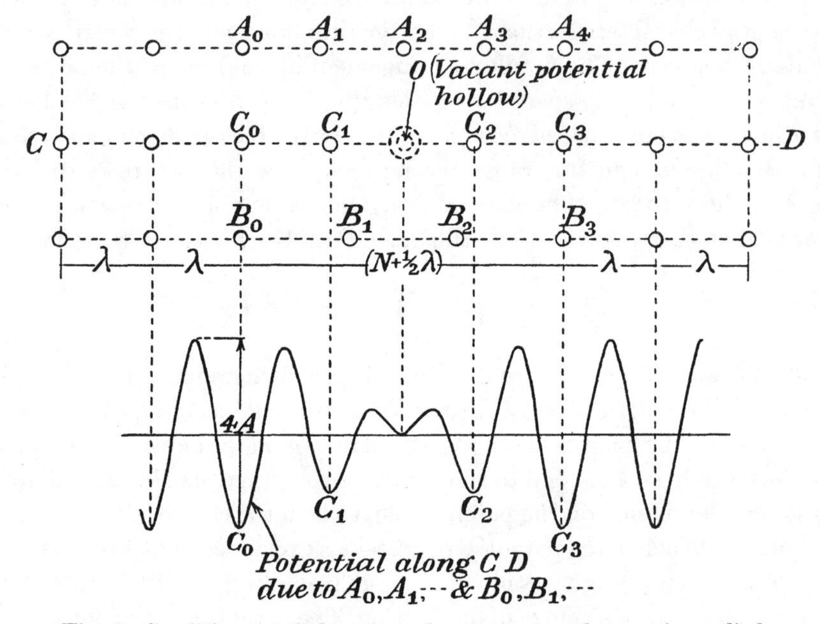


Fig. 6. Conditions at dislocation when no external stress is applied.

The foregoing picture is very incomplete because it takes no account of the mutual action of the atoms in the row along which the jumps occur, but the theory of the equilibrium of the atoms in *perfect* metallic crystals is so incomplete that it does not seem worth while to attempt any more complete discussion of equilibrium in a disturbed lattice.

## Effect of Applying an External Shear Stress

We are now in a position to examine, in the light of the foregoing rough representation of atomic structure, the effect of applying an external shear stress. We may imagine that a shear stress applied in the direction indicated by the arrows in fig. 7 gives the row of atoms $A_0, A_1, A_2, \ldots$, a small displacement to the right relative to the row $B_0, B_1, B_2, \ldots$. If this small displacement is $\delta$, the potential along the line $CD$ is now

$$P = -A\cos 2\pi\frac{x-\delta}{\lambda}\left(\frac{N+1}{N+\frac{1}{2}}\right) - A\cos 2\pi\frac{x}{\lambda}\left(\frac{N}{N+\frac{1}{2}}\right). \tag{4}$$

It will readily be seen that the effect of this is to lower the height of the hill which the atoms to the right of the central hollow must jump in order to enter the central hollow. The hill which the atoms to the left of the centre must jump is correspondingly raised. The actual potential curve for $N=3$, $\delta/\lambda=0{\cdot}1$ is shown in fig. 7. The positions of the atoms and of the vacant potential hollow or centre of the dislocation are also shown in fig. 7. Here height of the potential hill $h_1$ (fig. 7), which the atom $C_1$ to the left of the centre of dislocation has to jump in order to land in the central potential hollow, is 2·70 A. The atom $C_2$ to the right of the centre has only to jump a hill of height $h_2 = 0{\cdot}85$ A.

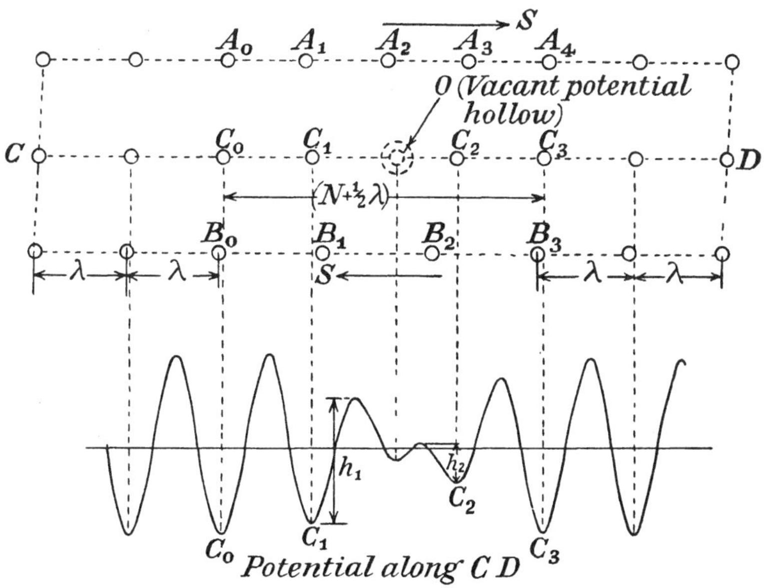


Fig. 7. Conditions at dislocation in crystal subjected to shear stress.

If the temperature is high enough to permit the jump $h_2$ but not high enough to permit the jump $h_1$, the only jump which can occur is to the left into the centre of dislocation from the potential hollow on its right. The actual centre, i.e. the vacant hollow, can therefore only move to the right. At higher temperatures, when the jumps $h_2$ and $h_1$ can both occur, the former will be the more frequent so that a positive centre of dislocation will necessarily migrate to the right under the action of a shear stress directed to the right. Similarly a negative dislocation will move to the left under the action of the same shear stress.

When the temperature is high enough to permit free movement of the centre of dislocation in the absence of applied shear stress the application of the smallest shear stress will cause the centre of dislocation to migrate along the slip-plane. Thus plastic strain will occur when the smallest stress is applied. At lower temperatures a finite shear stress may be necessary in order that the potential hill on one side of

23-2

the centre of dislocation may be reduced in height to such an extent that the neighbouring atom can jump into the central hollow.

In the foregoing discussion an atomic mechanism is roughly represented which would give metallic crystals the following properties:

(1) At temperatures above $T_D$ any stress, however small, applied parallel to the slip-plane in the positive direction (see fig. 7) will cause positive centres of dislocation to travel in the positive direction and negative centres to travel in the negative direction;

(2) At temperatures below $T_D$ the centres will not move till the shear stress attains some finite value.

In the work which follows these properties will be assumed and some possible consequences of their existence will be explored. It must be remembered that (1) is an accurate statement of what is to be expected of the atomic mechanism, but (2) is only a rough representation of some more complicated relationship involving temperature and rate of deformation which would result from a more complete consideration, from the statistical point of view, of the connection between temperature and the frequency of the stability interchanges, to which the motion of a dislocation is here ascribed.

## Distribution of Stress near a Unit Dislocation

The distribution of stress due to a unit dislocation cannot be calculated at points within a few lattice spacings of the centre of the disturbance, but at greater distances the theory of elasticity must apply. The general theory of dislocations in continuous isotropic elastic solids has been treated extensively by Volterra.* In the neighbourhood of a centre of dislocation the stresses become very large so that it is necessary to imagine that the material is cut away round the actual centre. We may therefore suppose that our material is initially unstrained and that it contains a hole $H$ (fig. 8). From some point $A$ on the boundary a centre of dislocation passes along a straight line $AB$ to the point $B$ which is on the surface of the hole $H$.

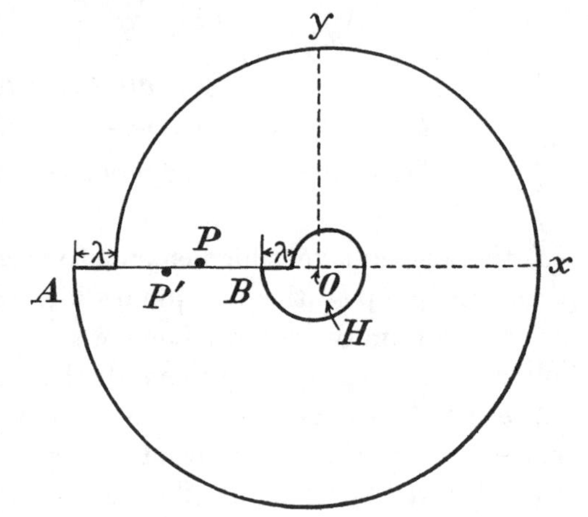

Fig. 8. Elastic cylinder with dislocation.

If $P$ is a particle just above $AB$ and $P'$ is just below it, they suffer a relative displacement $\lambda$ during the passage of the centre of dislocation between them.

The appearance of a circular ring which has suffered this kind of strain is shown in fig. 8. The stress distribution round such a dislocation has been calculated by

* *Ann. Éc. Norm.* XXIV (1907).

Timpe.* It depends on the surface tractions applied at the surface of the hole $H$ as well as on the amount of slip; but if the resultant force and couple due to these surface tractions are both zero the stresses due to the dislocation vary as $r^{-1}$, $r$ being the distance from the centre $O$, while the stresses due to the surface tractions vary as $r^{-2}$ or higher inverse powers.

At points far from the centre the components of stress due to a positive dislocation or slip of amount $\lambda$ on the positive axis of $x$ are

$$X_x = -Y_y = -\frac{\mu\lambda}{\pi}\frac{y}{x^2+y^2}, \quad X_y = \frac{\mu\lambda}{\pi}\frac{x}{x^2+y^2} \tag{5}$$

or in polar co-ordinates

$$\widehat{rr} = -\widehat{\theta\theta} = +\frac{\mu\lambda\sin\theta}{\pi r}, \quad \widehat{r\theta} = \frac{\mu\lambda\cos\theta}{\pi r}, \tag{6}$$

where $\mu$ is the modulus of rigidity.

In the particular instance when no traction is exerted on the surface of a circular hole $H$ of radius $a$, the complete expressions for the components are

$$\left.\begin{aligned} \widehat{rr} &= +\frac{\mu\lambda}{\pi r^3}\sin\theta\,(r^2-a^2), \\ \widehat{r\theta} &= \frac{\mu\lambda\cos\theta}{\pi r^3}(r^2-a^2), \\ \widehat{\theta\theta} &= -\frac{\mu\lambda}{\pi r^3}\sin\theta\,(r^2+a^2). \end{aligned}\right\} \tag{7}$$

These are identical with (6) when $a$ is small compared with $r$.

It appears that whatever may be the state of stress at distances from the dislocation centre comparable with $\lambda$ the stress at large distances is represented by (5). Thus, even though nothing is known about the exact conditions of equilibrium of atoms at a dislocation, the stress distribution at a distance of several lattice cells away is known, for it depends only on the amount of the slip. The application of the theory of dislocations in an elastic body to a crystal is, strictly speaking, only justifiable if the crystal is elastically isotropic, a condition which appears to be fulfilled for aluminium. In cubic crystals the error due to want of isotropy is likely to be small. It will be left out of consideration in the following pages.

## Mutual Action and Equilibrium of Centres of Dislocation

We may consider a single set of parallel planes as possible slip-planes. If these are taken as perpendicular to the axis of $y$, the application of a positive shear stress will cause a positive centre to move to the right or a negative centre to the left. For the present we may suppose that the temperature is above $T_D$ so that the dislocation moves, however small $X_y$ may be. Since each centre of dislocation produces a

* Göttingen Diss., Leipzig (1905).

distribution of shear stress in its neighbourhood, every centre must tend to cause all neighbouring centres to migrate along their slip-planes.

Referring to equation (5) it will be seen that in the neighbourhood of a positive centre of dislocation, $A$, $X_y$ is positive when $x$ is positive, if therefore there is another positive centre $B$ whose $x$ co-ordinate relative to $A$ is positive, $B$ will move to the right under the influence of $A$. That is, positive centres of dislocation repel one another. Similarly two negative centres repel one another. On the other hand, positive centres attract negative centres and *vice versa*. If there are only two centres in the whole field and they are both of the same sign, they will repel one another to an infinite distance (provided the temperature is above $T_D$). Centres of opposite sign will attract one another till the line joining them is perpendicular to the slip-plane, when they will rest in equilibrium under their mutual influence. This position is illustrated in fig. 9 (*a*).

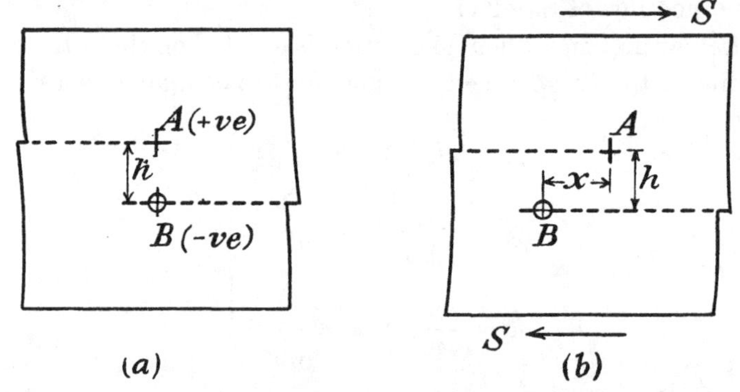


Fig. 9. Positive and negative dislocations in equilibrium. (*a*) Under no external stress, (*b*) when shear stress is applied.

If two centres of opposite sign lie on the same slip-plane, they will attract one another, but, since the elastic equations cannot be applied when they come within a few lattice distances of one another, it is not possible to say whether or not they will neutralize one another and leave a perfect crystal structure.

We may now consider the effect of applying a positive shear stress $S$ so that the upper face of a crystal block is subjected to a tangential stress acting towards the right. If only one dislocation is present, it will move across the block, to the right if it is positive and to the left if it is negative, however small $S$ may be. In either case the result is the same, namely, a shift of the upper portion of the block relative to the lower part through a distance $\lambda$. Apart from this shift the properties of the block are unchanged by the passage of the dislocation.

Conditions are very different when two dislocations of opposite sign are present. Suppose, for instance, that owing to the application of a very small shearing force a positive dislocation starts moving along a slip-plane from the left, and a negative dislocation starts to move from the right along another slip-plane, distant $h$ from the first. Even without application of the shear stress they would arrive under their mutual influence at the positions $A$, $B$ (fig. 9 ($a$)), where $AB$ is perpendicular to the slip-planes. On the other hand, in order to make them continue to move along their respective slip-planes beyond $AB$, a shear stress must be applied. A shear stress $S$ will cause the positive dislocation to move to the right and the negative one to the left till

$$S=\frac{\mu\lambda x}{\pi(x^2+h^2)}, \tag{8}$$

where $x$ is the projection on the slip-plane of the distance between them. In this position each dislocation is in equilibrium.

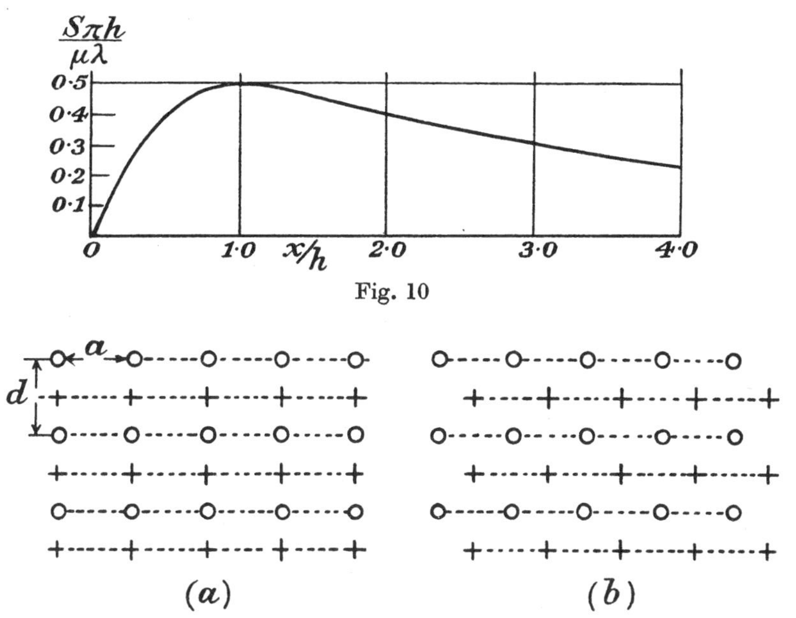


Fig. 10

Fig. 11. Equilibrium arrangements of dislocations, ($a$) stable, ($b$) unstable.

The connection between $S$ and $x$ is shown in fig. 10. It will be seen that as $x/h$ increases $S\pi h/\mu\lambda$ increases till the maximum value 0·5 is reached at $x=h$. The maximum possible value of $S$ consistent with equilibrium is therefore $\frac{1}{2}\frac{\mu\lambda}{\pi h}$ or $0{\cdot}16\,\mu\lambda/h$. If $S$ is less than this, the two centres of dislocation cannot escape from one another, whereas when $S>\mu\lambda/2\pi h$ they pass one another and escape from their mutual attraction. Thus plastic distortion or slipping can only proceed, so far as the movement of these two centres of dislocation are concerned, when $S>\mu\lambda/2\pi h$. This is remarkable in view of the fact that the strength of the material is zero when only one centre of dislocation is present. Since $S$ is inversely proportional to $h$ it appears that slipping will proceed on the slip-planes which are farthest from planes of slip already containing centres of dislocation.

We may now consider possible equilibrium arrangements of centres of dislocation. It has been seen that positive and negative centres cannot exist on the same slip-plane, for they would run together and neutralize one another. The simplest arrangements in which every centre is in equilibrium with its neighbours are those in which all the positive centres and all the negative centres lie on interpenetrating rectangular nets. Two possible equilibrium configurations are shown as (*a*) and (*b*), fig. 11, where the positive centres are represented by crosses and the negative ones by round spots. The application of a positive shear stress to a crystal containing such a system would move all the positive centres to the right and all the negative ones to the left along the slip-planes, represented by broken lines in fig. 11.

We shall calculate the displacement of the positive lattice relative to the negative when a given shear stress $S$ is applied to the crystal. For this purpose we must first calculate the distribution of $X_y$ due to the complete lattice of dislocations. The shear stress due to a row of positive dislocations regularly spaced at distance $d$ apart in a line perpendicular to the slip-planes is

$$\frac{\mu\lambda}{\pi}\sum_{n=-\infty}^{n=+\infty}\frac{x}{x^2+(y+nd)^2}, \tag{9}$$

where $x$ is the distance of the point considered from the plane of the dislocations and $y$ is its ordinate relative to an axis which passes through one of them. The sum of this series is known. It is

$$\frac{\mu\lambda}{d}\frac{\sinh(2\pi x/d)}{\cosh(2\pi x/d)-\cos(2\pi y/d)}; \tag{10}$$

when $y=0$ this becomes $$\mu\lambda d^{-1}\coth(\pi x/d); \tag{11}$$

when $y=\frac{1}{4}d$ it is $$\mu\lambda d^{-1}\tanh(2\pi x/d); \tag{12}$$

when $y=\frac{1}{2}d$ it is $$\mu\lambda d^{-1}\tanh(\pi x/d). \tag{13}$$

Starting from the equilibrium configuration (*a*) fig. 11, for which each negative centre is midway between two positive centres on a line perpendicular to the slip-planes, we shall give the positive lattice a displacement $\xi$ relative to the negative lattice. Taking one of the positive centres as origin, the co-ordinates of the negative centres are $(-\xi+ma,(n+\frac{1}{2})d)$ where $n$ and $m$ are integral numbers positive or negative, $a$ is the distance apart of centres of dislocation on a slip-plane. If $a=d$, the positive centres lie at the corners of a square lattice. As $\xi$ increases from zero the shear stress at the origin due to the negative centres changes, but that due to the positive centres, which remain fixed, is unaltered. If $S'$ is the shear stress at the origin due to all the centres of dislocation (except the positive centre which is situated there), then $S'=0$ when $\xi=0$. The contribution to $S'$ due to the displacement of the $m$th row of negative dislocations is (see (13))

$$\mu\lambda d^{-1}\{\tanh\pi d^{-1}(ma-\xi)-\tanh\pi d^{-1}ma\}. \tag{14}$$

As $m$ increases the value of this expression rapidly decreases, so that a convergent expression is obtained for $S'$, namely,

$$S'=\mu\lambda d^{-1}\sum_{m=-\infty}^{m=+\infty}\{\tanh\pi d^{-1}(ma-\xi)-\tanh\pi d^{-1}ma\}. \tag{15}$$

If the change in $S'$ is traced as $\xi$ increases, it will be found that each increase in $\xi$ by an amount $a$ gives rise to a change in $S'$ of amount $-2\mu\lambda d^{-1}$. This change may be regarded as being due to the separation of a layer of positive centres at the right-hand boundary of the crystal and a layer of negative centres on the left, thus it is a polarization effect arising from the distribution of dislocations on the boundary of the crystal. It will be noticed that the effect of this polarization is to increase the shear stress at all points in the field, not only at the lattice points, but an amount $-2\mu\lambda/d$ for each increase of amount $a$ in the displacement. Thus the mean shear stress over the whole crystal due to the presence of the centres of dislocation is increased by $-2\mu\lambda/d$.

Besides the effect of the relative displacement of the two lattices of centres of dislocation the conditions of stress at the boundary of the crystal itself have an unknown effect on the stress within it. If we suppose that the mean stress in any portion of the crystal is equal to $S$ the externally applied stress, i.e. if we suppose that the mean stress taken over a portion of the crystal which includes a large number of centres of dislocation is uniform throughout the volume of the crystal and equal to $S$, then the mean stress due to the combined effects of the distribution of centres of dislocation and to the conditions at the boundary must be zero. Thus to obtain the true shear stress at positive centres of dislocation, due to the displacement of the lattice of negative centres, we must subtract from the expression (15) the mean value of the shear stress in the crystal due to the whole body of the centres of dislocation.

## Mean Value of Shear Stress

The mean value of the shear stress due to the distribution of centres of dislocation may be found by taking the mean value over a strip parallel to the axis of $y$ and of breadth $a$. If this mean stress be represented by $S_M$, the change in $S_M$ which occurs when the displacement of the negative centres relative to the positive ones (which we suppose fixed) increases by $\delta\xi$ may be regarded as being due to the removal of a strip of thickness $\delta\xi$ on one side of the strip and replacing it by a strip also of the thickness $\delta\xi$ but located in a region where the shear stress has increased by $-2\mu\lambda/d$. The mean value of $S_M$ is therefore increased by $-2\mu\lambda\delta\xi/ad$. Hence

$$\left.\begin{aligned} \frac{dS_M}{d\xi} &= -\frac{2\mu\lambda}{ad}, \\ \text{so that} \qquad S_M &= -2\mu\lambda\xi/ad. \end{aligned}\right\} \tag{16}$$

Hence when the effects of the boundary have been taken into consideration in such a way that the mean stress throughout the body of the crystal is equal to the externally applied stress the resultant stress at a positive centre is $S+(S'-S_M)$.

If the temperature is greater than $T_D$ so that centres of dislocation move freely under the influence of the slightest shear stress, the equilibrium condition is therefore

$$S+S'-S_M=0, \tag{17}$$

or combining (15), (16) and (17)

$$S = \mu\lambda d^{-1}(F), \tag{18}$$

where
$$F = -\frac{2\xi}{a} + \sum_{m=-\infty}^{m=+\infty} \left\{ \tanh\frac{\pi m a}{d} - \tanh\frac{\pi}{d}(ma - \xi) \right\}. \tag{19}$$

$F$ is a function of $\xi/a$ and of $a/d$. The values shown graphically in the curves of fig. 12 have been calculated for values of $\xi/a$ ranging from 0 to 0·5 for each of the three lattices $a = d$, $a = 0{\cdot}8d$, $a = 0{\cdot}5d$. It will be seen that each curve rises from zero at $\xi = 0$ and falls to zero at $\xi/a = 0{\cdot}5$. This is to be expected because $\xi/a = 0{\cdot}5$ corresponds to the equilibrium arrangement represented by (*b*), fig. 11. The fact that

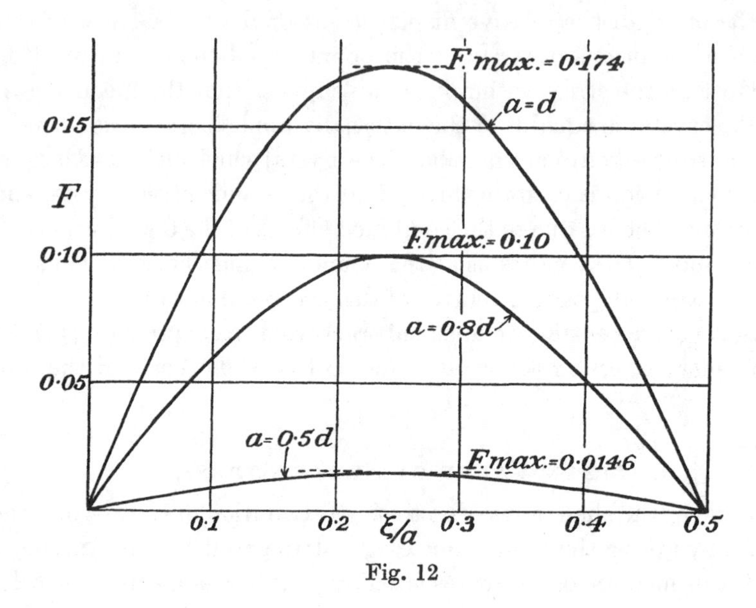


Fig. 12

$F$ is positive over the whole range from $\xi/a = 0$ to 0·5 shows that (*b*) is an unstable configuration while (*a*) is stable. In each case $F$ has a maximum value (see fig. 12). For $a = d$ it is $F_{\text{max.}} = 0{\cdot}174$. For $a = 0{\cdot}8d$ it is $F_{\text{max.}} = 0{\cdot}10$. For $a = 0{\cdot}5d$ it is $F_{\text{max.}} = 0{\cdot}0146$.

It will be seen, therefore, that when the externally applied shear stress $S$ is less than $\mu\lambda d^{-1}(F_{\text{max.}})$ a value of $\xi$ can be found such that every centre of dislocation, positive or negative, is in a position where the shear stress is zero, so that it can remain at rest. When $S$ rises above this value the positive centres will migrate steadily to the right and the negative ones to the left, and plastic distortion will set in. Thus the crystal has a finite shear strength

$$S = \mu\lambda d^{-1}(F_{\text{max.}}). \tag{20}$$

If the crystal is initially perfect, the first few dislocations migrate through it under the action of a shear stress which may be regarded as infinitesimally small. As the distortion proceeds, however, the number of dislocations will increase and the average value of $d$ will decrease so that the resistance to shear will increase with the amount of distortion. This explanation of hardening by cold work is identical in principle with one put forward in a qualitative form by the present writer some years ago,* but the introduction of the idea of a unit dislocation which gives rise to a calculable field of elastic stresses has now made it possible to present the theory in a more definite and constructive form than formerly.

## Diagonal Lattices of Dislocations

So far the only arrangement of lattices which has been discussed is that in which the two kinds of dislocation lie on two interpenetrating rectangular nets. Another equilibrium arrangement is that in which the positive and negative dislocations lie on interpenetrating lattices such that each cell is a rhombus with one of its diagonals parallel to the slip-planes. A diagonal arrangement of this kind is shown in fig. 13. If the dislocations are initially arranged as configuration ($a$), fig. 13, a relative displacement of the positive and negative lattices through $\frac{1}{2}a$ brings the system to configuration ($c$) which is the mirror image of ($a$) in the slip-planes. Thus if ($a$) is stable ($c$) will also be stable and vice versa.

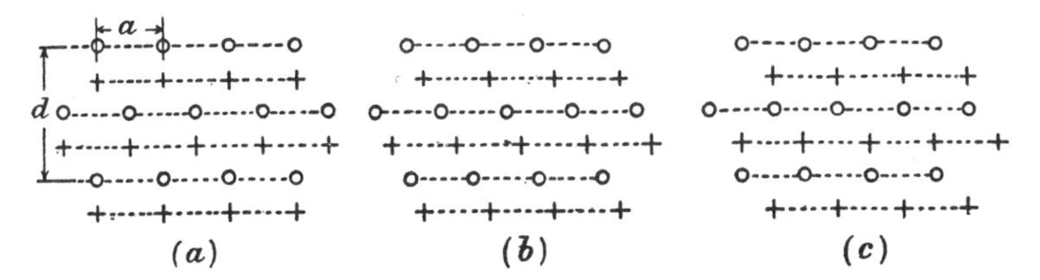


Fig. 13. Equilibrium arrangements of dislocations, ($a$) and ($c$) stable, ($b$) unstable.

The system must also be in equilibrium when the displacement is $0{\cdot}25a$ and $0{\cdot}75a$. This configuration is shown in ($b$), fig. 13.

If $a$ and $d$ represent the two diagonals of the lattice cell and $\xi$ the displacement of the positive relative to the negative lattice from the equilibrium positions ($a$), fig. 13, the shear stress at one of the positive centres due to a row of negative centres is (see equation (12))

$$\frac{\mu\lambda}{d}\tanh\frac{2\pi}{\lambda}\left(\frac{ma}{2}-\xi\right),$$

where $m$ takes all integral values. Proceeding on the same lines as in the case of the rectangular lattice the condition of equilibrium is

$$S' = \mu\lambda d^{-1}(F), \tag{21}$$

where

$$F = -\frac{4\xi}{a} + \sum_{m=-\infty}^{m=+\infty}\left\{\tanh\frac{2\pi}{d}\left(\frac{na}{2}\right) - \tanh\frac{2\pi}{d}\left(\frac{na}{2}-\xi\right)\right\}. \tag{22}$$

* *Trans. Faraday Soc.* xxiv (1928), 121; paper 15 above.

When $a = d$, i.e. when a square lattice is set diagonally relative to the slip-planes, the following values for $F$ were calculated from (22)

| $\xi/a$ | 0 | 0·05 | 0·1 | 0·15 | 0·20 | 0·25 |
|---|---|---|---|---|---|---|
| $F$ | 0 | 0·109 | 0·169 | 0·160 | 0·095 | 0 |

The maximum value is $F_{max.} = 0 \cdot 17$.

The fact that $F$ is positive in the range $\xi/a = 0$ to $\xi/a = 0 \cdot 5$ shows that the equilibrium configurations (*a*) and (*c*), fig. 13, are stable while (*b*) (for which $\xi/a = 0 \cdot 25$) is unstable.

When $d = a\sqrt{3}$ the centres of dislocation lie on a net consisting of equilateral triangles. The corresponding values of $F$ calculated from (22) are:

| $\xi/a$ | 0 | 0·05 | 0·10 | 0·15 | 0·20 | 0·25 |
|---|---|---|---|---|---|---|
| $F$ | 0 | 0·0179 | 0·0289 | 0·0283 | 0·0176 | 0 |

and $F_{max.} = 0 \cdot 030$.

## Relationship between Strain and the Motion of Centres of Dislocation

In the simple case when only one centre of dislocation is present the effect of moving it across the crystal from left to right has already been seen (fig. 4). The part of the crystal above its path is displaced through a distance $\lambda$ relative to the lower part. If the centre of dislocation does not travel across the crystal from side to side but merely moves through a distance $x$ from the outer surface, the displacement is $x\lambda/L$, where $L$ is the total length along the slip-plane through which a dislocation can move. $L$ might, as we have so far assumed, be the total width of the specimen or it might be the width of any subdivision of the crystal the boundary of which is impervious to centres of dislocation. The total shift due to centres regularly spaced at intervals $a$ along the slip-plane is therefore $L\lambda/2a$. The shift due to a row of positive centres, which have moved into the crystal from the left, is the same as that due to a row of negative centres which have moved in from the right.

Since in the rectangular arrangement (fig. 11) the operative slip-planes are spaced at distance $\frac{1}{2}d$ apart the relative shift of two slip-planes separated by a thickness $h$ is therefore $hL\lambda/ad$. By definition therefore the shear strain

$$s = L\lambda/ad. \tag{23}$$

For the diagonal arrangement of dislocations (fig. 13), the operative slip-planes are situated at distance $\frac{1}{4}d$ apart so that

$$s = 2L\lambda/ad. \tag{24}$$

### Theoretical Stress-strain Relationship

We have seen that the least shear stress at which plastic flow occurs when the temperature is higher than $T_D$ is $S=\mu\lambda d^{-1}F_{\text{max.}}$. Combining this with (23) or (24) $d$ can be eliminated. The resulting equation is

$$\frac{S}{\mu\sqrt{s}}=\kappa\sqrt{\frac{\lambda}{L}}, \tag{25}$$

where $\kappa$ is a constant depending on the particular arrangement of centres of dislocation. The $(S, s)$ curve is therefore parabolic. The values of $\kappa$ are given below.

$$\text{Rectangular lattice}\quad\begin{cases} a=d & \kappa=0{\cdot}174,\\ a=0{\cdot}8d & \kappa=0{\cdot}089,\\ a=0{\cdot}5d & \kappa=0{\cdot}010,\end{cases}$$

$$\text{Diagonal lattice}\quad\begin{cases} a=d & \kappa=0{\cdot}120,\\ a=d/\sqrt{3} & \kappa=0{\cdot}016.\end{cases}$$

If the temperature is below $T_D$, all the foregoing work is unchanged, but the condition of equilibrium of a centre of dislocation is now

$$S-S_T=\mu\lambda d^{-1}(F),$$

where $S_T$ is the finite stress which must be attained before a single dislocation can migrate in a perfect crystal. The resulting strain hardening or plastic stress-strain relationship is then

$$\frac{S-S_T}{\mu\sqrt{s}}=\kappa\sqrt{\frac{\lambda}{L}}. \tag{26}$$

### Choice of a Value for $\kappa$ for Comparison with Experiment

It will be seen that comparatively small variations in the arrangements of centres of dislocation produce very large changes in the value of $\kappa$. A crowding of positive centres of dislocation on one set of slip-planes increases the ease with which negative centres can migrate along the intermediate planes. Thus in the rectangular lattice of dislocations a decrease in the ratio $a/d$ corresponds with a much greater decrease in $\kappa$.

In trying to form some estimate as to which of these regular arrangements of dislocations is likely to represent most closely the true condition of a plastically strained crystal it is necessary to consider the manner in which the dislocations may arise at the boundary. Since in this theory the crystal is undisturbed by the passage of a dislocation, i.e. the crystal is just as perfect behind it as it is in front, the dislocations might be supposed to arise at the boundary or within the crystal owing to thermal agitation in a random manner. The hypothesis that the points of origin of the dislocations are distributed at random precludes the formation of regular lattices of dislocations of the type we have been considering. The kind of regularity which produces very low values of $\kappa$ is therefore unlikely to arise.

On the other hand, it is difficult to calculate the effect of a random production of dislocations because the mutual action of neighbouring centres of dislocation must itself tend to prevent the formation of a purely random distribution of centres inside the crystal. If, for instance, in the course of plastic distortion a positive and a negative centre approach one another from opposite sides of the crystal along planes which are separated by a very small distance $h$, they will not separate again till $S$ reaches the value $\mu\lambda/2\pi h$. Thus a pseudo regularity in which the distance of a dislocation from its nearest neighbour tends always to be about $\mu\lambda/2\pi S$ is likely to arise.

We might, perhaps, visualize the whole process of plastic deformation as follows. At points distributed at random in a crystal slipping begins, owing possibly to thermal agitation, by separation of a positive and negative dislocation. These move away from one another to an average distance apart $L$ under the action of the external shear stress $S$. Slipping through one lattice distance has therefore occurred along each line of length $L$ which joins a positive and negative pair of centres. If at any stage there are $N$ such pairs of dislocation lines per unit volume, the average distance of the nearest negative dislocation from any given positive one is proportional to $1/\sqrt{N}$. A very rough approximation is $1/2\sqrt{N}$.

If only the shear stress due to the nearest negative dislocation is considered, the shear stress necessary for its escape is of order of magnitude

$$S=\frac{\mu\lambda}{2\pi}(2\sqrt{N})=\mu\lambda\sqrt{N}/\pi.$$

The strain is evidently $s=N\lambda L$ so that

$$\frac{S}{\mu\sqrt{s}}=\frac{1}{\pi}\sqrt{\frac{\lambda}{L}}. \tag{27}$$

This expression is identical with (25) if $\kappa=\pi^{-1}=0{\cdot}32$, but it is likely to overestimate $S$ because it takes account only of the effect of the nearest dislocation of opposite sign.

In choosing a value to assume for $\kappa$ in order to compare the theory with observation we might take it that there are reasons for supposing that $\kappa=0{\cdot}1$ is too low, and $\kappa=0{\cdot}3$ is too high, accordingly we may use $\kappa=0{\cdot}2$ as the most probable value in the present very imperfect state of the theory. Calculations based on this value should at any rate give the correct order of magnitude for the strength of crystals.

# 22

# THE MECHANISM OF PLASTIC DEFORMATION OF CRYSTALS

# PART II. COMPARISON WITH OBSERVATIONS

REPRINTED FROM

*Proceedings of the Royal Society*, A, vol. CXLV (1934), pp. 388–404

## Comparison with Observed Plastic Stress-strain Relationships

According to the theory given in Part I the strain-hardening or plastic stress-strain curve for a pure metal should be a parabola. In figs. 1, 2 and 3, Part I, parabolas are drawn, the parameters being chosen so that they lie as close as possible to the points which represent actual observations. It will be seen that for aluminium and gold the agreement is good. For a single crystal of copper the agreement is not good, but, on the other hand, the plastic stress-strain curve for polycrystalline specimens of copper which is shown in fig. 1 is very nearly parabolic over a large range.

The observations for iron seem to show that there is a small finite elastic limit, i.e. $S_T$ may be finite. Parabolas corresponding with the existence of a small elastic limit and with no elastic limit have been drawn. It seems that the observed points lie rather closer to the former curve. In any case, the observed curves have the essential characteristic of the theoretical ones that they are very steep at small strains, but get less and less steep as the strain increases.

In some branches of physics the measure of agreement between theory and observation which is shown in figs. 1, 2 and 3, Part I, would perhaps not be considered encouraging. It must be remembered, however, that up to the present no theory of the strength of metals has been devised which is capable of being expressed in an analysable form. Also the plastic stress-strain curves of similar metals vary so much in detail that it would be impossible to devise any general theory which would account so accurately for the behaviour of every metal crystal that good agreement with actual individual observations could be attained. Another difficulty is that the large changes which very small amounts of impurity can cause in the stress-strain curves make it difficult to be certain that the observed curves are really those which would be found if the test were made with a perfect crystal of a metal containing no impurities.

### NECESSITY FOR A MOSAIC STRUCTURE OF SYSTEM OF FAULTING

The stress-strain relationship (25) Part I contains the ratio $\lambda/L$. It is clear that $L$ must not be interpreted as an external linear dimension of a single crystal specimen because the observed stress-strain relationship does not depend on the size of the specimen employed. The length $L$ must be a length connected with the structure of the crystal. It represents the distance through which a dislocation can travel freely along a slip-plane under the influence of a small shear stress before being held up by some fault in the perfection of the regular crystal structure, or surface of misfit. It is therefore a linear dimension connected with the spacing of such faults in the crystal. If the faults or surfaces of misfit are everywhere opaque to dislocations, $L$ would be the linear dimension of a cell of a superstructure or a mosaic.

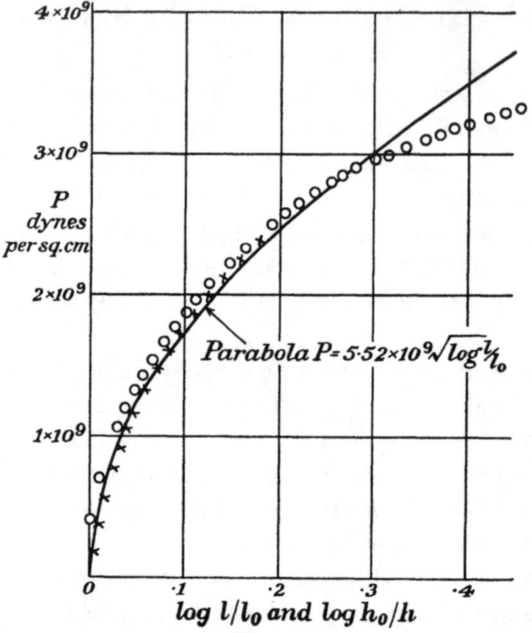


Fig. 1. Plastic stress-strain curve for polycrystalline copper in direct tension or compression. ×, tension, $\log l/l_0$; ○, compression, $\log h_0/h$.

From the parabolas which most nearly fit the observed plastic stress-strain curves the values of $S/\sqrt{s}$ may be found. Thus for aluminium the parabola shown in fig. 1, Part I, is $S=3{\cdot}8\times10^8\sqrt{s}$ dynes per sq.cm. $\mu$ may be taken as $2{\cdot}6\times10^{11}$ so that $S/\mu\sqrt{s}=1{\cdot}46\times10^{-3}$. For copper the parabola of fig. 2, Part I, is $S=8{\cdot}8\times10^8\sqrt{s}$ and $\mu=4{\cdot}55\times10^{11}$, so that $S/\mu\sqrt{s}=1{\cdot}94\times10^{-3}$. For gold the parabola of fig. 2, Part I, is $S=4{\cdot}5\times10^8\sqrt{s}$ and $\mu=2{\cdot}8\times10^{11}$, so that $S/\mu\sqrt{s}=1{\cdot}6\times10^{-3}$. For iron $\mu=8{\cdot}3\times10^{11}$, accordingly the parabola $S=2\times10^9\sqrt{s}$ of fig. 3, Part I, which corresponds with the assumption that iron has no finite elastic limit, gives $S/\mu\sqrt{s}=2{\cdot}4\times10^{-3}$. The parabola $S=1{\cdot}2\times10^8+1{\cdot}5\times10^9\sqrt{s}$, which corresponds with the assumption that iron crystals have an elastic limit of $1{\cdot}2\times10^8$ dynes per sq.cm., gives $(S-S_T)/\mu\sqrt{s}=1{\cdot}8\times10^{-3}$.

These values which are set forth in column 2 of table 1 are according to the present theory equal to $\kappa\sqrt{(\lambda/L)}$. In this expression $\lambda$ is the least distance through which the atoms above the slip-plane must be displaced relative to those below in order that a perfect structure may be re-formed. With face-centred cubic crystals like copper, aluminium, and gold, for which the possible directions of slip are parallel to the diagonals of faces of the cube, $\lambda = b/\sqrt{2}$, where $b$ is the length of the side of the unit cube. For iron which is body-centred, the direction of slip is parallel to a cube diameter and $\lambda = \frac{1}{2}b\sqrt{3}$. The values of $b$ are known from X-ray data and are given in column 3 of table 1, the corresponding values of $\lambda$ being given in column 4. Dividing the figures in column 4 by the squares of the figures given in column 2 values are found for $L\kappa^{-2}$; these are given in column 5.

It is only possible to proceed further by assuming a value for $\kappa$. Taking $\kappa = 0{\cdot}2$ the values given in table 1, column 6, are found for $L$.

Table 1

| (1) | (2) $S/\mu\sqrt{s}$ | (3) $b$ (cm.) | (4) $\lambda$ (cm.) | (5) $K\kappa^{-2}$ (cm.) | (6) $L$ (cm.) ($\kappa = 0{\cdot}2$) |
|---|---|---|---|---|---|
| Al | $1{\cdot}46 \times 10^{-3}$ | $4{\cdot}05 \times 10^{-8}$ | $2{\cdot}86 \times 10^{-8}$ | $1{\cdot}34 \times 10^{-2}$ | $5{\cdot}3 \times 10^{-4}$ |
| Cu | $1{\cdot}94 \times 10^{-3}$ | $3{\cdot}6 \times 10^{-8}$ | $2{\cdot}55 \times 10^{-8}$ | $0{\cdot}68 \times 10^{-2}$ | $2{\cdot}7 \times 10^{-4}$ |
| Au | $1{\cdot}61 \times 10^{-3}$ | $4{\cdot}08 \times 10^{-8}$ | $2{\cdot}87 \times 10^{-8}$ | $1{\cdot}11 \times 10^{-2}$ | $4{\cdot}4 \times 10^{-4}$ |
| Fe | $2{\cdot}4 \times 10^{-3}$ | $2{\cdot}86 \times 10^{-8}$ | $2{\cdot}47 \times 10^{-8}$ | $0{\cdot}43 \times 10^{-2}$ | $1{\cdot}7 \times 10^{-4}$ |

Observed spacing of surface marks in crystals:

| | |
|---|---|
| Bismuth | $1{\cdot}4 \times 10^{-4}$ cm. (Goetz) |
| Zinc | $0{\cdot}8 \times 10^{-4}$ cm. (Straumanis) |
| Iron | $0{\cdot}25 \times 10^{-4}$ cm. (Belaiew) |

## Comparison with Observation of Systems of Faulting and Lineage Systems in Metals

The length $L$ appears in this theory as the distance which a dislocation can travel before being stopped by some barrier such as a surface of misfit. The existence of such surfaces inside a crystal is essential to the theory, but so far they have been regarded as hypothetical. On the other hand, a large number of observations appear to show that surfaces of misfit do occur in crystals.

By suitable treatment Goetz has shown that it is possible to reveal on the surface of bismuth crystals some very closely spaced marks. These consist of three sets of lines spaced $1{\cdot}4 \times 10^{-4}$ cm. apart forming a triangular pattern.

When zinc or cadmium* are deposited from vapour they form a system of plates $0{\cdot}8 \times 10^{-4}$ cm. thick.

Etching pits in iron† seem to point to the existence of a block structure with sides of length $0{\cdot}25 \times 10^{-4}$ cm.

* Straumanis, *Z. Phys.* XIII (1931), 316; XIX (1932), 63.

† Belaiew, *Proc. Roy. Soc.* A, CVIII (1925), 295.

A micro-photograph* which is referred to by Zwicky shows a structure which is brought out by etching a copper crystal. This structure indicates a rectangular arrangement of surfaces of misfit (or according to Zwicky a superstructure) the fundamental linear dimension of which appears (judging from the photograph) to be about $1{\cdot}5 \times 10^{-4}$ cm. These figures are given at the end of table 1 in order to facilitate comparison with the values of $L$ which are necessary in the foregoing theory to account for the observed plastic stress-strain relationships.

It will be seen that the observed spacings are of the same order of magnitude as the values of $L$ given in column 6, table 1, namely, $10^{-4}$ cm., but $L$ is always rather greater than any of the smallest spacings which have been observed. This discrepancy might perhaps be due to an error in calculating the value of $\kappa$. If $\kappa$ were

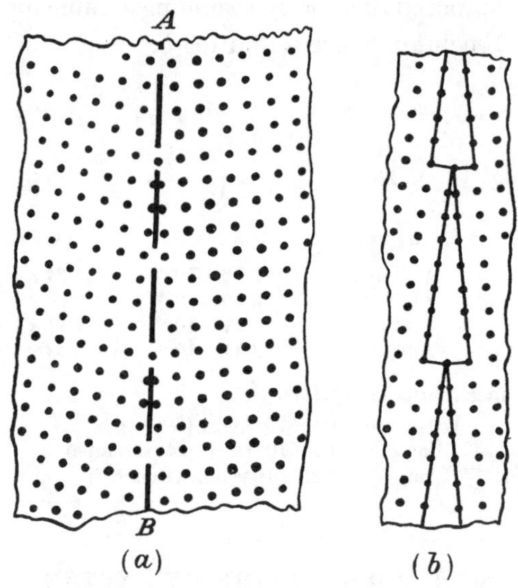


Fig. 2. Boundary of two crystals at slightly different orientations: (*a*) showing portions of surface of misfit opaque to dislocations; (*b*) showing boundaries of separate crystal portions.

0·1 instead of 0·2, all the figures in column 6 would be a quarter of the values there given, and the values of $L$ would be very close to the observed spacings, i.e. about $1{\cdot}0 \times 10^{-4}$ cm.

This explanation of the discrepancy would involve the assumption that the surfaces of misfit which are observed are completely opaque to the passage of dislocations over their whole area. It seems unlikely that such an hypothesis can be true. Whatever conception is held as to the nature of surfaces of misfit it seems that the disturbance of the lattice must be greater in some parts of them than in others. Consider, for instance, Darwin's type of mosaic in which the whole crystallite consists of blocks or bands in which the orientation of the crystal axes of any one block differs from that of its neighbours by several minutes of arc. Fig. 2 represents two neighbouring blocks, bands, or dendrites of a crystal with a cubic lattice. The

* Photograph, Plate 1, *Brown Boveri Rev.*, January 1929.

surface of misfit is shown in fig. 2 (*a*) as $AB$. If the positions of the atoms on the two sides of $AB$ be examined, it will be seen that in certain regions, some of which are represented in fig. 2 (*a*) by gaps in the line $AB$, the distance of atoms on one side from the nearest atoms on the other is the same as that which belongs to the perfect crystal structure. At these points therefore the disturbance of the lattice in passing from one block to the other is small. At the intermediate points the disturbance is a maximum. It is to be expected that as the temperature increases the proportion of the whole area of the surfaces of misfit which are transparent to dislocations will increase.

If $Z$ represents the proportion which is opaque to dislocations, it is to be expected that $Z$ will decrease with rising temperature. Thus at very low temperatures $Z$ might be 1·0 while at a sufficiently high temperature it might tend to 0. When $Z=1{\cdot}0$ and the surfaces of misfit are entirely opaque to dislocations the length $L$ is equal to $B$, the distance between surfaces of misfit. Thus we might expect $L$ to approximate to $B$ at low temperatures, but at higher temperatures $L$ should be greater than $B$.

## Variation in Plasticity with Temperature

It has been seen already that at room temperatures the calculated values of $L$ are of the same order as, but greater than, the observed spacings $B$ of surfaces of misfit. It remains to be seen whether as the temperature is reduced $L$ tends to become equal to $B$, as the present theory predicts it should.

A very complete set of experiments on the plasticity of a metal crystal over a wide range of temperatures has been made by Boas and Schmid,* who determined the $(S, s)$ curves of aluminium crystals over the range for $-185°$ to 600° C. Their results are shown in fig. 3, which is reproduced from their paper. The shaded areas cover all the observations made with many specimens at each temperature.

It will be seen that the $(S, s)$ curves are very similar at all temperatures and agree at 18° C. with the curve of fig. 1, Part I, but that as the temperature rises they become horizontal at rapidly decreasing values of $s$. In order to find the value of $S/\sqrt{s}$ from which $L$ must be calculated the values of $S$ have been taken from Boas and Schmid's curves at $s=0{\cdot}2$. These are given in column 2, table 2.

The values of $L$ calculated from the formula $\dfrac{S}{\mu\sqrt{s}}=\kappa\sqrt{\dfrac{\lambda}{L}}$, where $\kappa=0{\cdot}2$, are given in column 5, table 2. It will be seen that $L$ decreases with decreasing temperature till at $-185°$ C. it is $1{\cdot}8\times10^{-4}$ cm. This is very close indeed to the spacing of surfaces of misfit which have been observed (see data at base of table 1).

## Interpretation of Boas and Schmid's Results

According to the foregoing theory the decrease in $S/\sqrt{s}$ with increasing temperature is due to an increase in $L$. This increase in $L$ is due to a decrease in the fraction $Z$ of the surface of misfit which can act as barriers to the dislocations. If the surfaces of

* *Z. Phys.* LXXI (1931), 713.

24-2

misfit are uniformly spaced at distance $B$ apart, it is possible to calculate the mean value of $L$ as a function of $Z$. Suppose that each surface of misfit acts as an independent barrier to a fraction $Z$ of the centres of dislocation which approach it. We

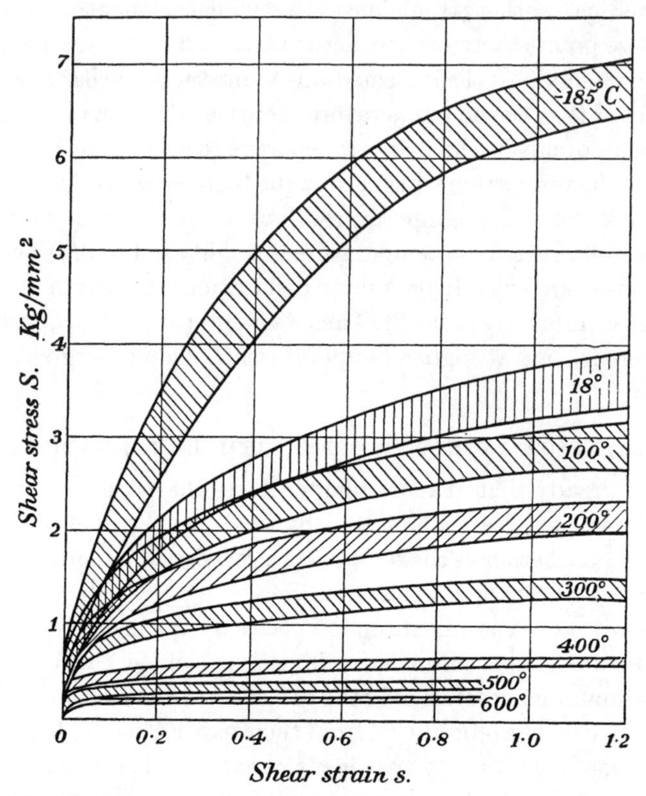


Fig. 3. ($S$, $s$) curves for aluminium crystals (Boas and Schmid).

Table 2

| Temperature (° C.) | $S$ (kg. mm.²) | $S/\mu\sqrt{s}$ $\times 10^3$ | $L\kappa^{-2}$ $\times 10^2$ | $L$ (cm.) ($\kappa=0{\cdot}2$) | $Z=\frac{2B}{L+B}$ $B=1{\cdot}5$ $\times 10^{-4}$ cm. | $B=1{\cdot}0$ $\times 10^{-4}$ cm. | $B=0{\cdot}5$ $\times 10^{-4}$ cm. |
|---|---|---|---|---|---|---|---|
| −185 | 3·0 | 2·53 | 0·477 | $1{\cdot}8\times10^{-4}$ | 0·91 | 0·71 | 0·54 |
| 18 | 1·95 | 1·64 | 1·06 | $4{\cdot}2\times10^{-4}$ | 0·53 | 0·38 | 0·21 |
| 100 | 1·72 | 1·45 | 1·36 | $5{\cdot}4\times10^{-4}$ | 0·43 | 0·31 | 0·17 |
| 200 | 1·37 | 1·156 | 2·14 | $8{\cdot}6\times10^{-4}$ | 0·30 | 0·21 | 0·11 |
| 300 | 0·97 | 0·818 | 4·28 | $1{\cdot}7\times10^{-3}$ | 0·16 | 0·11 | 0·057 |
| 400 | 0·50 | 0·472 | 12·84 | $5{\cdot}1\times10^{-3}$ | 0·06 | 0·038 | 0·019 |
| 500 | 0·34 | 0·287 | 34·8 | $1{\cdot}4\times10^{-2}$ | 0·02 | 0·014 | 0·007 |
| 600 | 0·20 | 0·169 | 100·0 | $4{\cdot}0\times10^{-2}$ | 0·01 | 0·005 | 0·002 |

may consider a large number $N$ of paths of centres of dislocation which cross a plane $AD$ (fig. 4) which is half-way between two surfaces of misfit, and we may calculate their total length. Taking first all the parts of the paths which lie to the left of the plane, $NZ$ of them end on the first barrier and the total length of these is $\frac{1}{2}BNZ$.

Of the $N(1-Z)$ paths which pass the first barrier, $NZ(1-Z)$ end on the second barrier and their total length is $NZ(1-Z)\frac{3}{2}B$. The total length of all the paths to the left of the central plane is therefore

$$NBZ\{\tfrac{1}{2}+\tfrac{3}{2}(1-Z)+\tfrac{5}{2}(1-Z)^2+\ldots\},$$

and since the total length of the parts of the paths to the right of the central plane is the same as that to the left, the total length of all $N$ paths is

$$NBZ\sum_0^\infty(2n+1)(1-Z)^n,$$

which on summation is found to be $NB\left(\frac{2-Z}{Z}\right)$. The average length $L$ is found by dividing this by $N$.

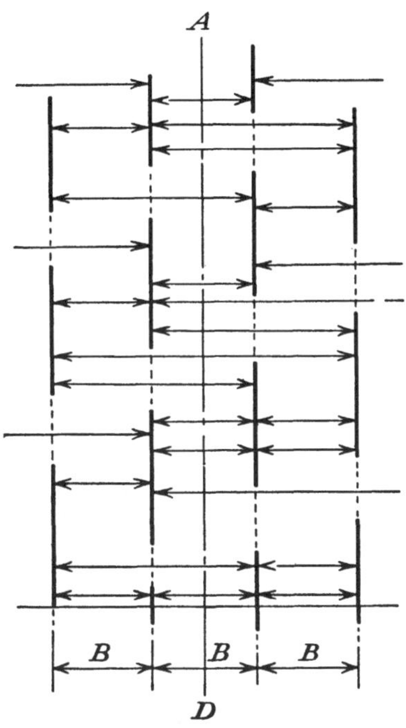


Fig. 4. Paths of dislocations ending on the opaque portions of surfaces of misfit.

Thus
$$L=B\left(\frac{2-Z}{Z}\right), \tag{28}$$

or
$$Z=\frac{2}{(1+L/B)}. \tag{29}$$

The expression (29) may be used to calculate $Z$ if $L$ and $B$ are known. No observations for aluminium indicating the spacing or even the existence of a mosaic system appear to have been made. On the other hand, the observations of surface markings which have been made on other crystals mostly indicate that spacings are usually between 0·5 and $2{\cdot}0\times10^{-4}$ cm. It is instructive to calculate the values of $Z$ which would correspond to surfaces of misfit separated by distances $B=1{\cdot}5\times10^{-4}$,

$1{\cdot}0\times10^{-4}$, and $0{\cdot}5\times10^{-4}$ cm. These are given in columns 6, 7 and 8 of table 2, and are shown in the curves of fig. 5.

Inspection of these curves reveals a very striking fact. Whatever value is chosen for $B$ the proportion $Z$ of the surfaces of misfit which is opaque to dislocations decreases in a regular manner till it reaches a very small value at 400° C. If no measurements had been made above 400° C., the curves would naturally be extrapolated through the small extension, shown dotted in fig. 5, which would have made them cut the axis $Z=0$ at about 480° C. Thus the prediction might have been made that at temperatures above 480° C. the crystal would have no permanent strength

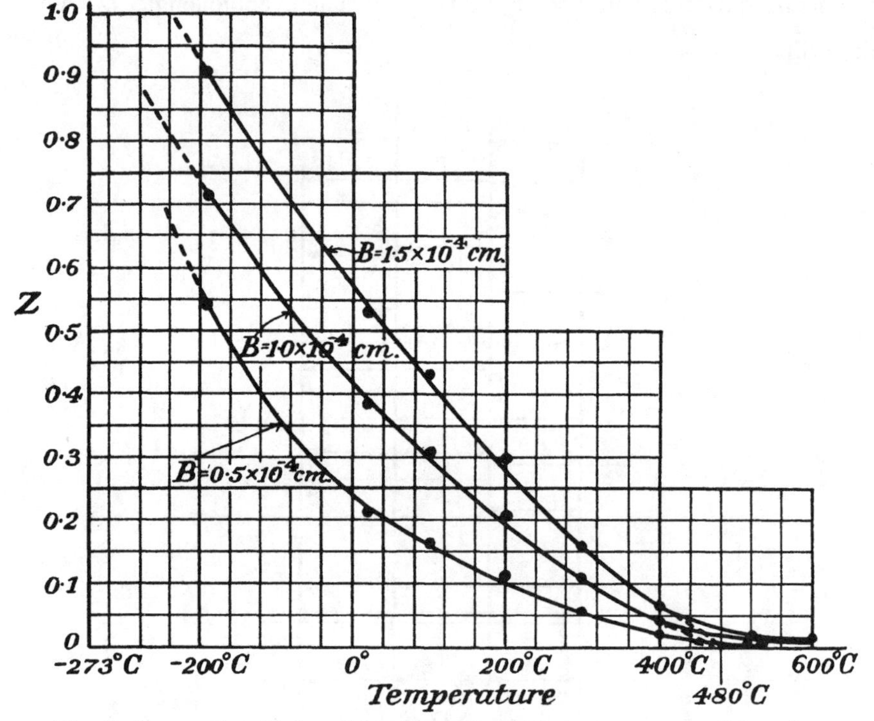


Fig. 5. Proportion $Z$ of surfaces of misfit which are opaque to dislocations as functions of temperature.

so that flow would occur even at the lowest stresses. The fact that recrystallization begins about this temperature indicates that thermal agitation is capable of causing switches among the atoms from one stable position to another even at the most unfavourable points, i.e. where the misfit between neighbouring crystals is most pronounced. It is to be expected therefore that at temperatures above 480° C. surfaces of misfit will not be able to act as barriers to dislocations in any part of their areas. This is equivalent to the statement that at the recrystallization temperature theory would lead one to expect that $Z=0$.

The fact that Boas and Schmid observed finite strengths at 500 and 600° C. might be explained either by the existence of a few foreign atoms as impurities which would still act as barriers to dislocations after the surfaces of misfit had ceased to do so, or alternatively it may be supposed that the strengths observed are not permanent strengths, but that the material is really flowing all the time and that if the velocity of deformation had been very much smaller, very much smaller strengths would have been observed. The latter view is confirmed by the fact that over the whole range of temperature from $-185$ to 400° C. the distortion was found by Boas and Schmid to be of the same nature as that originally found at room temperatures;* a slipping parallel to one plane of type (111) in direction of type [110]. At 500 and 600° C. the distortion was found to be of quite a different type. Test pieces pulled at these temperatures appeared to slip on several planes simultaneously.

The cross-section of a single crystal of aluminium originally circular becomes elliptical on being pulled. At temperatures below 400° C. the major axis of the ellipse remains almost unchanged during the elongation of the specimen, all the contraction occurring in the minor axis. At 500 and 600° C., on the other hand, Boas and Schmid found that both the major and minor axes contract as they would if there were no permanent resistance to shearing on any crystal plane. If, at temperatures above 500° C., slipping can occur on all crystal planes of type (111) it might be expected that the rate of slipping would be greatest in that direction and on that plane which would function alone as the slip-plane at temperatures below 400° C.

The final position of the axis of the specimens in relation to the crystal axes was found by Boas and Schmid by means of X-rays. Below 400° C. it lay in the direction of type [112] which necessarily results when slipping occurs only on the plane of type (111) for which the component of shear stress is greatest. Above 500° C. they found that the end position tended to the directions [111] or [100]. Either of these directions might result from slipping on all four planes of type (111), but in order to calculate the actual movement of the direction of the axis of the specimen in relation to the crystal axes it would be necessary to make some assumption about the relationship between the rate of slipping on any plane and the component of shear stress parallel to it.

## Plasticity at very Low Temperatures

According to the present theory the increase in strain hardening with decreasing temperature is due to a decrease in $L$. If we regard $L$ as being determined by the spacing of existing surfaces of misfit in the crystal, which do not depend on its temperature, the decrease in $L$ can only be attributed to increasing opacity of these surfaces with decreasing temperature to centres of dislocation, i.e. to an increase in $Z$.

As the temperature decreases towards absolute zero the strain hardening will increase, but the increase cannot proceed beyond the point at which $Z=1$, i.e. when $L$ is equal to $B$. In aluminium, for instance, the strain hardening relationship corresponding with Boas and Schmid's measurements at 185° C. is $S/\mu\sqrt{s}=2{\cdot}53\times10^{-3}$

* Taylor and Elam, *Proc. Roy. Soc.* A, CII (1923), 643; paper 5 above.

(see table 2), and this corresponds with $L = 1{\cdot}8 \times 10^{-4}$ cm. The maximum possible amount of strain hardening (corresponding with $Z = 1$) is:

$$\frac{S}{\mu\sqrt{s}} = \begin{cases} 2{\cdot}8 \times 10^{-3} & \text{if} \quad B = 1{\cdot}5 \times 10^{-4}\,\text{cm.,} \\ 3{\cdot}4 \times 10^{-3} & \text{if} \quad B = 1{\cdot}0 \times 10^{-4}\,\text{cm.,} \\ 4{\cdot}8 \times 10^{-3} & \text{if} \quad B = 0{\cdot}5 \times 10^{-5}\,\text{cm.} \end{cases}$$

Thus no very great increase in strain hardening is to be expected at very low temperatures. This indeed is found to hold, for experiments on cadmium crystals and on zinc crystals in liquid helium* have shown that neither the elastic limit nor the strain hardening relationship are appreciably different at 1·2° abs. from what they are at 4·2° abs., 12° or 20° abs. The maximum strength which the crystal can attain after considerable straining is, however, dependent on temperature.

It must be remembered that the theory accounts only for the increase in strength with increasing distortion, it is not explicitly concerned with the question whether or not the dislocations can migrate freely along slip-planes without the application of a finite shear stress.

The most careful observations all seem to indicate that a finite shear stress is necessary before plastic deformation begins, but experiments at low temperatures with aluminium, zinc, and cadmium indicate that this elastic limit only changes slightly with temperature. If the observed elastic limit were really the shear stress necessary to make a single dislocation migrate in a perfect crystal, a large variation with temperature might be expected.

On the other hand, if the centres of dislocation can move freely even at the absolute zero of temperature in the absence of shear stress, the existence of a finite elastic limit might be accounted for by supposing that the crystal is not initially in a stress-free state. If surfaces of misfit exist in a crystal, they must necessarily give rise to internal stresses in the body of the crystal and these stresses would, according to the present theory, prevent the free motion of centres of dislocation provided they were greater than the externally applied stress. These considerations, however, must be regarded as speculative until some method can be found for calculating the internal stresses which might be expected.

## Possible Explanation of the Elastic Limit

If the explanation of the elastic limit just given be accepted, it may be possible to calculate at any rate its order of magnitude. According to the present theory plastic distortion begins when centres of dislocation begin to move through the crystal. If the centres will migrate under the action of any shear stress, however small, which is directed parallel to the slip-planes, plastic distortion will begin when the applied shear stress is greater than the shear stresses, which occur in the body of the crystal owing to the conditions at the surfaces of misfit, for that is the condition that the centres can migrate through the initial 'stress barriers' which must fill the body of the crystal.

* Meissner, Polanyi and Schmid, *Z. Phys.* LXVI (1930), 477.

## CALCULATION OF STRESSES DUE TO SURFACE OF MISFIT

It is possible to form a rough estimate of the magnitude of the 'stress barriers' when the surface of misfit is conceived to be the surface dividing two perfect crystal fragments of which the crystallographic axes are inclined to one another at a small angle $\alpha$. Consider the two cubic arrays of atoms shown in fig. 2 and suppose that the surface of separation between them bisects the angle $\alpha$ between their crystallographic axes. Leaving out of consideration surface effects we may suppose that each fragment if separated from its neighbour would have a stepped surface, the 'steps' occurring at intervals $2b/\alpha$ along the interface. If the two fragments are brought into contact so that the edge of each step in the one meets a corresponding edge in the other, as shown in fig. 2 (*b*), we may think of each fragment as a continuous elastic solid and we may suppose that a pressure distribution is applied normally to each stepped surface so that the steps disappear, leaving a plane boundary. Symmetry ensures that the two portions will be in equilibrium under their mutual pressure at the boundary surface. The internal stresses are therefore those due to the normal pressure distributions which cause the stepped surface to become plane.

In the case illustrated in fig. 2 (*b*) the displacements along the surface of separation vary from $-\frac{1}{2}b$ to $+\frac{1}{2}b$. This arrangement, however, is not that which gives rise to the least elastic disturbance. By bringing the edges of opposite steps into contact it will be seen that a line of atoms which passes through the two edges of the steps contains two more atoms than the corresponding row just below these edges. If the fragments are arranged so that the steps of one alternate with the steps of the other, there is only one more atom in the row which passes through an edge than there is in the corresponding row just beneath it. The displacement necessary to make the two fragments fit now varies from $-\frac{1}{4}b$ to $+\frac{1}{4}b$, but the wave-length of the disturbance along the surface of separation is now only $b/\alpha$ instead of $2b/\alpha$.

Taking axes $ox$ in a plane boundary of an elastic solid, $oy$ perpendicular to it and directed normally inwards the elastic system which is required is that which corresponds with a displacement $v=\frac{1}{2}\alpha x$ over the range $x=-b/2\alpha$ to $x=+b/2\alpha$, this displacement being repeated periodically with wave-length $b/\alpha$.

Consider the stress function

$$\chi = F(1+my)\,m^{-2}\,e^{-my}\sin mx,$$

which satisfies the equation of equilibrium,* namely, $\nabla^4\chi=0$. The corresponding stress components are

$$\left.\begin{aligned} X_x &= \frac{\partial^2\chi}{\partial y^2} = -F(1-my)\,e^{-my}\sin mx, \\ Y_y &= -\frac{\partial^2\chi}{\partial x^2} = -F(1+my)\,e^{-my}\sin mx, \\ X_y &= -\frac{\partial^2\chi}{\partial x\,\partial y} = Fmy\,e^{-my}\cos mx. \end{aligned}\right\} \tag{30}$$

* See Love, *Mathematical Theory of Elasticity*, 4th ed. p. 204.

At $y=0, X_y=0$ so that the $\chi$ represents the effect of applying a normal stress $-F\sin mx$ over the surface $y=0$.

Using a method given by Love the displacements corresponding with (30) may be found. They are

$$\left.\begin{aligned} u&=-\frac{F}{2\mu m}\left(-\frac{\mu}{\lambda+\mu}+my\right)e^{-my}\cos mx,\\ v&=\frac{F}{2\mu m}\left(my+\frac{\lambda+2\mu}{\lambda+\mu}\right)e^{-my}\sin mx, \end{aligned}\right\} \tag{31}$$

where $\lambda$ and $\mu$ are the elastic constants.

Corresponding to the surface displacements

$$\left.\begin{aligned} u&=A_m\left(\frac{\mu}{\lambda+2\mu}\right)\cos mx,\\ v&=A_m\sin mx, \end{aligned}\right\} \tag{32}$$

the stress components are therefore

$$\left.\begin{aligned} X_x&=-2\mu mA_m\left(\frac{\lambda+\mu}{\lambda+2\mu}\right)(1-my)\,e^{-my}\sin mx,\\ Y_y&=-2\mu mA_m\left(\frac{\lambda+\mu}{\lambda+2\mu}\right)(1+my)\,e^{-my}\sin mx,\\ X_y&=2\mu m^2A_m\left(\frac{\lambda+\mu}{\lambda+2\mu}\right)y\,e^{-my}\cos mx. \end{aligned}\right\} \tag{33}$$

Putting $\xi=2\pi x\alpha/b$ the surface displacement is

$$\tfrac{1}{2}\alpha x=f(\xi)=\frac{b\xi}{4\pi},$$

in the range $-\pi<\xi<\pi$; now in this range

$$\tfrac{1}{2}\xi=\sin\xi-\tfrac{1}{2}\sin 2\xi+\tfrac{1}{3}\sin 3\xi\ldots$$

so that at $y=0$

$$v=\frac{b}{2\pi}\Sigma(-1)^{n+1}\frac{\sin n\xi}{n}. \tag{34}$$

Comparing (32) with (34) it will be seen that surface displacement $v$ may be expressed in the form $\Sigma A_m\sin mx$ provided $n\xi=mx$ and

$$A_m=(-1)^{n+1}b/2\pi n.$$

Hence from (33) the stress components are

$$\left.\begin{aligned} X_x&=2\mu\alpha\left(\frac{\lambda+\mu}{\lambda+2\mu}\right)\Sigma(-1)^n\,(1-n\eta)\,e^{-n\eta}\sin n\xi,\\ Y_y&=2\mu\alpha\left(\frac{\lambda+\mu}{\lambda+2\mu}\right)\Sigma(-1)^n\,(1+n\eta)\,e^{-n\eta}\sin n\xi,\\ X_y&=2\mu\alpha\left(\frac{\lambda+\mu}{\lambda+2\mu}\right)\Sigma(-1)^{n+1}\,n\eta\,e^{-n\eta}\cos n\xi, \end{aligned}\right\} \tag{35}$$

where $\eta=2\pi\alpha y/b$.

These expressions represent the distribution of stress which might be expected near a surface of misfit between two crystal fragments the crystallographic axes of which are inclined to one another at a small angle $\alpha$.

In order to visualize the nature of this stress distribution we may calculate the value of $X_y$ at various points.

Let $$f(\xi,\eta)=\Sigma(-1)^{n+1}n\eta\,e^{-n\eta}\cos n\xi, \tag{36}$$

so that $$X_y=2\mu\alpha\left(\frac{\lambda+\mu}{\lambda+2\mu}\right)f(\xi,\eta). \tag{37}$$

By summing the series (36) it can be shown that along the line $\xi=0$,

$$f(\xi,\eta)=\frac{\eta\,e^{-\eta}}{(1+e^{-\eta})^2};\quad \text{when}\quad \xi=\pi,\quad f(\xi,\eta)=\frac{-\eta\,e^{-\eta}}{(1-e^{-\eta})^2},$$

and when $\xi=\frac{1}{2}\pi$, $f(\xi,\eta)=\dfrac{2\eta\,e^{-2\eta}}{(1+e^{-2\eta})^2}$. These functions are shown graphically in fig. 6.

It will be seen that when $\eta>1$ the order of magnitude of the shear stresses in the interior of the crystal due to surfaces of misfit is

$$X_y=2\mu\alpha\left(\frac{\lambda+\mu}{\lambda+2\mu}\right)\eta\,e^{-\eta}.$$

To find numerical values for this expression it is necessary to assume values for $\alpha$ and $\eta$. If the surfaces of misfit are spaced at distances $L$ apart, the value of $\eta$ at the plan mid-way between two such surfaces is $\pi\alpha L/b$.

If $L$ is taken as $10^{-4}$ cm. and $b$ as $3\times10^{-8}$ cm., $\eta=10^4\alpha$. Taking $\alpha=1'=3\times10^{-4}$ radian, i.e. assuming that the crystal consists of a Darwin type of mosaic in which the orientation of the crystal axes of neighbouring fragments differs on the average by $1'$ of arc, $\eta=3$ and $\eta\,e^{-\eta}=0{\cdot}15$. Thus the order of magnitude of $X_y$ is

$$2\mu\,(3\times10^{-4})\,(0{\cdot}015)=10^{-4}\mu.$$

For aluminium $\mu=2{\cdot}6\times10^{11}$, so that

$$X_y=2{\cdot}6\times10^{7}\ \text{c.g.s., or }260\,\text{g.mm.}^2.$$

For copper $\mu=4{\cdot}5\times10^{11}$, so that $X_y=450\,\text{g.mm.}^2$.

These values are, in fact, of the same order of magnitude as the observed elastic limits of copper and aluminium. It seems, therefore, that it is justifiable to state that the existence of a finite elastic limit in a crystal is not inconsistent with the hypothesis that centres of dislocation could migrate freely through a *perfect* unstressed crystal of copper or aluminium in a stress-free state (if indeed such a crystal can exist).

*Slip lines.* In the preceding pages the distribution of centres of dislocation has been considered as statistically uniform. If, however, a slipping begins in a limited region of the crystal by separation of positive and negative centres of dislocation, the accumulation of positive centres on the right and negative centres on the left borders of this region will produce positive shear stresses outside the region and negative shear stress inside it. When a positive external shear stress is applied, therefore, the shear stress will be greatest to the right and left of the disturbed region. Thus the application of a positive external stress will tend to extend the disturbed

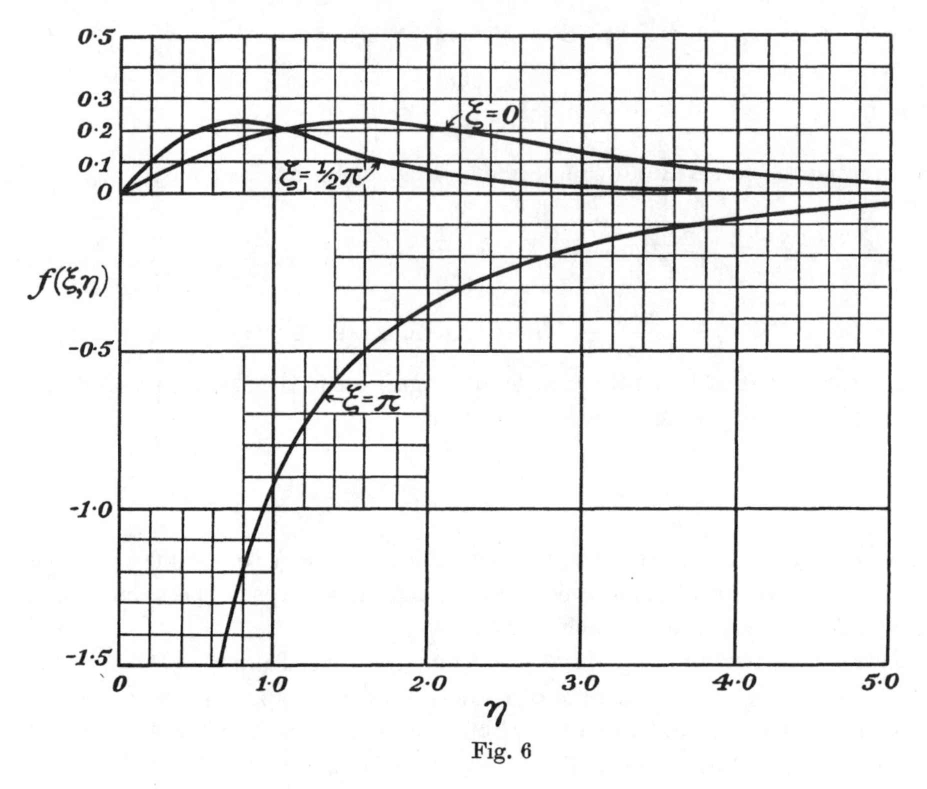

Fig. 6

area in the direction of the slip-lines. The extension of the disturbed region would thus be analogous to the extension of a crack. It seems, then, that the conception of plastic distortion here put forward leads to an expectation that regions of concentrated slipping would extend in the directions of the slip-planes. The existence of slip-lines is therefore consistent with the theory though it is inconsistent with the detailed conception of a regular lattice of dislocations which was used in deriving parabolic stress-strain relationship.

# 23

# THE STRENGTH OF ROCK SALT

REPRINTED FROM

*Proceedings of the Royal Society*, A, vol. CXLV (1934), pp. 405–15

The strength of rock salt has so far only been studied from the theoretical point of view without considering the effect of plastic strain. Experiment shows, however, that plastic strain is the main factor determining the strength of well-annealed crystals. A recent theory of the strength of metals is applied to rock salt and shown to lead to a parabolic relationship between tensile stress, $P$, and plastic strain $\log(l/l_0)$ where $l$ is the length of a stretched bar of rock salt whose initial length is $l_0$. This relationship is

$$P/\mu\sqrt{\{\log(l/l_0)\}} = 2\sqrt{2\kappa}\sqrt{(\lambda/L)},$$

where $\mu$ is the elastic modulus of rigidity, $\lambda$ is the atomic spacing along the plane of slip, $\kappa$ is a constant whose value can be calculated approximately, $L$ is the mean free path of centres of dislocation.

Comparing this with observations of plasticity in rock salt good agreement is found, and $L$ is shown to be of order of magnitude $10^{-4}$ cm. This is the order of magnitude of the distance which has been observed between faults in the structure of rock salt. It is concluded that the strain in rock salt occurs in the crystalline parts of the structure where the crystal order is perfect, and that the strength is determined by the mean free path $L$ of the centres of dislocation. $L$ is determined by the distance apart of the faults and by the temperature. The theory therefore assigns a definite function to the faults in determining the strength of crystals irrespective of their actual crystallographic or atomic nature.

The strength of rock salt has attracted a very large number of workers because the equilibrium of the crystal lattice appears to be well understood. The relationship between atomic forces and elastic constants has been calculated using Born's theory, and has been shown to agree with experiment. The force necessary to tear a crystal into two parts across a crystal plane has also been calculated using the same assumptions about atomic forces as those which account for the elastic properties. Very great disagreement, however, was found. Thus Zwicky* found that a rock salt crystal ought theoretically to stand a stress of 200,000 g.mm.$^2$ before breaking while the breaking stress actually observed† in samples tested at 'room' temperatures is only about 450 g.mm.$^2$.

The reason for this discrepancy has been discussed by Joffé and others who concluded that it was due to surface cracks. By pulling rock salt crystals at high temperatures Joffé found that they stretch plastically, and that strength increases with the amount of plastic stretching. He gives a curve in which the diminution in area of cross-section of a specimen is related to the increase in strength. With

* *Phys. Z.* XXV (1924), 223.

† Joffé, *Z. Phys.* XXII (1924), 286.

large amounts of stretching the strength could be raised to 5000 g. per mm.$^2$. Thus the strength is increased by plastic distortion just as it is for a metal.

Recently Piatti* has described experiments in which considerably higher strength has been attained in rock salt crystals which are stretched under controlled conditions in a bath of salt solution at room temperature. These strengths were associated with plastic extension, but the amount of plastic extension necessary to raise the strength to any given amount was very much less than that found by Joffé; thus Joffé gives the extension necessary to attain a strength of 5000 g.mm.$^2$ as that which reduces the area of cross-section to 1/30th of its original thickness. This corresponds with an elongation of 3000 %. In Piatti's experiments the crystals attained strengths from 10,000 to 20,000 g.mm.$^2$ with extension of about 10–20 %.

When a crystal of rock salt suffers plastic distortion the perfection of the crystal lattice is destroyed so that experiments designed to raise the strength of a rock salt by plastic distortion to the theoretical value calculated for a perfect crystal can only be successful when the crystal is no longer in the condition for which the calculation was made.

For this reason the theoretical interpretations which have hitherto been given concerning the results of experiments on the strength of rock salt seem to be misleading. For a pure and well-annealed rock salt crystal (natural or artificial) the principal factors which determine its strength seem to be the temperature and the amount of plastic distortion which it has undergone since it was in its undistorted condition. Any theoretical discussion of strength must necessarily involve both these factors. It appears, however, that no one has yet introduced the amount of distortion into theoretical discussion, and it is for this reason that I put forward in the following pages an analysis of experiments on the plasticity of rock salt which is based on a recent theory of the strain-hardening of metallic crystals.†

This theory is concerned with metallic crystals in which the slipping is parallel to one slip-plane. The slipping of a plane of atoms over the adjacent parallel plane is conceived as a jump which would leave the regularity of the crystal structure unaltered if slipping took place over the whole area of the slip-plane at once. The characteristic feature of the theory, however, is that the areas on the slip-plane over which the jump has occurred are considered to spread out from centres and to be propagated along the plane till they are stopped by some fault or irregularity in the crystal structure. The edges of the areas over which slipping has taken place are called dislocations and the slipping process is regarded as being determined entirely by the motion of these dislocations. The plastic properties of the crystal are determined by the distance $L$ through which the dislocation can move before its progress is stopped by a fault in the crystal structure.

If $S$ is the component of shear stress parallel to the slip-plane and $s$ is the total

* *Z. Phys.* LXXVII (1932), 401; *Nuovo Cim.* (1932), 102.

† 'The mechanism of plastic deformation of crystals. Parts I and II', *Proc. Roy. Soc.* A, CXLV (1934), 362–404; papers 21 and 22 above.

plastic shearing strain since the crystal was in a completely undeformed state the theoretical relationship between $S$ and $s$ to which this theory leads is

$$\frac{S}{\mu\sqrt{s}}=\kappa\sqrt{\frac{\lambda}{L}}, \tag{1}$$

where $\mu$ is the coefficient of elastic rigidity and $\lambda$ is the atomic spacing along the direction of slip, i.e. the least jump which will leave the crystal structure unaltered. $\kappa$ is a constant which depends on the arrangement of centres of dislocation in the crystal. Values $\kappa=0\cdot12$ and $\kappa=0\cdot17$ were found for two cubic arrangements.

In applying this theory to rock salt crystals the first step is to find the connection between the plastic stress-strain relations in a tensile test of a crystal of rock salt and those which would be found if the crystal could be caused to slip parallel to one slip-plane. It has frequently been remarked that when rock salt is stretched in the plastic condition the distortion does not appear to be due to slipping parallel to one plane. On the other hand, the distortion is definitely related to the crystal structure because in Joffé's experiments a round specimen is described as thinning down to a narrow band of elliptic cross-section. At first sight this seems to show a difference between the behaviour of metals and of rock salt. Further consideration, however, reveals the fact that if rock salt behaves *exactly* like a metal, slip occurring on that slip-plane for which the component of shear stress in the direction of shear is greatest, slipping would always be possible on two planes simultaneously instead of on one only. To prove this we may notice that the slip-planes are of type (110), and the direction of slip is of type [110]. The six possible slip-planes occur in three pairs which intersect on the three cubic axes of the crystal. Considering one such pair we may denote $\theta_1$ and $\theta_2$ as the angles between the normals to the planes and the direction of the axis of the specimen. If $\phi_1$ and $\phi_2$ are the directions which the projections of the longitudinal axis of the specimen on these planes make with the direction of slip, the component of shear stress parallel to the direction of slip is $S_1=P\cos\theta_1\sin\theta_1\cos\phi_1$ in one case and $S_2=P\cos\theta_2\sin\theta_2\cos\phi_2$ in the other. It is easy to prove that when the two slip-planes are perpendicular to one another and the direction of slip perpendicular to their line of intersection

$$\cos\theta_1\sin\theta_1\cos\phi_1=\cos\theta_2\sin\theta_2\cos\phi_2,$$

so that $S_1=S_2$. Hence the components of shear stress on the two members of a pair of slip-planes which intersect on a cubic axis are equal to one another. If therefore rock salt obeys the same law of distortion that has been found for aluminium, slipping on two planes simultaneously is always to be expected instead of the single slipping which occurs in general with metals of the cubic system.

The relationship between the extension of a specimen and the amount of slip has been discussed both for double and single slipping. For example, when the axis of the specimen is one of the cubic axes (as it was, for instance, in Piatti's experiments) we may suppose that during an extension of the specimen from $l$ to $l+\delta l$ an amount of slip $\frac{1}{4}\delta s$ has occurred on each of the four slip-planes which are at 45° to the axis. Then $\delta l = \frac{1}{2} l \delta s$, so that $s = 2 \log (l/l_0)$, where $l_0$ is the original length of the specimen. This relationship is still true if only two of the four slip-planes considered are operative. The tensile stress $P$ is then equal to $2S$. Hence

$$\frac{P}{\sqrt{\{\log (l/l_0)\}}} = 2\sqrt{2}\frac{S}{\sqrt{s}},$$

so that (1) becomes

$$\frac{P}{\sqrt{\{\log (l/l_0)\}}} = 2\sqrt{2}\kappa\mu\sqrt{\frac{\lambda}{L}}. \tag{2}$$

## Comparison with Observations

To compare this formula for the strength of rock salt with observation it is only possible to use data for crystals in which both $P$ and $l/l_0$ have been measured. Usually the extension has been measured, if at all, only at the breaking point, so that it is only possible to compare one stage in the process of plastic distortion of one specimen with another stage of the process in that of another specimen. The available data from Piatti's experiments are given in table 1. In that table $P$ is given in grammes per square millimetre in column 1. In column 2 the value of $l/l_0$ and in column 3 the value of $P/\sqrt{\{\log (l/l_0)\}}$ are also expressed in grammes per square millimetre. The table is arranged so that the figures in column 3 decrease from top to bottom of the column. It will be seen that different specimens vary very largely, the values of $P/\sqrt{\{\log (l/l_0)\}}$ varying from $75\cdot0 \times 10^3$ to $6\cdot5 \times 10^3$ g.mm.$^2$. In table 2 similar figures are taken from a paper by Smekal.* In this case the variation in $P/\sqrt{\{\log (l/l_0)\}}$ is not so large as in Piatti's experiments, but the lower limit is nearly the same, namely, $6 \times 10^3$ g.mm.$^2$

The great variability in the results given in tables 1 and 2 may be due to the presence of impurity for it has been shown that the introduction of small amounts of foreign material,† e.g. lead chloride and cuprous chloride into a rock salt crystal has a very great hardening effect. It may also be due to varying amounts of distortion or work-hardening in the natural crystals before starting the experiments.

In a more recent set of measurements Thiele‡ has obtained the load-extension or strain-hardening curves for several specimens of rock salt at a number of different temperatures from 20 to 600° C. Thiele's results also showed considerable variations, but by comparing the load-extension curves of different specimens he was able to distinguish between them in regard to their state of hardness in their natural condition. The yield point of one of his specimens (no. 13), for instance, was about 10 times as high as that of some others while the breaking stress was about seven times as high. By prolonged heating at 600° C. he was able to reduce this particular specimen to a condition similar to that of others. In this way it was found possible

* *Phys. Z.* XXXII (1931), 187.
† Metag, *Z. Phys.* LXXVIII (1932), 363.
‡ *Z. Phys.* LXXV (1932), 763.

to avoid the very great variability among individual measurements which characterized the results of Piatti, and to approximate to a definite standard characteristic of pure rock salt in a work-free* condition.

The complete results for one such specimen are shown in fig. 1, which is reproduced from Thiele's paper. In this diagram the maximum value of the extension was only 3·5 % so that ordinates which are expressed as extension per cent are very nearly proportional to log $(l/l_0)$. According to the formula (2), if $L$ depends only on the temperature and the disposition of pores, system of faulting or superstructure in the crystal, then $P/\sqrt{\{\log(l/l_0)\}}$ should be constant. The theoretical prediction is therefore that the curves shown in fig. 1 should be parabolas.

Table 1. *Piatti's measurements*

| $P$ (g. mm.²) | $l/l_0$ | $P/\sqrt{\{\log(l/l_0)\}}$ (g. mm.²) |
|---|---|---|
| 23150 | 1·100 | $75{\cdot}0 \times 10^3$ |
| 19530 | 1·215 | $43{\cdot}1 \times 10^3$ |
| 10775 | 1·093 | $36{\cdot}0 \times 10^3$ |
| 13890 | 1·187 | $33{\cdot}5 \times 10^3$ |
| 9540 | 1·085 | $33{\cdot}5 \times 10^3$ |
| 10290 | 1·120 | $31{\cdot}0 \times 10^3$ |
| 13370 | 1·215 | $30{\cdot}0 \times 10^3$ |
| 12650 | 1·215 | $28{\cdot}5 \times 10^3$ |
| 9920 | 1·128 | $28{\cdot}0 \times 10^3$ |
| 10685 | 1·190 | $26{\cdot}0 \times 10^3$ |
| 10160 | 1·207 | $23{\cdot}5 \times 10^3$ |
| 9260 | 1·270 | $18{\cdot}9 \times 10^3$ |
| 6945 | 1·155 | $18{\cdot}3 \times 10^3$ |
| 8960 | 1·320 | $16{\cdot}8 \times 10^3$ |
| 6250 | 1·240 | $13{\cdot}4 \times 10^3$ |
| 4720 | 1·140 | $13{\cdot}1 \times 10^3$ |
| 5785 | 1·228 | $12{\cdot}8 \times 10^3$ |
| 5340 | 1·200 | $12{\cdot}8 \times 10^3$ |
| 3654 | 1·097 | $12{\cdot}1 \times 10^3$ |
| 3150 | 1·122 | $10{\cdot}5 \times 10^3$ |
| 3470 | 1·129 | $9{\cdot}7 \times 10^3$ |
| 2045 | 1·102 | $6{\cdot}5 \times 10^3$ |

Table 2. *Measurements given by Smekal* (Phys. Z. XXXII (1931), 187)

| $P$ (g. mm.²) | $l/l_0$ | $P/\sqrt{\{\log(l/l_0)\}}$ (g. mm.²) |
|---|---|---|
| 4350 | 1·058 | $18{\cdot}0 \times 10^3$ |
| 2660 | 1·032 | $15{\cdot}0 \times 10^3$ |
| 3390 | 1·051 | $15{\cdot}0 \times 10^3$ |
| 2970 | 1·074 | $10{\cdot}9 \times 10^3$ |
| 3430 | 1·114 | $10{\cdot}2 \times 10^3$ |
| 2180 | 1·103 | $6{\cdot}8 \times 10^3$ |

To test this the points on Thiele's curves representing his actual observations were marked on a diagram (fig. 2) and parabolas drawn so as to pass as nearly as possible through the observed points. It will be seen that the agreement with theory is very good.

* 'Verformungsfrei.'

Thiele's results are tabulated in table 3. Values of $P$ and $l/l_0$ taken for his curves are given in columns 1 and 2 while in column 3 are given the values of $P/\sqrt{(\log l/l_0)}$ calculated from the figures of columns 1 and 2.

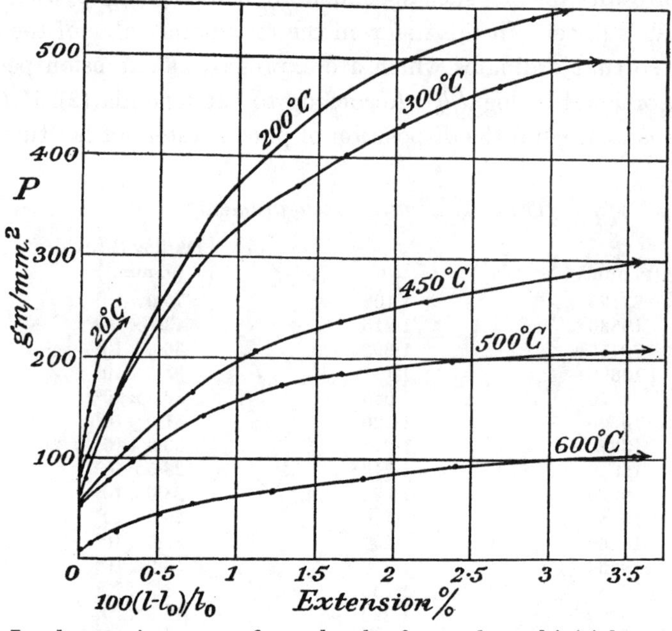


Fig. 1. Load-extension curves for rock salt after prolonged initial annealing (Thiele).

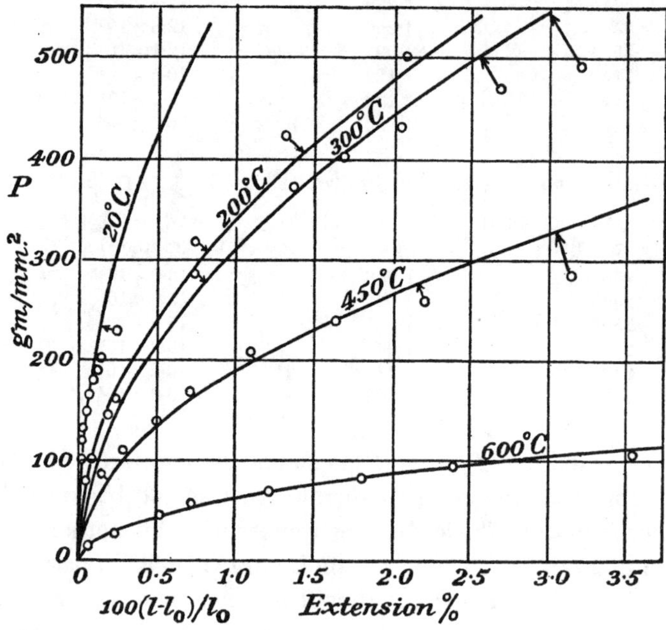


Fig. 2. Comparison of theory with experiment. The curves represent the theoretical relationship $P/\sqrt{(\log l/l_0)} = \text{constant}$.

It will be seen that at each temperature $P/\sqrt{\{\log (l/l_0)\}}$ is nearly constant. The average value of the constant for each observed temperature is also given in column 3.

It is worth noticing that at 20° C. the constant is $6 \times 10^3$ g.mm.$^2$, which is identical with the lowest values obtained both in Piatti's and Smekal's experiments. This fact may be significant since Piatti's and Smekal's experiments were carried out under water which allowed an extension of 10 % to take place. Thiele's experi-

Table 3. *Thiele's measurements (taken from his curves)*

| $P$ (g. mm.$^2$) | $l/l_0$ | $P/\sqrt{\{(\log (l/l_0)\}}$ (g. mm.$^2$) |
|---|---|---|
| | Temperature 20° C. | |
| 145 | 1·0005 | $6{\cdot}5 \times 10^3$ |
| 165 | 1·0007 | $6{\cdot}2 \times 10^3$ |
| 180 | 1·0010 | $5{\cdot}7 \times 10^3$ |
| 200 | 1·0013 | $5{\cdot}5 \times 10^3$ |
| | | Mean $6{\cdot}0 \times 10^3$ |
| | Temperature 200° C. | |
| 140 | 1·002 | $3{\cdot}1 \times 10^3$ |
| 320 | 1·0075 | $3{\cdot}7 \times 10^3$ |
| 420 | 1·013 | $3{\cdot}7 \times 10^3$ |
| 500 | 1·021 | $3{\cdot}4 \times 10^3$ |
| 540 | 1·029 | $3{\cdot}2 \times 10^3$ |
| | | Mean $3{\cdot}4 \times 10^3$ |
| | Temperature 300° C. | |
| 160 | 1·0028 | $3{\cdot}0 \times 10^3$ |
| 290 | 1·0075 | $3{\cdot}3 \times 10^3$ |
| 372 | 1·0140 | $3{\cdot}1 \times 10^3$ |
| 400 | 1·0167 | $3{\cdot}1 \times 10^3$ |
| 428 | 1·0205 | $3{\cdot}0 \times 10^3$ |
| 475 | 1·0269 | $2{\cdot}9 \times 10^3$ |
| 495 | 1·0320 | $2{\cdot}8 \times 10^3$ |
| | | Mean $3{\cdot}0 \times 10^3$ |

| $P$ (g. mm.$^2$) | $l/l_0$ | $P/\sqrt{\{\log (l/l_0)\}}$ (g. mm.$^2$) |
|---|---|---|
| | Temperature 450° C. | |
| 85 | 1·0015 | $2{\cdot}2 \times 10^3$ |
| 106 | 1·0029 | $2{\cdot}0 \times 10^3$ |
| 139 | 1·0050 | $2{\cdot}0 \times 10^3$ |
| 167 | 1·0073 | $2{\cdot}0 \times 10^3$ |
| 206 | 1·0115 | $1{\cdot}9 \times 10^3$ |
| 236 | 1·0165 | $1{\cdot}8 \times 10^3$ |
| 256 | 1·0221 | $1{\cdot}7 \times 10^3$ |
| 285 | 1·0313 | $1{\cdot}6 \times 10^3$ |
| | | Mean $1{\cdot}9 \times 10^3$ |
| | Temperature 600° C. | |
| 15 | 1·0006 | $0{\cdot}61 \times 10^3$ |
| 29 | 1·0024 | $0{\cdot}59 \times 10^3$ |
| 44 | 1·0051 | $0{\cdot}62 \times 10^3$ |
| 55 | 1·0072 | $0{\cdot}65 \times 10^3$ |
| 68 | 1·0122 | $0{\cdot}62 \times 10^3$ |
| 80 | 1·0180 | $0{\cdot}60 \times 10^3$ |
| 92 | 1·0240 | $0{\cdot}59 \times 10^3$ |
| 104 | 1·0355 | $0{\cdot}55 \times 10^3$ |
| | | Mean $0{\cdot}60 \times 10^3$ |

ments were not performed under water and the specimen seems to have stretched uniformly only up to about 0·2 % extension. This seems to suggest that the presence of water round the specimen may have no effect on the strain-hardening curve though it has a very large effect on the total extension which the specimen can suffer before breaking.

## Determination of $L$

The constancy of the observed values of $P/\sqrt{\{\log (l/l_0)\}}$ at any given temperature seems to confirm the accuracy of the conception here put forward of the mechanism of plastic deformation. Accordingly we may proceed with more confidence to use (2) for the determination of $L$. This formula may be rewritten

$$L = 8\mu^2\kappa^2\lambda[P/\sqrt{\{\log (l/l_0)\}}]^{-2}. \tag{3}$$

For rock salt $\mu$ may be taken as $1{\cdot}3 \times 10^{11}$ c.g.s. $\lambda$ is the least relative distance through which two parts of a crystal situated on opposite sides of a slip-plane

25-2

must move in order that the perfect crystal structure may be reformed. For the cubic lattice of NaCl

$$\lambda = \tfrac{1}{2}\sqrt{2} \times (\text{side of unit cube}) = \tfrac{1}{2}\sqrt{2}(5\cdot628 \times 10^{-8}) = 3\cdot98 \times 10^{-8} \text{ cm.}$$

The value of $\kappa$ depends on the manner in which the dislocations are distributed. Values $\kappa = 0\cdot12$ and $0\cdot17$ were calculated for special cases (i.e. square lattices of dislocations). In the absence of any definite method for estimating $\kappa$ it is only possible to point out that the various assumptions regarding the distribution of centres of dislocation lead to values which do not differ in order of magnitude and to calculate the values of $L$ which correspond with the various assumptions.

Setting $\kappa = 0\cdot12$ and $0\cdot17$ in (2) and using the observed values of $P/\sqrt{\{\log(l/l_0)\}}$ the two sets of values given in table 4 have been calculated for $L$.

Table 4. *Values of L calculated from formula* $L = 8\mu^2\kappa^2\lambda[P/\sqrt{\{\log(l/l_0)\}}]^{-2}$

| Temperature (°C.) | $P/\sqrt{\{(\log(l/l_0)\}}$ (g. mm.$^2$) | $L$ (cm.) $\kappa=0\cdot12$ | $L$ (cm.) $\kappa=0\cdot17$ |
|---|---|---|---|
| 20 | $6\cdot0 \times 10^3$ | $2\cdot2 \times 10^{-4}$ | $4\cdot4 \times 10^{-4}$ |
| 200 | $3\cdot4 \times 10^3$ | $6\cdot8 \times 10^{-4}$ | $1\cdot4 \times 10^{-3}$ |
| 300 | $3\cdot0 \times 10^3$ | $8\cdot8 \times 10^{-4}$ | $1\cdot9 \times 10^{-3}$ |
| 450 | $1\cdot9 \times 10^3$ | $2\cdot2 \times 10^{-3}$ | $4\cdot4 \times 10^{-3}$ |
| 600 | $0\cdot6 \times 10^3$ | $2\cdot2 \times 10^{-2}$ | $4\cdot4 \times 10^{-2}$ |

## Comparison between $L$ and the Average Size of an Element of a Mosaic Crystal Structure

It will be seen from table 4 that at 20° C. $L$ is of order of magnitude $10^{-4}$ cm. This is of the same order of magnitude as the estimates which have already been made of the average distance apart of pores or faults in the structure of rock salt. From X-ray data Darwin, Bragg and James* concluded that the linear dimensions of the elements of their mosaic are greater than $0\cdot5 \times 10^{-4}$ cm. From the work of Seidentopf† on the diffusion of the vapour of alkali metals into rock salt Smekal‡ considered that the average distance apart of 'Löckerstellen' or loose places in the structure is about $2 \times 10^{-4}$ cm.

It seems likely, therefore, that the 'Löckerstellen' of Smekal or the faults at the boundaries of the blocks of a mosaic structure do, in fact, act as barriers to the free movement of dislocations. If the present theory is accepted these faults control the plastic properties of the crystal by limiting the movement of dislocations, but the actual slipping occurs in the parts of the crystal where the atoms lie in perfect order. This conception of the mechanics of plastic deformation is the inverse of what has sometimes been proposed, namely, that the slipping occurs in the faults leaving the intervening blocks as fragments in perfect crystalline order.

* *Phil. Mag.* I (1926), 897.
† *Phys. Z.* VI (1905), 855.
‡ *Phys. Z.* XXVI (1925), 709.

At high temperatures the mean free path $L$ of the dislocations is many times as great as the mean size of an element of the mosaic. This might be accounted for by supposing that dislocations can cross the faults or boundaries between the elements of a mosaic at some places and not at others. At high temperatures a fault would act as a barrier only at points where the discontinuity between the crystal structure and neighbouring blocks was a maximum. At lower temperatures a greater proportion of the area of the fault surfaces would be opaque to dislocations. This suggestion has been partially analysed in a previous paper in connection with aluminium.

## The Elastic Limit

In the previous paper a suggestion was made that the elastic limit was caused by internal elastic stresses produced by the misfit at the boundaries of elements of a mosaic. This idea led to an approximate formula for determining the order of magnitude of the elastic limit. When the average angle between the orientation of the crystal axes of neighbouring blocks is 1 minute of arc and the average diameter of a block is $10^{-4}$ cm. the value found for the elastic limit was of order of magnitude $10^{-4}\mu$. For rock salt $\mu = 1{\cdot}3 \times 10^{11}$, so that the order of magnitude of the elastic limit which would result from a block structure of this type would be $1{\cdot}3 \times 10^7$ c.g.s. or 130 g.mm.$^2$.

Referring to fig. 1 it will be seen that the elastic limit of the specimen there referred to was about 90 g.mm.$^2$, at 20° C., and that it decreased to 50 g.mm.$^2$ at 500° C., and 10 g.mm.$^2$ at 600° C. The theory, therefore, gives at any rate the correct order of magnitude for the elastic limit.

# 24

# A THEORY OF THE PLASTICITY OF CRYSTALS

REPRINTED FROM

*Zeitschrift für Kristallographie*, A, vol. LXXXIX (1934), pp. 375–85

The fact that the macroscopic distortion of metallic crystals is a shear parallel to a crystal plane and in a crystal direction and the fact that this remains true even when the distortion is large shows that the plastic strain must be chiefly due to the sliding of one plane of atoms over its immediate neighbour in such a way that the perfect crystal structure is reformed after each atomic jump. It is supposed that slipping occurs over limited lengths $L$ of the slip-plane, and it is shown that this type of plastic strain necessarily gives rise to elastic stresses near the two dislocations which occur at the two ends of each of these lengths $L$.

It is then shown that the assumption that such dislocations will migrate through the crystal, owing perhaps to temperature agitation, under the influence of even the smallest shear stress leads to a definite picture of the mechanics of plastic distortion. This theory of strain-hardening is expressible in quantitative form and gives a parabolic relationship between stress and plastic strain, namely, $S/\mu\sqrt{s}=K\sqrt{(\lambda/L)}$. This expression is in good agreement with the results of experiment in the cases of metals which crystallize in the cubic system and in the case of rock salt.

The observed parabolic relationship is then used in connection with the formula to determine $L$ which is found, at room temperatures, to be of order of magnitude $10^{-4}$ cm. This is of the same order of magnitude as the observed spacings of faults in metals and rock salt. According to this theory the part played by the system of faulting or mosaic structure is to limit the free motion of centres of dislocation. The actual strain takes place inside the 'blocks' of the mosaic structure and the crystallographic nature of the faults, i.e. whether they are boundaries of dendrites, a superstructure or merely 'pores' is immaterial from the point of view of the theory.

It seems to be universally admitted that the strength and plastic properties of metallic crystals and of rock salt are structure-sensitive properties so that they probably depend in some way on the existence of faults or irregularities in crystal structure. Up to the present all attempts to find out what the connection is seem to have been directed to finding a theoretical value for the breaking strength or elastic limit and connecting it with some hypothetical kind of mosaic structure. When the experimental data are considered it is at once obvious that in many cases, particularly with cubic metals and rock salt, the elastic limit is very sensitive to small quantities of impurity and to the thermal and mechanical history of the crystal. When the crystals are well annealed the elastic limit is usually very small compared with the strength which they ultimately attain after considerable plastic distortion. The breaking stress of the most plastic metallic crystals is merely the stress at which a tensile specimen becomes unstable so that all stretching concentrates at any spot where it happens to be thinnest. It depends essentially on the amount of plastic distortion which has taken place.

Even in cases like that of rock salt in air at room temperature, plastic distortion occurs before breakage. Any search therefore for a direct connection, independent of the amount of plastic strain, between mosaic structure and strength must lead in a contrary direction to that indicated by the experimental evidence.

Examination of the experimentally determined plastic stress-strain relations for cubic metals like copper or aluminium seems to show that the elastic limit is very low indeed, but that very small amounts of distortion cause large increases in strength. As the amount of distortion increases the stress necessary to produce a further distortion goes on increasing, but the rate of increase in strength with plastic strain continually decreases. The plastic stress-strain curves in general resemble the parabolas $S = \text{constant} \times \sqrt{s}$, where $S$ is the stress and $s$ the total plastic strain, since the crystal was in a fully annealed condition. Recent experiments by Thiele* have shown that when rock-salt crystals are very thoroughly annealed the plastic stress-strain relations approach a standard form for any given temperature which is very similar to that obtained with metals, though unannealed crystals exhibit extreme variability in this respect. It seems to me therefore that experimental evidence points to the possibility of a direct connection between mosaic structure and the plastic stress-strain relationship, but not directly between mosaic structure and the strength.

Up to the present no one seems to have considered either the geometrical nature or the magnitude of the plastic strain in connection with the strength, so that it is not surprising that very little progress has been made towards understanding the strength of plastic crystals.

It is only possible to take account of the plastic strain in connection with some definite theory concerning the physical nature of a work-hardened or plastically strained crystal. Several qualitative theories have been proposed but none of them take any account of the factor which experiment shows to be the most important, namely the amount of strain. In the following pages an attempt is made to supply a theory giving a picture of a plastically strained crystal which is capable of quantitative interpretation.

The simplest type of plastic strain observed on a macroscopic scale consists of a translation or shear parallel to a crystal plane and in a crystal direction. The fact that this translation preserves this crystallographic direction even when the amount of shear is large can, I think, only be explained by supposing that the greater part at any rate of the deformation is due to a combination of simple jumps or stability interchanges in which one plane of atoms moves through one lattice spacing relative to its immediate neighbour so that the perfect crystalline order is reformed. In order that plastic distortion may occur, atoms must in some way jump the potential barriers which surround them when they form part of a perfect crystal structure. Unless they fall into the next available place along that crystallographic axis which is the line of slip on the macroscopic scale it is very difficult to understand why the

* W. Thiele, *Z. Phys.* LXXV (1932), 763.

relative displacement of two atoms which were originally neighbours should be parallel to this crystallographic axis.

Jumps of the kind just described would transform the block of atoms represented diagrammatically in (*a*), fig. 1, into the configuration (*c*). It is clear, however, that the whole strain cannot be regarded as compounded of movements of this kind because the transformations from (*a*) to (*c*) in fig. 1 leaves the crystal with its properties unchanged.

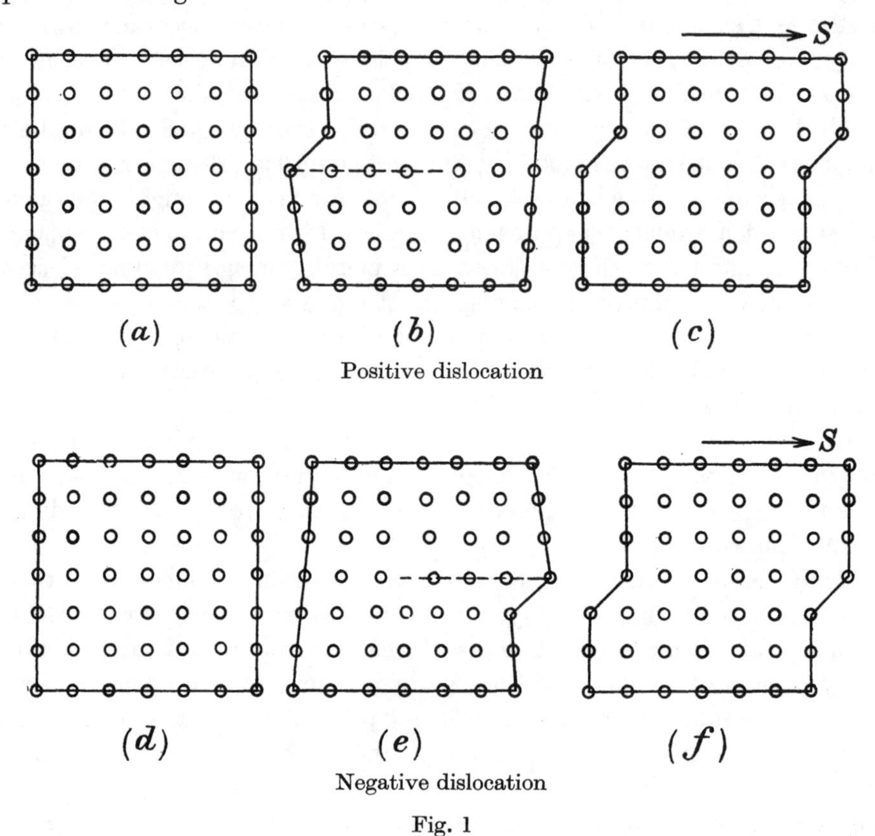


Positive dislocation

Negative dislocation

Fig. 1

If the jump is supposed to take place not over the whole but only over a portion of any plane, the macroscopic strain which is observed would still be accounted for as a combination of such translational movements over limited areas and now the strained crystal would be different from the original unstrained crystal because it would contain lines which will be called 'dislocations', where the portions of planes which have slipped are divided from those which have not. The diagram (*b*), fig. 1, represents the arrangement of atoms in such a dislocation. Here it is represented as two-dimensional, so that the dislocation is marked by a point instead of by a line on the plane of slip as it would be in three dimensions.

One of the chief difficulties in considering slipping as an atomic phenomenon occurring simultaneously over all atoms in a plane is that shearing forces comparable in order of magnitude with the elastic moduli of the crystal would be necessary in order to displace one plane relative to its neighbour to the position of unstable equilibrium from which the atoms would fall into their new stable positions. This difficulty does not arise when the area over which the translational glide has occurred is limited, because the process can then be regarded as a spreading of these areas, one atom at a time, owing to thermal agitation. The conception proposed is therefore that the centres of dislocation move along the slip-planes atom by atom in jumps which carry the atom over a very low potential barrier from one position to another. It may be supposed that in a perfect crystal structure under no external stress the centres of dislocation are freely mobile, provided the temperature is high enough, being able to jump back and forth along a crystallographic plane of slip. The application of even the smallest stress will then cause the jumps which relieve stress to occur more frequently than those in the reverse direction. Thus if the temperature is sufficiently high the centres of dislocation will travel along the slip-planes under the action of any shear stress however small it may be.

It will be observed that in two dimensions dislocations are of two types which will be called positive and negative respectively. They are represented as configurations (*b*) and (*e*) in fig. 1. In a positive dislocation (*b*) a row of atoms of given length immediately above the centre of the dislocation contains one more atom than the same length of a row below the centre. In a negative dislocation (*e*) the reverse is the case.

The action of a shear stress $S$ in the direction indicated in fig. 1 will cause a positive centre to migrate to the right and a negative centre to the left. The result of the passage of a dislocation across the block from side to side is identical in the two cases, namely the shift of the upper half of the crystal block one atomic spacing to the right relative to the lower half.

We may visualize the process of plastic strain as consisting in the generation of pairs of positive and negative dislocations which separate under the action of the applied shear stress, the positive ones moving to the right and the negative ones to the left or vice versa.

We may now examine the consequences of the assumptions (*a*) that the strain is of the nature described above, and (*b*) that the dislocations will migrate freely under the influence of any shear stress however small which has a component parallel to the plane of slip.

Each centre of dislocation is itself the centre of a field of elastic strain. To find in detail the stresses very close to the centre of the dislocation it would be necessary to determine the actual positions of the atoms from some assumed law of force between them, but at distances more than a few lattice spacings the theory of dislocations in an elastic medium must apply. If the plane of slip is taken as the axis of $x$, and if $\lambda$ is one lattice distance along the slip-plane, the three components of stress are*

$$X_x = -\frac{\mu\lambda y}{\pi(x^2+y^2)}, \quad Y_y = \frac{\mu\lambda y}{\pi(x^2+y^2)}, \quad X_y = \frac{\mu\lambda x}{\pi(x^2+y^2)},$$

* Timpe, Göttingen Diss., Leipzig (1905).

where $x$ and $y$ are the co-ordinates of the point considered, the centre of the dislocation being taken as origin, and $\mu$ is the coefficient of rigidity; thus on a plane parallel to the slip-plane the shear stress is $X_y$. Considering two centres, the shear stress at each due to the presence of the other will, according to assumption ($b$), cause them to migrate along their respective slip-planes. In this way we may see that two positive or two negative centres will repel one another while a positive and negative centre will migrate into such a position that the line joining them is perpendicular to the slip-planes.

If a shear stress $S$ is applied to a perfect crystal which contains only one dislocation this dislocation will migrate and will continue to migrate as long as $S$ is applied, even when $S$ is very small. On the other hand, if the crystal contains one positive and one negative dislocation in equilibrium on two slip-planes distant $h$ apart, the application of a positive shear stress $S$ will cause the positive dislocation to move to the right and the negative one to the left until

$$S = \mu\lambda d/\pi(d^2 + h^2), \tag{1}$$

where $d$ is the projection of the line joining the centres on the slip-plane. Since the maximum value of $d/(d^2+h^2)$, as $d$ varies, is $\frac{1}{2}h$, the maximum value of $S$ is $\mu\lambda/2\pi h$. If therefore $S < \mu\lambda/2\pi h$ the two dislocations will merely be displaced to new positions of equilibrium, but if $S > \mu\lambda/2\pi h$ the two centres will escape from their mutual reaction and will continue to migrate away from one another as long as the shear stress is maintained. In other words, plastic strain will only occur if $S > \mu\lambda/2\pi h$.

Consider now the whole process of slipping and suppose that at any stage $N$ pairs of centres of dislocation per unit area (i.e. $N$ lines of dislocation per unit volume) have been separated, migrating in opposite directions along slip-planes till the average distance apart of any pair is $L$. If the dislocations are distributed evenly through the volume of the crystal the distance of each from its nearest neighbour of opposite sign is proportional to $N^{-\frac{1}{2}}$. We may assume therefore that the application of a shear stress $S > \dfrac{c\mu\lambda\sqrt{N}}{2\pi}$ would cause the centres of dislocation to migrate past one another. Here $c$ is a numerical factor of order of magnitude unity. The shear stress necessary for plastic deformation is therefore

$$S = c\mu\lambda\sqrt{N}/2\pi. \tag{2}$$

It is not possible to calculate the actual value of $c$ without making special assumptions regarding the distribution of the centres of dislocation. I have calculated for instance the particular case of a cubic arrangement in which centres of dislocation of opposite sign lie on two interpenetrating cubic lattices. In this case I find

$$S = 0\cdot174\mu\lambda\sqrt{N},$$

so that

$$c = (0\cdot174)\,2\pi = 1\cdot1. \tag{3}$$

Next consider the plastic strain $s$. The motion of one dislocation across a cube whose sides are of unit length causes the upper part of the cube to shift through a distance $\lambda$ relative to the lower part. The shear strain $s$ is defined as the lateral translation of the top of a unit cube relative to its base. The separation of a single pair of dislocations in a unit cube to distance $L$ apart therefore gives rise to an element of strain $\delta_s = \lambda L$, and $N$ such pairs give rise to a strain

$$s = N\lambda L. \tag{4}$$

Combining (2) and (4) it will be seen that

$$\frac{S}{\mu\sqrt{s}} = \frac{c}{2\pi}\sqrt{\frac{\lambda}{L}} = 0{\cdot}16c\sqrt{\frac{\lambda}{L}}. \tag{5}$$

The remarkable formula (5) is derived from two considerations only: (1) the conception of the nature of strain which is forced on us by the observed macroscopic relationship between plastic strain and crystallographic axes and (2) the assumption that the special singularity which is called a dislocation is freely mobile in a perfect crystal on one crystallographic plane, called a slip-plane, provided the temperature is sufficiently high. The condition of free mobility seems to necessitate that the dislocation is situated in a region where the atoms are perfectly well ordered in a regular lattice. It seems justifiable to suppose therefore that the dislocations would be unable to pass through places in the crystal where the structure is not perfect. Such places might be pores (Smekal), a superstructure (Zwicky) or boundaries of the blocks of a Darwin type of mosaic or interlineage boundaries (Buerger). The exact nature of the faults is immaterial in the present theory; all that is essential is that the distances between the fault shall in some way determine the length $L$, which is a 'mean free path' for dislocations.

If $L$ is determined by the distance between faults in the structure it should not depend on the amount of the strain so that $L$ might be expected to be constant while $S$ and $s$ vary. In that case the relation (5) gives a theoretical relationship between $S$ and $s$, namely $S \propto \sqrt{s}$. The $(S, s)$ curves should therefore be parabolas.

Measurements have been made with single crystals of aluminium, copper, gold and rock salt. The metals slip on one plane so that the $(S, s)$ curves can be found, and in each case a parabola passes close to the observed points.* The values of $S/\mu\sqrt{s}$ corresponding with the nearest parabola to the observed points are given in column 3 of table 1. The theory therefore which was designed to account for the observed mode of distortion of metallic crystals also accounts for the plastic stress-strain relationship. So far as I know no other theory of the strength of crystals has yet been devised which accounts even approximately for the very large increase in resistance to distortion which occurs when a metal is deformed from an initially well annealed condition.

Rock salt slips in the direction [110] and on the planes (110) and it can be shown that the shear stress is always identical on two such planes so that double slipping is always to be expected. Taking the case of a rod of rock salt cut so that its

* In the case of copper single crystals the agreement is not good, but the load-extension curves for polycrystalline copper rods are very nearly parabolic.

longitudinal axis is in a direction [100] if $\delta\epsilon$ is a change in extension corresponding to slips of total amount $\delta s \dfrac{\delta\epsilon}{1+\epsilon} = \dfrac{\delta s}{\sqrt{2}}$ and $P = 2S$, where $P$ is the tensile stress in the specimen. $S/\sqrt{s}$ is therefore equal to $P/2\sqrt{\{2\log(1+\epsilon)\}}$, where $\epsilon = (l - l_0)/l_0$, $l$ being the length and $l_0$ the initial length of the specimen. Hence in order to apply (5) to rock salt it is necessary to express it in the form

$$\frac{P}{\mu\sqrt{\{\log(1+\epsilon)\}}} = \frac{2\sqrt{2c}}{2\pi}\sqrt{\frac{\lambda}{L}}. \tag{6}$$

To compare (6) with experiment it is necessary to use observations made with very thoroughly annealed rock salt. Such observations have been made by Thiele* at various temperatures and his results are shown in fig. 2 where values of $P$ are

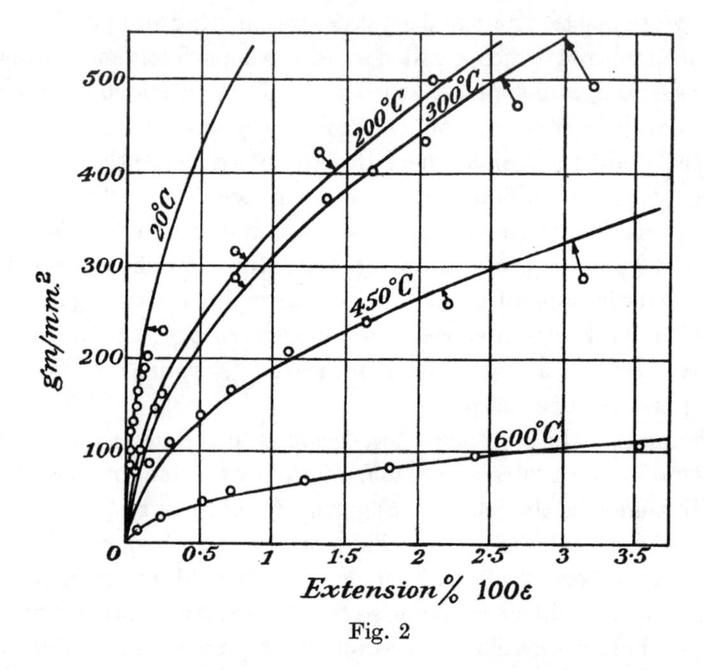


Fig. 2

shown for various values of $100\epsilon$. Since the maximum value of $\epsilon$ in these experiments was 0·035, $\log(1+\epsilon)$ is indistinguishable from $\epsilon$ so that if formula (6) is true Thiele's load-extension curves should be parabolas. Appropriate parabolas have been inserted in fig. 2 and it will be seen that the agreement with theory is remarkable.

## Calculation of $L$ and Comparison with Systems of Faulting in Crystals

The agreement between theory and experiment in regard to the form of the plastic stress-strain curve makes it possible to use the observed constant values of $S/\sqrt{s}$ or $P/\sqrt{\{\log(1+\epsilon)\}}$ obtained experimentally in order to find from (5) or (6) the value of $L$.

* W. Thiele, loc. cit.

In table 1 are given the data for Cu, Al, Au and Fe for all of which $(S, s)$ curves have been determined experimentally. The values of $\mu$ taken from physical tables are given in column 2 and the values of $S/\mu\sqrt{s}$ obtained by fitting the best parabola to the observed $(S, s)$ curve is given in column 3. The values of $\lambda$ given in column 4 are the appropriate atomic distances along the direction of slip and are obtained from X-ray data. The values of $Lc^{-2}$ calculated from (5) are given in column 4 and the values of $L$ found by taking $c = 1{\cdot}1$ are given in column 5.

Table 1. *Metals at room temperatures*

Formula (5) may be written $Lc^{-2} = \dfrac{\lambda}{4\pi^2}\left[\dfrac{S}{\mu\sqrt{s}}\right]^{-2}$

| | $\mu \times 10^{-11}$ | $S/\mu\sqrt{s}$ | $\lambda$ (cm.) | $Lc^{-2}$ (cm.) | $L$ (cm.) ($c = 1{\cdot}1$) |
|---|---|---|---|---|---|
| Cu | 4·5 | $1{\cdot}94 \times 10^{-3}$ | $2{\cdot}55 \times 10^{-8}$ | $1{\cdot}7 \times 10^{-4}$ | $2{\cdot}0 \times 10^{-4}$ |
| Al | 2·6 | $1{\cdot}46 \times 10^{-3}$ | $2{\cdot}86 \times 10^{-8}$ | $3{\cdot}1 \times 10^{-4}$ | $3{\cdot}8 \times 10^{-4}$ |
| Au | 2·8 | $1{\cdot}61 \times 10^{-3}$ | $2{\cdot}87 \times 10^{-8}$ | $2{\cdot}8 \times 10^{-4}$ | $3{\cdot}4 \times 10^{-4}$ |
| Fe | 8·3 | $2{\cdot}4 \times 10^{-3}$ | $2{\cdot}47 \times 10^{-8}$ | $1{\cdot}1 \times 10^{-4}$ | $1{\cdot}3 \times 10^{-4}$ |
| At $-185°$ C.: | | | | | |
| Al | 2·6 | $2{\cdot}53 \times 10^{-3}$ | $2{\cdot}86 \times 10^{-8}$ | $1{\cdot}1 \times 10^{-4}$ | $1{\cdot}3 \times 10^{-4}$ |

Observed spacing of faults: Bismuth $1{\cdot}4 \times 10^{-4}$ cm.
Zinc and cadmium $0{\cdot}8 \times 10^{-4}$ cm.
Iron $0{\cdot}25 \times 10^{-4}$ cm.
Copper $1{\cdot}5 \times 10^{-4}$ cm.

Table 2. *Rock salt;* $\mu = 1{\cdot}3 \times 10^{11}$, $\lambda = 3{\cdot}98 \times 10^{-8}$

(6) may be written $Lc^{-2} = 2\pi^{-2}\lambda\mu^2[P/\sqrt{\{\log(1+\epsilon)\}}]^{-2} = 1{\cdot}36 \times 10^{14}[P/\sqrt{\{\log\{(1+\epsilon)\}}]^{-2}$

| Temperature (° C.) | $P/\sqrt{[\log(1+\epsilon)]}$ (g. mm.$^2$) | $Lc^{-2}$ (cm.) | $L(c = 1{\cdot}1)$ (cm.) |
|---|---|---|---|
| 20 | $6{\cdot}0 \times 10^{-3}$ | $3{\cdot}9 \times 10^{-4}$ | $4{\cdot}7 \times 10^{-4}$ |
| 200 | $3{\cdot}4 \times 10^{-3}$ | $1{\cdot}17 \times 10^{-3}$ | $1{\cdot}4 \times 10^{-3}$ |
| 300 | $3{\cdot}0 \times 10^{-3}$ | $1{\cdot}5 \times 10^{-3}$ | $1{\cdot}8 \times 10^{-3}$ |
| 450 | $1{\cdot}9 \times 10^{-3}$ | $3{\cdot}8 \times 10^{-3}$ | $4{\cdot}6 \times 10^{-3}$ |
| 600 | $0{\cdot}6 \times 10^{-3}$ | $3{\cdot}8 \times 10^{-2}$ | $4{\cdot}6 \times 10^{-2}$ |

Observed spacing of faults, $2 \times 10^{-4}$ cm.

*Note.* In applying the formula it must be remembered that 1 g./mm.$^2 = 9{\cdot}81 \times 10^4$ c.g.s.

In table 2 similar data are given for rock salt. The values of $P/\sqrt{\{\log(1+\epsilon)\}}$ taken from the appropriate parabolas of fig. 2 are given in column 2. The values of $Lc^{-2}$ calculated from (6) are given in column 3 and the values of $L$ found by taking $c = 1{\cdot}1$ are given in column 4.

It will be seen from the two tables that in all these cases at room temperature $L$ is of order of magnitude $10^{-4}$ cm. We may now make a comparison with existing data concerning the spacing of faults in crystals. In the case of bismuth Goetz* found regular lines spaced $1{\cdot}4 \times 10^{-4}$ cm. apart. When zinc or cadmium† are deposited from vapour they form a system of plates $0{\cdot}8 \times 10^{-4}$ cm. Etching pits in iron‡

* Goetz, *Proc. Nat. Acad. Sci. U.S.A.* XVI (1930), 99.
† Straumanis, *Z. Phys. Chem.* B, XIII (1931), 316; XIX (1932), 63.
‡ Belaiew, *Proc. Roy. Soc.* A, CVIII (1925), 295.

seem to point to the existence of a block structure with sides of length $0{\cdot}25\times10^{-4}$ cm. A micro-photograph to which Zwicky* refers shows a structure in copper with spacing about $1{\cdot}5\times10^{-4}$ cm.

From X-ray data Darwin, Bragg and James† concluded that the linear dimensions of the elements of their mosaic are in the case of rock salt greater than $0{\cdot}5\times10^{-4}$ cm. From the work of Siedentopf‡ on the diffusion of vapour of alkali metals into rock salt Smekal§ considered that the average distance apart of 'Lockerstellen' or loose places in the structure of rock salt is about $2\times10^{-4}$ cm. These data are collected together for comparison with $L$ at the ends of tables 1 and 2.

It will be seen that at room temperatures the mean free path $L$ of the dislocations in a crystal which is undergoing plastic strain is of the same order of magnitude as the estimates which have been made by microscopic observation of the spacing of faults in the structure.

Though they are of the same order of magnitude the calculated values of $L$ at room temperatures are rather greater than the spacing of surface faults or marks which have been observed. This difference increases greatly as the temperature rises and gives rise to the possibility that all parts of the internal fault surfaces are not equally opaque to dislocations. At high temperatures the proportion of the fault surfaces which is opaque to the passage of dislocations might be expected to be less than at low temperatures and if this were the case the increase in $L$ would be explained. This question is subjected to analysis in a paper which I expect to publish shortly. In the case of aluminium for which the $(S, s)$ curves have been determined‖ down to the temperature $-185°$ C. formula (5) gives $L=1{\cdot}3\times10^{-4}$ cm. (see table 1) so that at this temperature $L$ is very close to the observed spacing of faults in copper, bismuth and zinc.

* Photograph plate 1 of *Brown Boveri Rev.* for Jan. 1929.

† Darwin, Bragg and James, *Phil. Mag.* (1926), 897.

‡ Siedentopf, *Phys. Z.* VI (1905), 855.

§ Smekal, *Phys. Z.* XXVI (1925), 709.

‖ Boas and Schmid, *Z. Phys.* LXXI (1931), 713.

# 25

# LATTICE DISTORTION AND LATENT HEAT OF COLD WORK IN COPPER

Paper written for the Aeronautical Research Committee (1935)

In a recent paper* Mr W. A. Wood has given measurements of the broadening of lines in the X-ray spectrum of copper owing to lattice distortion brought about by cold work. His experiment shows that there is a correspondence between the hardening effect of cold work and the change in the spacing of the reflecting planes.

In the course of his discussion he remarks that this 'would fall into line with theories elaborated by G. I. Taylor and others, provided that the dislocations, in those theories hypothetical, were identified as centres of lattice distortion'.

In this connection it seems worth while to attempt to make some more definite estimate on the basis of my theory of what X-ray analysis might be expected to reveal. The theory involves the idea that dislocations exist and that they can migrate, thus causing plastic distortion, provided that the applied external stress exceeds the maximum elastic stress, due to the neighbouring dislocations, along the crystallographic line on which the dislocation can move.

The elastic stress near a dislocation increases rapidly as its centre is approached, and if X-rays could reveal the corresponding distortions a very large though faint penumbra would be expected near a reflection line as soon as the first centre of dislocation appeared. This penumbra would darken without widening as the number and dislocations increased.

The regions near the dislocations where very high elastic stresses occur are, however, so small that it seems possible that they might not give rise to observable reflections. If the regions very close to the dislocations are supposed to give no reflection, and it is supposed that the X-ray beam is reflected chiefly from the greater part of the crystal which lies in between them, then the breadth of the X-ray line would indeed be related to the strength of the material through my theory, for the lattice distortion as measured by the broadening of the lines would then correspond with the stress barrier, which, in my theory, prevents the dislocations from migrating till sufficient external stress is applied.

Since the theory involves the idea that plastic distortion only begins when the externally applied stress is great enough to overcome these stress barriers, we are in a position to estimate the elastic distortion corresponding with the stresses at these barriers. In the case of copper stretched by direct end-load the maximum lattice

* W. A. Wood, 'Latent energy due to lattice distortion of cold-worked copper', *Phil. Mag.* XVIII (1934).

distortions which might be expected to be revealed by X-ray photographs would be equal to $P/E$, where $P$ is the maximum stress to which the copper was subjected and $E$ is Young's modulus.

The experimentally observed curve* of $P$ against $\log l/l_0$, where $l/l_0$ is the ratio of final to initial length of the copper, can therefore be used to predict the relationship between the lattice distortion† $\delta d/d$ and $l/l_0$.

Mr Wood's observations were made with rolled copper for which the relationship between $P$ and $l/l_0$ is rather different from that obtained in a pure load-elongation experiment; this difference, however, is not very large, so that it seems legitimate to compare Mr Wood's observed values of $\delta d/d$ with those calculated in the manner described above. In table 1 Mr Wood's observations are given in columns 1 and 2.

Table 1

| Observation | | Theory | | |
|---|---|---|---|---|
| 1 | 2 | 3 | 4 | 5 |
| Reduction in area (%) | $\delta d/d$ | $\log l/l_0$ corresponding to col. 1 | $P$ (dynes/sq.cm.) | $\frac{P}{E}\left(=\max. \frac{\delta d}{d}\right)$ |
| 5 | $5{\cdot}5\times 10^{-4}$ | 0·05 | 1·2 | $10\times 10^{-4}$ |
| 9·8 | 11·1 | 0·10 | 1·8 | 15 |
| 13·1 | 14·9 | 0·14 | 2·1 | 17 |
| 14·8 | 14·0 | 0·16 | 2·3 | 19 |
| 15·2 | 16·7 | 0·16 | 2·3 | 19 |
| 18·0 | 15·8 | 0·20 | 2·4 | 20 |
| 50 | 16·7 | 0·69 | 3·4 | 27 |
| 90 | 16·7 | 3·0 | 3·4 | 27 |

In column 3 is given the value of $\log(l/l_0)$ corresponding with the observed reduction of area (column 1). In column 4 the value of $P$ is taken from the observed load-extension curve for copper*. In column 5 are given the values of $P/E$ which, according to the theory described above, is roughly equal to the maximum value of $\delta d/d$ found by the X-rays. $E$ for copper is taken as $12{\cdot}3\times 10^{11}$ c.g.s.

On comparing columns 2 and 5 in table 1 it will be seen that the calculated maximum value of $\delta d/d$ is of the same order of magnitude as that observed. On the average the theoretical value is about 50 % higher than the observed lattice distortion. This is certainly quite as good an agreement as would be expected in view of the many elements of uncertainty involved.

## Latent Heat of Cold Working

Having measured the lattice distortion Mr Wood attempts to form an estimate of the latent energy stored up in the metal owing to the lattice distortion. As a result of his analysis of his X-ray photographs he comes to the conclusion that $\frac{15}{16}$ths of

* For this curve see fig. 1 in 'The mechanism of plastic deformation of crystals. Part II', *Proc. Roy. Soc.* A, CXLV (1934), 388; paper 22 above.

† Where $d$ is the spacing and $\delta d$ the change in spacing which measures the lattice distortion.

the material is strained or expanded by an amount $\delta d/d = 2{\cdot}8 \times 10^{-4}$, while $\frac{1}{16}$th is strained $\delta d/d = 16{\cdot}7 \times 10^{-4}$. The rises in temperature which would be necessary in order that the metal might expand these amounts by ordinary thermal expansion are 16·8 and 100° C. Mr Wood assumes that whatever the means by which this expansion against molecular attractions is effected, the same amount of energy would have to be supplied, so that the latent heat is the heat required to expand the body through the amounts observed by X-ray analysis. This is

$$\tfrac{15}{16} \times 16{\cdot}8 \times 0{\cdot}09 + \tfrac{1}{16} \times 100 \times 0{\cdot}09 = 1{\cdot}7 \text{ cal. per g.*}$$

A more direct method of performing the calculation would be to imagine the expansion performed not by raising the temperature but by applying a uniform tension over the surface. If $\kappa$ is the volume elasticity, the work done per c.c. in expanding through a change in volume $\delta v$ is $W = \frac{1}{2}\kappa\left(\frac{\delta v}{v}\right)^2 = \frac{9}{2}\kappa\left(\frac{\delta d}{d}\right)^2$. Taking the value $\delta d/d = 2{\cdot}8 \times 10^{-4}$, $\kappa = 13 \times 10^{11}$ for copper,

$$W = \tfrac{9}{2}(13)(10^{11})(2{\cdot}8)^2(10^{-8}) = 4{\cdot}5 \times 10^5 \text{ ergs per c.c.}$$

For the parts where $\delta d/d = 1{\cdot}67 \times 10^{-3}$, $W = 1{\cdot}6 \times 10^7$ ergs per c.c.

Using the same assumption as Mr Wood, the total latent energy would be $\frac{15}{16} \times 4{\cdot}5 \times 10^5 + \frac{1}{16} \times 1{\cdot}6 \times 10^7 = 1{\cdot}4 \times 10^6$ ergs per c.c. This is equivalent to

$$\frac{1{\cdot}4 \times 10^6}{4{\cdot}2 \times 10^7} = 0{\cdot}033 \text{ cal. per c.c.} \quad \text{or} \quad \frac{0{\cdot}033}{8{\cdot}9} = 0{\cdot}0037 \text{ cal. per g.}$$

Comparing this with Mr Wood's value, 1·7 cal. per g., it will be seen that there can be no justification for Mr Wood's assumption.

The fact that the energy due to elastic strains of the amount measured by X-ray methods is so very small compared with the observed latent energy of cold working shows that the greater part of the latent energy must be connected with elastic strains which are much greater than those observed by X-rays. It has already been pointed out that the stresses close to a dislocation must be very much greater than those detected by X-rays. It seems therefore that most of the latent energy due to cold work must be concentrated in regions very close to the dislocations.

This idea would naturally lead to the prediction that the latent energy should be proportional to the number of dislocations. This, however, is inconsistent with observation for it has been found that when a bar of copper is extended the latent energy of cold work is proportional to the cold work done. According to my theory of the strength of soft metal the stress $P$ is proportional to $\sqrt{N}$, where $N$ is the number of dislocations per unit area. The strain is proportional to $N$, so that the work done during the stretching of a bar from a soft state in which it is free from dislocations to a state in which there are $N$ dislocations per unit area is proportional to $\int_0^N \sqrt{N}/dN$ or $N^{\frac{3}{2}}$. If the theory is correct therefore the latent energy absorbed must be proportional to $N^{\frac{3}{2}}$ instead of $N$.

* 0·09 is the specific heat of copper.

26 BSP

# 26

# THE EMISSION OF THE LATENT ENERGY DUE TO PREVIOUS COLD WORKING WHEN A METAL IS HEATED*

REPRINTED FROM

*Proceedings of the Royal Society*, A, vol. CLXIII (1937), pp. 157–81

Four principal methods have been used for measuring the latent heats in metals due to change of phase or state, or to cold working:

(1) The total heat at any stage can be measured at a number of temperatures by means of a calorimeter.

(2) The rate of cooling of the specimen can be measured and the latent heat deduced on the assumption of Newton's Law of Cooling.

(3) The rate of heating can be measured when the specimen is placed in a furnace the temperature of which rises at a known uniform rate, Newton's Law being used to deduce rate of increase of thermal energy.

(4) The specimen can be heated electrically in such a way that all the electrical energy is used in heating the specimen.

Of these (1) and (4) involve direct measurements of heat energy, while in (2) and (3) the amount of absorption or emission of heat is deduced indirectly from the heating or cooling curves.

(1) and (4) have been used for measuring specific heats or latent heat of fusion where large quantities of heat are concerned, but when it is desired to measure small quantities of latent energy, such as that stored up in a metal during cold working, (2) and (3) have been used because they permit of greater sensitivity, especially when they are applied differentially.

The latent energy retained by a metal after it has been subjected to cold working can be found by measuring the heat evolved and the mechanical energy expended during the cold working. The difference between these lies latent in the metal, and if the metal is heated the latent energy must be released before the melting-point is reached.

Several attempts have been made to measure this latent energy. C. J. Smith† measured the difference between the heat evolved when a gramme of cold-worked and annealed copper wire were dissolved in a solution of bromine in potassium bromide. This method is insensitive because the heat evolution, due to dissolving the annealed copper, is large compared with that due to the latent energy of internal strains in the cold-worked material.

* With H. QUINNEY.

† *Proc. Roy. Soc.* A, CXXV (1929), 619.

The most sensitive method so far attempted is the differential method of Sato.* In this method two specimens were cut from the same bar of metal. One is cold worked and the other annealed. The external dimensions were identical and they were inserted in two equal holes in a block of silver which were heated at a uniform rate. They were insulated from the silver block by a silica lining. The difference $T$ in temperature between the two specimens was measured. In accordance with usual procedure in differential measurements of this kind Newton's Law of Cooling is assumed, and if the thermal resistance of the two silica linings and surfaces of the specimens are identical the difference between the initial heat content of the two specimens is proportional to $\int T\,dt$, where $t$ is the time. Sato found that it was not possible to get identical thermal resistance at the surface of the two specimens so that there was still a difference between the temperatures of the two specimens placed in his apparatus when they had received identical treatment.

To overcome this difficulty he first measured $T$, the difference in temperature between a strained and unstrained specimen, while raising their temperatures sufficiently high to remove all internal strain. After allowing the apparatus to cool, he reheated without moving the two specimens. If the difference in temperatures on the second heating is $T_0$, Sato takes the latent heat as proportional to $\int (T-T_0)\,dt$. As a qualitative method for finding the temperatures at which the release of latent energy occurs this method is admirable, but in order to make it yield quantitative results, several assumptions must be made which need considerable justification. If $1/K_1$ and $1/K_2$ are the thermal resistances between the silver block and the annealed and cold-worked specimens respectively, and if $H_1$ and $H_2$ are the differences in the energy content after heating to a given temperature, then $H_1-H_2=\int K_1(T_s-T_1)+K_2(T_s-T_2)\,dt$, where $T_s$ is the temperature of the silver block.

Similarly on the second heating, if $K_1$ and $K_2$ are unaltered, $H_1-H_2'$ the change in heat content on heating through the same range as before, then

$$H_1-H_2'=\int K_1(T_s-T_1)+K_2(T_s-T_2')\,dt,$$

and combining (1) and (2) we have

$$H_2-H_2'=K_2\int (T_2-T_2')\,dt.$$

In order to use the method for determining latent heat it is necessary to determine $K_2$. This was done by means of a separate experiment in which the latent energy of a specimen of brass, subjected to a definite amount of cold work, was determined by a colorimeter.

* *Sci. Rep. Tohoku Univ.* xx (1931), 140–77.

26-2

If the value of $K_2$, so obtained by dividing $H_2 - H_2'$ found by the calorimeter for brass by $\int (T_2 - T_2')\, dt$, can be assumed constant for all specimens which have received the same mechanical treatment and for all temperatures, the method can be used quantitatively for determining latent heats of cold work in specimens of the same material subjected to different amounts of cold work or in specimens of different materials.

Unfortunately no evidence is given in Sato's paper that this very drastic assumption is justified. It is difficult therefore to accept his results except in the case of brass subjected to the same amount of cold work as that with which the calorimeter experiments were made.

It may be significant that in the experiments to be described below, in which direct measurements of energy input were made, excellent agreement was found with Sato's results in this particular case of brass, but in most other cases there was wide disagreement between his measurements of the amount of latent heat and ours. The temperatures at which we found that the latent heat is liberated agree with those measured by Sato, though our method was too insensitive to allow of any very complete comparison in that respect.

The difficulty inherent in Sato's method occurs in practically all metallurgical investigations which are concerned with quantities of heat. So long as qualitative results only are desired, the heating and cooling curves of specimens contained in furnaces subjected to known rates of heating and cooling are useful. When quantitative results are needed, it is necessary to make so many assumptions regarding the transfer of heat through the surface of the specimens that results are frequently of little value.

In order to avoid the necessity for assuming, as Sato did, that the thermal resistances at the surfaces of specimens of different materials but of the same shape are the same, it is possible to devise a differential method in which only the lesser assumption is made that the thermal resistance of a specimen remains unchanged during two successive heatings.

In an apparatus sketched in fig. 1 two similar specimens, one cold worked and the other annealed placed in two arms of a silica U-tube, were heated by two furnaces, made to be as nearly as possible identical in all respects, and supplied by the same current. The temperature difference between the specimens was measured as the pair of furnaces was heated to a temperature higher than that necessary for annealing. After cooling, the temperatures of both furnaces were again raised, and the difference between the temperatures of the two specimens again measured. The results in the cases of copper and aluminium are shown in figs. 2 and 3, where the abscissae refer to difference in temperature between the two specimens. The fact that none of the curves actually coincides with the axis must be attributed to some accidental difference between the furnaces. If this accidental difference is assumed to remain unchanged when the specimen is heated a second or third time without removal from the furnace, the difference between the curves obtained on the first and those obtained on the second or third heating must be due to some change in the specimen which occurs during the first heating. This change will be assumed to consist in the evolution of the latent heat due to cold working.

On inspecting the curves of figs. 2 and 3 it will be seen that the curves for the second and third heating are identical, whereas the curve obtained during the first heating lies above them. In this respect the curves resemble those obtained by Sato, but they differ from them in an important particular. As the temperature

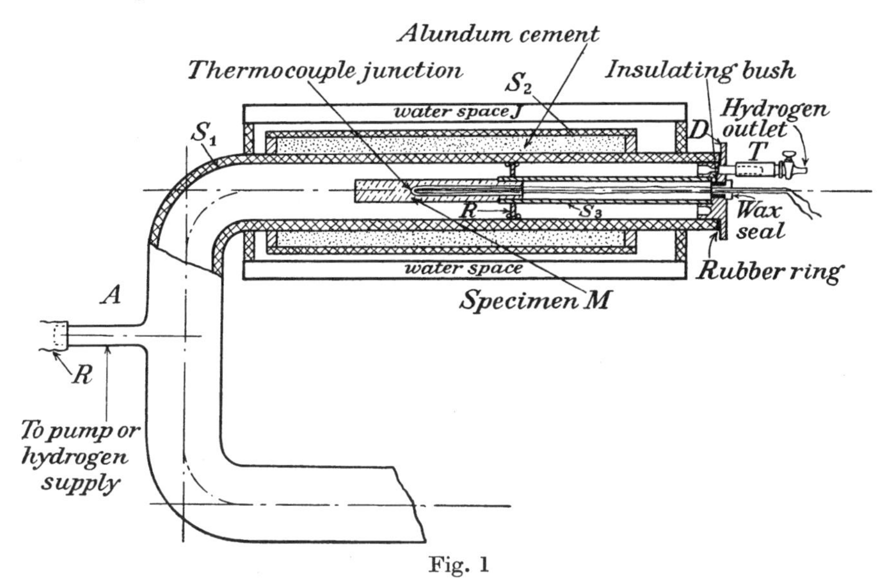


Fig. 1

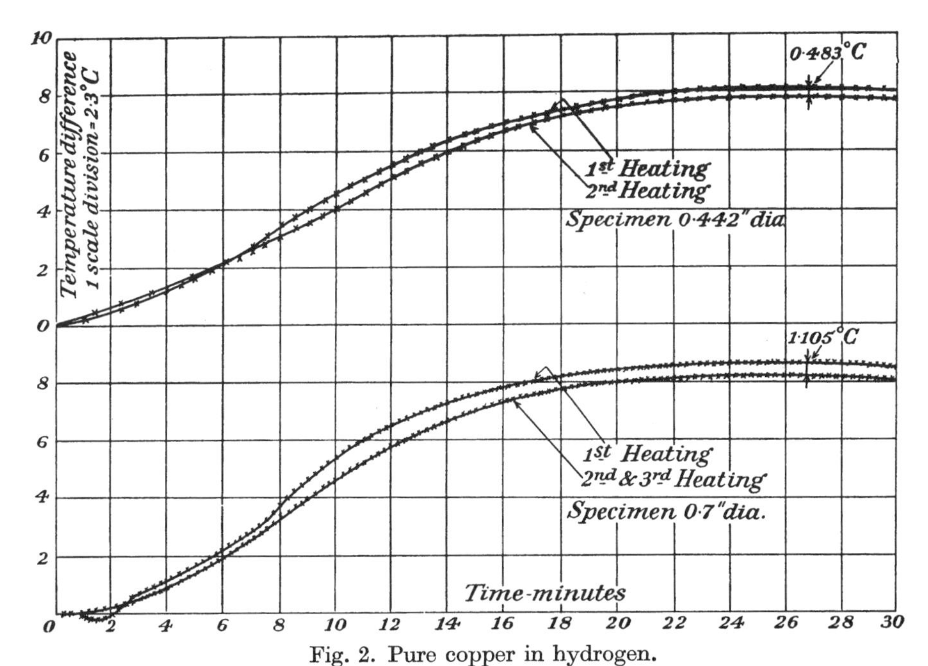


Fig. 2. Pure copper in hydrogen.

rises the evolution of latent heat in Sato's experiments produces a transitory difference between the temperatures of the two specimens which disappears after the heat has ceased being evolved.

As explained above, the latent heat is proportional to the area of the time-temperature difference curve, but the actual magnitude of the latent heat can only be estimated when the thermal resistance between the specimen and the cell which contains it has been measured by some independent method.

In the present experiments the curves show that the evolution of latent heat causes the two curves to separate as soon as the heat begins to be evolved. As the temperature rises above that at which latent heat is given out the separation of the two curves begins to decrease, but soon attains a value which remains practically constant for the rest of the experiment.

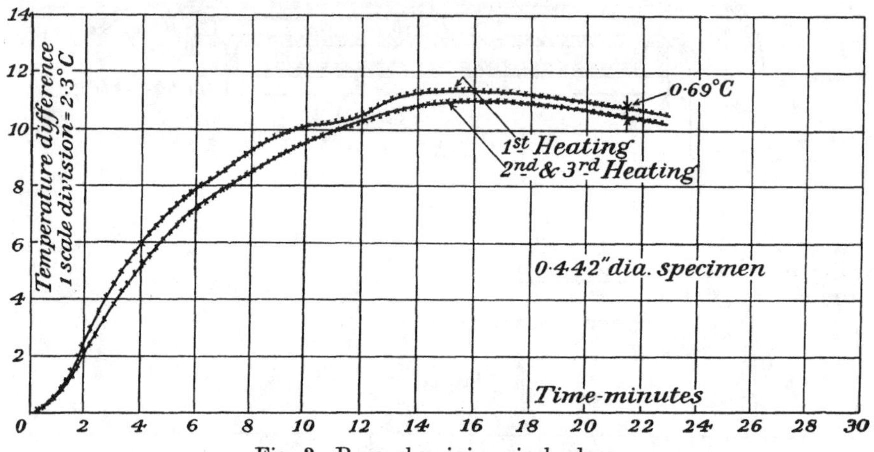


Fig. 3. Pure aluminium in hydrogen.

In fig. 3, which refers to aluminium, this constant difference was 0·69° C. Fig. 2, which refers to copper, contains two pairs of curves, one for a specimen 0·442 in. in diameter and 4 in. long, and the other for a specimen 0·7 in. diameter and 4 in. long. Both had been subjected to the same amount of cold work per gramme. For these two copper specimens the constant temperature differences after losing their latent energy were 0·48 and 1·10° C.

This constant value which the temperature difference attains when all the latent energy has been evolved does not depend on the thermal resistance between the furnace and specimen, but is evidently connected with the fact that part of the furnace near the specimen can be regarded as being in thermal contact with the specimen.

### Calculation of Latent Heat for Copper

If the specimens were very long in comparison with their diameter, the heat capacity of that part of the furnace which might be regarded as being in thermal contact with the specimen might be expected to be independent of the diameter of the specimen.

This assumption was verified by estimating directly the heat capacity per unit length of the furnace and comparing it with that deduced by the method which will now be described.

Assuming, therefore, that $h$ the relevant part of the heat capacity of the furnace is identical for two specimens of different diameters, but the same length, it is possible to calculate $h$ from the two measurements of the constant temperature differences, shown in fig. 2, which were made with copper specimens of the same length (4 in.) and diameter 0·442 and 0·70 in. respectively.*

The latent energy evolved by the smaller specimen was sufficient to raise the temperature of the specimen and furnace 0·483° C. If the evolution of latent energy due to cold work would have been sufficient to raise the temperature of the copper alone through $\theta°$ C.,

$$\frac{\theta}{0{\cdot}483}=\frac{h+45{\cdot}7\sigma}{45{\cdot}7\sigma}, \tag{1}$$

where $\sigma$ is the specific heat of copper and 45·7 g. is the weight of the smaller specimen. Similarly with the large specimen whose weight was 133 g. and temperature difference 1·105° C.

$$\frac{\theta}{1{\cdot}105}=\frac{h+133\sigma}{133\sigma}. \tag{2}$$

Since $\theta$ may be assumed the same with both specimens because the same amount of work per g. was done on each, $\theta$ and $h/\sigma$ can be determined by solving (1) and (2).

The result is

$$\theta=3{\cdot}37°\text{ C.}, \tag{3}$$

and

$$h/\sigma=273{\cdot}3. \tag{4}$$

In these specimens the work done in twisting was measured in the manner described in a previous paper†. In each case it was 4·6 cal./g. This would have been sufficient to raise the temperature 48·5° C. if the whole work had been turned into heat. The fraction of the work done which is absorbed and afterwards measured when it is released on reheating is therefore 3·37/48·5 = 0·07. 7 % of the cold work originally done on the specimen was therefore released in the form of heat during the first subsequent heating (to about 600° C.).

This is in close agreement with the results given in a previous paper† where the latent energy was measured during the process of absorption. Fig. 6 of that paper shows the proportion of the cold work done in twisting copper rods which remains latent. It will be seen that over a great part of the range this was $7\frac{1}{2}$ %. Referring to table 3† it will be seen that for copper this fraction remained practically constant until cold work equivalent to 10·5 cal./g. had been done. The present specimens which had 4·6 cal./g. are therefore within this range.

* This method has been used previously by Quinney; see *Proc. Roy. Soc.* A, CXXIV (1929), 591.

† Taylor and Quinney, *Proc. Roy. Soc.* A, CXLIII (1934), 307; paper 19 above.

Using the results obtained with copper specimens for determining the capacity for heat of the furnace, the latent energy contained in cold-worked aluminium and in 70/30 brass was measured.

### Aluminium

In the experiment represented in fig. 3 the observed final temperature difference between the annealed and the cold-worked specimen was 0·69° C. and the mass was 20 g., so that taking 0·22 as the specific heat the equation equivalent to (1) is

$$\frac{\theta}{0\cdot 69}=\frac{h+20\times 0\cdot 22}{20\times 0\cdot 22}. \tag{5}$$

Taking the specific heat of copper 0·1, it will be seen from (4) that $h=27\cdot 3$.

Thus from (5) $\theta=4\cdot 98°$ C.

The work done on the specimen during twisting was equivalent to 11·1 cal./g. which is equivalent to a rise in temperature of $11\cdot 1/0\cdot 22=51°$ C. The fraction of the work done which is emitted as heat when the specimen is reheated is therefore $4\cdot 98/51=0\cdot 098$, i.e. 9·8 %.

This may be compared with the proportion absorbed in Farren and Taylor's experiment* which was 8 %.

### Brass

Similar observations with 70/30 brass gave the proportion emitted as 15·8 %.

### Design of Calorimeter Furnace

The method described above though free from the greatest defects of the differential method, as applied by Sato, still leaves room for much improvement. In the first place, the heat capacity of the furnace is always so much greater than that of the specimen that the difference in temperature between the annealed and the cold-worked specimen due to the release of the heat is only a small fraction of what it would be if the heat capacity of the furnace could be reduced till it was smaller than that of the specimen.

In the second place it is necessary to assume that the heat capacity of the part of the furnace in thermal contact with the specimen is the same for all specimens of the same length, provided this length is several times as great as the diameter of the furnace. This assumption appears to be true, because the results are in good agreement with those obtained by the use of other methods, but nevertheless it seemed desirable to attempt the design of apparatus which would not suffer from these disadvantages. In doing so it was noticed that the advantages which appear at first sight to favour differential methods are largely illusory. Two furnaces cannot be constructed so that they are really identical. It is only by the use of

* Farren and Taylor, *Proc. Roy. Soc.* A, CVII (1925), 422; paper 6 above.

methods such as that of Sato in which both specimens are heated in the same furnace that the advantages of a differential method can be expected, and then, as we have seen, differences in the thermal resistance at the surface of different specimens may well produce far greater temperature differences than those due to cold work.

These considerations led us to abandon further attempts to use differential methods and to concentrate on devising a furnace in which fewer assumptions need be made about the nature of the thermal interchanges between the specimen and furnace.

In the apparatus now to be described the heat is supplied to the inner surface of a hollow specimen by means of a small electric furnace only 0·11 in. diameter and $1\frac{1}{2}$ in. long. The whole of the electric energy input is therefore transferred to the specimen itself except for the small amount which is necessary to raise the temperature of the furnace and its support.

The thermal resistance between the interior-heating furnace and the specimen does not affect the amount of heat which flows into the specimen, though it does determine the amount by which the temperature of the furnace must exceed that of the specimen in order that all the heat produced by the furnace may pass into the specimen. At the surface of the specimen heat can be lost by convection or radiation. Convection losses are avoided by using a high vacuum. Radiation losses are avoided by a guard-ring furnace which almost completely surrounds the specimen, and is maintained at the same temperature as the specimen itself.

If the guard ring were completely closed, and accurately at the temperature of the specimen, there would be no loss or gain of heat through the outer surface of the specimen whatever its condition or radiative properties might be.

It is clear that with this design the thermal resistance and radiating properties of the surface have no influence on the heat measurements—or at any rate have only a small secondary influence, depending on the necessary departures from the ideal conditions of an infinitely small interior furnace, and a completely closed guard-ring furnace maintained at exactly the same temperature as the specimen.

Fig. 4 shows a sectional elevation of the furnace constructed in accordance with these principles. In this figure the hollow specimen is shown in position as $C$ and the furnace as $D$.

Enlarged details of this part of the apparatus are shown inset.

The requirements which had to be satisfied were: ($a$) the specimen must be set on a support which is as good a non-conductor as possible, ($b$) the whole of the furnace must be inside the specimen, so that all heat which leaves it passes directly into the specimen, ($c$) the guard-ring furnace surrounding the specimen must be easily removable, so that the specimen and thermojunctions can be changed, ($d$) the apparatus must be vacuum tight and all heating and thermojunction leads must pass through vacuum-tight joints.

The first of these requirements ($a$) was attained by mounting the specimen on a silica post $B$, which was so designed that the cross-section near the specimen was as small as possible. This post, which was made by the Thermal Syndicate,* contained eight holes for wire leads. The two central ones conveyed the heating leads and also the two platinum potential leads for measuring the energy input to the furnace. These potential leads, which were themselves insulated by very small silica tubes, were fused to the two large gauge (0·018 in. diameter) platinum leads to the furnace. The small gauge (0·01 in. diameter) platinum wire of the furnace itself was fused to the upper end of these leads, and the lower ends were welded to the copper-heating leads which passed out of the furnace. By bringing the potential leads from a point which was actually inside the specimen (see inset fig. 4), it was possible to ensure that the measured energy input was wholly used inside the specimen.

The thermojunction $E$ on the specimen consisted of a platinum-rhodium disc to which the wires were welded. This disc was screwed by a platinum-rhodium screw to the specimen. The thermojunction leads passed through two of the holes in the silica post.

(*b*) The furnace which is shown inset as $D$, in fig. 4, consisted of a silica tube 0·11 in. diameter and $1\frac{1}{2}$ in. long containing six holes through which the platinum heating wire (0·01 in. diameter) was threaded. The cylindrical specimen was so formed that it rested on the top of the silica post, being thus electrically insulated from the furnace wire.

(*c*) The delicate nature of the electrical connections of the heating element $D$ made it necessary to construct a special slide so that the specimen could be moved into position and the thermocouple screwed on without risk of injury.

(*d*) After the specimen and thermojunctions had been placed in position it was necessary to bring the guard-ring furnace $L$, with all its electrical connections, and the casing $A$ into position and to make the joints vacuum tight. This operation was facilitated by attaching the guard-ring furnace to a silica post $B$ which passed through the top of the casing, and was made vacuum tight by wax which was poured into the cup $K$ (fig. 4).

The whole of the casing was made of copper and had a highly polished interior which was water jacketed in order to assist in maintaining repeatable temperature conditions inside the apparatus. It could be moved in a vertical slide and bedded down on to a vacuum-tight rubber ring $J$ recessed into the base $M$.

To ensure the maintenance of the vacuum at the point where the wires from the heating element passed out of the furnace, a flange was brazed to the tube $N$ which was brazed to the base in the casing $M$. This cylindrical flanged head which is drilled to take six $\frac{1}{4}$ in. diameter securing bolts carries a corresponding disc $O$ by means of which it is possible to collect and compress all the leads which are conveyed through the silica post, between two layers of rubber, coated with vacuum grease as shown in the figure.

The guard ring was also heated with platinum wires 0·01 in. diameter carried

* The authors are indebted to Mr Millar of the Thermal Syndicate for the preparation of this silica post.

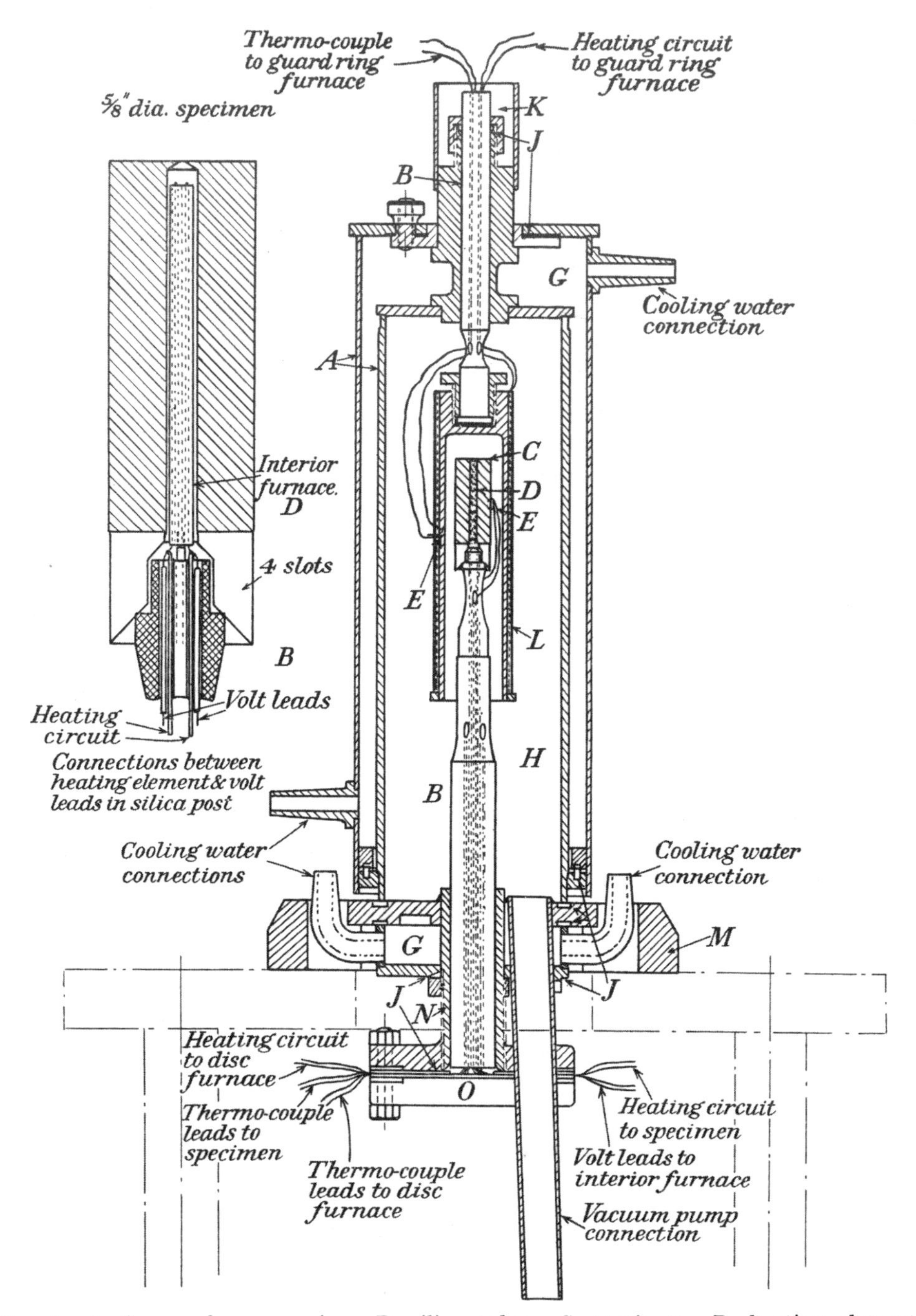


Fig. 4. *A*, Copper furnace-casing; *B*, silica tubes; *C*, specimen; *D*, heating element; *E*, thermo-couple points; *G*, water space; *H*, vacuum space; *J*, rubber joints; *K*, sealing wax; *L*, guard-ring furnace; *M*, base of casing; *N*, tube; *O*, disc.

in fifty fused alumina tubes kindly supplied for the purpose by Dr Desch of the N.P.L. These tubes were attached vertically to the guard ring by three bindings of tungsten wire. The guard-ring tube itself was of nickel and was painted outside with a solution of silica gel and graphite to facilitate the transfer of heat by radiation from the furnace wire to the metallic nickel. Round the outside of the heating wires of the guard ring was bound a sheet of pure nickel insulated with a thin sheet of mica to prevent radiation to the casing.

Most of the specimens were $\frac{5}{8}$ in. diameter and 2 in. long and contained a hole which originally fitted closely round the heating element, but in later experiments this hole was enlarged to facilitate the escape of any gas which might be given off by the specimen during heating.

The specimens after cold working in torsion were cut off to correct length, with a minimum of delay and the outside cylindrical surfaces were not machined after twisting. No grease or lubricant of any kind was used in drilling and machining the specimen, and the machining operations were done at a slow speed so as to avoid as far as possible any heating of the specimen. After washing in carbon tetrachloride and drying in ether the specimens were handled only by means of rubber finger stalls, to avoid any risk of introducing impurities. As soon as the thermocouple had been attached to the specimen the casing $A$ was lowered into position and the apparatus was exhausted.

To carry out a test three observers were required. A heating current of about 2 amp. was found suitable for the interior furnace or heating element and the p.d. recorded by the potential leads attached to a high resistance (12000 $\Omega$) voltmeter was generally about 6 V. The current in this interior furnace was controlled by hand during the whole of the test and maintained constant by one observer, who likewise recorded the voltmeter readings every $\frac{1}{2}$ min.

To maintain the guard ring at the same temperature as the specimen, about 2 amp. were required. This current was adjusted where necessary by a second observer who recorded the difference between the readings of the thermocouple attached to the specimen and those of a similar one screwed to the nickel of the guard ring. Any difference which the controlling observer had been unable to eliminate were recorded by him every $\frac{1}{2}$ min. A third observer recorded the temperature of the specimen by means of the potentiometer. The procedure was as follows: Having recorded every $\frac{1}{2}$ min. the temperature of the specimen, it was possible, by means of a double-pole change-over switch, to transfer the potentiometer to the guard-ring thermocouple, and the extent to which this was out of balance, if any, was recorded by the second observer who made the necessary adjustment of the guard-ring current to maintain balance. In this way five sets of readings were taken simultaneously, i.e. the current supplied to the interior furnace, potential drop between its potential leads, the temperature of specimen, that of the guard ring, and the guard-ring current.

In the early tests the vacuum was maintained throughout the test by having the whole apparatus attached directly to a Hyvac pump which was started some time before the beginning of the test. It was found, however, that in every case a considerable drop in potential at the potential leads of the interior furnace occurred soon after heating, and also occasionally during the first heating of a specimen. At the same time a small discharge tube attached to the apparatus indicated that this drop in potential was associated with a rise of pressure in the apparatus. This rise was attributed to a sudden evolution of gas which decreased the thermal resistance between the specimen and the heating element so that the heating wire became cooler and therefore suffered a corresponding drop in electrical resistance. Most metals at some stage appeared to give off gases which produced this effect, and it was only after a great many trials that methods were finally devised for overcoming this difficulty.

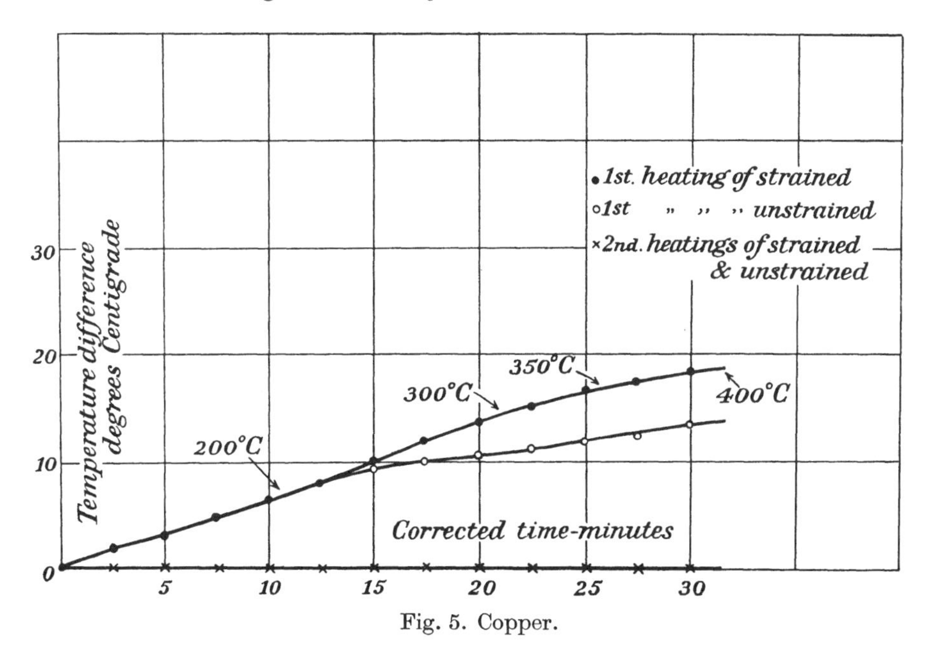


Fig. 5. Copper.

Some of our results were obtained before we had finally perfected these methods. In the case of copper for instance, in our early experiments we found a difference between the temperature-time curves of an annealed specimen obtained during the first and during subsequent heatings in the apparatus. We also found a difference between the first and subsequent heatings of a specimen which had been subjected to cold work after annealing. The temperature-time curves of the second and subsequent heatings of the annealed and of the cold-worked specimens were, however, found to be identical. It seemed therefore reasonable to suppose that the difference between these two differences was due entirely to the cold work. These differences are shown in fig. 5, and it will be seen that at temperatures above 350° C. they amounted to 13° C. for annealed and 18° C. for the cold-worked specimen. The difference, namely 5° C., was assumed to be due entirely to the latent energy due to cold work. This assumption, however, did not remain an unconfirmed assumption, because when methods for eliminating spurious effects had been perfected, the temperature-time curves of annealed and of cold-worked specimens were again compared, and it was then found that for the annealed specimen the curve obtained on the first heating was identical with that for

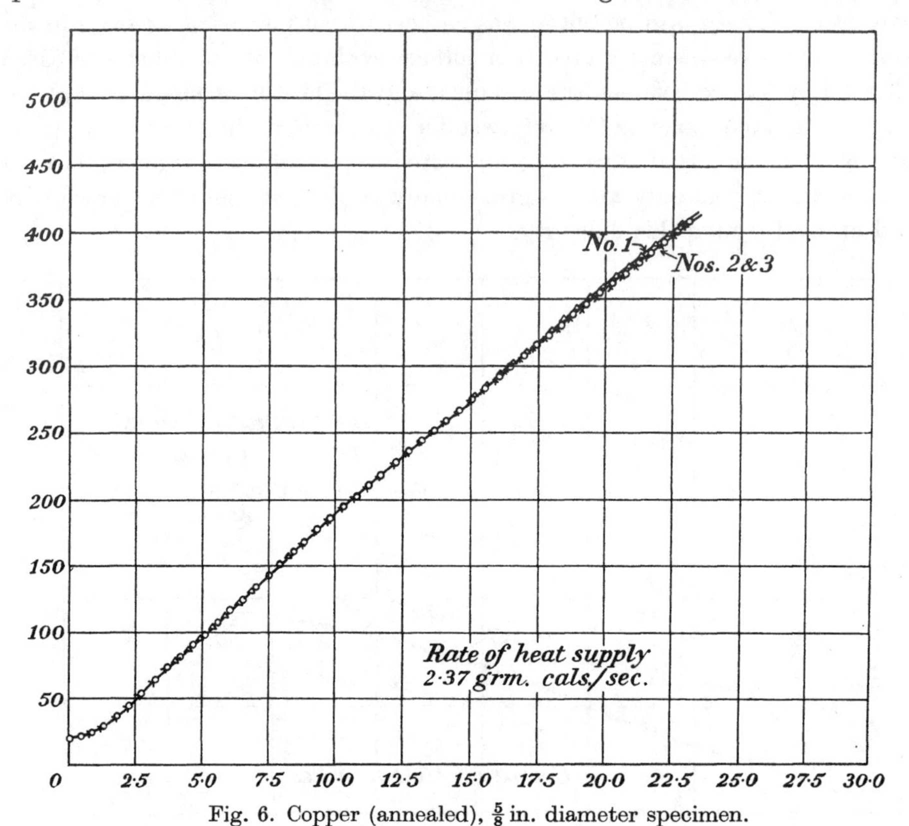


Fig. 6. Copper (annealed), $\frac{5}{8}$ in. diameter specimen.

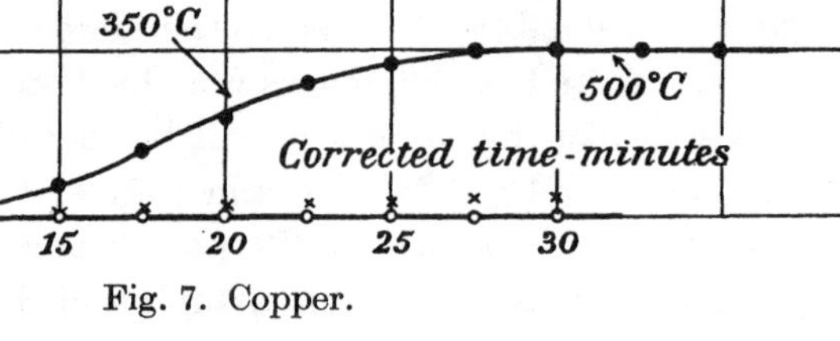


Fig. 7. Copper.

subsequent heatings (see fig. 6), while for the cold-worked specimen there was at temperatures above 500° C. a difference of 5° C. See fig. 7.

The cold-worked specimen in this case had received the same amount of cold work in the early experiments, so that the agreement between the latent heats

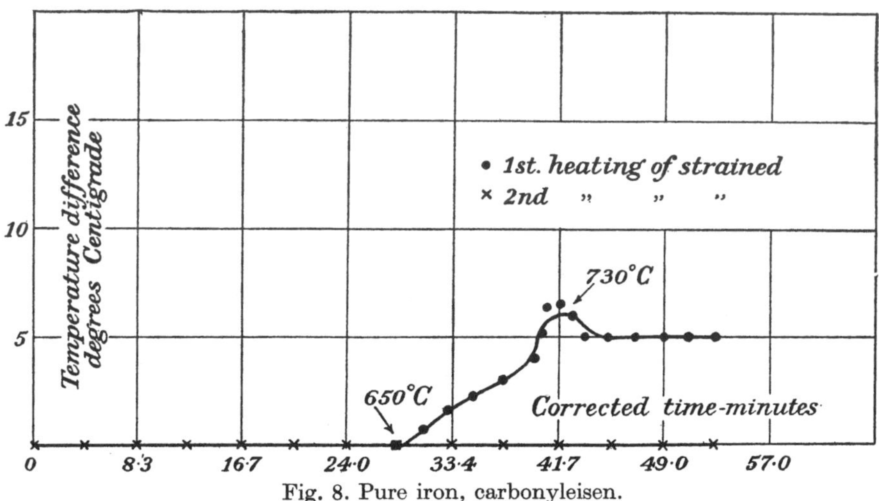


Fig. 8. Pure iron, carbonyleisen.

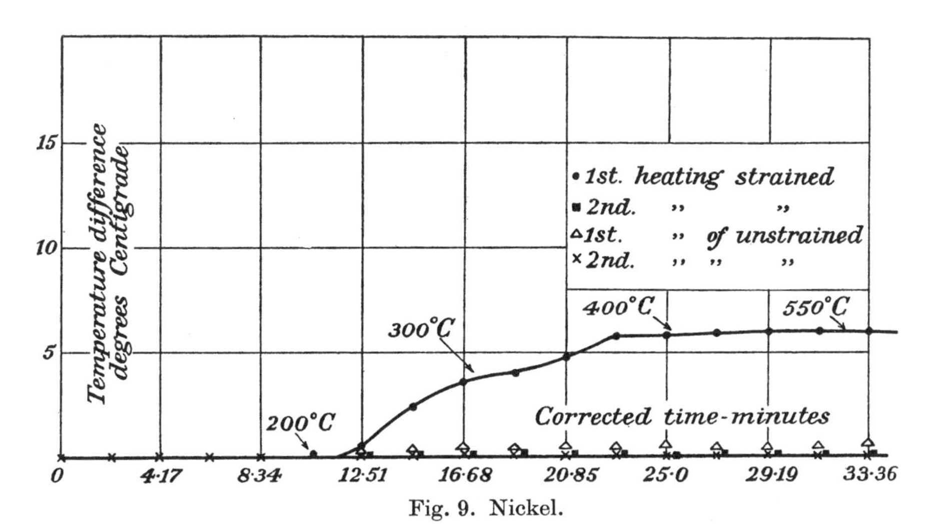


Fig. 9. Nickel.

found in the two cases (namely that required to raise the temperature 5° C.) is proof of the truth of the assumption that the difference between the differences of the temperature-time curves on first and on subsequent heatings is due to the latent heat of cold work.

The results shown in fig. 6 and onwards are those obtained after perfecting the apparatus in a number of respects enumerated below.

Instead of merely maintaining the vacuum for some time before commencing the test, the jackets were heated to a definite temperature well above the dew point, so that possible traces of moisture in the calorimeter were removed before commencing the test. During the whole of this time the calorimeter was exhausted by using a mercury vapour pump with the Hyvac pump as a backing pump.

The hole in the specimen into which the heating element fitted was enlarged and modified to allow any gases which might be evolved on heating to escape easily.

The exhaust passage in the calorimeter was considerably enlarged to facilitate the removal of the gases and the pump was set working and allowed to run in all cases for at least 1 hr. before testing. In some cases the pump was allowed to continue to 24 hr. before the test was made.

These and other modifications resulted in an almost complete elimination of the lowering in the potential drop through the heating coils which had previously been obtained during the first heating. The temperature-time curves for first, second and third heatings of an annealed specimen were now in very close though not always perfect agreement.

As may be seen in fig. 7, a maximum difference of about 0·5° C. was found for copper between the second and third heatings, while for iron (fig. 8) no measurable difference was observed.

With nickel (fig. 9) the maximum difference between the second heating of the strained and the second heating of the unstrained specimens was only a small fraction of a degree. The points in fig. 9 represent the differences between the various heating curves and those of the second heating of the *unstrained* specimen. It will be seen that the points representing the second heating of the *strained* specimen lie very nearly exactly on the base line.

## Correction of Observed Temperature-time Curves for Variation in Energy Supply

As already stated, the current in the interior furnace was maintained constant, and in view of the fact that the potential drop was subject to variation during the test, the rate of heat supply to the specimen was not quite constant.

With the perfect heat insulation which this apparatus ensures it seems certain that the rate of rise in temperature would be proportional to the heat input; accordingly it is possible to correct the observed temperature-time curves to an assumed standard constant heat input. The slope of these corrected temperature-time curves is then directly proportional to the specific heat of the metal.

The details of the method can be followed by referring to fig. 10, where the observed temperature-time and volt curves are shown. It will be seen that during the experiment the p.d. across the heating coil leads increased from 6·0 to 6·35 V. In this case the standard p.d. was taken as 6 V.

In taking the observations the potential difference between the ends of the heating coil was read every $\frac{1}{2}$ min. Since the current is kept constant the corrected time for constant heat input is therefore proportional to the sum of all the readings of the voltmeter which had been taken since starting the test. In practice the sum of these readings was used as a convenient abscissa for the temperature-time curves, and the corrected time for a standard heat input nearly equal to the mean actual input was afterwards calculated. In this way the corrected temperature-time curve shown in fig. 10 was obtained.

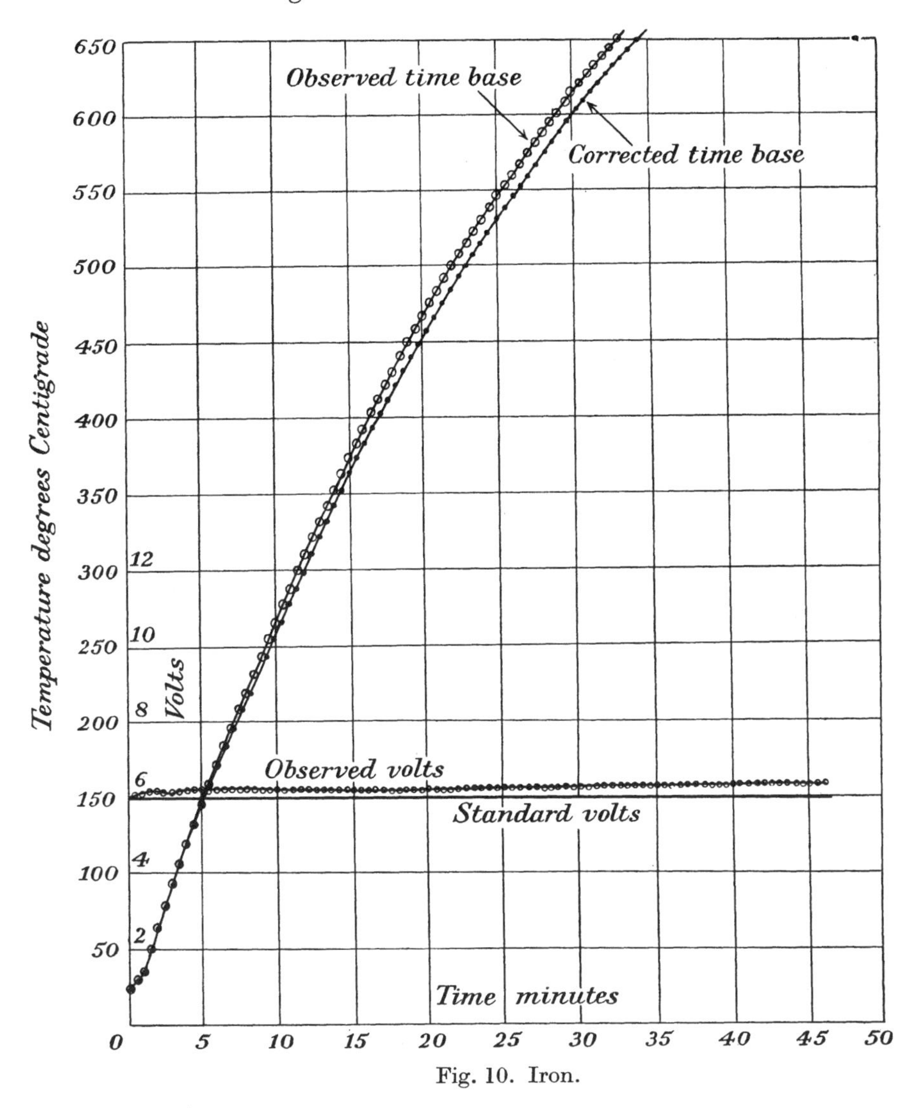


Fig. 10. Iron.

Fig. 11 shows the temperature curve for tool steel also plotted to a corrected time base. This curve is the mean of three independent tests, the separate points being indistinguishable. The heat absorption at the recalescence point is seen to be 68° C.

It would be difficult to reproduce the temperature-time curves similar to that of fig. 10 on a scale which would permit the difference due to cold work to be appreciated. Accordingly the principal results of the measurements are here

27 BSP

exhibited by means of curves showing the difference at any given time between the corrected temperature-time curves for the first and for subsequent heatings. Figs. 5, 7, 8 and 9, copper, iron and nickel for instance, which have already been described, are of this type.

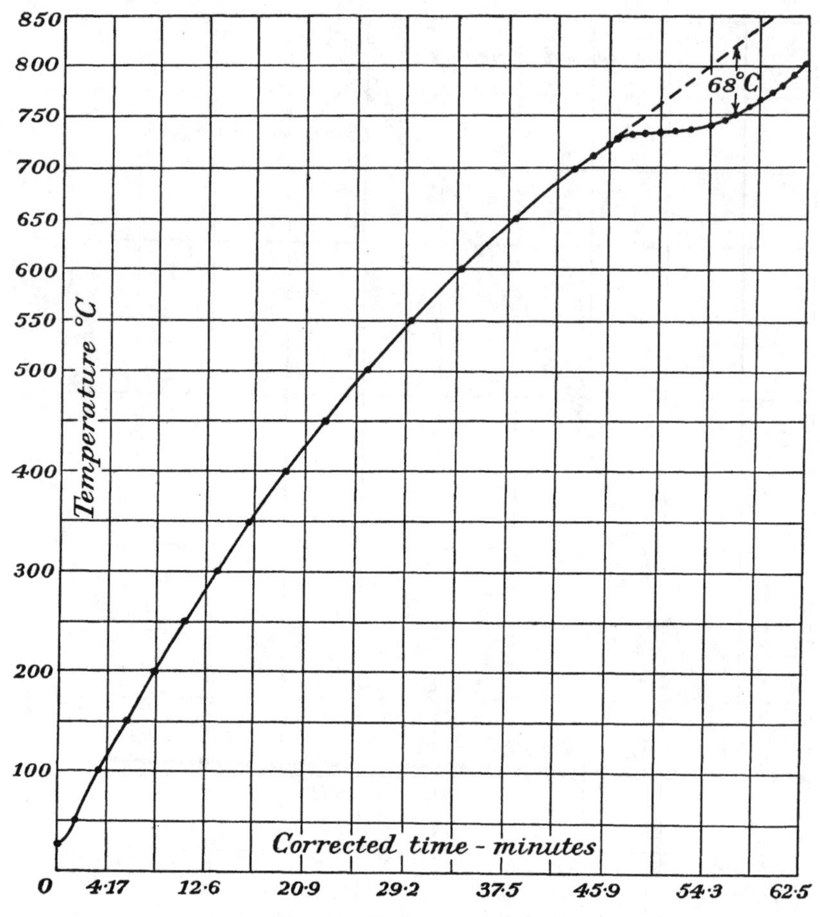


Fig. 11. Tool steel, 0·89% C.

## Results

### *Copper*

It has already been mentioned that at a temperature of 500° C. the difference in temperature attributed to the release of internal energy due to cold work was 5° C. Assuming the specific heat of copper at 500° C. to be 0·10 this is equivalent to 0·5 cal./g. The mechanical work done on the specimen during twisting was measured by apparatus previously described and was in each case 5·6 cal./g. A fraction 0·5/5·6 (i.e. 9 %) of the mechanical work done was then released on reheating. This is in good agreement with the value 9·8 % found by the differential apparatus and is in fair agreement with the difference (7·5 %) between the work done and heat emitted during the process of twisting.

*Nickel*

The work done in twisting the nickel specimen referred to in fig. 9 was 13·9 g.cal./g. while the latent heat of cold work produced a difference in temperature (see fig. 9) between the strained and unstrained specimens of 6° C. at 400° C. Taking the specific heat of nickel at 400° C. to be 0·13, the latent heat of cold work was $6 \times 0{\cdot}13 = 0{\cdot}78$ cal./g.

The proportion of the cold work originally done on the specimen which is released on reheating is therefore 0·78/13·9 or 5·6 %.

*Iron*

For iron the corresponding figures are: work done in twisting = 7·83 g.cal./g.; temperature difference due to release of cold work (see fig. 8), 5° C. at 730° C.

The specific heat of iron between 700° C. and 800° C. is very variable. Measurements as low as 0·19 and as high as 0·29 seem to have been made. Using the mean value 0·24 the heat energy evolved is $5 \times 0{\cdot}24 = 1{\cdot}2$ cal./g. The work done was 7·83 cal./g. so that a proportion 1·2/7·83, i.e. 15 % of the total work done on the specimen, was given out during the heating.

*Aluminium*

Measurements were made with aluminium specimens, but it was found that volatilization of the metal in the vacuum furnace made the results unreliable.

*Brass* 70/30

Work done in twisting = 2·93 g.cal./g. Temperature difference due to release of cold work = 5·5° C. and taking 0·088 as the specific heat of the specimen, the heat energy evolved was $5{\cdot}5 \times 0{\cdot}088 = 0{\cdot}485$ g.cal./g. That is, 14·7 % of the total work done on the specimen during twisting is given out on heating.

The results of Stott and Rosenhain* and also those of Smith† are not included in Table 1 as the cold work done in drawing the wires through the dies is not known.

*Specific Heats*

If the heat capacity of the interior furnace is negligible the slope of the corrected temperature-time curve is a measure of the specific heat of the metal at that temperature.

In order to find the true specific heat from the curves it was therefore necessary to estimate the heat capacity of the furnace. This was done by estimating the capacity for heat of the parts of the internal furnace and further checking same by comparing the rate of temperature of two specimens of pure copper of the same

* *Proc. Roy. Soc.* A, CXL (1933), 9.
† Ibid. CXXV (1929), 619.

27-2

length but different diameters. Calling $h$ the capacity for heat of the furnace the average value was 0·49 cal./°C. which is approximately 5 % of the capacity for heat of a nickel specimen. This value of $h$ was determined as follows:

Table 1

| Metal | Heat equivalent of work done (g.cal./g.) | Energy absorbed during cold working / Work done on specimen | Energy emitted on heating / Work done on specimen | Authority |
|---|---|---|---|---|
| Cu | 5·5 | — | 0·09 | Present results |
| Cu | 0·82 | 0·09 | — | Farren and Taylor, *Proc. Roy. Soc.* A, CVII (1925), 422; paper 6 above |
| Cu | 0·96 | 0·075 | | Taylor and Quinney, *Proc. Roy. Soc.* A, CXLIII (1934), 307; paper 19 above |
| Cu | 6·2 | 0·086 | — | |
| Cu | 14·1 | 0·081 | | |
| Cu | 18 | — | 0·018 | Sato, *Sci. Rep. Tohoku Univ.* XX (1931), 140 |
| Cu | 9·7 | — | 0·031 | Sato; Ibid. |
| Cu | 4·2 | — | 0·042 | Sato; Ibid. |
| Fe (pure) | 7·8 | — | 0·15 | Present results (vacuum calorimeter) |
| Fe (mild steel) | 1·1 | 0·13 | — | Farren and Taylor; Ibid. |
| Fe (decarburized mild steel) | 3·2 | 0·09 | — | Taylor and Quinney; Ibid. |
| | 9·6 | 0·07 | — | Taylor and Quinney; Ibid. |
| Fe | 16·1 | — | 0·008 | Sato; Ibid. |
| Fe | 10·4 | — | 0·009 | Sato; Ibid. |
| Al | 1·2 | 0·07 | — | Farren and Taylor; Ibid. |
| Al (single crystal) | 1·8 | 0·05 | — | Farren and Taylor; Ibid. |
| Al | 11·8 | — | 0·098 | Present results (differential method) |
| Al | 5·3 | — | 0·34 | Sato; Ibid. |
| 70/30 brass | 4·67 | — | 0·16 | Present results (differential method) |
| 70/30 brass | 2·93 | — | 0·15 | Present results (vacuum calorimeter) |
| 70/30 brass | 1·12 | — | 0·24 | Sato; Ibid. |
| 70/30 brass | 3·42 | — | 0·16 | Sato; Ibid. |
| 70/30 brass | 8·24 | — | 0·13 | Sato; Ibid. |
| Nickel | 13·9 | — | 0·056 | Present results (vacuum calorimeter) |

Copper specimens of the same length but different diameters weighing 82·07 and 24·80 g. respectively were used. From the corrected temperature-time curves the rates of rise of temperature in °C./sec. were found at a number of temperatures ranging from 490° C. down to 100° C.

If $\theta_1$ is the slope of the temperature-time curve for the large specimen, and $\theta_2$ that for the smaller one, and if $\sigma$ is the specific heat of the metal at any given temperature then, since the energy supplied was 10 W. in both cases,

$$(82{\cdot}07\sigma+h)=\frac{10}{4{\cdot}17}\times\theta_1$$
$$=H_1. \tag{6}$$

Similarly for the smaller specimen

$$(24{\cdot}80\sigma+h)=\frac{10}{4{\cdot}17}\times\theta_2 \tag{7}$$

$$=H_2,$$

where $h$ = capacity for heat of furnace and adjoining material. The present results together with comparable result of previous experiments are given in table 1. In this table column 3 contains measurements made during the *absorption* of energy and column 4 contains measurements made during the *release* of energy on reheating.

Table 2. *Values of $H_1$ and $H_2$*

| Temperature (° C.) | 490 | 400 | 350 | 250 | 200 | 150 | 100 |
|---|---|---|---|---|---|---|---|
| $82{\cdot}07\sigma+h=H_1=$ | 8·98 | 8·97 | 8·65 | 8·6 | 8·22 | 7·82 | 7·82 |
| $24{\cdot}80\sigma+h=H_2=$ | 3·145 | 3·052 | 3·01 | 2·95 | 2·86 | 2·72 | 2·61 |
| $57{\cdot}27\sigma=$ | 5·84 | 5·93 | 5·64 | 5·65 | 5·36 | 5·10 | 5·21 |
| $\sigma=$ | 0·102 | 0·1037 | 0·0985 | 0·0985 | 0·0936 | 0·0891 | 0·091 |
| Values of $h$ = | +3·145 −2·540 =0·61 | +3·052 −2·57 =0·48 | +3·01 −2·49 =0·52 | +2·95 −2·44 =0·51 | +2·86 −2·36 =0·50 | +2·72 −2·21 =0·51 | +2·61 −2·29 =0·32 |

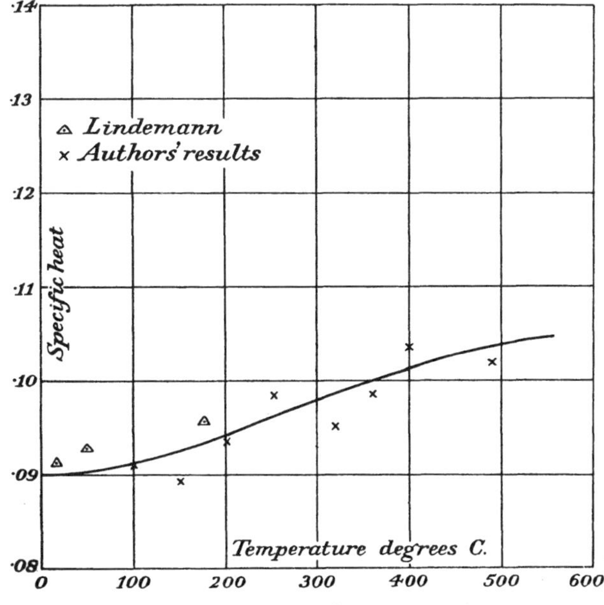


Fig. 12. Copper.

In table 2 the values of $H_1$ and $H_2$ are given for temperatures ranging from 490° C. downwards. Subtracting $H_2$ from $H_1$ the fourth line of the table gives 57·276, and dividing this by 57·27 the values of $\sigma$ are given in the fifth line. Using these values of $\sigma$ the value of $h$ can be found by substituting in (7).

The values of $h$ found in this way are given in line 6. It will be seen that except for the values at 100° C. and to a less extent at 490° C. the values are remarkably constant.

The mean of the series of values such as are shown in table 2 was

$$h = 0{\cdot}49\,\text{g.cal./}^\circ\text{C.}$$

The heat capacity of the furnace was therefore always a small fraction of the heat capacity of the specimen; for instance, the specimen with least heat capacity was aluminium 5·5 cal./° C. The capacity for heat of furnace was therefore 9 % of

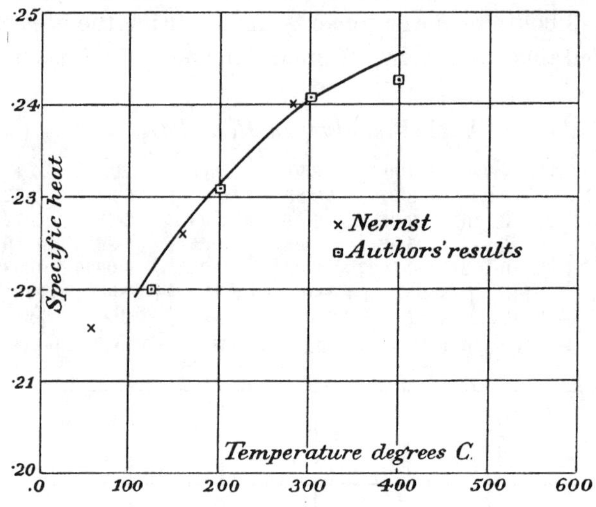


Fig. 13. Aluminium.

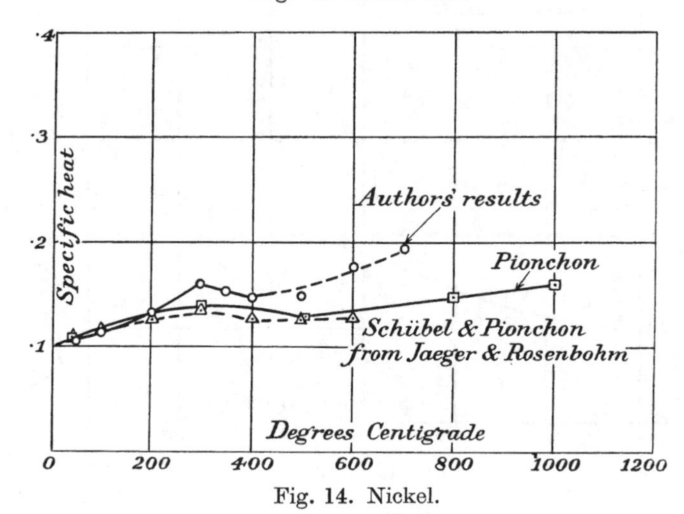


Fig. 14. Nickel.

specimen. The specimen with greatest capacity was 10·0 cal./° C. and the capacity for heat of the furnace was therefore 5 %. Errors in determining the heat capacity of the furnace give rise to only very small errors in the specific heats obtained.

From the corrected temperature-time curves, using this value (0·49 g.cal./° C.) for the heat capacity of the furnace, the specific heats for copper, aluminium, nickel and iron are shown in figs. 12, 13, 14 and 15. The results obtained by other

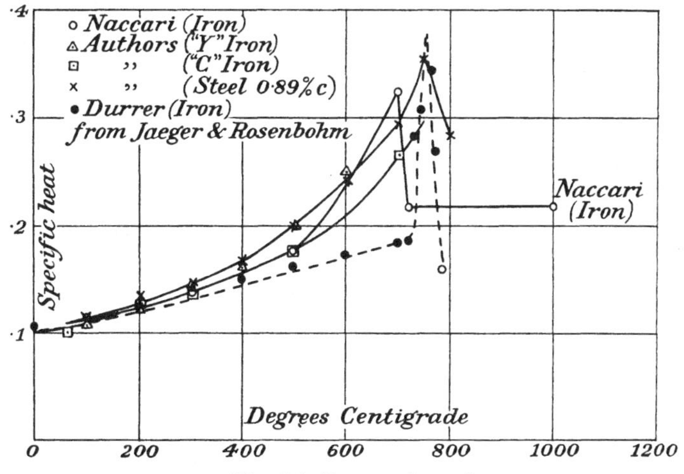


Fig. 15. Iron and steel.

investigations are also shown. It will be seen that for aluminium and copper the results are in good agreement with those of Nernst and Lindemann*. For iron, steel and nickel, our values appear higher at higher temperatures.

In conclusion, we wish to express our thanks to Professor C. E. Inglis, F.R.S., for permission to carry out the work in his laboratory.

* *S.B. preuss. Akad. Wiss.* (1911), p. 494; ibid. (1912), p. 1160.

# 27

# PLASTIC STRAIN IN METALS

(May Lecture to the Institute of Metals, delivered 4 May 1938)

REPRINTED FROM

*Journal of the Institute of Metals*, vol. LXII (1938), pp. 307–24

The work of the author with Dr Elam on the straining of metallic single crystals is described. The application of experimental results with single crystals to poly-crystalline aggregates is discussed.

In the May Lecture last year Professor Andrade* gave a very clear account of some of the main lines along which researches on metallic crystals have developed. I hope now to discuss some of the questions treated by Professor Andrade, but in greater detail than he was able to do in the time at his disposal. I propose also to put forward some thoughts about how our knowledge of metallic single crystals can help us to understand the mechanical properties of crystal aggregates.

I must begin by making the confession that I am not a metallurgist; I may say, however, that I have had the advantage of help from, and collaboration with, members of your Institute, whose names are a sure guarantee that the metals I have used were all right, even if my theories about them are all wrong. Perhaps I may be excused if I give an account of how I first came to have anything to do with metals. I was present at the Royal Society on the occasion when Sir Harold Carpenter described the fascinating series of researches which enabled him and Dr Elam to prepare very large single crystals of aluminium. He showed test-pieces which had been pulled in a testing machine with the result that lines originally scratched on them at right angles to their longitudinal axes had become oblique during the plastic straining. These lines, it seemed, must provide the clue to the relationship between the crystallographic axes and the plastic strain. At that time the existence of slip lines on the surface of strained metals was well known, and it was known also that they are the traces of crystallographic planes. It was freely stated that they became visible, owing to slipping of the metal, just as the edges of cards in a pack become visible when the top of the pack is pushed sideways. This, however, is a very different matter from stating that the total strain is identical with that which is produced in a pack when cards slide over one another. The surface markings may, for instance, develop quite independently of what goes on inside the crystal, because the surface is known to be in a different state from the interior.

* E. N. da C. Andrade, *J. Inst. Metals*, LX (1937), 427.

When the phenomena shown by Sir Harold Carpenter's strained crystals were regarded from the geometrical point of view, it was clear that one could completely determine the strain in the crystal: (*a*) if the strain was uniform, so that lines in the specimen which were originally parallel remained parallel when it was strained, and (*b*) if the extensions or contractions in six independent directions could be measured. We may, for instance, imagine that a square-sectioned bar is cut from a single crystal. In fig. 1, *ABCD* is the square section which we may suppose marked with scratches on the surface, and *AF*, *BE*, *CG* are edges of the bar. If now we measure the proportional extension during strain of the six lines, *BA*, *BC*, *BE*, *AC*, *AE*, *EC*, then the strained position of every particle is determined. If we measure only five, then the strain is not completely determined, unless some further assumption is made. We may find, for instance, that the density is unchanged by the strain; then five, and only five, independent extensions or components have to be measured to determine the strain. The six components of strain need not, of course, be those shown in fig. 1. One may, for instance, measure the extensions *BE*, *BA*, *BC*, and the angles *ABE*, *CBE*, and the angle between the faces of the specimen. Sir Harold Carpenter and Dr Elam's original specimens were not suitable for making accurate measurements of the six components of strain. Accordingly, encouraged by Sir Harold, Dr Elam and I collaborated in preparing and marking single crystals of such proportions that accurate strain measurements could be made. Fig. 2 shows one of our specimens after 70 % extension. The strain was very uniform, even after such great extension.

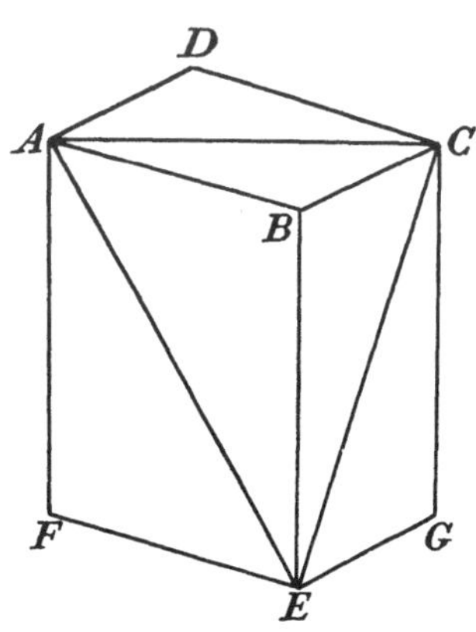

Fig. 1. Measurement of six lengths to determine strain.

The usual method by which strains are analysed is to find the positions and elongations of the axes of the strain ellipsoid, i.e. the strained shape and position of an originally spherical piece of material. In our case, however, we found instead the cone which passes through the intersection of the strain ellipsoid and the original sphere. This cone, which is shown in pl. 1*a* with the strain ellipsoid, evidently contains the strained positions of all directions which remain unstretched, and is therefore termed the unstretched cone. Our reason for adopting this procedure was that if the whole strain is, in fact, due to slipping parallel to one crystal plane, that crystal plane must form part of this cone, because slipping parallel to a plane gives rise to strain which leaves all directions in the plane of slipping unchanged in length. If part of the unstretched cone consists of a plane, it is a mathematical necessity that the whole cone must consist of two planes.

I will not trouble you with the method by which we calculated the position of the unstretched cone* in the strained specimen, but it is necessary for the

* Taylor and Elam, *Proc. Roy. Soc.* A, CII (1923), 643; paper 5 above.

argument that one should understand how this cone and the directions of the crystal axes were represented on plane diagrams. For this purpose we used the stereographic projection. Each direction in space can be regarded as marking a point on a sphere. The surface of this sphere is then projected on to a plane from a point on its circumference.

In the projection, great circles on the sphere, which contain all directions in space which lie in a plane, are projected into circles. Of these, the great circle which represents the plane parallel to the plane of projection is the smallest. I will call it the 'bounding circle'. The part of the projection which lies inside the bounding circle corresponds with a complete hemisphere, and if we are thinking about orientations in space, and are not concerned with the sense of directions on a straight line—i.e. if we do not consider whether a vertical line is pointing upwards or downwards, but only concern ourselves with the fact that it is vertical— then all orientations can be represented on a hemisphere, and so by the part of the stereographic projection which lies inside the bounding circle.

One property of the stereographic projection is that small circles on the sphere, which represent circular cones in space, also project into circles. Small circles can be distinguished from great circles, however, by the fact that projected great circles always cut the bounding circle in the projection at opposite ends of a diameter. Small circles never do so.

When we came to set out on a stereographic projection the points representing directions in our unstretched cone, calculated from the measurements made on our stretched single crystal specimen, we found that they did in fact lie on a circle which cut the bounding circle at opposite ends of a diameter.

Fig. 3 shows the stereographic projection of points on the unextended cone calculated from measurements made before and after a strain which extended a specimen cut from an aluminium crystal from 10 to 30 % elongation. The circles drawn most nearly through the calculated points are shown in the diagram. It will be seen that they do in fact cut the bounding circle at opposite ends of diameters. This is a proof that, in the case to

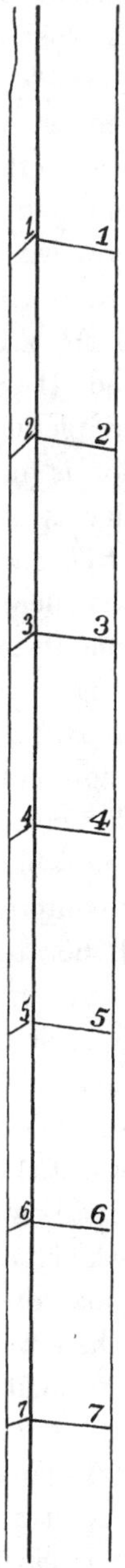

Fig. 2. Marked specimen cut from aluminium single crystal, after 70 % extension.

which this diagram refers, the unstretched cone really does degenerate into two planes, so that the total distortion can in fact be produced by slipping on either of these planes. It is impossible from two sets of external measurements made before and after straining to say which of these two planes is the plane on which slipping takes place, but we always found with aluminium that one of the two coincides with an octahedral crystal plane. This plane we took as the slip-plane, and we found that the direction of slip is the diagonal of a cube face or edge of the octahedron.

In fig. 3, $B'$ is the direction of slip calculated from the external measurements of lines on the specimen, and $B$ marks the orientation of a crystal axis represented by (101), i.e. it is in the direction of one of the diagonals of a cube face, or an edge

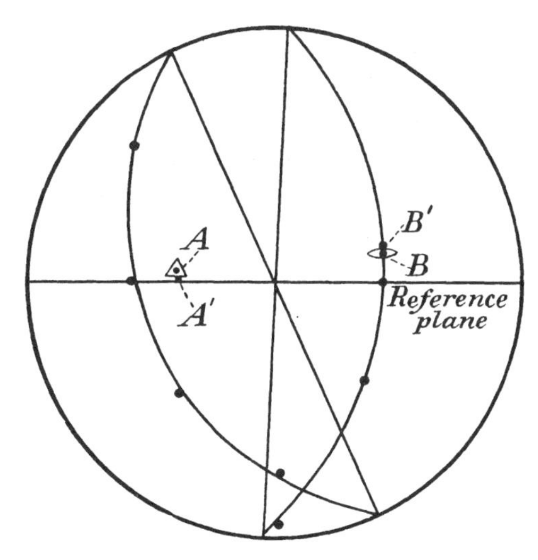


Fig. 3. Stereographic projection of unextended cone, with diameters marked to prove that cone is two planes.

of the octahedron corresponding with the cubic symmetry of the crystal. The point $A'$ represents the direction of the normal to the plane which is represented in fig. 3 by the circular arc containing $B'$. The point $A$ represents the crystal axis, determined by X-rays, which is the diagonal of the cube in cubic symmetry. The crystal axes $A$ and $B$ are evidently associated with one of the planes of the unextended cone, but not with the other. The plane has accordingly been taken as the plane of slip. It will be seen that the plane of slip is parallel to a face of the octahedron associated with the cubic symmetry of the crystal and the direction of slip is parallel to an edge of the octahedron.

This method is more complicated than that used by Professor Andrade and by practically all other workers in the field. They strain the crystal and observe marks on the surface, which they prove are the traces of a crystallographic plane. They then *assume without further proof* that the strain is of a simple type consisting of a shear parallel to the plane marked out by the surface markings, and then make the two angle or extension measurements which are necessary to determine the direction of slip if their initial assumption is true.

Professor Andrade offers the opinion, in one of his papers, that this simplified method is in some cases more accurate than the complete analysis that I have described. I do not agree with this contention, provided that *proper precautions are taken* to ensure that the specimen is strained uniformly. In some cases the simplified method is inapplicable, because the assumed strain by shearing parallel to a crystal plane does not in fact take place; in others the slip lines on which the method relies do not make their appearance.

Fig. 6 shows the stereographic diagram of one of the cases analysed. The unextended cone is nothing like two planes; in fact this diagram was proved to correspond with compound slipping on two octahedral planes. For cases to which it is applicable, the simple method provides a quick means of identifying slip-planes provided that one has made certain, by complete strain measurements, that the strain is due to shear parallel to a plane.

The accuracy of the complete analysis depends on the uniformity of the strain. If a tensile specimen slips unequally in different parts of its length, strain measurements on its surface will vary from place to place. For this reason, I developed a method of straining in which flat discs cut from a crystal were compressed between parallel steel faces. This method ensured that the compression at all points of the disc was the same, and thus secured uniformity in one, at any rate, of the components of strain.

Pl. 1*b* is a photograph of a circular disc cut from an aluminium crystal, before and after compression. In spite of reduction to half the original thickness, the scribed lines are still quite straight, and with this technique there is no 'barrelling' of the section perpendicular to the parallel faces.

The great uniformity of our compression specimens was obtained by a special technique. If one compresses a short cylinder of solid metal between parallel planes the friction at the top and bottom normally holds the ends from expanding laterally and the specimen assumes a shape like a barrel. If the compressing faces are ground and polished and then greased, the first thing that happens when a compressive load is applied is that the grease is squeezed out. This causes an outward tangential force due to viscous drag to act over the top and bottom of the specimen, that is, a force in the opposite direction to the friction which would act in the absence of grease. By compressing the specimen in very small stages one can in this way get far greater uniformity of strain than can be obtained in a tensile specimen.

When a single crystal is extended, the orientation of the crystal axes relative to the axis of extension varies as the straining proceeds. If the strain simply consists of sliding parallel to a crystal plane in a crystal direction, the cause and nature of this change in orientation becomes clear if we imagine the slip-plane as fixed, and the orientation of the axis of the specimen as changing. The specimen axis must rotate in a great circle towards the direction of slip. In fig. 4, 0 represents the initial position of a specimen axis in one of the triangles of cubic symmetry. The point (110) represents the crystal axis towards which slipping has occurred. The dotted line represents the great circle along which the specimen axis would move if the slipping were of the type contemplated, and the calculated positions of the specimen axes for extensions of 9, 41, 51 and 70 % are marked off on the dotted line. The positions of the specimen axis measured by X-rays are also shown. It will be seen that at 9 and 41 % extension there is good agreement, but that as soon as the representative point reaches the boundary between the two symmetry triangles it does not continue along the calculated path, but remains close to the boundary of the two triangles. This is because symmetry requires that when the representative point gets into the right-hand triangle slipping shall start in the direction of the

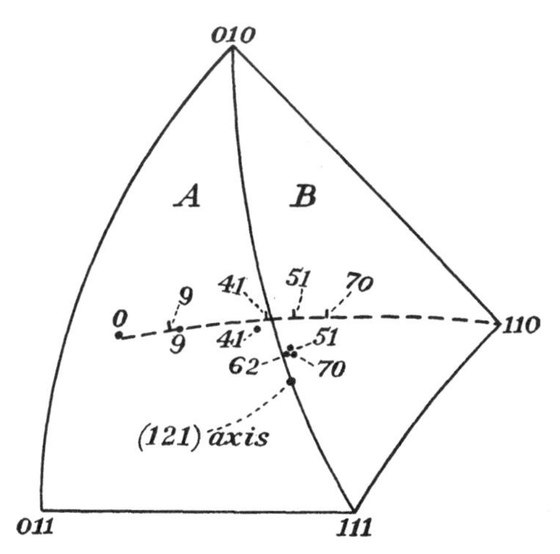


Fig. 4. Change in orientation of axis of specimen (see fig. 2) relative to crystal axes, during extension of 70 %. ●, observed by X-rays; ı, calculated.

axis represented in fig. 4 by (011). Slipping towards the direction (011) would move the representative point back to the boundary between the triangles. Thus slipping continues on two planes, and the representative point remains on the plane midway between the two directions of slip. Finally, it reaches a crystal axis shown by (121) in fig. 4, which is the point midway between the two directions of slip (110) and (001).

This, and the similar case of a single crystal under compression, are the only two cases in which a preferred orientation of crystal axes due to straining has been explained, though the phenomenon of preferred orientation has been found experimentally by means of X-rays in a large number of drawn, rolled, and otherwise worked metals.

I have mentioned that with aluminium the slipping is on an octahedral plane in the direction of its edges. An octahedron has eight faces, but pairs of them are parallel to one another so that there are four possible slip-planes, and on each of these there are three possible directions of slip, making twelve possible types of slipping in all. We have seen that, when a single crystal of aluminium is pulled, the strain is due to one only of these twelve. As Professor Andrade told you last year, we found that if the shear stress is resolved parallel to all the four possible slip-planes in each of the three possible directions of slip, the operative slip is that one of the twelve possibles for which the shear stress is greatest. We found, further, that this law of maximum shear stress determines the same slip-plane and direction for all possible positions of the specimen axis within one of the triangles into which the diagram of cubic symmetry is divided. When a single crystal of aluminium, or an aggregate of such crystals, is strained, the resistance to further straining increases as the plastic strain increases. When the slipping is on one crystal plane, the resistance to shear depends only on the amount of shear strain that has occurred since the crystal was in its original fully annealed state. Professor Andrade showed

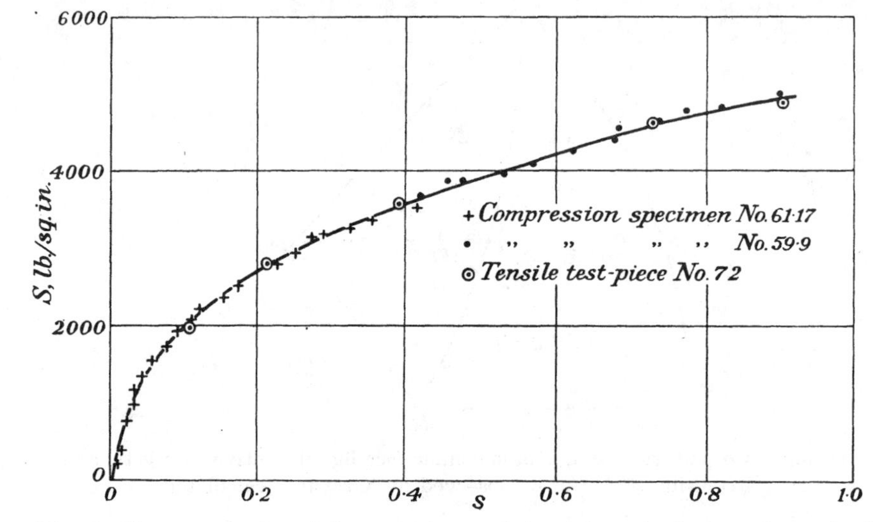


Fig. 5. Shear stress, $S$, and shear strain, $s$, of aluminium single crystals resolved on to slip-plane and direction. ●, +, in compression; ⊙, in tension.

some curves giving the relationship between the plastic shear strain $s$ and the shear stress $S$. Fig. 5 shows the relationship between $S$ and $s$, derived from experiments on the crystals of identical material in tension and in compression. It will be seen that the fact that though in one case there was compression perpendicular to the slip-plane, while in the other there was tension, no difference is observed in the $S$–$s$ relationship.

On the other hand, the resistance seems to increase rather more rapidly when double slipping occurs than when the whole strain is due to single slipping. One of our specimens had its crystal axes in the symmetrical position where double slipping might be expected to take place. The complete analysis of double slipping was carried out for various stages of compression. The unstretched cone was worked out from the measurements of the specimen, and also calculated on the assumption of equal slipping on each of the two possible slip-planes. The two cones are shown in fig. 6, and it will be seen that they are only very slightly different. Thus, the strain is in fact very nearly due to the type of double slipping which the symmetry and maximum shear stress rule prescribes. Fig. 4 shows the $S$–$s$ curve

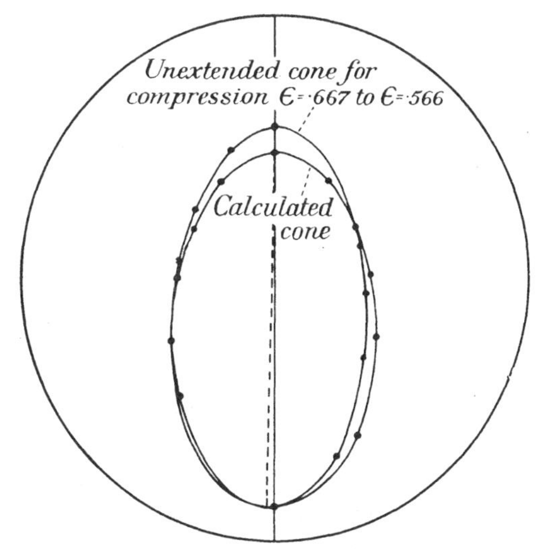


Fig. 6. Stereographic projections of calculated and observed cones for double slipping.

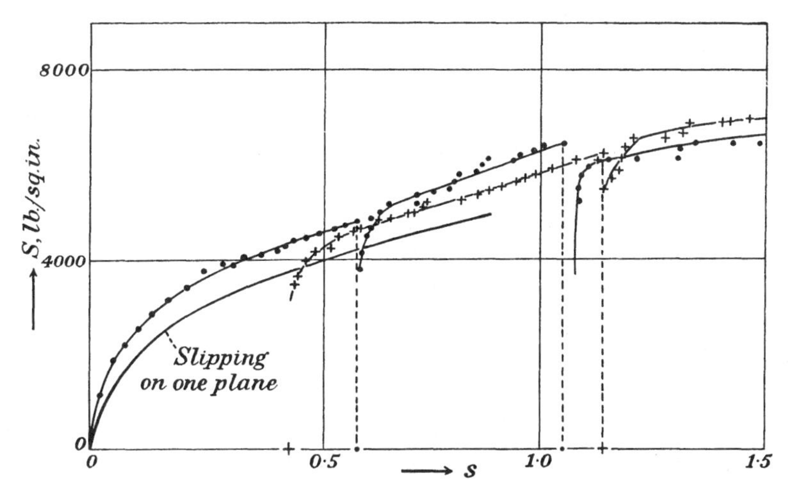


Fig. 7. Shear stress, $S$, and shear strain, $s$, for double slipping.

derived from the analysis of double slipping; it is of the same type as that for single slipping, but is rather higher.

Now we approach a complicated and difficult problem, namely, the analysis of stress and strain in an aggregate of crystals when the whole aggregate is strained plastically.

I think that I can say, without fear of contradiction, that no self-consistent or valid theory of plastic crystal aggregates has yet been put forward, though a number of invalid attempts have been made in this direction. The essential difficulty in connecting experimental results obtained with single crystals with those obtained in aggregates is to imagine how it is possible for slipping to go on inside crystals so that the boundaries of neighbouring crystal grains shall still be in contact after the slipping has taken place. All attempts made so far to correlate the mechanical properties of crystal aggregates with those of single crystals rest on the same fallacy, namely that each crystal grain can be treated as though its neighbours did not exist. The recent work of Cox and Sopwith,* for instance, visualizes a crystal aggregate as consisting of a large number of cylindrical single crystals combined together into a cylindrical aggregate. When the aggregate is extended parallel to the length of the cylinders each crystal extends just as a single crystal would if it were removed from its neighbours, and the total force required to extend the aggregate is the sum of the forces required to extend each crystal. If the crystals fitted together as a solid mass before straining they would certainly not, in this conception, fit together after straining, so that holes would be produced between the grains.

I have said that it is a fallacy to think of the grains in an aggregate as being independent of one another so far as strain is concerned, but it seems to me that a still more fundamental fallacy is involved in the existing way of thinking of stresses in the grains of plastic aggregate at all. When a cylindrical specimen cut from a single crystal is subjected to an end load, it is only possible to think about the stress at any point inside it because that stress can be assumed to be uniform. On the other hand, if two single crystal cylinders are stuck together along their length, and an end load is applied, the stress is quite indeterminate until they begin to stretch, because one may be subjected to an initial compressive load which is balanced by an equal tensile load in the other.

A crystal aggregate may be likened to a mechanical system in which each part bears on its neighbours with a frictional contact. A simple model which illustrates some of the properties of frictional systems is shown in fig. 8. It consists of a board lying in the angle made by vertical and horizontal boards. The stress in it is quite indeterminate, and might have any value between certain limits. If, for instance, one were to push on the vertical board, bending it slightly, one could increase the compression in the sloping board without making it slip. Now suppose we push the sloping board till it slips. Amonton's law of friction, according to which the ratio of the tangential to the normal force at a sliding contact is equal to the coefficient of friction, now makes the forces everywhere determinate. If, instead of obeying Amonton's law, the friction at both sliding surfaces were independent of normal force, the force system would again be determinate if the tangential force at each contact were known. To find $P$, the force with which the sloping board must be pushed in order that it may slide, we can solve the equations of equilibrium, calculating normal reactions at the points of contact. On the other hand, we can proceed more simply by what is called the principle of virtual work.

* H. L. Cox and D. G. Sopwith, *Proc. Phys. Soc.* XLIX (1937), 134.

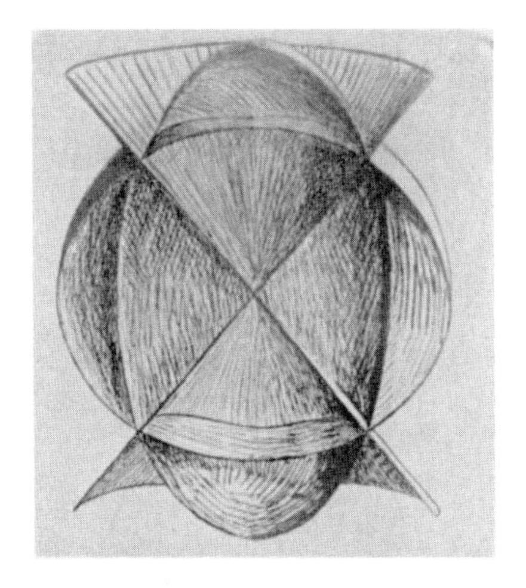

*a*, Unextended cone and strain ellipsoid.

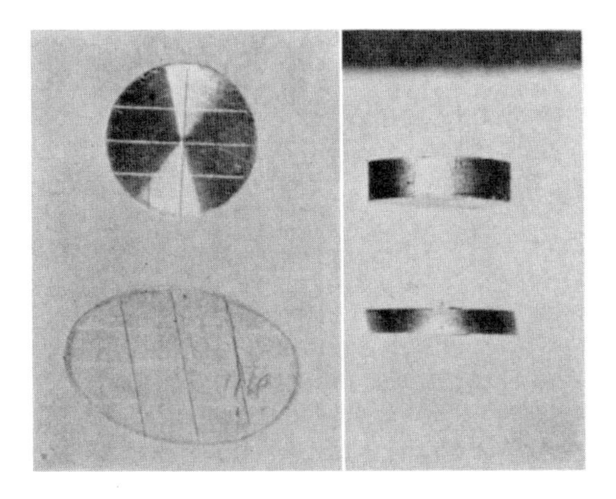

*b*, Front and side views of a compressed disc cut from a single crystal.

We imagine that the force $P$ pushes the system through a distance $x$ at its point of application. The work done is then $Px$. If $s_1$ and $s_2$ are the distances through which the ends of the sloping board slide, and the friction forces are $f_1$ and $f_2$, the energy wasted at the points of sliding contact is $s_1f_1+s_2f_2$. The principle of conservation of energy then gives

$$P=f_1\frac{s_1}{x}+f_2\frac{s_2}{x}.$$

The ratios $s_1/x$ and $s_2/x$ are determined by purely geometrical considerations. It will be seen that by this principle of virtual work we have determined the force $P$ without bringing in the conception of stress at all.

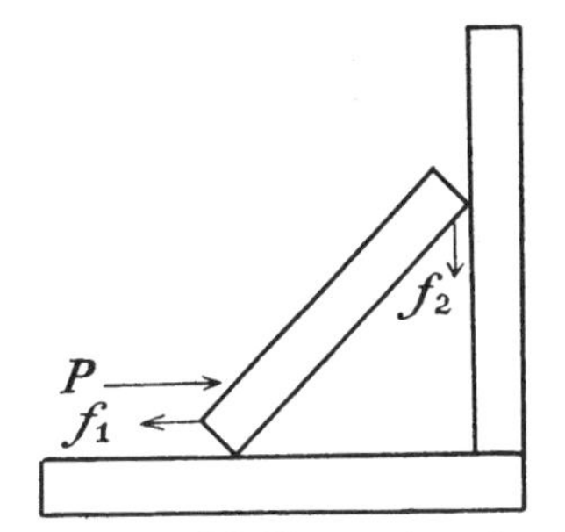


Fig. 8. Model illustrating simple system with friction

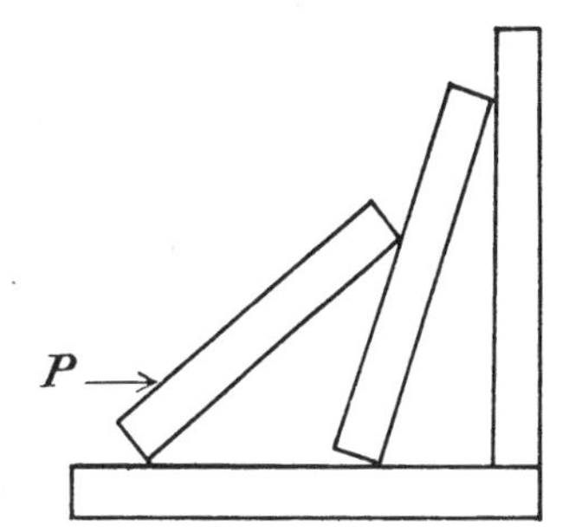


Fig. 9. Friction system with two possible modes of slip.

Now consider the more complicated system consisting of two boards, which is shown in fig. 9. When the outer sloping board is pushed, one of two things happens: either the inner sloping board remains fixed, the contacts at the two ends of the outer board slipping, or no slipping occurs at the contact of the two sloping boards, but all the remaining three contacts slide. One cannot arrange the boards so that all the possible contacts slide at once when the outer one is pushed. In systems like this the only general rule that can be given for determining the force necessary to cause motion is to assume that there is no slipping at as many of the frictional contacts as is possible in view of the geometrical constraints in the system. We can then calculate the force $P$ which corresponds with each of the motions which satisfy this condition, using the principle of virtual work. The motion which actually occurs will be that for which $P$ is least.

Now let us see how this principle can be used to determine the combination of shears or slips which will arise when any given strain is forced on a crystal by an external agency. Take first the case of a cubic single crystal extended in one direction, and able to expand or contract freely in all perpendicular directions. The single geometrical condition can be satisfied if one only of the twelve possible types of slipping is operative. In this case, the virtual work equation is $Ss=Px$, where $s$ is the amount of slip corresponding with extension, $x$, so that if the shear strength $S$ is the same for all the twelve possible types of slip, the principle of least possible energy dissipation for a given extension tells us that only the slip-plane is operative for which $s$ is least when $x$ is prescribed. It is a matter of simple geometry to show that with single slipping, $x/s$ is identical with the 'stress factor', i.e. it is equal to the ratio $S/P$, so that the condition that $s/x$ shall be the least possible is identical with the condition, derived from the conception of stress, that the operative slip is that for which $S/P$ is the greatest of the twelve possible values.

We are now in a position to see how one can determine the system of complex slipping which will occur when any given strain is produced in a crystal. A strain has, as I stated earlier, six components, but when the strain is composed of shear strains only, without volume expansion, this is reduced to five. If these five components of strain are given, we can combine five out of the twelve possible shears or modes of slipping to produce the required strain. We could, of course, combine six, seven, or more shears to produce the same strain, but our study of the mechanics of frictional systems shows that the least energy is wasted, or virtual work done, with a combination of five only. To choose the five, we can only try every combination of five out of the possible twelve, and see which corresponds with the least virtual work or energy dissipated.

At first sight, this seems a formidable task, because there are 792 ways of choosing five things from a group of twelve. We must remember, however, that the range of choice is much more restricted than that contemplated in this estimate. In the first place, the three directions of slip on any one plane are not independent, since the strain due to slipping in one direction can be produced by combining shears in the two other directions. Thus, the twelve shears are divided into four groups of three shears each and only two can be assigned to any one group. This reduces the number to 648. *Next it is found that the geometrical condition for a given strain cannot be satisfied if the five shears are chosen so that two are taken from one group, i.e. one slip-plane, and the remaining three are chosen one from each of the three remaining groups. This reduces the number of choices to 324, all of which must be chosen so that two shears occur on each of two planes, one on the third and none on the fourth.* Next it must be noticed that on a plane where there are two shears there are three ways of choosing the pair. Thus, if we work out any one of the 324 combinations, eight more can immediately be deduced without further analysis. This reduces the 324 in the ratio 9:1, i.e. to 36. Finally, it turns out that a further geometrical inconsistency rules out one-third of these 36, so that, finally, we are left with an irreducible number of 24 combinations of five shears.

Since the resistance to shear $S$ has been shown, experimentally, to be the same for all the twelve crystallographically similar shears, the energy dissipated in any combination of five shears is simply equal to $S$ multiplied by the sum of the five component shears. Thus, to find which of the 24 combinations is effective, we must take each of the 24 possible combinations of five shears and determine their five values so that they give rise to the given external strain (which is specified, of

* *Editor's note.* The statements between the two asterisks are not correct; see the note at the end of paper 28.

course, by five components) when combined together. In each case we then form the sum of the five shears, without regard to sign. The smallest of the 24 resulting sums is that which, by the principle of virtual work, or least energy dissipation,

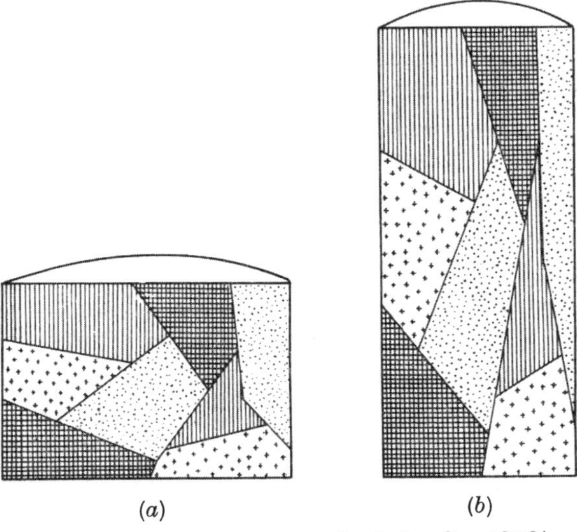


Fig. 10. Crystal aggregate. *a*, unstrained; *b*, after 125 % extension.

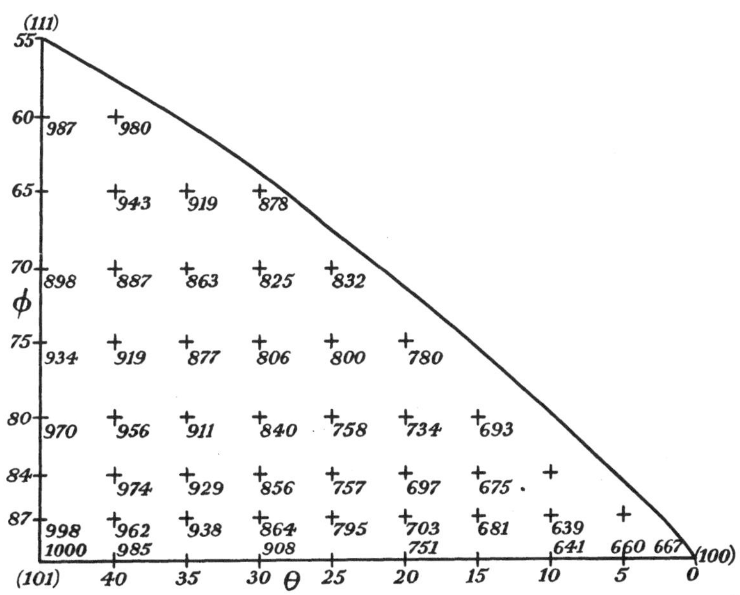


Fig. 11. Stereographic projection showing orientations for which complete calculations were made. Figures are sum of shears when extension per unit length of aggregate is 272. Crosses ( + ) show orientations for which complete calculations were made.

corresponds with the operative combination of five shears. All this sounds complicated, but the whole process involves only the simplest mathematical operations, repeated a great many times.

I have now described how the system of complex slipping which will occur on application of any given externally applied strain to a single crystal can be determined. It remains to apply the results to crystal aggregates. If you look at a microphotograph of the cross-section of a drawn wire, you will see that the crystal grains are all elongated in the direction of extension, and contracted in the perpendicular direction. Each grain, in fact, suffers exactly the same strain as the surrounding material in bulk. With this strain, all the grain boundaries necessarily remain in contact, no holes forming between them. I have therefore taken the case of an aggregate in which the grains take up all possible orientations, and have imagined that the aggregate is strained by extending it by a small amount in one direction, while at the same time contracting it by half the amount in all perpendicular directions, thus keeping the volume unchanged. The diagrams shown in fig. 10, which are drawn to scale, show on the left an imaginary section of a coarse-grained, round bar, on the right the same bar with the same grains when extended to $2\frac{1}{4}$ times its initial length.

I selected a number of orientations of the axis of extension in relation to the crystal axes, and have worked out by the method described above and with the help of Mallock's equation-solving machine, the particular combination of five shears which is effective at each orientation. The points representing the axis of extension are nearly uniformly distributed over the fundamental spherical triangle of cubic symmetry (see fig. 11) so as to represent random orientation in the aggregate.

## Results

The sum of the five shears necessary to give rise to an extension of the aggregate equal (in arbitrary units) to 272 is shown in fig. 11 (in the same units for each orientation). From this, the direct stress $P$ which must be applied to the aggregate in order that plastic strain by complex slipping may proceed, has been calculated. Then assuming, as is warranted by experiments on single crystals, that the increase in shear stress on a crystal plane in complex slipping depends on the sum of the shears in the same way that $S$ depends on $s$ in single slipping, one can deduce the load-extension curve for an aggregate from the $S$–$s$ curve of a single crystal. The result is shown in fig. 12. Fortunately, Dr Elam had measured the load-extension curve with a polycrystalline specimen of the same material from which the single crystals had been grown. Her observed points are marked in fig. 12.

The rotation of the crystal axes due to the five shears in each grain inside the aggregate was next calculated, in exactly the same way as the rotation due to single slipping. The rotation of the specimen axis relative to the crystal axis for an extension of 2·37 % is shown, in fig. 13, for each of the calculated orientations. This diagram is not a stereographic one; it shows in rectangular co-ordinates the co-latitude $\theta$ and longitude $\phi$ of the specimen axis referred to crystal axes placed with a cubic axis at the pole and a cubic plane as the meridian $\phi=0$. Comparison with fig. 11 shows that this diagram is only slightly distorted when compared with a true stereographic projection. The arrows in fig. 13 show the extent of the rotation; the orientation at the beginning of the extension is represented by the point from which the arrow springs, and the final position by the point of the arrow. You will see that over a large part of the total area of the triangle there are two arrows radiating from each point. This is because at points within those areas two different combinations of five shears correspond with exactly the same sum of the shears. Any combination of these two sets of five shears taken in varying proportions could equally well occur. I have accordingly filled in the angles

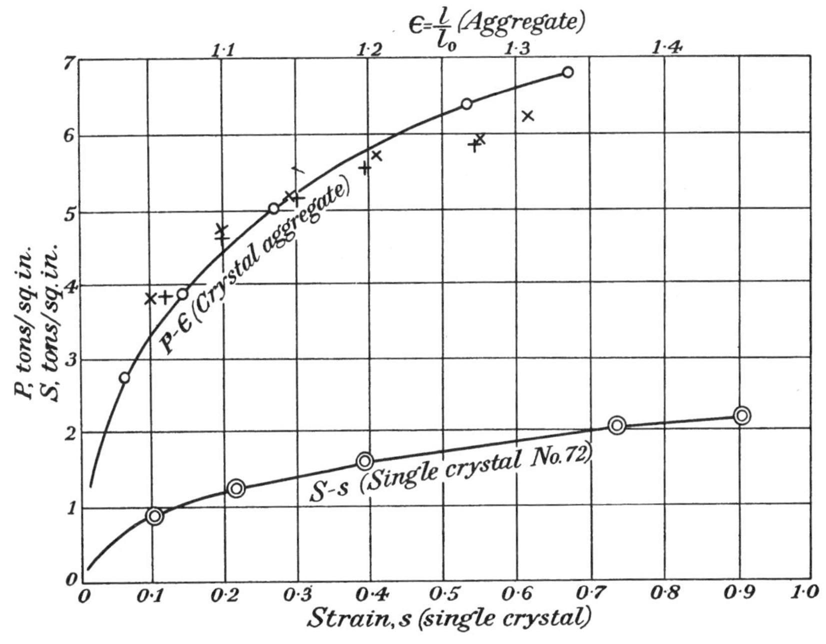


Fig. 12. Load-extension curve ($P$–$\epsilon$) for aggregate and stress-strain curve ($S$–$s$) for single crystal. ○, calculated from single crystal measurements. ×, +, aggregate, ◎, single crystal, measured by Dr Elam.

between the arrows to show the range of possible movement of the specimen axis relative to the crystal axes. The whole triangle is divided into areas within which one or two combinations of five shears are effective. Each operative combination of five shears is denoted by a letter in fig. 13. It will be seen that within the area *G* the axes of all grains rotate so that a (111) axis tends to come into line with the axis of extension of the aggregate. Moreover, the representative points of many grains which are in the area *EC* will move until they cross the boundary between *G* and *EC* and will then move along unambiguous paths towards the (111) axis. Grains whose representative points are in the neighbourhood of the (100) corner of the triangle will rotate until a cubic axis is parallel to the specimen axis. The crystal axes of grains which are near the (101) axis will tend to rotate towards either the (111) or the (100) axes. Thus, the aggregate will tend to attain a state in which the crystal axes of grains have either a (111) or a (100) axis, but no grains with a (101) axis in the direction of extension. For compression of a crystal aggregate, exactly the reverse rotation would occur; no grains would be expected with (111) or (100)

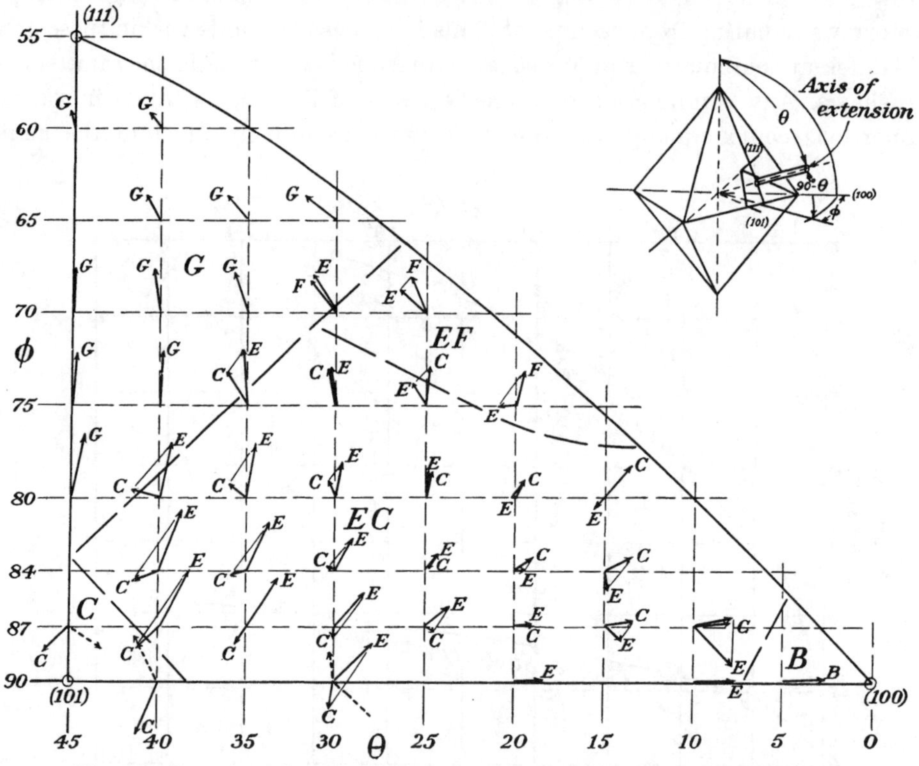


Fig. 13. Rotation of crystal axes in aggregate during extension of 2·37 %.

axes in the direction of compression, but the crystal axes of all grains would tend to rotate until a (101) axis was in or near the direction of compression. These preferred orientations are in fact exactly those found by X-rays.

# 28

# ANALYSIS OF PLASTIC STRAIN IN A CUBIC CRYSTAL

REPRINTED FROM
*Stephen Timoshenko 60th Anniversary Volume.* New York: Macmillan Co. (1938), pp. 218–24

When a single crystal of a metal in the form of a cylinder of uniform section is strained plastically by simple end loading it suffers a uniform shearing strain, planes of particles parallel to one of its crystallographic planes remaining unstrained while they move relative to one another in a crystallographic direction. With certain cubic crystals, such as copper and aluminium, the slip-plane which remains unstrained is an octahedral (111) plane and the direction of shear is parallel to one of the three edges (110) of the octahedron.

Among the twelve possible shears which this rule allows, only one is operative, namely, that one for which $\sin\theta\cos\phi$ is greatest, where $\theta$ is the angle between the plane of slip and the axis of the specimen and $\phi$ is the angle between this axis and the direction of slip.

If $P$ is the longitudinal stress due to end loading of the specimen $P\sin\theta\cos\phi$ is the component of shear stress resolved parallel to the shear planes and in the direction of their relative motion. The above statement may therefore be regarded as the mathematical expression of either of two equivalent principles, either (*a*) we may say that the operative shear is that one of the twelve possibles for which the resolved shear stress in the direction of strain is greatest, or (*b*) we may say that among the twelve possible shears that one is operative for which the shear strain necessary to extend the specimen a given amount is least.

This principle in either of its forms (*a*) or (*b*) suffices to determine the plastic strain when the stress is applied in such a way that there are no constraints or stresses in directions perpendicular to the axis of the specimen. These conditions have in fact been satisfied in all experiments with single crystals which have so far yielded analysable results. They are not satisfied, however, in some of the most interesting and important cases of plastic strain in crystals. The crystals in an aggregate of crystal grains, for instance, cannot all be strained by shearing parallel to one crystal plane, for such a strain would necessarily make them cease to fit together at their boundaries.

To discuss such problems it is necessary to extend the scope both of experiment and of theory so as to include cases where shears occur simultaneously on several crystal planes, but even before the necessity for this extension was realized experiments in which two shears occurred simultaneously had already been carried out. In these experiments the axis of the specimen was equally inclined to two octahedral planes so that the principle or rule, in either of its forms (*a*) or (*b*), leads to the conclusion that shearing might take place on either of the two planes. In these cases, however, it happened that if shearing occurred on only one of the two possible planes the strain automatically brought the axis of the specimen to such a position that the other plane was more favourably orientated than the one on which the shearing had already taken place. In these circumstances therefore the rule decrees that shearing on the first plane shall stop and that it shall begin on the second. Direct application of the rule therefore enables us to deduce that when the axis of the specimen comes into the symmetrical position equal slipping will occur on two planes and the crystal axes will remain symmetrically disposed with respect to the specimen axis as the slipping proceeds. It has in fact been found that this consequence of the rule is realized under experimental conditions.

With these experimental facts before us it seems that we are in a position to generalize the rule in its form (*b*) so as to make it applicable in cases where several shears must occur simultaneously in order that any assigned boundary conditions may be satisfied. The generalized rule in form (*b*) is: 'Of all the possible combinations of the twelve shears which could produce an assigned strain, only that combination is operative for which the sum of the absolute values of shears is least.' This statement, of course, might be regarded as the expression of the principle of virtual work to a mechanical system each component of which, if it moves at all, is subject to a frictional constraint. The words 'absolute values' convey the idea that the resistance to shear does not depend on the direction of slipping. The generalized rule (*b*) is consistent with the experimental results described above, and would discriminate between the cases where one shear or two shears are operative.

It does not seem possible to generalize the rule in its form (*a*), at any rate in such a way that the rule could be usefully applied.

The rule in its generalized form (*b*) makes it possible to determine the complete system of internal shear strains when a crystal is subjected to any assigned external strain, but the analysis is very laborious. The following sketch of the operations involved may perhaps be of interest.

It will be assumed that the strains are all small. Let $a_1 a_2 a_3$, $b_1 b_2 b_3$, $c_1 c_2 c_3$, $d_1 d_2 d_3$ be the twelve shear strains, three parallel to each of the four octahedral planes $a, b, c, d$, of a cubic crystal. Taking axes parallel to the cubic axes the octahedral planes are parallel to

$$x \pm y \pm z = 0. \tag{1}$$

The effect of the small strain $a_1$ parallel to the plane $x+y+z=0$ and in the direction of the octahedron edge whose direction cosines are $(0, -1/\sqrt{2}, +1/\sqrt{2})$ is to give the displacements $u, v, w$ to a particle at $xyz$ where

$$\left.\begin{aligned} u &= 0, \\ v &= -a_1(x+y+z)/\sqrt{6}, \\ w &= a_1(x+y+z)/\sqrt{6}. \end{aligned}\right\} \tag{2}$$

The six components of strain are then given by

$$\sqrt{6}\,e_{xy}=\sqrt{6}\left(\frac{\partial u}{\partial y}+\frac{\partial v}{\partial x}\right)=-a_1+a_2+b_1-b_2+c_1-c_2-d_1+d_2, \tag{3a}$$

$$\sqrt{6}\,e_{yz}=\sqrt{6}\left(\frac{\partial v}{\partial z}+\frac{\partial w}{\partial y}\right)=-a_2+a_3-b_2+b_3+c_2-c_3+d_2-d_3, \tag{3b}$$

$$\sqrt{6}\,e_{zx}=\sqrt{6}\left(\frac{\partial w}{\partial x}+\frac{\partial u}{\partial z}\right)=-a_3+a_1+b_3-b_1-c_3+c_1+d_3-d_1, \tag{3c}$$

$$\sqrt{6}\,e_{xx}=\sqrt{6}\frac{\partial u}{\partial x} \qquad =a_2-a_3+b_2-b_3+c_2-c_3+d_2-d_3, \tag{3d}$$

$$\sqrt{6}\,e_{yy}=\sqrt{6}\frac{\partial v}{\partial y} \qquad =a_3-a_1+b_3-b_1+c_3-c_1+d_3-d_1, \tag{3e}$$

$$\sqrt{6}\,e_{zz}=\sqrt{6}\frac{\partial w}{\partial z} \qquad =a_1-a_2+b_1-b_2+c_1-c_2+d_1-d_2. \tag{3f}$$

Equations (3) are not independent since shearing strains cannot alter the volume, so that $e_{xx}+e_{yy}+e_{zz}=0$. Thus there are five independent components of strain, and if these are given any combination of the twelve strains which satisfies equations (3) will give rise to the assigned strain.

So far only geometrical considerations have been involved. It remains to see how the rule in form (*b*) can be used to pick out which combination of the twelve shears is operative. Writing

$$S=|a_1|+|a_2|+|a_3|+|b_1|+\ldots+|d_3|, \tag{4}$$

where $|a_1|=a_1$ when $a_1$ is positive or $-a_1$ when $a_1$ is negative. The rule is expressed in the statement that the particular combination of shears is operative for which $S$ is the least among all those which satisfy equations (3).

To understand how the choice can be made it is convenient to consider first the simplified problem with only three variables. Suppose first we have to choose that combination of three variables $x_1, x_2, x_3$ which satisfies the linear equation

$$Ax_1+Bx_2+Cx_3=D, \tag{5}$$

and corresponds with the least possible value of $|x_1|+|x_2|+|x_3|$. Consider the surface whose equation in rectangular co-ordinates $x_1, x_2, x_3$ is

$$|x_1|+|x_2|+|x_3|=E, \tag{6}$$

where $E$ is positive. It is clear that it is in fact the surface of the octahedron whose eight faces are

$$\pm x_1 \pm x_2 \pm x_3=E.$$

The parts of these planes which are continuations of the faces beyond the octahedron itself are not included in (6) because at all points on those continuations $|x_1|+|x_2|+|x_3|$ is greater than $E$.

The condition equation (5) is represented in the $x_1, x_2, x_3$ system of co-ordinates by a plane. To find the least value of $|x_1|+|x_2|+|x_3|$ consistent with equation (5) we may imagine the increasing series of octahedra represented by increasing $E$ in equation (6) from $E=0$. At first the octahedra will have no point in common with the plane equation (5). In general, the first point of the increasing octahedron to come into contact with the plane equation (5) will be one of its vertices. The corresponding value of $E$ will be the least possible value of $|x_1|+|x_2|+|x_3|$ which is consistent with equation (5).

It will be seen, therefore, that when only one condition is to be satisfied, the combination of $x_1, x_2, x_3$ which give the least value of $|x_1|+|x_2|+|x_3|$ occurs when two of the three variables are zero. This is evidently the analogue in three variables of the problem which has already been considered.

A cylindrical crystal under end load could extend owing to a very large number of combinations of shears. The assigned extension, however, supplies only one linear relationship between them. The above analysis, extended to twelve variables, shows that when our postulate in form (*b*) is applied the 'operative combination of shears' is in fact only one shear, the remaining eleven shears being zero, as is observed in this case.

Returning to the three-variable problem, if the three variables are required to satisfy two linear relationships instead of only one, a straight line instead of a plane is defined in the $x_1 x_2 x_3$ space. Imagining as before that the octahedron, equation (6), expands from the origin it will first establish contact with this line at a point on one of its edges. Thus the least value of $|x_1|+|x_2|+|x_3|$ will be obtained when either $x_1=0$ or $x_2=0$, $x_3=0$. To complete the solution in this instance the three cases must be solved separately. Thus putting $x_1=0$, $x_2$ and $x_3$ are determined from the two given linear conditions. Similarly with $x_2=0$ and with $x_3=0$ the remaining pairs of variables are found. In this way 3 values of $E$ are determined. The least of the three is the least possible value of $|x_1|+|x_2|+|x_3|$ which can satisfy the two given linear conditions.

These examples show how we can proceed to choose the combination of shears which satisfies our initial principle postulate. Five shears are necessary in order to satisfy the five conditions of equations (3). The least value of $S$ will therefore correspond with a combination in which seven of the shears are taken as zero. It is necessary therefore to solve equations (3) using all possible combinations of five variables out of the twelve, and then to choose that combination for which $S$ is least.

At first sight this seems a formidable task because there are 792 ways in which five things can be chosen from twelve. It turns out, however, that the range of choice is much more limited than appears at first sight. In the first place, three directions of shear on one plane are not geometrically independent, in fact only the differences $a_1-a_2$, $b_2-b_3$, etc., appear in equations (3). Thus the twelve shears are divided into four groups from each of which only two may be selected. This reduces the 792 to 648. Next it is found that equations (3) cannot be satisfied if the five are chosen so that two come from one group and 1 from each of the remainder. The choice is limited therefore to cases where two come from each of two groups, one from one of the remaining groups, and none from the last. This reduces the number of choices to 324.*

Next it may be noted that on a plane on which two shears are chosen there are three ways of choosing the pair. If one pair is determined, then by adding the same amount to each of the three shears, an operation which leaves all the equations (3) unaltered, either of the two elements of the pair can be reduced to zero. Thus so far as equations (3) are concerned each of the three pairs are reduced to one pair, and since each combination of five that is to be considered contains two groups from each of which one of the three possible pairs must be taken, any given choice of the five variables gives identical results with eight other possible choices. Thus the total number of choices is reduced from 324 to one-ninth of 324, i.e. 36.

Finally it is found that a further geometrical inconsistency rules out twelve of the thirty-six, so that twenty-four choices are left, each one of which must be tried independently.

Having solved equations (3) with any one of these twenty-four combinations of five variables, it is necessary first to choose the pairs of variables from each of the two groups so that their sum is least. Take for instance the case when the five variables chosen are $a_1 a_2 b_1 b_2 c_1$. Suppose $a_1$ and $a_2$ are both positive and $a_2 > a_1$. Then by subtracting $a_1$ from all three members of the group $a_1$, $a_2$, $a_3$ we have a new choice of variables namely 0, $-a_1 + a_2$, $-a_1$. The contribution to $S$ of the original set is $a_1 + a_2$ but the contribution of the new choice is $|a_2 - a_1| + |-a_1| = a_2$. Thus if when equations (3) are solved the two variables of any one group are found to be of the same sign a new choice is possible in which the contribution to $S$ is smaller and is equal to the absolute value of the larger of the two variables.

If the two roots of equations (3) from one group are of opposite sign, the choice corresponding with the smallest contribution to $S$ has already been made. It is therefore possible to write down without further consideration the particular choice among the nine equivalent solutions of equations (3) which corresponds with the smallest contribution to $S$.

It remains to solve each of the twenty-four combinations and to choose the one which corresponds with the least value of $S$.

One example of this process has been worked out completely with results which are described in non-mathematical terms in the twenty-ninth May lecture to the Institute of Metals.†

It may now be noticed that when the particular choice of five out of twelve variables has been made in the manner described above, the rotation of the crystal axes relative to the outer surface of the specimen can be determined. In the preceding analysis the crystal axes have been chosen as axes of reference. If $\omega_1$, $\omega_2$, $\omega_3$

* *Editor's note.* The statements in the last three sentences of this paragraph ('Next it is found...choices to 324.') are not correct; see the note at the end of this paper.

† *J. Inst. Metals*, LXII (1938), 307; paper 27 above.

are the rotations of the lines of particles which lie in the principal axes of strain relative to the crystal axes, it is found that

$$\left.\begin{aligned}\omega_1\sqrt{6}&=-a_1+b_1-c_1+d_1+\tfrac{1}{2}(a_2+a_3-b_2-b_3+c_2+c_3-d_2-d_3),\\ \omega_2\sqrt{6}&=-a_2+b_2+c_2-d_2+\tfrac{1}{2}(a_1+a_3-b_1-b_3-c_1-c_3+d_1+d_3),\\ \omega_3\sqrt{6}&=-a_3-b_3+c_3+d_3+\tfrac{1}{2}(a_1+a_2+b_1+b_2-c_1-c_2-d_1-d_2).\end{aligned}\right\} \quad (7)$$

If the strain is carried out so that the particles in the principal axes of strain do not rotate, as for instance when the crystal forms part of a specimen which is extended slightly in one direction while contracting equally by the amount in all perpendicular directions, then the crystal axes rotate relative to the specimen through angles $-\omega_1$, $-\omega_2$, $-\omega_3$. In this way rotation of crystal axes in an aggregate subjected to uniform extension has been calculated and compared with the results of X-ray examinations.*

Finally it will be observed that the work done per unit volume making the plastic deformation is

$$W=SF, \quad (8)$$

where $F$ is the shear stress which would cause slipping on any one of the four possible slip-planes if applied parallel to one of its three possible slip directions.

The work done by external forces can be measured, thus for a crystal aggregate in the form of a cylinder under end load the work done per unit volume is $P\epsilon$ where $P$ is the longitudinal stress and $\epsilon$ the proportional extension.

Since $S$ is already expressed in terms of $\epsilon$ the virtual work equation

$$W=SF=P\epsilon \quad (9)$$

enables us to find $P/F$. It is therefore possible to calculate the strength of a crystal aggregate in terms of the shear strengths of the crystals of which it is composed. The calculation has been carried out in the case of aluminium. All crystal orientations were assumed equally probable and the load-extension curve for the aggregate was calculated from the 'shear-strain–shear-stress' relationship obtained by measurement, using a single crystal.* Good agreement was found with measurements or load-extension made with an aggregate of the same material as that from which the single crystals were prepared.

[*Note added by the author in September 1955 on the number of independent choices of combinations of five shears which are necessary to produce an arbitrary strain.*

The 648 combinations of five out of twelve shears of which only two may be on any one plane may be divided into two groups. Group I contains the 324 cases in which two of the five shears occur on one plane, two on another and one on one of the remaining slip-planes, the remaining plane being inactive. For the reason explained, each solution of equations (3) of paper 28 corresponds with nine choices so that $\frac{1}{9}\times 324=36$ independent solutions of (3) are required. Of these thirty-six, twenty-four correspond to cases where the fifth shear is in a direction at 60° to the line of intersection of the two planes on which two shears operate and in the

* *J. Inst. Metals*, LXII (1938), 307; paper 27 above.

remaining twelve this angle is 90°. This subgroup of twelve cases is inadmissible because such combinations cannot produce an arbitrary strain. Thus the 324 members of group I are reduced to twenty-four. These are the twenty-four on which the conclusions of papers 27 and 28 are based.

Group II contains the 324 cases in which double slipping occurs on one plane and single slipping on each of the others. In this case since double slipping occurs on only one plane the 324 cases can be reduced to $\frac{1}{3} \times 324 = 108$. The 108 can be divided into two subgroups. Subgroup II*a* has thirty-six members and contains all cases in which the directions of displacement in the three single shears are all parallel to one plane. Subgroup II*b* has seventy-two members and contains all combinations for which the directions of the three single shears are not coplanar. Members of II*b* are not capable of producing an arbitrary strain but members of II*a* are. Thus to obtain the combination of shears corresponding with the least possible virtual work it is necessary to obtain $24 + 72 = 96$ solutions of equations (3). The erroneous conclusion, given in both papers 27 and 28, that none of the choices in group II could give rise to an arbitrary strain was based on errors of sign in equations (3) of paper 28 which have now been corrected.

Since only twenty-four of the ninety-six choices were considered in papers 27 and 28 it might be expected that the results given in figs. 11, 12 and 13 of paper 27 would be wrong, and it was because some differences were noticed between fig. 13 and a similar diagram given by J. F. W. Bishop (*J. Mech. Phys. Solids*, III (1954), 136, fig. 5) that the error was found. Bishop's analysis was based on finding stress systems which could operate combinations of five shears simultaneously. He found that many different shear combinations are capable of producing the same minimum virtual work. Thus the spread of possible changes in the direction of the crystal axes owing to the shears corresponding with minimum virtual work is much greater than that shown in fig. 13 of paper 27. On the other hand the calculation by Bishop and Hill (*Phil. Mag.* XLII (1951), 1298) of the strength of an aggregate is almost identical with that shown in fig. 12 of paper 27. This may be taken to indicate that in most of the crystallographic positions of the axis of extension of the aggregate the maximum virtual work did correspond with one or more of the twenty-four cases which were considered as well as with several of the seventy-two cases which were neglected.]

# 29

# STRESS SYSTEMS IN AEOLOTROPIC PLATES PART I*

REPRINTED FROM

*Proceedings of the Royal Society*, A, vol. CLXXIII (1939), pp. 162–72

Equations which can be used for a system of generalized plane stress in a plate whose material has any kind of aeolotropy have been obtained recently by Huber.† When the material has two directions of symmetry at right angles in the plane of the plate the equation for the stress function takes a comparatively simple form. In the present paper solutions in polar co-ordinates of this equation are obtained which give single-valued expressions for the displacements, and these solutions are applied to the problem of an isolated force acting at an internal point of an infinite aeolotropic plate. The stress distributions due to such a force acting at a point of certain highly aeolotropic materials such as oak and spruce are represented by polar diagrams.

## INTRODUCTION

1. The distribution of stress in sheets of uniform aeolotropic material has been discussed by Michell.‡ In particular, he gave the solution of the elastic equations for an isolated force acting at the edge of an aeolotropic plate and at the vertex of an angle cut from such a plate. In these cases the solution is very simple because the stress turns out to be purely radial. When expressed in polar co-ordinates the stress components $\widehat{r\theta}$ and $\widehat{\theta\theta}$ are zero, and this is true whether the material is isotropic or aeolotropic.

Several examples of the stress systems due to loads distributed along the edges of aeolotropic plates and disks have recently been published by Wolf,§ Okubo‖ and Sen.¶ The method of solution which is used in these examples differs from that used by Michell in the paper referred to above but is related to that used by him in a subsequent paper** in which he shows that the differential equations in three dimensions for stress and strain in an aeolotropic material possessing elastic symmetry equivalent to that of a crystal of the hexagonal system, can be expressed in the form

$$\left(\frac{\partial^2}{\partial x^2}+\frac{\partial^2}{\partial z^2}+k_1\frac{\partial^2}{\partial y^2}\right)V_1=0,$$

$$\left(\frac{\partial^2}{\partial x^2}+\frac{\partial^2}{\partial z^2}+k_2\frac{\partial^2}{\partial y^2}\right)V_2=0,$$

* With A. E. GREEN. † *Stephen Timoshenko 60th Anniversary Volume* (1938), 89.

‡ *Proc. Lond. Math. Soc.* XXXII (1900), 35.

§ *Z. angew. Math. Mech.* XV (1935), 249.

‖ *Sci. Rep. Tôhoku Univ.* (ser. 1), XXV (1937), 1110; *Phil. Mag.* XXVII (1939), 508.

¶ *Phil. Mag.* XXVII (1939), 596. ** *Proc. Lond. Math. Soc.* XXXII (1900), 247.

where $k_1$ and $k_2$ are functions of the elastic constants, $V_1$ and $V_2$ are functions of the components of strain, and the body is elastically symmetrical around lines parallel to the axis of $y$. A particular case which is included in Michell's analysis is that of an aeolotropic sheet parallel to the plane $z=0$ when it is subjected to a system of generalized plane stress. His equations can then be reduced* to a single equation

$$\left(\frac{\partial^2}{\partial x^2}+\alpha_1\frac{\partial^2}{\partial y^2}\right)\left(\frac{\partial^2}{\partial x^2}+\alpha_2\frac{\partial^2}{\partial y^2}\right)\chi=0,$$

for the stress function $\chi$, where $\alpha_1$ and $\alpha_2$ are functions of the four elastic constants for this type of symmetry. Wolf†, Okubo‡ and Sen§ use an equation of this type but limit their analysis‖ by assuming that the elastic constants of the material with which they are dealing are so related that either $\alpha_1$ or $\alpha_2$ is equal to 1.

The above stress distributions in aeolotropic sheets can be discussed without reference to the displacements. When a force is applied at an internal point a purely radial distribution of stress corresponds with a distribution of displacement which is not single-valued. Michell¶ shows how to find the particular distribution of stress which corresponds with single-valued displacements but he only completes the solution in the case of an isotropic sheet.

Equations which can be used for a system of generalized plane stress in a plate whose material has any kind of aeolotropy have been obtained recently by Huber**. When the material has two directions of symmetry at right angles in the plane of the plate the equation for the stress function reduces to the form given above. In the present paper solutions in polar co-ordinates of this equation are obtained which give single-valued expressions for the displacements, and these solutions are applied to complete the discussion of Michell's problem of an isolated force acting at an internal point of an aeolotropic plate. The corresponding stress distributions in certain highly aeolotropic materials such as oak and spruce are represented by polar diagrams.

* The algebra required for the reduction is long and it is easier to obtain the equation for $\chi$ independently.

† Loc. cit.

‡ Loc. cit.

§ Loc. cit.

‖ An even more drastic assumption is made by Westergaard (*Stephen Timoshenko 60th Anniversary Volume* (1938), 268) in extending Boussinesq's three-dimensional investigation of the stress in an isotropic material caused by a vertical point force acting on a horizontal plane boundary to a special aeolotropic material in which horizontal planes are inextensible. Michell (*Proc. Lond. Math. Soc.* XXXII (1900), 247) gives the more general solution of Boussinesq's problem for an aeolotropic material which is isotropic in horizontal planes but not necessarily inextensible.

¶ Loc. cit. p. 35.

** Loc. cit.

FUNDAMENTAL EQUATIONS

2. The analytical expressions of Hooke's law in an aeolotropic solid body* will be taken, in matrix notation, to be

$$(e_{xx}, e_{yy}, e_{zz}, e_{yz}, e_{zx}, e_{xy}) = \begin{vmatrix} s_{11} & s_{12} & s_{13} & s_{14} & s_{15} & s_{16} \\ s_{21} & s_{22} & s_{23} & s_{24} & s_{25} & s_{26} \\ s_{31} & s_{32} & s_{33} & s_{34} & s_{35} & s_{36} \\ s_{41} & s_{42} & s_{43} & s_{44} & s_{45} & s_{46} \\ s_{51} & s_{52} & s_{53} & s_{54} & s_{55} & s_{56} \\ s_{61} & s_{62} & s_{63} & s_{64} & s_{65} & s_{66} \end{vmatrix} (\widehat{xx}, \widehat{yy}, \widehat{zz}, \widehat{yz}, \widehat{zx}, \widehat{xy}), \tag{1}$$

where $s_{rt} = s_{tr} (r, t = 1, 2, \ldots, 6)$.

Attention is now confined to a plate of uniform thickness in the $(x, y)$ plane, the normal to the plate being in the direction of the $z$-axis. The plate is assumed to be in a state of generalized plane stress in which the stress component $\widehat{zz}$ vanishes everywhere, and the components $\widehat{xz}$ and $\widehat{yz}$ are zero at the two surfaces of the plate. Then, from (1), the relations between the *mean* values, taken through the thickness of the plate, of the stresses and strains in the $(x, y)$ plane, are

$$\left.\begin{aligned} e_{xx} &= s_{11}\widehat{xx} + s_{12}\widehat{yy} + s_{16}\widehat{xy}, \\ e_{yy} &= s_{12}\widehat{xx} + s_{22}\widehat{yy} + s_{26}\widehat{xy}, \\ e_{xy} &= s_{16}\widehat{xx} + s_{26}\widehat{yy} + s_{66}\widehat{xy}, \end{aligned}\right\} \tag{2}$$

and the *mean* stresses can be expressed in the form

$$\widehat{xx} = \frac{\partial^2 \chi}{\partial y^2}, \quad \widehat{yy} = \frac{\partial^2 \chi}{\partial x^2}, \quad \widehat{xy} = -\frac{\partial^2 \chi}{\partial x \partial y}, \tag{3}$$

where $\chi$ is a stress function. By inserting the expressions (2) and (3) in the identical relation

$$\frac{\partial^2 e_{yy}}{\partial x^2} + \frac{\partial^2 e_{xx}}{\partial y^2} = \frac{\partial^2 e_{xy}}{\partial x \partial y},$$

the following equation is obtained for $\chi$:

$$s_{22}\frac{\partial^4 \chi}{\partial x^4} - 2s_{26}\frac{\partial^4 \chi}{\partial x^3 \partial y} + (2s_{12} + s_{66})\frac{\partial^4 \chi}{\partial x^2 \partial y^2} - 2s_{16}\frac{\partial^4 \chi}{\partial x \partial y^3} + s_{11}\frac{\partial^4 \chi}{\partial y^4} = 0. \tag{4}$$

This is equivalent to the equation given by Huber.†

When the material of the plate has two directions of symmetry at right angles in the $(x, y)$ plane and these are taken to be parallel to the directions of $x$ and $y$, then $s_{16} = s_{26} = 0$ and equation (4) becomes

$$\left(\frac{\partial^2}{\partial x^2} + \alpha_1 \frac{\partial^2}{\partial y^2}\right)\left(\frac{\partial^2}{\partial x^2} + \alpha_2 \frac{\partial^2}{\partial y^2}\right)\chi = 0, \tag{5}$$

where $\qquad \alpha_1 \alpha_2 = s_{11}/s_{22}, \quad \alpha_1 + \alpha_2 = (s_{66} + 2s_{12})/s_{22}. \qquad (6)$

* A. E. H. Love, *Mathematical Theory of Elasticity* (1927, 4th ed.), § 72. † Loc. cit.

The constants $\alpha_1$ and $\alpha_2$ may take real or imaginary values. It will be supposed here that $\alpha_1$ and $\alpha_2$ are real and positive as this is the case for a large number of materials including those materials for which numerical results are given in this paper.

The mean displacements $\bar{u}$ and $\bar{v}$ are found by a similar method to that used for an isotropic material* and are given by

$$\left.\begin{aligned}\bar{u}&=(s_{12}-s_{11})\frac{\partial\chi}{\partial x}+s_{11}\frac{\partial\psi}{\partial y},\\ \bar{v}&=(s_{12}-s_{22})\frac{\partial\chi}{\partial y}+s_{22}\frac{\partial\psi}{\partial x},\end{aligned}\right\} \tag{7}$$

while $\psi$ is to be found from the equation

$$\frac{\partial^2\psi}{\partial x\,\partial y}=\nabla_1^2\chi, \tag{8}$$

with the additional restriction that

$$\frac{\partial^2\psi}{\partial x^2}+\alpha_1\alpha_2\frac{\partial^2\psi}{\partial y^2}=(1-\alpha_1)(1-\alpha_2)\frac{\partial^2\chi}{\partial x\,\partial y}. \tag{9}$$

## Fundamental Stress Functions

3. Three sets of plane polar co-ordinates are now introduced by the equations

$$x+iy=re^{i\theta},\quad x+iy/\alpha_1^{\frac{1}{2}}=r_1e^{i\theta_1},\quad x+iy/\alpha_2^{\frac{1}{2}}=r_2e^{i\theta_2}. \tag{10}$$

Then the fundamental solutions of (5) which are of order $n$ are

$$r_1^{\pm n}\cos n\theta_1,\quad r_1^{\pm n}\sin n\theta_1,\quad r_2^{\pm n}\cos n\theta_2,\quad r_2^{\pm n}\sin n\theta_2. \tag{11}$$

In order to find the displacements which correspond to these stress functions it is convenient to deal with the stress function

$$\chi=(x+iy/\alpha^{\frac{1}{2}})^{\pm n}, \tag{12}$$

where $\alpha$ stands for either $\alpha_1$ or $\alpha_2$. The real and imaginary parts of $\chi$ give the stress functions (11). From (8) and (9) the corresponding value of $\psi$ is found to be

$$\psi=i(\alpha^{-\frac{1}{2}}-\alpha^{\frac{1}{2}})(x+iy/\alpha^{\frac{1}{2}})^{\pm n}, \tag{13}$$

and so the expressions (7) for the displacements are single-valued when $n$ is an integer.

If $n=1$, the only surviving distinct and significant solutions are

$$\frac{\cos\theta_1}{r_1},\ \frac{\sin\theta_1}{r_1},\ \frac{\cos\theta_2}{r_2},\ \frac{\sin\theta_2}{r_2}, \tag{14}$$

* E. G. Coker and L. N. G. Filon, *A Treatise on Photo-elasticity* (1931), §2·25.

the solutions $r_1 \cos\theta_1, r_1 \sin\theta_1, r_2 \cos\theta_2, r_2 \sin\theta_2$ being trivial as they give zero stresses. It can easily be seen, however, that further solutions of (5) are given by

$$\chi' = \alpha_1^{\frac{1}{2}} x \log r_1 - y\theta_1, \quad \chi'' = y \log r_1 + \alpha_1^{\frac{1}{2}} x\theta_1, \tag{15}$$

with similar expressions containing the suffix 2. The values $\psi'$ and $\psi''$ of $\psi$ which correspond respectively to $\chi'$ and $\chi''$ are found to be

$$\psi' = (\alpha_1^{\frac{1}{2}} - \alpha_1^{-\frac{1}{2}})\chi'', \quad \psi'' = -(\alpha_1^{\frac{1}{2}} - \alpha_1^{-\frac{1}{2}})\chi', \tag{16}$$

and hence if $u'$, $v'$ and $u''$, $v''$ are the displacements corresponding to $\chi'$ and $\chi''$ respectively then $u'$ and $v''$ are single-valued but $u''$, $v'$ are many-valued and of the form

$$u'' = \alpha_1^{\frac{1}{2}}(s_{12} - \alpha_2 s_{22})\theta_1, \quad v' = -(s_{12} - \alpha_1 s_{22})\theta_1. \tag{17}$$

If stress functions are restricted to those which give single-valued stresses and displacements, then by combining functions of the type given in (15) the following stress functions of order 1 are obtained, in addition to those in (14):

$$(s_{12} - \alpha_2 s_{22})(\alpha_1^{\frac{1}{2}} x \log r_1 - y\theta_1) - (s_{12} - \alpha_1 s_{22})(\alpha_2^{\frac{1}{2}} x \log r_2 - y\theta_2), \tag{18}$$

and

$$(s_{12} - \alpha_2 s_{22})(\alpha_2^{-\frac{1}{2}} y \log r_2 + x\theta_2) - (s_{12} - \alpha_1 s_{22})(\alpha_1^{-\frac{1}{2}} y \log r_1 + x\theta_1). \tag{19}$$

The only significant functions of zero order which give single-valued stresses and displacements are

$$\log r_1, \quad \log r_2, \quad \theta_1, \quad \theta_2, \quad r_1^2, \quad r_2^2. \tag{20}$$

## Expansions in Polar Co-ordinates

4. In a future paper it is hoped to apply the above stress functions to problems of stress systems in plates containing circular holes, and for this purpose it is necessary to express the functions in terms of the plane polar co-ordinates $r$, $\theta$. The required expressions will be given here for reference.

After a little straightforward reduction, the equations (10) give

$$\frac{e^{i\theta_1}}{r_1} = \frac{(1+\gamma_1)e^{i\theta}}{r(1+\gamma_1 e^{2i\theta})}, \quad r_1 e^{i\theta_1} = \frac{r(e^{2i\theta} + \gamma_1)}{(1+\gamma_1)e^{i\theta}}, \tag{21}$$

where

$$\left.\begin{aligned} \beta_1 = \frac{\alpha_1 + 1}{\alpha_1 - 1}, \quad \gamma_1 = \beta_1 - (\beta_1^2 - 1)^{\frac{1}{2}} \quad (\beta > 1), \\ \gamma_1 = \beta_1 + (\beta_1^2 - 1)^{\frac{1}{2}} \quad (\beta < -1), \end{aligned}\right\} \tag{22}$$

with similar expressions which are obtained by changing the suffix 1 into 2.

Hence

$$\frac{e^{in\theta_1}}{r_1^n} = \frac{e^{in\theta}}{r^n}(1+\gamma_1)^n \sum_{s=0}^{\infty} \binom{n+s-1}{s} \gamma_1^s (-)^s e^{2is\theta}, \tag{23}$$

since $|\gamma_1| < 1$, and

$$\left.\begin{aligned}(1+\gamma_1)^n r_1^n e^{in\theta_1} &= r^n \Big\{ e^{in\theta} + \gamma_1^n e^{-in\theta} + n(\gamma_1 e^{i(n-2)\theta} + \gamma_1^{n-1} e^{-i(n-2)\theta}) \\ &\quad + \ldots + \binom{n}{\frac{1}{2}n-1} (\gamma_1^{\frac{1}{2}n-1} e^{2i\theta} + \gamma_1^{\frac{1}{2}n+1} e^{-2i\theta}) + \binom{n}{\frac{1}{2}n} \gamma_1^{\frac{1}{2}n} \Big\} \quad (n \text{ even}), \\ &= r^n \Big\{ e^{in\theta} + \gamma_1^n e^{-in\theta} + n(\gamma_1 e^{i(n-2)\theta} + \gamma_1^{n-1} e^{-i(n-2)\theta}) \\ &\quad + \ldots + \binom{n}{\frac{1}{2}(n-1)} (\gamma_1^{\frac{1}{2}(n-1)} e^{i\theta} + \gamma_1^{\frac{1}{2}(n+1)} e^{-i\theta}) \Big\} \quad (n \text{ odd}),\end{aligned}\right\} \tag{24}$$

where $\binom{n}{s}$ is the binomial coefficient $\dfrac{n!}{s!(n-s)!}$. Also

$$\log r_1 - i\theta_1 = \log r - i\theta - \log(1+\gamma_1) + \sum_{s=1}^{\infty} \frac{(-)^{s-1}\gamma_1^s}{s} e^{2is\theta}. \tag{25}$$

Expansions for the fundamental stress functions in terms of $r$ and $\theta$ are obtained by separating the real and imaginary parts of these expressions. The stresses are then obtained from the usual formulae

$$\widehat{rr} = \frac{1}{r^2}\frac{\partial^2 \chi}{\partial \theta^2} + \frac{1}{r}\frac{\partial \chi}{\partial r}, \quad \widehat{r\theta} = -\frac{\partial}{\partial r}\left(\frac{1}{r}\frac{\partial \chi}{\partial \theta}\right), \quad \widehat{\theta\theta} = \frac{\partial^2 \chi}{\partial r^2}. \tag{26}$$

## Isolated Force in an Infinite Plate

5. Consider the stress function

$$\chi = \frac{P(s_{12}-\alpha_2 s_{22})}{2\pi(\alpha_1-\alpha_2)\,s_{22}} \{\alpha_1^{\frac{1}{2}} x \log r_1 - y\theta_1\} - \frac{P(s_{12}-\alpha_1 s_{22})}{2\pi(\alpha_1-\alpha_2)\,s_{22}} \{\alpha_2^{\frac{1}{2}} x \log r_2 - y\theta_2\}, \tag{27}$$

which has been shown to give single-valued stresses and displacements although it is not itself single-valued. The corresponding mean stresses are given by

$$\left.\begin{aligned}2\pi(\alpha_1-\alpha_2)\,s_{22}\widehat{xx} &= -\frac{(s_{12}-\alpha_2 s_{22})\,Px}{\alpha_1^{\frac{1}{2}}(x^2+y^2/\alpha_1)} + \frac{(s_{12}-\alpha_1 s_{22})\,Px}{\alpha_2^{\frac{1}{2}}(x^2+y^2/\alpha_2)}, \\ 2\pi(\alpha_1-\alpha_2)\,s_{22}\widehat{xy} &= -\frac{(s_{12}-\alpha_2 s_{22})\,Py}{\alpha_1^{\frac{1}{2}}(x^2+y^2/\alpha_1)} + \frac{(s_{12}-\alpha_1 s_{22})\,Py}{\alpha_2^{\frac{1}{2}}(x^2+y^2/\alpha_2)}, \\ 2\pi(\alpha_1-\alpha_2)\,s_{22}\widehat{yy} &= \frac{\alpha_1^{\frac{1}{2}}(s_{12}-\alpha_2 s_{22})\,Px}{x^2+y^2/\alpha_1} - \frac{\alpha_2^{\frac{1}{2}}(s_{12}-\alpha_1 s_{22})\,Px}{x^2+y^2/\alpha_2}.\end{aligned}\right\} \tag{28}$$

Resolving the forces on a circular boundary $r=$ constant in the directions of the $x$- and $y$-axes gives

$$\widehat{rx} = \widehat{xx}\cos\theta + \widehat{xy}\sin\theta, \quad \widehat{ry} = \widehat{xy}\cos\theta + \widehat{yy}\sin\theta. \tag{29}$$

If then $X$, $Y$ are the total average forces per unit thickness applied to this circular boundary by the material outside,

$$X = \int_0^{2\pi} \widehat{rx}\, r\, d\theta = -P, \quad Y = \int_0^{2\pi} \widehat{ry}\, r\, d\theta = 0. \tag{30}$$

Also, this set of forces has no moment about the origin.

If these forces are reversed it will be seen that the material inside the circle applies to the material outside the circle a force $P$ in the positive direction of the $x$-axis.

In polar co-ordinates the stresses $\widehat{rr}$ and $\widehat{r\theta}$ are found to be

$$\widehat{rr}=\frac{P\cos\theta}{2\pi(\alpha_1-\alpha_2)s_{22}r}\left\{\frac{\alpha_1^{\frac{1}{2}}(s_{12}-\alpha_2 s_{22})\{\alpha_1-3-(\alpha_1-1)\cos 2\theta\}}{\alpha_1+1+(\alpha_1-1)\cos 2\theta}-\frac{\alpha_2^{\frac{1}{2}}(s_{12}-\alpha_1 s_{22})\{\alpha_2-3-(\alpha_2-1)\cos 2\theta\}}{\alpha_2+1+(\alpha_2-1)\cos 2\theta}\right\}, \tag{31}$$

and

$$\widehat{r\theta}=\frac{P\sin\theta}{2\pi(\alpha_1-\alpha_2)s_{22}r}\{\alpha_1^{\frac{1}{2}}(s_{12}-\alpha_2 s_{22})-\alpha_2^{\frac{1}{2}}(s_{12}-\alpha_1 s_{22})\}. \tag{32}$$

6. Similarly, it may be shown that a force $P$ in the positive direction of the $y$-axis is given by a stress function

$$\chi=\frac{P(s_{12}-\alpha_2 s_{22})}{2\pi(\alpha_1-\alpha_2)s_{22}}\{\alpha_2^{-\frac{1}{2}}y\log r_2+x\theta_2\}-\frac{P(s_{12}-\alpha_1 s_{22})}{2\pi(\alpha_1-\alpha_2)s_{22}}\{\alpha_1^{-\frac{1}{2}}y\log r_1+x\theta_1\}, \tag{33}$$

which produces single-valued stresses and displacements. The corresponding values of $\widehat{rr}$ and $\widehat{r\theta}$ are

$$\widehat{rr}=\frac{P\sin\theta}{2\pi(\alpha_1-\alpha_2)s_{22}r}\left\{\frac{\alpha_1^{-\frac{1}{2}}(s_{12}-\alpha_1 s_{22})\{3\alpha_1-1+(\alpha_1-1)\cos 2\theta\}}{\alpha_1+1+(\alpha_1-1)\cos 2\theta}-\frac{\alpha_2^{-\frac{1}{2}}(s_{12}-\alpha_2 s_{22})\{3\alpha_2-1+(\alpha_2-1)\cos 2\theta\}}{\alpha_2+1+(\alpha_2-1)\cos 2\theta}\right\}, \tag{34}$$

$$\widehat{r\theta}=\frac{P\cos\theta}{2\pi(\alpha_1-\alpha_2)s_{22}r}\{\alpha_1^{-\frac{1}{2}}(s_{12}-\alpha_1 s_{22})-\alpha_2^{-\frac{1}{2}}(s_{12}-\alpha_2 s_{22})\}. \tag{35}$$

The stresses due to an isolated force in an isotropic plate can be obtained from the above results by taking the limit as $\alpha_1\to\alpha_2\to 1$.

7. As examples, the stress distributions due to an isolated force at an internal point of a sheet or wide board of oak and of spruce have been evaluated numerically. These materials were chosen because they possess a very high degree of aeolotropy. In the case of spruce, for instance, Young's modulus for the direction of the fibres is 26·4 times as great as for the transverse direction while for oak this ratio is 5·9. The sheets are in each case cut parallel to the grain of the wood. The direction of the fibres (which are parallel to the core of the original tree trunk) is taken as the axis of $y$. With this orientation of the saw cuts the axis of $x$ is in the direction which is perpendicular to the fibres and also to all lines which pass through the central core of the tree. The values of the elastic constants are taken from a paper by Hörig* and they are reproduced in table 1, together with the corresponding values of $\alpha_1$ and $\alpha_2$. In the case of spruce Hörig uses the measurements of H. Carrington and in the case of oak those of J. Stamer and H. Sieglerschmidt. Complete references to these are given in Hörig's paper. The inverses of the constants $s_{11},\ldots,s_{66}$ are

* *Ingen. Arch.* VI (1935), 8.

measured in kg./sq. mm., $1/s_{22}$ and $1/s_{11}$ being Young's moduli along and transverse to the grain of the wood.

The formulae (31) and (32) give the stress distribution due to a force which is perpendicular to the grain, and the formulae (34) and (35) give the stress distribution due to a force which is parallel to the grain. The values of $r\widehat{rr}/P$ are shown in table 2

Table 1

| | $s_{11}$ | $s_{22}$ | $s_{12}$ | $s_{66}$ | $\alpha_1$ | $\alpha_2$ |
|---|---|---|---|---|---|---|
| Oak | 10·15 | 1·72 | −0·87 | 12·8 | 5·321 | 1·109 |
| Spruce | 15·5 | 0·587 | −0·33 | 11·5 | 16·91 | 1·56 |

Table 2. *Values of* $(r/P)\widehat{rr}$

| | Oak | | Spruce | |
|---|---|---|---|---|
| $\theta°$ | $P$ parallel to grain | $P$ perpendicular to grain | $P$ parallel to grain | $P$ perpendicular to grain |
| 0 | 0·443 | 0·183 | 0·718 | 0·140 |
| 10 | 0·390 | 0·182 | 0·483 | 0·139 |
| 20 | 0·287 | 0·182 | 0·246 | 0·138 |
| 30 | 0·199 | 0·183 | 0·137 | 0·137 |
| 40 | 0·137 | 0·183 | 0·0850 | 0·137 |
| 50 | 0·0953 | 0·183 | 0·0558 | 0·143 |
| 60 | 0·0644 | 0·179 | 0·0368 | 0·157 |
| 70 | 0·0401 | 0·161 | 0·0227 | 0·178 |
| 80 | 0·0193 | 0·105 | 0·0109 | 0·171 |
| 85 | — | 0·0570 | — | 0·113 |
| 88 | — | 0·0234 | — | 0·0496 |
| 90 | 0·0 | 0·0 | 0·0 | 0·0 |

For an isotropic material whose Poisson's ratio is 0·25,

$$r\widehat{rr}/P = 0{\cdot}259 \cos\theta,$$

and

$$r\widehat{r\theta}/P = (0{\cdot}0597)\sin\theta.$$

The values of $\dfrac{r\widehat{r\theta}}{P\sin\theta}$, where $\theta$ is measured from the direction of the force, are

0·0375 when the force is parallel to the grain } for oak,
0·0910 when the force is perpendicular to the grain }

0·0264 when the force is parallel to the grain } for spruce.
0·136 when the force is perpendicular to the grain }

for values of $\theta$ between 0° and 90°, $\theta$ now being measured in all cases from the direction of the force. Owing to the symmetry of the problem the stress distribution in the other three quadrants may be obtained from that in the first quadrant.

The stress distributions due to an isolated force in a sheet of oak and a sheet of spruce are shown in fig. 1.

## Discussion of Stress Distributions in Wood Sheets

8. A force acting at a point of a sheet of spruce, in the direction of the grain, produces a high stress along the radii which are nearly parallel to the grain but very little radial stress in directions which differ from this by more than about 30°. At the same time the radial-tangential stress $\widehat{r\theta}$ is extremely small, its maximum value

being only 0·0368 of the maximum radial stress. This may be compared with the corresponding ratio 0·231 for an isotropic material whose Poisson's ratio is 0·25. When the force is applied perpendicularly to the grain the radial stress $\widehat{rr}$ is not a maximum in the direction of the applied force, being greatest at an angle of approximately 75° with this direction and therefore more nearly in the direction of the grain. In this case the ratio of the maximum shear stress to the maximum radial stress is very much greater than in the previous case, being about 0·764.

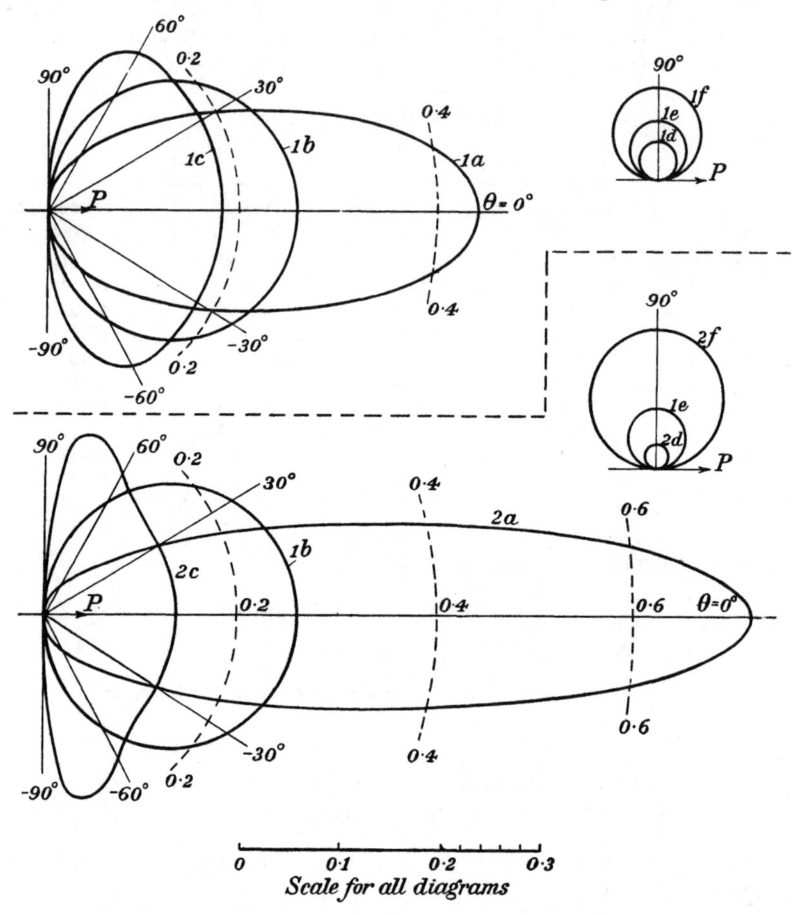


Fig. 1. Polar diagrams showing radial stress $\widehat{rr}$ and shear stress $\widehat{r\theta}$. Radial distances give

(1$a$) $(r/P)\,\widehat{rr}$, $P$ acting parallel to the grain } oak,
(1$d$) $(r/P)\,\widehat{r\theta}$, $P$ acting parallel to the grain }

(1$c$) $(r/P)\,\widehat{rr}$, $P$ acting perpendicularly to the grain } oak,
(1$f$) $(r/P)\,\widehat{r\theta}$, $P$ acting perpendicularly to the grain }

(1$b$) $(r/P)\,\widehat{rr}$, isotropic material ($\sigma = 0{\cdot}25$).

(1$e$) $(r/P)\,\widehat{r\theta}$, isotropic material ($\sigma = 0{\cdot}25$).

The diagrams 2$a$, 2$c$, 2$d$, 2$f$ represent the corresponding stresses for spruce. To facilitate comparison the diagrams for an isotropic material are included with both those for oak and those for spruce.

The results for oak are of a similar nature but as oak is not so highly aeolotropic as spruce the effect of the grain is less striking.

As may be seen by comparison of the constants for oak and spruce in table 1 the main element of aeolotropy in these materials lies in the great difference between $s_{11}$ and $s_{22}$. The ratio $\dfrac{\text{Young's modulus along grain}}{\text{Young's modulus across the grain}}$ which is $s_{11}/s_{22}$, and is therefore 1·0 for isotropic materials, is 26·4 in the case of spruce, and 5·9 in the case of oak.

# 30

# PROPAGATION OF EARTH WAVES FROM AN EXPLOSION

Paper written for the Civil Defence Research Committee, Ministry of Home Security (1940)

Though the plastic properties of materials have been studied statically and, to a limited extent, the effect of rate loading on plastic stress-strain relations has also been measured, no dynamical study of the propagation of waves in a non-elastic material appears to have been made. The present work, which is limited to propagation in one dimension as in a bar of plastic material, presupposes that observations have shown that a pressure suddenly applied at one end of a plastic bar gives rise to a progressive pressure wave within which the velocities of particles increase at a rate which has been observed. The analysis determines the plastic stress-strain relationship which the material must obey in order that the velocity-time relationship observed at a fixed station may be dynamically possible.

The calculation is carried out so as to correspond with observations made at 20 ft. from a certain buried bomb. At this station the earth began to move 0·027 sec. after the explosion and at 0·13 sec. it had acquired a velocity of 4 ft. per sec. Within these limits it is possible to assume a variety of time distributions of acceleration. Two were chosen, namely, Case B, a uniform acceleration of the ground from 0 to 4 ft. per sec., and Case A, in which the acceleration was greatest at the first onset of the earth wave and afterwards decreased. In Case A the maximum stress was calculated to be 75 lb. per sq.in. In Case B it was 51 lb. per sq.in. The pressure which might be expected to act on a rigid immovable wall buried in earth which had the plastic stress-strain relationship shown in fig. 1 would be rather greater, but not very much greater, than these values.

This analysis was intended to illustrate the connection between the plastic properties of a material and the transformation of a suddenly applied pressure at one point into a gradually increasing pressure in other parts of the material.

The velocity with which compressional elastic waves of small amplitude are propagated from an explosion in the ground seems usually to be less than, but of the same order of magnitude as $\sqrt{(E/\rho)}$, $E$ being a mean of the elastic constants of the earth as measured statically and $\rho$ the density.

Velocities of order 4000 or 5000 ft. per sec. are measured as compared with 4600 ft. per sec. for water and 17,000 for steel. In seismic work distortional waves with velocities of about half the velocity of the compressional waves are also observed. This also is in accordance with what would be expected from the theory of elasticity.

The comparatively large disturbances which are propagated outwards near the explosive source move with a velocity which is much smaller than anything that can be accounted for by pure elastic theory. Any complete calculation of the dynamics of earth movement near an explosion must take account of the plastic stress-strain relations in earth as well as the changes of compactness with depth, and the complexity of such work would prevent any useful result from being obtained by mathematicians in a finite time, even if the physical data on which calculations might be based were known. A complete frontal attack on the problem being impossible at present, it seems worth while to explore some simpler problems in which the physical causes of the main observable phenomena are introduced in such a way that the mathematician has a chance of being effective. The principles which have guided me in the choice of the particular problem which is described below are:

(1) The velocity with which spherical waves of small amplitude are propagated outwards in an elastic solid is identical with the velocity of propagation of plane waves, though the amplitude decreases rapidly with distance from the source in the spherical case, and does not do so with plane waves. For this reason the solution of the problem in the propagation of disturbances in one dimension may be expected to reveal the connection between the physical properties of a material and rates of propagation of non-elastic waves, whereas it would not be expected to reveal much about the decrease in disturbance with distance from the explosive source.

(2) The low velocity with which the large disturbances are propagated indicates that they are being controlled by the plastic rigidity or resistance to shear in the material rather than by its compressibility.

(3) The comparatively large permanent plastic strain or set of the earth after the explosion indicates either that the mean density of the ground through which the disturbance has passed has increased during its passage or that the surface of the ground has risen. The latter alternative seems from the measurement of vertical earth movement to be the correct one.

(4) From the above considerations it may be expected that the passage of a longitudinal wave down a bar of plastic material might exhibit the same kind of displacement-time and pressure-time characteristics as those observed in the earth near an explosion, provided that the wave was produced by a similar explosion and the bar had the same physical properties as the earth. In this connection it may be noted that the plastic bar would expand laterally during the passage of the wave just as the level of the earth rises during the passage of the earth shock from an explosion.

## Propagation of a Compression Wave in a Plastic Bar

If $a$ is the area of cross-section of the bar (originally uniform), $\rho$ its density and $p$ the stress normal to a cross-section, the equations of motion and continuity are

$$\left.\begin{aligned} \frac{\partial u}{\partial t}+u\frac{\partial u}{\partial x} &= -\frac{1}{\rho a}\frac{d(pa)}{d(\rho a)}\frac{\partial(\rho a)}{\partial x}, \\ \frac{\partial(\rho a)}{\partial t}+u\frac{\partial(\rho a)}{\partial x} &= -\rho a\frac{\partial u}{\partial x}, \end{aligned}\right\} \tag{1}$$

where $u$ is the material velocity.

Analytically these are identical with the equations of motion in one dimension of a fluid of density $\rho' = \rho a$, and pressure $p' = pa$. The solution of these equations has been given by Riemann, and in a less general manner by Earnshaw, in the case when $p'$ is a function of $\rho'$ only. Riemann and Earnshaw showed that it is possible for a disturbance to be propagated in one direction only and that in such a progressive wave

$$u = f(\rho'), \tag{2}$$

where

$$\frac{dp'}{d\rho'} = [\rho' f'(\rho')]^2 \tag{3}$$

and $f'(\rho')$ is written for $\frac{d}{d\rho'}[f(\rho')]$.

Riemann and Earnshaw further showed that this progressive wave has the property that $p'$, $\rho'$ and $u$ are constant at a point which moves forward with velocity

$$\frac{dx}{dt} = u + \left(\frac{dp'}{d\rho'}\right)^{\frac{1}{2}}, \tag{4}$$

i.e. with a velocity which is the sum of the velocity of the fluid and the local velocity of sound waves.

In so far as $pa$ can be regarded as a function of $\rho a$ only, Riemann's solution can be applied directly to the propagation of a progressive disturbance in a plastic bar. This limitation is severe, but cases where the stress, starting from zero, continually increases may satisfy this requirement, though it will certainly not be satisfied in any part of the solid after the maximum stress has been attained there, because plastic stress-strain curves are not reversible.

The part of the disturbance produced by an explosion where the above condition is likely to be satisfied is the front of the advancing compression wave where the stress is increasing and the earth is accelerating outwards. In a plastic bar such a region can be produced by suddenly applying a force $F$ at the end of the bar. If this force is maintained constant a compression wave will travel along the bar. If $U$ is the velocity with which the end of the bar is moved, the whole of the bar behind a transition region will move with velocity $U$. The forward end of this transition region will move with velocity equal to the value of $\sqrt{\{d(pa)/d(\rho a)\}}$ when $p = 0$, and the rear end will move with velocity equal to $U + \left[\frac{d(pa)}{d(\rho a)}\right]_{pa=F}$. The relationship between $U$ and $F$ as well as the sequence of events in the transition region depends on the form of the plastic stress-strain curve. As an example of the application of these ideas we may take a special case which is so constructed that it may be compared with some observations of ground movement made at a station 20 ft. from an exploding bomb.* In this case, instruments at 20 ft. from the bomb recorded first a small shock, presumably an elastic wave, which began about 0·004 sec. after the explosion. No big movement began till 0·027 sec. after the explosion when the earth began to move outward from the explosive source with an acceleration comparable with $g$. The outward velocity continued to increase, or at any rate did not begin to decrease, till 0·13 sec. after the explosion.

* 'Survey test on open trench after bombing', by Stewartby; paper written for the Civil Defence Research Committee, 1940.

To simplify the analysis we shall imagine that the earth at 20 ft. from a sudden explosion remains at rest for 0·027 sec., then accelerates uniformly, till at 0·13 sec. it has attained a velocity of 4 ft. per sec. (this is approximately the measured value). The question which the analyst can answer is: What plastic stress-strain relationship in the earth can give rise to this velocity-time relationship at a distance of 20 ft. from the explosive source?

The characteristic of the Riemann solution is that the points at which $pa$, $\rho a$ and $u$ are constant move outwards with constant velocities. If the movement $y$ is due to the sudden application of force $F$ at the end of a bar and subsequent maintenance of this force at constant value, then if the velocity-time curve is known at one point it is known at all points of the bar and at all times. Two examples will be worked out showing how the plastic stress-strain relationship can be deduced from observations of the velocity-time relationship observed at a fixed station.

*Case A*. The rate of propagation of the disturbance, namely, $c+u$, is a linear function of $u$.

*Case B*. The time rate of increase in $u$ at a fixed station is constant. When the material velocity $u$ is small compared with the velocity of propagation $c+u$, this assumption implies that the plastic material accelerates uniformly during the interval of time considered.

*Case A*

In this case
$$c+u=c_1-\frac{u}{U}[c_1-(c_2+U)], \tag{5}$$
where
$$c_1^2=\left[\frac{d(pa)}{d(\rho a)}\right]_{p=0} \quad \text{and} \quad c_2^2=\left[\frac{d(pa)}{d(\rho a)}\right]_{pa=F}$$
Comparing this with (3), (4) and (2) it will be seen that in a progressive disturbance
$$\left[\frac{d(pa)}{d(\rho a)}\right]^{\frac{1}{2}}=c_1-\frac{c_1-c_2}{U}u=\rho af'(\rho a) \tag{6}$$
and
$$u=f(\rho a). \tag{7}$$
Hence
$$\rho a\frac{du}{d(\rho a)}=c_1-\frac{c_1-c_2}{U}u. \tag{8}$$
The solution of (8) is
$$u=\frac{Uc_1}{c_1-c_2}\left[1-\left(\frac{\rho_0 a_0}{\rho a}\right)^{(c_1-c_2)/U}\right], \tag{9}$$
where $\rho_0$ and $a_0$ are the initial density and area of cross-section of the bar. Inserting this value of $u$ in (6) the equation for $pa$ is found to be
$$\left[\frac{d(pa)}{d(\rho a)}\right]^{\frac{1}{2}}=c_1\left(\frac{\rho_0 a_0}{\rho a}\right)^{(c_1-c_2)/U}. \tag{10}$$
The solution of (10) is
$$pa=\frac{c_1^2(\rho_0 a_0)\,U}{(2c_1-2c_2-U)}\left[1-\left(\frac{\rho_0 a_0}{\rho a}\right)^{(2c_1-2c_2-U)/U}\right]. \tag{11}$$
This equation gives $pa$ in terms of $\rho a$ and the velocities $c_1$, $c_2$ and $U$. Though $\rho$ and $a$ cannot be separated without further knowledge, it is unnecessary for our purpose

to do so because $\rho a$ is directly related to the plastic strain. If $l$ is the length of a small portion of the bar, the initial length of which was $l_0$, the plastic compressive $s$ is defined by

$$s=\frac{l_0-l}{l_0}, \tag{12}$$

and since

$$a\rho l=a_0\rho_0 l_0, \tag{13}$$

$$\frac{a\rho}{a_0\rho_0}=\frac{1}{1-s}. \tag{14}$$

Since $pa$ is the total compressive force acting over the cross-section, (11) might be regarded as a relationship between load and compressive strain of the material during a compression test. In cases where the change in density and the total strain $s$ is small, compared with $l$, (11) may be written

$$p=\frac{c_1^2\rho_0 U}{2c_1-2c_2-U}\left[1-(1-s)^{(2c_1-2c_2-U)/U}\right], \tag{15}$$

or approximately

$$p=\frac{c_1^2\rho_0 U}{2c_1-2c_2-U}\left[1-\exp\left(-\frac{(2c_1-2c_2-U)}{U}s\right)\right]. \tag{16}$$

When $U$ is small compared with $2c_1-2c_2$, as it is in the case under consideration, $(2c_1-2c_2-U)/U$ is large; in fact, the values assumed above were $U=4$ ft. per sec., $c_1=20$ ft./0·027 sec. $=740$ ft. per sec., $c_2+U=20$ ft./0·13 sec. $=153$ ft. per sec., so that

$$\frac{2c_1-2c_2-U}{U}=286 \tag{17}$$

and

$$\frac{c_1-c_2}{U}=143{\cdot}5. \tag{18}$$

Taking the density of the material to be 3 so as to correspond with earth, the values of $p$ calculated from (16) in lb. per sq.in. for a range of values of $s$ are shown in curve $A$, fig. 1. It will be seen that for a compression of 1 % the stress is 74 lb. per sq.in. The velocity-time relationship observed at a point 20 ft. from the explosion at the end of the bar is given by

$$\frac{20\text{ ft.}}{t\text{ sec.}}=c+u=c_1-\left(\frac{c_1-c_2-U}{U}\right)u, \tag{19}$$

$t$ being time from the explosion or sudden application of force at the end of the bar.

This relationship is shown in curve $A$ of the inset at the top of fig. 1.

The expression for $u$ corresponding with (16) when changes in $\rho a$ are small is

$$u=\frac{Uc_1}{c_1-c_2}\left[1-\exp\left(\frac{-c_1-c_2}{U}s\right)\right]. \tag{20}$$

The strains and stresses corresponding with $u=1$, 2, 3 and 4 ft. per sec. are marked in fig. 1 on curve $A$.

*Case B (uniform acceleration)*

In this case it may be assumed that $t=A+Bu$, where $A$ and $B$ must be determined so that the velocities of propagation correspond with $c_1$ at the beginning of the disturbance and $c_2+U$ at the end. If $X$ represents the distance from the end of the bar (20 ft. in the present case)

$$u+c=\frac{X}{t}=\frac{X}{A+Bu}. \tag{21}$$

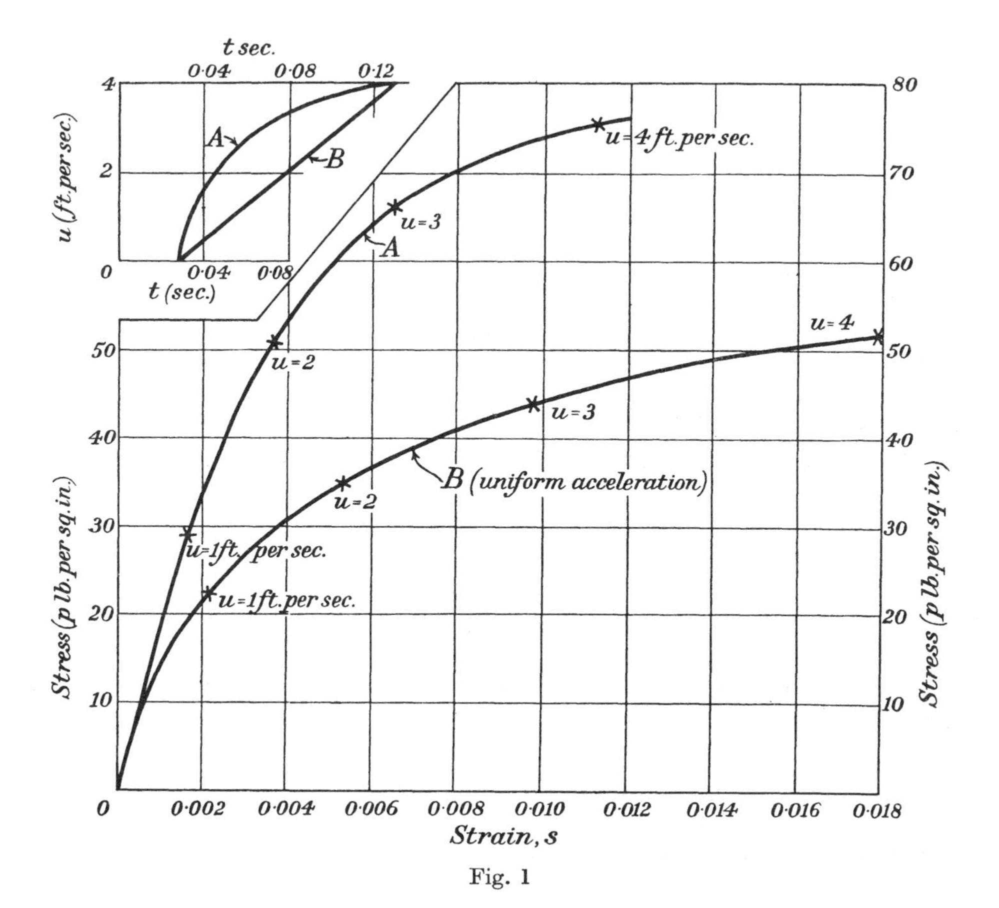


Fig. 1

The conditions determining $A$ and $B$ are

$$c_1=X/A, \tag{22}$$

$$U+c_2=\frac{X}{A+BU}. \tag{23}$$

Solving (22) and (23) for $A/X$ and $B/X$, (21) may be written

$$u+c=\frac{c_1}{1+\dfrac{u}{U}\left(\dfrac{c_1-c_2-U}{U+c_2}\right)}; \tag{24}$$

hence, writing $u' = u/U$,

$$\frac{c}{U} = \frac{\dfrac{c_1(c_2+U)}{(c_1-c_2-U)\,U} - \left(\dfrac{c_2+U}{c_1-c_2-U}\right)u' - u'^2}{\dfrac{c_2+U}{c_1-c_2-U} + u'}, \tag{25}$$

where $u'$ varies from 0 to 1 in the interval under consideration.

The Riemann equation (3) may be written

$$c = \rho a \frac{du}{d(\rho a)}, \tag{26}$$

which may be integrated to give

$$\log\left(\frac{\rho a}{\rho_0 a_0}\right) = \int_0^u \frac{du}{c}. \tag{27}$$

If the changes in $\rho a$ are small

$$\log\left(\frac{\rho a}{\rho_0 a_0}\right) = s, \tag{28}$$

so that

$$s = \int_0^u \frac{du}{c}. \tag{29}$$

Substituting for $c$ from (25), (29) may be integrated.

In the case under consideration where $U = 4$, $c_1 = 740$, $c_2 + U = 153$ ft. per sec.,

$$\frac{c_1(c_2+U)}{(c_1-c_2-U)\,U} = 48{\cdot}3, \quad \frac{c_2+U}{c_1-c_2-U} = 0{\cdot}261,$$

so that

$$s = \int_0^{u'} \frac{(u'+0{\cdot}261)\,du'}{48{\cdot}3 - 0{\cdot}261u' - u'^2}. \tag{30}$$

The terms $0{\cdot}261u'$ and $u'^2$ are small compared with $48{\cdot}3$, so that (30) may be integrated approximately, thus

$$s = \frac{1}{48{\cdot}3}\left(\tfrac{1}{2}u'^2 + 0{\cdot}261u'\right). \tag{31}$$

The stress $p$ is found by integrating (3)

$$pa = \int\left[\rho a \frac{du}{d(\rho a)}\right]^2 d(\rho a) = \int(\rho a)\,c\,du. \tag{32}$$

If the plastic strain $s$ is small the changes in $\rho a$ in the interval of integration may be neglected so that (32) becomes

$$p = \rho U^2 \int_0^{u'} \frac{c}{U}\,du'. \tag{33}$$

Neglecting $u'^2$ and $0{\cdot}261u'$ compared with $48{\cdot}3$, this gives in the case under consideration

$$p = 48{\cdot}3\rho U^2 \log\left(1 + \frac{u'}{0{\cdot}261}\right). \tag{34}$$

Taking $\rho = 3$, as in Case A, the following values have been calculated from (31) and (34):

| $u'$ | $u$ (ft./sec.) | $p$ (lb./sq. in.) | $s$ |
|---|---|---|---|
| 0 | 0 | 0 | 0 |
| 0·2 | 0·8 | 18·5 | 0·0015 |
| 0·4 | 1·6 | 30·0 | 0·0038 |
| 0·6 | 2·4 | 38·8 | 0·0070 |
| 0·8 | 3·2 | 45·5 | 0·0110 |
| 1·0 | 4·0 | 51·0 | 0·0158 |

These values are shown in fig. 1 as curve $B$. The corresponding linear velocity-time curve observed at $X = 20$ ft. is shown as $B$ in the inset to fig. 1.

# 31

# CALCULATION OF STRESS DISTRIBUTION IN AN AUTOFRETTAGED TUBE FROM MEASUREMENTS OF STRESS RINGS

Paper written for the Advisory Council on Scientific Research and Technical Development, Ministry of Supply (1941)

The measurement of residual stress in an 'autofrettaged cylinder', that is, a cylinder which is so constructed that it is in a state of internal stress, is sometimes determined by cutting disks from it and marking circles on their plane ends. The material is then cut away except for a thin ring containing the marked circle. The radial expansion of the ring, $\delta r$, produced by the removal of the surrounding material, is measured and the quantity

$$\phi = -E\,\delta r/r \tag{1}$$

is defined as the 'residual stress'. Here $E$ is Young's modulus. The present note shows how far measurements of $\phi$ can be used to find the distribution of internal stress that existed before the rings were isolated from the disks.

If $p$, $t$ and $l$ were the radial, tangential and longitudinal stresses in the cylinder

$$\frac{\phi}{E} = -\frac{\delta r}{r} = \frac{1}{E}\{t - m(p+l)\}, \tag{2}$$

where $m$ is Poisson's ratio. The equation of equilibrium is

$$r\frac{dp}{dr} + p - t = 0. \tag{3}$$

Combining (3) with (2),
$$r\frac{dp}{dr} + (l-m)\,p - \phi - ml = 0. \tag{4}$$

The solution of (4) is
$$p = r^{-l+m}\left[c + \int r^{-m}(\phi + ml)\,dr\right], \tag{5}$$

and since $p = 0$ at the inner radius $r = r_1$, the constant $C$ must be chosen to satisfy this requirement; thus

$$p = r^{-l+m}\int_{r_1}^{r} r^{-m}(\phi + ml)\,dr. \tag{6}$$

(6) would give the radial stress if both $\phi$ and $l$ could be measured. The cutting of the disk, however, releases the longitudinal stress so that $l$ may be taken as zero and (6) becomes

$$p = r^{-l+m}\int_{r_1}^{r} r^{-m}\phi\,dr. \tag{7}$$

Since $p$ must vanish at the outer radius $r = r_2$, evidently

$$\int_{r_1}^{r_2} r^{-m}\phi\,dr = 0. \tag{8}$$

If this condition is not satisfied when the measured values of $\phi$ are set in (8) the method of measurement is faulty. This equation is therefore a test of the stress ring method for measuring residual stresses.

The tangential stress is, from (2),

$$t=\phi+\frac{m}{r^{l-m}}\left[\int_{r_1}^{r} r^{-m}\phi\,dr\right]. \tag{9}$$

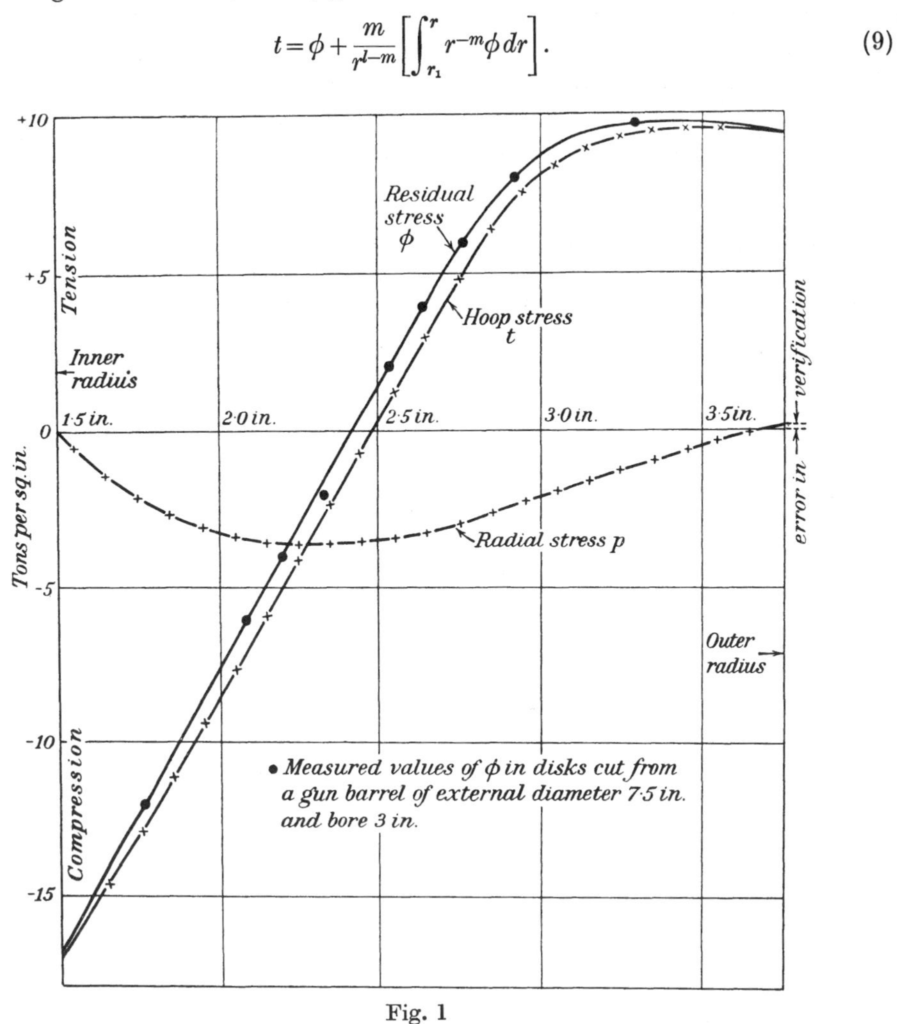


Fig. 1

### VERIFICATION OF OVERALL CONDITION OF EQUILIBRIUM

The equilibrium of half the tube under the action of hoop stress acting on an axial plane is represented by the equation

$$\int_{r_1}^{r_2} t\,dr=0.$$

This equation can be verified by partial integration of (9), when it is found to be true provided (8) is satisfied.

30 BSP

## Numerical Calculation for Autofrettaged Cylinder

The attached diagram (fig. 1) shows the measured values* of $\phi$ at all radii of a disk cut from a gun from the inner radius 1·5 in. to the outer radius 3·75 in.* The values of $t$ and $p$ calculated using formulae (7) and (9) (and assuming Poisson's ratio $m = 0{\cdot}29$) are shown. The fact that $p$ so nearly vanishes at the outer radius† shows that (8) is very nearly verified and thus confirms the accuracy of the stress ring measurements. The fact that the $\phi$ curve is so close to the true hoop stress ($t$) curve shows that the definition of residual stress as $\phi$ (see equation (1)) is a sound rough approximation.

* Measurements made under the direction of Dr Greaves at Woolwich Arsenal.

† The calculated value of $p$ was $+0{\cdot}206$ ton/sq.in. This may be compared with the maximum radial (compressive) stress $p = -3{\cdot}58$ tons/sq.in.

# 32

# THE PLASTIC WAVE IN A WIRE EXTENDED BY AN IMPACT LOAD

Paper written for the Civil Defence Research Committee, Ministry of Home Security (1942)

The analysis of waves of stress in a plastic wire recently given by von Kármán is compared with a previous one given by the author in a note (paper 30 above) to which von Kármán has not had access. The only difference between them is that von Kármán's work applies only when the strain is small,* a restriction which does not apply to the earlier analysis.

In making experiments to test von Kármán's theory, Duwez assumes that the plastic wire is brought instantaneously to rest when the load is released. This assumption leads to an apparent discrepancy between theory and observation. The present paper contains an approximate analysis of the actual state of the rod during the period while it is coming to rest after the stressing has stopped. Good agreement is found with Duwez's observations of the residual strain distribution when the wire is examined after coming to rest.

A recent note† by von Kármán analyses the plastic wave which is propagated down a wire when the end is suddenly given a uniform velocity. Von Kármán's analysis is similar to a previous one first given in my paper 'Propagation of earth waves from an explosion' (paper 30 above), and subsequently used by me to calculate the plastic wave in a wire struck by a bullet. Von Kármán's formulae are, however, only approximations based on the assumption that the plastic extension of any element of the wire is small, and since this assumption leads to no significant simplification, it seems worth while to reproduce my original analysis (which does not make this assumption) for comparison, and to apply it to the experimental results obtained by Dr Duwez.‡

The following symbols will be used:

$m$ = mass of wire per unit length before stretching (not variable).

$a$ = area of the wire at any section.

$a_0$ = initial area of section.

$\sigma$ = true stress.

$L = a\sigma$ = total force over any cross-section.

$L_0 = a_0\sigma_0$ = force applied during impact.

* *Footnote added by the author in 1955.* It was pointed out to me by Dr Bohlenblust that von Kármán's analysis is correct for finite strains if the analysis and all the symbols used in it are regarded as defined in a Lagrangian system of co-ordinates instead of the Eulerian system here used. This point, however, was not appreciated either by Dr Duwez or by me.

† 'On the propagation of plastic deformation in solids', National Defense Research Council Report, no. A 29 (1942).

‡ 'Preliminary experiments on the propagation of plastic deformation', National Defense Research Council Report, no. A 33 (1942).

30-2

$u = f(\epsilon)$ = velocity of any element of the wire.

$u_0$ = velocity of end of wire during impact.

$1+\epsilon$ = (length of element of wire)/(original length of same element).

$t$ = time from first application of load.

$t_0$ = duration of load when finite time of application is considered.

$C^2 = \frac{(\epsilon+1)^2}{m}\frac{dL}{d\epsilon}$.

$x$ = co-ordinate of any point measured downwards from the initial position of the end of the wire.

$X$ = distance of the point where the wave of relief of stress stops the plastic flow from the original position of the end of the wire, i.e. co-ordinate of this point is $x = -X$ and $X$ may be regarded as a function of $t$ when $t > t_0$.

$X'$ = initial unstretched length of the portion of the wire below $x = -X$.

$\epsilon_1$ = value of $\epsilon$ at $x = -X$.

$L_1$ = force at $x = -X$.

$u_1$ = velocity of the wire below $x = -X$.

$d$ = distance of lower end of specimen from any point where $\epsilon$ is measured after specimen has come to rest.

$d'$ = distance of lower end of specimen from point where strain is $\epsilon$ at time $t = t_0$.

$D$ = depth of end of wire below its initial position at any time after the release of load, i.e. when $t > t_0$.

$D_0$ = measured extension of wire, i.e. value of $D$ when wire comes to rest.

$t_D$ = Duwez's calculated time of impact $= D_0/u_0$.

$H$ = extension of wire at instant of release of load, i.e. at $t = t_0$.

$c_0$ = value of $c$ corresponding with $\epsilon = \epsilon_0$, $u = u_0$.

$c_{el.}$ = velocity of elastic waves.

It is assumed that $L/m$ (and consequently $\sigma/\rho$) is a function of $\epsilon$ only.

The equation of motion then takes the form

$$\frac{m}{1+\epsilon}\left(\frac{\partial u}{\partial t} + u\frac{\partial u}{\partial x}\right) = \frac{\partial L}{\partial x} = \frac{dL}{d\epsilon}\frac{\partial \epsilon}{\partial x}, \tag{1}$$

and the equation of continuity is

$$\frac{\partial \epsilon}{\partial t} + u\frac{\partial \epsilon}{\partial x} = \frac{\partial u}{\partial x}(1+\epsilon). \tag{2}$$

These equations are identical with the equation of Earnshaw and Riemann for the propagation of a wave of finite amplitude in a compressible fluid. Earnshaw's method of solution was to assume that

$$u = f(\epsilon). \tag{3}$$

Then (1) becomes

$$\frac{m}{\epsilon+1}\frac{df}{d\epsilon}\left(\frac{\partial \epsilon}{\partial t} + u\frac{\partial \epsilon}{\partial x}\right) = \frac{dL}{d\epsilon}\frac{\partial \epsilon}{\partial x}, \tag{4}$$

and (2) gives

$$\frac{\partial \epsilon}{\partial t} + u\frac{\partial \epsilon}{\partial x} = \frac{df}{d\epsilon}\frac{\partial \epsilon}{\partial x}(1+\epsilon). \tag{5}$$

Eliminating $\frac{\partial \epsilon}{\partial t}+u\frac{\partial \epsilon}{\partial x}$ between (4) and (5) gives $\left(\frac{df}{d\epsilon}\right)^2=\frac{1}{m}\frac{dL}{d\epsilon}$. Hence if $u$ is measured relative to the part of the wire which is at rest and $\epsilon=0$,

$$u=\int_0^{\epsilon}\sqrt{\left(\frac{1}{m}\frac{dL}{d\epsilon}\right)}\,d\epsilon. \tag{6}$$

Substituting for $df/d\epsilon$ in (5),

$$\frac{\partial \epsilon}{\partial t}+\left\{u-(1+\epsilon)\sqrt{\left(\frac{1}{m}\frac{dL}{d\epsilon}\right)}\right\}\frac{\partial \epsilon}{\partial x}=0. \tag{7}$$

Writing
$$c=(1+\epsilon)\sqrt{\left(\frac{1}{m}\frac{dL}{d\epsilon}\right)},$$

$$\frac{\partial \epsilon}{\partial t}+(u-c)\frac{\partial \epsilon}{\partial x}=0. \tag{8}$$

Thus at points which move with velocity $u-c$, $\epsilon$, and consequently $L$, are invariable.

To complete the solution when the end of a very long wire is given any assigned variation of velocity or load with time, one can apply directly the known method for determining the motion of a compressible fluid contained in a tube when a piston at one end is moved in any given manner.

## Case when the Plastic Wave is caused by a Constant Load suddenly applied at Time $t=0$

In this case the solution is very simple. All values of $u$ from $u=0$ to $u=u_0$ (and consequently all values of $\epsilon$ from $\epsilon=0$ to $\epsilon=\epsilon_0$) may be regarded as starting instantaneously at time $t=0$ from $x=0$ and each is propagated with velocity $u-c$. Thus at time $t$ and distance $x$

$$\frac{x}{t}=u-c. \tag{9}$$

Since both $u$ and $c$ depend only on $\epsilon$ and can be evaluated using (6) and (7), $u$ and $\epsilon$ are in this case functions of $x/t$ only. In general, it is found that the plastic stress-strain relationship is such that $c-u$ is positive and decreases continuously with increasing $\epsilon$. Thus (9) applies in the range of $(-x/t)$ from $[c]_{\epsilon=0}$ down to $c_0-u_0$, where $c_0$ is the value of $c$ corresponding with $u_0$. For values of $(-x/t)$ less than $(c_0-u_0)$, $u$, $c$ and $\epsilon$ are constant and equal to $u_0$, $c_0$ and $\epsilon_0$.

## Comparison with von Kármán's Formulae

Von Kármán's expression equivalent to (6) is

$$u=\int_0^{\epsilon}\sqrt{\left(\frac{1}{\rho}\frac{d\sigma}{d\epsilon}\right)}\,d\epsilon. \tag{10}$$

When (6) is expressed in terms of $\sigma$ and $a$

$$u=\int_0^{\epsilon}\left[\frac{d}{d\epsilon}\left\{\frac{\sigma}{\rho(1+\epsilon)}\right\}\right]^{\frac{1}{2}}d\epsilon. \tag{11}$$

(11) is true even when $\rho$ as well as $\sigma$ is a function of $\epsilon$, but if $\rho$ is supposed constant (11) may be written

$$u=\int_0^\epsilon\left[\frac{1}{\rho(1+\epsilon)}\left\{\frac{d\sigma}{d\epsilon}-\frac{\sigma}{1+\epsilon}\right\}\right]^{\frac{1}{2}}d\epsilon=\int_0^\epsilon\frac{c\,d\epsilon}{1+\epsilon}. \tag{12}$$

(12) is identical with (10) only when $\sigma/(1+\epsilon)$ is neglected compared with $d\sigma/d\epsilon$ and $\epsilon$ is neglected compared with 1·0. It will be noticed that the limiting value of $u$ occurs when $d\sigma/d\epsilon=\sigma/1+\epsilon$, which is, of course, the condition for the onset of instability in static stretching.

In the approximation used by Duwez for the special problem of constant end-load starting suddenly at $t=0$, $u$ is neglected compared with $c$ so that (9) becomes

$$x/t=-c. \tag{13}$$

If, as in Duwez's comparison, $x$ is taken as co-ordinate relative to the end of the wire rather than to the upper (fixed) part, the correct formula for comparison would be

$$x-\int_0^t u\,dt=(u-c)\,t. \tag{13a}$$

## Determination of Motion after Removal of Load

The determination of the motion of plastic bars or wires in cases where hysteresis is important is in general a difficult problem. Mr F. G. Friedlander has for some months been working on it and he has recently found a general method for dealing with certain classes of problem of this kind. In the particular case considered here the analysis can be carried to a conclusion. The case of a cylinder of plastic material projected at high speed normally at a fixed plane target is not included among the cases where a complete solution has been found, but much progress has been made by approximate methods. The nature of the present problem can be appreciated by working out a specially simple case.

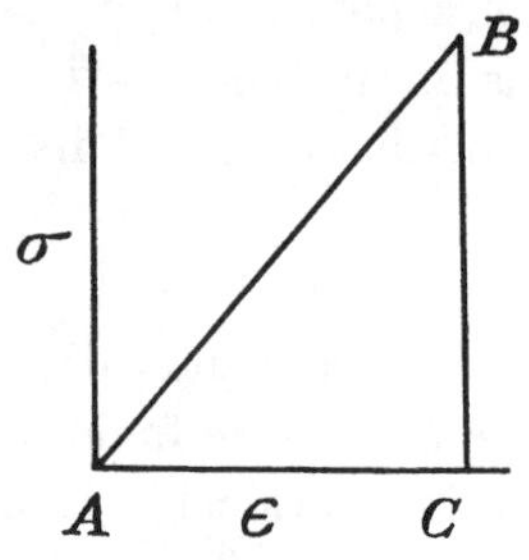

Fig. 1

Using von Kármán's assumption that the plastic strain is small, we may take the case of a material for which the plastic stress-strain curve is linear, though with a much smaller slope than the linear elastic stress-strain curve applicable on removal of the load or reloading to a stress less than the maximum load first applied. This is represented by the sketch (fig. 1), where $AB$ represents the plastic stress-strain curve and $BC$ the elastic condition which applies on relieving the load or reloading to a stress less than that represented by the point $B$.

With this stress-strain curve the plastic wave produced by suddenly applying a stress $\sigma_0$ and maintaining it is a sharp-fronted wave. The particle velocity $u_0$, stress $\sigma_0$ and strain $\epsilon_0$ are shown in fig. $2a$. The front travels with velocity $c_1$, where $c_1^2=\sigma_0/\epsilon_0\rho$.

If at the stage represented by fig. $2a$ the load is released, an elastic wave of compression travels back in the same direction as the plastic wave but at the much higher speed, $c_2$, of elastic waves. Fig. $2b$ represents the condition after this wave has passed half-way towards the plastic wave front. Fig. $2c$ shows the condition at the instant when this elastic wave catches up the plastic wave. At this stage the stress is everywhere zero but the velocity of the wire behind the wave is not zero. It is in fact $u_{11}=u_0(1-c_1/c_2)$ and the extension is $\epsilon_{11}=\epsilon_0(1-c_1^2/c_2^2)$. This differs from $\epsilon_0$ by a quantity of the second order, i.e. by a term containing $c_1^2/c_2^2$. At this stage a wave of elastic extension is reflected back towards the free end of the wire. Fig. $2d$ shows the condition when this elastic wave has gone half-way back towards the free end.

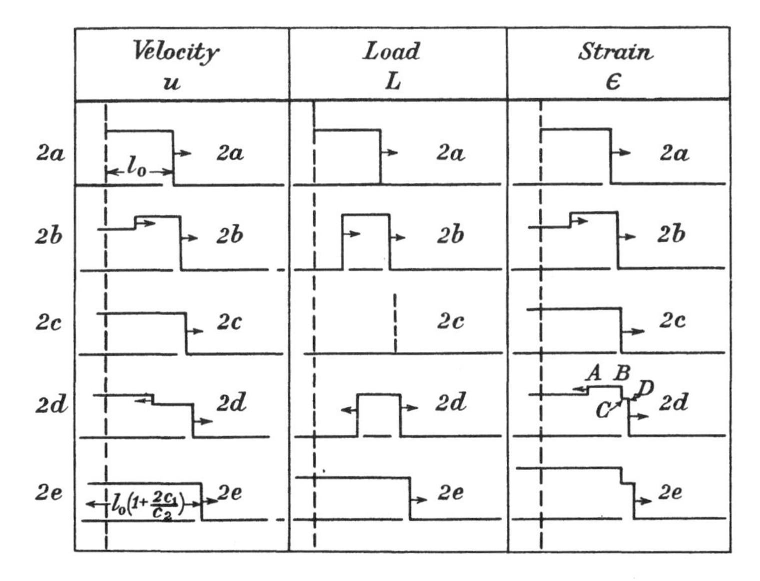


*Note u is always from right to left. The plastic wave front moves from left to right, the elastic wave fronts in both directions.*

Fig. 2

Assuming that $c_1$ is small compared with $c_2$ the stress in this wave is

$$\sigma_1=\sigma_0(1-2c_1/c_2)$$

and the velocity is $u_0(1-2c_1/c_2)$. The strain, however, is not the same at all points. In the range represented by $AB$ in the right-hand diagram of fig. $2d$, the strain is that caused by a stress $\sigma_0(1-2c_1/c_2)$ acting on a piece of wire which has already been strained plastically by a load $\sigma_0$. It is, in fact, $\epsilon_{12}=\epsilon_{11}+\epsilon_0(1-2c_1/c_2)\,(c_1^2/c_2^2)$, which differs from $\epsilon_0$ by $2\epsilon_0(c_1/c_2)^3$, a quantity of the third order. The strain in the remaining part of the wave where the plastic wave is penetrating into material not previously stressed (shown as $CD$ in fig. $2d$) is the plastic strain corresponding with a wave in which the particle velocity is $u_0(1-2c_1/c_2)$, i.e. it is $\epsilon_{13}=\dfrac{\sigma_0(1-2c_1/c_2)}{\rho c_1^2}$, so that $\epsilon_{13}/\epsilon_0=1-2c_1/c_2$.

When the reflected tension wave has reached the free end the distribution of velocity and stress shown in fig. 2*e* are exactly the same as they were when the stress was first released (i.e. as in fig. 2*a*) except that

(1) the velocity of the wire behind the wave front is reduced in the ratio $1-2c_1/c_2$,

(2) the stress is reduced in the ratio $1-2c_1/c_2$,

(3) if $l_0$ is the length of wire which has been stretched when the load is released the length at the end of the sequence described above is $l_0(1+2c_1/c_2)$,

(4) the strain is to the first order of small quantities $\epsilon_0$ over a length $l_0$ and $\epsilon_0(1-2c_1/c_2)$ over the remaining length $2l_0c_1/c_2$.

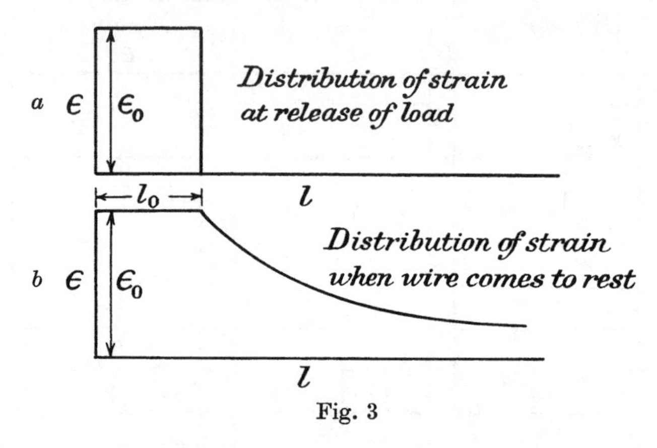


Fig. 3

The whole sequence of reflections just described can now be repeated again and again till the whole motion dies out. At each stage the length of the part of the wire which has stretched is increased in the ratio $(1+2c_1/c_2):1$, while the stress $\sigma_1$ immediately behind the plastic wave front has decreased in the ratio $(1-2c_1/c_2):1$. The extension $\epsilon_1$ of the part of the wire which has been newly affected by the stress waves is less than that in the previous series of reflections in the ratio $(1-2c_1/c_2):1$. In this way one finds for the changes in $\epsilon_1$, $u_1$ and $l$ which occur during one sequence of reflections

$$\frac{\delta\epsilon_1}{\epsilon_1}=\frac{\delta u_1}{u_1}=-\frac{\delta l}{l}=-\frac{2c_1}{c_2}. \tag{14}$$

It appears therefore that the strain which is ultimately 'fixed' at distance $l$ from the end of the wire can be found by integrating (14). Thus

$$l\epsilon_1=l_0\epsilon_0. \tag{15}$$

The strain $\epsilon_0$ which exists in the length $l_0$ at the instant when the end-load is removed remains unchanged, so that the distribution of strain which is fixed is that shown in fig. 3*b*. This is very different from the strain shown in fig. 3*a* which existed in the wire at the moment of release of load.

### Alternative Discussion for the Case when the Velocity of the Elastic Wave is Large compared with that of the Plastic Wave

The above discussion considers the effect of elastic waves which are continually reflected backwards and forwards between the free end of the wire and the advancing plastic wave. Exactly the same equations could have been deduced by assuming that the whole of the wire behind the plastic wave front is rigid, but that the plastic wave continues to progress till the whole wire comes to rest. A similar assumption has given useful results in the problem of the deceleration of a plastic cylinder on striking a rigid plate. The equation for the deceleration of the rigid part of the wire is

$$\sigma = -\rho l \frac{du}{dt}, \tag{16}$$

and the linear stress-strain law (exhibited in fig. 1) gives

$$\sigma = \rho c_1 u = \rho \epsilon_1 c_1^2.$$

This, together with the kinematic equation $dl/dt = c_1$, gives

$$\sigma = \rho u \, dl/dt. \tag{17}$$

Combining (16) and (17) $$l\frac{du}{dt} + u\frac{dl}{dt} = 0,$$

so that $$lu = \text{constant} = l_0 u_0,$$

and hence $$l\epsilon_1 = \text{constant} = l_0 \epsilon_0, \tag{18}$$

which is identical with (15).

### Analysis of 'Fixing' of Plastic Strain with Arbitrary Stress-strain Relationship

The method just described is general and can be applied to a wire with any load-extension curve *even when the strain is not small.* The simple special case discussed above derived its simplicity from the fact that when the plastic stress-strain curve is linear, as shown in fig. 1, the plastic boundary moves with uniform speed and there is no motion ahead of it. In the general case this simplification does not apply. If $X$ is the distance *from the original position of the end of the wire* of the boundary between the part of the wire which is still stretching plastically at time $t$ and the part which has ceased to extend, the problem is to determine $X$ as a function of $t$. This must be done by ensuring that at $-x = X$ the stress in the advancing plastic wave will give rise to the force necessary to decelerate the 'rigid' portion of the wire (i.e. the part in which plastic flow has ceased), at the same time ensuring continuity of velocity at $-x = X$. Continuity of stress at $-x = X$ gives

$$L_1 = -M \, du_1/dt, \tag{19}$$

where $u_1$ is the velocity in the rigid part of the wire, $L_1$ is the total force at the boundary, and $M$ the mass of the rigid portion. The plastic wave equations (6) and (7) give

$$u_1 = \int_0^{\epsilon_1} \frac{c_1}{1+\epsilon} d\epsilon \quad \text{and} \quad c_1 = (1+\epsilon_1)\left[\frac{1}{m}\left(\frac{dL}{d\epsilon}\right)_\epsilon\right]^{\frac{1}{2}}. \tag{20}$$

If $X'$ is the initial unstretched length of the portion of the wire bounded by $x = -X$ then

$$mX' = M.$$

Since $$X/t = c - u \quad \text{(see (9))}, \tag{9a}$$

both $X/t$ and consequently $X'/t$ are functions of $\epsilon$ only. The relationship between $X$ and $X'$ is therefore expressed by the formula

$$\frac{X}{t} - \frac{X'}{t} = \int_0^{\epsilon_1} \epsilon \frac{d}{d\epsilon}\left(\frac{X'}{t}\right) d\epsilon, \tag{21}$$

so that $$\frac{d}{d\epsilon}\left(\frac{X}{t}\right) = (1+\epsilon)\frac{d}{d\epsilon}\left(\frac{X'}{t}\right),$$

hence $$\frac{d}{d\epsilon}\left(\frac{X'}{t}\right) = \frac{1}{1+\epsilon}\frac{d}{d\epsilon}\left(\frac{X}{t}\right) = \frac{1}{(1+\epsilon)}\frac{d}{d\epsilon}(c-u), \tag{22}$$

and since, from (6) and (7), $$\frac{du}{d\epsilon} = \frac{c}{1+\epsilon},$$

(22) becomes $$\frac{d}{d\epsilon}\left(\frac{X'}{t}\right) = \frac{1}{1+\epsilon}\left(\frac{dc}{d\epsilon} - \frac{c}{1+\epsilon}\right) = \frac{d}{d\epsilon}\left(\frac{c}{1+\epsilon}\right),$$

so that $$\frac{X'}{t} = \frac{c}{1+\epsilon} = \frac{du}{d\epsilon}. \tag{23}$$

(19) may now be written $$\frac{L_1}{m} = -X'\frac{du_1}{d\epsilon_1}\frac{d\epsilon_1}{dt}, \tag{24}$$

and, substituting for $X'$ from (23), (24) becomes

$$\frac{L_1}{m} = -t\left(\frac{du_1}{d\epsilon_1}\right)^2\frac{d\epsilon_1}{dt} = -t\frac{c_1^2}{(1+\epsilon_1)^2}\frac{d\epsilon_1}{dt}. \tag{25}$$

Hence, from (7), $$\frac{L_1}{m} = -t\frac{d\epsilon_1}{dt}\left(\frac{1}{m}\frac{dL_1}{d\epsilon_1}\right).$$

This equation may be integrated, giving

$$L_1 t = \text{constant} = L_0 t_0. \tag{26}$$

This relationship is due to Mr Friedlander. To find $X$ as a function of $t$ it is necessary first to calculate $u$ and $c$ by numerical integration of (6) and (7) using the observed relationship between $L$ and $\epsilon$. In this way corresponding values of $\epsilon_1$, $L_1$, $u_1$ and $c_1$ are found. $L_0$ is the value of $L$ corresponding with the velocity $u_0$ at which the load was applied, so that $t/t_0$ can be found from (26). $X$ is then found by multiplying the value of $t$ so found by $c-u$.

## Comparison with Experiments of Dr Pol Duwez

*Values of c.* Dr Pol Duwez has made experiments with copper wire in order to test von Kármán's theory. His measurements are expressed in terms of $\sigma$ instead of $L/m$ so that for verification (12) may be used. It is found in the case of copper that for a given value of $\epsilon$ the first approximation used by Duwez for $c$ is slightly higher than that deduced by the more accurate solution. In fact if the value for $c$ used by Duwez is multiplied by the factor $(1+\epsilon)^{\frac{1}{2}}\left(1-\frac{\sigma}{1+\epsilon}\frac{d\epsilon}{d\sigma}\right)^{\frac{1}{2}}$, the correct value is found for comparison with experiments. For $\epsilon_0=0{\cdot}066$ Duwez's approximation gave $c_0=1100$ ft. per sec. The above correcting factor is found to be 0·98, so that the true calculated value of $c_0$ for $\epsilon_0=0{\cdot}066$ is

$$c_0=1080\,\text{ft. per sec.} \tag{27}$$

The quantity which Duwez deduces as the observed value of $c_0$ is 910 ft. per sec., but for the reasons which follow I do not think that Duwez's method for deducing $c_0$ from the facts of observation is correct.

Duwez measured the distance $d$ from the lower end of his specimen to the point where the extension is $\epsilon_1$. He then took $c=d/t_D$, where $t_D$ is what he describes as the 'calculated time of impact'. This calculation was based on the assumption that the motion of the wire stops instantly when the load is removed. Thus he assumes that $t_D$ is the same quantity as $t_0$, the actual time of impact, and he calculates its value by dividing $D_0$, the measured total extension of the wire, by $u_0$, the measured velocity of the end of the wire during the impact. Duwez therefore uses the formula

$$c=d/t_D=u_0d/D_0. \tag{28}$$

It is clear from the analysis here given that this method for finding $c$ is incorrect. If the velocity $c_{\text{el.}}$ of the elastic wave reflected from the free end of the wire at the moment the load is released is assumed infinite, the correct formula for the velocity $c_0$ of the back of the plastic wave relative to the free end is

$$c_0=d_0/t_0=u_0d_0/H, \tag{29}$$

where $H$ is the distance moved by the end of the wire while the load is acting (i.e. in time $t_0$), $u_0$ is the velocity of the end while the load is acting, and $d_0$ is the value of $d$ corresponding with $\epsilon=\epsilon_0$.

Formula (29) differs from Duwez's (28) by an amount which is not small even when his approximation is applicable, because the end of the wire would continue to move downward after the removal of the load if it were free to do so. It is true that in Duwez's apparatus the position, in space, of the end of the wire was apparently fixed at the end of the impact. This would, if the wire were constrained to keep straight, give rise to a plastic compression wave, but it is certain that in a wire of the thickness used by Duwez this compression would immediately be released by lateral instability so that the wire would in fact behave as though it could go on stretching vertically after the removal of the load until the whole motion was stopped by the tension in the plastic tension wave. The distance through which the end of the wire moves after the removal of the load must always be comparable with $H$ because only a portion of the work done by the load acting through the distance $H$ has been used in stretching the wire. The remainder is in the form of kinetic energy which can only be dissipated by further stretching of the wire. Thus, if the velocity of the elastic wave can be assumed as very large compared with that of the plastic wave, the ratio $D_0/H$ is always greater than 1·0 and does not in general tend to 1·0 for small plastic extensions of the wire.

It will be seen later that for the case of a wire stretched at 92 ft. per sec. $D_0/H$ is calculated to be 1·36 so that the method adopted by Duwez for finding $t_0$ from his observations makes him overestimate it by 36 %. On the other hand, the assumption that the velocity of the reflected elastic wave which fixes the end of the uniformly stretched portion of the wire is infinite causes an error in the opposite direction. The velocity of elastic waves in copper may be taken as $c_{\text{el.}} = 3{\cdot}7 \times 10^5$ cm. per sec. The condition which determines the length $d_0$ is evidently

$$\frac{d_0}{c_0} - \frac{d_0}{c_{\text{el.}}} = t_0 \quad \text{or} \quad c_0 = \frac{d_0}{t_0}\left(1 - \frac{c_0}{c_{\text{el.}}}\right). \tag{30}$$

If $c_{\text{el.}}$ is much larger than $c_0$ (30) may be written

$$c_0 = \frac{d_0}{t_0}\left(1 - \frac{d_0}{t_0 c_{\text{el.}}}\right).$$

Thus for comparison with theory the 'observed' value of $c_0$ must be taken as

$$c_0 = 1{\cdot}36\frac{d_0}{t_D}\left\{1 - \frac{1{\cdot}36\, d_0}{(c_{\text{el.}})(t_D)}\right\}, \tag{31}$$

or if only things actually measured are included

$$c_0 = 1{\cdot}36\frac{d_0 u_0}{D_0}\left\{1 - 1{\cdot}36\frac{d_0}{D_0}\frac{u_0}{c_{\text{el.}}}\right\}. \tag{32}$$

The measured value of $d_0 u_0/D_0$ was 910 ft. per sec. or $2{\cdot}77 \times 10^4$ cm. per sec., so that the value of $c_0$ which would be deduced from observations is, from (32),

$$(1{\cdot}36)(2{\cdot}77 \times 10^4)\left(1 - 1{\cdot}36 \times \frac{2{\cdot}77}{3{\cdot}7}\right) = 3{\cdot}38 \times 10^4 \text{ cm. per sec.} = 1110 \text{ ft. per sec.}$$

This must be compared with 1080 ft. per sec. deduced theoretically (see (27)) by applying formula (7) to the stress-strain data measured in static loading tests. It is no doubt accidental that these two figures agree so very closely.

## Distribution of Plastic Strain after Test

In the absence of any analysis, Dr Duwez assumes that the strains observed when the wire has come to rest are identical with those which existed at the moment when the load was removed. It is clear from the analysis here given that this hypothesis could not be expected to give results agreeing with observation.

In order to make a comparison with Dr Duwez's observations, the distribution of strain has been calculated for the case where the velocity of the impact load was 92·5 ft. per sec. It will be seen from equations (9$a$) and (26) that $X/t$ and $t/t_0$ can be calculated directly from the load-extension curve obtained in a static loading test. The only dynamic measurement necessary in order to compare the observed and calculated distribution of plastic strain in the wire is that of $t_0$.

It has already been seen that Duwez's deduction of $t_0$ from the facts of observation is incorrect, but that a more accurate value of $t_0$ may perhaps be inferred. On the other hand, if the true value of $t_0$ is used the value of $X$ calculated on the assumption that $c_{\text{el.}}$ is infinite will be too small. For comparison of the observed distribution of residual strain down the wire with that calculated a suitable value to take for $t_0$ is $(X_0+D_0)/c_0$ where $D_0+X_0$ is the observed distance along the wire of the point where the variable part of the plastic wave begins and $c_0$ is the calculated value $c$ for extension $\epsilon_0$.

This method necessarily makes one point of the observed and calculated distributions coincide and the comparison is then concerned with the rest of the distribution.

The results of the calculation are given in table 1. This table is expressed in terms of c.g.s. units. In column 2 are given the values of $\sigma$ in dynes per sq.cm. taken from fig. 2 of Duwez's paper. In column 3 are given values of $c$ taken from fig. 3 of Duwez's paper and in column 4 the values which he would have obtained if he had used (7) instead of von Kármán's approximate formula. Column 5 gives the values of $u$ calculated from (6). Column 6 gives $X/t=c-u$. Column 7 gives the values of $t/t_0$ given by the formula (26) in its alternative form

$$\frac{t}{t_0}=\frac{\sigma_0}{\sigma}\left(\frac{1+\epsilon}{1+\epsilon_0}\right) \tag{33}$$

and $$\sigma_0=1{\cdot}94\times10^9 \quad \epsilon_0=0{\cdot}066.$$

Column 8 gives $X/t_0$ and is found by multiplying the figures in columns 6 and 7. Values of $\dfrac{D-H}{t_0}=\displaystyle\int_1^{t/t_0} u\,d\left(\frac{t_0}{t}\right)$ obtained by numerical integration from the figures in columns 5 and 7 are given in column 9.

$D_0$, the final value of $D$ when the wire comes to rest, is seen from column 9 to be given approximately by $(D_0-H)/t_0=9{\cdot}7\times10^{-2}$ cm. per sec. Since the corresponding value of $u_0$ is, from column 5, $27{\cdot}0\times10^2 t_0$, hence $H=27{\cdot}0\times10^2 t_0$, so that

$$\frac{D_0}{H}=\frac{27{\cdot}0+9{\cdot}7}{27{\cdot}0}=1{\cdot}36. \tag{34}$$

The measured value of $d_0=X_0+D_0$ for one experiment at $u_0=92{\cdot}5$ ft. per sec. is given in the text of Duwez's paper (in his table 1) as 9·45 in. Measurement of the corresponding curve in fig. 12 gives it as about 9·7 in.

Table 1

| 1 | 2 | 3 | 4 | 5 | 6 | 7 | 8 | 9 | 10 | 11 | 12 |
|---|---|---|---|---|---|---|---|---|---|---|---|
| $\epsilon$ | $\sigma \times 10^{-9}$ (dynes/sq.cm.) | $c \times 10^{-4}$ (cm./sec.) (Duwez) | $c \times 10^{-4}$ (cm./sec.) (corrected) | $u \times 10^{-3}$ (cm./sec.) | $X/t \times 10^{-4}$ (cm./sec.) | $t/t_0$ | $X/t_0 \times 10^{-4}$ (cm./sec.) | $(D-H)/t_0 \times 10^{-2}$ (cm./sec.) | $X+D$ (in.) | $(c-u+u_0)\, t_0$ (in.) | $d$ (in.) (measured: Duwez) |
| 0 | — | — | — | — | — | — | — | 9·7 | — | — | — |
| 0·005 | 1·10 | 5·30 | 5·08 | 0·55 | 5·02 | 1·66 | 8·33 | 8·7 | 26·3 | 16·0 | — |
| 0·010 | 1·21 | 4·14 | 3·98 | 0·78 | 3·96 | 1·52 | 6·40 | 7·7 | 20·4 | 12·6 | 21·0 |
| 0·015 | 1·33 | 3·99 | 3·87 | 0·97 | 3·77 | 1·38 | 5·22 | 6·5 | 16·8 | 12·2 | 17·6 |
| 0·020 | 1·40 | 3·87 | 3·77 | 1·16 | 3·65 | 1·32 | 4·83 | 5·9 | 15·6 | 11·7 | 15·5 |
| 0·030 | 1·53 | 3·71 | 3·63 | 1·52 | 3·48 | 1·22 | 4·25 | 4·5 | 13·8 | 11·3 | 13·5 |
| 0·040 | 1·66 | 3·59 | 3·53 | 1·87 | 3·34 | 1·14 | 3·79 | 3·1 | 12·4 | 10·9 | 12·1 |
| 0·050 | 1·76 | 3·44 | 3·39 | 2·20 | 3·17 | 1·08 | 3·43 | 2·2 | 11·3 | 10·4 | 10·9 |
| 0·066 | 1·94 | 3·27 | 3·21 | 2·70 | 2·94 | 1·00 | 2·94 | 0 | 9·7 | 9·7 | 9·7 |

Taking the latter value the appropriate value of $t_0$ for comparison between the observed and calculated distributions of $\epsilon$ is

$$t_0 = \frac{9\cdot7 \times 2\cdot54}{3\cdot21 \times 10^4} = 0\cdot000768 \text{ sec.}$$

Multiplying the figures in columns 8, table 1, by 0·000768, the calculated values of $X$ are obtained in centimetres. For comparison with the measured values of $d = X + D$, which are given in column 12 in inches, these figures are divided by 2·54 and $D$, calculated from column 9, is added. The resulting calculated values of $X + D$ are given in column 10.

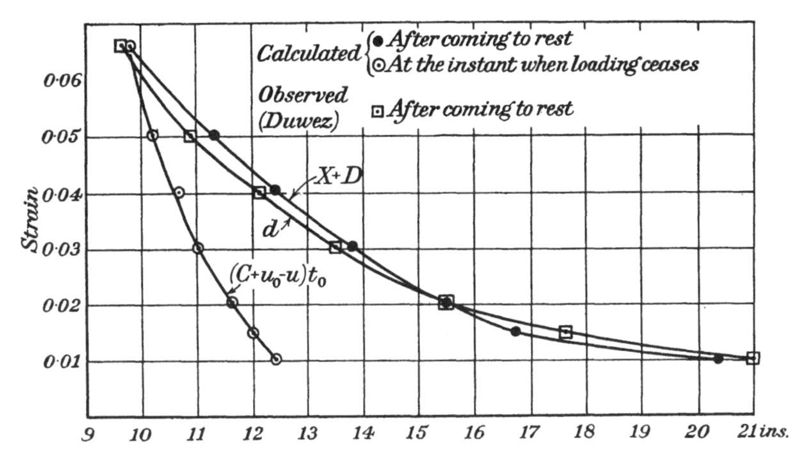


Fig. 4. Distribution of strain in wire after loading at 92·5 ft. per sec. for time $t_0 = 0\cdot00077$ sec.

The distribution of strain along the wire at the moment when the load is released, i.e. $t = t_0$, is given by

$$d' = H + (c - u)\,t_0 \quad \text{or} \quad d = (c - u + u_0)\,t_0, \tag{35}$$

where $d'$ is the distance from the end of the wire at time $t = t_0$.

This is the calculated value of the strain which Duwez compares with the observed figures given in column 10. Using $t_0 = 0\cdot000768$ the resulting values of $d'$ are given in inches in column 11.

The results are compared in fig. 4. It will be seen that the present calculation is in very good agreement with the observation of Duwez.

# 33

# THE MECHANICAL PROPERTIES OF CORDITE DURING IMPACT STRESSING*

Paper written for the Advisory Council on Scientific Research and Technical Development, Ministry of Supply (1942)

This report deals with the relationships between nominal stress, true stress, longitudinal strain and lateral strain for cordite under the very rapid compression stressing which occurs when cordite cylinders, fired from a pneumatic gun, are projected against the end-surface of a long steel pressure-bar. These relationships are deduced from measurements of the velocity of projection of the cordite cylinders, the maximum pressure in the elastic wave produced in the steel bar by the impact, and the maximum area of contact between the cordite cylinders and the end of the pressure bar. The velocity of projection of the cylinders is found from their mass and their dimensions and from the pressure which accelerates them along the barrel of the gun. Two modifications—one electrical, the other mechanical—of the Hopkinson bar are used to determine the pressure in the elastic wave in the bar, whilst the area of contact is found from measurement of the circular spot formed on the end of the pressure bar when the impacting end of the cordite cylinder is smeared with grease.

The relationships between the stresses and strains are deduced from the experimental results by means of a general theory of the propagation of stress waves of finite displacement in a medium obeying a non-linear stress-strain law.

The results show that, under the conditions of these experiments, there is a large increase in strength compared with the strength under static conditions, the majority of the specimens showing a sevenfold increase in strength for short periods of stressing. These stresses are accompanied by correspondingly large strains, amounting in some cases to a decrease of over 50 % in the length of the specimens and to an increase of over 20 % in their diameter. The stress-strain relationship is similar in form to that of a soft metal, such as copper.

## 1. Introduction and Theory of the Experiments

A method† has recently been developed for determining the strength of steel and other materials at very high rates of stress in which cylindrical specimens were fired from a rifle or smooth bore gun at a hard steel target. The length which remains undistorted was measured and also the final length of the specimen. From these data and the velocity of the projectile a minimum value can be assigned for the compressive strength. It was thought that the same method might be applied to determine the greatest load which can be applied to cordite for a very short time without breaking it. This direct application, however, proves impossible because the above-described method involves the assumption that the velocity of the

* With R. M. Davies.

† G. I. Taylor, 'Stress-strain relationship in impact'; paper written for Civil Defence Research Committee (1939). See also 'The use of flat-ended projectiles for determining dynamic yield stress. I. Theoretical considerations', *Proc. Roy. Soc.* A, CXCIV (1948), 289; paper 39 below.

projectile is small compared with the velocity of the elastic waves in it. The coefficient of elasticity of cordite, however, is so small that the breaking stress can only be reached in a cylindrical projectile of this material if the velocity of impact is comparable to the velocity of elastic waves.

Though this prohibits the use of this purely ballistic method for determining the stress, we have been able to use two developments—one electrical, the other mechanical—of the Hopkinson bar method* to determine the pressure exerted by a cylindrical projectile of cordite when fired directly at the end of the bar.

(1) *The electrical method*

This method consists in measuring electrically the variation with time of the displacement of the far end of a steel bar due to the longitudinal elastic waves set up by the pressure of the cordite projectile.

The pressure in a longitudinal elastic wave in steel is

$$p = \rho c v, \tag{1}$$

where $\rho$ is the density of steel, $c$ the wave velocity in steel, and $v$ the particle velocity in the wave. The displacement $\xi$ of the far end of the bar is due to the superposition of a wave of compression moving away from the point of impact and an equal wave of tension moving in the opposite direction. Thus

$$d\xi/dt = 2p/\rho c, \tag{2}$$

and $d\xi/dt$ is found by graphical differentiation of the measured displacement-time curve.

In all cases the pressure of the cordite was found to rise to a maximum value within a few (5–15) microseconds of the moment of impact. This small apparent delay is probably partly instrumental, so that we can only state that the maximum pressure is attained within a period of 5–15 $\mu$sec.

The maximum pressure, attained almost instantaneously, remained constant for a period which corresponds with that required for a stress wave to travel twice the length of the specimen. The pressure then decreases and finally disappears as the specimen rebounds.

The specimens were examined as soon as possible after impact, and no change in their external dimensions could be detected even when, as sometimes happened, a small crack was visible as an opaque plane within the translucent material. In such cases the pressure-time curve showed a sudden drop followed by a rise which, however, never brought the pressure up to the initial maximum. This drop we associate with the moment of breakdown.

(2) *The mechanical method*

The displacement-time curve obtained using the electrical method showed that the duration of impact never exceeded 120 $\mu$sec. and was therefore much less than

* B. Hopkinson, *Phil. Trans.* A, CCXIII (1914), 437; R. Robertson, *Trans. Chem. Soc.* CXIX (1921); J. W. Landon and H. Quinney, *Proc. Roy. Soc.* A, CIII (1923), 622.

the time taken for the elastic wave to travel 4 ft. in steel. For this reason it is clear that the detached end-portion of a 2 ft. bar (as described in Hopkinson's original paper*) would not be overtaken by the bar itself, and could therefore be used to determine the maximum pressure. Some time ago we developed a method for finding the maximum pressure, replacing Hopkinson's cylindrical detached portions by a few steel balls of suitably chosen diameters. From measurements of the velocities with which balls of different diameters are projected from the end-surface of the bar remote from the impact, it is possible to calculate the velocity of this surface due to the impact, and, from equation (2), the maximum pressure developed during the impact.

We have verified the accuracy of the two methods of experiment by measuring the pressure developed when a bullet strikes the end of the pressure bar, and the pressure in a detonation wave in an explosive mixture of equal volumes of oxygen and acetylene. In the first case, the relation between pressure and time can be calculated* to a fair degree of accuracy by assuming that the lead bullet is devoid of strength and behaves as a fluid at impact. In the second case, the pressure in the initial part of the detonation wave can be calculated from thermochemical data. In both cases, the experimental results were in agreement with those obtained by calculation.

*Theory of the method.* Projecting at a range of velocities from 100 to 450 ft. per sec. a number of determinations of maximum pressure were made by the electrical and mechanical methods. From these measurements it is possible to determine the stress-strain relationship for cordite at high rates of stressing by means of the theory of waves in plastic bars or wires. This theory,† first given in connection with the propagation of stress waves in earth, was used later to analyse the plastic waves in stretched wires when suddenly loaded at one end.‡ Though the theory is described as one of plastic strains, it applies equally well to finite elastic stresses and strains.

In the present case, where a compression load is suddenly applied to one end of the cordite cylinder, a compression wave starts from this end and travels through the cordite. The feature which distinguishes 'plastic' waves in bars of material for which the stress-strain relationship is non-linear from those considered in elastic theory for which the relationship is linear, is that the 'plastic' waves change their form as they pass down the bar, whereas elastic waves preserve their form. In fact, the point in a progressive plastic wave where the stress or strain has a given value moves down the cylinder with velocity $c_1$ relative to the material, which itself is moving with velocity $u$ relative to the unstrained part; the velocity of propagation

* See references in footnote, p. 481.

† G. I. Taylor, 'Propagation of earth waves from an explosion' (1940), paper 30 above; and 'The plastic wave in a wire extended by an impact load' (1942), paper 32 above.

‡ T. von Kármán, 'On the propagation of plastic deformation in solids', National Defense Research Council Report, no. A 29 (1942); Pol Duwez, 'Preliminary experiments on the propagation of plastic deformation', National Defense Research Council Report, no. A 33 (1942).

relative to the unstrained part is thus $c_1+u$. $c_1$ is a function of the strain provided that the stress $S$ is a function of $\epsilon$ only.

We may assume that within the range of these experiments, the variation in the stress-strain relationship with rate of stressing is not appreciable.

The relevant formulae* of the plastic wave theory applied to the case of compression are then:

$$c_1^2=\frac{(1-\epsilon)^2}{\rho_0}\frac{dS}{d\epsilon}, \tag{3}$$

$$u=\int_0^{\epsilon}\sqrt{\left(\frac{1}{\rho_0}\frac{dS}{d\epsilon}\right)}\,d\epsilon, \tag{4}$$

where $F$ = total force over a cross-section at any instant,

$a$ = cross-sectional area of an element at any stage of compression,

$a_0$ = original cross-sectional area of the element,

$S$ = 'nominal' compressive stress $=F/a_0$,

$\sigma=F/a$ = true compressive stress,

$1-\epsilon$ = (length of an element at any stage of compression)/(original length of the same element),

$\rho_0$ = original density of the material,

$c_1$ = velocity of propagation of strain $\epsilon$ relative to the material,

$u$ = velocity of an element relative to the unstrained material.

Let $U$ = velocity of projection of the cordite cylinder, $\sigma_1$, $S_1$, $\epsilon_1$, $a_1$ = values of $\sigma$, $S$, $\epsilon$ and $a$ at the point of impact. The formulae (3) and (4) apply to all parts of the plastic wave. The measured velocity $U$ of the cordite projectile differs from the value of $u$ at the impact end of the cylinder by the very small velocity of the end of the steel bar. This difference will be neglected, and considering the point of impact, equation (4) may be written

$$U=\int_0^{\epsilon_1}\sqrt{\left(\frac{1}{\rho_0}\frac{dS}{d\epsilon}\right)}\,d\epsilon. \tag{5}$$

The observations give the relationship between $S_1$ and $U$. Equation (5) can therefore be used for determining the corresponding values of $\epsilon_1$. Writing this equation in the form

$$\frac{1}{\rho_0}\frac{dS_1}{d\epsilon_1}=\left(\frac{dU}{d\epsilon_1}\right)^2$$

or

$$\frac{dU}{d\epsilon_1}=\frac{1}{\rho_0}\frac{dS_1}{d\epsilon_1}\frac{d\epsilon_1}{dU}=\frac{1}{\rho_0}\frac{dS_1}{dU},$$

$$\epsilon_1=\rho_0\int_0^U\frac{dU}{dS_1}\,dU. \tag{6}$$

Using the experimentally determined relationship between $S_1$ and $U$, the relationship between $S_1$ and $\epsilon_1$ can be determined by integrating equation (6) numerically.

* See references in footnote †, p. 482.

31-2

To determine the true stress, $\sigma$, it is necessary also to know the area of contact, $a_1$, between the specimen and the pressure bar during impact. This was measured by smearing the end of the cordite projectile with grease; during the impact the grease is pressed between the cordite and the steel, and a well-defined circular patch of grease is left on the steel surface. The area of this patch can be determined with good accuracy by measuring its diameter with a low-power measuring microscope.

As we have previously stated, the quantity which is measured in the experiments is the pressure $p$ in the elastic wave in the bar due to the impact of the cordite projectiles; the value of the nominal stress $S_1$ is derived from $p$ by the relationship

$$S_1 = pA/a_0, \tag{7}$$

where $A$ is the area of cross-section of the pressure bar.

## 2. Description of the Apparatus and Experimental Procedure

### (1) *The electrical method*

Fig. 1 is a sketch in elevation of the general arrangement of the apparatus for the electrical method. It consists of

($A$) pneumatic gun with a smooth barrel,

($B$) directing cone to ensure that the cordite cylinder fired from ($A$) strikes the centre of the pressure bar ($C$) normally,

($C$) steel pressure bar suspended ballistically by two bifilar suspensions, each 7 ft. long,

($D$) throw indicator to measure the ballistic throw of the pressure bar,

($E$) condenser unit, consisting of an insulated brass disc, mounted elastically, and forming a parallel plate condenser with the end of the bar,

($F$) ring switch, which starts the time sweep and intensifying circuits when the switch is broken by the arrival of the elastic wave.

The recording of the displacement-time curve of the end of the bar is carried out photographically with a cathode-ray oscillograph and as shown schematically in the diagram, the oscillograph is connected to

(i) the condenser unit $E$, through the amplifier and the feed unit,

(ii) the ring switch $F$, through the sweep unit,

(iii) a radio-frequency oscillator which gives a timing wave of known frequency.

The following sections describe the various parts of the apparatus in greater detail.

#### (A) *The pneumatic gun*

The cordite cylinders are projected from the pneumatic gun $A$ whose air chamber $G$ consists of a brass tube, 4 in. diameter, $\frac{1}{32}$ in. wall, and $6\frac{1}{2}$ in. long, into which are soldered two discs of brass; the forward disc is fitted with six $\frac{1}{4}$ in. bolts and a central hole, $1\frac{1}{4}$ in. diameter, is bored in it. The barrel $H$ is a brass tube $\frac{1}{2}$ in. internal diameter, and about 12 in. long; it is soldered to a brass cylinder $I$, $1\frac{1}{2}$ in. long and $1\frac{3}{4}$ in.

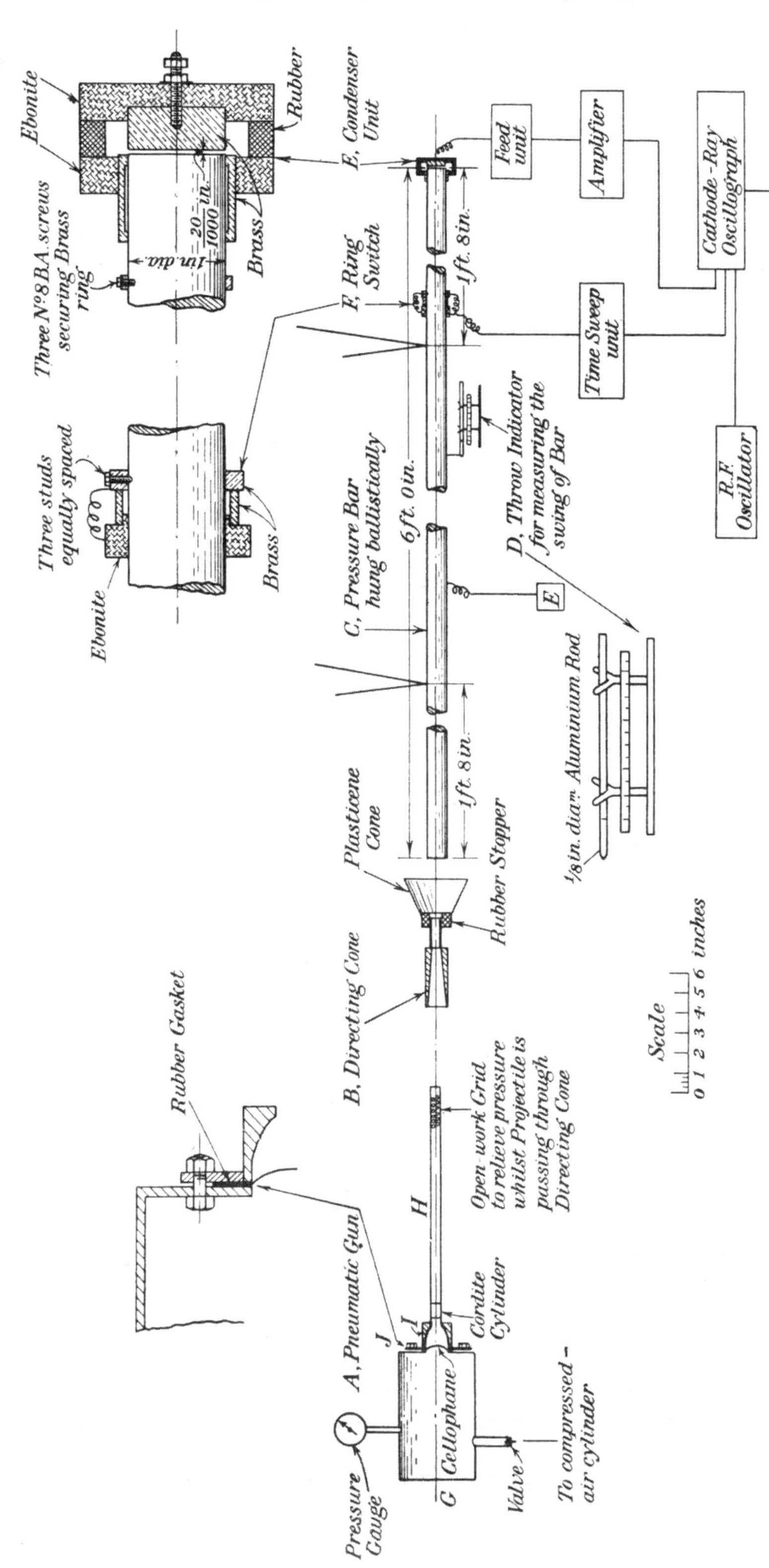


Fig. 1. Apparatus used in the electrical method.

external diameter, which is in turn soldered to a brass disc, $J$, 4 in. diameter, with a central hole, $1\frac{1}{4}$ in. diameter. As the diagram shows, the internal surface of $I$ is faired from the internal diameter of the barrel to the diameter of the hole in $J$.

A thin rubber gasket, 3 in. diameter with a central hole $1\frac{1}{2}$ in. diameter, and one or more circular pieces of cellophane sheet, $2\frac{1}{2}$ in. diameter and about 0·001 in. thick, are placed between the forward end-plane of the air chamber and the disc $J$; these are then clamped together by nuts on the bolts fitted on the end-plate so as to form an air-tight joint.

The air chamber is connected by means of a valve and reinforced rubber tubing to a cylinder containing compressed air; the pressure of the air in the chamber at any instant can be read on a Bourdon gauge.

The cordite cylinders which are used as projectiles are carefully turned so that their diameter is uniform and is about 0·003 in. less than the diameter of the barrel. In order to project a cylinder, it is placed in the position shown in the diagram and air is admitted to the air chamber by opening the valve connecting the chamber to the reservoir. As the pressure in the chamber increases, the cellophane sheet distends and then bursts suddenly at a pressure determined by the thickness of the cellophane and the number of pieces used; the air valve is shut immediately the cellophane bursts. The cordite cylinder is accelerated along the barrel under the action of the air in the chamber until it reaches the open grid-work near the forward end of the barrel. The function of this grid is to allow the air behind the projectile to expand laterally and so prevent any disturbance of the motion of the projectile in its later stages.

Let $P$ = gauge pressure (i.e. the difference between the actual pressure and atmospheric pressure) of the air in the chamber when the cellophane diaphragm bursts,

$M$ = mass of the projectile,

$L$ = the effective length of the gun barrel, i.e. the distance through which the projectile moves under the action of the pressure $P$,

$a_0$ = the area of cross-section of the projectile.

Assuming that there is no friction between the projectile and the barrel, and that there is no appreciable leakage of air past the projectile, the acceleration of the projectile is $Pa_0/M$, and the velocity $U$ with which it leaves the barrel is thus

$$U = \sqrt{(2Pa_0L/M)}. \tag{8}$$

Experiment shows (see section (B) below) that this equation holds to within 10 % in the range of velocities covered in the present experiments.

The value of $U$ in a typical case is derived thus: in the majority of the experiments, $L = 9·78$ in. $= 24·8$ cm., $a_0 = 1·25$ sq.cm.; for a diaphragm consisting of a single thickness of cellophane, $P$ is about 27 lb. per sq.in., i.e. $1·86 \times 10^6$ dynes per sq.cm., with a cordite cylinder about 3·57 cm. long, weighing 7 g., $U$ will therefore be

$$\sqrt{(2 \times 1·86 \times 10^6 \times 24·8/7)} = 3650 \text{ cm. per sec.} = 119 \text{ ft. per sec.}$$

For cylindrical projectiles of a given material, $U$ can be altered by varying either (i) $P$ (by using cellophane sheets of different thicknesses, or by varying the number of sheets used), or (ii) $M/a_0$ (by altering the length of the cylinder), or (iii) $L$ (by replacing the barrel by one of different length).

In our experiments, all these methods were used for varying $U$, and values lying between 100 and 450 ft. per sec. were obtained.

(B) *The directing cone*

When the projectile leaves the barrel of the pneumatic gun, it is followed by a certain amount of air blast even when the forward end is pierced with vent holes. On the one hand, this blast may upset the motion of the projectile, causing it to strike the end of the pressure bar inaccurately; on the other hand, the blast will falsify determinations of momentum made by using the pressure bar ballistically. These two sources of error are avoided by placing the directing cone, $B$, shown in fig. 1, between the gun and the pressure bar, and then carefully aligning the apparatus.

This cone consists of two brass cylinders which are soldered together and machined so that the internal surface consists of a truncated cone followed by a cylinder whose internal diameter is equal to the internal diameter of the gun barrel ($\frac{1}{2}$ in.). The axial length of the truncated cone is 3 in., the diameter of the larger end is 1 in., and the diameter of the smaller end is equal to the internal diameter of the cylindrical portion; the length of the cylindrical portion is $1\frac{1}{2}$ in. The rubber stopper and the cone of plasticene shown in the diagram are used to avoid damage to the cordite cylinder when it rebounds after striking the bar.

Table 1. *Area of cross-section of the gun barrel* = 1·25 *sq.cm.*

| | | | |
|---|---|---|---|
| Effective length of gun barrel ($L$) (cm.) | 50·5 | 50·5 | 24·8 |
| Mass of projectile ($M$) (cm.) | 2·78 | 2·78 | 5·58 |
| Bursting pressure ($P$) (lb. per sq.in.) | 50 | 42 | 29·5 |
| Velocity of projectile ($U$): | | | |
| (1) by ballistic exp. (cm. per sec.) | $1{\cdot}23\times10^4$ | $0{\cdot}98\times10^4$ | $0{\cdot}51\times10^4$ cm/sec. |
| (2) from eq. (1) (cm. per sec.) | $1{\cdot}25\times10^4$ | $1{\cdot}14\times10^4$ | $0{\cdot}47\times10^4$ cm/sec. |

The cylindrical portion of the directing cone ensures that the projectile strikes the end-surface of the pressure bar accurately, and experiment shows that when the apparatus is carefully set up, the centre of the area of impact of the cylinder is within less than 0·05 in. of the centre of the surface of the pressure bar. The cone also helps to reduce the disturbing effect of the air blast on the ballistic measurements, since it allows the gun to be placed at a different distance from the pressure bar to enable the blast to expand laterally so as to give no forward momentum to the bar.

In the course of our work, a number of proof experiments were carried out in which the velocity of various cordite projectiles was calculated by means of equation (8) and determined by using the pressure bar as a ballistic pendulum. In these experiments, the projectile was caught in a piece of plasticene which was attached to the near end of the bar. The results of some of these experiments are shown in table 1.

These results show that the combined effect of frictional losses, etc., in the gun barrel and directing cone, of air blast from the gun, and of imperfect alignment of the apparatus, is to introduce a discrepancy of about 10 % between the values of $U$ calculated from equation (8) and those determined ballistically. A discrepancy of the same order was found in earlier experiments in which the projectile—in this case, a steel ball—was fired directly into a ballistic pendulum, the effect of blast being eliminated by firing the ball through a paper disc mounted on the axle of an electric motor revolving at 3000 r.p.m. In this arrangement, the resistance of the paper to the passage of the ball was negligibly small, and at the same time, the revolving paper, on account of its dynamical rigidity, prevented the blast from disturbing the pendulum.

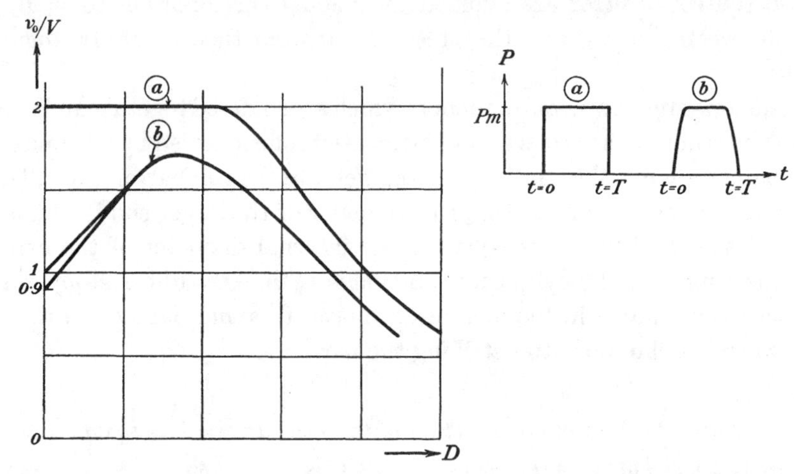


Fig. 2. The variation of the ratio of the velocity of projection $v_0$ of the indicator balls to the maximum velocity $V$ of the surface with the diameter $D$ of the balls. Inset: the relationship between pressure $p$ and time $t$.

(C), (D), (E), (F) *The steel pressure-bar and associated apparatus**

(2) *The mechanical method*

In experiments made by this method, a steel pressure-bar, 1 in. diameter, 2 ft. 2 in. long, is supported vertically, using rubber to prevent transmission of elastic waves from the bar to its supports. The two end-surfaces of the bar are ground flat so that they lie at right angles to the axis of the bar. The pneumatic gun, described in connection with the electrical method, is supported so that cordite cylinders are fired vertically at the centre of the lower surface of the pressure bar. No momentum measurements are made; the directing cone is therefore omitted from the arrange-

* *Editor's note.* The pressure bar will not be described here, since a full account can be found in the paper by R. M. Davies, 'A critical study of the Hopkinson pressure bar', *Phil. Trans.* A, CCXL (1948), 375–457.

ment, accurate aiming being secured by placing the upper end of the gun barrel as near as possible to the lower end of the bar.

The upper horizontal surface of the pressure bar is lapped and four steel balls, of the type used in ball bearings, are arranged on a diameter of the surface; the balls and the surface are carefully cleaned, first by washing with carbon tetrachloride to remove grease and finally polishing with Selvyt cloth. The four steel indicator balls and the upper portion of the bar are illuminated by a source of light, suitably placed.

When the cordite cylinder strikes the lower end of the bar, an elastic wave travels up the bar and produces a motion of the upper surface of the bar; the steel balls are then projected upwards like the short cylindrical end-pieces in the Hopkinson bar experiment. The velocity of projection of the balls is determined from the height of their first jump, which is found from photographs taken with a stationary plate-camera focused on the balls. The camera shutter is opened just before the cordite projectile is fired and closed about half a second afterwards; this exposure is usually sufficient to allow the balls to jump up and rebound several times.

Some time ago we investigated the factors which determine the velocity of projection of the indicator balls in an experiment of this type. It was found that, for a given impact and a given diameter of the indicator balls, the velocity of projection was independent of the position of the balls on the upper surface and of the length of the pressure bar, provided that this length exceeded twice the length of the elastic wave produced in the bar by the impact. When this condition is satisfied the relation between the diameter $D$ and the velocity of projection, $v_0$, of the indicator balls depends on the relation between pressure $p$ and time $t$ in the elastic wave in the bar. Two distinct cases have to be considered:

*Case* ($a$). The value of $p$ increases instantaneously from zero to a constant value $p_m$ and then after a time $T$ decreases instantaneously to zero, as shown in inset ($a$) of fig. 2. The relationship between $v_0$ and $D$ is shown in the main curve ($a$) of this figure, in which $D$ is plotted as abscissa and the ratio of $v_0$ to the maximum velocity $V$ of the surface as ordinate ($V = 2p_m/\rho c$). For small values of $D$, $v_0/V$ is constant and equal to 2; when $D$ increases beyond a certain value, $v_0/V$ decreases with increasing $D$.

*Case* ($b$). If, as in the impact of a lead bullet, $p$ takes a finite time to rise from zero to its maximum value $p_m$ and then to decrease to zero, as shown in inset ($b$) of fig. 2, then the relation between $v_0/V$ and $D$ is of the form shown in the main curve ($b$) of the figure. For small values of $D$, $v_0/V$ increases linearly with $D$; for larger values of $D$, $v_0/V$ increases more slowly, then attains a maximum value, and finally decreases with increasing $D$. The slope of the initial portion of the curve and the maximum value of $v_0/V$, depend on the time taken by the pressure to reach its maximum value. As this time decreases and the limiting case of an instantaneous rise is approached, the slope of the initial portion of the curve increases whilst the maximum becomes flatter and the height of the maximum ordinate approaches the value two.

Let $v_e/V$ be the ordinate of the point of intersection of the initial portion of the $(v_0/V, D)$ curve with the axis of $v_0/V$; in the cases investigated, calculation and experiments with lead bullets show that $v_e/V$ lies between 1 and 0·9, the exact value depending on the shape of the initial, rising part of the $(p, t)$ curve. Thus in case (*b*) plotting the observed values of the velocity of projection $v_0$ against ball diameter $D$ and extrapolating to $D=0$ gives the velocity $v_e$ and, $V=v_e$, subject to what must be regarded as an unknown error of order 10 %.

The shape of the $(v_0, D)$ curve obtained in an experiment with indicator balls enables us to distinguish immediately between the two cases which have been considered. In experiments with gaseous detonation waves, balls ranging in diameter between $\frac{1}{16}$ and $\frac{5}{32}$ in. jumped to the same height (cf. curve (*a*), fig. 5), showing that in this case, the $(p, t)$ curve had a very sharp front. Experiments with lead bullets gave $(v_0, D)$ curves which were of the same form as curve (*b*) of fig. 5. In both cases, the value of $V$ can be determined by extrapolation to $D=0$; in case (*a*), $V=\frac{1}{2}v_e$, and in case (*b*), $V=v_e$. The value of the maximum pressure, $p_m$, in the elastic wave in the bar is then found from the relation

$$p_m = \tfrac{1}{2}\rho c V. \tag{9}$$

In the present experiments with cordite projectiles, the $(v_0, D)$ curves were of type (*b*). It was found that balls ranging in diameter from $\frac{1}{16}$ to $\frac{1}{8}$ in. gave points on the initial, linear portion of the $(v_0, D)$ curve. In one case the values of $v_0$ for $D=\frac{1}{8}$, $\frac{3}{32}$ and $\frac{1}{16}$ in. were 115, 99·5 and 83·3 cm. per sec. respectively, the figure for the $\frac{1}{16}$ in. balls being the average of the velocities (82·7, 83·9 cm. per sec.) for the two balls. By extrapolation, $v_e = 51{\cdot}5$ cm. per sec., and since $\rho_1 c$ for this particular pressure bar was $4{\cdot}10 \times 10^6$ g. cm.$^{-2}$ sec.$^{-1}$, $p_m = \frac{1}{2} \times 4{\cdot}10 \times 10^6 \times 51{\cdot}5 = 1{\cdot}06 \times 10^8$ dynes per sq. cm.

## 3. Experimental Results

Cordite is extruded in the form of long cylinders, and cylinders $1\frac{1}{2}$ in. long and $\frac{1}{2}$ in. diameter were supplied. These had been cut so that some had their axes parallel with the original axes of the material and some at right angles to them.

No great difference in the stress-strain relationship was found between the different batches. One of them (RNCF 3573, cut at right angles to the axis), however, withstood a nominal stress (load/original area) of over 5 tons per sq.in. without suffering any visible damage. The others developed small cracks at nominal stresses of the order 3 tons per sq.in. and broke in pieces longitudinally at stresses exceeding 5 tons per sq.in. This behaviour may be compared with a static test in which it was found that complete disintegration of the specimen occurred at a nominal stress of 0·67 tons per sq.in.

The experimental results obtained by both the electrical and the mechanical methods are summarized in table 2 and shown graphically in fig. 3; in this diagram the velocity of projection, $U$, of the cylinders is plotted as abscissa and the maximum nominal stress, $S_1$, in the specimen as ordinate, no distinction being made between the four batches of cordite.

In the cases where the specimens were damaged, the value of $S_1$ given in table 2 and shown in fig. 3 is the maximum value derived from the experimental data. This implies that these nominal stresses are stresses which, in many cases, can be sustained by cordite for short periods without causing damage; when applied for longer periods, these stresses may give rise to damage in varying degrees.

In fig. 3 the horizontal and vertical lines, drawn through the experimental points represent to scale the variation which would be expected on account of errors of

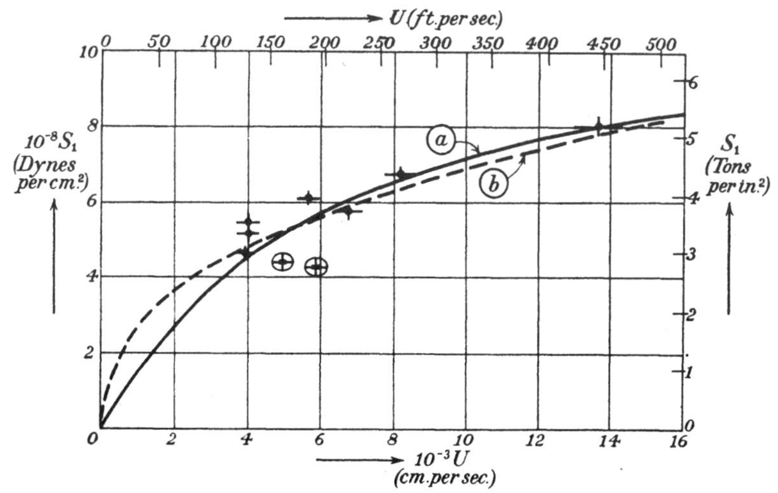


Fig. 3. The relation between the velocity of impact $U$ of the cordite projectiles and the maximum nominal stress $S_1$ in the projectile during impact. ✦, points obtained in the electrical experiments. ⊕, points obtained in the mechanical experiments. Curve $a$, $S_1 = \frac{10^8 U}{550 + 0{\cdot}085U}$. Curve $b$, $S_1 = 1{\cdot}57 \times 10^7\ U^{0{\cdot}41}$. ($S_1$ in dynes per cm²., $U$ in cm. per sec.)

Table 2

| Method | Specimen | $U$ (cm. per sec.) | $S_1$ (dynes per cm.²) | Damage |
|---|---|---|---|---|
| Electrical | C | $3{\cdot}97 \times 10^3$ | $4{\cdot}60 \times 10^8$ | Small crack |
| | A | 4·01 | 5·11 | Small crack |
| | B | 4·02 | 5·41 | Very small crack |
| | C | 5·65 | 6·09 | Small crack |
| | D | 6·76 | 5·79 | Undamaged |
| | A | 8·16 | 6·71 | Small crack |
| | B | 13·64 | 8·00 | Large crack |
| Mechanical | D | 4·98 | 4·38 | Undamaged |
| | D | 5·85 | 4·26 | Undamaged |

*Note.* A = 3468 cut parallel to the axis. B = 3468 cut at right angles to the axis. C = 3573 cut parallel to the axis. D = 3573 cut at right angles to the axis.

experiment, namely, ± 5 % in the velocity of projection $U$, ± 2 % in the nominal stress $S_1$ when measured electrically and ± 5 % when measured by the indicator-ball method; these estimates of the experimental error are based, for $U$, on the data given in table 1, and, for $S_1$, on the results of our previous experiments with gaseous explosions and lead bullets.

The scatter of the points in the diagram is greater than one would expect from our estimates of the experimental errors, and it may be due to variability in the cordite, or perhaps to temperature effects, since no special steps were taken to control temperature during the experiments, which were carried out in a laboratory where the average room temperature was 18° C. over the period of experiments. The points given by the mechanical method are lower than those deduced from the electrical experiments; this may be due to the uncertainty in the value of the ratio of $v_e$ to $V$ (*vide* § 2) which we have assumed to be unity.

This scatter prevents us from drawing a well-determined curve through the experimental points in order to find the stress-strain relationship by numerical integration of equation (6). We have therefore taken two different empirical expressions for $S_1$ in terms of $U$, which might equally well represent our results, and

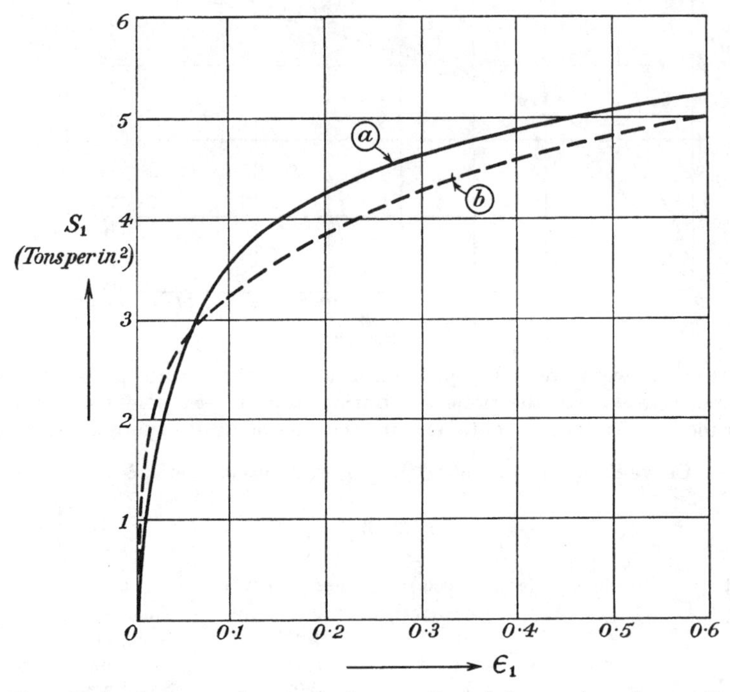


Fig. 4. The relation between the nominal stress $S_1$ and the strain $\epsilon_1$ for cordite during impact stressing. Curve $a$, derived from $S_1 = \dfrac{10^8 U}{550 + 0{\cdot}085U}$. Curve $b$, derived from $S_1 = 1{\cdot}57 \times 10^7 U^{0\cdot41}$.

we have deduced the $(S_1, \epsilon_1)$ relationship by applying equation (6) to each in turn. When $S_1$ is expressed in dynes per sq.cm. and $U$ in cm. per sec., the empirical relations are

$$S_1 = \frac{U}{b + cU}, \quad \text{where} \quad b = 5{\cdot}5 \times 10^{-6}, \quad c = 8{\cdot}5 \times 10^{-10},$$

$$= 10^8 U/(550 + 0{\cdot}085U), \tag{10}$$

and

$$S_1 = \beta U^\gamma, \quad \text{where} \quad \beta = 1{\cdot}57 \times 10^7, \quad \gamma = 0{\cdot}41$$

$$= 1{\cdot}57 \times 10^7 U^{0\cdot41}. \tag{11}$$

The relation (10) is shown by the continuous curve ($a$) in fig. 3 and the relation (11) by the broken-line curve ($b$).

Since
$$\epsilon_1 = \rho_0 \int_0^U \frac{dU}{dS_1} dU,$$

equation (10) gives
$$\epsilon_1 = \rho_0 \frac{b^2}{3c}\left[\left(1+\frac{cU}{b}\right)^3 - 1\right] \tag{12a}$$

or
$$\epsilon_1 = \rho_0 \frac{b^2}{3c}\left[\left(1+\frac{cS_1}{1-cS_1}\right)^3 - 1\right], \tag{12b}$$

and equation (11) gives
$$\epsilon_1 = \frac{\rho_0}{\beta\gamma(2-\gamma)} U^{2-\gamma}, \tag{13a}$$

or
$$\epsilon_1 = \frac{\rho_0}{\gamma(2-\gamma)} \frac{S_1^{(2-\gamma)/\gamma}}{\beta^{2/\gamma}}. \tag{13b}$$

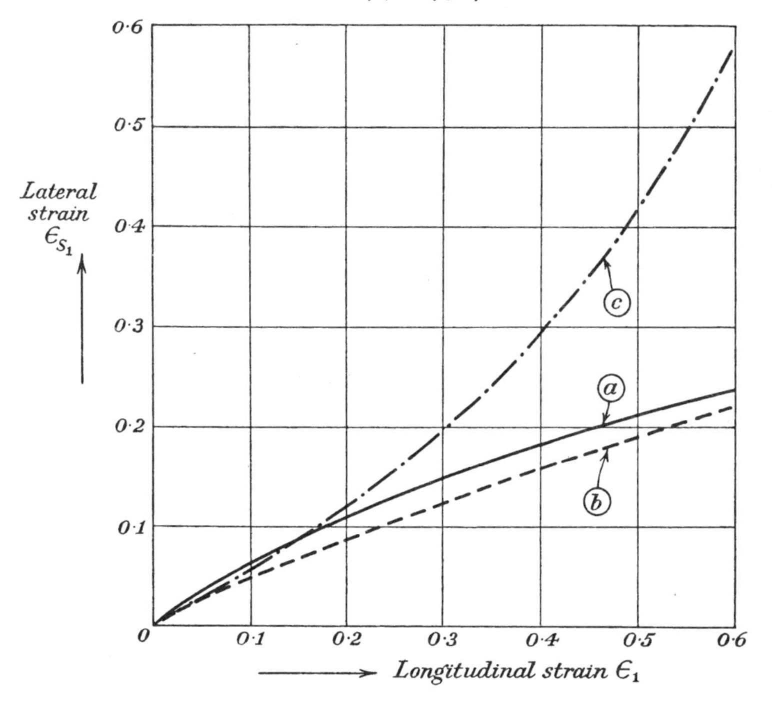


Fig. 5. The relation between lateral strain $\epsilon_{s_1}$ and longitudinal strain $\epsilon_1$. Curve $a$, experimental curve, assuming that $S_1 = \dfrac{10^8\, U}{550 + 0{\cdot}085U}$. Curve $b$, experimental curve, assuming that $S_1 = 1{\cdot}57 \times 10^7 U^{0{\cdot}41}$. Curve $c$, calculated curve for an incompressible material.

The average value of the density $\rho_0$ of the cordite used in our experiments was found to be 1·57 g. per c.c.; using this value of $\rho_0$ and the values of the constants $b$, $c$, $\beta$ and $\gamma$ given above, the stress-strain relationships derived from equations (12$b$) and (13$b$) are shown in fig. 4, the former by the continuous-line curve ($a$) and the latter by the broken-line curve ($b$).

It will be noticed that though the empirical relationships (10) and (11) differ considerably in form, there is not much difference between the final results as far as the $(S_1, \epsilon_1)$ curve is concerned. This curve is of the same form as that obtained with soft metals, such as copper; the scale of the curves in the two cases is, however, very different.

In order to obtain the true stress $\sigma_1$ (load/actual area), cordite cylinders were fired at a steel anvil and the maximum area of contact, $a_1$, or the radius, $r_1$, of the circle of contact of the cylinders and the anvil was measured as described in §1. It was found more convenient with the pressure bars used in this work to do this in a separate experiment; it can, however, be done at the same time as the main experiments if the cordite cylinders are fired, not at the end of the pressure bar, but at the front surface of a steel anvil whose back surface is ground flat and lapped, and wrung with a slight smear of grease to the surface of the pressure bar, which is similarly ground plane and lapped.

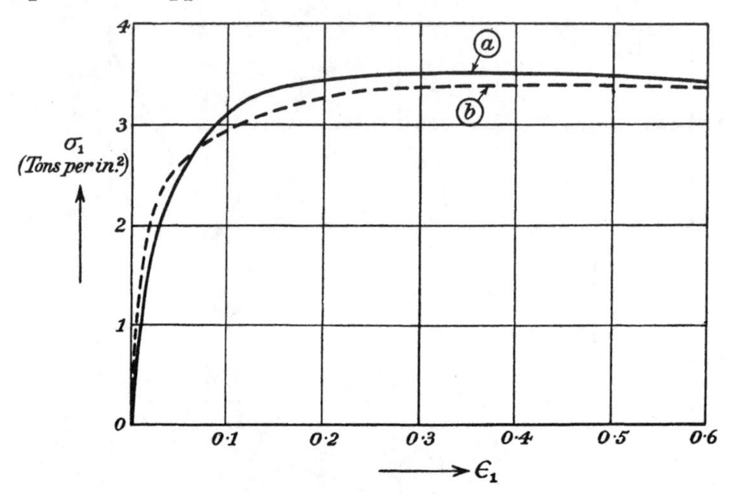


Fig. 6. The relation between the true stress $\sigma_1$ and the strain $\epsilon_1$ for cordite during impact stressing. Curve $a$, derived from $S_1 = \dfrac{10^8 U}{550 + 0\cdot085U}$. Curve $b$, derived from $S_1 = 1\cdot57 \times 10^7 U^{0\cdot41}$.

Let $a_0$, $r_0$ = original area and radius respectively of the cordite cylinders,

$$\epsilon_{s_1} = \frac{r_1}{r_0} - 1 = \text{lateral strain at the point of impact.}$$

These experiments give $r_1$ or $\epsilon_{s_1}$ as a function of $U$, whilst the longitudinal strain, $\epsilon_1$, is given in terms of $U$ by equations (12$a$) and (13$a$). We can therefore derive the relationship between the lateral and longitudinal strains; this is shown graphically in fig. 5, the continuous curve ($a$) being derived from equation (10) and the broken-line curve ($b$) from equation (11). For incompressible material, the relation between $\epsilon_{s_1}$ and $\epsilon_1$ is

$$\epsilon_{s_1} = \frac{1}{\sqrt{(1-\epsilon_1)}} - 1, \tag{14}$$

which is shown as a chain-dotted line in fig. 5 for purposes of comparison. It is seen that within the limits of experimental error, cordite behaves as an incompressible material for small strains and as a compressible material for larger strains; in this respect, its behaviour is similar to that of rubber.

For any given value of the strain $\epsilon_1$, the value of the nominal stress $S_1$ is given by equation (12*b*) and (13*b*) and by fig. 4, whilst the corresponding value of the lateral strain $\epsilon_{s_1}$ is given in fig. 5. Since $\sigma_1 = S_1 a_0/a_1$ and $a_1 = a_0(\epsilon_{s_1}+1)^2$, the value of $\sigma_1$ for any given strain $\epsilon_1$ can be calculated. The relationship between $\epsilon_1$ and $\sigma_1$ is shown in fig. 6 in which the continuous curve (*a*) gives the value of $\sigma_1$ derived from equation (10), and the broken-line curve from equation (11). It will be noticed that when the longitudinal strain exceeds 15 %, $\sigma_1$ attains a constant limiting value of about 3·4 tons per sq.in.

In impact experiments of this type, the maximum rate of strain of the projectiles, occurring at the impacting end at the moment of impact, is theoretically infinite. The mean rate of strain is equal to the ratio of the strain to half the total time of impact; the average value of this quantity in these experiments was about $3 \times 10^3$ per sec.

The most striking features of our results are: (1) the large impact stresses which cordite can sustain for short periods of the order $2 \times 10^{-5}$ sec. without breakdown, sometimes, in the case of one cylinder cut at right angles to the axis, for a period exceeding $10^{-4}$ sec.; these stresses are as much as seven times the static strength; (2) these stresses are accompanied by correspondingly large strains; for example, when a cordite projectile, $\frac{1}{2}$ in. long and $\frac{1}{2}$ in. diameter, is projected at 450 ft. per sec., the maximum nominal stress is about 5·2 tons per sq.in. and the maximum true stress about 3·5 tons per sq.in.; at the instant of maximum compression, the longitudinal and lateral strains are about 0·55 and 0·225, which means that the length of the specimen decreases from 0·5 to 0·225 in., whilst the diameter increases from 0·5 to 0·613 in.; (3) though the external dimensions, when measured as soon as practicable (say, 30 sec.) after impact, are indistinguishable from the original dimensions, yet the velocity of rebound is much less than that of impact.

# 34

# THE DISTORTION UNDER PRESSURE OF AN ELLIPTIC DIAPHRAGM WHICH IS CLAMPED ALONG ITS EDGE

Paper written for the Interdepartmental Co-ordinating Committee on Shockwaves, Ministry of Home Security (1942)

It will be assumed that the plastic material has the ideal property that it deforms by a negligible amount until the yield point is reached and that it subsequently flows plastically without further increase in stress. In other words, it will be assumed that no strain hardening occurs.

In general, the criterion for plastic flow is that some function of the stress components shall exceed a certain limit. Two such criteria have been extensively used, namely (i) that of Mohr, according to which the material flows when the maximum stress difference exceeds a certain limit, (ii) that of von Mises, according to which flow occurs when the sum of the squares of the principal stress differences exceeds a certain limit. If $\sigma_1$ and $\sigma_2$ are the principal stress components parallel to the surface of the diaphragm and both are positive (i.e. tensions), and if both are large compared with the stress perpendicular to the diaphragm, then the two criteria for plastic flow are:

$$\text{Mohr:}\quad \left.\begin{aligned} \sigma_1 &= P \quad \text{when} \quad \sigma_1 > \sigma_2, \\ \sigma_2 &= P \quad \text{when} \quad \sigma_2 > \sigma_1, \end{aligned}\right\} \tag{1}$$

$$\text{Von Mises:}\quad \sigma_1^2 + \sigma_2^2 - \sigma_1\sigma_2 = P^2, \tag{2}$$

where $P$ is the tensile strength as measured in a testing machine.

## Circular Diaphragm

If $\sigma_1$ is the radial stress and $\sigma_2$ the tangential stress, the equation of equilibrium of an element in the radial direction is

$$r\frac{d\sigma_1}{dr} + \sigma_1 - \sigma_2 = 0, \tag{3}$$

where $r$ is the radial co-ordinate. In the centre of the diaphragm symmetry alone requires that $\sigma_1 = \sigma_2 = P$. If Mohr's criterion is accepted three alternatives are possible, either

(i) $\sigma_2 = P$, $\sigma_1 < P$. This must be rejected because it is inconsistent with (3).

(ii) $\sigma_2 < P$, $\sigma_1 = P$. In this case (3) gives $\sigma_1 = \sigma_2$.

(iii) $\sigma_1 = \sigma_2 = P$.

Thus only the third alternative is possible. Similar reasoning gives the same result if von Mises's criterion is accepted so that in either case $\sigma_1 = \sigma_2$ and the stress distribution is like that of a soap film or stretched membrane. The sheet therefore assumes a spherical form under the action of a uniform pressure $p$. If $h_0$ is the displacement of the centre the displacement perpendicular to the plane of the edge at radius $r$ is

$$z = h_0(1 - r^2/r_1^2), \tag{4}$$

where $r_1$ is the radius of the diaphragm. The pressure necessary to produce the displacement is

$$p = 4Pth_0/r_1^2, \tag{5}$$

where $t$ is the thickness of the sheet.

The work $\delta W$ done on the material by the pressure during the displacement of the centre from $h_0$ to $h_0 + \delta h_0$ is

$$\delta W = \int_0^r 2\pi pr \left[\frac{dz}{dh}\delta h_0\right] dr. \tag{6}$$

Substituting for $p$ and $z$ from (4) and (5) and integrating with respect to $r$ and $h_0$ the total work done is

$$W = \pi Pth_0^2. \tag{7}$$

Since $\pi r_1^2$ is the area of the diaphragm the average work done per c.c. of the material of the plate is

$$E_p = Ph_0^2/r_1^2. \tag{8}$$

## Distribution of Plastic Strain in the Diaphragm

Though (8) represents the average amount of work done over the whole area of the diaphragm the distribution of strain, and therefore the work done per c.c., is far from uniform. If $\zeta$ is the radial component of displacement parallel to the initial plane of the diaphragm the two components of strain are

$$\left.\begin{aligned} &\text{radial:} && \epsilon_1 = \frac{d\zeta}{dr} + \frac{1}{2}\left(\frac{dz}{dr}\right)^2, \\ &\text{tangential:} && \epsilon_2 = \zeta/r. \end{aligned}\right\} \tag{9}$$

Since $\sigma_1 = \sigma_2$ symmetry ensures that $\epsilon_1 = \epsilon_2$. The equation for $\zeta$ is therefore

$$\frac{d\zeta}{dr} + \frac{1}{2}\left(\frac{dz}{dr}\right)^2 = \frac{\zeta}{r}, \tag{10}$$

and substituting for $z$ from (4) this becomes

$$\frac{d\zeta}{dr} - \frac{\zeta}{r} + \frac{2h_0^2 r^2}{r_1^4} = 0. \tag{11}$$

The solution of (11) which corresponds with $\zeta = 0$ at $r = r_1$ is

$$\zeta = h_0^2\left(\frac{r}{r_1^2} - \frac{r^3}{r_1^4}\right). \tag{12}$$

The principal strains $\epsilon_1$ and $\epsilon_2$ are each equal to

$$\frac{\zeta}{r} = \frac{h_0^2}{r_1^2}\left(1 - \frac{r^2}{r_1^2}\right). \tag{13}$$

It appears therefore that the strain is zero at the edge and equal to $h_0^2/r_1^2$ at the centre. The strain at any point in the surface of the sheet is proportional to the displacement of that point from its initial position in a direction perpendicular to the sheet, at any rate so far as any one state of the diaphragm is concerned. When the strains at a given point in the diaphragm are compared at various stages of displacement they are proportional to $h_0^2$.

The mean displacement is

$$\bar{z}=\frac{\text{volume, } V\text{, between initial and final positions of diaphragm}}{\text{area, } A\text{, of diaphragm}}.$$

For a circular diaphragm and small displacement

$$\bar{z}=\tfrac{1}{2}h_0. \tag{14}$$

## Elliptical and other Non-circular Diaphragms

Though the distortion of a non-circular diaphragm cannot be treated so simply as that of a circular one, an approximation to the displacement might be made by assuming that the diaphragm assumes the same form as a flat membrane of soap film when displaced by a uniform pressure perpendicular to its plane. There is, however, one non-circular shape for which the complete calculation of stress, strain and displacement can be made, namely, for elliptical diaphragms. This calculation is here carried through in order to estimate the error that may be expected if the assumption is made that the diaphragm is displaced into the same form as a membrane with uniform tension in all directions.

Taking the equation for the edge of the elliptic diaphragm or plate as

$$\frac{x^2}{a^2}+\frac{y^2}{b^2}=1, \tag{15}$$

where $2a$ is the major axis, it will be assumed that the equation for the displaced sheet is

$$\frac{z}{h_0}=1-\frac{x^2}{a^2}-\frac{y^2}{b^2}, \tag{16}$$

and it will be shown that this assumption is consistent with the satisfaction of all the required plastic stress, strain and equilibrium conditions. It will be assumed provisionally that the stress is uniform at all points but not isotropic, thus $\sigma_x$ and $\sigma_y$ will be taken as the components of stress parallel to the surface of the plate. The condition of equilibrium in direction normal to the surface is

$$p=t\left(\frac{\sigma_x}{\rho_x}+\frac{\sigma_y}{\rho_y}\right), \tag{17}$$

where $t$ is the thickness of the plate and $\rho_x$, $\rho_y$ are the principal radii of curvature. For a sheet of the form (16) $1/\rho_x=2h_0/a^2$, $1/\rho_y=2h_0/b^2$, so that

$$p=2h_0t\left(\frac{\sigma_x}{a^2}+\frac{\sigma_y}{b^2}\right). \tag{18}$$

Since (18) does not contain $x$ or $y$ it can be satisfied if, as has been assumed, $\sigma_x$ and $\sigma_y$ are independent of $x$ and $y$. It will be noticed that (16) shows that the displaced plastic sheet is in fact of the same shape as a displaced soap film. The fact that $\sigma_x = \sigma_y$ in the soap film, but not in the present case, makes no difference to the form of (16).

It now remains to find out whether a distribution of displacements and strains in the surface of the sheet can be found which is consistent with the normal displacement (16) and at the same time allows the plastic stress-strain relations to be satisfied.

## Plastic Stress-strain Relationships

Assuming that $\sigma_x/\sigma_y$ is constant over the ellipse and equal to $\alpha$, it seems clear that, whatever the stress-strain relationships may be, the ratio of the strain components in the surface of the plate, namely, $\epsilon_x$ and $\epsilon_y$, must also be constant. Taking $\epsilon_x/\epsilon_y = \beta$ the experimental law of plasticity will define the relationship between $\alpha$ and $\beta$. Investigations by Lode* and by Taylor and Quinney† by different experimental methods give results which are in good agreement. They have been expressed in terms of Lode's variables

$$\left.\begin{aligned} \mu &= 2\left(\frac{\sigma_2-\sigma_3}{\sigma_1-\sigma_3}\right)-1, \\ \nu &= 2\left(\frac{\epsilon_2-\epsilon_3}{\epsilon_1-\epsilon_3}\right)-1. \end{aligned}\right\} \tag{19}$$

where $\sigma_1, \sigma_2, \sigma_3, \epsilon_1, \epsilon_2, \epsilon_3$ are the principal stresses and strains and $\sigma_1 > \sigma_2 > \sigma_3$. The experimental relationship between $\mu$ and $\nu$ is shown in fig. 10 of Taylor and Quinney's paper.

In the present case where the strains $\sigma_1$ and $\sigma_2$ are in the plane of the sheet we may take $\sigma_3 = 0$. Since it will be found that $\sigma_y > \sigma_x$ it is necessary to take $\sigma_x$ and $\sigma_2$ and $\sigma_y$ as $\sigma_1$. In this case therefore

$$\mu = 2\alpha - 1. \tag{20}$$

Since the material may be taken as incompressible to the degree of approximation here required

$$\epsilon_1 + \epsilon_2 + \epsilon_3 = 0, \tag{21}$$

and substituting this in Lode's variable $\nu$

$$\nu = 3\beta/(\beta+2). \tag{22}$$

Some values of $\mu$ and the corresponding experimental values of $\nu$ taken from fig. 10 of Taylor and Quinney's paper are given in columns 1 and 3 of table 1. The values of $\alpha$ from (20) are given in column 2. Values of $\beta$, found by using the values of $\nu$ given in column 3 in (22), are given in column 4. This experimental relationship between $\alpha$ and $\beta$ is shown in fig. 1.

* Lode, *Z. Phys.* XXXVI (1926), 913.

† Taylor and Quinney, 'The plastic distortion of metals', *Phil. Trans.* A, CCXXX (1930), 323; paper 16 above.

32-2

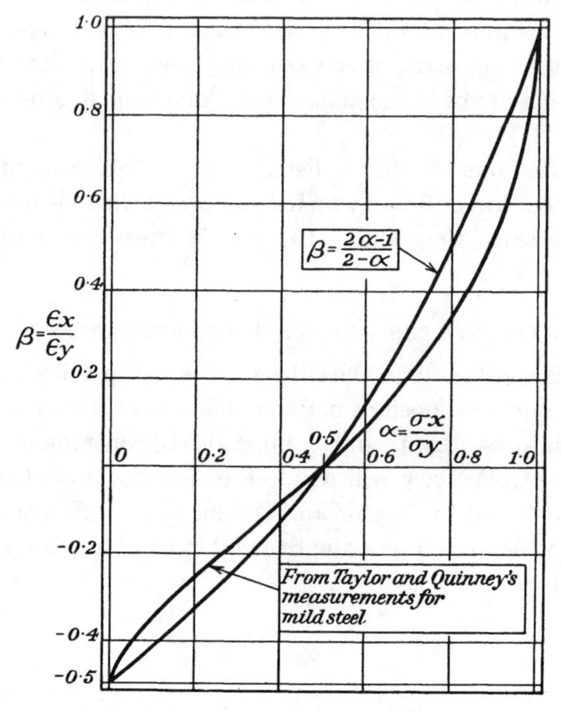


Fig. 1

Table 1

| 1 | 2 | 3 | 4 |
|---|---|---|---|
| $\mu$ | $\alpha$ | $\nu$ (obs.) | $\beta$ (from col. 3 using (22)) |
| +1·0 | 1·00 | +1·000 | +1·000 |
| +0·9 | 0·95 | +0·770 | +0·691 |
| +0·8 | 0·90 | +0·635 | +0·536 |
| +0·7 | 0·85 | +0·535 | +0·434 |
| +0·6 | 0·80 | +0·440 | +0·343 |
| +0·5 | 0·75 | +0·360 | +0·273 |
| +0·4 | 0·70 | +0·280 | +0·206 |
| +0·3 | 0·65 | +0·195 | +0·139 |
| +0·2 | 0·60 | +0·120 | +0·083 |
| +0·1 | 0·55 | +0·050 | +0·034 |
| 0 | 0·50 | 0 | 0 |
| −0·1 | 0·45 | −0·050 | −0·033 |
| −0·2 | 0·40 | −0·120 | −0·077 |
| −0·3 | 0·35 | −0·195 | −0·122 |
| −0·4 | 0·30 | −0·280 | −0·171 |
| −0·5 | 0·25 | −0·360 | −0·214 |
| −0·6 | 0·20 | −0·440 | −0·256 |
| −0·7 | 0·15 | −0·535 | −0·302 |
| −0·8 | 0·10 | −0·635 | −0·347 |
| −0·9 | 0·05 | −0·770 | −0·408 |
| −1·0 | 0·00 | −1·000 | −0·500 |

The only relevant *a priori* theory on this subject is that the ratios of the stress components in a plastic body are related to the ratios of strain components by the same relationship that applies to viscous fluids, namely $\mu = \nu$. From (20) and (22) it will be seen that, in terms of $\alpha$ and $\beta$, this would give

$$\beta = (2\alpha - 1)/(2 - \alpha). \tag{23}$$

This relationship is also shown in fig. 1. It has been used by von Mises and others in deriving theoretical solutions of problems in plasticity.

## Strain and Displacement of Particles

The form of the expression (12) for the radial displacement of particles in a circular diaphragm suggests the possibility that a solution may be found by assuming for the two components of displacement $\xi$, $\eta$ parallel to the axes $x$ and $y$ the forms

$$\left.\begin{aligned} \xi &= Ax(1 - x^2/a^2 - y^2/b^2), \\ \eta &= By(1 - x^2/a^2 - y^2/b^2). \end{aligned}\right\} \tag{24}$$

This assumption satisfies the condition that $\xi = \eta = 0$ at the edge of the plate. The components of strain in the surface of the distorted plate are then

$$\left.\begin{aligned} \epsilon_x &= \frac{\partial \xi}{\partial x} + \frac{1}{2}\left(\frac{\partial z}{\partial x}\right)^2 = A\left(1 - \frac{3x^2}{a^2} - \frac{y^2}{b^2}\right) + \frac{2h_0^2 x^2}{a^4}, \\ \epsilon_y &= \frac{\partial \eta}{\partial y} + \frac{1}{2}\left(\frac{\partial z}{\partial y}\right)^2 = B\left(1 - \frac{x^2}{a^2} - \frac{3y^2}{b^2}\right) + \frac{2h_0^2 y^2}{b^4}. \end{aligned}\right\} \tag{25}$$

Since it has been assumed that $\sigma_x/\sigma_y = \alpha$ is independent of $x$ and $y$, the unique experimental relationship between $\alpha$ and $\beta$ ensures that $\beta = \epsilon_x/\epsilon_y$ shall be independent of $x$ and $y$. Substituting for $\epsilon_x$ and $\epsilon_y$ from (25) in the equation $\epsilon_x = \beta\epsilon_y$, the condition that $\beta$ may be independent of $x$ and $y$ is found by equating the constant terms and the coefficients of $x^2$ and $y^2$. Hence

$$\left.\begin{aligned} &A = \beta B, \\ &B(-3/a^2 + 1/a^2) + 2h_0^2/\beta a^4 = 0, \\ &B(-1/b^2 + 3/b^2) - 2h_0^2/b^4 = 0, \end{aligned}\right\} \tag{26}$$

and therefore

$$\left.\begin{aligned} B &= h_0^2/b^2, \\ A &= h_0^2/a^2, \\ \beta &= b^2/a^2. \end{aligned}\right\} \tag{27}$$

Having determined $\beta$ from (27) and hence $\alpha$ from the curve in fig. 1, $\sigma_x$ and $\sigma_y$ can be connected with $P$, the testing machine measured yield stress, by the formulae

$$\sigma_x = P\alpha, \quad \sigma_y = P \quad \text{if Mohr's theory is used} \tag{28}$$

or

$$\sigma_x = P\alpha/\surd(1 - \alpha + \alpha^2), \quad \sigma_y = P/\surd(1 - \alpha + \alpha^2) \quad \text{according to von Mises's theory.} \tag{29}$$

This completes the solution of the problem, all the necessary conditions being satisfied. Substituting from (28) or (29) in (18) the following expressions are derived connecting the pressure, the maximum displacement $h_0$ and the yield stress:

$$p=\frac{h_0tP}{a^2}(\alpha+1/\beta) \quad \text{according to Mohr's theory,} \tag{30}$$

or

$$p=\frac{h_0tP}{a^2}(\alpha+1/\beta)(1-\alpha+\alpha^2)^{-\frac{1}{2}} \quad \text{according to von Mises's theory.} \tag{31}$$

If the stress is assumed uniform, as in a soap film, $\sigma_x=\sigma_y=P$ so that

$$p=\frac{h_0tP}{a^2}(1+1/\beta). \tag{32}$$

For the case when $b/a=\frac{2}{3}$, $\beta=\frac{4}{9}$ and the curve of fig. 1 gives $\alpha=0{\cdot}86$. In this case the factors multiplying $h_0tP/a^2$ in (30), (31) and (32) are 3·11, 3·31 and 3·25 respectively. The assumption that the plate behaves like a membrane with uniform tension equal to $P$ in all directions therefore *overestimates* the work which is necessary to produce a given displacement by 4·5 % if Mohr's criterion is accepted. On the other hand, this assumption *underestimates* the work by 1·8 % if von Mises's strain-energy criterion is used.

As the length of the ellipse increases the error increases if von Mises's criterion is used. It is 6 % when $a/b=2$, 10 % when $a/b=3$ and 15 % when $a$ becomes infinite.

# 35

# STRESS SYSTEMS IN AEOLOTROPIC PLATES PART III*

REPRINTED FROM

*Proceedings of the Royal Society*, A, vol. CLXXXIV (1945), pp. 181–95

The stress distribution round a circular hole in an infinite aeolotropic plate subjected to tension in one direction is found theoretically in the case when the material of the plate has two directions of symmetry at right angles to one another. Numerical work is carried out using the elastic constants found in experiments made with specimens cut from the highly aeolotropic woods spruce and oak. An attempt is made to apply the calculated stress concentrations in conjunction with measurements of ultimate strength to determine the kind of failure that might be expected near a hole in a highly stressed wood plate.

## INTRODUCTION

1. In previous papers,† fundamental stress functions were obtained for problems of generalized plane stress in a plate of aeolotropic material which has two directions of symmetry at right angles, and some of these functions were used to find the stresses produced by isolated forces acting in the plane of the plate. Numerical work was carried out for the highly aeolotropic materials oak and spruce, and the results were found to be in striking contrast with those for isotropic materials.

In the present paper the fundamental stress functions are used to find the stress distribution in an infinite aeolotropic tension member which contains a circular hole. These stress functions which satisfy the equations of equilibrium and which produce single-valued expressions for the corresponding stresses and displacements, are combined in an infinite series so as to satisfy the boundary conditions. The resulting formal solution is not completely satisfactory owing to difficulties of convergence in some parts of the plate. It may, however, be modified and expressed in a finite form which represents the stress distribution in the whole of the plate. As in previous papers, numerical work is carried out for certain specimens of oak and spruce.

## FUNDAMENTAL EQUATIONS

2. Consider an elastic aeolotropic plate whose material has two directions of symmetry at right angles in the plane of the plate. The axes of $x$ and $y$ are taken to be parallel to these directions and the $z$-axis will then be normal to the plate. The plate is imagined to be in a state of generalized plane stress in which the stress

* With A. E. GREEN.

† A. E. Green and G. I. Taylor, *Proc. Roy. Soc.* A, CLXXIII (1939), 162, paper 29 above; A. E. Green, *Proc. Roy. Soc.* A, CLXXIII (1939), 173.

component $\widehat{zz}$ vanishes everywhere and the components $\widehat{xz}$ and $\widehat{yz}$ are zero at the surfaces of the plate. Then the *mean* stresses are given by

$$\widehat{xx}=\frac{\partial^2\chi}{\partial y^2}, \quad \widehat{yy}=\frac{\partial^2\chi}{\partial x^2}, \quad \widehat{xy}=-\frac{\partial^2\chi}{\partial x\,\partial y}, \tag{2·1}$$

or, in polar co-ordinates,

$$\widehat{rr}=\frac{1}{r^2}\frac{\partial^2\chi}{\partial\theta^2}+\frac{1}{r}\frac{\partial\chi}{\partial r}, \quad \widehat{r\theta}=-\frac{\partial}{\partial r}\left(\frac{1}{r}\frac{\partial\chi}{\partial\theta}\right), \quad \widehat{\theta\theta}=\frac{\partial^2\chi}{\partial r^2}, \tag{2·2}$$

and the *mean* values of the stresses and strains are related by the equations

$$e_{xx}=s_{11}\widehat{xx}+s_{12}\widehat{yy}, \quad e_{yy}=s_{12}\widehat{xx}+s_{22}\widehat{yy}, \quad e_{xy}=s_{66}\widehat{xy}. \tag{2·3}$$

As shown in a previous paper,* the stress function $\chi$ satisfies the boundary conditions and the equation

$$\left(\frac{\partial^2}{\partial x^2}+\alpha_1\frac{\partial^2}{\partial y^2}\right)\left(\frac{\partial^2}{\partial x^2}+\alpha_2\frac{\partial^2}{\partial y^2}\right)\chi=0, \tag{2·4}$$

where $$\alpha_1\alpha_2=s_{11}/s_{22}, \quad \alpha_1+\alpha_2=(s_{66}+2s_{12})/s_{22}, \tag{2·5}$$

and the analysis is confined to cases where $\alpha_1$ and $\alpha_2$ are real and positive.

Three sets of plane polar co-ordinates are now introduced by the equations

$$x+iy=re^{i\theta}, \quad x+iy/\alpha_1^{\frac{1}{2}}=r_1e^{i\theta_1}, \quad x+iy/\alpha_2^{\frac{1}{2}}=r_2e^{i\theta_2}. \tag{2·6}$$

Then $$r_1^{-2n}\cos 2n\theta_1, \quad r_2^{-2n}\cos 2n\theta_2, \quad \log r_1, \quad \log r_2, \tag{2·7}$$

where $n$ is a positive integer, are solutions of equation (2·4) which are symmetrical about both co-ordinate axes, and they give single-valued expressions for the corresponding stresses and displacements.* Expansions of these functions in terms of $r$ and $\theta$ will be needed and are*

$$\left.\begin{aligned}\frac{\cos 2n\theta_1}{r_1^{2n}}&=\frac{(1+\gamma_1)^{2n}}{r^{2n}}\sum_{s=n}^{\infty}\binom{n+s-1}{s-n}(-)^{s-n}\gamma_1^{s-n}\cos 2s\theta,\\ \log r_1&=\log r+\sum_{s=1}^{\infty}\frac{(-)^{s-1}\gamma_1^s}{s}\cos 2s\theta-\log(1+\gamma_1),\end{aligned}\right\} \tag{2·8}$$

where $$\gamma_1=\frac{\alpha_1^{\frac{1}{2}}-1}{\alpha_1^{\frac{1}{2}}+1}, \tag{2·9}$$

with similar expansions for $r_2^{-2n}\cos 2n\theta_2$ and $\log r_2$ which are got by changing the suffix 1 into 2. By using the formulae (2·2) the corresponding stresses can be obtained as series which are absolutely and uniformly convergent.

## The Tension Problem

3. The plate is now supposed to contain a circular hole of radius $a$ and the origin of co-ordinates is taken at the centre of the hole. A uniform tension $T$ is then applied to the plate at infinity parallel to the $x$-axis. If the hole were absent the stresses would be derived from a stress function

$$\chi_0=\tfrac{1}{2}Ty^2=\tfrac{1}{4}Tr^2(1-\cos 2\theta). \tag{3·1}$$

* Green and Taylor, loc. cit.

In order to allow for the effect of the hole suitable stress functions which give zero stresses at infinity are added to $\chi_0$ so that the complete stress function gives zero normal and shear tractions over the boundary of the hole. The stress system must be single-valued, symmetrical about both the co-ordinate axes, and must give single-valued expressions for the corresponding displacements. Thus, it is assumed that the complete stress system may be derived from the stress function

$$\chi=\chi_0+A_0\log r_1+B_0\log r_2+\sum_{n=1}^{\infty}\left\{\frac{A_{2n}\cos 2n\theta_1}{(1+\gamma_1)^{2n}r_1^{2n}}+\frac{B_{2n}\cos 2n\theta_2}{(1+\gamma_2)^{2n}r_2^{2n}}\right\}, \tag{3·2}$$

where $A_0, A_1, \ldots, B_0, B_1, \ldots$ are constants which are to be found from the boundary conditions at the edge of the hole. Using (2·8) the stress function (3·2) may be expressed in terms of $r$ and $\theta$ and then the corresponding stresses may be derived with the help of (2·2). The radial and tangential stresses are

$$\widehat{rr}=\tfrac{1}{2}T(1+\cos 2\theta)+\frac{A_0+B_0}{r^2}+\frac{4}{r^2}\sum_{s=1}^{\infty}(-)^s s(A_0\gamma_1^s+B_0\gamma_2^s)\cos 2s\theta$$
$$-\sum_{n=1}^{\infty}\sum_{s=n}^{\infty}\binom{n+s-1}{s-n}(4s^2+2n)(-)^{s-n}(A_{2n}\gamma_1^{s-n}+B_{2n}\gamma_2^{s-n})\frac{\cos 2s\theta}{r^{2n+2}}, \tag{3·3}$$

$$\widehat{r\theta}=-\tfrac{1}{2}T\sin 2\theta+\frac{2}{r^2}\sum_{s=1}^{\infty}(-)^s(A_0\gamma_1^s+B_0\gamma_2^s)\sin 2s\theta$$
$$-\sum_{n=1}^{\infty}\sum_{s=n}^{\infty}\binom{n+s-1}{s-n}(4sn+2s)(-)^{s-n}(A_{2n}\gamma_1^{s-n}+B_{2n}\gamma_2^{s-n})\frac{\sin 2s\theta}{r^{2n+2}}. \tag{3·4}$$

If it is assumed that the expression (3·3) for $\widehat{rr}$ may be written as a cosine series and that the expression (3·4) for $\widehat{r\theta}$ may be written as a sine series then the conditions for $\widehat{rr}$ and $\widehat{r\theta}$ to be zero when $r=a$ are

$$\left.\begin{aligned}&\tfrac{1}{2}T+\frac{A_0+B_0}{a^2}=0,\\&\tfrac{1}{2}T-\frac{4}{a^2}(A_0\gamma_1+B_0\gamma_2)-\frac{6}{a^4}(A_2+B_2)=0,\\&-\tfrac{1}{2}T-\frac{2}{a^2}(A_0\gamma_1+B_0\gamma_2)-\frac{6}{a^4}(A_2+B_2)=0,\end{aligned}\right\} \tag{3·5}$$

and

$$\left.\begin{aligned}&\frac{4}{a^2}(-)^s s(A_0\gamma_1^s+B_0\gamma_2^s)-\sum_{n=1}^{s}\frac{(-)^{s-n}}{a^{2n+2}}(4s^2+2n)\binom{n+s-1}{s-n}(A_{2n}\gamma_1^{s-n}+B_{2n}\gamma_2^{s-n})=0,\\&\frac{2}{a^2}(-)^s(A_0\gamma_1^s+B_0\gamma_2^s)-\sum_{n=1}^{s}\frac{(-)^{s-n}}{a^{2n+2}}2s(2n+1)\binom{n+s-1}{s-n}(A_{2n}\gamma_1^{s-n}+B_{2n}\gamma_2^{s-n})=0,\end{aligned}\right\} \tag{3·6}$$

for $s\geqslant 2$. These equations are satisfied by

$$\left.\begin{aligned}&A_{2n}=-\frac{Ta^{2n+2}\gamma_1^n(1+\gamma_2)}{2(\gamma_1-\gamma_2)}\frac{(2n-1)!}{n!\,(n+1)!},\quad B_{2n}=\frac{Ta^{2n+2}\gamma_2^n(1+\gamma_1)}{2(\gamma_1-\gamma_2)}\frac{(2n-1)!}{n!\,(n+1)!}\quad(n\geqslant 1),\\&A_0=\frac{Ta^2(1+\gamma_2)}{2(\gamma_1-\gamma_2)},\quad B_0=-\frac{Ta^2(1+\gamma_1)}{2(\gamma_1-\gamma_2)}.\end{aligned}\right\} \tag{3·7}$$

FURTHER DISCUSSION OF THE SOLUTION

4. A formal solution of the problem is given in the previous paragraph but it cannot be regarded as completely satisfactory because it is found that when $r=a$ the series (3·2) and the corresponding series (3·3) and (3·4) for the stresses do not converge for all values of $\theta$, and for all values of $\gamma_1$ and $\gamma_2$ which lie between 0 and 1. Before writing down the boundary conditions the order of summations in (3·3) and (3·4) was reversed, and closer investigation shows that the resulting cosine series for $\widehat{rr}$ and sine series for $\widehat{r\theta}$ do actually converge for all values of $\theta$, for all $r \geqslant a$ and for all values of $\gamma_1$ and $\gamma_2$ between 0 and 1. The series (3·2), when it converges, may be summed in finite terms and it will be seen that this sum gives a stress function which represents the complete solution of the problem for all the values of $\theta$ and $r$ that are needed. Moreover, this sum may be expanded as a convergent cosine series from which may be derived a convergent cosine series for $\widehat{rr}$ and a convergent sine series for $\widehat{r\theta}$, these being the same as those used above but which were there obtained by a method which was only valid for some values of $\theta$ and $r$.

Instead of summing the series (3·2) it is easier to sum the series for the stresses which are derived from (3·2). It will then be found that the stresses are the *real* parts of the following expressions:

$$\widehat{rr}=\frac{(1+\gamma_1)(1+\gamma_2)}{4\gamma_1\gamma_2}T\cos 2\theta + \frac{T}{4(\gamma_1-\gamma_2)}\left\{\frac{(1+\gamma_2)(1-\gamma_1e^{2i\theta})^2}{\gamma_1e^{2i\theta}\left[1-\dfrac{4a^2\gamma_1e^{2i\theta}}{r^2(1+\gamma_1e^{2i\theta})^2}\right]^{\frac{1}{2}}}-\frac{(1+\gamma_1)(1-\gamma_2e^{2i\theta})^2}{\gamma_2e^{2i\theta}\left[1-\dfrac{4a^2\gamma_2e^{2i\theta}}{r^2(1+\gamma_2e^{2i\theta})^2}\right]^{\frac{1}{2}}}\right\}, \quad (4·1)$$

$$\widehat{r\theta}=-\frac{(1+\gamma_1)(1+\gamma_2)}{4\gamma_1\gamma_2}T\sin 2\theta - \frac{iT}{4(\gamma_1-\gamma_2)}\left\{\frac{(1+\gamma_2)(1-\gamma_1^2e^{4i\theta})}{\gamma_1e^{2i\theta}\left[1-\dfrac{4a^2\gamma_1e^{2i\theta}}{r^2(1+\gamma_1e^{2i\theta})^2}\right]^{\frac{1}{2}}}-\frac{(1+\gamma_1)(1-\gamma_1^2e^{4i\theta})}{\gamma_2e^{2i\theta}\left[1-\dfrac{4a^2\gamma_2e^{2i\theta}}{r^2(1+\gamma_2e^{2i\theta})^2}\right]^{\frac{1}{2}}}\right\}, \quad (4·2)$$

$$\widehat{\theta\theta}=-\frac{(1+\gamma_1)(1+\gamma_2)}{4\gamma_1\gamma_2}T\cos 2\theta - \frac{T}{4(\gamma_1-\gamma_2)}\left\{\frac{(1+\gamma_2)(1+\gamma_1e^{2i\theta})^2}{\gamma_1e^{2i\theta}\left[1-\dfrac{4a^2\gamma_1e^{2i\theta}}{r^2(1+\gamma_1e^{2i\theta})^2}\right]^{\frac{1}{2}}}-\frac{(1+\gamma_1)(1+\gamma_2e^{2i\theta})^2}{\gamma_2e^{2i\theta}\left[1-\dfrac{4a^2\gamma_2e^{2i\theta}}{r^2(1+\gamma_2e^{2i\theta})^2}\right]^{\frac{1}{2}}}\right\}. \quad (4·3)$$

Referred to Cartesian axes the components of stress are the *real* parts of

$$\widehat{xx}=\frac{(1+\gamma_1)(1+\gamma_2)T}{4\gamma_1\gamma_2} + \frac{T}{4(\gamma_1-\gamma_2)}\left\{\frac{(1+\gamma_2)(1-\gamma_1)^2}{\gamma_1\left[1-\dfrac{4a^2\gamma_1e^{2i\theta}}{r^2(1+\gamma_1e^{2i\theta})^2}\right]^{\frac{1}{2}}}-\frac{(1+\gamma_1)(1-\gamma_2)^2}{\gamma_2\left[1-\dfrac{4a^2\gamma_2e^{2i\theta}}{r^2(1+\gamma_2e^{2i\theta})^2}\right]^{\frac{1}{2}}}\right\}, \quad (4·4)$$

$$\widehat{yy} = -\frac{(1+\gamma_1)(1+\gamma_2)T}{4\gamma_1\gamma_2}$$

$$-\frac{T}{4(\gamma_1-\gamma_2)}\left\{\frac{(1+\gamma_2)(1+\gamma_1)^2}{\gamma_1\left[1-\frac{4a^2\gamma_1 e^{2i\theta}}{r^2(1+\gamma_1 e^{2i\theta})^2}\right]^{\frac{1}{2}}} - \frac{(1+\gamma_1)(1+\gamma_2)^2}{\gamma_2\left[1-\frac{4a^2\gamma_2 e^{2i\theta}}{r^2(1+\gamma_2 e^{2i\theta})^2}\right]^{\frac{1}{2}}}\right\}, \quad (4\cdot5)$$

$$\widehat{xy} = -\frac{iT}{4(\gamma_1-\gamma_2)}\left\{\frac{(1+\gamma_2)(1-\gamma_1^2)}{\gamma_1\left[1-\frac{4a^2\gamma_1 e^{2i\theta}}{r^2(1+\gamma_1 e^{2i\theta})^2}\right]^{\frac{1}{2}}} - \frac{1(+\gamma_1)(1-\gamma_2^2)}{\gamma_2\left[1-\frac{4a^2\gamma_2 e^{2i\theta}}{r^2(1+\gamma_2 e^{2i\theta})^2}\right]^{\frac{1}{2}}}\right\}. \quad (4\cdot6)$$

The square roots in the above formulae are to be evaluated by the following rule—if

$$R_1 e^{i\phi_1} = 1-\frac{4a^2\gamma_1 e^{2i\theta}}{r^2(1+\gamma_1 e^{2i\theta})^2}, \quad R_2 e^{i\phi_2} = 1-\frac{4a^2\gamma_2 e^{2i\theta}}{r^2(1+\gamma_2 e^{2i\theta})^2},$$

then $\phi_1$ and $\phi_2$ are to be chosen so that they lie between $-\pi$ and $\pi$, and $R_1$ and $R_2$ are positive. The square roots are then $R_1^{\frac{1}{2}} e^{\frac{1}{2}i\phi_1}$, $R_2^{\frac{1}{2}} e^{\frac{1}{2}i\phi_2}$.

It may easily be verified that this stress system satisfies the equations of equilibrium, gives zero values for $\widehat{rr}$ and $\widehat{r\theta}$ at the edge of the hole and reduces to a uniform tension $T$ at infinity parallel to the $x$-axis. It therefore represents the complete solution of the problem, the form (3·2) being a valid expansion of the stress function for only a restricted range of values of $\theta$ and $r$. The stresses in an isotropic tension member containing a circular hole may be deduced by finding the limit of the above results as $\gamma_1 \to \gamma_2 \to 0$.

The stresses at the edge of the circular hole are of special interest and they take a comparatively simple form. Thus, when $r=a$,

$$\widehat{\theta\theta} = \frac{T(1+\gamma_1)(1+\gamma_2)(1+\gamma_1+\gamma_2-\gamma_1\gamma_2-2\cos 2\theta)}{(1+\gamma_1^2-2\gamma_1\cos 2\theta)(1+\gamma_2^2-2\gamma_2\cos 2\theta)}, \quad (4\cdot7)$$

$$\widehat{xy} = -\tfrac{1}{2}\widehat{\theta\theta}\sin 2\theta, \quad \widehat{xx} = \widehat{\theta\theta}\sin^2\theta, \quad \widehat{yy} = \widehat{\theta\theta}\cos^2\theta. \quad (4\cdot8)$$

## Numerical Discussion

5. In the first paper of this series numerical work was carried out for the problem of an isolated force acting at a point in a wide board of oak and of spruce cut so that the annular layers are parallel to the plane of the board. The elastic constants which

Table 1

| | $s_{11}$ | $s_{22}$ | $s_{12}$ | $s_{66}$ | $\alpha_1$ | $\alpha_2$ | $\gamma_1$ | $\gamma_2$ |
|---|---|---|---|---|---|---|---|---|
| Oak | 10·15 | 1·72 | −0·87 | 12·8 | 5·321 | 1·109 | 0·395 | 0·026 |
| Spruce | 15·5 | 0·587 | −0·33 | 11·5 | 16·91 | 1·56 | 0·608 | 0·111 |

were used for this purpose are reproduced in table 1, and these values are used for numerical work in the present paper. The grain of the wood will then be parallel to the $y$-axis or perpendicular to the tension. If the tension is applied parallel to the

Table 2. *Values of stresses on the edge of the circle: oak*

| $\theta°$ | $\widehat{xy}/T$ | $\widehat{xx}/T$ | $\widehat{yy}/T$ | $\widehat{\theta\theta}/T$ |
|---|---|---|---|---|
| | | Tension parallel to grain | | |
| 0 | 0 | 0 | −0·412 | −0·412 |
| 10 | 0·0664 | −0·0117 | −0·376 | −0·388 |
| 20 | 0·0996 | −0·0363 | −0·274 | −0·310 |
| 30 | 0·0691 | −0·0399 | −0·120 | −0·160 |
| 40 | −0·0492 | 0·0413 | 0·0587 | 0·100 |
| 50 | −0·263 | 0·313 | 0·221 | 0·534 |
| 60 | −0·540 | 0·935 | 0·312 | 1·25 |
| 70 | −0·752 | 2·07 | 0·274 | 2·34 |
| 75 | −0·752 | 2·81 | 0·201 | 3·01 |
| 80 | −0·626 | 3·55 | 0·110 | 3·66 |
| 90 | 0 | 4·36 | 0 | 4·36 |
| | | Tension perpendicular to grain | | |
| 0 | 0 | 0 | −2·43 | −2·43 |
| 10 | 0·291 | −0·0513 | −1·65 | −1·70 |
| 20 | 0·105 | −0·0382 | −0·289 | −0·327 |
| 30 | −0·343 | 0·198 | 0·595 | 0·793 |
| 40 | −0·742 | 0·622 | 0·884 | 1·51 |
| 45 | −0·873 | 0·873 | 0·873 | 1·75 |
| 50 | −0·950 | 1·13 | 0·797 | 1·93 |
| 60 | −0·940 | 1·63 | 0·543 | 2·17 |
| 70 | −0·739 | 2·03 | 0·269 | 2·30 |
| 80 | −0·404 | 2·29 | 0·0712 | 2·36 |
| 90 | 0 | 2·38 | 0 | 2·38 |

Table 3. *Values of stresses on the edge of the circle: spruce*

| $\theta°$ | $\widehat{xy}/T$ | $\widehat{xx}/T$ | $\widehat{yy}/T$ | $\widehat{\theta\theta}/T$ |
|---|---|---|---|---|
| | | Tension parallel to grain | | |
| 0 | 0 | 0 | −0·195 | −0·195 |
| 10 | 0·0323 | −0·0057 | −0·183 | −0·189 |
| 20 | 0·0543 | −0·0198 | −0·149 | −0·169 |
| 30 | 0·0533 | −0·0308 | −0·0922 | −0·123 |
| 40 | 0·0137 | −0·0115 | −0·0164 | −0·0279 |
| 50 | −0·0852 | 0·102 | 0·0715 | 0·173 |
| 60 | −0·266 | 0·461 | 0·154 | 0·615 |
| 70 | −0·529 | 1·45 | 0·193 | 1·65 |
| 75 | −0·653 | 2·44 | 0·175 | 2·61 |
| 80 | −0·683 | 3·87 | 0·120 | 3·99 |
| 85 | −0·483 | 5·52 | 0·0423 | 5·57 |
| 90 | 0 | 6·37 | 0 | 6·37 |
| | | Tension perpendicular to grain | | |
| 0 | 0 | 0 | −5·14 | −5·14 |
| 10 | 0·380 | −0·0669 | −2·15 | −2·22 |
| 20 | −0·187 | 0·0680 | 0·513 | 0·581 |
| 30 | −0·735 | 0·424 | 1·27 | 1·70 |
| 40 | −1·02 | 0·854 | 1·21 | 2·07 |
| 45 | −1·06 | 1·06 | 1·06 | 2·13 |
| 50 | −1·06 | 1·26 | 0·888 | 2·15 |
| 60 | −0·922 | 1·60 | 0·533 | 2·13 |
| 70 | −0·671 | 1·85 | 0·244 | 2·09 |
| 80 | −0·352 | 2·00 | 0·0621 | 2·06 |
| 90 | 0 | 2·04 | 0 | 2·04 |

grain it is only necessary to interchange the values of $s_{11}$ and $s_{22}$ which will change the signs of $\gamma_1$ and $\gamma_2$. The inverses of the constants $s_{11}, \ldots, s_{66}$ are the Young's moduli. They are measured in kg./sq.mm.

The values of the stresses on the edge of the circular hole have been evaluated by using the formulae (4·7) and (4·8). The results are given in tables 2 and 3. The stresses

Table 4. *Values of stresses on axis perpendicular to tension*

| | Spruce | | Oak | |
|---|---|---|---|---|
| $r/a$ | $\widehat{xx}/T$ | $\widehat{yy}/T$ | $\widehat{xx}/T$ | $\widehat{yy}/T$ |
| | | Tension parallel to grain | | |
| 1·0 | 6·37 | 0 | 4·36 | 0 |
| 1·05 | 3·58 | 0·126 | 3·28 | 0·139 |
| 1·1 | 2·70 | 0·145 | 2·70 | 0·196 |
| 1·2 | 1·99 | 0·139 | 2·07 | 0·226 |
| 1·3 | 1·67 | 0·123 | 1·75 | 0·220 |
| 1·4 | 1·50 | 0·107 | 1·56 | 0·204 |
| 1·5 | 1·39 | 0·0937 | 1·44 | 0·186 |
| 1·6 | 1·32 | 0·0824 | 1·35 | 0·168 |
| 1·7 | 1·26 | 0·0730 | 1·29 | 0·152 |
| 1·8 | 1·22 | 0·0650 | 1·25 | 0·138 |
| 1·9 | 1·19 | 0·0583 | 1·21 | 0·125 |
| 2·0 | 1·17 | 0·0526 | 1·18 | 0·114 |
| 2·5 | 1·09 | 0·0337 | 1·10 | 0·0751 |
| | | Tension perpendicular to grain | | |
| 1·0 | 2·04 | 0 | 2·38 | 0 |
| 1·05 | 1·97 | 0·096 | 2·24 | 0·116 |
| 1·1 | 1·90 | 0·182 | 2·11 | 0·205 |
| 1·2 | 1·77 | 0·325 | 1·91 | 0·340 |
| 1·3 | 1·67 | 0·436 | 1·76 | 0·431 |
| 1·4 | 1·59 | 0·522 | 1·64 | 0·491 |
| 1·5 | 1·52 | 0·589 | 1·54 | 0·527 |
| 1·6 | 1·45 | 0·640 | 1·46 | 0·548 |
| 1·7 | 1·40 | 0·678 | 1·39 | 0·557 |
| 1·8 | 1·36 | 0·705 | 1·34 | 0·557 |
| 1·9 | 1·32 | 0·724 | 1·30 | 0·552 |
| 2·0 | 1·29 | 0·737 | 1·26 | 0·543 |
| 2·1 | 1·26 | 0·743 | — | — |
| 2·2 | 1·23 | 0·745 | — | — |
| 2·3 | 1·21 | 0·744 | — | — |
| 2·4 | 1·19 | 0·740 | — | — |
| 2·5 | 1·18 | 0·733 | 1·14 | 0·466 |
| 3·0 | 1·11 | 0·677 | — | — |

on the lines $\theta = 0$ and $\theta = \frac{1}{2}\pi$ which may be obtained from (4·4) and (4·5) have also been evaluated and these results are reproduced in tables 4 and 5. In all the tables $\theta = 0$ is a line through the centre of the circular hole parallel to the applied tension.

The distribution of $\widehat{\theta\theta}$ over one quadrant of the hole is shown in fig. 1 for the case where the tension is parallel to the grain and in fig. 2 for the case when it is perpendicular to the grain. The distribution of $\widehat{\theta\theta}$ for a hole in an isotropic sheet, namely,

$$\widehat{\theta\theta} = T(1 - 2\cos 2\theta),$$

is also shown in figs. 1 and 2 for comparison. In these figures the sheet is supposed to be in a state of tension in the direction $\theta = 0$ and $\widehat{\theta\theta}$ is positive where there is tension and negative where the stress is compressive. It will be seen that with the highly aeolotropic spruce the maximum stress rises to $6{\cdot}37T$ when the tension is applied parallel to the grain but that the tensile stress rises only to $2{\cdot}04T$ when the

Table 5. *Values of stresses on axis parallel to tension*

| | Spruce | | Oak | |
|---|---|---|---|---|
| $r/a$ | $\widehat{xx}/T$ | $\widehat{yy}/T$ | $\widehat{xx}/T$ | $\widehat{yy}/T$ |
| | | Tension parallel to grain | | |
| 1·0 | 0 | −0·195 | 0 | −0·412 |
| 1·05 | −0·0093 | −0·176 | −0·0160 | −0·357 |
| 1·1 | −0·015 | −0·160 | −0·0252 | −0·310 |
| 1·2 | −0·021 | −0·132 | −0·0270 | −0·233 |
| 1·3 | −0·019 | −0·109 | −0·0128 | −0·176 |
| 1·4 | −0·012 | −0·0896 | 0·0116 | −0·132 |
| 1·5 | −0·0005 | −0·0738 | 0·0425 | −0·0985 |
| 1·6 | 0·014 | −0·0606 | 0·0772 | −0·0723 |
| 1·7 | 0·0309 | −0·0496 | 0·114 | −0·0521 |
| 1·8 | 0·0496 | −0·0403 | 0·152 | −0·0362 |
| 1·9 | 0·0698 | −0·0325 | 0·190 | −0·0237 |
| 2·0 | 0·091 | −0·026 | 0·227 | −0·0139 |
| 2·5 | 0·201 | −0·0052 | 0·393 | 0·0113 |
| 3·0 | 0·307 | | | |
| 5·0 | 0·611 | | | |
| 10·0 | 0·871 | | | |
| | | Tension perpendicular to grain | | |
| 1·0 | 0 | −5·14 | 0 | −2·43 |
| 1·05 | −0·0311 | −1·86 | −0·0537 | −1·40 |
| 1·1 | 0·0459 | −0·926 | −0·0281 | −0·875 |
| 1·2 | 0·207 | −0·270 | 0·0771 | −0·370 |
| 1·3 | 0·337 | −0·0498 | 0·188 | −0·150 |
| 1·4 | 0·439 | 0·0399 | 0·287 | −0·0407 |
| 1·5 | 0·518 | 0·0789 | 0·372 | 0·0165 |
| 1·6 | 0·582 | 0·0950 | 0·443 | 0·0477 |
| 1·7 | 0·634 | 0·100 | 0·504 | 0·0640 |
| 1·8 | 0·676 | 0·0998 | 0·555 | 0·0723 |
| 1·9 | 0·711 | 0·0967 | 0·599 | 0·0761 |
| 2·0 | 0·741 | 0·0920 | 0·637 | 0·0770 |
| 2·5 | 0·837 | 0·0679 | 0·764 | 0·0663 |
| 3·0 | 0·889 | | | |

tension is applied perpendicular to the grain. On the other hand, in the latter case there is a region where the compressive stress rises to $5{\cdot}14T$. The regions of high stress concentration extend only over a small area however. It will be seen in fig. 1 that the region where the stress exceeds that for an isotropic sheet extends only 14° from the position of maximum stress concentration. Similarly this region extends radially only 0·15 time the radius from the edge of the hole into the material. The distribution of $\widehat{\theta\theta}$ (or $\widehat{xx}$) along the $y$-axis is shown in fig. 3.

It will be seen that with highly aeolotropic materials like wood the region of high stress concentration is limited to a small area where the fibres which have been cut in making the hole lie close to the uncut fibres.

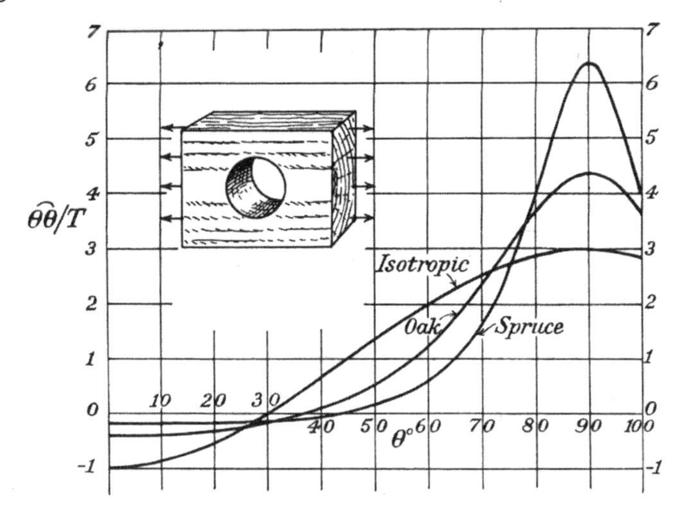


Fig. 1. Stress distribution round the edge of a hole. Stress applied parallel to the grain, parallel to $\theta=0$.

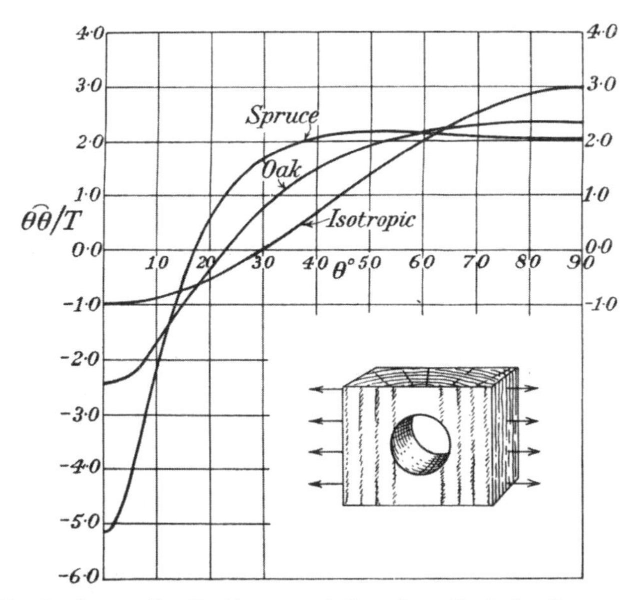


Fig. 2. Stress distribution round the edge of a hole. Stress applied perpendicular to the grain, parallel to $\theta=0$.

The distribution of $\widehat{xy}$ round the edge of the hole is shown in fig. 4.

The high stresses which occur near a hole in a stressed sheet have a technical interest for they may cause failure of a material which would otherwise have withstood the stress. In an isotropic sheet subject to tension in one direction a circular hole increases stress at the point of maximum stress concentration by a factor 3. Since the stress at the edge of the hole is a pure tension or compression in the circumferential direction, and the maximum stress occurs in the part of the circumference where the stress is tensile, failure of the material might be expected to be of

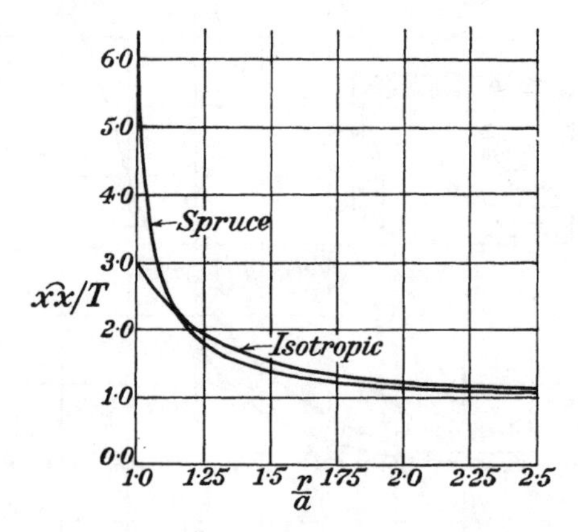


Fig. 3. Distribution of $\widehat{xx}$ along $\theta = \frac{1}{2}\pi$ when stress is applied parallel to the grain and to $\theta = 0$.

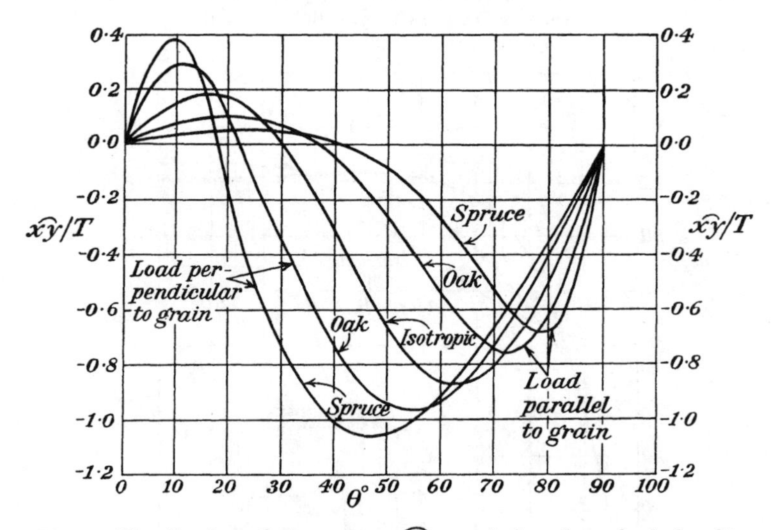


Fig. 4. Distribution of shear stress $\widehat{xy}$ round the edge of a hole. Stress applied parallel to $\theta = 0$.

the same type as that which occurs in the absence of the hole, but at a load only one-third as great. In actual practice the material usually withstands greater loads than this, a fact which has been explained by supposing that the stress at the point of maximum stress concentration does not reach its calculated value owing to slight plastic yielding there.

In aeolotropic materials the conditions of failure under complex stresses have not been fully investigated, but the ultimate stresses for direct tensile and compressive loads applied parallel and perpendicular to the grain of certain woods as well as the ultimate shear stresses for shears applied in these directions have been measured. Measurements of this type are included in a report by C. F. Jenkin* and the relevant figures† for spruce taken from this report are given in table 6.

It is of interest to compare the loads at which failure of the five types contemplated in table 6 might be expected to take place in the neighbourhood of a hole in a spruce plank under tensile or compressive loads. For this purpose the value of the externally applied tensile or compressive stress, $T$, which would theoretically raise the stress near the hole to its ultimate value has been calculated. Four cases are considered:

(i) tension applied parallel to the grain,
(ii) compression applied parallel to the grain,
(iii) tension applied perpendicular to the grain,
(iv) compression applied perpendicular to the grain,

Table 6. *Ultimate yield stresses for spruce*

| | lb./sq.in. |
|---|---|
| Tension parallel to the longitudinal fibres | 18,000 |
| Compression parallel to the fibres | 5,000 |
| (Corresponding in the present work with $\widehat{xx}$ when the external load is applied along the grain and $\widehat{yy}$ when it is perpendicular to it) | |
| Tension acting in the tangential direction, i.e. perpendicular to a vertical plane through the centre of the tree when standing vertically | 400 to 800 |
| Compression in this direction | |
| (Corresponding with $\widehat{yy}$ when the load is applied along the grain and $\widehat{xx}$ when applied perpendicular to it) | 700 |
| Shear stress across a plane perpendicular to the tangential direction | 1,100 |
| (Corresponding with $\widehat{xy}$) | |

and for each case the value of $T$ is calculated for each of the five types of failure referred to in table 6, namely,

(*a*) rupture of the longitudinal fibres.
(*b*) breakdown of longitudinal fibres in compression,
(*c*) rupture perpendicular to the fibres,
(*d*) breakdown by compression perpendicular to the fibres,
(*e*) shearing rupture parallel to the fibres.

Taking case (i), (*a*) the stress maximum $\widehat{xx}$ occurs at $\theta = 90°$ and is equal to $6{\cdot}37T$. From table 6 this corresponds with 18,000 lb. so that the tensile load necessary to cause breakdown of type (*a*) in case (i) is

$$T = 18{,}000/6{\cdot}37 = 2800\,\text{lb./sq.in.}$$

* *Report on Materials of Construction used in Aircraft and Aircraft Engines.* London: H.M. Stationery Office (1920).

† These figures, like those given in table 1 for the elastic constants, are average figures taken from results of experiments with a large number of specimens which exhibit considerable variations among themselves.

In each case the appropriate maximum positive or negative values of $\widehat{xx}$, $\widehat{yy}$, $\widehat{xy}$ have been taken from table 3. The results are given in table 7.

The type of failure which might be expected to occur corresponds with the least of the values given in table 7 for each of the four cases. These are indicated in

Table 7. *Calculated stresses for failure near a hole in a spruce plank*

| Failure type lb./sq.in. | Case (i) | Case (ii) |
|---|---|---|
| *a* | $\frac{18,000}{6\cdot37}=2800$ | $\frac{18,000}{0\cdot0308}=580,000$ |
| *b* | $\frac{5000}{0\cdot0308}=160,000$ | $\frac{5000}{6\cdot37}=780$ |
| *c* | $\frac{400\text{–}800}{0\cdot193}=2000\text{–}4000$ | $\frac{400\text{–}800}{0\cdot195}=2000\text{–}4000$ |
| *d* | $\frac{700}{0\cdot195}=3600$ | $\frac{700}{0\cdot193}=3600$ |
| *e* | $\frac{1100}{0\cdot683}=1600$ | $\frac{1100}{0\cdot683}=1600$ |

| Failure type lb./sq.in. | Case (iii) | Case (iv) |
|---|---|---|
| *a* | $\frac{18,000}{1\cdot27}=14,000$ | $\frac{18,000}{5\cdot14}=3500$ |
| *b* | $\frac{5000}{5\cdot14}=970$ | $\frac{5000}{1\cdot27}=3900$ |
| *c* | $\frac{400\text{–}800}{2\cdot04}=200\text{–}400$ | $\frac{400\text{–}800}{0\cdot067}=6000\text{–}12,000$ |
| *d* | $\frac{700}{0\cdot067}=10,000$ | $\frac{700}{2\cdot04}=350$ |
| *e* | $\frac{1100}{1\cdot06}=1000$ | $\frac{1100}{1\cdot06}=1100$ |

Fig. 5*a*. Shear crack when stress is applied parallel to the grain.

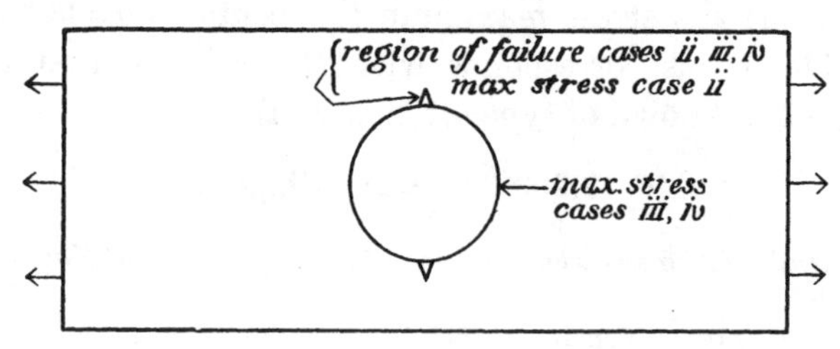


Fig. 5*b*. Position of failure when compression is applied parallel to grain or when tension or compression is applied perpendicular to grain.

table 7 by lines enclosing the four calculated values. It will be noticed that in cases (ii), (iii) and (iv) failure may be expected at the point on the circumference of the hole which lies on the diameter perpendicular to the direction of the applied stress. This is the position of maximum stress when the stressed sheet is isotropic. In case (ii) this position corresponds with the point of maximum stress concentration, but in cases (iii) and (iv) greater stresses occur in other parts of the field. Case (i) is more interesting. Here the type of breakdown by shearing parallel to the grain is quite different from that in an isotropic sheet. The maximum value of $\widehat{xy}$ occurs at $\theta = 78°$ (see fig. 4). The calculated position of the shear crack is shown in fig. 5*a*. The position of the region where breakdown is to be expected in cases (ii), (iii) and (iv) is shown in figure 5*b*.

33-2

# 36

# THE TESTING OF MATERIALS AT HIGH RATES OF LOADING

(James Forrest Lecture to the Institution of Civil Engineers, delivered 21 May 1946)

REPRINTED FROM

*Journal of the Institution of Civil Engineers*, vol. XXVI (1946), pp. 486–518

CONTENTS



## Introduction

The properties of materials subjected to impact loads during which the stresses rise and fall with great rapidity has always excited interest among engineers. During the years immediately preceding the war, a number of workers designed machines for carrying out high-speed tests of materials under conditions where both the stress and strain could be measured at every instant of the test. The design of such machines presents very considerable difficulties because, as the rate of straining increases, the inertia of such parts as the grips which hold the test specimen, the levers by which the load is applied, and so forth, give rise to forces which are comparable with those which produce permanent deformation or breakage in a slow-speed test. By careful designing, machines can be made in which the effects of inertia can be reduced or allowed for, provided they are not too large. The higher the rate of straining, however, the more difficult it becomes to separate the effects due to inertia of the machine from real physical changes in the strength properties of material. Finally, at very high speeds a condition is reached in which the inertia of the specimen itself gives rise to changes of stress along its length, which must be taken into account when seeking to interpret experimental results.

## Methods of Testing

*Dr John Hopkinson's experiment.* I thought it might interest you to hear an account of some of the methods used in attempting to make tests at the highest possible rates of stressing. The earliest work of this kind with which I am acquainted was done in 1872, by Dr John Hopkinson.* He hung a 27 ft. length of iron wire vertically from an iron block. A ball-shaped iron weight pierced by a hole was threaded on the wire, and was dropped down the wire from a measured height so that it struck an iron clamp fixed to the wire near its lower end. The experiments consisted in determining the heights required to break the wire when weights ranging from 7 to 41 lb. were used. The wire would support $3\frac{1}{2}$ cwt. dead weight. Dr Hopkinson found that all the weights would break the wire somewhere, usually near the bottom if dropped 5 ft., whilst a fall of only 2 ft. would break the wire at the top in all cases except that of the 7 lb. weight. This result is very different from that obtained with an ordinary impact machine like the Izod. In such machines the energy absorbed in breaking a specimen is found to depend very little on the speed; in fact, when the machine is designed like that of Southwell so as to eliminate the risk of energy escaping down the supports of the pendulum, it is found that the energy used in breaking a specimen is, in general, equal to that which would be required to break it statically. The use of ordinary impact machines is merely a convenient way of making a measurement which is essentially concerned with static conditions.

Dr Hopkinson expressed his remarkable result in the following words: 'In problems of this kind it has been assumed by some that two blows were equivalent when their *vis vivas* were equal,† by others when the momenta were equal; my result is that they are equivalent when the velocities or heights of fall are equal.'

Although Dr Hopkinson's experimental results certainly showed that with his apparatus the *vis viva*, or energy of the blow, was not the thing which determined the behaviour of the wire, I must confess that if my knowledge of the subject was confined to these experiments I should not have felt that they justified the statement that 'blows are equivalent when their velocities are equal'. I suspect that Dr Hopkinson would have agreed on this point, but his experiments are presented in his paper as being a rough confirmation of a theory of the propagation of stress waves in a wire and the result quoted is really a theoretical one based on the assumption that the wire is perfectly elastic till it breaks.

Dr Hopkinson's theory is very simple. Suppose that the end of a long vertical wire of mass $m$ per unit length is suddenly moved downwards with velocity $u_0$. An elastic wave travels upwards with the velocity $a$ of elastic waves. In this way the velocity of the wire is $u_0$ over the whole height from the bottom to the wave-front. The rate at which downward momentum is increasing in the wire is $mau_0$ and the equation of motion is therefore

$$p_0 = mau_0, \tag{1}$$

where $p_0$ denotes the tension in the wire. The strain in the wire is found from the consideration that after time $t$ a length $at$ has increased by an amount $u_0 t$. Thus the strain $\epsilon$ is $u_0/a$. Finally, the elastic equation is

$$\text{Modulus of elasticity of the wire} = \frac{p_0}{\epsilon} = \frac{mau_0}{(u_0/a)} = ma^2;$$

* See *Collected Scientific Papers*, vol. II, p. 316.

† As is in fact found to be the case with the Izod and other slow-speed impact machines.

and, since the modulus of elasticity of a wire is $E \times$ cross-section, where $E$ is Young's modulus, and $m = \rho \times$ cross-section, where $\rho$ denotes density,

$$a^2 = \frac{E}{\rho}. \tag{2}$$

So long as the velocity of the lower end of the wire is maintained at $u_0$ the tension $p_0$ remains constant; if, however, the motion is caused by the impact of a weight of mass $M$ this tension will produce an upward acceleration in $M$ which will reduce the downward velocity and hence the tension of the wire. This is represented mathematically by the equation

$$-\frac{M\,du_1}{dt} = mau_1, \tag{3}$$

where $u_1$ denotes the velocity of $M$ at time $t$. The solution of equation (3) is

$$u_1 = u_0 e^{-mat/M}.$$

At time $t$ there is no tension in the wire at heights greater than $at$, but at lower heights the tension is

$$p = mau_0 e^{m(x-at)/M}. \tag{5}$$

This represents a wave with a sharp front where the tension is $mau_0$, but behind this point the tension decays, so that at distance $M/m$ below the front the tension is only $1/e$, that is, 0·37 of its greatest value $mau_0$.

The length $M/m$, which is simply the length of wire which weighs as much as the dropped weight, is therefore a characteristic length. In Dr Hopkinson's experiments this was 270 ft. even when his smallest (7 lb.) weight was being used. Since his wire was only 27 ft. long, the elastic wave must have reached the top and been reflected down to the bottom again before the tension there had decreased appreciably owing to the slowing down of the mass $M$.

It is a pretty exercise for a mathematician to work out the whole story of the way in which the successive reflections of the wave slow down and finally stop the falling weight. I have carried out this calculation as far as the fifth reflection. I find that the results can be expressed in terms of the non-dimensional ratio

$$\alpha = \frac{mL}{M} = \frac{\text{(length of wire)}}{\text{(characteristic length } M/m)}. \tag{6}$$

To demonstrate the result, I have shown in fig. 1 the way in which the tension at the top and at the bottom of the wire would vary with time in Dr Hopkinson's case ($\alpha = \frac{27}{270} = 0{\cdot}1$) if the top fixing of his wire were rigid and the mass $M$ were also rigid. In this diagram the ordinate represents $p/mau_0$, where $p$ denotes the tension; that is, it represents the ratio of the stress at any time to the maximum stress in the wave before the first reflection. The horizontal co-ordinate represents $at/L$, so that reflections occur at the top when $at/L = 1$, 3 and 5 and at the bottom when $at/L = 2$, 4 and 6.

If the mass of the wire had been neglected, the tension in it would have been the same at the top and the bottom and would have varied harmonically. It turns out that in this case the stress can also be expressed in terms of $\alpha$ and the non-dimensional co-ordinates used in fig. 1. The formula is

$$\frac{p}{mau_0} = \frac{1}{\sqrt{\alpha}} \sin\left(\sqrt{\alpha}\frac{at}{L}\right). \tag{7}$$

This relationship is shown in fig. 1. It will be seen that when $\alpha = 0{\cdot}1$ the stress at each end varies above and below the stress represented by equation (7) by an amount approximately equal to $mau_0$, and that the maximum stress is therefore approximately $(1+1/\sqrt{\alpha})\, mau_0$, which for $\alpha = 0{\cdot}1$ is $4{\cdot}2\, mau_0$.

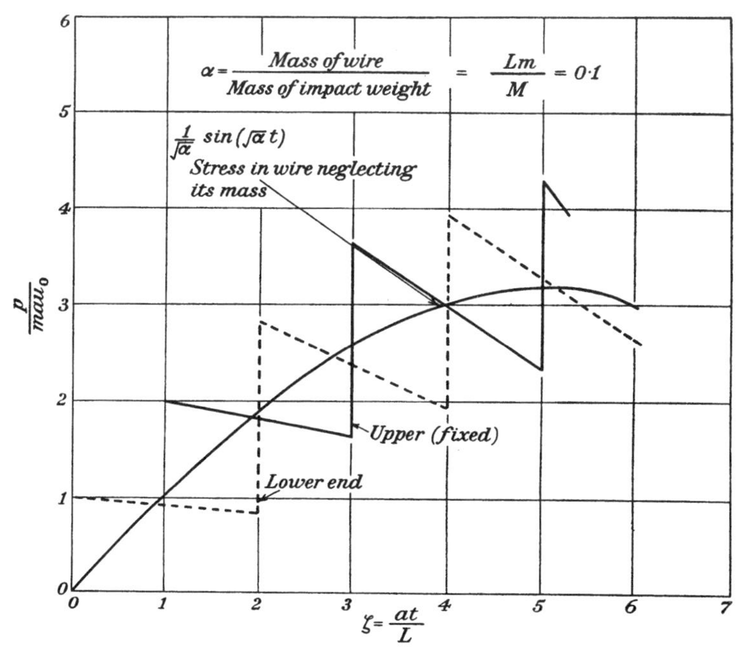


Fig. 1. Dr J. Hopkinson's experiment.

*B. Hopkinson's experiment.* Thirty years after the publication of Dr John Hopkinson's work, Bertram Hopkinson* repeated his father's experiment with improved technique, so that he could measure the maximum strain and the residual permanent strain near the top of the wire. His object was to find out whether the wire could bear for a very short period a load exceeding the static elastic limit without yielding. He assumed that the maximum load occurred on the first reflection at the fixed end of the wire and that, apart from a correction for the load initially hung on the wire to keep it straight, this maximum was twice that in the initial wave. It is clear from fig. 1 that if he had used a weight as heavy as even the lightest used by his father, his assumption that the maximum stress occurred at the first reflection would have been very far out. He evidently appreciated this fact, for he used only a

* 'The effects of momentary stresses in metals', *Proc. Roy. Soc.* A, LXXIV (1905), 498.

1 lb. weight, impacting on the lower end of 30 ft. of no. 10 gauge wire weighing 1·3 lb. I have repeated the calculation of the stresses at the ends of the wire for this case, namely, $\alpha = 1{\cdot}3$. The results are shown in fig. 2, and the values of the stresses immediately after the successive impacts are given for both $\alpha = 0{\cdot}1$ and $\alpha = 1{\cdot}3$ in table 1. It will be seen that in B. Hopkinson's experiment the maximum stress occurs at the third reflection, that is, the second reflection at the top. The value of this, however, is only $2{\cdot}15\,mau_0$, instead of $2{\cdot}0\,mau_0$, as was assumed (apart from the above-mentioned corrections). This error only strengthens the validity of B. Hopkinson's result, namely, that his steel wire had withstood without yielding a load nearly twice as great as the statically determined yield stress.

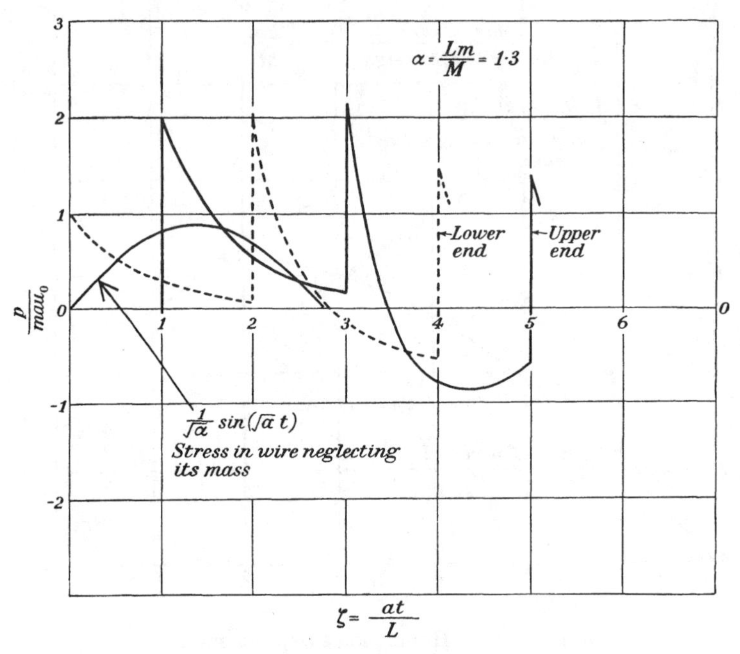


Fig. 2. B. Hopkinson's experiment.

It seems that every impact must have given rise to some permanent distortion in Dr J. Hopkinson's experiment, for the stresses calculated by the method outlined above were, in all cases, far higher than those at which yield began to occur in B. Hopkinson's experiment. Even a small amount of plastic stretching renders these calculations invalid, and the great merit of B. Hopkinson's experiment was that he measured the maximum elastic extension when there was no permanent extension as well as when there was.

*Duration of stress.* Referring to fig. 2, the maximum stress is, theoretically, only attained instantaneously, but a stress exceeding 90 % of this maximum was maintained without producing measurable permanent set for a period of about 0·06 of the time taken for the wave to travel 30 ft., that is, for $\frac{1}{10000}$ sec. The velocity of the weight was, in this case, 17 ft. per sec., and $2\rho au_0$ is then 31 tons per sq.in. Thus

B. Hopkinson's experiment shows that a stress greater than $0{\cdot}90 \times 31 = 28$ tons per sq.in. was maintained for $\frac{1}{10000}$ sec. without producing permanent set. Hopkinson gives the static yield stress of his wire as 18 tons per sq.in.

Table 1. *Values of $p/mau_0$ attained at successive reflections*

| | No. of reflection | Initial | 1st | 2nd | 3rd | 4th | 5th |
|---|---|---|---|---|---|---|---|
| General formulae | Bottom | 1·0 | — | $2+e^{-2\alpha}$ | — | $2+2e^{-2\alpha}+e^{-4\alpha}-4\alpha e^{-2\alpha}$ | — |
| | Top | 0 | 2·0 | — | $2(1+e^{-2\alpha})$ | — | $2(1+e^{-2\alpha}+e^{-4\alpha}-4\alpha^{-2\alpha})$ |
| $\alpha=0{\cdot}1$, J. Hopkinson's experiment | Bottom | 1·0 | — | 2·82 | — | 3·92 | — |
| | Top | 0 | 2·0 | — | 3·64 | — | 4·33 |
| $\alpha=1{\cdot}3$, B. Hopkinson's experiment | Bottom | 1·0 | — | 2·07 | — | 1·45 | — |
| | Top | 0 | 2·0 | — | 2·15 | — | 1·39 |

## Bouncing-ball Experiments

Recently Dr R. M. Davies and I measured the minimum load which is required to indent a steel surface with a steel ball statically. We also found the least height from which a steel ball must be dropped to produce an indentation on the same surface. According to Hertz's theory of impact, the distribution of stress near the circle of contact in the dynamic case is similar to that in the static case. The ratio of the magnitude of the stresses at similar points is the same all over the field, but the scale of the field of stress is determined by the radius of the circle of contact, which is determined by the pressure between the two surfaces in the two cases. The ratio (dynamic yield stress)/(static yield stress) is, therefore, the same as the ratio of these stresses at any pair of corresponding points in the two cases, provided that the yield first occurs in the same part of the field in the two cases. One might, for instance, take the normal pressure at the mid-point of the circle of contact in the two cases, or the stress at the point in the material where the shear stress is a maximum; the result will be the same.

When an elastic ball of radius $R$ falls on a plane surface of a material with the same elastic constants, Hertz's theory gives, for the normal pressure $p_d$ at the mid-point of the circle of contact,

$$p_d = \frac{1}{\pi}\left(\tfrac{5}{2}\pi\rho v^2\right)^{\frac{1}{5}}\left(\frac{E}{1-\sigma^2}\right)^{\frac{4}{5}}, \tag{8}$$

where $E$ denotes Young's modulus, taken as $2{\cdot}0 \times 10^{12}$ dynes per sq.cm.;

$\sigma$ denotes Poisson's ratio, taken as 0·286;

$v$ denotes the velocity of the ball before impact, in cm. per sec.;

$\rho$ denotes the density, taken as 7·9.

For static load, with force $P$ dynes, Hertz gives the stress at the mid-point of the circle of contact as

$$p_s = \frac{1\cdot 145}{\pi}\left(\frac{P}{R^2}\right)^{\frac{1}{3}}\left(\frac{E}{1-\sigma^2}\right)^{\frac{2}{3}}, \tag{9}$$

so that

$$\frac{p_d}{p_s} = \frac{(2\cdot 5\pi\rho)^{\frac{1}{5}}}{1\cdot 145}\frac{v^{\frac{2}{5}}R^{\frac{2}{3}}}{P^{\frac{1}{3}}}\left(\frac{E}{1-\sigma^2}\right)^{\frac{2}{15}}. \tag{10}$$

If the static load is produced by gravity acting on a mass $M$ grams, and the dynamic load by a fall through $h$ cm. then $v^2 = 2gh$ and $P = Mg$. Taking $g = 981$, it will be found that for steel equation (10) reduces to

$$\frac{p_d}{p_s} = 40\cdot 4 h^{\frac{1}{5}} R^{\frac{2}{3}} M^{-\frac{1}{3}}. \tag{11}$$

*Experimental.* Small polished areas were prepared on the materials to be tested. These were cleaned carefully to remove grease and set either horizontally or vertically. Steel balls were dropped on the horizontal surface or hung on a fine thread and swung as pendulums against a vertical surface. After the impact the test surface was examined to see whether the ball had made any impression. Our first method for doing this was to lay a small piece of optically flat glass over the area and by illuminating with monochromatic light, we could easily see any impression which was as deep as one-quarter of the light wave-length. Later we found an arrangement of lighting in the field of view of a microscope which made use of the fact that the depression was a small concave mirror of about the same radius as the ball which produced it. When correctly illuminated and focused, an indentation showed up as a brilliant spot on a dark ground. We found that as we decreased the height of fall of the ball the indentation remained easily identifiable down to a certain value of $h$, at which it suddenly disappeared altogether. For $h$ we took the mean of the least height at which an indentation was visible and the greatest height at which no indentation could be observed.

Great care was taken in the static tests to load the balls so that no instantaneous dynamic load would occur. The mean of the greatest mass which would just fail to produce an indentation and the least mass which would just do so was taken as $M$. Some experimental results are shown in table 2. The results of applying equation (11) to these are shown in column 10 of table 2. It will be noticed that three values are determined for one of the steels, namely, $p_d/p_s = 1\cdot 03$, $1\cdot 18$ and $1\cdot 01$. This is the largest variation that we found. It seems, however, that the polishing probably alters the surface layers of the steel to a depth comparable with the radius of contact, and that considerable variations may be expected in any case from a test which involves only metal in the surface layers. It must be remembered that the static and dynamic measurements never apply to exactly the same spot. In other investigations it has been found that the dynamic is nearly the same as the static yield point when steel with a high static yield is used. Our results have been plotted in fig. 3 to show the values obtained by the ball-dropping test for steels with varying yield

Table 2. *Ball indentation tests*

| Steel | Static: Brinell no. | Static: Least mass for indentation (g.) | Static: Greatest mass for no indentation | Static: $M$ (g.) | Dynamic: Least height for no indentation | Dynamic: Greatest height for no indentation | $h$ (cm.) | $R$ (cm.) | $p_d/p_s$ (from equation (11)) | Yield stress in tensile test (tons/sq.in.) |
|---|---|---|---|---|---|---|---|---|---|---|
| WTM | 351 | 4510 | 3610 | 4060 | 1·1 | 0·9 | 1·0 | 0·317 | 1·18 | 74 |
| WTN | 321 | 8020 | 6330 | 7180 | 0·4 | 0·3 | 0·35 | 0·476 | 1·03 | 69 |
| WTN | 321 | 2400 | 1910 | 2150 | 0·4 | 0·3 | 0·35 | 0·317 | 1·18 | 69 |
| WTN | 321 | 1100 | 890 | 1000 | 0·5 | 0·4 | 0·45 | 0·159 | 1·01 | 69 |

stress. It will be seen that there is a tendency for $p_d/p_s$ to rise as the static yield stress decreases.

Equation (9) refers to the normal component of pressure at the mid-point of the circle of contact; the maximum difference between the principal stresses occurs within the metal at a point 0·48 time the radius of the circle of contact below the

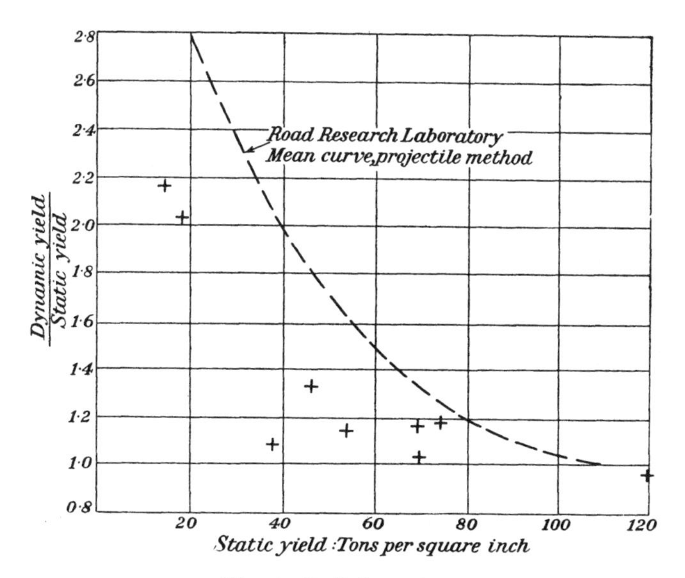


Fig. 3. Ball-dropping test.

surface, and its value is 0·63 time the above-mentioned normal stress. The maximum stress difference is $0{\cdot}63\left(\frac{1{\cdot}145}{\pi}\right)\left(\frac{P}{R^2}\right)^{\frac{1}{3}}\left(\frac{E}{1-\sigma^2}\right)^{\frac{2}{3}}$. When these values are worked out, inserting the observed values of $P$ and $R$, it is found that the maximum stress difference so calculated is greater than the yield stress determined in a tensile test. As Mrs Tipper has pointed out to me, it is nearer to the 'ultimate stress'. Some determinations are shown in table 3.

Table 3

| | Brinell no. | | | |
|---|---|---|---|---|
| | 217 | 269 | 302 | 115 |
| Yield stress (tons per sq.in.) | 38 | 46 | 54 | 14 |
| Ultimate stress (tons per sq.in.) | 48 | 56 | 64 | 25 |
| Maximum stress-difference calculated from static ball-indentation test (tons per sq.in.) | 42 | 50 | 67 | 24 |

## Tests in which the Yield Point is Passed

In the tests so far described the loads are increased until the first sign of plastic deformation appears, and the stress at that point is calculated from the elastic equations. In tests where the yield stress is exceeded, this method cannot be used and it is necessary to adopt some means for measuring stress and strain independently. Various experimenters have devised apparatus for achieving this end. Brown and

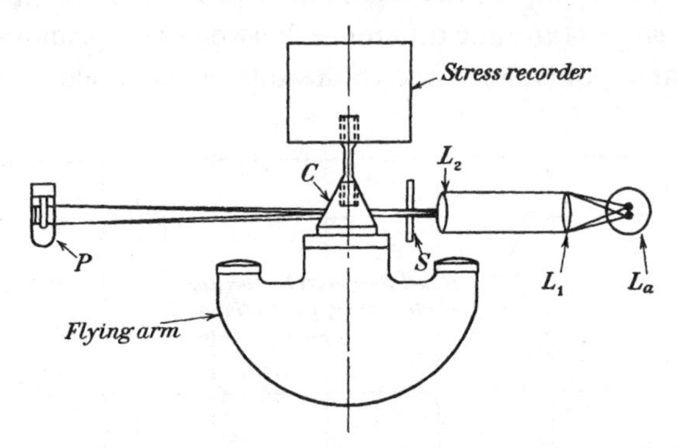


Fig. 4. Brown and Vincent's apparatus.

Vincent*, at the National Physical Laboratory, made the unit shown in fig. 4, which could be mounted on the anvil of a gravity-operated impact machine. The apparatus was constructed so that a yoke attached to the end of the specimen could be accelerated by two arms of the hammer striking simultaneously. The load at the other end of the specimen was transferred through a stress-measuring crystal to the anvil. The strain was measured by passing light through a slit whose width measured the extension of the specimen. The light fell on a photo-electric cell and the resulting electric disturbance was used to deflect the beam of an oscillograph. The piezo-electric current from the stress unit deflected the beam in a perpendicular direction so that stress and strain were recorded simultaneously.

Figs. 5 and 6, which are reproduced from Brown and Vincent's paper, show their measurements of stress strain during tests in which the hammer speeds were 10, 20 and 30 ft. per sec. The static values are shown for comparison. Fig. 5

* A. F. C. Brown and N. D. G. Vincent, 'The relationship between stress and strain in the tensile impact test', *Proc. Instn Mech. Engrs*, CXLV (1941), 126. See also D. S. Clark and G. Dätwyler, *Proc. Amer. Soc. Test. Mat.* XXXVIII (1938), 98.

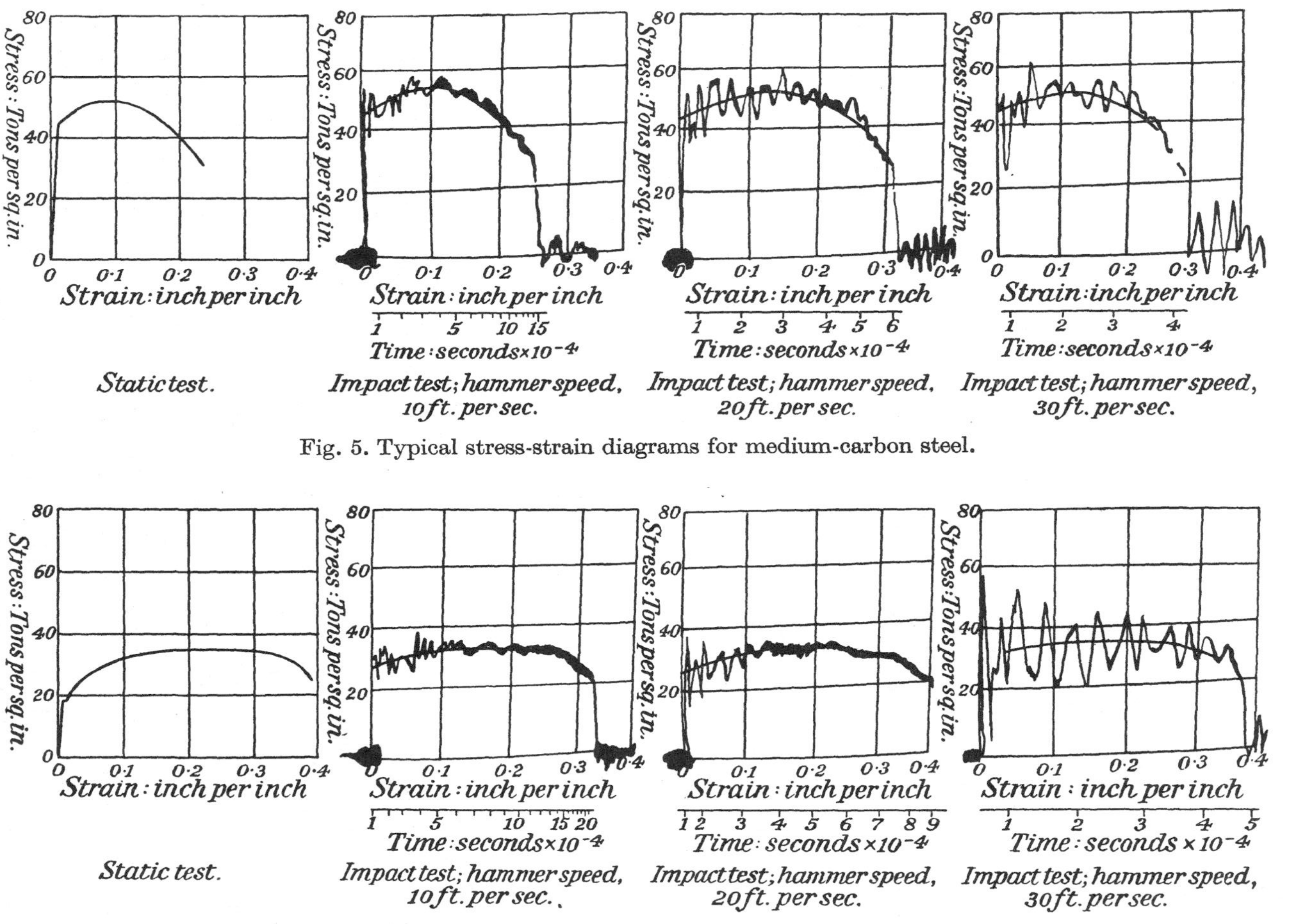


Fig. 5. Typical stress-strain diagrams for medium-carbon steel.

Fig. 6. Typical stress-strain diagrams for low-carbon manganese steel.

refers to a mild steel, whilst fig. 6 refers to a stronger alloy steel. It will be seen that they obtain the same result as Hopkinson and others, namely, that with mild steel the yield point is considerably increased when the duration of the test is very small. They also agree with our result using a bouncing ball, that there is very little increase in yield stress due to this cause when a steel of higher yield stress is used. Two further points, however, are shown in these figures. The first is that the increase in stress at the yield due to the short duration of the test is still maintained during the period of plastic extension which follows. The second is that when the speed of one end of the specimen is suddenly increased to anything in excess of about 30 ft. per sec., large oscillations occur which make the interpretation of the record difficult.

Another method was developed by the late Mr H. Quinney and myself during the years 1937–39. We had expected that vibration troubles would make it very difficult to measure stresses during tests of very short duration by means of apparatus attached to the end of the specimen. These troubles would be accentuated

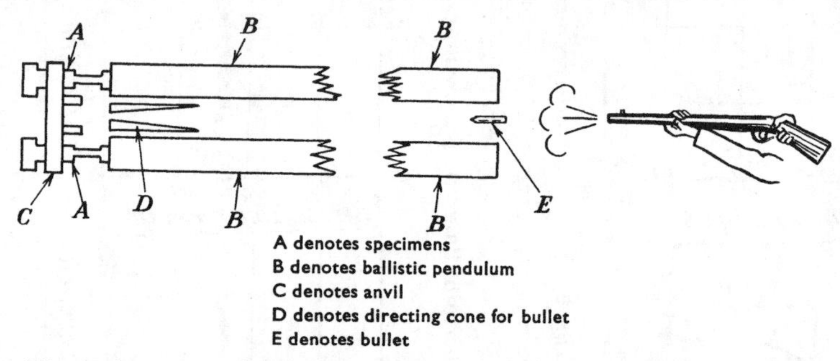


Fig. 7. Apparatus for independent measurement of stress and strain.

by the fact that if a heavy attachment to the end of the specimen was accelerated by a blow from an arm attached to a heavy body moving at high speed, vibrations of the attachment and the impacting body might be violent and associated with more energy than would be dissipated in stretching the specimen. Some of this might go into the specimen and falsify the results. For this reason, we pursued the opposite policy to that afterwards followed by Messrs Brown and Vincent. We used two equal specimens connected together by a loosely fitting hardened steel yoke or anvil. The arrangement is shown in fig. 7. The anvil was light—the heaviest we used weighed only 15 g. Lead bullets weighing from 1·8 to 2·7 g. were fired from a 0·22 in. rifle at the mid-point of the anvil. We found that the lead behaved like a fluid, spreading out in a thin film over the plane surface of the anvil. We measured the force exerted by the bullet and found it to be, in fact, just what a fluid jet of diameter 0·22 in. and the density of lead would exert. We did not, however, rely on this. We set the anvil so that it did not engage with the ends of the specimen until the bullet had completely disintegrated and lost all forward momentum. The specimens were screwed into the ends of two parallel steel bars which were hung as a ballistic pendulum. Thus the momentum of the anvil could be measured if the specimen stretched, but did not break, by observing the swing of these bars. In fact, we chose specimens of such dimensions that they would stretch slightly, but not break, when a single shot was fired at the yoke. By marking the specimens with very light marks, and measuring them between each shot, we were able to measure the small plastic extension produced by the shot. Since the mass of the pendulum was very large in comparison with that of the anvil, the kinetic energy of the anvil was practically entirely absorbed by the plastic extensions of the specimen. The mean tension during this extension was therefore taken as being given by

$$\tfrac{1}{2}MV^2 = 2p\epsilon, \tag{12}$$

where $\epsilon$ denotes the plastic extension of each specimen, $M$ the mass of the yoke, and $V$ the velocity of the anvil. The factor 2 is introduced because the two specimens stretched equally. If there was any appreciable difference between them they were rejected, and the experiment was started again. To get even stretching it was necessary to hit the centre of the anvil within about $\frac{1}{50}$ in. This accuracy cannot be obtained with a rifle fired from a distance of 15 ft., so we made a hardened steel directing cone of small cone angle (fig. 7), so that the bullets which were slightly off centre would be pushed back as quietly as possible towards the centre of the target. The anvil had two baffles to keep the lead spray from the specimen.

The results for a particular kind of mild steel are shown in fig. 8 and for copper in fig. 9. Each point represents the mean stress and mean strain during impact. It will be seen that although the method of experiment prevents an accurate determination of the yield point from being made, the results for mild steel are very similar to those of Brown and Vincent. The mean stress in the first interval of stretching is considerably above the yield point measured statically. The experiments on pure annealed copper, on the other hand (fig. 9), show only a small increase in stress due to rate of straining.

## Rate of Strain

The rate of strain has little meaning in the tests we first considered. The factor which can be understood and roughly measured or calculated is the length of time during which the stress exceeds some assigned value. In the experiments of Brown and Vincent and those of Quinney and myself, where the plastic stretching subsequent to elastic failure was measured, the rate of strain is a measurable quantity. In the case of experiments of Brown and Vincent it is simply

$$\left(\frac{\text{velocity of the end of the specimen}}{\text{length of parallel portion of specimen}}\right).$$

It is expressed as inches per inch per second, and has dimension $(\text{sec.})^{-1}$. In our experiments, where the velocity of the anvil decreases from its impact velocity to zero, one can take

$$\text{mean rate of strain} = \frac{1}{2}\left(\frac{\text{velocity of anvil}}{\text{length of specimen}}\right).$$

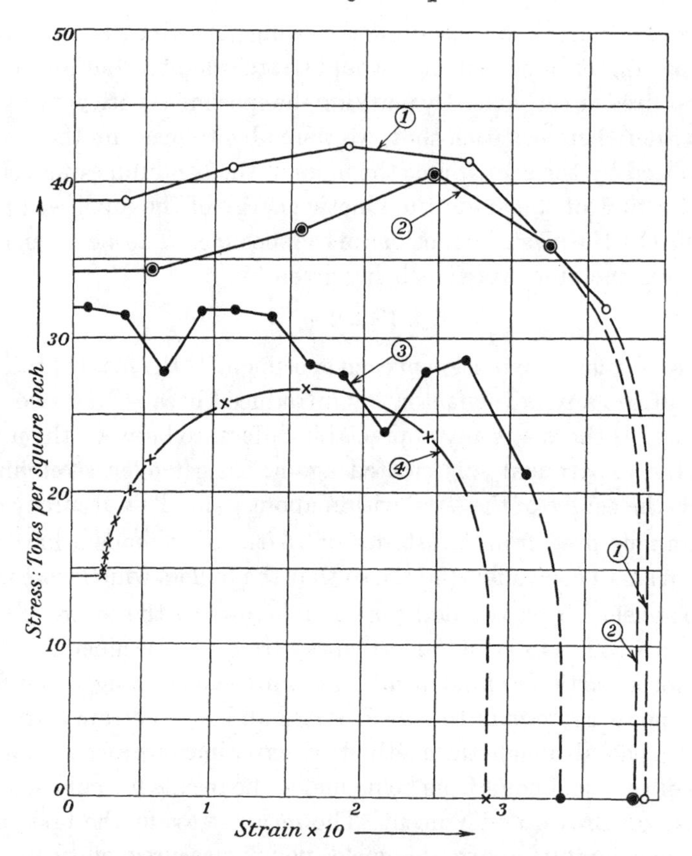


Fig. 8. Mild steel specimens, vacuum annealed at 760° C. for 2 hr., analysis: C 0·15, Si 0·25, Mn 0·71, S 0·03, P 0·04, Ni 0·05, Cr 0·01.

(1) Bullet speed: 1,400 ft. per sec.
Velocity of anvil during test (mean): 150 ft. per sec.

$$\frac{\text{Relative velocity of ends of specimen (ft. per sec.)}}{\text{Length of specimen (ft.)}} = 3000\ \text{sec.}^{-1}$$

(2) Bullet speed: 1200 ft. per sec.
Velocity of anvil during test (mean) = 127 ft. per sec.
Mean rate of strain: 2860 sec.$^{-1}$.

(3) Bullet speed: 70 ft. per sec.
Velocity of anvil during test (mean): 54 ft. per sec.
Mean rate of strain = 1530 sec.$^{-1}$.

(4) Static test.

To find the yield stress, or the stress at which the plastic strain is less than 0·001, the curves of fig. 8 are carried back to the stress axis. The values of yield stress so found are 2·2, 2·4 and 2·6 times the static value at mean rates of straining of 1240, 1530 and 3000 sec.$^{-1}$. These values are shown in fig. 10 for comparison with the results of Brown and Vincent, and also with some results at still higher rates of straining made by analysing the plastic distortion of cylindrical projectiles when striking a hard steel armour plate.

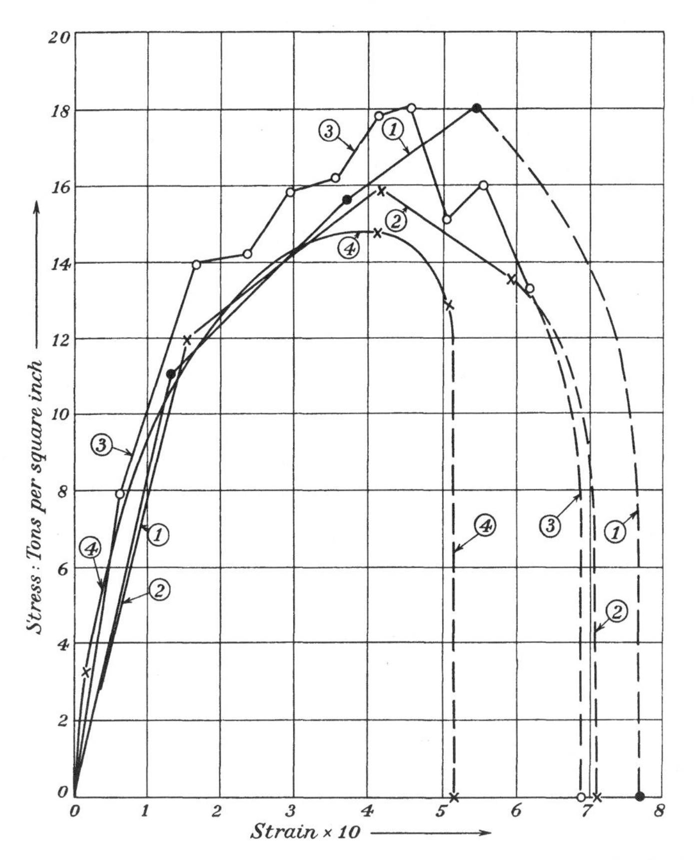


Fig. 9. Pure electrolytic copper, vacuum annealed at 450° C. for half an hour.

(1) Bullet speed: 1400 ft. per sec.
Velocity of anvil during test (mean): 150 ft. per sec.
Mean rate of strain:

$$\frac{\text{Relative velocity of ends of specimen (ft. per sec.)}}{\text{Length of specimen (ft.)}} = 3000 \text{ sec.}^{-1}.$$

(2) Bullet speed: 1200 ft. per sec.
Velocity of anvil during test (mean): 127 ft. per sec.
Mean rate of strain = 2860 sec.$^{-1}$.

(3) Bullet speed: 700 ft. per sec.
Velocity of anvil during test (mean): 54 ft. per sec.
Mean rate of strain = 1530 sec.$^{-1}$.

(4) Static test.

## Impact at very High Speeds

In the experiments so far described the analysis of the results has been possible because either (*a*) the conditions before failure are known from the theory of elasticity, so that the stresses can be calculated up to the moment the material ceases to behave elastically, or (*b*) the experiments have been conducted so that over the significant part of the range of testing the plastic strain is uniform over the portion of the specimen where the strain occurs. When attempts are made to increase the rate of strain by increasing the relative velocity of the two ends of a specimen beyond a certain limit, the inertia forces within its own length produce variations in stress which are of the same order as those applied at the end. The velocity at which this occurs might be expected to be of the order of magnitude of $\sqrt{(p/\rho)}$, where $p$ denotes the yield stress and $\rho$ the density of the material. With steel of density 7·9 and yield stress 20 tons per sq.in., $\sqrt{(p/\rho)} = 660$ ft. per sec.; with lead of yield, say,

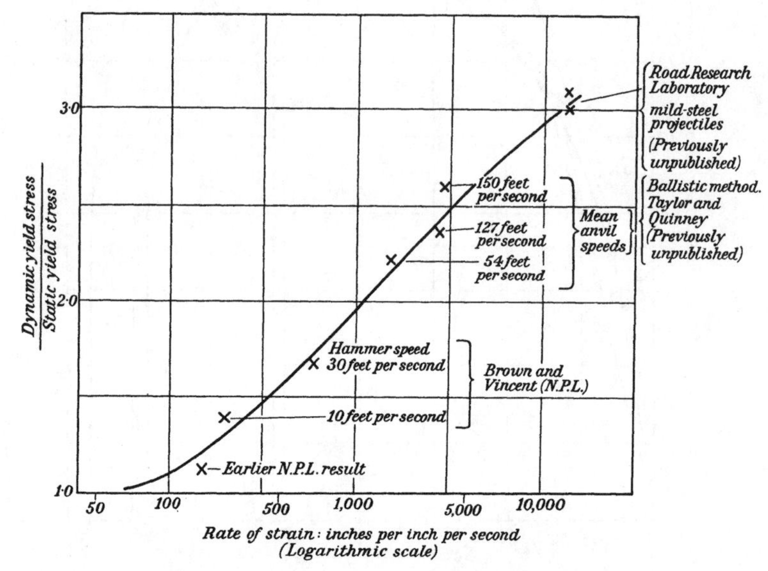


Fig. 10. Effect of rate of strain on dynamic yield stress.

1 ton per sq.in. and density 11, it is about 120 ft. per sec. Any attempt to force a relative velocity much greater than $\sqrt{(p/\rho)}$ is likely to make the metal flow like a liquid; indeed, I have already mentioned that a lead bullet fired at a speed of about 1000 ft. per sec. produces a thin spray of lead on striking a steel target, just as a jet of water would. To understand in a quantitative way what happens when impact occurs at such speeds would, in general, be beyond the power of our present analytical methods, even if we knew beforehand the stress-strain characteristics of the material. The inverse problem of finding the stress-strain characteristics from the observed behaviour of materials is even more difficult unless the experiments are very carefully designed.

An obvious and very simple way in which rapid loading can be applied is to project a cylindrical specimen at a hard steel plate. This has the advantage that any measurements that are made must necessarily be made without attaching measuring apparatus to the specimen. It has the disadvantage that the motion of the metal at the impact end is difficult—one might, perhaps, say impossible—to analyse with our present mathematical and experimental technique. On the other hand, there is no longitudinal stress (that is, no pressure acts) at the rear end of the specimen during its deceleration. If one could think of the specimen as long in comparison with its diameter, the stress near the rear end would be due to a series of longitudinal waves passing backwards and forwards in the portion which had not yet yielded. Since the velocity of elastic waves would be far greater than that of the projectile, a very large number of such waves would pass backwards and forwards in the rear end of the projectile before it was brought to rest. Under these conditions the stress in the rear part of the projectile may be regarded as being the same as that which would occur in a rigid cylindrical body subjected to the deceleration of the rear end. The stress would, in fact, increase linearly in accordance with the equation

$$\text{Stress} = (\text{density}) \times (\text{deceleration}) \times (\text{distance from rear end}). \tag{13}$$

If instantaneous measurements of the deceleration and of the position of the plastic-elastic boundary could be made, the equation (13) would enable us to measure the yield stress. Although experiments have been made during the war in both the United States and Great Britain to determine the deceleration of the back of a projectile, no such experiments had been made in the autumn of 1938, when Mr Quinney and I had already obtained the results shown in fig. 8, and I was thinking how we could obtain still higher rates of strain. It seemed to me that, even without measuring the deceleration, we should know that at some period it must have been greater than that of a particle moving with the velocity of impact $v_0$ and stopped at uniform deceleration in a distance equal to the difference between the initial and final lengths of the specimen. Thus the deceleration must have been greater than $\dfrac{v_0^2}{2(L-l_1)}$, where $L$ denotes the initial and $l_1$ the final length of the specimen. Since the distance between the rear end and the plastic boundary can never have been less than that which could be found by measuring the specimen after the impact, it will be seen from equation (13) that the minimum possible value of the yield stress is

$$S = S_2 = \frac{\rho v_0^2 x_1}{2(L-l_1)}, \tag{14}$$

where $x_1$ denotes the distance of the plastic-elastic boundary from the rear end after the impact.

To test this, I constructed a small catapult capable of projecting cylinders of paraffin wax at 175 ft. per sec. at a heavy metal plate hung as a ballistic pendulum. Paraffin wax has the property that it remains transparent until a sudden collapse occurs at a definite breakdown stress. The edge of the region in which breakdown had occurred in my cylinders was clearly defined and could be measured to within 0·5 mm. Some results are shown in table 4. The ballistically measured minimum value of its breakdown stress $S_2$ is given in column 6 and the statically measured stress $S_3$ in column 7. It will be seen that $S_2$ is considerably greater than $S_3$, and when a less rough method of calculation is applied the ratio $S_2:S_3$ is more than 2:1, instead of the minimum value of approximately 1·3:1 which is found from the figures in table 4. This result seemed encouraging, but about this time scientific

34-2

activities connected with the possibility that we should find ourselves at war made me drop this work. During the war, however, the Road Research Laboratory took up the work and fired a large number of cylindrical projectiles at heavy armour plates.

Fig. 11 shows a set of 1 in. mild-steel cylinders of diameter 0·3, after being fired at speeds ranging from 810 to 2120 ft. per sec. It will be seen that a portion of each of them appears to be unstrained, whilst the impact end in each case has spread laterally on the target. The specimens were measured with micrometers and the

Table 4. *Yield stress for paraffin wax*

| $L$ (cm.) | $l_1$ (cm.) | $\rho v_0^2$ (dynes per sq.cm.) | $x_1$ (cm.) | $S_2 = \frac{x_1}{L-l_1}(\frac{1}{2}\rho v_0^2)$ (dynes per sq.cm.) | $S_2$ measured ballistically (lb. per sq.in.) | $S_3$ measured statically (lb. per sq.in.) |
|---|---|---|---|---|---|---|
| 1·774 | 1·635 | $1{\cdot}25\times10^7$ | 0·95 | $4{\cdot}25\times10^7$ | 638 | |
| 1·757 | 1·625 | $1{\cdot}31\times10^7$ | 0·95 | $5{\cdot}1\times10^7$ | 760 | 485 |
| 1·779 | 1·625 | $1{\cdot}39\times10^7$ | 0·90 | $4{\cdot}05\times10^7$ | 604 | |

distance $x_1$ at which the lateral plastic expansion first became as great as 0·002 was determined. The lengths before and after impact were also measured, as well as the striking velocity of the cylinder. Plate 1 shows two projectiles of very different sizes, but of the same material, which had been fired at the same speed (575 ft. per sec.). It will be seen that they appear to be almost identical in shape.

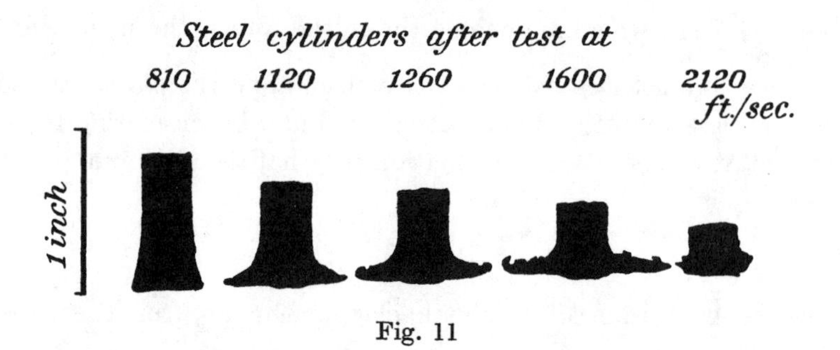


Fig. 11

To analyse experiments of this kind completely would have been possibly only if the whole of the successive positions of each section of the cylinder had been measured. Since only the final condition was measurable, it was necessary to make some kind of assumption about how the material reached its final state. The first assumption I made was that the yield stress $S$ has the same value throughout the test. With such an assumption ordinary dynamical reasoning yields the following equation connecting the distance $x$ of the plastic boundary from the rear of the specimen with the distance $z$ of the rear of the specimen from the target plate:

$$x\frac{d^2z}{dt^2} = -\frac{S}{\rho}, \tag{15}$$

where $t$ denotes the time since the moment of impact. To get the second equation, which is necessary in order to determine both $x$ and $z$ as functions of $t$, it would be necessary either to have a mathematically formulated conception of the way in which material mushrooms out between the plastic boundary and the target, or to have some measurements of the actual variation of $z$ with $t$. Failing these, the best that can be done is to assume some arbitrary variation in $x$ with $t$ consistent with its initial value $L$ and its final measured value $x_1$. Fortunately, the final results do not seem sensitive to the possible range of variations of $x$ within these limits. Accordingly, I assumed that the plastic boundary moves out with constant velocity $c$ during the impact from the target surface to the measured distance $l_1 - x_1$. Thus I assumed that

$$x = z - ct, \tag{16}$$

where

$$cT = l_1 - x_1, \tag{17}$$

and $T$ denotes the total duration of the impact. The equation of motion is now

$$(z - ct)\frac{d^2z}{dt^2} = -\frac{S}{\rho}. \tag{18}$$

This equation can be solved, giving the result

$$(u_0 + c)^2 - \left(\frac{dx}{dt}\right)^2 = \frac{2S}{\rho}\log_e\left(\frac{L}{x}\right). \tag{19}$$

At the end of the impact $x = x_1 = l_1 - cT$ and $dz/dt = 0$, so that $dx/dt = -c$. Hence equation (19) becomes

$$S = \rho\left(\frac{u_0^2 + 2u_0 c}{2\log_e(L/x_1)}\right). \tag{20}$$

This expression contains $c$, which is unknown and must be determined through the consideration that when equation (19) is integrated to give the actual connection between $x$ and $t$, the value $T$ which corresponds with $x_1$ must also satisfy the equation (17). I have found a method for carrying out this operation, but as it involves mathematical technique which it would be out of place to describe in this lecture, and as it also turns out that a simple approximation can be made which appears to agree with the more elaborate method to within a small percentage, I confine myself to the simple theory. If the deceleration of the rear portion of the cylinder had been uniform, the time of impact $T$ would have been given by

$$L - l_1 = \tfrac{1}{2}u_0 T = \tfrac{1}{2}u_0\left(\frac{l_1 - x_1}{c}\right),$$

so that

$$\frac{2c}{u_0} = \left(\frac{l_1 - x_1}{L - l_1}\right).$$

Thus equation (20) becomes

$$S = \frac{\rho u_0^2}{2\log_e(L/x_1)}\left(\frac{L - x_1}{L - l_1}\right). \tag{21}$$

This formula assumes that the cylinder does not chip out a hole in the target plate. If it does, in fact, chip out a hole of depth $d$, the distance of the rear of the projectile from the target plate at the end of the impact is $l_1 - d$, instead of $l_1$, so that equation (21) should be changed to

$$S = \frac{\rho u_0^2}{2\log_e(L/x_1)}\left(\frac{L-x_1}{L-l_1+d}\right). \tag{22}$$

This is the formula which has been used by Mr Whiffin, of the Road Research Laboratory, for deducing the dynamic yield stress from his measurement of the cylinders after impact. He made many experiments to find out whether the total size or the ratio of diameter to length of the cylinders affected his result. Figs. 12 and 13 show some results for two mild steels. In fig. 13 it will be seen that as the striking velocity increases from 400 to 2800 ft. per sec. there appears to be a small rise in yield strength; but this is very small and may be due to errors of measurement. Between these values and those determined statically there is a great difference, the dynamic value being more than twice the static value. Fig. 14 shows that this

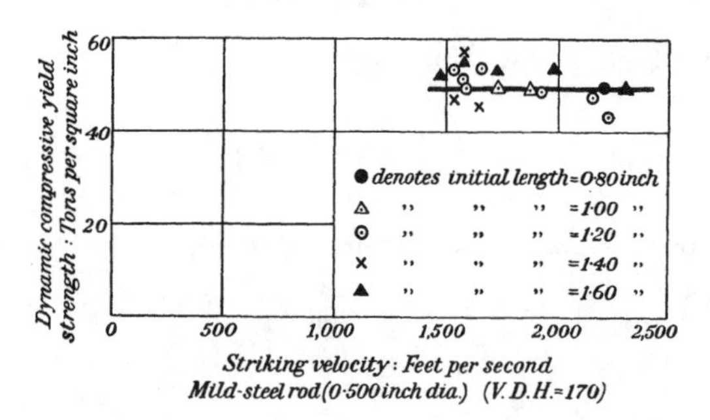


Fig. 12. Effect of variation of striking velocity on the dynamic compressive yield strength of cylindrical projectiles of mild steel.

characteristic of mild steel is also shared by duralumin, although the increase in yield strength is not so great as in the case of mild steel of the same static strength.

The fact that the values are so uniform in spite of very large variation in the values $u_0$, $x_1/L$ and $l_1/L$ is a rather striking testimony to the adequacy of the theoretical analysis, which it must be remembered is incomplete in that it replaces our ignorance of the complex actual flow of metal near the target by the *assumption* that the plastic-elastic boundary moves uniformly from the target out to its final measured position.

Even more convincing evidence of this point is provided by the results of Mr Whiffin's measurements of dynamic yield stress in steel and duralumin of varying static strength. Fig. 15 shows the ratio (dynamic yield strength)/(static yield strength) for steels of varying static strengths. It will be seen that as the static yield increases this ratio decreases until at a static strength of 120 tons per sq.in.*

the dynamic yield is not appreciably different from the static value. This is in agreement with our results obtained in the ball-dropping tests and also with those of Brown and Vincent. The fact that the ratio dynamic/static strength for a given

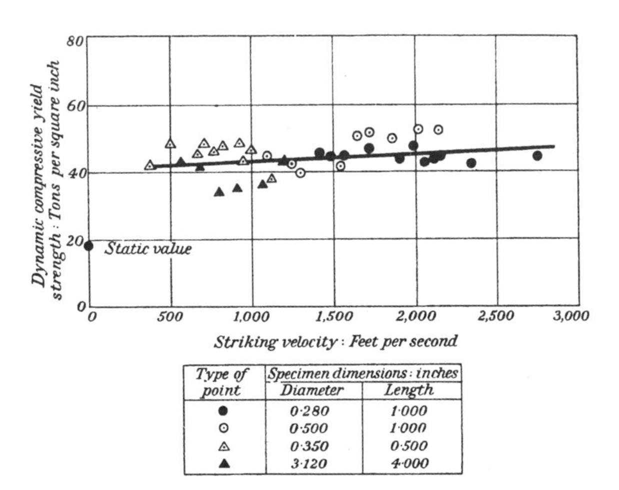


| Type of point | Specimen dimensions: inches | |
|---|---|---|
| | Diameter | Length |
| ● | 0·280 | 1·000 |
| ⊙ | 0·500 | 1·000 |
| △ | 0·350 | 0·500 |
| ▲ | 3·120 | 4·000 |

Fig. 13. Effect of variation of striking velocity on the dynamic compressive yield strength of cylindrical projectiles of mild steel (V.D.H. 120).

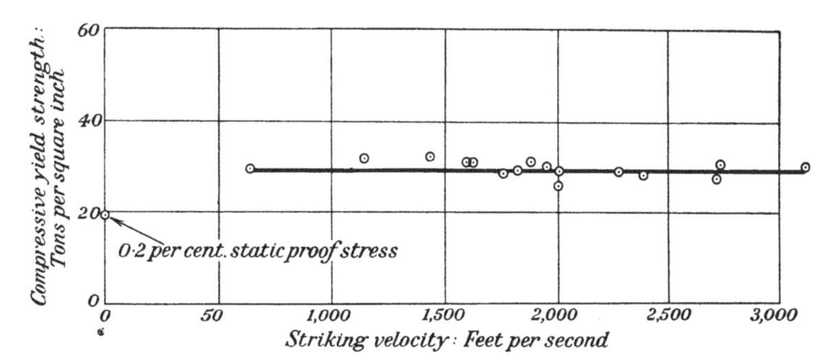


Fig. 14. Compressive yield strength of cylindrical projectiles made of duralumin of 'normal' strength.

yield stress is greater in these experiments than in those of Brown and Vincent or those obtained by our ballistic method, is probably due to the higher rate of straining or the shorter duration of the impact in the Road Research Laboratory experiments. At first sight one might expect that the high-velocity impacts would be of much shorter duration than the low-velocity ones, and that this would make the dynamic yield at high striking velocities greater than those obtained at lower

* The results for 120 lb. per sq.in. static refer to experiments with two types of cylinders of the same material (nickel-chrome steel). One kind gave perfect coherent pellets and a dynamic strength of 125 tons per sq.in. The other partially shattered and gave 92 tons per sq.in., so should not have been included.

speeds. The analysis indicates, however, that over the whole range of velocities used there is comparatively little change in duration of impact, provided the initial length is kept constant. This may well be the reason why so little change in dynamic yield is found when the impact velocity is varied.

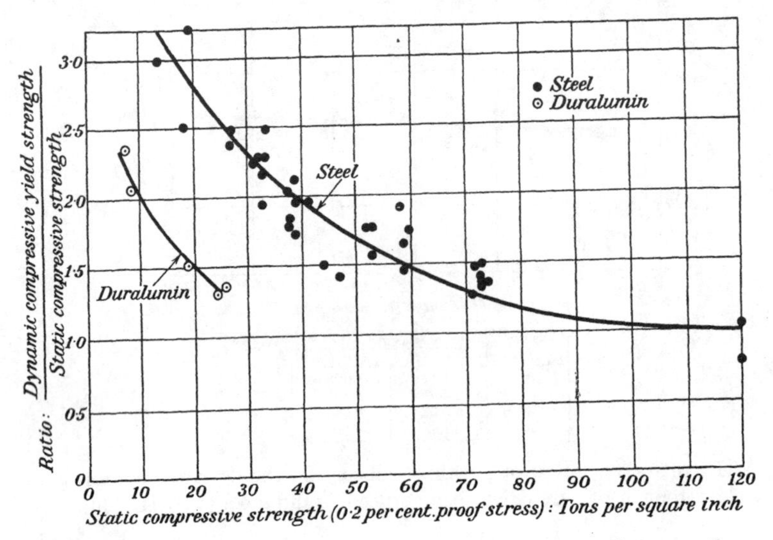


Fig. 15. Variation of dynamic strength with static strength.

Table 5. *Rate of strain and duration $T$ of impact in one set of measurements with mild steel projectiles*

| $u_0$ (ft. per sec.) | $l_1$ (in.) | $x_1$ (in.) | From (22) $S$ (tons per sq.in.) | More accurate analysis $S$ (tons per sq.in.) | Rate of strain (in. per in. per sec. sec.$^{-1}$) | Duration of impact $T$ (sec.) |
|---|---|---|---|---|---|---|
| 1310 | 0·57 | 0·41 | 62 | 64 | $1{\cdot}3\times10^4$ | $5{\cdot}5\times10^{-5}$ |
| 1395 | 0·57 | 0·36 | 66 | 56 | $1{\cdot}3\times10^4$ | $5{\cdot}1\times10^{-5}$ |
| 1445 | 0·49 | 0·33 | 58 | 60 | $1{\cdot}3\times10^4$ | $5{\cdot}9\times10^{-5}$ |
| 1510 | 0·46 | 0·28 | 56 | 59 | $1{\cdot}2\times10^4$ | $6{\cdot}0\times10^{-5}$ |
| 1600 | 0·41 | 0·25 | 55 | 58 | $1{\cdot}3\times10^4$ | $6{\cdot}2\times10^{-5}$ |
| 1876 | 0·30 | 0·16 | 56 | 57 | $1{\cdot}3\times10^4$ | $6{\cdot}2\times10^{-5}$ |
| 1890 | 0·40 | 0·22 | 70 | 76 | $1{\cdot}4\times10^4$ | $5{\cdot}3\times10^{-5}$ |
| 1900 | 0·30 | 0·17 | 60 | 60 | $1{\cdot}4\times10^4$ | $6{\cdot}1\times10^{-5}$ |
| 2300 | 0·22 | 0·09 | 60 | 64 | $1{\cdot}5\times10^4$ | $5{\cdot}6\times10^{-5}$ |
| 2400 | 0·21 | 0·08 | 64 | 69 | $1{\cdot}6\times10^4$ | $5{\cdot}4\times10^{-5}$ |

It is difficult to assign a value for the rate of strain in these experiments, because the strain is not uniform over the length of the specimens. As a rough approximation, one may take the mean strain in the plastic portion and divide it by the time of impact. The mean strain is

$$1-\left(\frac{\text{length of plastically strained portion}}{\text{initial length of this portion}}\right)=1-\frac{l_1-x_1}{L-x_1}=\frac{L-l_1}{L-x_1}.$$

The time of impact is of the order $T=\frac{2(L-l_1)}{u_0}$, so that the mean rate of strain in the plastic part is $\frac{u_0}{2(L-x_1)}$. The values of this are given in column 6 of table 5, and the values of the time of impact in column 7. It will be seen that there is little variation in the rate of strain or the time of impact as the impact velocity changes from 1310 to 2440 ft. per sec. The average rate of strain is 13,500 sec.$^{-1}$ and, since the mild steel in this experiment was comparable with those used in the experiments whose results are shown in fig. 10, it is of interest to set down Mr Whiffin's results in fig. 10, although it must be remembered that the two sets of results are not strictly comparable. It appears from fig. 10 that all the results for mild steel with yield point between 15 and 20 tons per sq.in. form a series which shows increasing yield strength with increase in the rate of strain.

## Stress Waves in a Plastic Material

As soon as the speed of testing becomes so high that inertia stress in the specimen is comparable with the stress associated with plastic strain, a complete picture of the course of events can be obtained only by considering plastic waves, just as the stress associated with elastic waves was considered in Hopkinson's experiments. It is true that, by a fortunate accident, it was possible to deduce the dynamic yield stress in the Road Research Laboratory experiments because the lack of knowledge of the exact motion in the mushrooming part of the projectile within the limits imposed by the final measurements did not seem to have much effect on the results, but this will not, in general, be true. Until recently no discussion of any case of plastic wave propagation has been given.

In 1940, my attention was drawn to a confidential paper on what happens to a wire when a bullet hits it. The results of experiments were excellently analysed by its author, using the theory of elasticity, and he came to the conclusion that there should be a critical velocity of impact, depending on the elasticity and on the strength of the wire, beyond which the bullet would break the wire. Experiments were carred out with many types of wire, in some of which plastic flow certainly occurred, and a critical velocity was always found. It was clear to me that this critical transverse velocity depended on there being a critical longitudinal velocity associated with the plastic waves, and I gave the theory of such waves to the department concerned; but, whilst I did not write a formal account of them, I did use a version of the theory in a report to the Ministry of Home Security.* Some time later the same treatment of the longitudinal plastic wave, produced by suddenly moving the end of a long wire at constant speed, was given independently in a confidential American report by von Kármán. Following von Kármán's paper much work was done on the subject both in the United States and in Great Britain. Since von Kármán and I had only written accounts of this work in confidential reports, I might have had some difficulty in referring to it in this lecture, but fortunately

* See papers 30 and 32 above.

exactly the same idea occurred a little later to a Russian (Rakhmatulin), who published it openly, thus absolving us from further responsibility in the matter.*

If one end of a long elastic wire is suddenly pulled longitudinally with a small velocity, $u_1$, and if the pull is maintained constant, a wave of tension equal to $mau_1$ (see equation (1)) travels along the wire. Now suppose that after this wave has been established, tension of the wire is suddenly increased still further by moving the end with additional velocity $u_2$; the tension is now $ma(u_1+u_2)$. If the velocity of the wave moving into the unstretched wire is $a_1$ and if $a_2$ denotes the velocity of the second wave *relative* to the wire that has already been stretched by the first pull, then the second wave will move with velocity in space equal to $a_2-u_1$. Thus, unless $a_2-u_1=a_1$, the wave fronts will separate or come together as they are propagated down the wire. In an elastic wire $u_1$ is always small in comparison with $a_1$ or $a_2$, which are equal if the wire obeys Hooke's law. Elastic waves in a wire are therefore propagated practically without change of form.

When the wire is plastic, or if its elasticity is not such that stress is proportional to strain, the velocity $C$ with which a small increment in velocity is propagated is

$$C=(1+\epsilon)\sqrt{\left(\frac{1}{m}\frac{dp}{d\epsilon}\right)}, \tag{23}$$

where $\epsilon$ denotes the extension at any point per unit length,

$p$ denotes the corresponding tension, and

$m$ denotes the mass per unit length.

In a plastic material like copper, for which the stress-strain curve is roughly parabolic, $dp/d\epsilon$ decreases as $\epsilon$ increases.

Therefore the velocity with which a disturbance is propagated decreases as the plastic strain increases. The wave produced by suddenly moving the end of a plastic wire and maintaining a constant tension at the end has a front which travels with the speed of an elastic wave, but the rear part, where the tension is greatest, travels much more slowly, so that the length of the wave continually increases as it passes down the wire.

These ideas can be expressed mathematically. If the end of a long plastic wire is moved with increasing or constant velocity a disturbance is propagated along it with velocity relative to the wire

$$c=(1+\epsilon)\sqrt{\left(\frac{1}{m}\frac{dp}{d\epsilon}\right)} \quad \text{or} \quad (1+\epsilon)\sqrt{\left(\frac{1}{\rho}\frac{d\sigma_\epsilon}{d\epsilon}\right)}, \tag{24}$$

where $\sigma_\epsilon$ denotes the 'nominal stress' per unit area.

The disturbance at any point consists of a portion of the wire which is stretched to strain $\epsilon$ and is moving in the direction of the pull with particle velocity

$$u=\int_0^\epsilon \sqrt{\left(\frac{1}{\rho}\frac{d\sigma_\epsilon}{d\epsilon}\right)}\,d\epsilon. \tag{25}$$

* K. A. Rakhmatulin, 'Propagation of a wave of unloading', *Appl. Math. Mech.* (Leningr.), IX (1945), 91.

The wave velocity in space of a point at which $u$, $\epsilon$, and $\sigma_\epsilon$ are constant is then $c-u$.

If a vertical wire is suddenly jerked downward, with velocity $u$ at time $t=0$, and the motion is maintained constant, all the states of velocity and strain are initially concentrated at the jerked end. Each state then travels down the wire at its appropriate speed $c-u$, so that the values of $\sigma_\epsilon$ and $\epsilon$ appropriate to the value of $u$ are found after time $t$ at distance

$$x=(c-u)\,t \tag{26}$$

from the initial position of the end of the wire.

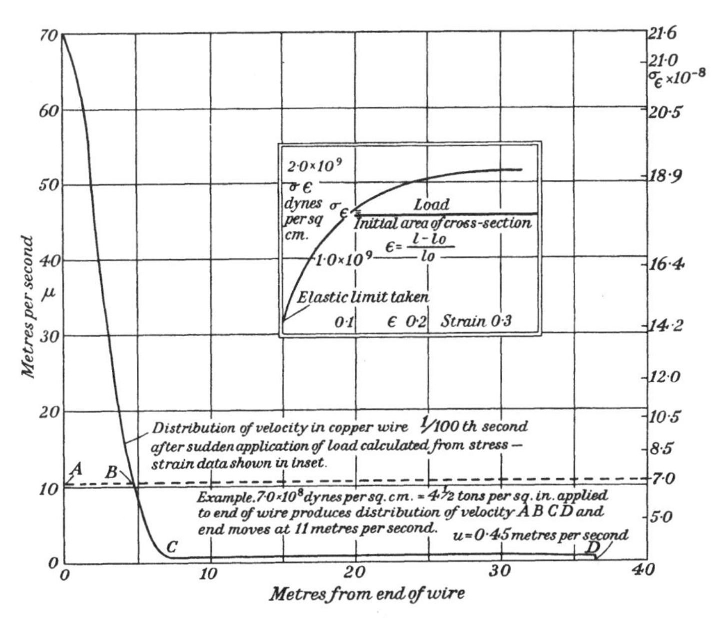

Fig. 16. Plastic wave in copper $\frac{1}{100}$ second after applying load.

The mechanics of these plastic waves can be understood most easily by working out an example in detail. For this purpose I have applied equations (24), (25) and (26) to an annealed copper wire, and the results are shown in fig. 16. The load-extension curve used in the calculations is shown inset, and the form of the wave $\frac{1}{100}$ sec. after the load had been applied is shown. The diagram gives the results for all velocities of applications of load. The stresses corresponding are shown on the right-hand side of fig. 16. It will be seen, for instance, that if a load of $7\times10^8$ dynes per sq.cm. (or $4\frac{1}{2}$ tons per sq.in.) were applied, it would move the end of the wire at 11 m. per sec. and at $\frac{1}{100}$ sec. the whole of the last 5 m. of wire would be under that stress. After a transition-zone extending from about 5 to 7 m. the elastic wave with stress at the elastic limit would stretch from 7 to 36 m.

It will be noticed, in fig. 16, that if the wire had been jerked at any speed greater than 70 m. per sec. the plastic wave would not have been able to relieve the stress, and the wire would immediately have broken off short. This brings us to a severe limitation of machines designed for high-speed tensile tests. A nicely designed machine of this type has been described by Mann,* and is shown in fig. 17. A heavy flywheel was rotated at high speed by a motor. Two pivoted flying arms, called horns, were released electrically from a point on the circumference and flew out under the influence of centrifugal force so as to strike an anvil attached to one end of the test specimen. The other end was attached to a heavy pendulum. It is simple to show that if the tension is the same at the two ends of the specimen, the swing of the pendulum is directly proportional to the work done in breaking it, and inversely, to the speed of the flywheel. Mann found that up to a certain speed the

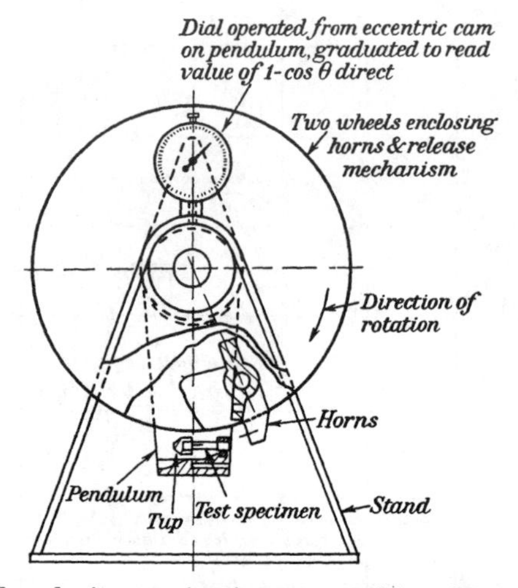


Fig. 17. High-velocity tension impact machine (Mann).

work done in breaking the specimen is constant, as is found with ordinary impact machines. At a certain critical speed, however, the energy, as determined by this method, decreased, whilst at the same time the elongation of the specimen remained constant or even increased. Fig. 18 shows one of Mann's sets of results.

It is, in general, very difficult to analyse the sequence of plastic waves reflected backwards and forwards in a plastic rod. The analysis of a plastic wave, started in a long wire from a longitudinal blow, shows, however, that there is a critical speed of impact beyond which the impacting body cannot communicate any energy to the wire. It seems likely that what Mann observed was a dynamical phenomenon of this type, though no doubt reflected waves, as well as the initial wave propagated from the 'horns' towards the pendulum, play a part. It is curious that the phenomenon announced by Dr J. Hopkinson in 1872 on theoretical grounds without adequate experimental evidence should be described in 1936 as a new experimental discovery of which no explanation had been suggested.

The analysis of plastic waves in finite metal rods or wires has been developed both in England and America. It is extremely complex owing to the fact that the plastic material concerned distorts only when the stress is increasing. If a part of the wave-

* H. C. Mann, 'High-velocity tension-impact test', *Proc. Amer. Soc. Test. Mat.* XXXVI (1936), 85.

system involves a decrease in stress the plastic flow ceases, and, in one case at least, which was worked out by Mr E. H. Lee, of the Armament Research Department, flow in parts of the specimen stopped, only to start up again at a later stage of the impact. In general, the final state of a specimen subject to impact can be calculated, with difficulty, when the stress-strain relation is known, but it is more difficult to reverse the process and find the stress-strain relationship by making measurements of the specimen after the test. For this reason, it is important to use every method available to find out how far a simplified conception like that underlying equation (22) can be verified experimentally.

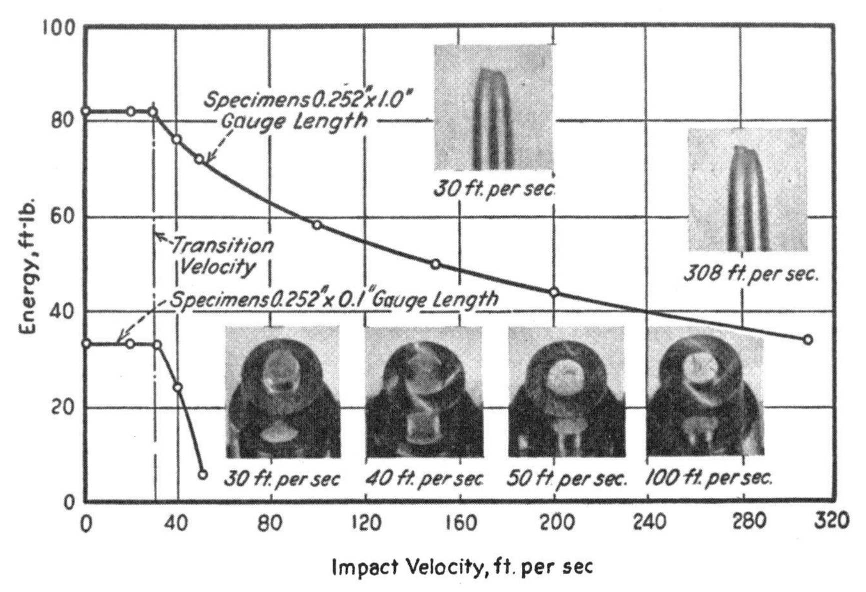


Fig. 18. Set of results using Mann's high-velocity tension impact machine.

In one case which has been mentioned it was found that nickel-chrome steel cylinders with static yield (0·2 % proof load) of 120 tons per sq.in., gave a dynamic strength of 125 tons per sq.in. when equation (22) was applied. In other experiments it was found that dynamic and static yield stresses tend to be the same when the static value is high. When it is remembered how completely the methods of measurement differ in the static and dynamic cases, this fact gives some confidence in the validity of the approximate equation (22).

The kind of experiment that might be expected to give the most complete results would be one in which the specimen was marked at regular intervals along a parallel portion of its length and the motion of these marks was followed by a drum camera, but there are considerable technical difficulties in doing work of this kind.

## Materials in which the Plastic Distortion is not Permanent

When the stress is released from a metal undergoing plastic deformation, it springs back only to the extent which would correspond with the elastic stress of the same magnitude. With materials which go by the name of plastics there is a much greater recovery; moreover, the recovery is, in general, determined not only by the amount of plastic deformation, but also by the duration of the load. This subject has been studied extensively in recent years by rheologists, but their experiments have not been carried to very high rates of straining. Dr E. Volterra and I have recently been making experiments in which short cylinders of various plastic materials have been placed on the plane end of a steel bar hung as a ballistic pendulum. Another bar, also hung as a ballistic pendulum, has been swung against it. The stress in the plastic cylinder has been deduced from the movement of the steel bar, which started from rest under the pressure exerted by the plastic specimen during the impact. The strain was deduced by photographing the gap between the steel bars with a rotating-drum camera. In this way stress-time and strain-time curves were found, and from these the stress-strain curve during an impact was constructed. Fig. 19 shows some of the results obtained with 'Polythene'. The stress-strain curve obtained in static compression tests, using specimens with lubricated ends, is also shown in the same figure.

In attempting to analyse the results we were driven, as the rheologists are, to think of some theoretical model which would behave in the same way. We adopted the model suggested many years ago by Boltzmann and others, consisting of a spring with the arbitrary static stress-strain relationship

$$\sigma = f(\epsilon), \tag{27}$$

and a series of springs and dashpots all connected in parallel. This system may be represented by the equation:

$$\sigma = f(\epsilon) + \int_0^t \phi(t-\tau)\frac{d\epsilon}{d\tau}\,d\tau, \tag{28}$$

where $\phi$ is an arbitrary function, $t$ denotes the time when the strain was $\epsilon$, and $\tau$ is a variable representing all times up to the time $t$. This formula takes account of the whole strain history of the plastic up to the time $t$, but it assumes that the stress at any moment is the sum of the effects of the strain when it existed at times $t-\tau$ before the instant concerned. A change in strain $\delta\epsilon$ is supposed to produce instantaneously a change in stress of amount $\phi(0)\,\delta\epsilon$, and after time $(t-\tau)$ it contributes an element $\phi(t-\tau)\,\delta\epsilon$ to the stress.

When the results of the experiments on polythene at room temperature were analysed, using this assumption, it was found that $\phi(t-\tau)$ was fairly represented over the range of our experiments by the formula

$$\phi(t-\tau) = Ae^{(\tau-t)/t_0} = 2{\cdot}0\times 10^9\, e^{600(\tau-t)} \text{ dynes per sq.cm.} \tag{29}$$

The values of $\phi$ found by analysing the experiments and also the values given by equation (29) are shown in fig. 20. This formula is the one which would be obtained with the model described above if only one dashpot were used and the spring attached to it had elasticity obeying Hooke's law. The time $t_0$ is described as the relaxation time. It must be remembered that the whole duration of the impact never exceeded 17 msec., so that the value of $\phi$ is not determined in these experiments beyond that time. It will be seen, however, that all the effects observed in these experiments are explained by the existence of an element in the material which has a relaxation time of about 1·7 msec.

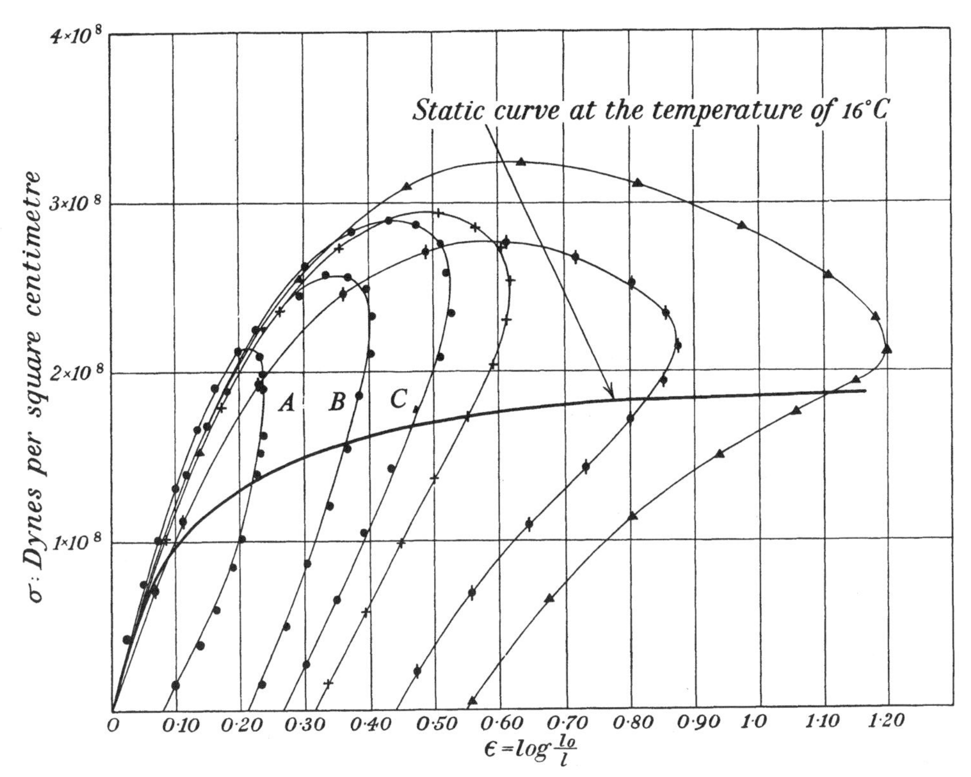


Fig. 19. Plastic stress/strain curves for polythene in direct compression. Comparison between static and dynamic tests at constant temperature and different initial speeds. (Experimental results.)

$\sigma = f(\epsilon)$ (static stress/strain relation).

$\sigma = f(\epsilon) + \int_0^t \phi(t-\tau) \frac{d\epsilon(\tau)}{d\tau} d\tau$ (dynamic stress/strain relation).

$\phi = \phi(t)$ (hereditary function).

Mass of the bar = 56·965 g. Length of specimens = 0·80 cm.

Diameter of specimens = 0·80 cm.

*Dynamic tests*

| Film no. | Date | Mark | Initial speed (cm. per sec.) | Temperature (° C.) | Time of impact ($10^{-3}$ sec.) |
|---|---|---|---|---|---|
| 1 | 5. i. 45 | ● (*A*) | 31·32 | 16 | 18·48 |
| 2 | 5. i. 45 | ● (*B*) | 44·29 | 16 | 16·48 |
| 4 | 8. i. 45 | ● (*C*) | 54·25 | 16·5 | 16·00 |
| 3 | 5. i. 45 | + | 62·64 | 16·5 | 16·48 |
| 3 | 8. i. 45 | ⧫ | 76·72 | 15·5 | 16·95 |
| 1 | 8. i. 45 | ▲ | 93·96 | 15 | 16·65 |

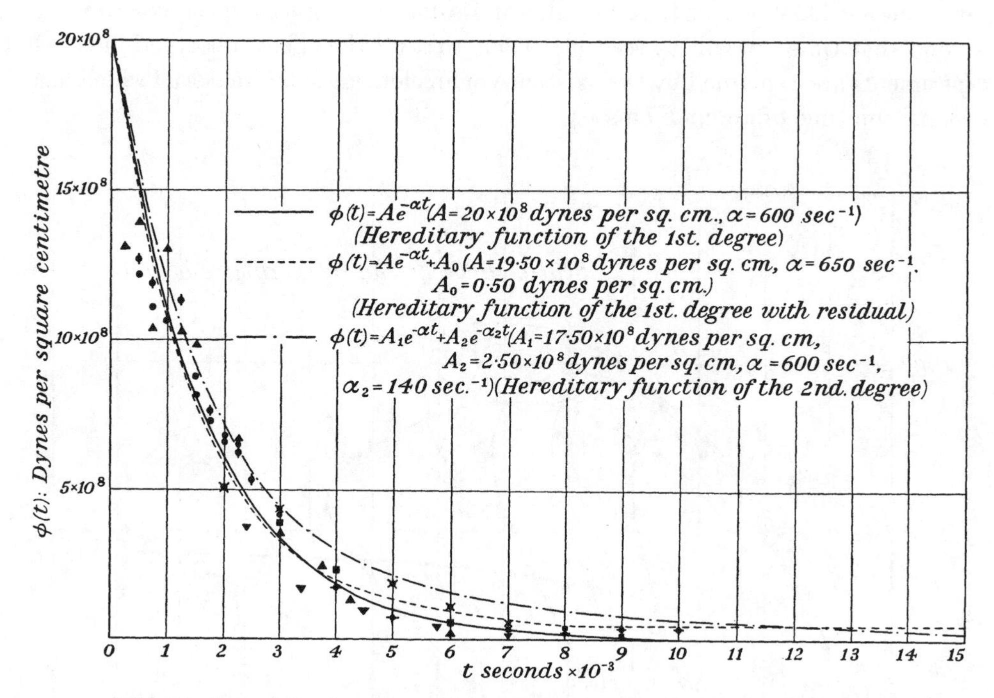


Fig. 20. Determination from the experimental results of the function $\phi(t)$ for polythene

| Film no. | Date | Mark | No. of bars | Mass of the bars (g.) | Initial speed (cm. per sec.) | Time of impact (sec. × $10^{-3}$) | Tempera-ture (°C.) |
|---|---|---|---|---|---|---|---|
| 2 | 1. v. 44 | ● | 1 | 6,000 | 99·00 | 5·00 | 16 |
| 1 | 1. v. 44 | ▲ | 1 | 6,000 | 140·00 | 5·10 | 16 |
| 1 | 14. viii. 44 | ⧫ | 2 | 6,000 | 88·50 | 5·35 | 19 |
| 3 | 5. i. 45 | ✦ | 1 | 56,965 | 62·64 | 16·48 | 16·5 |
| 4 | 8. i. 45 | ■ | 1 | 56,965 | 54·25 | 16·00 | 16·5 |
| 2 | 5. i. 45 | ✖ | 1 | 56,965 | 44·29 | 16·48 | 16 |
| 3 | 8. i. 45 | ▼ | 1 | 56,965 | 76·72 | 16·95 | 15·5 |

From the general relation $\sigma = f(t) + \int_0^t \phi(t-\tau)\,\frac{d\epsilon(\tau)}{d\tau}\,d\tau$ (dynamic stress/strain relation):

$$[\sigma - f(\epsilon)]_{t=0} = \left[\phi(0)\,\frac{d\epsilon(0)}{dt}\right]\delta t$$

$$[\sigma - f(\epsilon)]_{t=1} = \left[\phi(0)\,\frac{d\epsilon(1)}{dt} + \phi(1)\,\frac{d\epsilon(0)}{dt}\right]\delta t$$

$$[\sigma - f(\epsilon)]_{t=2} = \left[\phi(0)\,\frac{d\epsilon(2)}{dt} + \phi(1)\,\frac{d\epsilon(1)}{dt} + \phi(2)\,\frac{d\epsilon(0)}{dt}\right]\delta t$$

$$[\sigma - f(\epsilon)]_{t=3} = \left[\phi(0)\,\frac{d\epsilon(3)}{dt} + \phi(1)\,\frac{d\epsilon(2)}{dt} + \phi(2)\,\frac{d\epsilon(1)}{dt} + \phi(3)\,\frac{d\epsilon(0)}{dt}\right]\delta t$$

PLATE 1

Mild steel cylinders after test at 575 feet per second.

## Conclusion

In this lecture I have tried to describe some of the ways in which the strength properties of materials at high rates of strain can be deduced from results of dynamic experiments. It seems that the difficulties in the way of testing at high rates of straining are of quite a different order of magnitude from those at ordinary laboratory speeds. The use of modern techniques makes it possible in some cases to disentangle the effects of inertia from real changes due to speed of loading in the strength properties of the material itself. In the case of plastics, where the memory which the material has for its past strain history is important, a start has been made, but much more work is necessary before any real understanding of such complex phenomena is likely to be attained.

In conclusion, I must express my thanks to the Director of the Road Research Laboratory and to Mr Whiffin for permission to publish the results of their experiments.

# 37

# A CONNECTION BETWEEN THE CRITERION OF YIELD AND THE STRAIN RATIO RELATIONSHIP IN PLASTIC SOLIDS

REPRINTED FROM

*Proceedings of the Royal Society*, A, vol. CXCI (1947), pp. 441–6

The assumption that the work done during a small plastic strain is a maximum as the yield-stress criterion is varied is shown to give rise to a connection between the yield-stress and the strain-ratio relationship. The strain-ratio relationship is that which exists between the ratios of principal stress differences and the ratios of the corresponding strain differences. It is common to assume that this relationship is one of simple proportionality. Experiments, however, show that this assumption is not true in metals. The observed strain-ratio relationship is used in conjunction with the assumption of maximum work during a given strain to calculate the criterion of yield. It is found that this is very close to, but not identical with, the Mises-Hencky criterion.

## Introduction

The three principal properties which define the plastic behaviour of an isotropic material and must be known or assumed before the solution of any problem in plastic flow can be attempted are:

(*a*) The criterion of yield (Mises-Hencky, maximum stress difference, etc.).

(*b*) The relationship between the ratios of small strains parallel to the principal axes and the ratios of principal stress differences. This relationship may be called the strain-ratio relationship.

(*c*) The amount of strain hardening after yielding.

When problems concerning small strains are considered it is frequently a good enough approximation to assume that there is no strain hardening, so that the criterion of yield is independent of the amount of small strain considered. On the other hand, both (*a*) and (*b*) must be known before any problems in which the stress depends on the strain can be solved.

It has usually been considered that (*a*) and (*b*) are quite unrelated, and from the purely mathematical point of view this is true. Problems are solvable mathematically, for instance, if either the Mises-Hencky or the maximum-stress difference hypothesis is assumed in conjunction either with the strain-ratio relationship observed with viscous fluids or with other strain-ratio assumptions. When the physical nature of a plastic material is considered, however, it seems that a certain relationship between (*a*) and (*b*) must exist. A relationship is here deduced using an assumption that the energy dissipation for a given strain is a maximum. The theoretical result is compared with available observations.

## Representation of Plasticity Conditions

### (a) *Criterion of yield*

If $\sigma_1$, $\sigma_2$ and $\sigma_3$ are the three principal stresses it may be assumed that the yield criterion depends only on the stress differences $\sigma-\sigma_2$, $\sigma_2-\sigma_3$, $\sigma_3-\sigma_1$ and is independent of the mean pressure $\bar{p}=-\frac{1}{3}(\sigma_1+\sigma_2+\sigma_3)$. It is convenient to express results in terms of the three residual stresses $\sigma_1'$, $\sigma_2'$, $\sigma_3'$, which result from subtracting the mean pressure from the principal stress, so that

$$\sigma_1'=\tfrac{2}{3}\sigma_1-\tfrac{1}{3}\sigma_2-\tfrac{1}{3}\sigma_3,\text{ etc.,} \quad \text{and} \quad \sigma_1'+\sigma_2'+\sigma_3'=0. \tag{1}$$

In these co-ordinates the Mises-Hencky criterion of yield is

$$(\sigma_1')^2+(\sigma_2')^2+(\sigma_3')^2=\tfrac{2}{3}Y^2, \tag{2}$$

and the maximum shear stress criterion is

$$\sigma_1'-\sigma_3'=Y, \tag{3}$$

where $Y$ is the yield stress in direct tensile loading and

$$\sigma_1'\geqslant\sigma_2'\geqslant\sigma_3'. \tag{4}$$

### (b) *Strain-ratio relationship*

It will be assumed that the compressibility of the material is unchanged when plastic flow sets in. If $e_1, e_2, e_3$ are the principal strains (assumed small) it is convenient to use the residual principal strains

$$e_1'=e_1-\tfrac{1}{3}(e_1+e_2+e_3),\text{ etc.,} \tag{5}$$

because

$$e_1'+e_2'+e_3'=0, \tag{6}$$

so that if

$$e_1'\geqslant e_2'\geqslant e_3', \tag{7}$$

all possible strain ratios have been considered when $e_3'/e_1'$ varies from $-\frac{1}{2}$ to $-2$.

In considering all possible relationships between stress ratios and strain ratios it is clear that any one of them could be represented by a curve in a diagram with $\sigma_3'/\sigma_1'$ as abscissa and $e_3'/e_1'$ as ordinate, the full range extending from $-\frac{1}{2}$ to $-2$ in each case. This diagram is unsymmetrical in the sense that if the yield is the same in compression as in tension the symmetrical diagram would show the compression part of the curve as similar to the part representing extension. The former extends from $\sigma_3'/\sigma_1'=-\frac{1}{2}$ to $-1$, while the latter extends from $-1$ to $-2$, so that the diagram is unsymmetrical. To avoid this difficulty Lode (1926) introduced the symmetrical variables

$$\left.\begin{aligned}\mu&=\frac{2\sigma_2'-\sigma_1'-\sigma_3'}{\sigma_1'-\sigma_3'} \quad \text{or} \quad -3\frac{\sigma_1'+\sigma_3'}{\sigma_1'-\sigma_3'},\\ \nu&=\frac{2e_2'-e_1'-e_3'}{e_1'-e_3'} \quad \text{or} \quad -3\frac{e_1'+e_3'}{e_1'-e_3'}.\end{aligned}\right\} \tag{8}$$

35-2

The results of experiments on all possible combinations of $\sigma_1', \sigma_2', \sigma_3', e_1', e_2', e_3'$ can be represented on a square diagram showing $\mu$ as abscissa and $\nu$ as ordinate and covering the range $-1<\mu<+1$, $-1<\nu<+1$. It will be seen that when $\nu$ is positive, $e_2'$ is positive. This corresponds with cases where the residual principal strain which has the greatest absolute value is compressive; the point $\nu=+1$ for instance corresponds with a uniaxial compressive load parallel to $e_3'$. When $e_2'$ is negative the residual strain which has the greatest absolute value is positive and $\nu=-1$ corresponds with uniaxial extension parallel to $e_1'$. In an isotropic medium the $\mu$, $\nu$ curve must pass through the points $\mu=\nu=-1$, $\mu=\nu=0$, $\mu=\nu=+1$. Otherwise no limitations on the form of the $\mu$, $\nu$ curve are imposed by considerations of symmetry

Table 1

| 1 | 2 | 3 | 4 | 5 | 6 | 7 |
|---|---|---|---|---|---|---|
| $\mu$ | $\nu$ | $-\sigma_3'/\sigma_1'$ | $\chi$ | $\frac{1}{\chi-\sigma_3'/\sigma_1'}$ | $m$ | $\frac{\frac{1}{2}m}{1+\sigma_3'/\sigma_1'}$ |
| −0·021 | −0·011 | 0·9860 | 1·022 | 0·498 | 0·025 | 0·907 |
| −0·242 | −0·171 | 0·8507 | 1·412 | 0·442 | 0·280 | 0·885 |
| −0·471 | −0·393 | 0·7286 | 2·295 | 0·331 | 0·515 | 0·946 |
| −0·597 | −0·502 | 0·6680 | 3·016 | 0·271 | 0·68 | 0·978 |
| −0·652 | −0·554 | 0·6429 | 3·484 | 0·240 | 0·70 | 0·980 |
| −0·773 | −0·659 | 0·5902 | 4·865 | 0·183 | 0·80 | 0·978 |
| −0·874 | −0·775 | 0·5487 | 7·889 | 0·118 | 0·90 | 0·997 |
| −0·954 | −0·876 | 0·5174 | 15·13 | 0·064 | 0·95 | 0·983 |

alone. The results of experiments in which complex stresses were produced in a tube (*a*) by simultaneous end-load and internal pressure,* and (*b*) by simultaneous end-load and torque are given in a paper by the present author and the late Mr H. Quinney.† It was found that with most of the metals used, namely, mild steel, decarburized mild steel, copper, nickel and aluminium, the $\mu$, $\nu$ curve lies below the line $\mu=\nu$ when $\mu$ and $\nu$ are positive (i.e. $\mu>\nu>0$), and above this line when $\mu$ and $\nu$ are negative (i.e. $\mu<\nu<0$). With very soft metals like lead and cadmium this was still true, but the experimental $\mu$, $\nu$ curves lie nearer to the line $\mu=\nu$ than with the above-mentioned metals. Experiments made with glass heated till it began to flow gave points lying very closely on the line $\mu=\nu$. This is to be expected in view of the fact that $\mu=\nu$ for viscous fluids.

Values of $\mu$ and $\nu$ found in experiments with annealed tubes of pure copper are given in columns 1 and 2 of table 1, and the corresponding values of $\sigma_3'/\sigma_1'=(\mu+3)/(\mu-3)$ are given in column 3.

## Possible Maximum Energy Dissipation Assumption

In a paper published recently‡ it is shown that if the strain-ratio relationship expressed by the equation

$$\mu=\nu \tag{9}$$

* W. Lode, *Z. Phys.* XXXVI (1926), 913.

† G. I. Taylor and H. Quinney, *Phil. Trans.* A, CCXXX (1931), 323; paper 16 above.

‡ R. Hill, E. H. Lee and S. J. Tupper, *Proc. Roy. Soc.* A, CXCI (1947), 287.

holds, the work done during a given small strain is a maximum if the Mises-Hencky criterion of flow applies. Since experiments with most metals reveal a marked divergence from the ideal relationship (9), it is of interest to find out what criterion of flow would correspond with the maximum of work during a small strain when the experimentally observed relationship between $\mu$ and $\nu$ is used instead of (9). The work done during a small strain is

$$W = \sigma_1' e_1' + \sigma_2' e_2' + \sigma_3' e_3'$$

or

$$W = \sigma_1'(2e_1' + e_3') + \sigma_3'(e_1' + 2e_3'). \tag{10}$$

A stationary value of $W$ for a given strain occurs when

$$\frac{d\sigma_3'}{d\sigma_1'} = -\frac{2e_1' + e_3'}{e_1' + 2e_3'}. \tag{11}$$

Table 2

| 1 | 2 | 3 | 4 (Mises-Hencky) | 5 (Mohr) |
|---|---|---|---|---|
| $\sigma_3'/\sigma_1'$ | $I$ | $\frac{3\sigma_1'}{2Y}$ | $\frac{\sqrt{3}}{2}\left\{1+\frac{\sigma_3'}{\sigma_1'}+\left(\frac{\sigma_3'}{\sigma_1'}\right)^2\right\}^{-\frac{1}{2}}$ | $\frac{3}{2}\left(1-\frac{\sigma_3'}{\sigma_1'}\right)^{-1}$ |
| −1 | 0 | 0·849 | 0·866 | 0·750 |
| −0·95 | 0·0247 | 0·870 | 0·888 | 0·770 |
| −0·90 | 0·0487 | 0·891 | 0·908 | 0·789 |
| −0·85 | 0·0716 | 0·912 | 0·927 | 0·810 |
| −0·80 | 0·0926 | 0·931 | 0·945 | 0·834 |
| −0·75 | 0·1114 | 0·949 | 0·960 | 0·856 |
| −0·70 | 0·1278 | 0·964 | 0·974 | 0·882 |
| −0·65 | 0·1416 | 0·978 | 0·985 | 0·909 |
| −0·60 | 0·1526 | 0·989 | 0·992 | 0·938 |
| −0·55 | 0·1605 | 0·996 | 0·997 | 0·969 |
| −0·525 | 0·1629 | 0·999 | 0·999 | 0·985 |
| −0·50 | 0·1640 | 1·00 | 1·00 | 1·00 |

It has been seen that experiments on the strain-ratio relationship give $\sigma_3'/\sigma_1'$ in terms of $e_3'/e_1'$, so that (11) will enable the criterion of yield to be determined by numerical integration. Alternatively, if the relationship between $\sigma_3'$ and $\sigma_1'$ is known by experiment, (11) gives directly the strain-ratio relationship. To determine the criterion of yield from the observed relationship between $\sigma_3'/\sigma_1'$ and $e_3'/e_1'$ (11) may be written

$$\frac{d\sigma_3'}{d\sigma_1'} = \sigma_1' \frac{d}{d\sigma_1'}\left(\frac{\sigma_3'}{\sigma_1'}\right) + \frac{\sigma_3'}{\sigma_1'} = -\frac{(2 + e_3'/e_1')}{(1 + 2e_3'/e_1')}. \tag{12}$$

Taking the experimental value of $e_3'/e_1'$ corresponding with each value of $\sigma_3'/\sigma_1'$, the corresponding experimental value of $-\frac{2 + (e_3'/e_1')}{1 + (2e_3'/e_1')}$ can be found. Calling this $\chi$, (11) can be integrated, giving

$$\text{constant} + \log_e \sigma_1' = \int \frac{d(\sigma_3'/\sigma_1')}{\chi - (\sigma_3'/\sigma_1')} = I. \tag{13}$$

Expressed in terms of $\nu$, $\chi=\dfrac{1-\nu}{1+\nu}$. Values of this for copper are given in column 4, table 1, and values of the integrand of (12) in column 5. To perform the integration the values given in column 5, table 1, were plotted against $\sigma_3'/\sigma_1'$. The values of the integral $I$ were then determined from this diagram for values of $\sigma_3'/\sigma_1'$ ranging from $-1$ to $-0{\cdot}5$. These are given in column 2, table 2. The value of the constant in (13) is determined by the fact that when $\sigma_3'/\sigma_1'=-0{\cdot}5$, $\sigma_1'=\frac{2}{3}Y$. In this way the values given in column 3 of table 2 were found for $3\sigma_1'/2Y$.

## Comparison with Mises-Hencky and Maximum Stress Difference Criteria

The Mises-Hencky criterion (2) may be written

$$\sigma_1'^2+\sigma_1'\sigma_3'+\sigma_3'^2=\tfrac{1}{3}Y^2, \tag{14}$$

so that

$$\frac{3}{2}\frac{\sigma_1'}{Y}=\frac{\sqrt{3}}{2}\left\{1+\frac{\sigma_3'}{\sigma_1'}+\left(\frac{\sigma_3'}{\sigma_1'}\right)^2\right\}^{-\frac{1}{2}}. \tag{15}$$

Values of this function are given in column 4 of table 2

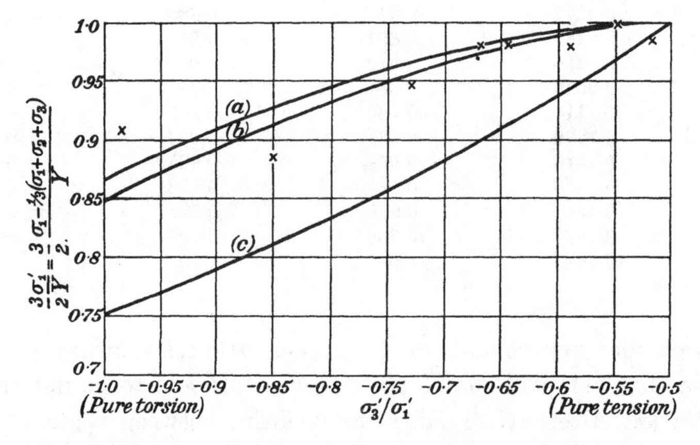


Fig. 1. (*a*) Mises-Hencky. (*b*) Calculated by stationary energy dissipation hypothesis. (*c*) Maximum stress difference (Mohr). × × Observed points (Taylor and Quinney*).

The maximum stress difference criterion is

$$\sigma_1'-\sigma_3'=Y \quad \text{or} \quad \frac{\sigma_1'}{Y}=\frac{1}{1-\sigma_3'/\sigma_1'}. \tag{16}$$

For comparison the values of $\dfrac{3}{2(1-\sigma_3'/\sigma_1')}$ are shown in column 5 of table 2. The values given in columns 3, 4 and 5 are shown graphically in fig. 1.

* Loc. cit.

## Comparison with Observed Criterion

The observations made with copper tubes subjected to combined end-load and torsion* were carried out by first loading the tube till it stretched, say, 0·5 %.

If the longitudinal stress for this pure extension was $Y$ the load was reduced to $P=mY(0<m<1)$ and torque applied till plastic flow occurred. It was found that though the beginning of inelastic displacement was not easy to measure the torque at which steady plastic flow occurred was definite. The shear stress corresponding with this condition being $S$, the angle $\theta$ between the axis of the tube and the axis of the greatest principal stress is given by

$$\tan 2\theta = \frac{2S}{P}. \tag{17}$$

The principal stresses are

$$\sigma_1 = \tfrac{1}{2}P(1+\sec 2\theta), \quad \sigma_2 = 0, \quad \sigma_3 = \tfrac{1}{2}P(1-\sec 2\theta), \tag{18}$$

so that

$$\sigma_1'/Y = \tfrac{1}{2}m(1+\sec 2\theta) - \tfrac{1}{3}m, \tag{19}$$

and

$$\frac{\sigma_1'}{\sigma_3'} = \frac{\frac{1}{3}+\sec 2\theta}{\frac{1}{3}-\sec 2\theta} \quad \text{or} \quad \cos 2\theta = 3\frac{\sigma_3'+\sigma_1'}{\sigma_1'-\sigma_3'}. \tag{20}$$

Hence

$$\frac{3}{2}\frac{\sigma_1'}{Y} = \frac{1}{2}\frac{m}{1+(\sigma_3'/\sigma_1')}. \tag{21}$$

The experimental values of $m$ and $\sigma_3'/\sigma_1'$ are given in columns 6 and 3 of table 1. The values of $3\sigma_1'/2Y$ found by inserting those in (21) for values of $m>0·28$ are given in column 7 of table 1. When $m$ is very small (21) tends to the form 0/0, so that another expression for $\sigma_1'$ is more accurate. The form used when $m=0·025$ and $m=0·28$ was

$$\frac{3}{2}\frac{\sigma_1'}{Y} = \frac{P}{Y}\left\{\frac{1}{4}+\frac{3}{4}\sqrt{\left(1+\frac{4S^2}{P^2}\right)}\right\}. \tag{22}$$

The experimental points are plotted in figure 1, where it will be seen that the material behaves very nearly in accordance with the prediction based on the assumed stationary energy principle and the measured strain ratio relationship. Both the observed points and those calculated by the minimum energy relationship lie close to the Mises-Hencky line.

* Taylor and Quinney, loc. cit.

## 38

# THE FORMATION AND ENLARGEMENT OF A CIRCULAR HOLE IN A THIN PLASTIC SHEET

REPRINTED FROM

*Quarterly Journal of Mechanics and Applied Mathematics*, vol. I (1948), pp. 103–24

When a circular hole is made in a flat sheet by a conical-headed bullet or by outward radial pressure on its edge, the metal near the hole piles up into a thickened crater. The mechanics of this deformation is discussed. The interest of the problem lies in the fact that the complete strain history of each element of the sheet has to be calculated. This is because the ratios of the principal stresses at each element of the sheet vary as the deformation proceeds, so that there is no relationship between the stress and total deformation but only between stress and strain increments occurring during a small expansion of the hole.

If $b$ is the radius of the hole at any time, the strain is found to be elastic at points where $r$, the radius, is $> 3\cdot64b$. In the annulus $2\cdot21b < r < 3\cdot64b$ the strain is plastic, but comparable in magnitude with the small elastic strain when $r > 3\cdot64b$. In the annulus $b < r < 2\cdot21b$ there is finite strain. At the edge of the hole the sheet has thickened to 2·61 times the thickness of the sheet.

Experiments made with lead show that the symmetrical deformation contemplated in this analysis does not occur; but an alternative unsymmetrical deformation is produced which calculation shows to require less work, in the ratio 2·6 to 1·0, than the symmetrical mode.

## INTRODUCTION

It is a comparatively easy matter to formulate simple laws of plasticity, but it is difficult to solve even the simplest special problems. Perhaps the most difficult feature in solving such problems is to connect the stress with the total deformation when the ratios of the principal stresses at each element of the material are not constant during the straining process. Most of the problems to which correct solutions have been given are concerned with plane strain of incompressible material, and in that case this difficulty does not arise. The chief interest in the solution of the special problem here discussed is that the ratios of the principal stresses change during the straining so that the whole history of the straining of each element has to be followed from the beginning to the end of the straining period.

In only one previously published case has this difficulty been faced.* The present solution was obtained before that work was started, and as its authors pointed out their recognition of the fundamental difficulty in this kind of work was prompted

* R. Hill, E. H. Lee, S. J. Tupper, 'The theory of combined plastic and elastic deformation with particular reference to a thick tube under internal pressure', *Proc. Roy. Soc.* A, CXCI (1947), 278.

by some unpublished work of mine which they had seen. The solution which follows is part of that work. The remainder will not be published as it has been superseded by the much more complete work of Hill, Lee and Tupper.

## Enlargement of a Hole in a Sheet

When a hole is enlarged in a sheet of metal by radial pressure, as it is when pierced by a pointed conical broach or a pointed bullet, the metal near the edge of the hole is thickened over an area extending out to radius, say, $r_2$. Beyond this radius the metal suffers only small thickening, and outside some radius $r_1$ it suffers only elastic strain.

A simplified analysis illustrating the main features of this state of affairs was given in a private communication to the author by H. A. Bethe. With his permission a shortened version of this analysis is given below.

### *Bethe's analysis*

The following assumptions were made:

(1) The sheet is thin and the radius of the hole is large compared with the thickness. The sheet is of infinite extent.

(2) Outside a radius $r_1$, the sheet is in a state of elastic stress due to radial pressure at $r_1$.

(3) The stress $\sigma_z$ perpendicular to the sheet is zero.

(4) In a region $r_2 < r < r_1$ the plastic strain is small so that the stress distribution adjusts itself till a flow condition

$$\sigma_\theta - \sigma_r = Y \tag{1}$$

is fulfilled. Here $Y$ is the yield stress.

(5) Mohr's condition of plastic flow is satisfied so that the maximum difference in absolute value between the principal stresses is equal to $Y$. It is further assumed that $Y$ is constant for large as well as small strains.

Using the equations of equilibrium and elasticity with assumptions (2) and (3), the following expressions for the stresses in the elastic region $r > r_1$ are derived:

$$\left.\begin{array}{ll}\text{radial stress} & \sigma_r = -A/r^2 \\ \text{tangential stress} & \sigma_\theta = +A/r^2 \\ \text{stress normal to sheet} & \sigma_z = 0\end{array}\right\} \quad (r > r_1). \tag{2}$$

Here $A$ is a constant for any given state of stress but varies as the straining proceeds.

Using the equation of equilibrium and the plasticity condition (4)

$$\left.\begin{array}{l}\sigma_r = Y\left(\log\dfrac{r}{r_1} - \dfrac{1}{2}\right) \\ \sigma_\theta = Y\left(\log\dfrac{r}{r_1} + \dfrac{1}{2}\right) \\ \sigma_z = 0\end{array}\right\} \quad (r_1 > r > r_2). \tag{3}$$

The condition that $\sigma_r$ is continuous at $r_1$ requires

$$\frac{A}{r_1^2}=\tfrac{1}{2}Y. \tag{4}$$

Bethe pointed out that, if Mohr's condition (5) is assumed, equations (1) and (3) are only applicable so long as $\sigma_\theta$ is positive ($\sigma_r$ is necessarily negative), for it is only then that the maximum difference between the principal stresses is that between $\sigma_r$ and $\sigma_\theta$. When $\ln(r/r_1)<-\tfrac{1}{2}$ the maximum difference is between $\sigma_r$ and $\sigma_z$. The radius $r_2$ at which $\sigma_\theta=0$ is therefore

$$r_2=r_1e^{-\frac{1}{2}}=0{\cdot}606r_1,$$

or

$$r_1=1{\cdot}65r_2. \tag{5}$$

Within the radius $r_2$ the condition of maximum stress difference could still be satisfied if $\sigma_\theta=0$ and $\sigma_r=-Y$. The equilibrium equation applicable to sheets of uniform thickness, namely,

$$\frac{d\sigma_r}{dr}+\frac{\sigma_r-\sigma_\theta}{r}=0, \tag{6}$$

would not be satisfied. On the other hand, if the material were able to flow within the radius $r_2$ in such a way that the thickness $h$ were equal to $h_0r_2/r$, $h_0$ being the original thickness, the equation of equilibrium of a thin sheet of variable thickness, namely,

$$\frac{d}{dr}(h\sigma_r)+\frac{h}{r}(\sigma_r-\sigma_\theta)=0, \tag{7}$$

would be satisfied.

If the hole is expanded from a pin-hole to radius $b$ and if the compressibility of the material be neglected, the volume of the metal contained between $r=b$ and $r=r_2$ must be equal to $\pi r_2^2h_0$. Thus

$$\pi r_2^2h_0=\int_b^{r_2}\frac{h_0r_2}{r}(2\pi r\,dr),$$

which yields

$$b=\tfrac{1}{2}r_2. \tag{8}$$

Hence

$$[h]_{r=b}=2h_0. \tag{9}$$

Bethe's model indicates a reason for the main feature of the observed plastic strains round the hole made in a sheet by a conical-headed punch or bullet, namely, the thickening of the plate close to the hole. When, however, actual plastic strain near the hole is considered in detail it will be seen that the arbitrary assumption that $\sigma_\theta=0$ and $\sigma_r=-Y$ is inconsistent with any theory of plasticity in which the ratios of increments of strain are related to corresponding ratios of stress differences. If $\sigma_\theta=\sigma_z=0$ and $\sigma_r=-Y$ through the range $b<r<r_2$, any theory of stress and strain in an isotropic plastic material would require that the small increments in strain occurring while the radius of the hole was increased from $b$ to $b+\delta b$ would be such that the radial strain $e_r$ is equal to the normal strain $e_z$. It is clear that this is not

consistent with the actual strain involved in the equation $h=h_0 r_2/r$. In fact, for a small increment $\delta b$ in radius of the hole, Bethe's model involves radial displacement $\delta u$, where

$$\delta u=\left(2-\frac{b}{r}\right)\delta b,$$

and the components of strain increment are

$$\left.\begin{aligned}\delta e_\theta&=\frac{2b-r}{r}\frac{\delta b}{b},\\ \delta e_z&=\frac{\delta h}{h}=\frac{4b}{r}\left(1-\frac{b}{r}\right)\frac{\delta b}{b},\\ \delta e_r&=\left(1-\frac{6b}{r}+\frac{4b^2}{r^2}\right)\frac{\delta b}{b}.\end{aligned}\right\}\tag{10}$$

Thus in Bethe's analysis the ratios $\delta e_r:\delta e_\theta:\delta e_r$ vary with radius while the ratios of the principal stresses are constant.

## Problems in Plasticity when the Ratios of Principal Stress Differences are not Constant at each Point during the Straining Process

The incompatibility, in any rational theory of plasticity, of Bethe's hypothesis $(\sigma_\theta=\sigma_z=0)$ with the strains represented by (10) must mean that this hypothesis is incorrect. To improve the analysis it is necessary to use some hypothetical assumption or experimental result relating the stress-difference ratios to the strain ratios. For small strains the simplest is that of Mises, namely,

$$\frac{\sigma_r-\sigma_\theta}{e_r-e_\theta}=\frac{\sigma_\theta-\sigma_z}{e_\theta-e_z}=\frac{\sigma_z-\sigma_r}{e_z-e_r}.\tag{11}$$

This, though not quite an accurate representation of experimental results, is the hypothesis most frequently used in discussing problems in plasticity. It will be used in the work which follows.

If the stress-difference ratios, namely, $\sigma_r-\sigma_\theta:\sigma_\theta-\sigma_z:\sigma_z-\sigma_r$, are constant at each element of the material during the whole straining process, the corresponding strain-difference ratios $e_r-e_\theta:e_\theta-e_z:e_z-e_r$ are constant and (11) will apply if $e_r$, $e_\theta$ and $e_z$ are the total strains. If, however, the stress-difference ratios change in the course of the deformation, (11) can only apply if $e_r$, $e_\theta$ and $e_z$ are small increments of strain occurring during a portion of the deformation in which there is small change in the stress distribution. Problems of plastic flow in which the stress ratios vary during the deformation can therefore only be solved by following the strain history of each element.

Eminent authorities have not always appreciated this point and have consequently published erroneous solutions of problems in which the stresses and *total* plastic strains have been related as though the stress distribution had been constant during the deformation when in fact it had not. The solutions of the problem of the straining of a thick-walled tube beyond the elastic limit which are given by Nadai* and by Sokolowsky† seem to me to be defective for this reason.

In the following pages the solution is given to the problem presented by a hole in a sheet of plastic material which is expanded from a pin-hole. The solution involves tracing the complete strain-history of each element of the sheet, but the analysis is much simplified by considerations of symmetry and similarity.

## Analysis of Strain round an Expanding Radial Hole in a Sheet

When a hole is enlarged the finite strain at any stage is made up of infinitesimal elements of strain which vary as the enlargement proceeds. Thus, when a small pin-hole in a plate is enlarged we must study the small strain produced in an element of the sheet which was originally at radius $s$ from the pin-hole, when the hole enlarges from radius $b$ to radius $b+\delta b$. In the more general case when the initial radius of the hole in the unstretched sheet is not zero this is very difficult to analyse, but when the expansion starts from a small pin-hole it may be expected that the configuration when the hole has radius $b_2$ will be similar to that round the hole when its radius is $b_1$ except that the radii where any given thickness occurs will be changed in the ratio $b_2/b_1$. Thus, if $h$ is the thickness and $u$ the radial displacement, it may be assumed that $h/h_0$ and $u/b$ and also the stresses are functions of $s/b$ only, where $h_0$ is the initial thickness of the sheet, and $s$ the initial radial distance of a particle.

To simplify matters I have assumed that the compressibility is so small that it may be neglected and the material taken as incompressible. The relationship between the small strain which occurs at any radius during the expansion of the hole through a small increase in radius from $b$ to $b+\delta b$ can be understood by referring to fig. 1. Here the ordinates represent $u$ and the abscissae $r$.

The initial radial distance of the element, which at a subsequent stage in the opening-out of the hole is at radius $r$, is related to $u$ by the equation

$$r=s+u. \tag{12}$$

In fig. 1, therefore, the displacement of a particle from its initial radius $s$ is represented by a line drawn at 45° to the axes. In particular the displacement of the particles which were initially at the pin-point where the hole began is represented by the 45°-line $OP_0P_1$. The curved line $P_0AQ_0$ represents the relationship between $r$ and $u$ which it is the object of the analysis to calculate. At a subsequent stage of the expansion, when the hole has expanded from radius $b$ to radius $b+\delta b$, the curve $P_1BCQ_1$ representing displacement is similar to $P_0AQ_0$, but with its linear dimensions increased in the ratio $b+\delta b:b$; thus in fig. 1,

$$\frac{P_1P_0}{OP_0}=\frac{AC}{AO}=\frac{AD}{r}=\frac{\delta b}{b},$$

so that

$$AD=r\delta b/b. \tag{13}$$

* *Plasticity* (McGraw-Hill, 1931), pp. 196–9.

† *The Theory of Plasticity* (in Russian) (Moscow, 1946), chapter III.

If $\delta r$ is the change in $r$ *for a given particle of material* when the hole expands from $b$ to $b+\delta b$, $\delta r$ is found by drawing the line $AB$ at 45° to the axes to meet the curve $P_1BCQ_1$ in $B$. If $\delta b/b$ is small enough, the arc $CB$ may be taken as straight so that, if $\pi-\alpha$ is the slope of $CB$ to the axis,

$$\frac{\partial u}{\partial r} = -\tan\alpha. \tag{14}$$

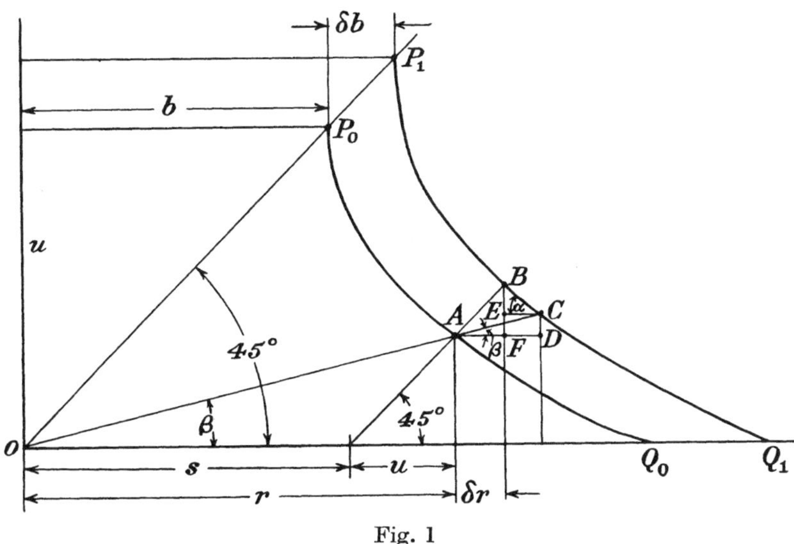


Fig. 1

If $\beta$ is the angle $AOQ_0$, $\tan\beta = u/r$. From the geometry of the figure $ABCD$ (fig. 1)

$$\delta r = AF = BF = CE\tan\alpha + DA\tan\beta = (DA-\delta r)\tan\alpha + DA\tan\beta. \tag{15}$$

Hence

$$\delta r = \left(\frac{\tan\alpha+\tan\beta}{1+\tan\alpha}\right) DA, \tag{16}$$

and, from (13),

$$\delta r = \left(\frac{u/r - \partial u/\partial r}{1-\partial u/\partial r}\right) r\frac{\delta b}{b}. \tag{17}$$

The radial strain component during the expansion of the hole from $b$ to $b+\delta b$ is $\dfrac{\partial}{\partial r}(\delta r)$, and differentiating (17) with respect to $r$, keeping $\delta b$ constant,

$$\frac{\partial}{\partial r}(\delta r) = -\left[\frac{(1-u/r)\,r}{(1-\partial u/\partial r)^2}\frac{\partial^2 u}{\partial r^2}\right]\frac{\delta b}{b}. \tag{18}$$

Since the strain during expansion of the hole from $b$ to $b+\delta b$ is proportional to $\delta b/b$, it is convenient to define strain components $\epsilon_r$, $\epsilon_\theta$, and $\epsilon_z$ so that strains during the small enlargement $\delta b$ are $\epsilon_r\delta b/b$, $\epsilon_\theta\delta b/b$, $\epsilon_z\delta b/b$. With this definition

$$\epsilon_r = -\left[\frac{r-u}{(1-\partial u/\partial r)^2}\right]\frac{\partial^2 u}{\partial r^2}. \tag{19}$$

The tangential strain is simply

$$\epsilon_\theta = \frac{b}{\delta b}\frac{\delta r}{r} = \frac{u/r - \partial u/\partial r}{1 - \partial u/\partial r}, \tag{20}$$

and the strain perpendicular to the sheet is

$$\epsilon_z = -\epsilon_r - \epsilon_\theta. \tag{21}$$

The thickness $h$ at any stage can be found simply from the equation of continuity; it is given by

$$\frac{h}{h_0} = \left(1 - \frac{u}{r}\right)\left(1 - \frac{\partial u}{\partial r}\right). \tag{22}$$

It can be verified that (21) is consistent with (22)

These expressions for strain take simple forms when expressed in terms of a new independent variable $\xi = r^2$ and a new dependent variable $\eta = s^2 = (r-u)^2$. Making these transformations and writing

$$p = \frac{d\eta}{d\xi}, \quad q = \frac{d^2\eta}{d\xi^2}, \tag{23}$$

(19), (20) and (21) become

$$\epsilon_r = -1 + \frac{2\eta q}{p^2} + \frac{\eta}{\xi p}, \tag{24}$$

$$\epsilon_\theta = 1 - \frac{\eta}{\xi p}, \tag{25}$$

$$\epsilon_z = -\frac{2\eta q}{p^2}, \tag{26}$$

while (22) reduces to the simple form

$$h/h_0 = p. \tag{27}$$

(26) can be deduced directly from (27).

The stress-equilibrium equation for a thin sheet is

$$\frac{\partial}{\partial r}(h\sigma_r) + \frac{h(\sigma_r - \sigma_\theta)}{r} = 0. \tag{28}$$

Two possible alternative forms for the strength condition might be considered:

(*a*) Mohr's stress criterion which may be written

$$\sigma_\theta - \sigma_r = Y \quad \text{if } \sigma_\theta \text{ is positive, i.e. tensile,}$$

or

$$-\sigma_r = Y \quad \text{if } \sigma_\theta \text{ is negative, i.e. compressive.} \tag{29}$$

(*b*) Mises's condition which may be written, when $\sigma_z = 0$,

$$\sigma_r^2 + \sigma_\theta^2 - \sigma_\theta\sigma_r = \text{constant}. \tag{30}$$

This reduces to $-\sigma_r = \text{constant}$ if $\sigma_\theta = 0$, and so is identical with Mohr's in that case.

If Bethe's assumption that $\sigma_\theta = 0$ combined with $\sigma_r = \text{constant}$ is used, (28) leads to

$$hr = \text{constant} = h_0 r_2', \tag{31}$$

where $r_2'$ is the outer boundary of the region of finite plastic strain. Substituting in (22)

$$\frac{r_2'}{r}=\left(1-\frac{u}{r}\right)\left(1-\frac{\partial u}{\partial r}\right), \tag{32}$$

which gives on integration $\quad \frac{1}{2}(r-u)^2=rr_2'+\text{constant}. \quad (33)$

Since $u=0$ when $r=r_2'$, the constant is $-\frac{1}{2}(r_2')^2$ and

$$u=r-\surd\{(2r-r_2')\,r_2'\}. \tag{34}$$

The inner boundary is where $b=r=u$, so that from (34)

$$b=\tfrac{1}{2}r_2', \tag{35}$$

which is Bethe's result if $r_2'$ is identified with his $r_2$.

## Introduction of the Strain-ratio Relationship

Since the strain-ratio relationship (11) involves only differences between the principal stresses, we may assume without loss of generality that $\sigma_z=0$. We shall assume further that the compressibility of the material is small enough to be neglected so that $\epsilon_r+\epsilon_\theta+\epsilon_z=0$. In these circumstances (11) may be written

$$\frac{\sigma_\theta}{\sigma_r}=\frac{\epsilon_\theta-\epsilon_z}{\epsilon_r-\epsilon_z}=\frac{\epsilon_r+2\epsilon_\theta}{2\epsilon_r+\epsilon_\theta}. \tag{36}$$

Substituting (27) and (36) in (28) the equilibrium condition reduces to

$$2\frac{d}{d\xi}(p\sigma_r)+\frac{p\sigma_r}{\xi}\left(\frac{\epsilon_r-\epsilon_\theta}{2\epsilon_r+\epsilon_\theta}\right)=0. \tag{37}$$

This equation must be used in conjunction with a strength criterion. Mohr's criterion (*a*) will be used. In this case (37) assumes two different forms according as $\sigma_\theta$ is negative (i.e. compressive) or positive (i.e. tangential tension). These are:
$\sigma_\theta$ negative, $\sigma_r=-Y$, so that (37) becomes

$$2q+\frac{p}{\xi}\left(\frac{\epsilon_r-\epsilon_\theta}{2\epsilon_r+\epsilon_\theta}\right)=0; \tag{38}$$

$\sigma_\theta$ positive, $\sigma_r-\sigma_\theta=-Y$, so that, from (36),

$$\sigma_r=\frac{2\epsilon_r+\epsilon_\theta}{\epsilon_r-\epsilon_\theta}(-Y),$$

and hence $\quad 2\frac{d}{d\xi}\left[p\left(\frac{2\epsilon_r+\epsilon_\theta}{\epsilon_r-\epsilon_\theta}\right)\right]+\frac{p}{\xi}=0. \quad (39)$

Substituting for $\epsilon_r$ and $\epsilon_\theta$ from (24) and (25) the resulting equations may be written:
$\sigma_\theta$ *negative (tangential compression)*

$$q^2\left(\frac{4\eta}{p^2}\right)+q\left(-1+\frac{2\eta}{\xi p}\right)+\frac{p}{\xi}\left(-1+\frac{\eta}{\xi p}\right)=0. \tag{40}$$

In this case, from (38), $\quad \frac{\epsilon_\theta}{\epsilon_r}=\frac{4q+p/\xi}{-2q+p/\xi}, \quad \frac{\sigma_\theta}{\sigma_r}=1-\frac{2q\xi}{p}, \quad (41)$

and in terms of Mohr's strength criterion the stresses are

$$\sigma_r = -Y, \quad \sigma_\theta = -Y\left(\frac{\sigma_\theta}{\sigma_r}\right). \tag{42}$$

*$\sigma_\theta$ positive (tangential tension)*

$$3w\left(\frac{\eta}{p^2}\right)\left(p-\frac{\eta}{\xi}\right) = \frac{4\eta^2 q^3}{p^4} + q^2\left(\frac{3\eta^2}{\xi p^3} + \frac{\eta}{p^2}\right) + 2q\left(-1 + \frac{3\eta^2}{\xi^2 p^2} - \frac{2\eta}{\xi p}\right) + \frac{p}{\xi} - \frac{2\eta}{\xi^2} + \frac{\eta^2}{\xi^3 p}, \tag{43}$$

where $w$ is written for $dq/d\xi$, i.e. $d^3\eta/d\xi^3$.

The expressions for $\epsilon_\theta/\epsilon_r$ and $\sigma_\theta/\sigma_r$ cannot be simplified by using the equation of equilibrium, and the full expressions derived from (24), (25) and (36) must be used, namely,

$$\frac{\epsilon_\theta}{\epsilon_r} = \frac{1-\eta/\xi p}{-1+2\eta q/p^2+\eta/\xi p}, \quad \frac{\sigma_\theta}{\sigma_r} = \frac{6\eta q}{4\eta q + \eta p/\xi - p^2} - 1, \tag{44}$$

and in terms of Mohr's condition $\sigma_r - \sigma_\theta = -Y$ the stresses are now

$$\sigma_r = -Y/(1-\sigma_\theta/\sigma_r), \quad \sigma_\theta = -Y\left(\frac{\sigma_\theta}{\sigma_r}\right)\Big/\left(1-\frac{\sigma_\theta}{\sigma_r}\right). \tag{45}$$

It will be seen that (43) is an ordinary differential equation of the third order and first degree while (40) is of the second order and second degree. The reason for this difference lies in the form of Mohr's strength condition. When $\sigma_\theta$ is positive three boundary conditions can be assigned at any given value of $\xi$ (i.e. of $r$). These might, for instance, be $u/r$, $h/h_0$ and $\sigma_r$ which can be transformed directly in assigned values of $q$, $p$ and $\eta$. When $\sigma_\theta$ is negative $\sigma_r$ cannot be assigned arbitrarily; it is in fact constant. Thus only $p$ and $\eta$ can be assigned arbitrarily.

### Boundary Condition at the Elastic-plastic Boundary

The elastic stresses due to radial displacement in an infinite sheet are

$$-\sigma_r = \sigma_\theta = \tfrac{1}{2} Y r_1^2/r^2, \tag{46}$$

where $r_1$ is the radius at which $\sigma_r - \sigma_\theta = -Y$. The corresponding small radial displacement is

$$u = \frac{1+m}{2}\left(\frac{Y}{E}\right)\frac{r_1^2}{r}, \tag{47}$$

where $E$ is Young's modulus and $m$ is Poisson's ratio. In the present investigation compressibility will be neglected and we will take $m=\frac{1}{2}$. In the elastic region, therefore, where $u/r$ is small compared with unity,

$$\eta = r^2\left(1-\frac{u}{r}\right)^2 = \xi\left(1-\frac{3}{2}\frac{Y}{E}\frac{r_1^2}{\xi}\right). \tag{48}$$

At the inner boundary of the elastic region therefore

$$p = \frac{d\eta}{d\xi} = 1, \quad q = 0, \quad \eta = \xi\left(1-\frac{3Y}{2E}\right). \tag{49}$$

PLATE 1

At the outer boundary of the plastic region, since $\sigma_r$ is positive, Mohr's criterion ensures that $\sigma_r - \sigma_\theta = -Y$. Since $\sigma_r$ is necessarily continuous through $r = r_1$, and it is assumed that $\sigma_r - \sigma_\theta = -Y$ at the elastic limit in the elastic region, $\sigma_\theta$ must be continuous through $r = r_1$ and equal to $\frac{1}{2}Y$. It is important to notice the reason why $\sigma_\theta$ is continuous at the plastic boundary in this case, because it is not necessary in general that $\sigma_\theta$ shall be continuous when Mohr's criterion is used. It will be shown in fact that $\sigma_\theta$ becomes discontinuous on the circle $r = r_2$ within the plastic region at the point where $\sigma_\theta = 0$.

### Strains and Displacements when $r_1 > r > r_2$

In the region within the circle $r = r_1$, where $\sigma_\theta$ is positive, it will be found that the strains are small, being of order $Y/E$. Assuming that

$$\eta = \xi(1 - \alpha\eta_1) \quad \text{and} \quad p = 1 + \alpha p_1,$$

where

$$\alpha = 3Y/2E,$$

$$p = \frac{d\eta}{d\xi} = 1 - \alpha\eta_1 - \alpha\xi\frac{d\eta_1}{d\xi}, \quad \text{so that} \quad p_1 = -\eta_1 - \xi\frac{d\eta_1}{d\xi}, \tag{50}$$

and

$$q = \frac{dp}{d\xi} = \alpha\frac{dp_1}{d\xi} = -\alpha\left(2\frac{d\eta_1}{d\xi} + \xi\frac{d^2\eta_1}{d\xi^2}\right). \tag{51}$$

When $\sigma_\theta$ is positive

$$\sigma_r = -Y\left(\frac{2\epsilon_r + \epsilon_\theta}{\epsilon_r - \epsilon_\theta}\right) = -\tfrac{1}{2}Y\left(\frac{-p + \eta/\xi + 4\eta q/p}{-p + \eta/\xi + \eta q/p}\right). \tag{52}$$

Substituting from (50) and (51) in (52) and neglecting terms in $\alpha^2$ compared with those containing $\alpha$

$$\left.\begin{aligned} \sigma_r &= -\tfrac{1}{2}Y\left(4 + \frac{3\psi}{\psi + \xi\psi'}\right), \\ \text{where} \quad \psi &= \frac{d\eta_1}{d\xi} \quad \text{and} \quad \psi' = \frac{d\psi}{d\xi}. \end{aligned}\right\} \tag{53}$$

Substituting this in (39),

$$\frac{d}{d\xi}\left(\frac{4p}{3} + \frac{p\psi}{\psi + \xi\psi'}\right) + \frac{p}{3\xi} = 0. \tag{54}$$

Neglecting terms which contain $\alpha$ as a factor compared with those that do not, $p$ may be taken as 1·0 and (54) may then be integrated giving

$$\frac{\psi}{\psi + \xi\psi'} + \tfrac{1}{3}\log\xi = \text{constant}. \tag{55}$$

The boundary condition at $r = r_1$ is $\sigma_r = -\sigma_\theta = -\frac{1}{2}Y$ so that from (53)

$$\psi/(\psi + \xi\psi') = -1.$$

The constant in (55) is therefore $-1 + \frac{1}{3}\log r_1^2$. Writing $\zeta$ for $\log(\xi/r_1^2)$, (55) becomes

$$\frac{d\psi}{d\zeta} + \psi\left(\frac{6 + \zeta}{3 + \zeta}\right) = 0. \tag{56}$$

The integral of (56) is

$$\log\psi + \zeta + 3\log(3 + \zeta) = \text{constant}. \tag{57}$$

Throughout the whole of the elastic region $-\sigma_r = \sigma_\theta$ and $\sigma_z = 0$, so that, at the elastic boundary, $p = 1$; hence, from (50),

$$\left[\eta_1 + \xi \frac{d\eta_1}{d\xi}\right]_{r=r_1} = 0, \tag{58}$$

and from the definition of $\eta_1$, (49) shows that $[\eta_1]_{r=r_1} = 1$. Hence from (58),

$$[\xi\psi]_{r=r_1} = -1;$$

(57) may therefore be written

$$\psi\xi(1 + \tfrac{1}{3}\ln \xi/r_1^2)^3 = -1, \tag{59}$$

and since $\psi = \dfrac{d\eta_1}{d\xi}$ and $\zeta = \log \xi r_1^{-2}$ this becomes

$$\frac{d\eta_1}{d\zeta} = -\frac{1}{(1 + \frac{1}{3}\zeta)^3}. \tag{60}$$

(60) may be integrated, giving

$$\eta_1 = \frac{3}{2(1 + \frac{1}{3}\zeta)^2} - \frac{1}{2}. \tag{61}$$

The equation for the thickness is

$$\frac{h}{h_0} = \frac{d\eta}{d\xi} = 1 - \alpha\left(\eta_1 + \frac{d\eta_1}{d\xi}\right) = 1 + \alpha\left[\frac{1}{2} - \frac{1+\zeta}{2(1 + \frac{1}{3}\zeta)^3}\right]. \tag{62}$$

The displacement is

$$u = \xi^{\frac{1}{2}} - \eta^{\frac{1}{2}} = \tfrac{1}{2}\alpha\eta_1 r = \tfrac{1}{2}\alpha r[\tfrac{3}{2}(1 + \tfrac{1}{3}\zeta)^{-2} - \tfrac{1}{2}]. \tag{63}$$

The stress can be found by substituting, from (56), $-\dfrac{6+\zeta}{3+\zeta}$ for $\dfrac{1}{\psi}\dfrac{d\psi}{d\zeta}$, i.e. for $\dfrac{\xi\psi'}{\psi}$, in (53). It is found that

$$\sigma_r = -\tfrac{1}{2}Y(1-\zeta) = -\tfrac{1}{2}Y\{1 - 2\log(r/r_1)\}, \tag{64}$$

and hence

$$\sigma_\theta = Y - \sigma_r = \tfrac{1}{2}Y\left(1 + 2\log\frac{r}{r_1}\right). \tag{65}$$

This is the well-known result which can be obtained without considering the strains and displacements, if it is assumed that the thickness of the plate does not vary.

### VALUES WHEN $\sigma_\theta = 0$

The radius $r$ at which $\sigma_\theta = 0$ is from (65) $r_2 = r_1 e^{-\frac{1}{2}} = 0{\cdot}606 r_1$ and corresponds to $\zeta = -1$.

Though the stress distribution in the range $r_1 > r > r_2$ is identical with that found in Bethe's investigation, the displacements and strains are not the same. For the case when Poisson's ratio is $\frac{1}{2}$, Bethe's method would give for the displacement when $\sigma_\theta = 0$

$$u_2 = \tfrac{1}{2}\alpha\frac{r_1^2}{r_2} = \tfrac{1}{2}\alpha r_2(2{\cdot}718). \tag{66}$$

Putting $\zeta = -1$ in (63) the displacement according to the present strain hypothesis is

$$u_2 = \frac{\alpha r_2}{2}\left\{\frac{3}{2(\frac{2}{3})^2} - \frac{1}{2}\right\} = \tfrac{1}{2}\alpha r_2\left(\frac{23}{8}\right) = \tfrac{1}{2}\alpha r_2(2{\cdot}875). \tag{67}$$

The displacement is in fact about 6% greater than that calculated on Bethe's strain hypothesis.

Putting $\zeta = -1$ in (62) the value of $h/h_0$ at $r = r_2$ is $1 + \frac{1}{2}\alpha$, and from (51) and (56)

$$q = -\alpha\left(2\psi + \frac{d\psi}{d\zeta}\right) = -\alpha\psi\left(2 - \frac{6+\zeta}{3+\zeta}\right) = -\alpha\psi\left(\frac{\zeta}{3+\zeta}\right);$$

hence, from (59),

$$q\xi = \frac{\alpha\zeta}{(3+\zeta)(1+\frac{1}{3}\zeta)^3},$$

so that when $\zeta = -1$,

$$q\xi = -\tfrac{27}{16}\alpha. \tag{68}$$

### Boundary Values at $\sigma_\theta = 0$

At the circle $r = r_2$, where $\sigma_\theta = 0$, $p$ and $\eta$ are continuous. Just inside the circle therefore, where $\sigma_\theta$ is negative,

$$\frac{\eta}{\xi} = 1 - \alpha\eta_1 = 1 - \tfrac{23}{8}\alpha \quad \text{and} \quad p = \frac{h}{h_0} = 1 + \tfrac{1}{2}\alpha. \tag{69}$$

When $\sigma_\theta$ is negative $q$ is determined by (40) when $\eta/\xi$ and $p$ are given. Substituting in (40) from (69), the values of $q\xi$ found by solving the resulting quadratic equation are (neglecting terms in $\alpha^2$)

$$q\xi = -\tfrac{1}{4} - \tfrac{85}{32}\alpha \quad \text{and} \quad q\xi = +\tfrac{27}{8}\alpha. \tag{70}$$

Neither of these values is the same as $q = -27\alpha/16$, the value just outside the boundary, so that $q$ is *not continuous* at $r = r_2$.

The two possible values of $q$ just inside the radius $r = r_2$ are therefore of opposite signs. Since

$$q = \frac{dp}{d\xi} = \frac{1}{h_0}\frac{dh}{d\xi} = \frac{1}{2h_0 r}\frac{dh}{dr},$$

a positive value of $q$ would correspond to a condition in which the plate gets thinner towards the centre, a configuration which does not appear to have physical significance. The negative value of $q$ for which $q\xi = -\frac{1}{4} - \frac{85}{32}\alpha$ will therefore be taken as the boundary value to be used in continuing the solution inward from $r = r_2$. It will be noticed that $\epsilon_r$, and consequently $\sigma_\theta$, are discontinuous as well as $q$. This discontinuity arises from the form of Mohr's criterion. It would not occur if von Mises's criterion had been used.

36-2

DISCONTINUITY IN $\epsilon_r$ AND $\epsilon_\theta$

Substituting $$q\xi = -\tfrac{1}{4} - \tfrac{85}{32}\alpha, \quad p = 1 + \tfrac{1}{2}\alpha, \quad \frac{\eta}{\xi} = 1 - \tfrac{33}{8}\alpha$$

in (24) and (25), it is found that

$$\epsilon_\theta = \tfrac{27}{8}\alpha, \quad \epsilon_r = -\tfrac{1}{2} - \tfrac{27}{4}\alpha,$$

and, substituting these in (36), $$\frac{\sigma_\theta}{\sigma_r} = \tfrac{1}{2} - \tfrac{81}{16}\alpha.$$

When $\alpha$ is small, i.e. when $E/Y$ is small, we may neglect $\alpha$ and take as the boundary condition at $r = r_2$ for calculating the stresses and displacements when $r < r_2$ the values

$$p = 1, \quad \eta/\xi = 1, \quad q\xi = -\tfrac{1}{4} \tag{71}$$

and the stresses are $\sigma_r = -Y$, $\sigma_\theta = -\tfrac{1}{2}Y$.

Thus the stress $\sigma_\theta$ suddenly changes from 0 to a *compressive stress* of $-\tfrac{1}{2}Y$ at the radius $r = r_2$.

CALCULATION OF STRESS AND STRAIN WHEN $r < r_2$

To calculate the distribution of stress and plastic strain inside the radius $r = r_2$, (40) must be solved step by step using the boundary values (71).

If $\delta\xi$ is the magnitude of a small step, the corresponding changes $p$ and $\eta$ may be taken as

$$\left.\begin{aligned} \delta\eta &= p\,\delta\xi + \tfrac{1}{2}q(\delta\xi)^2, \\ \delta p &= q\,\delta\xi. \end{aligned}\right\} \tag{72}$$

After calculating the values of $p$ and $\eta$ at the end of each step these values are inserted in (40) and the resulting quadratic for $q$ is solved, the root which derives by continuous variation of $\eta$, and $p$ from $qr_2^2 = -\tfrac{1}{4}$, being chosen in each case.

The results of applying this process are given in table 1 and are shown graphically in fig. 2. Values of the principal variables $\xi/r_2^2$ and $\eta/r_2^2$ are given in columns 1 and 2, table 1. Values of $p$ and $-qr_2^2$ are given in columns 3 and 4. Using these values of $p$, $q$ and $\xi$, values of $\sigma_\theta/\sigma_r$ calculated from (41) are given in column 8, and the corresponding values of $\sigma_r/Y$ and $\sigma_\theta/Y$ in columns 9 and 10. It will be seen that $\sigma_\theta$, which begins as a compressive stress equal to half the radial stress at the outer limit of the region of finite plastic flow, rapidly decreases till when $\xi/r_2^2 = 0{\cdot}35$ it becomes zero, and if the process is carried farther, using (41), $\sigma_\theta$ becomes a tension. When $\xi/r^2 = 0{\cdot}30$, for instance, the calculated value of $\sigma_\theta/\sigma_r$ is $-0{\cdot}124$. For values of $\xi/r_2^2$ less than $0{\cdot}35$, therefore, the alternative form (43) of the equilibrium equation must be used.

Since $\sigma_r$ is continuous and equal to $-Y$ at $\xi/r_2^2 = 0{\cdot}35$ and $\sigma_\theta = 0$ when $\xi/r^2$ is just greater than $0{\cdot}35$, while Mohr's criterion ensures that

$$\sigma_r - \sigma_\theta = -Y$$

when $\sigma_\theta$ is positive, it seems that $\sigma_\theta = 0$ when $\xi/r_2^2$ is just less than 0·35. Since both $\sigma_r$ and $\sigma_\theta$ are therefore in this case continuous through the radius where $\sigma_\theta$ changes sign, $\epsilon_r$ and $\epsilon_\theta$ are also continuous. Hence from (24) $q$ is continuous. The values of $\eta, p$, and $q$ at $\xi/r_2^2 = 0{\cdot}35$ can therefore be inserted in (43) and the value of $w = d^3\eta/d\xi^3$ at $\xi/r_2^2 = 0{\cdot}35$ determined.

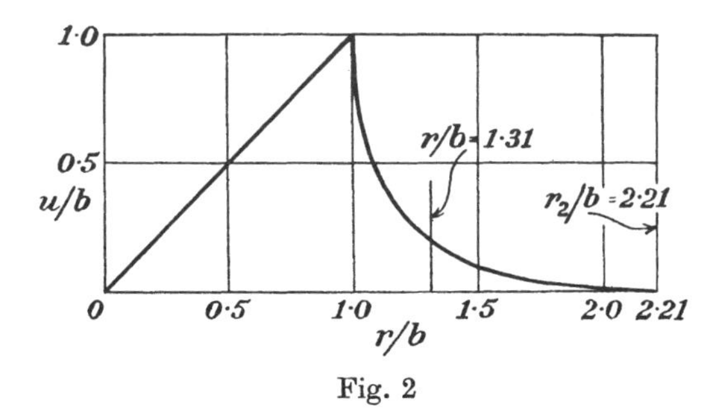


Fig. 2

Table 1

| 1 | 2 | 3 | 4 | 5 | 6 | 7 | 8 | 9 | 10 |
|---|---|---|---|---|---|---|---|---|---|
| $\xi/r_2^2$ | $\eta/r_2^2$ | $p$ | $-qr_r^2$ | $wr_2^4$ | $r/b$ | $u/b$ | $\sigma_\theta/\sigma_r$ | $\sigma_r/Y$ | $\sigma_\theta/Y$ |
| | | | | From eqn. (40), $\sigma_\theta$ negative | | | | | |
| 1·0 | 1·0 | 1·0 | 0·25 | — | 2·21 | 0 | +0·50 | −1·0 | −0·50 |
| 0·90 | 0·899 | 1·025 | 0·305 | — | 2·096 | 0·001 | — | −1·0 | −0·47 |
| 0·80 | 0·795 | 1·055 | 0·381 | — | 1·978 | 0·008 | +0·440 | −1·0 | −0·44 |
| 0·75 | 0·741 | 1·075 | 0·431 | — | 1·915 | 0·012 | +0·397 | −1·0 | −0·40 |
| 0·70 | 0·687 | 1·096 | 0·493 | — | 1·850 | 0·019 | +0·370 | −1·0 | −0·37 |
| 0·65 | 0·632 | 1·121 | 0·566 | — | 1·782 | 0·024 | +0·343 | −1·0 | −0·34 |
| 0·60 | 0·575 | 1·149 | 0·660 | — | 1·712 | 0·035 | +0·310 | −1·0 | −0·31 |
| 0·50 | 0·457 | 1·257 | 0·998 | — | 1·563 | 0·070 | +0·212 | −1·0 | −0·21 |
| 0·45 | 0·392 | 1·317 | 1·240 | — | 1·483 | 0·100 | +0·152 | −1·0 | −0·15 |
| 0·40 | 0·325 | 1·379 | 1·583 | — | 1·398 | 0·138 | +0·082 | −1·0 | −0·08 |
| 0·35 | 0·254 | 1·450 | 2·070 | — | 1·307 | 0·194 | +0·000 | −1·0 | 0 |
| (0·30) | (0·179) | (1·554) | (2·910) | — | (1·210) | (0·276) | (−0·124) | (−1·0) | (+0·12) |
| | | | | From eqn. (43), $\sigma_\theta$ positive | | | | | |
| 0·35 | 0·254 | 1·450 | 2·070 | 26·5 | 1·308 | 0·194 | — | — | — |
| 0·30 | 0·178 | 1·587 | 3·397 | 57·3 | 1·210 | 0·278 | −0·092 | −0·916 | +0·084 |
| 0·27 | 0·129 | 1·715 | 5·117 | 108·0 | 1·149 | 0·356 | −0·168 | −0·857 | +0·143 |
| 0·24 | 0·075 | 1·917 | 8·357 | 281·0 | 1·083 | 0·478 | −0·328 | −0·753 | +0·247 |
| 0·22 | 0·034 | 2·140 | 13·98 | 934·0 | 1·037 | 0·630 | −0·569 | −0·638 | +0·362 |
| 0·21 | 0·012 | 2·326 | 23·22 | 5190·0 | 1·013 | 0·771 | −0·739 | −0·576 | +0·424 |
| 0·205 | 0 | 2·61 | — | — | 1·00 | 1·00 | −1·000 | −0·500 | +0·500 |

The changes in $\eta$, $p$ and $q$ during the first step $\delta\xi$ in the new region are calculated using the formulae

$$\left.\begin{aligned} \delta\eta &= p\,\delta\xi + \tfrac{1}{2}q(\delta\xi)^2 + \tfrac{1}{6}w(\delta\xi)^3, \\ \delta p &= q\,\delta\xi + \tfrac{1}{2}w(\delta\xi)^2, \\ \delta q &= w\,\delta\xi. \end{aligned}\right\} \qquad (73)$$

Values of $\eta/r_2^2$, $p$, $-qr_2^2$, and $wr_2^4$ found in this way are given in the lower part of table 1, corresponding to $0{\cdot}35 > \xi/r_2^2 > 0{\cdot}205$. Values of $\sigma_\theta/\sigma_r$, $\sigma_r/Y$, and $\sigma_\theta/Y$ from (44) and (45) are given in columns 8, 9, 10 of table 1.

## Conditions at Edge of Hole

It will be seen in table 1 that as $\xi/r_2^2$ decreases to 0·21, $-qr_2^2$ and $wr_2^4$ are rising very rapidly. A study of the values of the terms in (43) reveals that by the time $\xi/r_2^2=0{\cdot}21$ is reached, one term on the right-hand side of the equation and one on the left-hand side are larger than any other terms. The limiting form of the equation when $\eta$ is small is in fact

$$\frac{3w\eta}{p}=-2q. \tag{74}$$

This equation can be integrated twice, thus

$$\left.\begin{aligned} -q&=A\eta^{-\frac{2}{3}} \\ \text{and}\qquad p^2&=B-6A\eta^{\frac{1}{3}}, \end{aligned}\right\} \tag{75}$$

$A$ and $B$ being the two constants of integration. To determine the constants the values of $\eta/r_2^2=0{\cdot}012$, $p=2{\cdot}326$, $qr_2^2=-23{\cdot}22$ can be used at $\xi=0{\cdot}21$. The resulting values of $A$ and $B$ are

$$A=1{\cdot}217r_2^{-\frac{2}{3}},\quad B=7{\cdot}07. \tag{76}$$

The limiting value of $p$ when $\eta=0$ is therefore

$$p=\sqrt{7{\cdot}07}=2{\cdot}66. \tag{77}$$

This is the limiting value of $h/h_0$ at the edge of the hole and may be compared with Bethe's value 2·0. It is not very different from the value at $\xi/r_2^2=0{\cdot}21$; to find the limiting value of $\xi$ therefore it is sufficient to take $p$ as constant and equal to 2·61 in the interval during which $\eta$ decreases from 0·012 to 0. Thus the limiting value of $\xi/r_2^2$ corresponding to the edge of the hole is

$$\xi_{\eta=0}r_2^{-2}=0{\cdot}21-\frac{0{\cdot}012}{2{\cdot}66}=0{\cdot}205. \tag{78}$$

The ratio $\left(\dfrac{\text{radius of finite plastic deformation}}{\text{radius of hole}}\right)$ is

$$\frac{r_2}{b}=\frac{1}{\sqrt{0{\cdot}205}}=2{\cdot}21. \tag{79}$$

This may be compared with Bethe's value 2·0.

Substituting the approximate limiting forms of $p$ and $q$ from (75) in (44), the limiting form for $\sigma_\theta/\sigma_r$ is

$$\lim_{\eta\to 0}\left(\frac{\sigma_\theta}{\sigma_r}\right)=\lim_{\eta\to 0}\frac{7{\cdot}30\eta^{\frac{1}{3}}r_2^{-\frac{2}{3}}}{7{\cdot}07-2{\cdot}43\eta^{\frac{1}{3}}r_2^{-\frac{2}{3}}}-1. \tag{80}$$

This tends to the value $-1$ as indicated in the last figure of column 8, and the corresponding values of $\sigma_r$ and $\sigma_\theta$ are therefore $-0{\cdot}5Y$ and $+0{\cdot}5Y$.

It will be noticed that the stress at the internal boundary could have been predicted *a priori* if it had been possible to assume that $h/h_0$ is finite at $r=b$, because clearly the total amounts of strain in the tangential and radial directions are both infinite at a hole which has been enlarged from a pin-hole. Thus the state of strain at the hole is such that symmetry alone must ensure that $\sigma_z$ is exactly half-way between $\sigma_r$ and $\sigma_\theta$. Since $\sigma_z=0$, $\sigma_r=-\sigma_\theta$. Similar considerations can be used to understand why the stress at points just inside the boundary $r=r_2$ corresponds with $(\sigma_\theta/\sigma_r)=+0{\cdot}5$, for at the edge of the region of finite plastic displacement, where the radial displacement is zero, $\epsilon_\theta=0$. Thus $\epsilon_r=-\epsilon_z$ and $\sigma_\theta$ must therefore be exactly half-way between $\sigma_r$ and $\sigma_z$. Hence, since $\sigma_z=0$, $\sigma_\theta=\frac{1}{2}\sigma_r$.

## Expressions in Terms of Radius of Hole

The radial variable is expressed in terms of the radius of the plastic region. To express the results in terms of $b$, it is necessary to tabulate $r/b=2{\cdot}21\sqrt{\xi}/r_2$. These values are given in column 6, table 1. The displacements $u/b=2{\cdot}21(\sqrt{\xi}-\sqrt{\eta})\,r_2^{-1}$ are tabulated in column 7.

The radial displacement is shown graphically in fig. 2 which may be compared with the diagrammatic sketch, fig. 1.

## Comparison with Bethe's Results

The thickness ratio $h/h_0=p$ is shown in fig. 3 and Bethe's values, namely, $h/h_0=2b/r$, are also shown. It will be seen that the main differences are that the present calculation shows the 'crater' extending farther radially than Bethe's and at the same time the 'crater' is much steeper close to the hole. From (75) it will be seen that $q\to-\infty$ as $\eta\to 0$ and, since $q=\dfrac{1}{2h_0 r}\dfrac{dh}{dr}$, the equations indicate that the edge of the 'crater' is a thin knife-edge. This deduction, however, is based on the assumption that the strains and stresses are uniform through the thickness of the plate. It is evident that this assumption ceases to be a good approximation when $dh/dr\to-\infty$. At first sight it might be thought that the extra thickness at $r=b$ above Bethe's $2h$ means that the work done in expanding the hole is greater according to the present calculations than in Bethe's calculations, but this is the reverse of the truth, for the radial stress at the hole is only $-\frac{1}{2}Y$ instead of Bethe's $-Y$. In fact the work done in expanding to a given radius is

$$\pi b^2 h_0(\tfrac{1}{2}Y)(2{\cdot}66)=1{\cdot}33\pi b^2 h_0 Y, \tag{81}$$

while if Bethe's strain assumption is used it is $2\pi b^2 h_0 Y$.

Fig. 4 shows the distribution of stress. This is of course very different from Bethe's, the most striking difference being that the present calculations predict a state of tangential tension in a ring which extends to 30 % of the radius of the hole from its edge and a tangential compression from that point to the edge of the region of large plastic distortion. In the plastic region $r_1>r>r_2$, where small strains comparable with the elastic strains occur, the stress is as calculated by Bethe, i.e. there is a tangential tension. In this connection it may be noticed that in comparing calculations of this kind with the behaviour of real materials, a metal which experiences considerable hardening with cold work might give results differing widely from the above theory.

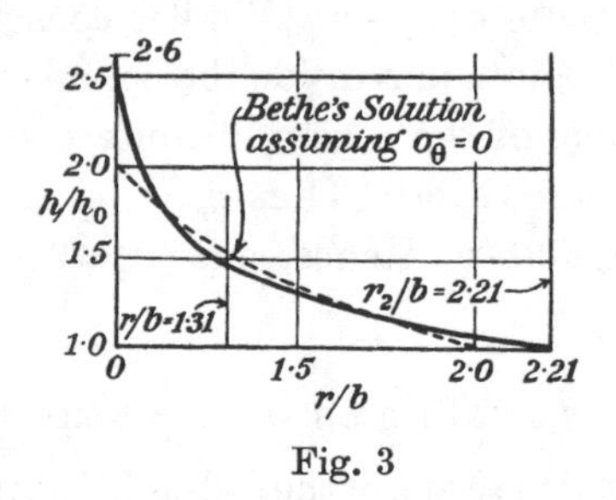


Fig. 3

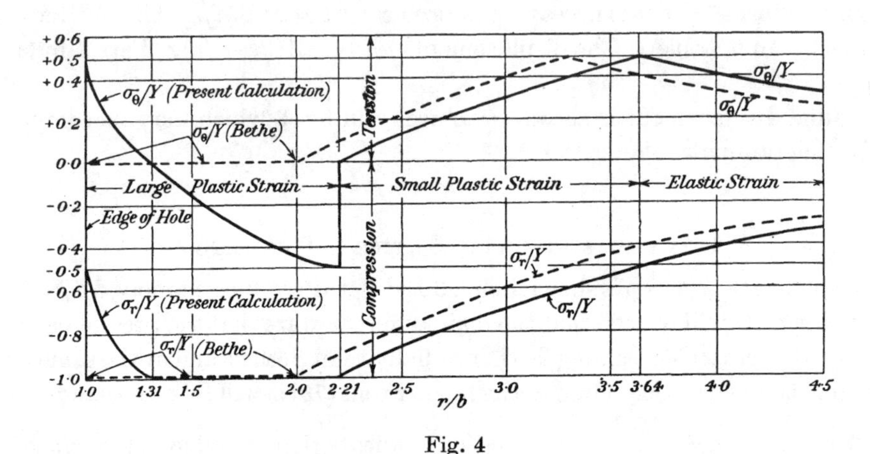


Fig. 4

## Experimental Work

An attempt was made to produce the distribution of strain contemplated in the foregoing analysis. A series of tapering sections of a cone of semi-vertical angle $1\frac{1}{2}°$ were made in mild steel. The smallest of these, which terminated in the vertex, was mounted in the chuck of a lathe and a flat sheet of lead was mounted on a slide rest so that its plane was perpendicular to the lathe bed. A small hole was bored in the lead sheet and the slide rest was moved parallel to the axis of rotation of the cone so that the vertex entered the hole and exerted radial pressure on its circumference. The cone was rotated while the hole was being enlarged so that the direction of the friction should be in the tangential direction rather than perpendicular to the plane of the lead sheet. 'Oildag' was used as a high-pressure lubricant between the sheet and the cone.

On reaching the base of the smallest section the lead sheet was withdrawn and the next largest section of the cone was fitted in the chuck. The hole was then enlarged to the largest diameter of this section. This process was repeated with successive sections, or broaching tools, till the hole was 2·45 cm. in diameter.

It was found that the hole could be enlarged symmetrically without bending the plate till its diameter was between 7 and 10 times the thickness of the sheet, but that at about this stage the sheet always bent out of its plane. This was not due to a sideways pressure, for the sheet sometimes bent towards the thick end of the broaching tool and sometimes away from it. It appears that the sheet is unstable and that an alternative form of deformation occurs in which the 'crater' develops into a short length of tube on one side or other of the sheet.

This tube is joined to the flat sheet by a region where the sheet is bent into the form of a curved fillet. The wall of the tube varies in thickness from that of the sheet at the fillet to zero at the outer edge. This edge contains the particles which were originally at the point where the vertex of the cone entered the sheet.

Sections of deformed sheets are shown in plate 1. These photographs were obtained by sawing the sheets in two in planes passing through their axes of symmetry and grinding down the surfaces so obtained till the true sections were revealed. The upper photograph shows the hole when its radius is 2·7 times the thickness of the sheet, while in the lower photograph the hole is about 10 times the thickness. It is clear that in the unsymmetrical case (lower photograph, plate 1) the material of the tube-shaped crater is expanding under the influence of a tangential tension $Y$ and that the stress perpendicular to the plane of the sheet is zero. The radial stress varies from zero at the outer surface to approximately $Yt/b$ at its inner surface. Here $b$ is the radius of the hole and $t$ the thickness of the wall of the tubular crater. Since $t$ is less than $h_0$ the radial stress is negligible when $b/h_0$ is large. The stress is therefore approximately uni-axial. The contraction of the material in the radial direction should therefore be equal to the contraction in the direction perpendicular to the sheet. This type of deformation would give rise to a distribution of thickness $t$ in the wall of the tubular crater for which $t=h_0\sqrt{(y/D)}$, where $y$ is the distance from the plane of the lip of the crater and $D$ is the height of this plane above the plane of the sheet. The initial radius, $s$, of the ring of particles which are at the point in the crater defined by $y$ is given by

$$\frac{s}{b}=\sqrt{\frac{y}{D}}. \tag{82}$$

Since the material is assumed to be incompressible

$$\pi b^2 h_0 = 2\pi b\int_0^D \sqrt{\frac{y}{D}}\,dy, \tag{83}$$

so that

$$D=\tfrac{3}{4}b \tag{84}$$

and

$$t=h_0\sqrt{\frac{y}{0{\cdot}75b}}=1{\cdot}15h_0\sqrt{\frac{y}{b}}. \tag{85}$$

The main features of the approximate theory represented by (84) and (85) are found, at any rate qualitatively, in the experiment represented in the lower photograph of plate 1. Since this unsymmetrical mode of deformation and also the symmetrical deformation are both theoretically possible it is of interest to compare the work done in opening out the hole to a given radius $b$ by the two modes.

The work done in expanding the ring of material which was originally contained between radii $s$ and $s+\delta s$ is

$$2\pi Y h_0 s\,\delta s \log\frac{b}{s},$$

so that the total work done is

$$w = 2\pi h_0 Y \int_0^b s \log\frac{b}{s}\,ds = \tfrac{1}{2}\pi b^2 h_0 Y. \qquad (86)$$

Comparing (86) with (81), it will be seen that the symmetrical mode requires 2·6 times as much work as the unsymmetrical mode for the same radius of hole. This fact seems to explain why it is the unsymmetrical type of deformation which actually occurs.

39

# THE USE OF FLAT-ENDED PROJECTILES FOR DETERMINING DYNAMIC YIELD STRESS PART I. THEORETICAL CONSIDERATIONS

REPRINTED FROM

*Proceedings of the Royal Society*, A, vol. CXCIV (1948), pp. 289–99

It has long been known that metals may be subjected momentarily to stresses far exceeding their static yield stress without suffering plastic strain. One of the simplest methods for subjecting a metal to a high stress for a short time is to form it into a cylindrical specimen and fire this at a steel target. The front part of this projectile crumples up, but the rear part is left undeformed. If the target is rigid the distance which this portion travels while it is being brought to rest may be taken as the difference between the initial length and the length of the deformed specimen after impact. Knowing the velocity of impact, a minimum possible value can be assigned to the maximum acceleration of the material, and from this a minimum value for the yield stress can be calculated. The actual yield stress is considerably greater than this minimum, and methods are given for calculating a more probable value.

## INTRODUCTION

When a cylindrical projectile strikes perpendicularly on a flat rigid target, the stress at the impact end immediately rises to the elastic limit, and an elastic compression wave travels towards the rear end. The stress in this wave is equal to the elastic limit. If the material is one in which the stress rises when the strain exceeds that corresponding with the elastic limit, the elastic wave is followed by a plastic one. On reaching the rear end of the projectile, the elastic wave is reflected as a wave of tension which is superposed on the compression wave. At this stage the velocity of the material in the part of the projectile which the reflected wave has not yet reached is $U-S/\rho c$, where $S$ is the yield stress of the material, $\rho$ its density, $c$ the velocity of elastic waves, and $U$ the velocity of impact. The stress in this portion is $S$. In the reflected wave extending from the wave front to the rear end of the projectile the velocity is $U-2S/\rho c$ and the stress is zero.

The reflected elastic wave runs forward along the projectile until it meets the front of the plastic wave advancing from the target plate. In this plastic wave, the stress will not rise appreciably above the yield stress at any point close to the plastic-elastic boundary, but the velocity may be nearly zero, or, at any rate, will be very different from that in the rear part where plastic flow has not taken place. At the moment when the reflected wave has just reached the plastic boundary, the part of the specimen which lies behind this is stress-free, and is moving as a solid body with velocity $U-2S/\rho c$. It is, therefore, in the same condition as the projectile at the moment of impact, except that its speed is $U-2S/\rho c$ instead of $U$, and its length is less than the original length $L$.

The length $x$ of this portion which has not yet suffered plastic strain will depend on the speed of the projectile, the speed of elastic waves in it, and the velocity with which the plastic-elastic boundary moves away from the target plate. Under the conditions of all the experiments which are here considered, $c$ is much greater than $U$, the velocity of the projectile, or $v$, the velocity of the plastic-elastic boundary, and $S$ is comparable with $\rho U^2$, but small compared with $\rho c^2$.

If $h$ is the distance of the plastic boundary from the target plate at any time, $x$ the length of the portion which has not yet been plastically compressed, and $u$ the velocity of this rear portion, the above considerations lead to the following equations for the small changes in $u$, $h$ and $x$ during one passage of an elastic wave from the plastic boundary and back to it. The duration of this double passage is

$$dt=\frac{2x}{c}, \tag{1}$$

so that

$$dh=v\frac{2x}{c}, \tag{2}$$

$$dx=-(u+v)\frac{2x}{c}, \tag{3}$$

$$du=-\frac{2S}{\rho c}. \tag{4}$$

Eliminating $c$, equations (1) to (4) reduce to

$$\frac{dh}{dt}=v, \tag{5}$$

$$\frac{dx}{dt}=-(u+v), \tag{6}$$

$$\frac{du}{dt}=-\frac{2S}{2x\rho}=-\frac{S}{\rho x}. \tag{7}$$

It will be noticed that (5), (6) and (7) are the equations which would be derived if the rear portions of the projectile were regarded as rigid, and all the quantities as continuously varying.

The equations (6) and (7) are not sufficient to determine the motion. In fact, the velocity of the plastic boundary, $v$, is determined by the plastic flow between this boundary and the target. To analyse the dynamics completely, it would be necessary to know all the intermediate states of the projectile between the instant of impact and the time when it comes to rest, or leaves the target plate. The object of the present work is to extract as much information as possible from measurements of the projectile recovered after impact. In the absence of measurements made during the impact, it is necessary to make some assumption about how the plastic boundary moves from the surface of the target to the final position in which it is measured after the projectile has come to rest.

## Simple Theoretical Model

In order to obtain a simplified picture of the phenomenon to serve as a framework for thinking about the motion, the simplest possible assumption about the plastic stress-strain relationship was made, namely, that the stress in the part of the projectile where the material is yielding is constant and equal to the yield stress $S$. Further, the radial inertia is neglected, so that the stress can be considered as constant over any cross-section. It is not possible to discuss the plastic flow between the plastic boundary and the target without analysing the complete problem of plastic waves. It is possible, however, to imagine a state in which the material which has just passed through the plastic boundary is brought to rest in a very short length.

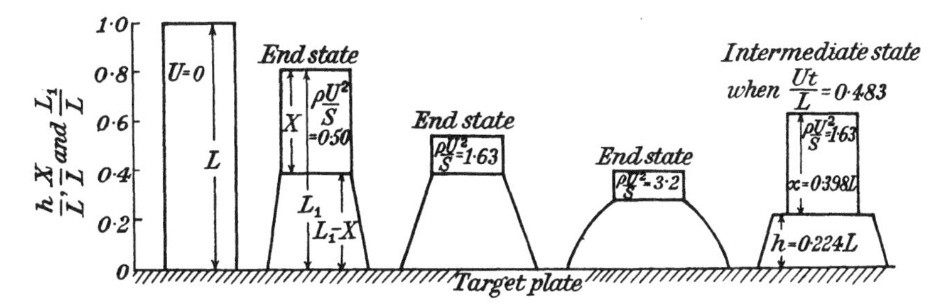


Fig. 1. Simple theoretical model of flat-ended projectile fired at speed $U$ at flat target.

For this to be possible, the material must spread out very rapidly. The appearance of the theoretical model at a time when the rear end is still moving is shown at the right-hand side of fig. 1. If $A_0$ is the cross-section of the projectile before it has been compressed plastically, and $A$ the area at the point where the material is brought to rest, the continuity equation is

$$A_0(u+v)=Av, \tag{8}$$

and if the stress is $S$ on both sides of the thin region where the change in area occurs, the momentum equation is

$$\rho A_0(u+v)\,u=S(A-A_0). \tag{9}$$

The longitudinal compressive strain at any point may be defined as

$$e=1-\frac{A_0}{A}. \tag{10}$$

Eliminating $v$ from (8) and (9), and employing (10) to eliminate $A$ and $A_0$, we obtain

$$\frac{\rho u^2}{S}=\frac{e^2}{1-e}. \tag{11}$$

From (6), (7), (8) and (10)
$$\frac{dx}{du}=\frac{(u+v)\rho x}{S}=\frac{\rho ux}{Se}. \tag{12}$$

Integrating (12)

$$\log_e(x^2) = \int \frac{1}{e} d\left(\frac{e^2}{1-e}\right) = \frac{1}{1-e} - \log_e(1-e) + \text{constant}. \tag{13}$$

At the moment of impact $x = L$, and $e = e_1$, say, and from (11)

$$\frac{\rho U^2}{S} = \frac{e_1^2}{1-e_1}. \tag{14}$$

Table 1. *Results of calculation based on simple theoretical model*

| $e_1$ | 0 | 0·1 | 0·2 | 0·3 | 0·4 | 0·5 | 0·6 | 0·7 | 0·8 | 0·9 | 1·0 |
|---|---|---|---|---|---|---|---|---|---|---|---|
| $X/L$ | 1·0 | 0·897 | 0·789 | 0·675 | 0·555 | 0·430 | 0·299 | 0·171 | 0·061 | 0·003 | 0 |
| $\rho U^2/S$ | 0 | 0·011 | 0·050 | 0·128 | 0·267 | 0·500 | 0·900 | 1·633 | 3·200 | 8·10 | — |
| $h/L$ | — | — | — | — | — | 0·382 | — | 0·376 | 0·288 | — | — |
| $L_1/L$ | — | — | — | — | — | 0·812 | — | 0·547 | 0·349 | — | — |

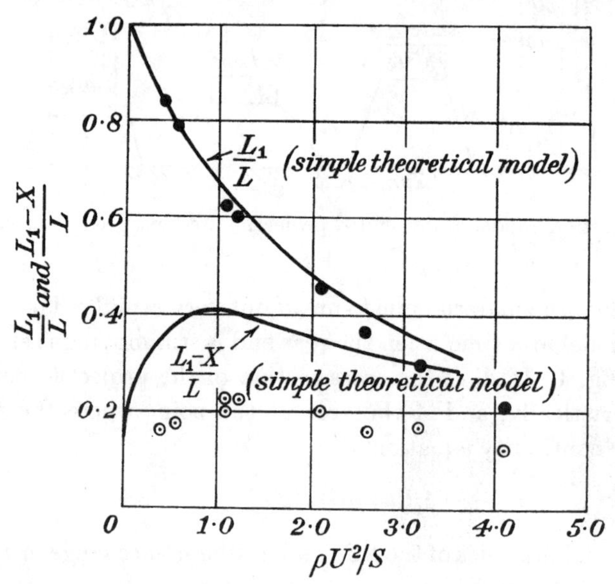


Fig. 2. Results of calculation based on a simple theoretical model. ● measured values $L_1/L$; ⊙ measured values $(L_1 - X)/L$.

When the projectile is brought to rest, $x = X$ and $e = 0$. $X$ is one of the lengths which can be measured, hence

$$\log_e\left(\frac{x}{L}\right)^2 = \frac{1}{1-e} - \log_e(1-e) - \frac{1}{1-e_1} + \log_e(1-e_1) \tag{15}$$

and

$$\log_e\left(\frac{X}{L}\right)^2 = 1 - \frac{1}{1-e_1} + \log_e(1-e_1). \tag{16}$$

Eliminating $e_1$ between (14) and (16) gives $X/L$ as a function of $\rho U^2/U$. The results are given in table 1 and shown graphically in figure 1.

To find the shape of the projectile after impact, we obtain from (5) and (6)

$$\frac{dh}{dx} = \frac{-v}{u+v} = -1 + e,$$

so that

$$h = -\int_L^x (1-e)\,dx. \tag{17}$$

This integration was performed numerically in three cases, $e_1 = 0{\cdot}5$, $0{\cdot}7$ and $0{\cdot}8$. Taking a given value of $e_1$, $\rho U^2/S$ was taken from table 1, and $(x/L)$ calculated from (15) for the range of values of $e$ from 0 to $e_1$. The resulting values of $h$ for $e_1 = 0{\cdot}5$ and $e_1 = 0{\cdot}7$ are given in table 2. The corresponding cross-section is $A = A_0/(1-e)$, so

Table 2. *Calculations for simple theoretical model of projectile for two impact velocities*

$e_1 = 0{\cdot}70$, $\rho U^2/S = 1{\cdot}633$

| $e$ | 0 | 0·05 | 0·10 | 0·15 | 0·20 | 0·25 | 0·30 | 0·35 |
|---|---|---|---|---|---|---|---|---|
| $x/L$ | 0·171 | 0·180 | 0·190 | 0·202 | 0·216 | 0·233 | 0·253 | 0·277 |
| $h/L$ | 0·376 | 0·368 | 0·358 | 0·348 | 0·336 | 0·323 | 0·309 | 0·293 |
| $Ut/L$ | 0·720 | 0·708 | 0·695 | 0·681 | 0·665 | 0·646 | 0·625 | 0·599 |
| $d/d_0$ | 1·00 | 1·026 | 1·054 | 1·087 | 1·118 | 1·155 | 1·196 | 1·241 |

| $e$ | 0·40 | 0·45 | 0·50 | 0·55 | 0·60 | 0·65 | 0·70 |
|---|---|---|---|---|---|---|---|
| $x/L$ | 0·307 | 0·346 | 0·398 | 0·468 | 0·571 | 0·730 | 1·000 |
| $h/L$ | 0·274 | 0·251 | 0·224 | 0·191 | 0·147 | 0·088 | 0 |
| $Ut/L$ | 0·568 | 0·531 | 0·483 | 0·420 | 0·333 | 0·207 | 0 |
| $d/d_0$ | 1·292 | 1·349 | 1·414 | 1·492 | 1·583 | 1·690 | 1·827 |

$e_1 = 0{\cdot}50$, $\rho U^2/S = 0{\cdot}50$

| $e$ | 0 | 0·05 | 0·10 | 0·15 | 0·20 | 0·25 | 0·30 | 0·35 | 0·40 | 0·45 | 0·50 |
|---|---|---|---|---|---|---|---|---|---|---|---|
| $x/L$ | 0·430 | 0·452 | 0·478 | 0·508 | 0·543 | 0·585 | 0·635 | 0·696 | 0·773 | 0·871 | 1·000 |
| $h/L$ | 0·382 | 0·361 | 0·337 | 0·311 | 0·282 | 0·250 | 0·213 | 0·172 | 0·124 | 0·067 | 0 |
| $Ut/L$ | 0·332 | 0·315 | 0·298 | 0·277 | 0·255 | 0·229 | 0·199 | 0·163 | 0·119 | 0·067 | 0 |
| $d/d_0$ | 1·000 | 1·026 | 1·054 | 1·087 | 1·118 | 1·155 | 1·196 | 1·241 | 1·292 | 1·349 | 1·414 |

that the diameter of the projectile at this point is $d = d_0\sqrt{\{1/(1-e)\}}$, where $d_0$ is the diameter of the projectile before firing. The values of $d/d_0$ and $h/L$ are given in table 2. Similar calculations were made for $e_1 = 0{\cdot}8$.

The shapes of the projectile in the three cases are shown in figure 1 for the case where the diameter was initially 0·3 of the height. Comparing these with the profiles of the steel slugs shown in fig. 1 of Mr Whiffin's paper (see part II), it will be seen that the calculated shape for $\rho U^2/S = 0{\cdot}5$ is very similar to that of the slug fired at 810 ft./sec. The shapes of the slugs fired at greater speeds do not resemble at all closely those calculated for $\rho U^2/S = 1{\cdot}63$ or $3{\cdot}2$. The plastically strained parts of the slugs fired at speeds greater than 810 ft./sec. have a concave profile, due, apparently, to the high radial velocity imparted to the material near the target which was neglected in the analysis.

As previously stated, the object of the calculations was to form the basis for an appropriate rough assumption for the rate at which the plastic boundary reaches its final position. To determine how $h$ varies with $t$, the time since the beginning of the impact, equation (7) was integrated numerically. Using equations (7) and (11)

$$t=-\int\frac{\rho x}{S}du=-\int\frac{\rho x}{S}\sqrt{\frac{S}{\rho}}\,d\left(\frac{e}{\sqrt{(1-e)}}\right)=L\sqrt{\frac{\rho}{S}}\int\frac{x}{L}\frac{(1-\frac{1}{2}e)}{(1-e)}de,$$

and, using (14), this may be written

$$\frac{Ut}{L}=\frac{e_1}{\sqrt{(1-e_1)}}\int_e^{e_1}\frac{x}{L}\frac{(1-\frac{1}{2}e)}{(1-e)^{\frac{3}{2}}}de. \tag{18}$$

The integration of (18) was performed numerically in the cases when $e_1=0{\cdot}5$ and $e_1=0{\cdot}7$. The results are given in table 2, together with the values of $h$. The relationship between $Ut/L$ and $h/L$ for these two values of $e_1$ are shown in fig. 3. It will be seen that $h$ increases nearly uniformly with $Ut/L$, so that the velocity of the plastic boundary, as it moves away from the target, is nearly uniform.

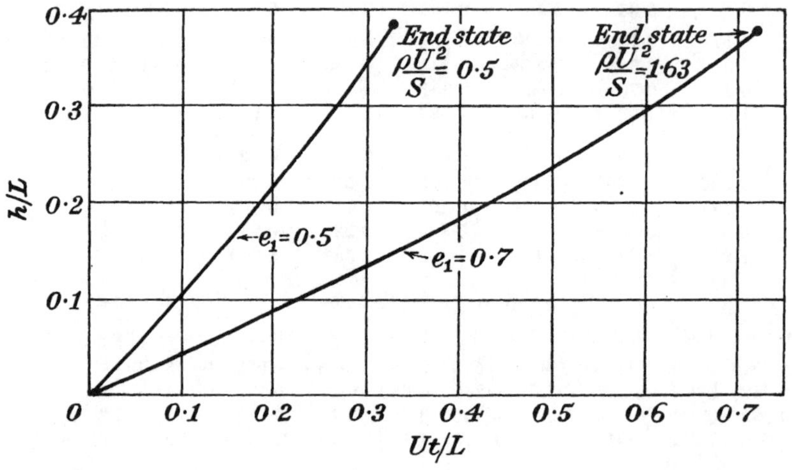


Fig. 3. Propagation of plastic boundary in a simple theoretical model.

The complete system of plastic waves occurring during an impact has been analysed by Messrs E. H. Lee and S. J. Tupper in a case where the plastic stress-strain relationship was assumed known. Good agreement was found with the strains calculated using the simple theoretical model here described. The main difference is that the strains calculated by the complete plastic-wave theory change in discontinuous jumps along the length of the projectile. This, as the authors point out in a paper which is not yet published, is due to the neglect in their analysis of the radial inertia of the plastic material as it spreads out near the target plate. The present calculations also neglect this radial inertia, but the discontinuities of strain disappear from the calculations when the finite difference equations (1), (2) and (3) are replaced by the differential equations (5), (6) and (7).

It is worth pointing out that though the plastic wave system can be calculated with considerable difficulty, when the plastic stress-strain relationship is known, this relationship can only be measured at low rates of straining. It is not possible to use these calculations directly to determine the stress-strain relationships from the shape of a projectile measured after an impact. It appears that the most which can be done at present is to determine approximately the yield stress using simplifying assumptions as to the general nature of the plastic stress waves. This is the line of attack developed in the present work.

## APPROXIMATE FORMULAE FOR ESTIMATING YIELD POINT FROM MEASUREMENTS OF SLUGS AFTER IMPACT

In developing a simple formula for estimating the yield point from measurements of the position of the yield boundary after impact, it will be assumed that the plastic-elastic boundary moves outwards at a uniform velocity from the impact end to its final position. This, together with the assumption that the yield boundary represents the position where a definite compressive stress $S$ is reached, is sufficient to fix the whole history of the deceleration of the rear part of the projectile.

If $C$ is written for the constant velocity of the plastic boundary, equations (6) and (7) give

$$\frac{du}{dx}=\frac{S}{\rho x(u+C)}, \tag{19}$$

which may be integrated to give

$$\frac{S}{\rho}\log_e\left(\frac{x}{L}\right)=\tfrac{1}{2}u^2+Cu-\tfrac{1}{2}U^2-CU. \tag{20}$$

When $u=0$, $x=X$, so that

$$\frac{S}{\rho}\log_e\left(\frac{X}{L}\right)=-\tfrac{1}{2}U^2-CU. \tag{21}$$

If the deceleration of the rear of the projectile were exactly uniform, $C/U$ could be determined simply from the fact that the time of deceleration $T$ is equal to $(L_1-X)/C$ (where $L_1$ = overall length of the projectile after test), and is also equal to $2(L-L_1)/U$, so that $C/U=\tfrac{1}{2}(L_1-X)/(L-L_1)$, and equation (21) would then assume the form

$$\frac{S_1}{\rho U^2}=\frac{(L-X)}{2(L-L_1)}\frac{1}{\log_e(L/X)}, \tag{22}$$

where $S_1$ has been used instead of $S$ to distinguish it from the value calculated by more exact methods. In fact, the deceleration is not uniform, so that (22) is only approximate. It is, however, the formula used with success by Mr Whiffin to interpret his experiments (see part II). In Mr Whiffin's experiments, slugs fired at very different speeds gave different values of $X/L$ and $L_1/L$, but, when the measured values were inserted in (22), the values of $S$ so deduced were found to be nearly independent of $U$. This fact affords a strong confirmation that (22) is an approximate equation which can be relied on to give the values of $S$ from measurements of $U$,

$X/L$ and $L_1/L$. In practice, it is sometimes found that the slug makes a depression in the target. In this case, equation (22) can still be applied, but if $L_1$ is the total measured final length, and $d$ the depth of the hole into which it fitted at the end of the impact, $L_1$ in (22) must be replaced by $(L_1-d)$.

## More Exact Calculation

The fact that the deceleration of the rear of the projectile is not uniform under the conditions assumed introduces an error which can be calculated. Since

$$\frac{dx}{dt}=-(u+C),$$

equation (20) may be written

$$\left(\frac{dx}{dt}\right)^2=\frac{2S}{\rho}\log_e\left(\frac{x}{L}\right)+(U+C)^2, \tag{23}$$

and when $u=0$, $x=X$, so that

$$\frac{2S}{\rho}\log_e\frac{X}{L}=C^2-(U+C)^2. \tag{24}$$

Writing $2S/\rho=a^2$, $K=(U+C)/a$, $x_1=x/L$, $t_1=at/L$, and $T_1=aT/L$, where $T$ is the duration of the impact, the non-dimensional form of (23) is obtained:

$$\frac{dx_1}{dt_1}=\sqrt{(K^2+\log_e x_1)},$$

so that

$$T_1=\int_1^{e^{-K^2}}\frac{dx_1}{\sqrt{(K^2+\log_e x_1)}}, \tag{25}$$

and, putting $K^2+\log_e x_1=z^2$,

$$T_1=2e^{-K^2}\int_{C/a}^{K}e^{z^2}dz. \tag{26}$$

Values of $F(K)=e^{-K^2}\int_0^K e^{z^2}dz$ have been tabulated * in the range $K=0$ to $K=12$. To use these tables, (26) can be expressed as

$$T_1=2[F(K)-\exp\{-K^2+(C/a)^2\}F(C/a)]. \tag{27}$$

Since the plastic-elastic boundary moves with velocity $C$, $CT=L_1-X$, which, in non-dimensional form, becomes

$$\frac{C}{a}T_1=\frac{L_1}{L}-\frac{X}{L}, \tag{28}$$

while (24) becomes

$$\log_e(L/X)=K^2-(C/a)^2. \tag{29}$$

Equations (27), (28) and (29), together with

$$K=U/a+C/a, \tag{30}$$

are sufficient to determine $C/a$, $U/a$, $K$ and $T$, when $X/L$ and $L_1/L$ are known.

* W. L. Miller and A. R. Gordon, *J. Phys. Chem.* xxxv (1931), 2878.

To estimate the error arising from the use of equation (22) instead of the more accurate equations (27) to (30) in any particular case where $L_1$ and $X$ are measured, it is necessary to solve transcendental equations. This laborious calculation was carried out by Mr Whiffin in the case of several of the experimentally determined values of $L_1$ and $X$. It was found that, in all cases, equation (22) underestimated the yield stress as compared with the equations (27) to (30). Subsequently, the corrections were calculated systematically in the form of a correcting factor $S/S_1$, which, when applied to $S_1$, as calculated from (22), gives the value which would have been obtained for $S$ if (27) to (30) had been solved.

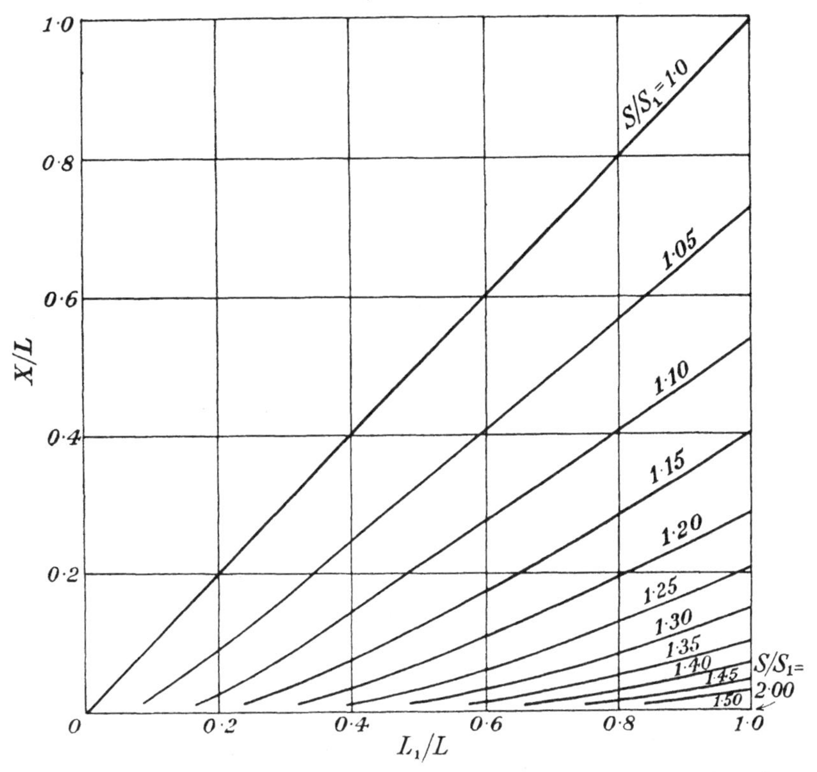


Fig. 4. Contours of ratio of yield value $S$, computed by exact formula, to value $S_1$ obtained by approximate formula.

Combining equations (27), (28) and (29)

$$\frac{L_1}{L}=2\frac{C}{a}F(K)-\left[2\frac{C}{a}F(C/a)-1\right]\frac{X}{L}. \tag{31}$$

Since
$$\frac{2S}{\rho U^2}=\frac{a^2}{U^2}=\frac{1}{(K-C/a)^2}$$

the correcting factor is
$$\frac{S}{S_1}=\frac{(L-L_1)\log_e(L/X)}{(L-X)\,(K-C/a)^2}. \tag{32}$$

37-2

In performing the calculation, a value of $X/L$ was taken, and a series of values of $C/a$ covering the range from 0 to infinity were used to calculate the corresponding values of $L_1/L$ and $S/S_1$. These were plotted on a diagram with ordinates $S/S_1$ and abscissae $L_1/L$. A series of curves were thus obtained, each corresponding with a given value of $X/L$. The values of $L_1/L$ corresponding with definite values of $S/S_1$, namely, 1·0, 1·05, ..., 2·0, were taken from these curves. In this way, and with the help of scattered values calculated by Mr Whiffin, the diagram of fig. 4 was constructed. Here the ordinates are $X/L$ and the abscissae $L_1/L$. The curves on the diagram are contours of equal correction factors. It will be seen that the curvature of these contours is very slight except near the axis $X/L=0$.

It will be noticed that the contours cut the line $L_1/L=1\cdot0$ without any apparent singularity. The limit points on $L_1/L=1\cdot0$ were calculated using the asymptotic form of $F(x)$ as $x\to\infty$, namely

$$\lim_{x\to\infty} F(x)=\frac{1}{2x}+\frac{1}{4x^3}+\dots. \tag{33}$$

In this way it is found that, when $C/a$ and $K$ are infinite,

$$L_1/L=1\cdot0, \tag{34}$$

and

$$\frac{S}{S_1}=2\left[\frac{1}{1-X/L}-\frac{1}{\log_e(L/X)}\right]. \tag{35}$$

The limiting points on the line $L_1/L=1\cdot0$ in fig. 4 were calculated from equation (35).

## Experiments with Paraffin Wax

The method described above was first applied to find the dynamic yield stress of paraffin wax. Transparent cylinders of this material were cut from cast blocks, and these were projected by means of a catapult, capable of giving them a speed of 125 ft. per sec., at a heavy anvil hung as a ballistic pendulum. The cylinders were 1·75 cm. long. After the impact they were found to be shorter, but they remained coherent. Paraffin wax has the property of remaining transparent under compressive stressing until a sudden collapse occurs. The material which has yielded is full of small cracks, which give it an opaque white appearance. The cylinders which had been projected were found to be opaque at the impact end, but they remained transparent at the rear end. The yield point was taken to correspond with the boundary between the transparent and opaque portions. The length of the transparent portion was taken as $X$. Static tests were also made by compressing paraffin wax cylinders between polished plates in a parallel-motion compression machine, and it was found that sudden and catastrophic breakdown occurred at a certain load, the wax remaining transparent up to the instant of breakdown.

Some results are given in table 3. It will be seen that the dynamic yield stress varied from 840 to 930 lb. per sq.in., while the static yield stress was only 485 lb. per sq.in. The ratio (dynamic yield stress)/(static yield stress) was therefore about 1·8.

Table 3. *Experiments with cylinders of paraffin wax*

| $L$ (cm.) | $U$ (ft. per sec.) | $L_1$ (cm.) | $X$ (cm.) | $S$ dynamic yield stress from eqn. (22) (lb. per sq. in.) | $S$ static yield stress (mean of 6 observations) (lb. per sq. in.) |
|---|---|---|---|---|---|
| 1·774 | 126 | 1·635 | 0·95 | 854 | — |
| 1·757 | 128 | 1·625 | 0·95 | 930 | 485 |
| 1·779 | 132 | 1·625 | 0·95 | 840 | — |

## Interpretation of Results

The measurements made by Mr Whiffin show that his specimens maintained the stresses, which are here called dynamic yield stresses, instantaneously without suffering strain greater than 0·2 %. When considering the results of mechanical tests at comparatively low speeds, in which the elastic limit is passed and plastic strain occurs, the rate of strain is definable in terms of the experimentally measured quantities. In fact, if $l$ is the length of the specimen at time $t$, and $L$ its original length, one definition of rate of strain is simply $1/L.dl/dt$. In experiments of the type here considered, it is possible to define the mean rate of strain of the plastically distorted portion of the projectile. If the approximation represented by (22) is adopted, the mean rate of strain is $U/2(L-X)$, because the reduction in length is $(L-L_1)$, and this is entirely confined to the material whose initial length was $(L-X)$. The total strain of the portion which has yielded is therefore $(L-L_1)/(L-X)$. Since the deceleration is assumed uniform, its duration is $2(L-L_1)/U$, so that the mean rate of strain is

$$\frac{(L-L_1)}{(L-X)}\frac{\frac{1}{2}U}{(L-L_1)}=\frac{\frac{1}{2}U}{(L-X)}. \tag{36}$$

Values of this mean rate of strain are given in Mr Whiffin's paper because they are the only rates of strain which can be deducéd from his measurements. It must, however, be remembered that, at all stages of the impact, the analysis refers only to the part of the projectile which has not yet suffered plastic compression, so that the connection between the rate of strain just defined and the yield stress can only be an indirect one.

It seems impossible to derive further information about the physical factors which determine the dynamic yield stress without making more complete measurements of the successive states of the projectile during impact. It is, however, worth noticing that, if the deceleration of the rear portion of the projectile is continuous, as is contemplated in equations (5), (6) and (7), the maximum stress $S$ at a distance $X$ from the rear end is only attained instantaneously at the end of the impact. At the beginning of the impact, the stress in this part of the projectile is $XS/L$. If the stress at distance $X$ rises uniformly from the value $XS/L$ at the beginning to $S$ at the end, the stress at time $t$ would be

$$S\left[\frac{X}{L}+\frac{t}{T}\frac{(L-X)}{L}\right].$$

The duration of stress greater than, say, $S(1-y)$, where $y$ is small, is $L/(L-X).Ty$. To a rough approximation, therefore, one may expect that, at the place where the yield point is found, the stress has exceeded, say, 99 % of $S$ for a duration of the order of (0·01) $L/(L-X.T)$, and, according to the simple theory of equation (22), $T=2(L-L_1)/U$. In one group of Mr Whiffin's experiments, recorded in his table 2, the estimated value of $T$ lies between 5·2 and $7{\cdot}0\times10^{-5}$ sec., so that there is very little variation in the duration of the impact, although there is a great variation in $U$ and $(L-L_1)$. It seems possible that the constancy of $S$, which is found for varying velocities of the projectile, is due to the constancy of the duration of impact.

# 40

# DISTRIBUTION OF STRESS WHEN A SPHERICAL COMPRESSION PULSE IS REFLECTED AT A FREE SURFACE

(Appendix to a paper by W. M. Evans, 'Deformation and fractions produced by intense stress pulses in steel')

REPRINTED FROM

*Research*, vol. v (1952), pp. 508–9

The stress produced in steel when a detonating explosive is fired at a point on its surface is no doubt very complicated. Though it is not to be expected that the actual stresses will resemble very closely the stress system when a spherical sound pulse is reflected at a free surface, it is perhaps worth while to work out some distributions of this type in order to compare the shape of the area where the maximum tensile stress occurs with that which is observed in actual experiments and as is shown in pl. 1 (taken from the paper by Mr Evans). The compression pulse will be assumed to originate at a point distant $d$ from the boundary and to have the form

$$p=\frac{A}{r}\exp\left(\frac{r-ct}{\lambda}\right)$$

where $\lambda$ is the pulse length, $t$ the time from its inception, $c$ the velocity of the sound wave in steel and $r$ the distance from the explosive source. The pulse is assumed sharp-fronted so that before the pulse reaches the surface, i.e. when $t<d/c$,

$$p=\frac{A}{r}\exp\left(\frac{r-ct}{\lambda}\right) \quad \text{when} \quad r<ct,$$

$$p=0 \quad \text{when} \quad r>ct.$$

After the reflection the stress distribution is

$$p=A\left\{\frac{1}{r}\exp\left(\frac{r-ct}{\lambda}\right)-\frac{1}{r'}\exp\left(\frac{r'-ct}{\lambda}\right)\right\} \quad \text{when} \quad r<r'<ct, \tag{1}$$

where $r'$ is the distance from the image of the source in the free surface.

Since $\exp(-ct)$ is a factor of both terms in (1) the absolute value of the stress at a fixed point will decrease with time so long as $r<r'<ct$, and since $r'>r$ at all points in the steel the maximum tensile stress will be attained at the instant that the second term appears, i.e. when $r'-ct=0$.

Thus the maximum tensile stress is

$$-\frac{A}{\lambda}\left\{\frac{1}{x}\exp(x-x')-\frac{1}{x'}\right\}, \quad \text{where} \quad x=r/\lambda,\ x'=r'/\lambda.$$

Writing
$$F=\frac{1}{x'}-\frac{1}{x}\exp(x-x'), \tag{2}$$

we may take a given value of $F$ and calculate the value of $x$ from (2) for a new series of values of $x'$. Such calculations are made and the contours for $F=0{\cdot}204$, $0{\cdot}20$, $0{\cdot}18$, $0{\cdot}15$, $0{\cdot}125$, $0{\cdot}10$ are shown in fig. 1 where the ordinates are values of $x$ and the abscissae are values of $x'$. To plot these contours in space it is necessary to know $d$,

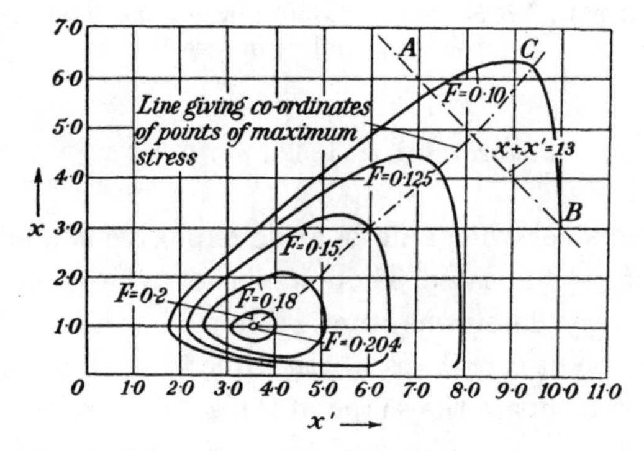


Fig. 1. Contour of $F=1/x'-\exp(x-x')/x$. (Broken lines used in construction of fig. 2.)

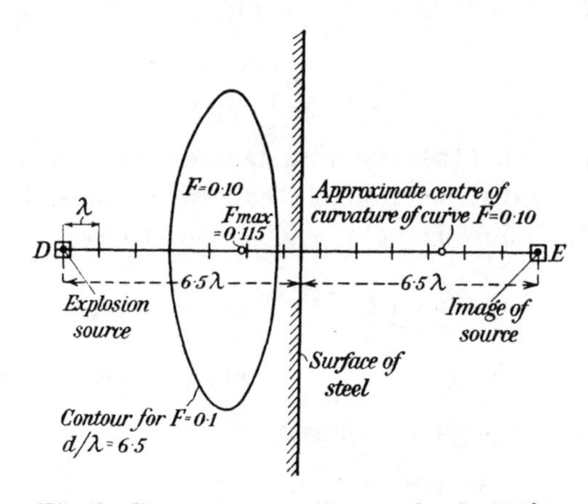


Fig. 2. Stress contour diagram for $d=6{\cdot}5\lambda$.

and since the minimum value of $r+r'$ is $2d$ the minimum value of $x+x'$ is $2d/\lambda$. If therefore the line $x+x'=2d/\lambda$ is described on fig. 2 the whole of any contour of constant tension in the steel corresponds with the part of the corresponding curve in fig. 2 which lies above the line $x+x'=2d/\lambda$.

Fig. 2 shows an example of the way in which fig. 1 can be used. There the thickness of the metal slab is taken as $6{\cdot}5\lambda$, i.e. 6·5 times the pulse length and the line $x+x'=13$ is shown as $AB$ in fig. 1. Successive points are selected on the portions $ACB$ of the contour for $F=0{\cdot}10$ and the corresponding values of $x$ and $x'$ are used as radii for

PLATE 1

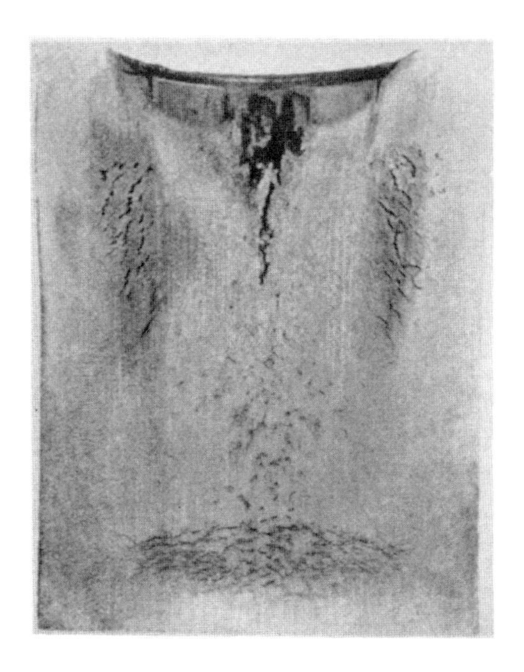

Etching of a specimen reveals a region of tension fractures near the base.

arcs of circles about the centres $D$ and $E$ which represent the source and image in fig. 2. The intersections of these pairs of circles represent successive points on the contours.

The position of the points of maximum stress is found by joining the points where lines $x+x'=$ constant, touch the contours of constant $F$ in fig. 1. This line is marked on fig. 1 and it will be seen that it cuts the line $x+x'=13$, in the point $x=4\cdot9:x'=8\cdot1$ corresponding with a value $F_{\text{max.}}=0\cdot115$. The position of $F_{\text{max.}}$ is marked in fig. 2. It is distant $6\cdot5\lambda-4\cdot9\lambda=1\cdot6\lambda$ from the reflecting free surface of the steel.

# 41

# STRAINS IN CRYSTALLINE AGGREGATES

REPRINTED FROM

*Proceedings of the Colloquium on Deformation and Flow of Solids (Madrid, 1955).*
Berlin: Springer (1956), pp. 3–12

Much study has been devoted in many countries to the mechanism of plastic strain in crystals by the formation and movement of dislocations. There has also been much experimental work on the macroscopic relationships between stress, strain and the crystal axes of single crystals. With this background of knowledge the man who contemplates a crystal aggregate and tries to understand how it behaves when forced to flow, is in much the same position as one who knows about the kinetic theories of gases and liquids, understands how the statistical effects of molecular collisions are averaged to yield the laws of viscosity, is familiar with Poisson's equations for the flow of viscous fluids, and then sets out to study turbulence. Dr Cottrell has used this analogy in a similar case. Even with the very simple laws summarized in Poisson's equations it has proved impossible so far to make any complete representation of turbulent fluid flow. Indeed, some people working on that subject have felt so discouraged at their lack of success in finding significant results based on Poisson's equations that they have concluded that these equations do not apply to turbulent flow and have tried to start again, as in the kinetic theory of gases, from known properties of molecules. Such people have had no success in these despairing efforts; nor are they likely to succeed, because Poisson's mathematical expression for the stress-rate of strain relations in a fluid represent the simplest possible averaging process for molecular interactions. The sole reason why we have not succeeded in giving a complete description of turbulent flow is not because we do not know enough about fluids, but because we are not clever enough to find the appropriate solutions of Poisson's equations.

Though this analogy between turbulent flow of fluids and the flow of an aggregate is not an exact one, it does illustrate the reason why workers on aggregates are forced to start with idealized macroscopic stress-strain relationships determined experimentally with single crystals rather than attempt to start with dislocations. Such studies will not take account, in the first instance, of the actual physical conditions at the grain boundaries, regarding them merely as surfaces at which the displacement and three out of the six components of stress are continuous. Ultimately, if a self-consistent description of an aggregate of crystals with idealized properties can be built up, it may be possible to go further and apply knowledge gained by the molecular theorists and experimenters on the physical conditions at crystal boundaries to aggregates, but that is merely a guess at future developments.

Experiments in which single crystals of certain metals, particularly those with the face-centred cubic arrangement atoms, are subjected to uniaxial stress, have established that deformation occurs by slipping in one direction parallel to one crystal plane and that of all crystallographically similar planes the slipping occurs only on the one for which the shear stress in the direction of the slip is greatest. The slipping produces a rotation of the axes of the specimen relative to the axes of the crystal. This rotation ultimately brings a second possible slip-plane into the most favourable position for slipping. When this occurs the slipping begins on the second plane also. These observational laws of macroscopic strain in certain single crystals have been deduced entirely by measurements of strain without any reference to surface 'slip lines', but these 'slip lines' do, in fact, mark out the intersections of the plane of slip, determined geometrically from measurements on the surface of the specimen.

In attempting to apply these laws of observation for single crystal grains to deduce the laws of plasticity of a polycrystalline mass, regarded as a continuum, some workers have simply imagined each crystal grain as being subjected to sufficient uniaxial stress in the direction of the macroscopic stress to make it yield. They then imagined the axes of grains to be orientated at random and each grain to be unaffected by its neighbours. With this model the sum of the forces due to each grain over any section of the specimen is equal to the load on it and the stress is $m\tau$, where $\tau$ is the yield stress for single slipping of each crystal. The value of $m$ so calculated by Sachs (1928) is $m = 2{\cdot}238$. This model has little relationship to poly-crystals because neither the stress nor the strain is continuous at the grain boundaries and grains which originally fitted together will not do so after straining. It has, however, one advantage, it does definitely give a lower bound to the value of $m$ because the crystal is essentially a frictional system and it is a general mechanical principle that removal of constraints anywhere in such a system decreases the work necessary to make any given strain in it.

To make irregular grains fit together, a combination of small elements of slipping is necessary at all points. If the strain of a grain boundary is prescribed and corresponds with a surface of particles in a continuum subjected to a uniform strain, then the material inside the boundary can itself be strained uniformly, but the conditions of stress at the boundary will not be satisfied. It is clear therefore that in the actual polycrystal the strain in an irregular grain would not be uniform even if the strain of its boundary were. On the other hand, it is possible to maintain macroscopic continuity within the material by appropriate combinations of slipping on crystal planes whatever the distribution of strain may be inside the grains. In general, the slip combinations will vary from point to point in the grain, and since each such combination will cause the crystal axis to rotate relative to the axes of strain the grains must necessarily cease to be true crystals and exhibit a spread of crystal directions through their volume. The extreme complexity of the problem makes me doubt whether even a valid statistical picture of the strain in polycrystals in which the stress and strain conditions are satisfied both at the boundaries and within the substance, will ever be attained. Certainly none has yet been proposed.

Failing a complete description it seemed to me some years ago * that one could calculate an upper bound to the stress in an aggregate of face-centred cubic crystals using the principle of virtual work. There is an infinity of ways in which combinations of slips on crystal faces can give rise to a given external strain of an aggregate and satisfy the conditions of continuity of displacement at crystal boundaries. If any one of these occurs the energy wasted is $\Sigma \tau\, ds$, where $\tau$ is the shear stress in the direction of slipping and $ds$ is an element of shear strain. This must be equal to the work done by the external forces. For an aggregate extending in one direction with longitudinal strain $d\epsilon$ under a longitudinal stress $P$, the work done is $P\, d\epsilon$. Hence in the restricted case where $\tau$ is constant,

$$\frac{P}{\tau}=\frac{\Sigma\,|\,ds\,|}{d\epsilon}.$$

It seemed to me* that if we could determine the combination of slips which produce the imposed external strain $d\epsilon$ and had the minimum value of $\Sigma\,|\,ds\,|$, we should have solved the problem. All other combinations of slips which satisfy the conditions of continuity at the boundaries of grains would require a higher value of $P$ than $(\tau/d\epsilon)\times$(minimum value of $\Sigma\,|\,ds\,|$) to produce deformation. Thus an upper bound to $P$ might be obtained by finding any combination of slips which satisfied the external strain condition and the condition of continuity of displacement at the grain boundaries. I did not consider the question of whether the stress on all active slip-planes could be the same because I imagined that the small elements of slip might occur successively rather than simultaneously. As the stress in the polycrystal increases the shear stress on one plane might rise to the value at which a small slip would occur. This would alter the stress distribution in the same crystal and raise the stress on another plane to the necessary value for slipping. Thus I supposed one could discuss the possibility that during an arbitrary strain an appropriate combination of slips might occur without thinking whether a stress system could occur in which the stress was raised to the slipping value on all active slip-planes *simultaneously*. It will be seen later that this question has been investigated much more thoroughly.†

The only kind of strain satisfying continuity of displacement at grain boundaries which I saw any hope of discussing was that in which the strain of each grain was identical with that of the aggregate. I therefore attempted to find the minimum value of $\Sigma\,|\,ds\,|$ for a crystal whose crystal axes relative to the strain axes of the aggregate was given. In a face-centred cubic crystal there are assumed to be 12 possible slips, three on each of the (111) planes in the directions of the diagonals of square faces of the fundamental cube.

* G. I. Taylor, *J. Inst. Metals*, LXII (1938), 307, paper 27 above; *Stephen Timoshenko 60th Anniversary Volume* (1938), p. 218, paper 28 above.

† J. F. W. Bishop and R. Hill, *Phil. Mag.* XLII (1951), 414 and 1298.

Taking the four planes as (111), $a$; $(1\bar{1}\bar{1})$, $b$; $(\bar{1}1\bar{1})$, $c$; $(\bar{1}\bar{1}1)$, $d$; and the directions of slip numbered with suffixes 1, 2, 3 according as they lie in the planes (100), (010), (001) the 12 slips are defined as $a_1, a_2, a_3, b_1, b_2, b_3, c_1, c_2, c_3, d_1, d_2, d_3$, and the components of strain relative to the cubic axes are

$$\left.\begin{aligned}
e_{xx} &= \frac{\partial u}{\partial x} = a_2 - a_3 + b_2 - b_3 + c_2 - c_3 + d_2 - d_3, \\
e_{yy} &= \frac{\partial v}{\partial y} = -a_1 + a_3 - b_1 + b_3 - c_1 + c_3 - d_1 + d_3, \\
e_{zz} &= \frac{\partial w}{\partial z} = a_1 - a_2 + b_1 - b_2 + c_1 - c_2 + d_1 - d_2, \\
2e_{yz} &= \frac{\partial w}{\partial y} + \frac{\partial v}{\partial z} = -a_2 + a_3 - b_2 + b_3 + c_2 - c_3 + d_2 - d_3, \\
2e_{zx} &= \frac{\partial u}{\partial z} + \frac{\partial w}{\partial x} = a_1 - a_3 - b_1 + b_3 + c_1 - c_3 - d_1 + d_3, \\
2e_{xy} &= \frac{\partial v}{\partial x} + \frac{\partial u}{\partial y} = -a_1 + a_2 + b_1 - b_2 + c_1 - c_2 - d_1 + d_2.
\end{aligned}\right\} \quad (1)$$

The object of my work was to find the minimum value of the sum of the absolute values of the twelve slips, subject to the condition (1). It was first shown that this minimum must be one of the cases in which only 5 of the 12 are active, but the number of combinations of 12 things taken 5 at a time is 792 so that the task of selecting the correct choice is likely to be very laborious. This number is considerably reduced by various considerations. First, since any direction in a plane can be compounded out of two components at 120° apart, only two slip components are needed from any one slip-plane, and corresponding with any given magnitude for an arbitrary shear on, say, the plane $a$ either the vector sum of $a_1 + a_2$ or $a_2 + a_3$ or $a_3 + a_1$ can be used, but if the same shear be added to each of $a_1$, $a_2$ and $a_3$ there are three ways in which the arbitrary shear can be composed of two only, and one can choose the pair which is least of $|a_1| + |a_2|$ or $|a_2| + |a_3|$ or $|a_3| + |a_1|$ without affecting any of the choices of the remaining 3 of the combinations of 5 shears. This reduces very greatly the number of choices for which (1) must be solved, in fact there turn out to be 24 independent choices in which the five shears are chosen so that 2, 2, 1, 0 are on each of the four slip-planes respectively. If the choices are made so that 2 are on one plane and one on each of the remaining planes there are 108 of them but 36 of these are not capable of combining into an arbitrarily chosen strain tensor, so that 72 is the correct number, making 98 in all. In my papers on the subject* I unfortunately made a mistake in signs in equation (1) and this had the effect of making me draw the false conclusion that choices in which two shears occur on only one of the planes are unable to combine to form an arbitrarily chosen strain tensor. It appears, however,† that I did in fact in most cases find the least value of $\Sigma|ds|$. In most

* G. I. Taylor, loc. cit. † J. F. W. Bishop and R. Hill, loc. cit.

orientations of crystal axes I found two different combinations giving the least value. Taking crystal axes with approximately uniform distribution of orientations I obtained a number approximately $m=3{\cdot}06$ as an upper bound for the ratio of tensile stress of the aggregate to the shear resistance of a single crystal. Surprisingly this upper bound is very close to the ratio of the observed strength of an aggregate of aluminium to that of a single crystal of the same material.

Though the theory predicted the strength of an aggregate it was incapable of predicting anything about the deformation texture (i.e. the way in which the distribution of orientations of crystal axes is altered by straining) because when the same minimum virtual work is found for several different slip combinations each of these would give rise to different rotations of the crystal axes relative to the specimen. Unless some method can be devised for distinguishing between these combinations or unless the directions of rotation of axes due to the different minimum combinations are not very different from one another no predictions could be made.

Later some theories were put forward (by Calman and Clews,* and by Kochendörfer†) to account for the deformation texture of strained aggregates; none of them could be taken seriously by a mathematically minded person because they all involved strains in which the grains would not fit together after the strain had happened. These authors recognized this, but it did not seem important to them. Their arguments roughly were that there are great irregularities near the grain boundaries, so one may perhaps assume that some of the material there flows like a fluid so as to permit single or double slipping to occur without loss of continuity. Such theories, to my mind, have little value. They do not give a true lower bound to the value of $m$ as Sachs's does. They do not give a true upper bound as my theory and Bishop and Hill's theory (which ultimately lead to the same result) do.

This was the position till J. F. W. Bishop and R. Hill‡ pointed out that the equations which relate the components of shear stress on the crystal planes in the directions of possible slipping to the six components of stress, referred to the cubic axes, are exactly the same as equations (1) when the six components of strain are replaced by those of stress and the elements of slip by the components of shear stress parallel to the slip directions.

Using this theorem and assuming that the shear stress on each of the active planes were the same, they were able to define 56 states of stress in each of which the components of shear stress were the same on 5 slip-planes. They used the principle that the one among the 56 which corresponds with the *greatest* virtual work consistent with a given externally applied strain is the one which operates. Taking as an example of their method the case when the applied strain is an extension in one direction and contractions in two directions perpendicular to it, they were able to divide the fundamental triangle of the face-centred cube into regions such that

* E. A. Calman and C. T. B. Clews, *Phil. Mag.* XLI (1950), 1085.

† A. Kochendörfer, *Plastische Eigenschaften von Kristallen.* Berlin: Springer (1941), p. 22.

‡ Loc. cit.

when a point representing the direction of extension fell in one region the particular member of the group of 56 was known.

This very interesting method of analysis yields the same physical result as mine. It uses the same physical assumption that the strain of every grain is the same as that of the aggregate. It gives the same upper bound for $m$, namely 3·06. It would give the same limits of uncertainty in the predicted deformation texture. In fact the deformation texture diagram given by Bishop* shows a greater amount of dispersion than mine† but this is due to the fact that I missed some of the possible combinations of 5 slips. The physical picture presented is much clearer than mine because the distribution of stress among the possible slip-planes is considered. Continuity of strain at the boundaries of grains is preserved in both but continuity of stress is not preserved in either.

A further development in Bishop and Hill's studies was the calculation of the deviation of the plastic properties of a crystal aggregate regarded as a continuum from the simplest law of ideal plastic continua, namely, Mises's assumption that the stress tensor and the rate of strain tensor are identical. In this they were not quite so successful as they and I were in predicting the value of $m$, but they showed that a crystal aggregate should differ from the ideal Mises plastic body in the way that both copper and aluminium have been observed to do.‡

With regard to deformation texture the theories of Bishop, Hill and myself are not capable of predicting anything quantitative because they do not contain any clue as to which of many equivalent combinations of 5 slips operates. Bishop § has analysed the effect of supposing that the choice is determined by differential hardening of active and passive slip-planes, but it can hardly be said that anything very decisive has yet come out of this suggestion.

## Polycrystalline Iron

Experiments with single crystals of iron have established the following facts: ‖

(1) When a unidirectional tension is applied the strain is due to slipping on one plane.

(2) The direction of slip is one of the four [111] axes.

(3) The plane of slip, as determined by external measurements, is not a crystallographic plane. It is that one of all the planes which pass through the crystallographic [111] axes on which the component of shear stress is a maximum.

This last statement has been criticized by people who have observed the slip-lines on surfaces of strained single crystals and identified them as being the traces of various crystallographic planes, of type (101), (121), (231) which contain axes

* J. F. W. Bishop, *J. Mech. Phys. Solids*, III (1954), 130 and 259.

† G. I. Taylor, loc. cit.

‡ G. I. Taylor and H. Quinney, *Phil. Trans.* A, CCXXX (1931), 323, paper 16 above; and E. Lode, *Z. Phys.* XXXVI (1926), 913.

§ J. F. W. Bishop, loc. cit.

‖ G. I. Taylor and C. F. Elam, *Proc. Roy. Soc.* A, CXII (1926), 337; paper 10 above.

normal to the (111)-type planes. This may well be the case but it has no relevance to the question, since the statement depends only on the results of careful measurements of external marks made for the purpose of analysing the total macroscopic strain. How this strain is attained by internal slippings, the motion of dislocations, etc., is an interesting physical question but it is unnecessary to have a complete picture of the microscopic internal condition when formulating laws which represent the results of experiments on macroscopic stress and strain relationships. The experimental result (3) must be interpreted in the sense that the shear stress required to operate a shear strain parallel to any arbitrary plane through the direction [111] is very nearly independent of the orientation of that plane. This result is not generally true for all body-centred cubic metals, in particular it is not true for $\beta$-brass, but it is true to the limit of accuracy of my measurements for iron. The result (3) is not inconsistent with the principle that slippings proceed over crystal planes. Slipping over an arbitrary plane passing through a given [111] axis can occur owing to an appropriate combination of the two nearest crystallographic slip-planes.

The maximum angle between the possible neighbouring slip-planes [101], [121], [123] is $19° 6'$ so that if the resistance to slip on each of those planes were the same, the maximum possible variation in the stress required to operate an arbitrary plane passing through the [111] direction is $1-\cos 19° 33' = 1-0{\cdot}986 = 0{\cdot}014$ or $1{\cdot}4\,\%$. Thus the experimental law (3) is so slightly at variance with the hypothesis that the shear on an arbitrary plane is due to an appropriate combination of two slips on crystal planes that no existing technique would be capable of distinguishing between them. The advantage in analysis of using the experimental law (3) instead of considering all possible slips on crystal planes is very great. C. S. Barrett * for instance remarks that with four [111] slip directions and 12 possible slip-planes passing through each of them, the number of independent slip systems is 48 and since the number of combinations of 48 things taken 5 at a time is 1,712,304 he suggests that the equations connecting compatibility of an arbitrary strain with 5 slip systems would have to be solved this number of times. If, however, the experimental law of slip in single crystals is used, only four independent slips, say $s_1$, $s_2$, $s_3$, $s_4$, on planes of arbitrary orientation need be considered. Thus the 48 variables are reduced to 8, namely, $s_1$, $s_2$, $s_3$, $s_4$ together with four variables $\alpha_1$, $\alpha_2$, $\alpha_3$, $\alpha_4$, specifying the orientations of the slip-planes.

The first step in analysing the problem is to set forth the strain equations connecting an arbitrary strain, expressed in terms of the cubic axes, with the four shears $s_1$, $s_2$, $s_3$, $s_4$ parallel to the cube diagonal axes $[111]$, $[1\bar{1}\bar{1}]$, $[\bar{1}1\bar{1}]$, $[\bar{1}\bar{1}1]$. The equations of planes through these are

$$\left.\begin{array}{ll} 0=x+\alpha_1 y-(1+\alpha_1)z & \text{passes through} \quad x=y=z, \\ 0=x-\alpha_2 y+(1+\alpha_2)z & \text{passes through} \ +x=-y=-z, \\ 0=-x+\alpha_3 y+(1+\alpha_3)z & \text{passes through} \ -x=y=-z, \\ 0=-x-\alpha_4 y-(1+\alpha_4)z & \text{passes through} \ -x=-y=z. \end{array}\right\} \quad (2)$$

* *Structure of Metals.* New York: McGraw-Hill (1952).

The strain equations become, after some reduction,

$$\left.\begin{aligned} e_{xx}&=\frac{\partial u}{\partial x}=S_1+S_2+S_3+S_4,\\ e_{yy}&=\frac{\partial v}{\partial y}=\alpha_1 S_1+\alpha_2 S_2+\alpha_3 S_3+\alpha_4 S_4',\\ e_{yz}&=\frac{\partial v}{\partial z}+\frac{\partial w}{\partial y}=-S_1-S_2+S_3+S_4,\\ e_{zx}&=\frac{\partial w}{\partial x}+\frac{\partial u}{\partial z}=-\alpha_1 S_1+\alpha_2 S_2-\alpha_3 S_3+\alpha_4 S_4,\\ e_{xy}&=\frac{\partial u}{\partial y}+\frac{\partial v}{\partial x}=(1+\alpha_1)S_1-(1+\alpha_2)S_2-(1+\alpha_3)S_3+(1+\alpha_4)S_4,\end{aligned}\right\}\tag{3}$$

where
$$S_n=\frac{s_n}{\sqrt{\{6(1+\alpha_n+\alpha_n^2)\}}}\quad(n=1,2,3,4).\tag{4}$$

The virtual work is
$$W=\tau\sum_n|s_n|=\tau\sum_n|S_n\sqrt{\{6(1+\alpha+\alpha^2)\}}|,\tag{5}$$

where $\tau$ is the shear stress necessary to operate the shear on any one of the planes (2).

If each crystal is forced by external constraints to assume the strain of the aggregate the work done by stresses over its surface is fully accounted for by the stresses taken into account in the expression for virtual work on the aggregate, so that, for instance, if the total strain of the aggregate is that due to a small extension $\epsilon$ along one direction and contractions $\frac{1}{2}\epsilon$ in all perpendicular directions. The stress $P$ which is necessary to extend the specimen is given by

$$P\epsilon=T\Sigma|s|.\tag{6}$$

For the same reasons that held in the case of the face-centred cubic metals the value of $P$ found from (6) is greater than the true value so that an upper bound to $P$ is found by choosing the smallest value of $\Sigma|s|$ consistent with (3). In the face-centred case this minimum always occurs when the minimum number, 5, of strains consistent with the strain equations (1) is operative. This is a direct consequence of the fact that the strain equations (1) are linear. The five equations (3) are not linear in all the 8 variables, so that the minimum value of $\Sigma|s|$ does not necessarily occur when, say, one of the four strains $s_1, s_2, s_3, s_4$ is zero.

On the other hand, if 5 variables, say $S_1$, $S_2$, $S_3$, $S_4$ and $\alpha_4 S_4$ are regarded as independent, (3) can be solved as linear equations to obtain their values in terms of $\alpha_1, \alpha_2, \alpha_3$, and the given components of strain. If these be substituted in (5) the problem reduces to that of finding the minimum value of an expression involving $\alpha_1, \alpha_2$ and $\alpha_3$ as they are varied.

The problem is therefore that of determining a true minimum value as three parameters vary continuously. This is quite different from that of the face-centred cubic metals in which the least value was required for an expression containing 5 variables selected from 12.

◎

# 编辑手记

本书是一部版权引进自英国剑桥大学出版社的英文版学术著作,中文书名为《杰弗里·英格拉姆·泰勒科学论文集.第1卷,固体力学》.杰弗里·英格拉姆·泰勒(1886—1975)是世界著名物理学家、数学家,同时也是流体动力学和波理论的专家,他被认为是20世纪最伟大的物理学家之一.从1958年到1971年出版的这四卷书中,巴彻勒共收集了杰弗里·英格拉姆·泰勒的近200篇论文.前三卷的论文大致按主题分组,第四卷整理了许多有关流体力学的各种论文.这些内容加在一起,可以让读者彻底了解泰勒爵士在流体动力学领域的广泛且多样的兴趣.在第四卷的结尾,巴彻勒为读者提供了按时间顺序列出的所有四卷论文的清单,以及泰勒爵士发表的其他文章的清单,从而完成了这项真正宝贵的研究和参考工作.

本书的主编是G. K.巴彻勒,他是剑桥大学三一学院的教授.他在本书的前言中指出:

> 除了一些重复的论文和一些被认为不再令人感兴趣的未发表的笔记以外,这四卷包含了我和杰弗里·泰勒爵士所能找到的所有科学论文.本卷中内容与其原始内容相比改动较小.第一次出版中的印刷错误和小的数学错误已被纠正,本卷中没有任何评论,但与最初出版的文本相比,论文中任何实质性的变化已在脚注中标明.所有的图形都以统一的样式

重绘. 为咨询委员会等撰写的几篇论文在这里首次发表;杰弗里·泰勒爵士浏览了这些论文,并做了一些小的修改,以满足更普遍的图书发行的需要. 本卷中未包括准备作为演讲或集体作品发表的纯粹说明性文章.

尽管第二卷、第三卷和第四卷中根据流体力学的不同方面对论文所进行的划分并不清晰,但整体来说,这四卷仍根据主题将论文分组. 在每卷中,论文都是按时间顺序排列的,每卷都有基于该顺序的自己的编号体系. 在第四卷末尾将提供所有四卷中的论文的完整列表.

本系列图书的出版是从供应部(通过流体运动小组委员会和它的上级机构,航空研究委员会)获得的财政支持. 我个人要感谢军备研究与发展机构的 J. W. 麦科尔博士和他的工作人员,他们提供了一份极为宝贵的清单,其中包括杰弗里·泰勒爵士已发表和未发表的几乎所有论文.



本书的版权编辑李丹女士为我们翻译了本书的目录:

20. 在材料屈服于剪切应力而保持其体积弹性时的断层
21. 晶体塑性变形的机理. 第一部分:理论
22. 晶体塑性变形的机理. 第二部分:比较与观察
23. 岩盐的强度
24. 晶体的可塑性理论
25. 铜的晶格畸变和冷加工潜热
26. 金属受热时由于先前的冷加工而产生的潜在能量的释放(与H. 奎尼合著)
27. 金属塑性应变
28. 立方晶体塑性应变分析
29. 各向异性板块中的应力系统. 第一部分(与 A. E. 格林合著)
30. 爆炸产生的地震波的传播
31. 通过应力环测量计算自粘性管中的应力分布
32. 冲击载荷作用下导线中的塑性波
33. 冲击应力条件下的力学性能(与 R. M. 戴维斯合著)
34. 沿边缘夹紧的椭圆膜片在压力作用下的变形
35. 各向异性板块中的应力系统. 第三部分(与 A. E. 格林合著)
36. 材料在高载荷率下的试验
37. 塑性体中屈服准则与应变比关系的联系
38. 薄塑料板中圆孔的形成和扩大
39. 用平板弹体确定动态屈服应力. 第一部分:理论考虑
40. 球形压缩脉冲在自由表面反射时的应力分布
41. 结晶聚合物应变

本书的主要内容是在固体力学领域. 由于本书的编辑出版时间较早,这几十年来固体力学又有了非常迅猛的发展,所以我们要用历史的观点来看待这些成果,为此笔者特意找到了一本1965年中国科学院干部局为了帮助在科学研究机关从事业务组织管理工作、政治工作和行政组织工作的同志学习自然科学知识,更好地为科学研究工作服务而编写的《自然科学学科简介》(科学出版社. 1965年. 北京),书中对固体力学的介绍是这样的:

> 固体力学和流体力学都是连续介质力学中的组成部分. 由于工程技术的发展,人们迫切要求知道在静态(在这种情况下外力和由外力引起的固体的变形都是不随时间而变的)和动态(例如撞击)情况下固体材料的各部分(严格说来是各点)的受力情况,变形,位移和破坏的情况,从而有效地利用固体材料做成我们所需要的机器部件和建筑构件.

固体的受力与变形情况基本上是受两类因素所决定的:(1)是材料固有的强度性质,其中最主要的是应力应变关系.研究上述问题的科学我们叫它为材料的强度理论;(2)材料的形状、尺寸及外力分布情况.研究这类问题的是结构强度理论.

当材料所受外力不大,相应的变形也不大,这时候固体材料是处于弹性变形状态,亦即外力撤走以后材料可以恢复原有形状.应力与应变是成正比的(虎克定律),从这个前提出发科学家推导了几个重要的基本方程,在不同条件下(这些条件相应于不同的几何形状,外力分布等)可以求得不同的解答.这些解答告诉我们固体材料内部受力及变形的情况.弹性力学中的静力学问题有很多已得到解决,并且在工程中被广泛地采用了.研究弹性波的传播问题也有许多结果,并且用来说明例如地震等现象.

当材料进入塑性变形阶段,这时候应力应变的关系就不能用简单的虎克定律来说明,而遵循更复杂的规律.也可以用与弹性力学类似的办法求出在塑性变形情况下固体材料的受力变形情况,这是因为材料在塑性情况下应力应变关系还待进一步研究,而从一些已有应力应变关系出发来求受力变形情况的问题也较弹性力学中同类问题要复杂一些.塑性力学的发展对于金属压力加工、高压容器的设计等许多工业中的重要问题关系也很密切.

固体力学中的实验,其内容包括:对固体材料应力应变关系的研究,应力分析的研究是固体力学中重要的一个组成部分,其中如光弹性法,电阻应变及脆性涂漆技术在工业中也得到广泛的应用.

固体的材料强度基本上是由两个因素决定的:(1)原子间的作用力;(2)原子的排列.尤其是后者是个非常复杂的问题,原子不仅有各种式样的排列,而且受温度、缺陷等影响很大.因此固体材料在不同温度,不同受力范围下,各种不同种类的材料,就呈现各种不同的而且是十分复杂的强度性质,用虎克定律进行概括是一种较粗糙的做法,适用范围比较小,当研究固体在反复受力情况下的长期强度,或高温强度时这种理论与实际上的差别往往是十分巨大的.随着工程技术(如火箭技术,近代的动力工业等)的飞跃发展,对于固体材料的特殊强度性质如蠕变、疲劳等的研究已经成为当前固体力学发展的主要方向了.

## 一、弹性力学与实验应力分析

弹性力学是一门与工程技术有紧密联系的学科,由于18世纪末及19世纪初叶火车及机械工业的飞速发展,弹性力学才开始有广泛的应用和发展;近代火箭技术、航空技术、土木水利工程等的发展,提出了很多弹性力学问题,因此就大大地丰富和发展了弹性力学.

简单说来,弹性力学是研究弹性体在力的作用下物体内部力的分布、变形和各点的位移的学科. 它是各种工程上受力构件(例如机械零件、飞机舱壁、水坝等)设计的依据. 具体一点说,它主要研究下列问题:

(1)对韧性材料(如钢)构成的构件,在什么样的外力下开始出现永久变形.

(2)对脆性材料(如生铁)构成的构件,在什么样的外力下出现最大限度的变形.

(3)结构在外力作用下的最大弹性变形.

(4)结构在静力或动力作用下,发生"不稳定"情况(即发生突然而来的新的大变形)的条件的研究.

(5)结构的振动.

(6)"应力集中"现象的研究,即研究各种构件的薄弱部分处的内力分布.

(7)物体在高速变形条件下,变形与内部的传播过程,即所谓"弹性波的研究".

(8)物体在受热时发生内力,内力的分布研究.

上述的各种问题都是在工程发展时所碰到的重要问题,而提炼成为弹性力学的研究内容.

弹性力学是"数学物理"的一个重要部分,它建筑在简单的实验基础——虎克定律上(即物体受力在一定限度内变形与所受力成正比),然后通过数学做普遍和精确的研究,因此弹性力学问题往往又是数学问题,由于在数学上的困难,很多问题得不到精确的解答. 于是弹性力学中就发展了很多近似计算方法,其中以"数值分析"用得最多,尤其在电子计算机出现之后,很多复杂问题就有了得出解答的可能.

为了解决实际问题,越过数学上的困难,除近似计算外,人们还发展了多种实验方法来解决弹性力学问题,这些方法往往统称为"实验应力分析",主要方法有下列几种:

电测法

电测法主要是利用电阻丝在变形时电阻发生变化的特点(或其他材料做成的"感受元件"在受力时或变形时发生的电学性质的变化),把电阻丝的变形与所研究结构的变形互相联系起来,通过电的测量推算出物体中的内力和变形. 电测法以它的精确性和简单得到最广泛的应用,但是它需要较复杂的设备.

光测法

光测法是利用某些材料(如塑料)在受力后发生光学性质的变化,而通过光学的测量换算得出物体中的内力分布. 光测法一般即指"光弹性"

方法.

光弹性学——这是利用一种光学灵敏度高的塑料做成机械零件的模型,在偏光仪中来测定在一定外力作用下的应力分布和大小,是一种直观性很强的应力测定方法,已广泛应用于机构强度的研究上.近年来,由于可以“冻结”应力的塑料的试制成功,这一方法已由求解平面的应力问题发展至能求解立体的应力问题,此外还可以用薄片的光弹性塑料粘贴于机械上,用可携带的偏光仪直接测出该机械部位上的应力分布及大小.这些新的发展以及塑料性能的不断改进已大大扩大了光弹性学在机械设计和水工建筑等方面的应用范围.我国近几年来已掌握了这一先进的应力分析技术,并能自制光弹性材料和光弹性仪,已为今后的发展打下了良好的基础.

这方面的研究课题有新的光弹性方法如“冻结法”“贴片法”“动态光弹性学”等的寻求和改善,新的塑料的试制以及更完善的光弹性仪的设计和制造等.

最近还有不少的尝试以应用此法于研究塑性力学,即研究在塑性变形下的应力分布和大小的问题.这门学科叫“光塑性”,其采用的设备与光弹性法基本上是相同的,但是塑料则有相当大的差别.这一方法可以说尚在开始和发展阶段,还是很不成熟的,但是可以看到有很大的发展前途.

其他

例如用机械方法测定受力物体某点的变形,从而算得其内力,这就是“变形仪”方法,又如“脆膜法”,利用脆漆涂在试件表面上,试件受力后脆漆发生裂纹,通过裂纹的研究可算出物体内的内力.

这里所谈到的是最常见的几种,其他还有很多就不一一列举了.

## 二、塑性力学

前面已经讲过关于弹性变形和弹性力学.什么是塑性变形和塑性力学呢?简单地说就是当作用在物体上的外力取消之后,物体的变形没有完全恢复,而产生了一部分永久的或暂时的残余变形,这种变形我们称为塑性变形.研究这种变形和作用力之间的关系,变形后物体内部应力分布的这门学科我们称它为塑性力学.从图1低碳钢简单拉伸实验图来看,*BC*这一段就是物体的塑性区域,研究塑性力学就是在这个范围内研究物体的应力和变形的问题.

塑性力学和弹性力学的根本区别在于,弹性力学是以应力应变成线性关系的广义虎克定律为基础的.而在塑性的范围中,应力应变已成非线性变化了(见图1中*BC*段),因此塑性力学中就没有像广义虎克定律那样统一的基本变形定律,它在不同条件下有它的特殊形式.

研究塑性力学内容之一是研究物体在塑性区应力和应变之间的关系，常称为基本规律的研究．在常温下，塑性变形和时间没有关系，因此应力应变关系用数学式子来表示，其中就没有时间这一因素．而在高温下，塑性变形随时间增长而增加，一般就称为蠕变，这时候应力应变关系中就明显地出现了时间的因子．不管常温高温，目前根据不同的假定有很多类型的应力应变关系的数学形式．因此用大量精确的实验来验证这些理论，哪一个是最符合客观规律，这就很重要了．这方面已做了不少工作，但是尚未出现被大家一致公认的塑性理论．

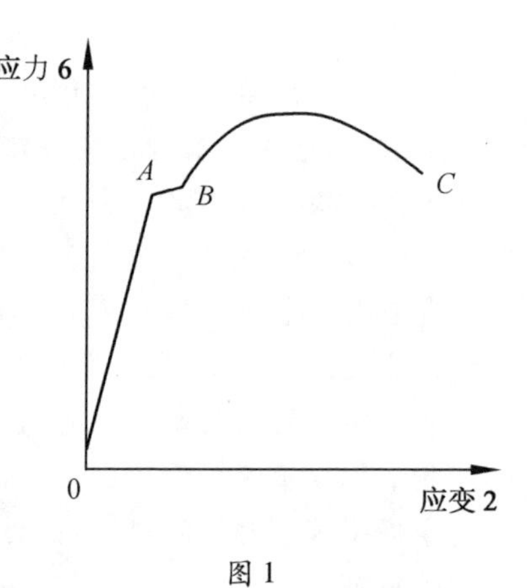

图 1

研究塑性力学内容之二就是利用基本规律来讨论材料发生塑性变形之后内部应力重新分布的情况，以便对各种机器零件做出最合理的强度设计．这方面由于数学上的困难，有些比较简单的问题已经解决，但是形状复杂的零件就很难解决．因此在塑性力学中，利用塑性问题中的某些特点，克服数学上的困难，各种研究方法的发展有重要的意义．

塑性力学在工程实践中有极重要的用途．各种机器的结构中许多强度设计都是以物体到达塑性阶段（开始屈服）作为依据的．实际上物体到达塑性还没有破坏，它还有能力继续工作，因此设计者往往把零件设计到部分发生塑性，还有部分保持弹性的状态．这样可以大量的节省材料．还有一种所谓极限设计，即结构每部分在某一条件下都同时破坏，这在军事上有很大用处，它的设计就以塑性力学为基础．高温情况下如汽轮机、燃气轮机、高速飞行器的考虑蠕变问题也已经越来越明显了．另外，在金属工艺上，如金属的压延、拉丝、锻造、冲压、切削等都是塑性变形的过程，把这些工艺现象提高到理论阶段，从而又进一步地去指导实践，对加速我国国民经济的发展具有重大的意义．近年来，利用塑性理论研究岩石压力的规律也正在发展．

## 三、材料的基本力学性质（材料力学）

要设计一台既好又省的机器或者设计一座实用经济的建筑物的时候，开始的阶段，我们从生产实践所积累的经验中，设想了所要设计的东西的要求、工作条件以及可能的各种结构型式；而后，运用了理论力学、材料力学、弹塑性力学、流体力学以及其他学科的知识进行了结构的受力情况分析（应力分析）．但最终还必须知道材料在这种受力条件下到底能否承受得住？应该怎样合理选择和使用合乎我们要求的材料？研究材料的力学性质亦就是为了探求这方面的规律性.当然,上面提及的各方面都是相互联系着、影响着.掌握了材料的性质,更加促进建立新的力学理论,寻找高强度、高性能的材料.

大家知道,金属材料可分为塑性材料(如低碳钢)及脆性材料(如铸铁).它们的性质都密切与金属的晶体结构以及微观原子结构有关,与工作条件有关.至今,积累了大量数据,但由于材料微观结构的复杂性,引起很大的计算困难.因而,对现有的力学现象和性质,大都停留于定性解释.所以目前工程应用中,主要是模拟结构的实际工作情况,来进行材料实验.把所得的力学性质作为强度设计依据.

通常表达金属材料力学性质有下列几个基本参数(它是从静拉伸图中得到):强度极限、屈服极限、延伸率、硬度.再根据需要来附加其他的性质,如铝合金应考虑弹性模量,合金钢应考虑冲击韧性等.随着新的高强度材料的发展,基本数据仍然起标志性的作用.

目前,由于动力机械的发展,广泛研究材料的动力性能.主要研究有:(1)材料在周期及非周期的载荷作用下的性质,即材料的疲劳.这个问题很重要,它直接影响结构的寿命和安全.其性质通常用疲劳极限或应力——载荷重复次数图来表达;(2)对可能发生振动的零件,如汽轮机叶片等,需研究材料的内耗,亦就是当强迫振动的力一消失,振动逐步消失的性能(一般用振动振幅衰减率表示);(3)结合穿甲、锻压及在爆破载荷作用下结构等的需要,研究在高速高能冲击载荷作用下材料的特性.

随着航空、火箭事业的发展,目前大量地在研究高温条件下高速飞行器的结构以及推进器零件的力学性质——蠕变.另外,结合高压高温管道联结紧密性研究,进行了防止零件丧失正常工作能力,零件破坏的材料应力松弛性质的研究.

上面只是粗略提及材料力学性质的几个方面.材料的力学性质在不同的受力情况下(拉、压、弯、扭、变合载荷、不规则的动力载荷等)都不完全一样,受到结构的几何尺寸、材料的化学成分、热处理情况、周围介质以及加工情况等而异.实际结构零件工作情况很复杂,必须根据具体的受力情况,做必要的简化,来做试验.用宏观和微观,化学物理和力学的分析方法研究材料本身性能的普遍规律性.

之所以不惜重金引进这样大部头(四卷)的科学家论文集,最根本的原因是因为泰勒爵士不仅是一位物理学家,同时还是世界著名的应用数学家.而我们数学工作室一直都在致力提升国人数学素养,特别是对中国近代数学为什么落后了这个问题感到困惑,他山之石可以攻玉,多看看西方近代数学发展的状况可能会有些帮助.

数学作为社会与人类文明这个巨大系统中的一个子系统,它的兴盛与衰微绝不是一个简单的问题.要回答这个难解之谜,不仅要考虑数学子系统本身的特点及运行方式,更要考虑到整个系统的运行,此外还要考查各子系统间的相互作用与相互影响,所以我们先从文明的结构谈起.

文明,从广泛的意义上讲指的是人类的文化.

文明,包括精神文明和物质文明.物质文明是精神文明的物质载体,同时又对精神文明的发展起到重要作用.精神文明,在一定意义上是人类思维活动的产物,没有人类的思维,没有人类的思维能力,是不会有什么文明的.

文明是有结构的,这指的是:

(1)宗教伦理、数学和自然科学、文学艺术、工程技术、哲学、政治、经济、法律等文明组成部分.

(2)其各个组成部分在整个文明中各有其特殊的、具体的地位.

(3)其各个组成部分之间的相互关系也是具体的.这些“组成”“地位”和“关系”就是文明的结构.文明的结构又是不断变化的.正因为如此,文明既有其民族性,又有其时代性.

数学的素质和数学在某文明中的地位会影响该文明在世界文明中的地位.从泰勒斯到毕达哥拉斯,再到柏拉图,都充分认识到数学的作用.柏拉图认为:作为一个统治者,为了很好地认识自己在所生活的那个多变的现象世界中的处境,应该学习数学.这是对数学在文明中的地位的最高评价.希腊数学是很有特色的,有一位数学史专家把希腊数学与印度数学相比较后,指出:

(1)在希腊人那里,数学取得了独立的地位,并且是为了数学本身的发展而被研究,而印度的数学却只不过是天文学的侍女;

(2)希腊数学是大众的,而印度数学是少数精英的,在希腊,数学的大门是对任何一个认真地研究它的人敞开的,在印度,数学教育几乎完全属于僧侣;

(3)希腊数学是理论的,而印度数学是实践的,希腊人对几何学有卓越的贡献,但对计算工作则不大认真,印度人却是有才能的计算家和拙劣的几何学者;

(4)希腊数学是逻辑的,而印度数学是直觉的,希腊人的著作在表述上力求清楚和合乎逻辑,印度人的著作却常被模糊不清和神秘的语言所笼罩;

(5)希腊数学是严格的,而印度数学是经验的,希腊数学的一个显著特征是主张严格证明,而印度数学则或多或少是经验的,很少给出证明和推导;

(6)希腊数学具有美学标准,而印度数学缺少这种标准,希腊人好像具备区别优劣的天性,印度数学却很不平衡,优秀的和拙劣的数学往往同时出现.

综上所述,充分体现了这位数学史专家对于希腊数学的深刻了解.同时也可以看到中国数学发展的一些特点.

现在让我们从另一个角度去看待中国古代数学发展落后这一问题,即解决问题与社会的关系.

# 编辑手记

回顾数学发展的历史,解决问题在数学的发展中起着重要的作用.没有解决的数学问题总是像没有被攀登的山峰一样耸立在数学家的面前,激励他们的征服欲望,这种欲望是不会因为征途的艰险和多次受到挫折而有所减弱的.

数学问题可以分为两类.

第一类是现实问题的抽象,主要是由于社会发展的需要而提出来的.由于解决这类问题的迫切需要,会使大量的数学家同时从不同的途径去攻克它.而这些问题的解决往往向数学界引入新的概念和新的方法,由此构造起新的数学体系,开辟新的数学领域.数学发展的一些重要变革,如微积分的创立、常微分方程、偏微分方程以及非线性微分方程的理论的提出都是采用这种发展形势.这个过程与社会发展有着密切的联系,正是天文学的迫切要求促使了中国和巴比伦数学的发展.

第二类问题则是数学家们的自由创造.其中有些问题虽然也有具体问题的原型,但是数学家的兴趣显然不在于作为原型的实际问题的需要,而是数学问题本身的数学结构,是攻克难题的乐趣和欲望.例如,哥尼斯堡七桥问题、四色问题,还有许多数学问题则根本没有什么具体的模型.例如,希尔伯特在 1900 年巴黎国际数学会议上代表大会的讲演中所提出的那些著名问题.这些问题纯粹是数学家们借助于抽象思维、逻辑推理,运用了数学技巧对概念进行巧妙的分析和抽象才提出来的.20 世纪数学的发展就是受到希尔伯特那个报告的强烈影响.

因此,数学的发展主要有两种方式.一种方式是根本性概念的变革导致新的数学领域的开拓.另一种方式则是使开始建立的新理论日臻完善.如果我们把前一种方式比作是对一个新阵地的突破,那么后一种方式就是扩大已经突破阵地的战果和巩固这块阵地.数学家们就是这样使自己的疆域不断向前推进的.

数学作为一种文化的子系统与经济生产、社会制度、文化传统、哲学思想等都有密切联系.现实世界提出的问题首先要经过数学家的抽象才能变成使用数学语言的数学问题.数学问题激励数学家发展数学方法去解决它们,由此积累了数学知识.要从数学知识的积累进一步上升为有系统的有内部逻辑结构的理论体系则需要更进一步的抽象.反过来,从数学理论的体系中可以用推理方法获得新的知识和方法,这些方法可以用来更有效地解决数学问题,也可以在现实世界获得应用.同时从数学理论的体系中也会产生新的数学问题,数学在实际中的应用促进各方面的发展也会产生新的实际问题.

数学的发展依赖于社会的发展所提供的新问题,它反过来也促进社会的发展.但是,促进社会发展的原因是很多的,而数学对社会发展所起的作用却不是直接的,特别是在数学没有受到充分重视,没有有效利用数学知识的手段和文化模式中尤其是这样.相反,社会的动荡和停滞却会严重地阻碍数学的发展.要有社会发展带来的大量新问题,又要为数学家提供稳定的工作环境,并保证他们的成果能迅速获得有效的应用.在中国历史上占有重要地位的大部分著作,主要是由一些问题和解法构成的.例如,中国算学的经典著作,《九章算术》包括 264 个数学问题以及解决这些问题的答案,但是对于与此有关的数学理论却没有给以足够的重视. 刘徽虽然在《九章算术注》中整理了解题方法的系统,而且创立了许多新解法,阐述了一些解法的原理,甚至是极为重要的思想,但是也没有形成欧几里得几何那样的数学范式.

至于其他的著作如《孙子算经》《夏侯阳算经》《张邱建算经》,包括《缀术》这样重要的数学著作都更像是一本数学问题集. 这充分反映了中国数学重实践的传统. 在中国的文化模式中,数学是一种"术",一种济世之术. 而在西方,数学是一门"学",是成体系的学问.

当然这并不是说中国的数学没有理论,没有结构. 在中国古代数学中,用面积证明几何定理,刘徽在《九章算术注》中引入的极限方法,中国数学家致力于解方程发展起来的代数方法都是极为有价值的数学理论,但是没有形成公理化的结构. 就是说还不注重概念的严格定义,没有求证的传统,也没有理论体系完备性的概念.

华裔菲尔兹奖得主丘成桐先生 1997 年 6 月在北京清华大学高等研究中心开幕式上有一个演讲,他说:"我们要谈中国数学的未来,先看一下我们的过去,现在中国人习惯上讲自己很了不起. 事实上,中国古代数学主要贡献在计算及其实用化方面,我们算圆周率算得位数很高,但是对数学理论没有系统化的讲究,基本上抗拒几何学的逻辑结构和发现抽象代数. 在我看来,它们在中国从来没有生过根,我们对传统的科学有不合理的热爱,结果不能接受新的观念,也不能对应用数学做出贡献."

纵观人类几千年文明史,可以说公理化思想是推动科学技术发展和社会进步的最深刻的思想源泉.

在古希腊诞生奥林匹克运动会与公理化思想不无关系. 最初的奥林匹克运动只有跑步一项比赛,参加比赛者不分种族和社会地位,站在同一起跑线上,人们平等竞争,民主评定. 谁先到达终点,雕塑家就用大理石塑造其健美雄姿,诗人作家描写其技巧和勇敢的伟绩. 这就是公理化思想在运动场上的体现,这就是奥林匹克精神之所在.

公理化思想在社会生活中的体现就是社会的法制化,在法律面前,人人平等. 公理化思想在技术领域的体现就是专利法的实行,社会保护发明者的利益,有偿推广发明者的成果,向全社会公开.

中华民族从来不乏能工巧匠和智力优秀者,但由于没有公理化思想的支持和相应的社会保护,发明只能保密,知识只能私有,科学技术主要靠祖传延续. 祖传常常导致失传,失传后人们再从零开始努力. 祖传的技术很难借鉴别人的成果,更不会推广普及.

祖冲之的数学著作《缀术》的失传是一个典型的例子. 唐朝科举制度分科取士,其中"明算科"规定《算经十书》之一的《缀术》做教科书,学习 4 年,这就足以见得它在当时受重视的程度和其内容的博大精深. 但是,这门课程后来被腐败的官僚取消,因为有许多官员不懂《缀术》,设立这门课程有损他们的威信.更有甚者,《缀术》被进一步污蔑为妖术,实行查禁,以致最后失传.为此,至今人们不能明确祖冲之的圆周率究竟是用什么方法计算的.可以设想,如果在一个公理化思想普及的社会,专权者能如此恣意横行吗?

另外中国学术研究有一种明显的复古主义倾向,这与重要科学技术常常失传有关.以中国历史上影响最为深远的一部数学书《九章算术》为例,它是汉朝初年的著作.这本书记述的数学内容,对当时的世界是了不起的贡献.但从汉朝一直到清朝,将近2 000年,中国数学相当部分是在解释《九章算术》,似乎越古越好.人们普遍认为,最古老的东西一定有无尽的内涵.造成这种情况的原因,也是由于没有建立公理体系和实行公理化的研究方法.

15~16世纪,公理化思想在欧洲逐渐传播开来.17世纪初,这种思想由传教士传到中国.最早进行这项活动的是意大利耶稣会教士利玛窦(Matteo Ricei,1552—1610),他与徐光启合译《几何原本》前6卷.200多年后的1580年,李善兰与英国人伟烈亚利(Veileyarli)合译《几何原本》后9卷和《代微积拾级》等书.这些用公理方法研究数学的著作和新思维没有引起封建统治者的重视,在学术界和民间也没有产生强烈的反响.但与此相反,稍后传入日本的这些著作和研究方法,受到日本政界的重视,明治皇帝带头维新,改革教育,推行公理化思想,由此日本焕然一新,跻身世界强国之列.

为什么西方科技从15世纪以后大大超越中国?因为他们普及公理化思想,把教育建立在公理化思想的基础上.公理化思想可保证社会平等竞争,保证科学技术连续发展,不会失传,保证每一代研究者都能站在科学的最前沿进行再创造,保证后人一定超过前人,今人一定胜过古人.中国科技知识的祖传方法,基本属于个体的智力劳动,而公理化的研究方法是社会性的劳动,个人的力量怎么能胜过全社会的力量呢?因此,15世纪西方在普及公理化思想后胜过中国就成为必然的了.其他诸多原因与这一原因相比,都不具有根本的性质.中国长治久安并保持繁荣的根本在于发展教育事业,发展教育事业的关键是大力宣扬公理化思想,加强全民族的公理意识.

从客观上说,数学没有在中国发展只是方法不对头,并不是不重视.

我们的祖先自认为对数学是重视的,有皇帝亲自学数学的,有给数学家立传的(如《畴人传》),还有用考数学来委任官吏的.有一则古代故事,它是记载在《唐阙史》里的.故事称颂"青州杨尚书损,政令颇肃".说唐朝有一位高级官员叫杨损的,执政认真严肃,任人唯贤."郡人戎校缺,必采于舆论而升陟之,缕及细胥贱卒,率用斯道."他提拔行政官吏和权衡功过是非,不按个人喜恶行事.有一次,要在两个办事人员中提升一个,"众推合授","较其岁月、职次、功绩、违法无少差异者."负责提升的人十分为难,便去请示上司杨损.杨损想了想说:"我得之矣.为吏之最,孰先于书算耶.姑听吾言:有夕道于丛林间者,聆群跖评窃贿之数.且曰,人六匹则长五匹,人七匹则短八匹,不知几人复几匹."并说"先达者胜."少顷,一吏果以状先,于是给他先提升了.

我们暂且不讲杨损公而善断,也不论算题难易与否,就其用书算来考核官吏一事,已是一大创造.但这些都是我们自己看自己有时难免美化,俗话说旁观者清,那就让我们看看"洋人"是如何看待中国古代科学落后的原因吧.韩琦①先生对此早有研究,他写道:

> 来自欧洲人所写的著作在法国科学家中也曾产生过影响.在1688年洪若翰、白晋等首批法国耶稣会士到达北京之前,法国出版了利玛窦、曾德昭、卫匡国等人撰写的有关中国的著作,值得注意的是,由于来华欧洲人的宣传,给法国人的印象是:中国皇帝非常重视科学,因此法国应该向中国学习.

1671年,法国耶稣会士科学家帕迪斯(G. G. Pardies,1636—1673)在所著*Elemens de Geometrie*(此书后来被译为中文,名《几何原本》,与徐光启、利玛窦根据Glavius, *Euclidis Elementorum Libri XV*所译的《几何原本》同名,收入康熙御制《数理精蕴》中)一书中给法国科学院先生们的信,曾这样写道:

> 先生们:
>
> 我的目的不仅仅是向各位呈献此书(就像呈献给强大的保护者一样),而是把此书赠给你们,把你们看作最高的仲裁者.事实上,在法国我们尚没有如此的仲裁机构,而在中国早已有之,它是由一批博学的数学家所组成,以终审所有的数学问题,这是中国最为重要的国家事务之一.尽管我国的法律一点都没有赋予这个仲裁权,而你们凭借专业特长,是有资格担当此任的.并且考虑到你们皇家科学院的成员构成,我们可以说它不仅仅是欧洲一个拥有最杰出人物的团体,而且是一个最高的仲裁机构,它的评判在学者中可以作为定论,当我们看到如此宏伟的建筑(当指法国皇家学会)拔地而起的时候,我们只能说它是一个建成来用做新的官办机构(Tribunal)的宫殿(Palacs),我们也只能说在建立皇家学会方面远胜中国皇帝的法国国王也许是从打算模仿中国的政策而建立起这个新的皇家学会的.先生们,你们知道,中国的钦天监通常设立两个观象台,它们都位于中国两个城市的附近,那些向我们描述中国的人说:无论是从地点的壮观,还是从青铜仪器的宏伟来说,我们在欧洲都找不到与此媲美的观象

---

① 韩琦:关于17,18世纪欧洲人对中国科学落后原因的论述.

> 台;这些仪器在700年前即已建成,数百年来陈设于宏伟的观象台的平台上,现在仍很完整,而且整洁如新,就像刚从冶炉中铸出一样;仪器刻度很精确,摆放得非常适合于观测,整个工作非常精致.

从这段话可看出,法国皇家科学院的建立,与中国的制度有一定关系,帕迪斯显然是在阅读了欧洲人所写的有关中国的著作后做出上述评论的.法国耶稣会士李明在《中国现状新志》一书中也曾引用了上述关于中国天文仪器的论述.

德国科学家莱布尼兹对中国的科学状况也产生了很大的兴趣,他通过与耶稣会士闵明我、白晋的会面和通信,对中国有了更多的了解.在《中国近事》(*Novissima Sinica*,1697)一书中,莱布尼兹曾认为中国的手工艺技能与欧洲的相比不分上下,而在思辨科学方面欧洲要略胜一筹,但在实践哲学方面,即在生活与伦理、治国方面,中国远远要比欧洲进步.在讲到中国数学的状况时,利玛窦曾这样写道:"但是当时最受中国人欢迎的书莫过于欧几里得的《几何原本》,这可能是因为世界上没有一个民族如同中国人一样重视数学.虽然数学的方法不同,他们发现了许多数学上的问题,只是得不到证明.在这种光景下,每个人都可以发现数学的问题,施展他的想象力,而找不到任何的答案."莱布尼兹关于中国数学没有证明的看法很可能来自利玛窦的著作.

在莱布尼兹看来,"研究数学不应看作只是工匠们的事情,而应作为哲学家的事务."而"中国人尽管几千年来发展着自己的学问,并奇迹般地用于实际应用,他们的学者可以得到很高的奖赏,然而他们在科学方面并没有达到极高的造诣.简单的原因是,他们缺少欧洲人的慧眼之一,即数学."也就是说,莱布尼兹认为中国科学落后的一个原因是中国人对数学的研究只注意于实用,而缺乏证明,妨碍了对其他科学的深入研究.

莱布尼兹所处的时代,正值康熙皇帝对耶稣会士表示欢迎的时期.他认为东方和西方的关系是具有统一世界的重要媒介,并期望中西交流能促进科学的发展.在致闵明我的信中,莱布尼兹写道:"欧洲应当感谢您开始了这方面的交流.您把我们的数学传授给中国人,反过来,中国人通过您将他们经过长期观察而得到的自然界的奥秘传授给我们.物理学更多地以实际观察为基础,而数学恰恰相反,则以理智的沉思为根基,后者乃我们欧洲之特长.""他们善于观察,而我们长于思考."莱布尼兹的宏伟计划是,以西方的数学特长来弥补中国数学的落后状态,同时输入中国长期观察的物理学,互为补充.莱布尼兹孜孜不倦地致力于他的东方研究计划的各种准备工作,他为创设柏林科学院做了很大的努力,以打开中国门户,而使中国与欧洲的文化互相交流.

白晋等人来华后,法国耶稣会士接踵来华,正是他们的频繁通信,改变了欧洲科学界对中国科学状况的看法,其中最早对中国科学落后原因提出较为全面的看法的当推巴多明.

1698 年,巴多明随白晋来华,在中国期间他参与了为康熙皇帝翻译满文《解剖学》的工作. 由于他精通满汉文字,因此在俄国使节来华时,经常应皇帝的请求担任翻译工作. 1723 年雍正皇帝驱逐传教士以后,使他有更多的时间从事于中国科学的历史和现状的研究. 最为可贵的是他与法国科学院院长梅兰(Dortous de Mairan)频繁通信. 1728 年至 1740 年梅兰向巴多明寄出了许多询问有关中国历史、天文学、中国文明的信,这些信在 1759 年发表,而巴多明的信大多发表在《耶稣会士通信集》(*Lettres Edifiantes et Curieuses*)中.

在致梅兰的信(1730 年 8 月 13 日写于北京)中,巴多明写道:

> (前略)先生,这也许使你莫名其妙.‘中国人很久以来就致力于所谓纯理论性的科学,却无一人深入其中’. 我和你一样,都认为这是难以令人置信的;我不归咎于中国人的心灵,如说他们缺少格物致知的光明及活力,他们在别的学科中的成就所需要的才华及干劲,不减于天文及几何,许多原因凑合而成,致使科学至今不能得到应有的进步,只要这些原因继续存在,仍是前进中的绊脚石.

这里巴多明提出了中国科学落后的问题,他的看法是:首先,凡是想一试身手的人得不到任何报酬. 从历史上来看,数学家的失误受到重罚,无人见到他们的勤劳受到奖赏,他们观察天象,免不了受冻挨饿……钦天监假如是一位饱学之士,热爱科学,努力完成科研;如果有意精益求精,或超过别人,加紧观察,或改进操作方法,在监内同僚之中就立刻引起轩然大波,大家是要坚持按部就班的……这种情况势必是一种阻力,以致北京观象台无人再使用望远镜去发现肉眼所看不见的东西,也不用座钟去计算精确的时刻. 皇宫内原来配备得很好,仪器都是出自欧洲的能工巧匠之手(指耶稣会士汤若望等人),康熙皇帝又加以改革,并把这些好的仪器都放在观象台内,他虽知道这些望远镜和座钟对准确观象是多么重要,但没有人叫数学家利用这些东西. 无疑地,还有人大反特反这些发明,他们抱残守缺,墨守成规,只顾私利……

巴多明认为:“使科学停滞不前的第二个原因,就是里里外外没有刺激与竞争. 假如中国邻邦有一个独立的王国,它研究科学,它的学者足以揭露中国人在天文学中的错误,中国人也许可以如大梦初醒,皇帝变得谨慎、追求进步.”

“(清代)建国初期,就有天文学家及几何学家,他们经过严格考试之后才被分配到钦天监,但是以后经过考验及功绩,他们当了省官或京官,假如数学及数学家更为尊荣的话,我们今日就有长期的观察记录,对我们大有用处,使我们少走许多弯路.”

从这封信可看出,巴多明对中国科学的现状的分析是很有见地的. 他认为国家必须从长远打算,重视知识分子,特别是要重视那些进行具体观测的钦天监的科学家;要有竞争体制,并暗示了向别国学习先进的东西.巴多明认为解决的办法之一是:“要使这些科学在中国兴旺发达,一个皇帝不够,而是要许多皇帝连续地优待勤学苦练、有创造发明之士;建立稳固的基金,以奖励有功人员;提供差旅费及必要的工具;解决数学家落魄穷困之扰,不致受不学无术者之折磨.”这些话至今仍有其现实意义.直到1998年世界著名数学家陈省身先生在接受杰克逊(Allyn Jackson)采访时还说:“我想在中国,数学进步的主要障碍是工资太低.”在观察中国古代科学发展状态的诸多欧洲人中,伏尔泰是其中最重要的一位,有人曾风趣地称:“当伏尔泰用中国的茶碗喝着阿拉伯的咖啡时,他感觉到他的历史视野扩大了.”伏尔泰对中国发生兴趣的原因是多方面的,他从小(1704~1711)在耶稣会士创办的路易大帝学校学习过,而这个学校与来华耶稣会士有较为密切的关系.伏尔泰不仅对中国古史、哲学、音乐、戏剧、道德感兴趣,也对中国科学非常关注.有人曾评论伏尔泰,说他不仅在形而上学方面有很深的造诣,并且在科学史上也有一定涵养.他关于中国科学的论述主要表现在对中国天文学史、中国文明史上的重要发明,以及对长城与大运河的评论.如在《哲学辞典》论中国的条目中,即以赞赏的口气论及中国天文观测的可靠:“中国古代编年史计算出来的32次日蚀,其中有28次是已经被欧洲数学家证实是准确无误的.”更有价值的是,伏尔泰对中国科学落后的原因做了较多的论述.伏尔泰有关中国科学的认识来自纳尔特(D. F. de Navarrete,1618—1686)主教、门多萨、亨宁斯(Henningius)、古斯曼(Louis de Gusman)主教、曾德昭、巴多明、宋君荣以及其他耶稣会士的一些著作.他认为中国科学与艺术落后的原因,主要表现在道德、教育和语言等方面.在《路易十四时代》一书中,伏尔泰认为:对祖先的崇拜导致了中国人缺乏胆识;并把中国人对祖先的崇拜与欧洲人对亚里士多德的崇拜结合起来.在《哲学辞典》中更直截了当地认为崇敬祖先在中国阻碍了物理学、几何学和天文学的进步.在给Pauw的第四封信中,伏尔泰认为中国没有好的数学家,这与教育制度很有关系.在《风俗论》中,伏尔泰写道:“试问中国人为什么这样落后,这样停滞不前?为什么天文学在他们那里古已有之却故步自封,这些人好像天生如此,和我们的性格不同,他们只想发明机器,有利民生,一劳永逸,不愿越雷池一步.很多艺术、科学源远流长,连绵不断,而进步是这样少,也许有两个原因:其一就是这个民族敬祖敬得出奇,在他们眼中看来,凡是老祖宗传下来的都是圆满无缺的;其二就是语言的性质,语言为一切知识之本.”他认为中国语言是所有认识的首要原则,因此他把科学落后与语言相联系,并把中国汉字的规范化以及字母化看作是促进科学进步的一个条件.

在历史领域,伏尔泰最先把中国置于世界中来考察.在《世界史》一书中,伏尔泰在首章论述了中国历史的悠久,指出:中国人的本性的力量使他们能够很早就有许多创造发明,诸如火药、印刷术和指南针等.但伏尔泰认为中国人没有能力进一步发展自己的发明创造,当他们发明了某种东西以后,就止步不前了,而西方人却设法把中国人的发明发展到一个新的高度,并扩展到了一个新的领域;尽管欧洲人在科学的洞察力方面比中国人差一些,但他们却能迅速地使科学上的发明创造趋于完善;这种推进事物发展的独特能力就使西方处于主动地位.这种观点与莱布尼兹在《中国近事》中所阐述的几乎一致.

法国重农学派代表人物奎奈(F. Quesnay,1694—1774)认为造成中国缺乏抽象思考及逻辑观念的原因是中国传统的实用性,"虽然中国人很好学,且很容易在所有的学问上成功,但是他们在思辨上很少进步,因为他们重视实利,所以他们在天文、地理、自然哲学、物理学及很多实用的学科上有很好的构想,他们的研究倾向应用科学、文法、伦理、历史、法律、政治等看来有益于指导人类行为及增进社会福利的学问."这与莱布尼兹的看法相一致.

1741年和1742年,休谟(David Hume,1711—1776)出版了他的《论文集》.此书收集了近50篇论文,其中不少谈到了中国.他把中国的科学与文化同贸易联系在一起,他认为没有什么能比若干邻近而独立的国家通过贸易和政策联合在一起更有利于提高教养和学问了.中国恰恰在这一方面有很大的缺陷,从而使原来可能发展出更完美和完备的教养和科学,在许多世纪的进程中,收获甚微.从外部来说,其原因在于没有更多的外贸对象,但从内部来说,是由于中国处于大一统的状态之下,说一种语言,在一种法律统治下,赞成相同的生活方式;对权威的宣传和敬畏,造成了勇气的丧失.休谟实际上以自己的见解回答了为什么在那个非凡的帝国,科学只取得了如此缓慢的进展这一问题.

狄德罗曾分析过中国没有出现像欧洲那些天才的原因,主要在于东方精神的束缚,在他看来,东方精神趋于安宁、怠惰,只囿于最切身的利益,对成俗不敢逾越,对于新事物缺乏热烈的渴求.而这一切恰恰与科学发展所需要的探索精神格格不入.他说:"虽然中国人的历史比我们悠久,但我们却远远走在了他们的前面."狄德罗是从中国文化的角度看待中国的问题,而莱布尼兹是从一个科学家的角度去看待中国科学,是比较求实的态度,这与一些旅行家的偏见是不可等量齐观的.

巴多明与法国思想家的看法,后来在一些法国科学著作中也有体现,如法国数学家蒙塔克莱(Montucla)在其所著《数学史》的第四卷中曾专门论述了中国科学,认为语言、敬祖、对数学的不重视导致了中国科学的落后.18世纪末19世纪初,也就是当英国大使马嘎尔尼出使中国会见乾隆皇帝的前后,也有许多英国人提到中国科学的现状与落后的原因.由于英国当时正经历着工业革命,科学得到了迅猛的发展,随之而产生的价值观念的变化使许多英国人在评论时对中国社会的缺陷进行了批评,这时对中国科学的评论则带有明显的轻蔑倾向.

所幸的是近年来随着综合国力的增强,国家对数学与科学也逐渐开始重视起来.我们来看一则消息:

> 2020年9月6日,2020未来科学大奖在北京揭晓.
>
> "物质科学奖"授予中国科学院院士、中国科学院金属研究所研究员、

沈阳材料科学国家研究中心主任卢柯,以奖励他开创性地发现和利用纳米孪晶结构及梯度纳米结构,实现铜金属的高强度、高韧性和高导电性.

他在接受《中国科学报》专访时表示,一个聚焦“未来”的大奖颁发给一个“古老”的学科,这给传统领域的科研人员带来了巨大鼓舞.

**《中国科学报》:**“未来”科学大奖颁给了一个“古老”学科,是否意味着传统学科其实也充满了突破的潜力?

**卢柯:**我想是的. 未来科学大奖应该是面向未来的. 而我相信材料科学这个传统领域依然有着广阔的未来.

我从大领域来讲是做金属材料的,从小领域来说是做纳米金属材料的. 金属材料的结构到了纳米尺度会表现出什么新性能,会出现什么新现象? 这就是传统材料科学领域中出现的新问题.

从这个意义上讲,把未来科学大奖颁给我,不是颁给我一个人,而是颁给了我们这个孕育着希望的领域. 很多同事在给我发来的微信、短信中,都说这是我们这个领域的喜讯. 事实上,未来科学大奖给我们这些从事传统领域研究的人一个巨大的鼓舞.

**《中国科学报》:**目前您所在领域面临的核心问题是什么?

**卢柯:**我们最初面临的问题就是怎么把微观结构缩小到纳米量级,所以我们做了二三十年的主要工作其实就是制备.

“制备”听起来是个简单的技术问题,实则不然. 一个材料制备不出来,除了我们的工艺技术水平不够,更重要的是我们对纳米尺度下材料的很多规律,比如结构形成演化规律、结构稳定性规律、结构性能关系规律都不清楚——这是一个综合性的问题.

所以,最后我们的功夫是花在怎么能够把结构演化到纳米尺度,怎么能稳定住它. 最终,我们发现的纳米孪晶结构和梯度纳米结构都是比较稳定的结构,而且它们表现出了很好的性能.

**《中国科学报》:**您曾说纳米孪晶结构的发现过程是很偶然的,这其中有什么故事吗?

**卢柯:**我们当时并没有想要做出纳米孪晶,只是想做出晶粒尺寸很小的铜. 结果却发现铜的晶粒尺寸并没有那么小,反而出来了一堆副产品(亚结构),就是现在我们所说的纳米孪晶结构. 但是,这个副产品的性能特别好.

过去我们不知道有这样一种纳米结构,于是花了很大精力来理解:它为什么会形成这个结构,这个结构为什么会表现出这样的性能? 但当我们把论文投出去后,送一次被退一次,送一次被退一次. 大家不认为这是一种典型的纳米结构,不是一种大家所认同的纳米结构. 因为纳米孪晶的晶粒尺寸不在纳米量级,只是它的层片厚度在纳米量级.

那段时期我们被拒稿折磨得快要受不了了,以至于都想放弃了.就在决定做最后一次努力的时候,《科学》接收了它,那是在2004年.

**《中国科学报》:**纳米孪晶这个结构有什么特点,有什么用处?

**卢柯:**孪晶之所以叫"孪"晶,是因为它是一个非常规则的晶界,晶界两边的晶格是对称的,所以它非常稳定——这是大家已经知道的.

但当我们把孪晶界之间的间距做到特别小,小到纳米级别,结果发现它的性能特别好,这是我们万万没想到的.

如果我们把普通铜的强度通过合金化或晶粒细化提高几倍,它就变得特别脆,导电性也会损失.而做成纳米孪晶铜以后强度可以提高到10倍,比一般钢的强度都高,同时它还有塑性,可以变形、加工,导电性还不会损失.也就是说,它在保持金属出色的导电性能的同时,又大幅提升了强度.

相比奖励重金,对后续发展好处更大的似乎是出版这些优秀科学家的文集.

比如北京大学出版社2004年出版的《黄昆文集》,当然还有一大批大师的文集,如北京大学出版社还出版过《周培源文集》《王选文集》.

本书从规模上与《苏步青数学论文全集》(共3卷,2001年由高等教育出版社出版)相当,后者是对科学共同体的一次重要的积累与贡献,值得我们效仿!

刘培杰

**2021年6月15日**

**于哈工大**

# 刘培杰物理工作室<br>已出版(即将出版)图书目录

| 序号 | 书　名 | 出版时间 | 定　价 |
|---|---|---|---|
| 1 | 物理学中的几何方法 | 2017—06 | 88.00 |
| 2 | 量子力学原理.上 | 2016—01 | 38.00 |
| 3 | 时标动力学方程的指数型二分性与周期解 | 2016—04 | 48.00 |
| 4 | 重刚体绕不动点运动方程的积分法 | 2016—05 | 68.00 |
| 5 | 水轮机水力稳定性 | 2016—05 | 48.00 |
| 6 | Lévy 噪音驱动的传染病模型的动力学行为 | 2016—05 | 48.00 |
| 7 | 铣加工动力学系统稳定性研究的数学方法 | 2016—11 | 28.00 |
| 8 | 粒子图像测速仪实用指南:第二版 | 2017—08 | 78.00 |
| 9 | 锥形波入射粗糙表面反散射问题理论与算法 | 2018—03 | 68.00 |
| 10 | 混沌动力学:分形、平铺、代换 | 2019—09 | 48.00 |
| 11 | 从开普勒到阿诺德——三体问题的历史 | 2014—05 | 298.00 |
| 12 | 数学物理大百科全书.第 1 卷 | 2016—01 | 418.00 |
| 13 | 数学物理大百科全书.第 2 卷 | 2016—01 | 408.00 |
| 14 | 数学物理大百科全书.第 3 卷 | 2016—01 | 396.00 |
| 15 | 数学物理大百科全书.第 4 卷 | 2016—01 | 408.00 |
| 16 | 数学物理大百科全书.第 5 卷 | 2016—01 | 368.00 |
| 17 | 量子机器学习中数据挖掘的量子计算方法 | 2016—01 | 98.00 |
| 18 | 量子物理的非常规方法 | 2016—01 | 118.00 |
| 19 | 运输过程的统一非局部理论:广义波尔兹曼物理动力学,第 2 版 | 2016—01 | 198.00 |
| 20 | 量子力学与经典力学之间的联系在原子、分子及电动力学系统建模中的应用 | 2016—01 | 58.00 |
| 21 | 动力系统与统计力学:英文 | 2018—09 | 118.00 |
| 22 | 表示论与动力系统:英文 | 2018—09 | 118.00 |
| 23 | 工程师与科学家微分方程用书:第 4 版 | 2019—07 | 58.00 |
| 24 | 工程师与科学家统计学:第 4 版 | 2019—06 | 58.00 |
| 25 | 通往天文学的途径:第 5 版 | 2019—05 | 58.00 |
| 26 | 量子世界中的蝴蝶:最迷人的量子分形故事 | 2020—06 | 118.00 |
| 27 | 走进量子力学 | 2020—06 | 118.00 |
| 28 | 计算物理学概论 | 2020—06 | 48.00 |
| 29 | 物质,空间和时间的理论:量子理论 | 2020—10 | 48.00 |
| 30 | 物质,空间和时间的理论:经典理论 | 2020—10 | 48.00 |
| 31 | 量子场理论:解释世界的神秘背景 | 2020—07 | 38.00 |
| 32 | 计算物理学概论 | 2020—06 | 48.00 |
| 33 | 行星状星云 | 2020—10 | 38.00 |

# 刘培杰物理工作室
# 已出版(即将出版)图书目录

| 序号 | 书　　名 | 出版时间 | 定　价 |
|---|---|---|---|
| 34 | 基本宇宙学:从亚里士多德的宇宙到大爆炸 | 2020—08 | 58.00 |
| 35 | 数学磁流体力学 | 2020—07 | 58.00 |
| 36 | 高考物理解题金典(第2版) | 2019—05 | 68.00 |
| 37 | 高考物理压轴题全解 | 2017—04 | 48.00 |
| 38 | 高中物理经典问题25讲 | 2017—05 | 28.00 |
| 39 | 高中物理教学讲义 | 2018—01 | 48.00 |
| 40 | 1000个国外中学物理好题 | 2012—04 | 48.00 |
| 41 | 数学解题中的物理方法 | 2011—06 | 28.00 |
| 42 | 力学在几何中的一些应用 | 2013—01 | 38.00 |
| 43 | 物理奥林匹克竞赛大题典——力学卷 | 2014—11 | 48.00 |
| 44 | 物理奥林匹克竞赛大题典——热学卷 | 2014—04 | 28.00 |
| 45 | 物理奥林匹克竞赛大题典——电磁学卷 | 2015—07 | 48.00 |
| 46 | 物理奥林匹克竞赛大题典——光学与近代物理卷 | 2014—06 | 28.00 |
| 47 | 电磁理论 | 2020—08 | 48.00 |
| 48 | 连续介质力学中的非线性问题 | 2020—09 | 78.00 |
| 49 | 力学若干基本问题的发展概论 | 2020—11 | 48.00 |


**联系地址:**哈尔滨市南岗区复华四道街10号　哈尔滨工业大学出版社刘培杰物理工作室

**网　　址:**http://lpj.hit.edu.cn/

**邮　　编:**150006

**联系电话:**0451—86281378　　13904613167

E-mail:lpj1378@163.com